Natural Products from Actinomycetes

Ravishankar V. Rai • Jamuna A. Bai
Editors

Natural Products from Actinomycetes

Diversity, Ecology and Drug Discovery

Springer

Editors
Ravishankar V. Rai
Department of Studies in Microbiology
University of Mysore
Mysore, Karnataka, India

Jamuna A. Bai
Department of Microbiology, School of Life Sciences
JSS AHER
Mysore, Karnataka, India

ISBN 978-981-16-6131-0 ISBN 978-981-16-6132-7 (eBook)
https://doi.org/10.1007/978-981-16-6132-7

This Springer imprint is published by the registered company Springer Nature Singapore Pte Ltd.
The registered company address is: 152 Beach Road, #21-01/04 Gateway East, Singapore 189721, Singapore

Preface

Actinomycetes are still one of the most important sources of chemical diversity and a reservoir to mine for novel structures that are requiring the integration of diverse disciplines. The decline in the number of new chemical scaffolds and the rediscovery problem of old known molecules has become a hurdle for industrial natural products discovery programmes; new actinomycetes compounds and leads have continued to be discovered and developed to the preclinical stages. These can range from novel strategies to isolate species previously not cultivated, innovative whole-cell screening approaches and on-site analytical detection and dereplication tools for novel compounds, to *in silico* biosynthetic predictions from whole-gene sequences and novel-engineered heterologous expression that have inspired the isolation of new NPs and shown their potential application in the discovery of novel antibiotics. The book is on the biology and the application of actinomycetes. The special features of the book focus on the Diversity, Chemical biology and Ecology of Actinomycetes and the application of modern genomic platforms for the discovery of antibiotics, anti-infectives and anticancer drugs from actinomycetes.

The edited book *Natural Products from Actinomycetes* is comprehensive with two sections consisting of 18 chapters. The first section is on "Diversity, Chemical biology and Ecology of Actinomycetes; and the Discovery of Natural Products" and covers topics on natural products from the soil, endophytic and marine-derived actinomycetes including new natural products discovery, chemical biology, new methods for discovering secondary metabolites, structure elucidation, biosynthetic research of natural products and new biological activities. Studying silent biosynthetic gene clusters in the genomes of actinomycetes that code for secondary metabolites which usually go undetected under standard fermentation conditions is also emphasized. The topics in this section focus on the effects of biological and chemical elicitation at the molecular level on secondary metabolism in actinomycetes. The diversity of natural products discovered by these approaches from the actinomycetes will be exemplified. The second section of the book is on "Genomic and synthetic biology approaches in Actinomycetes drug discovery". In this section, the topics are on the application of metabolic engineering to optimize natural product synthesis and the use of omics data and the engineering of regulatory genes. The advanced tools of synthetic biology and metabolic engineering including cluster assembly, CRISPR/Cas9 technologies and chassis strain development for natural product overproduction in actinomycetes are covered. Also, the use of bioinformatics tools for reprogramming of biosynthetic pathways through polyketide synthase and non-ribosomal peptide synthetase engineering will be highlighted. These new and advanced genomic and molecular tools are expected to accelerate the discovery and development of new natural products from actinomycetes with medicinal and other industrial applications.

The book includes contributions by both internationally recognized and well-established scientists. It is a valuable source of information for young researchers and students who are keen to unfold the natural product repository hidden in the Actinomycetes class.

Mysore, Karnataka, India

Ravishankar V. Rai
Jamuna A. Bai

Contents

1 **Actinobacteria from Marine Environments: A Unique Source of Natural Products** 1
Mariana Girão, Inês Ribeiro, and Maria de Fátima Carvalho

2 **Exploration of Insects and Mollusks for New Secondary Metabolites from Actinobacteria** 47
Chandra Risdian, Nasim Safaei, Michael Steinert, and Joachim Wink

3 **Polar Actinobacteria: A Source of Biosynthetic Diversity** 71
Adriana Rego, Maria de Fátima Carvalho, Pedro Leão, and Catarina Magalhães

4 **Actinobacteria from Arid Environments and Their Biotechnological Applications** 91
Feiyang Xie, Pharada Rangseekaew, and Wasu Pathom-aree

5 **Endophytic Actinobacteria in Biosynthesis of Bioactive Metabolites and Their Application in Improving Crop Yield and Sustainable Agriculture** 119
Yadollah Bahrami, Yaser Delbari, Kimia Rafiei Buzhani, Elham Kakaei, Yaser Mohassel, Sasan Bouk, and Christopher M. M. Franco

6 **An Overview on Natural Product from Endophytic Actinomycetes** 151
Govindan Nadar Rajivgandhi, R. T. V. Vimala, G. Ramachandran, Chelliah Chenthis Kanisha, N. Manoharan, and Wen-Jun Li

7 **Regulation of Secondary Metabolites Through Signaling Molecules in *Streptomyces*** 167
Kenji Arakawa and Toshihiro Suzuki

8 **New Strategies to Activate Secondary Metabolism in *Streptomyces*** 185
Paula Yagüe, Nathaly Gonzalez-Quiñonez, Gemma Fernández-García, Sergio Alonso-Fernández, and Angel Manteca

9 Novel Agroactive Secondary Metabolites from Actinomycetes in the Past Two Decades with Focus on Screening Strategies and Discovery 199
Kaimei Wang, Shaoyong Ke, Wei Fang, Zhaoyuan Wu, and Yani Zhang

10 Quorum Sensing and Quorum Quenching Metabolites in Actinomycetes 223
Jamuna A. Bai and Ravishankar V. Rai

11 Metabolic Engineering of Actinomycetes for Natural Product Discovery 267
Erin E. Drufva, Tien T. Sword, and Constance B. Bailey

12 Application of CRISPR/Cas9 Editing for Production of Secondary Metabolites in Actinomycetes 309
Jamuna A. Bai and Ravishankar V. Rai

13 Synthetic Biology in Actinomycetes for Natural Product Discovery 331
Shuqing Ning, Tingting Wu, Yushuang Ren, and Yunzi Luo

14 Endophytic Actinomycetes: Secondary Metabolites and Genomic Approaches 363
Nattakorn Kuncharoen and Somboon Tanasupawat

15 Mining for NRPS and PKS Genes in *Actinobacteria* Using Whole-Genome Sequencing and Bioinformatic Tools 393
Heidi El-Gawahergy, Dina H. Amin, and Alaa F. Elsayed

16 Glycopeptide Antibiotics: Genetics, Chemistry, and New Screening Approaches 411
Oleksandr Yushchuk and Bohdan Ostash

17 Biotechnological Aspects of Siderophore Biosynthesis by Actinobacteria 445
Artur Maier, Carolin Mügge, and Dirk Tischler

18 An Overview of Biomedical, Biotechnological, and Industrial Applications of *Actinomycetes* 475
H. A. D. Ruwandeepika, G. C. P. Fernando, and T. S. P. Jayaweera

Editors and Contributors

About the Editors

Ravishankar V. Rai received his MSc (1980) and PhD (1989) from the University of Mysore, India. Currently, he is working in the Department of Studies in Microbiology, University of Mysore, Mysore. His current research and publications in food microbiology, microbial quorum sensing, microbial influenced corrosion and nanotechnology have been well received by the international scientific committee. He edited books with reputed publishers such as CRC Press and Wiley publications. Prof. Rai has received awards from UNESCO Biotechnology Action Council Programme (Visiting Fellow, 1996), UGC Indo-Israel Cultural Exchange Programme (1998), DBT Overseas Fellowship (2008), Indo-Hungarian Educational Exchange Programme Fellowship (2011) and INSA—bilateral exchange fellowship (2015), Incoming Fellowship (2017) from Cardiff University in the UK, and invitation from Mauritius Research Council, Mauritius (2018), to conduct collaborative research with renowned scientists from international universities. He visited Massey University, New Zealand, under Eric Ojala fund International Travel Fellowship in 2019.

Jamuna A. Bai is working as an Assistant Professor in JSS Academy of Higher Education & Research, Mysore. She completed her MSc and PhD in Microbiology from the University of Mysore, India. She pursued her PhD under the guidance of Dr. Ravishankar V. Rai. Her doctoral research was funded by NMPB, Govt of India, and Indian Council of Medical Research. As an ICMR Senior Research Fellow and for her doctoral research, she carried out studies on food safety, role of quorum sensing and biofilms in food-related bacteria and developed phytochemical based quorum-sensing inhibitors and anti-biofilm agents. She completed her postdoctoral research in UGC sponsored University with Potential Excellence Project on synthesis and biological application of functionalized nanoparticles from the University of Mysore. Her research work included application of functionalized nanomaterials and peptides as antimicrobials, biofilm inhibitors and anticancer agents. She has authored 30 research articles, reviews and book chapters and co-edited 6 books for CRC Press, USA. She has received grants and awards from IAFP, USA, and ICFMH to present her research findings at international conferences. She has been awarded the

Interstellar Initiative Early Career Investigator Funding jointly by the Japan Agency for Medical Research and Development (AMED) and the New York Academy of Sciences (NYAS) in 2021.

Contributors

Sergio Alonso-Fernández Área de Microbiología, Departamento de Biología Funcional, IUOPA, ISPA, Facultad de Medicina, Universidad de Oviedo, Oviedo, Spain

Dina H. Amin Department of Microbiology, Faculty of Science, Ain Shams University, Cairo, Egypt

Kenji Arakawa Unit of Biotechnology, Graduate School of Integrated Sciences for Life, Hiroshima University, Hiroshima, Japan

Yadollah Bahrami Medical Biology Research Center, Faculty of Medicine, Kermanshah University of Medical Sciences, Kermanshah, Iran
Department of Medical Biotechnology, School of Medicine, College of Medicine and Public Health, Flinders University, Adelaide, SA, Australia
Department of Medical Biotechnology, Faculty of Medicine, Kermanshah University of Medical Sciences, Kermanshah, Iran

Jamuna A. Bai Department of Microbiology, School of Life Sciences, JSS AHER, Mysore, Karnataka, India

Constance B. Bailey Department of Chemistry, University of Tennessee-Knoxville, Knoxville, TN, USA

Sasan Bouk Department of Medical Biotechnology, Faculty of Medicine, Kermanshah University of Medical Sciences, Kermanshah, Iran

Kimia Rafiei Buzhani Department of Medical Biotechnology, Faculty of Medicine, Kermanshah University of Medical Sciences, Kermanshah, Iran

Maria F. Carvalho CIIMAR—Interdisciplinary Centre of Marine and Environmental Research, University of Porto, Terminal de Cruzeiros do Porto de Leixões, Matosinhos, Portugal
ICBAS Institute of Biomedical Sciences Abel Salazar, University of Porto, Porto, Portugal

Yaser Delbari Department of Medical Biotechnology, Faculty of Medicine, Kermanshah University of Medical Sciences, Kermanshah, Iran

Erin E. Drufva Department of Chemistry, University of Tennessee-Knoxville, Knoxville, TN, USA

Heidi El-Gawahergy Department of Microbiology, Egyptian Drug Authority (EDA), Cairo, Egypt

Alaa F. Elsayed Department of Microbiology, Faculty of Science, Ain Shams University, Cairo, Egypt

Wei Fang Hubei Biopesticide Engineering Research Centre, Hubei Academy of Agricultural Sciences, Wuhan, China
National Biopesticide Engineering Research Centre, Wuhan, China

Maria de Fátima Carvalho Interdisciplinary Centre of Marine and Environmental Research (CIIMAR/CIMAR) University of Porto, Matosinhos, Portugal
ICBAS Institute of Biomedical Sciences Abel Salazar, University of Porto, Porto, Portugal

Gemma Fernández-García Área de Microbiología, Departamento de Biología Funcional, IUOPA, ISPA, Facultad de Medicina, Universidad de Oviedo, Oviedo, Spain

G. C. P. Fernando Department of Livestock Production, Faculty of Agricultural Sciences, Sabaragamuwa University of Sri Lanka, Belihuloya, Sri Lanka

Christopher M. M. Franco Department of Medical Biotechnology, School of Medicine, College of Medicine and Public Health, Flinders University, Adelaide, SA, Australia

Mariana Girão CIIMAR—Interdisciplinary Centre of Marine and Environmental Research, University of Porto, Terminal de Cruzeiros do Porto de Leixões, Matosinhos, Portugal
ICBAS Institute of Biomedical Sciences Abel Salazar, University of Porto, Porto, Portugal

Nathaly Gonzalez-Quiñonez Área de Microbiología, Departamento de Biología Funcional, IUOPA, ISPA, Facultad de Medicina, Universidad de Oviedo, Oviedo, Spain

T. S. P. Jayaweera Department of Livestock Production, Faculty of Agricultural Sciences, Sabaragamuwa University of Sri Lanka, Belihuloya, Sri Lanka

Elham Kakaei Department of Medical Biotechnology, Faculty of Medicine, Kermanshah University of Medical Sciences, Kermanshah, Iran

Chelliah Chenthis Kanisha Department of Nanotechnology, Noorul Islam Centre for Higher Education, Kumaracoil, Tamil Nadu, India

Shaoyong Ke Hubei Biopesticide Engineering Research Centre, Hubei Academy of Agricultural Sciences, Wuhan, China
National Biopesticide Engineering Research Centre, Wuhan, China

Nattakorn Kuncharoen Department of Plant Pathology, Faculty of Agriculture, Kasetsart University, Bangkok, Thailand

Pedro Leão Interdisciplinary Centre of Marine and Environmental Research (CIIMAR/CIMAR) University of Porto, Matosinhos, Portugal

Wen-Jun Li State Key Laboratory of Biocontrol, Guangdong Provincial Key Laboratory of Plant Resources and Southern Marine Science and Engineering Guangdong Laboratory (Zhuhai), School of Life Sciences, Sun Yat-Sen University, Guangzhou, People's Republic of China
State Key Laboratory of Desert and Oasis Ecology, Xinjiang Institute of Ecology and Geograph, Chinese Academy of Sciences, Urumqi, People's Republic of China

Yunzi Luo Frontier Science Center for Synthetic Biology and Key Laboratory of Systems Bioengineering (Ministry of Education), School of Chemical Engineering and Technology, Tianjin University, Tianjin, China
Department of Gastroenterology, State Key Laboratory of Biotherapy, West China Hospital, Sichuan University and Collaborative Innovation Center of Biotherapy, Chengdu, China
Collaborative Innovation Center of Chemical Science and Engineering (Tianjin), Tianjin University, Tianjin, China

Catarina Magalhães Interdisciplinary Centre of Marine and Environmental Research (CIIMAR/CIMAR) University of Porto, Matosinhos, Portugal
Department of Biology, Faculty of Sciences, University of Porto, Porto, Portugal

Artur Maier Microbial Biotechnology, Ruhr-Universität Bochum, Bochum, Germany

N. Manoharan Department of Biotechnology, Bharathidasan University, Tiruchirappalli, Tamilnadu, India

Angel Manteca Área de Microbiología, Departamento de Biología Funcional, IUOPA, ISPA, Facultad de Medicina, Universidad de Oviedo, Oviedo, Spain

Yaser Mohassel Department of Clinical Biochemistry, Faculty of Medicine, Kermanshah University of Medical Sciences, Kermanshah, Iran

Carolin Mügge Microbial Biotechnology, Ruhr-Universität Bochum, Bochum, Germany

Shuqing Ning Frontier Science Center for Synthetic Biology and Key Laboratory of Systems Bioengineering (Ministry of Education), School of Chemical Engineering and Technology, Tianjin University, Tianjin, China

Bohdan Ostash Department of Genetics and Biotechnology, Ivan Franko National University of Lviv, Lviv, Ukraine

Wasu Pathom-aree Department of Biology, Faculty of Science, Chiang Mai University, Chiang Mai, Thailand
Research Center in Microbial Diversity and Sustainable Utilization, Faculty of Science, Chiang Mai University, Chiang Mai, Thailand

Ravishankar V. Rai Department of Studies in Microbiology, University of Mysore, Mysore, Karnataka, India

Govindan Nadar Rajivgandhi State Key Laboratory of Biocontrol, Guangdong Provincial Key Laboratory of Plant Resources and Southern Marine Science and Engineering Guangdong Laboratory (Zhuhai), School of Life Sciences, Sun Yat-Sen University, Guangzhou, People's Republic of China
Department of Marine Science, Bharathidasan University, Tiruchirappalli, Tamil Nadu, India

G. Ramachandran Department of Marine Science, Bharathidasan University, Tiruchirappalli, Tamil Nadu, India

Pharada Rangseekaew Department of Biology, Faculty of Science, Chiang Mai University, Chiang Mai, Thailand

Adriana Rego Interdisciplinary Centre of Marine and Environmental Research (CIIMAR/CIMAR) University of Porto, Matosinhos, Portugal
ICBAS Institute of Biomedical Sciences Abel Salazar, University of Porto, Porto, Portugal

Yushuang Ren Department of Gastroenterology, State Key Laboratory of Biotherapy, West China Hospital, Sichuan University and Collaborative Innovation Center of Biotherapy, Chengdu, China

Inês Ribeiro CIIMAR—Interdisciplinary Centre of Marine and Environmental Research, University of Porto, Terminal de Cruzeiros do Porto de Leixões, Matosinhos, Portugal
ICBAS Institute of Biomedical Sciences Abel Salazar, University of Porto, Porto, Portugal

Chandra Risdian Microbial Strain Collection (MISG), Helmholtz Centre for Infection Research (HZI), Braunschweig, Germany
Research Unit for Clean Technology, National Research and Innovation Agency Republic of Indonesia (BRIN), Bandung, Indonesia

H. A. D. Ruwandeepika Department of Livestock Production, Faculty of Agricultural Sciences, Sabaragamuwa University of Sri Lanka, Belihuloya, Sri Lanka

Nasim Safaei Microbial Strain Collection (MISG), Helmholtz Centre for Infection Research (HZI), Braunschweig, Germany

Michael Steinert Institut für Mikrobiologie, Technische Universität Braunschweig, Braunschweig, Germany

Toshihiro Suzuki Department of Fermentation Science, Faculty of Applied Biosciences, Tokyo University of Agriculture, Tokyo, Japan

Tien T. Sword Department of Chemistry, University of Tennessee-Knoxville, Knoxville, TN, USA

Somboon Tanasupawat Department of Biochemistry and Microbiology, Faculty of Pharmaceutical Sciences, Chulalongkorn University, Bangkok, Thailand

Dirk Tischler Microbial Biotechnology, Ruhr-Universität Bochum, Bochum, Germany

R. T. V. Vimala Department of Biotechnology, Bharathidasan University, Tiruchirappalli, Tamilnadu, India

Kaimei Wang Hubei Biopesticide Engineering Research Centre, Hubei Academy of Agricultural Sciences, Wuhan, China
National Biopesticide Engineering Research Centre, Wuhan, China

Joachim Wink Microbial Strain Collection (MISG), Helmholtz Centre for Infection Research (HZI), Braunschweig, Germany

Tingting Wu Department of Gastroenterology, State Key Laboratory of Biotherapy, West China Hospital, Sichuan University and Collaborative Innovation Center of Biotherapy, Chengdu, China

Zhaoyuan Wu Hubei Biopesticide Engineering Research Centre, Hubei Academy of Agricultural Sciences, Wuhan, China
National Biopesticide Engineering Research Centre, Wuhan, China

Feiyang Xie PhD's Degree Program in Applied Microbiology, Faculty of Science, Chiang Mai University, Under the CMU Presidential Scholarship, Chiang Mai, Thailand
Department of Biology, Faculty of Science, Chiang Mai University, Chiang Mai, Thailand

Paula Yagüe Área de Microbiología, Departamento de Biología Funcional, IUOPA, ISPA, Facultad de Medicina, Universidad de Oviedo, Oviedo, Spain

Oleksandr Yushchuk Department of Genetics and Biotechnology, Ivan Franko National University of Lviv, Lviv, Ukraine

Yani Zhang Hubei Biopesticide Engineering Research Centre, Hubei Academy of Agricultural Sciences, Wuhan, China
National Biopesticide Engineering Research Centre, Wuhan, China

Actinobacteria from Marine Environments: A Unique Source of Natural Products

1

Mariana Girão, Inês Ribeiro, and Maria de Fátima Carvalho

Abstract

Nature is considered the primary source of new biotechnologically relevant molecules with application in different fields. The production of natural products by microorganisms is largely associated with the phylum Actinobacteria, which is gifted with a rich metabolic machinery that allows the production of a vast panoply of bioactive compounds. The demand for new natural products led to a recent focus on the marine environment. In fact, ocean bioprospection has opened new perspectives for the discovery of new drugs and other useful chemical structures due to the enormous unexplored biodiversity found in these environments.

This chapter focus on current knowledge on the diversity and distribution of Actinobacteria in marine environments and how adaptation of these microorganisms to the ocean can be favorable for the discovery of new bioactive molecules. New natural products discovered since 2017 from marine Actinobacteria as well as their chemical diversity and biotechnological potential are also highlighted. The combination of different isolation and cultivation methods, along with genetic approaches and in silico tools, is also mentioned here as beneficial strategies for obtaining marine actinobacterial strains and unveiling their full biotechnological potential.

Mariana Girão and Inês Ribeiro contributed equally to this work.

M. Girão · I. Ribeiro · M. de F. Carvalho (✉)
CIIMAR—Interdisciplinary Centre of Marine and Environmental Research, University of Porto, Terminal de Cruzeiros do Porto de Leixões, Matosinhos, Portugal

ICBAS Institute of Biomedical Sciences Abel Salazar, University of Porto, Porto, Portugal
e-mail: mariana.martins@ciimar.up.pt; iribeiro@ciimar.up.pt; mcarvalho@ciimar.up.pt

R. V. Rai, J. A. Bai (eds.), *Natural Products from Actinomycetes*, https://doi.org/10.1007/978-981-16-6132-7_1

Keywords

Marine Actinobacteria · Bioactive compounds · Corals · Sponges · Macroalgae · Bioprospection · Pharmaceuticals · Antifouling compounds · Biosurfactants · Genome mining · Biosynthetic gene clusters

1.1 Introduction

Nature has always been the answer to the constant and vital pursuit for novel biotechnologically relevant molecules, to feed both the pharmaceutical and industrial fields. Natural products (NPs) embody a large family of diverse chemical entities with a wide assortment of bioactivities. Due to their complexity and unique chemical structures and biosynthetic pathways, NPs and their derivatives have been an unmatched supplier of pharmaceutically and industrially relevant agents, representing an unrivalled source of key scaffolds for chemical molecules' development (Andersson et al. 2020). These small molecules can be produced by any living organism, being usually associated to secondary metabolism products. If in earlier times most of these substances were obtained from plants and animals, today, microbial NPs represent the backbone of most currently used compounds with biotechnological, pharmaceutical, and therapeutic applications, derived or inspired from nature (Busi and Pattnaik 2018). Microbes are indeed talented chemists: over the past few decades, more than 200,000 NPs have been reported (Busi and Pattnaik 2018), many mined from microbial secondary metabolism and exhibiting a wide spectrum of bioactivities. Within the bacterial domain, the ability to produce a broad range of biotechnologically relevant NPs is particularly noteworthy for the phylum Actinobacteria, and so more for the members of the order Actinomycetales, commonly referred to as actinomycetes. The enormous potential of these microorganisms can be illustrated by the fact that about two in each three antibiotics in current clinical use, and compounds with anticancer, antifungal, or antiparasitic properties, are Actinobacteria sourced (Carroll et al. 2020), with a particularly rich contribution of the *Streptomyces* genus that, by itself, is responsible for the production of more than 50% of these compounds (Carroll et al. 2020).

Despite the numerous NPs already described and characterized, the pursuit for novel bioactive compounds is still—now, more than ever—a major scientific goal. Aware of the unrivalled features of Actinobacteria as true antibiotic factories, during decades, drug bioprospection programs have centered their efforts on exploiting these bacteria from classic terrestrial sources: the so-called golden days of antibiotic discovery. Years of intensive exploration of terrestrial Actinobacteria translated in an increase in the re-isolation of known compounds, perhaps the major bottleneck in the field, and in the slowdown in the discovery of novel bioactive molecules (Andersson et al. 2020). This struggle for finding novel chemistry awakened the scientific community to the need of mining less explored sources, such as the marine environment (Bister et al. 2004). In fact, bioprospection of oceans has opened up new perspectives for the discovery of novel drugs and other chemical structures, due to the enormous unexplored biodiversity found in marine environments (Sayed et al. 2020). This unique biodiversity is a result of the high contrasting conditions encountered between terrestrial and marine environments, mainly in terms of temperature, light, pressure, salinity, oxygen, and nutrients availability, creating new opportunities for the generation of new chemical diversity (Lozupone and Knight 2007). The presence of Actinobacteria as true dwellers of marine environments was only recognized about 30 years ago with the discovery of *Rhodococcus marinonascens*, the first marine Actinobacteria to be characterized. These microorganisms have been shown to be physiologically adapted to live in marine environments as well as metabolically active for the production of relevant bioactive compounds (Jensen et al. 1991; Mincer et al. 2002).

Recent trends in the exploration of the diversity of marine Actinobacteria and their secondary metabolites are based on culture-independent and metagenomic approaches. Nonetheless, culture-dependent studies are still essential for studying the metabolic potential of Actinobacteria by allowing a direct and non-manipulated enzymatic expression and production of bioactive compounds (Vester et al. 2015).

This chapter describes the diversity and bioactive potential of marine Actinobacteria and also highlights the best and most used culture-dependent methods and in silico tools to study these microorganisms.

1.2 Marine Actinobacteria: Where Can We Find Them?

Actinobacteria are a phylum of Gram-positive bacteria widely distributed throughout our planet, representing one of the largest bacterial taxonomic units, both in number and diversity (Carroll et al. 2020). The presence and diversity of Actinobacteria in marine environments are still poorly explored; however, in recent years, this has been the subject of several studies (Bredholdt et al. 2007; Claverias et al. 2015; Duncan et al. 2014; Jose and Jha 2017; Liu et al. 2019c; Ribeiro et al. 2020). Microorganisms from this phylum are present in both pelagic and benthic environments, and they are usually found in symbiotic association with a wide range of marine organisms, such as sponges, corals, fish, and macroalgae, among others (Nathani et al. 2020; Penesyan et al. 2010). Furthermore, Actinobacteria have also been identified in various extreme marine ecosystems, including deep-sea regions as the Mariana Trench, hydrothermal vents, and polar marine waters (Bull and Goodfellow 2019; Abdel-Mageed et al. 2020; Zhang et al. 2020a; Sivasankar et al. 2018). Only relatively recently were Actinobacteria recognized to be indigenous of marine environments (Mincer et al. 2002). Several new species, genera, and even families of Actinobacteria have been isolated from marine sediments, seawater, and marine fauna (Subramani and Aalbersberg 2013; Subramani and Sipkema 2019). Studies on the diversity of Actinobacteria in marine environments allowed identifying six genera of obligate marine Actinobacteria: *Euzebya*, *Marihabitans*, *Marinitenerispora*, *Paraoerskovia*, *Salinispora*, and *Spinactinospora* (Chang et al.

2011; Kageyama et al. 2008; Khan et al. 2009; Kurahashi et al. 2010; Maldonado et al. 2005; Ng et al. 2019) (Fig. 1.1a). In addition, it is noteworthy the discovery of a new subclass, "*Candidatus Actinomarinales*," exclusively characterized by marine obligate Actinobacteria from the epipelagic zone (López-Pérez et al. 2020).

Marine Actinobacteria have been found in several marine matrices, with sediments, sponges, corals, and seawater being the sources with the greatest abundance and diversity of these microorganisms (Fig. 1.1b). Despite having a lower abundance of marine Actinobacteria, marine flora and organisms like tunicates, mollusks, marine fish, and sea cucumbers have been shown to host new genera of Actinobacteria not found in other matrices. However, this lower actinobacterial abundance may just be a reflection of the great underexploration of these marine sources (Fig. 1.1a).

1.2.1 Sediments and Seawater

Marine sediments are composed of organic and inorganic matter and support complex microbial communities with different ecological functions, including Actinobacteria (Patin et al. 2017). In fact, the isolation of Actinobacteria from marine sediments, including intertidal, subtidal, and deep-sea sediments, from diverse world regions, has been frequently reported (Jose and Jha 2017; Solano et al. 2009; Meena et al. 2019). For example, in a study conducted by us with a coastal marine sediment collected in northern Portugal, we were able to isolate 52 actinobacterial strains belonging to the genera *Arthrobacter*, *Actinomadura*, *Herbiconiux*, *Micromonospora*, *Nocardiopsis*, *Polymorphospora*, and *Streptomyces* (Ribeiro et al. 2020). Twenty-eight actinobacterial isolates affiliated to the genera *Streptomyces*, *Salinispora*, *Nocardiopsis*, *Verrucosispora*, *Micromonospora*, *Prauserella*, and *Promicromonospora* were recovered from sediments collected in the South China Sea (Yang and Song 2018). In a study conducted by Gozari et al. (2019), 168 actinobacterial strains were isolated from 14 sediment samples collected from the northern part of the Oman Sea, with these strains being distributed among four main families: Streptomycetaceae, Micromonosporaceae, Nocardiaceae, and Pseudonocardiaceae.

Several new Actinobacteria genera have also been isolated from marine sediments, like the genera *Actinomarinicola*, *Actinotalea*, *Aestuariimicrobium*, *Arenivirga*, *Demequina*, *Flaviflexus*, *Haloactinomyces*, *Halopolyspora*, *Litorihabitans*, *Marinactinospora*, *Mariniluteicoccus*, *Marinitenerispora*, *Marisediminicola*, *Miniimonas*, *Paraoerskovia*, *Sciscionella*, *Sediminihabitans*, *Sediminivirga*, and *Spinactinospora* (He et al. 2020; Tian et al. 2009a, b; Khan et al. 2009; Chang et al. 2011; Ng et al. 2019; Hamada et al. 2012, 2017, 2019; Zhang et al. 2014, 2016a; Du et al. 2013; Lai et al. 2014, 2017; Li et al. 2010; Ue et al. 2011; Jung et al. 2007; Yi et al. 2007).

Culture-dependent and culture-independent studies have also demonstrated the presence of Actinobacteria in seawater, including deep-sea water (Wang et al. 2017b, 2020; Nimnoi and Pongsilp 2020; Sheikh et al. 2019). In a study conducted

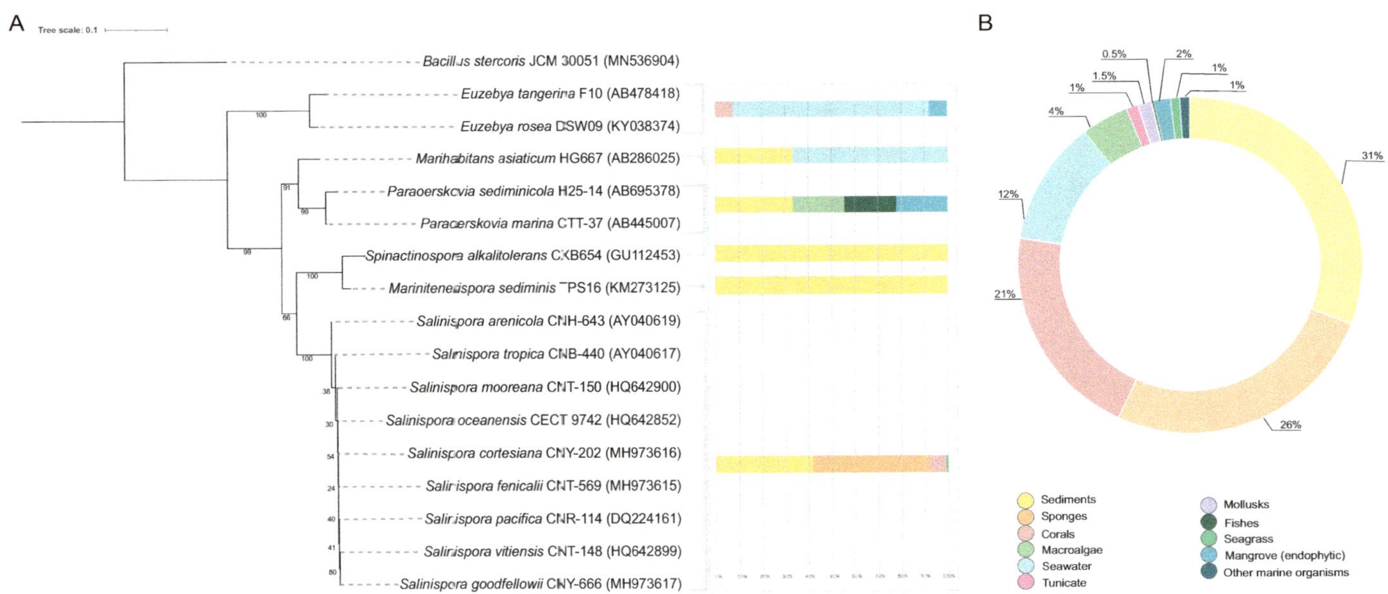

Fig. 1.1 Actinobacteria in the marine environment. (**a**) Phylogenetic tree of obligate marine Actinobacteria strains and their abundance in various marine sources. In total, 17 sequences were aligned using the MUSCLE alignment tool from Geneious software package (version 11.1.4, Biomatters Ltd, Auckland, New Zealand). The phylogenetic tree was built using the maximum likelihood method with 1000 bootstraps, using the MEGA software (Molecular Evolutionary Genetics Analysis) version 11.0. The most suitable nucleotide evolutionary model was calculated based on the lowest Akaike Information Criterion (AIC) using MEGA Version 11.0, in which the Tamura-Nei model revealed to be the most suitable one for the selected data. *Bacillus stercoris* JCM 30051 was used as an outgroup. Scale bar corresponds to 0.1 substitutions per nucleotide position. (**b**) Distribution of Actinobacteria by various marine sources based on 10298 16S rRNA gene sequences retrieved from GenBank (16S ribosomal RNA sequences database). To perform a selective search, it was used as keywords "16S ribosomal RNA and Actinobacteria" combined with other keywords related with the marine environment, like "marine sediment," "sponge," "coral," "macroalgae," "algae," "seaweed," "seagrass," "endophytic mangrove," "marine fish," "mollusk," "tunicate," "ascidian," "sea cucumber," and "seawater"

by Chen et al. (2016), the genera *Brevibacterium*, *Cellulomonas*, *Dietzia*, *Kocuria*, *Microbacterium*, *Micrococcus*, *Mycobacterium*, *Nocardiopsis*, *Rhodococcus*, *Saccharopolyspora*, *Streptomyces*, *Tsukamurella*, and *Verrucosispora* were isolated from deep-sea water samples collected along the Southwest Indian Ridge (Chen et al. 2016). Kai et al. (2017) studied the bacterial diversity along the water column of the South Atlantic Ocean in 17 water samples, showing the presence of Actinobacteria of the genera *Gordonia*, *Microbacterium*, *Micrococcus*, and *Mycobacterium* (Kai et al. 2017). Some studies also identified new genera of Actinobacteria in seawater, like the genera *Marihabitans*, *Oceanitalea*, *Ornithinibacter*, *Ponticoccus*, *Salinibacterium*, *Serinicoccus*, and *Thalassiella* (Fu et al. 2012; Han et al. 2003; Kageyama et al. 2008; Lee et al. 2016a; Lee and Lee 2008; Xiao et al. 2011; Yi et al. 2004).

1.2.2 Marine Fauna

1.2.2.1 Corals

Corals are marine invertebrates belonging to the phylum Cnidaria that typically live in colonies with several identical individuals. Coral reefs hold a valuable biological diversity of prokaryotic communities distributed by coral mucus, tissues, and the calcium carbonate skeleton (Mahmoud and Kalendar 2016). Actinobacteria have been identified as part of the bacterial communities associated with diverse corals and/or their mucus (Zhang et al. 2013; Lampert et al. 2008; Nithyanand and Pandian 2009). A study conducted by Lampert et al. (2006) revealed that Actinobacteria comprised 23% of the cultivable bacterial community associated with the mucus of the Red Sea coral *Fungia scutaria*. Culture-independent studies showed that Actinobacteria were one of the most dominant phyla found in the black corals *Antipathes ceylonensis* and *Antipathes dichotoma* from the South China Sea (Liu et al. 2018). New actinobacterial species have been isolated from diverse corals. As example, the species *Corynebacterium maris*, *Glutamicibacter mishrai*, *Janibacter corallicola*, *Janibacter alkaliphilus*, *Myceligenerans cantabricum*, *Nocardiopsis coralliicola*, *Prauserella coralliicola*, *Pseudokineococcus galaxeicola*, *Rhodococcus electrodiphilus*, and *Saccharopolyspora coralli* were isolated from the corals *Fungia granulosa*, *Favia veroni*, *Anthogorgia* sp., *Acropora gemmifera*, *Caryophyllidae* sp., *Menella praelonga*, *Galaxea* sp., *Galaxea fascicularis*, and *Porites* sp., respectively (Ben-Dov et al. 2009; Das et al. 2020; Kageyama et al. 2007; Li et al. 2012a, b, 2020; Ramaprasad et al. 2018; Wu et al. 2014; Zhou et al. 2020; Sarmiento et al. 2015).

1.2.2.2 Sponges

Marine sponges are the oldest multicellular organisms in the world and are sessile invertebrates that feed on planktonic organisms by filtration (Hentschel et al. 2012; Van Soest et al. 2012). Sponges establish symbiotic relationships with many microorganisms in such a way that microbial biomass may constitute up to 40% of the sponge's total volume (Webster and Taylor 2012). Within the bacterial realm associated with marine sponges, Actinobacteria are one of the most common phyla living in symbiosis with these organisms (Schmitt et al. 2012; Abdelmohsen et al. 2014a; Dat et al. 2018; Najafi et al. 2018). Several studies on the characterization of the microbiome of marine sponges have demonstrated the wide abundance and diversity of Actinobacteria associated with them. For example, the marine sponges *Haliclona* sp., *Callyspongia* sp., and *Desmacella* sp. revealed to be excellent hosts for Actinobacteria, harboring 21 actinobacterial genera, of which *Microbacterium*, *Pseudonocardia*, *Streptomyces*, *Kocuria*, *Aeromicrobium*, *Brachybacterium*, and *Nocardiopsis* represented about 82% of the total actinobacterial strains (Liu et al. 2019d). A study on the phylogenetic diversity of Actinobacteria associated with the marine sponge *Hymeniacidon perlevis*, collected in the South China Sea, identified several strains of the genera *Gordonia*, *Mycobacterium*, *Nocardia*, *Rhodococcus*, and *Streptomyces*, having also reported for the first time the genus *Salinispora* in a marine sponge (Sun et al. 2010). Studies on the bacterial diversity of various marine sponges led to the isolation of 25 new actinobacterial strains, affiliated with 14 genera (Kämpfer et al. 2014, 2015; Khan et al. 2010; Li et al. 2019b, d, e, f, g; de Menezes et al. 2015, 2017, 2019; Nouioui et al. 2017; Olson et al. 2007; Pimentel-Elardo et al. 2008, 2009; Silva et al. 2016; Souza et al. 2017; Supong et al. 2013a, b; Thawai et al. 2017; Zhang et al. 2006b, 2012). Among these, the species *Spongiactinospora rosea*, recovered from the marine sponge *Craniella* sp. obtained from the South China Sea, was found to represent a new genus and species within the family Streptosporangiaceae (Li et al. 2019b).

1.2.2.3 Tunicates, Mollusks, Marine Fishes, and Other Organisms

Actinobacteria are known to establish symbiotic relationships with various marine invertebrates, like tunicates and mollusks, where they contribute to the health and defense of these organisms by enlarging their chemical arsenal (Peraud et al. 2009).

Tunicate organisms are divided in three classes: Ascidiacea, Appendicularia, and Thaliacea (Holland 2016). The class Ascidiacea is the most diverse and abundant one, with over 3000 species reported (Chen et al. 2018b). It integrates organisms commonly known as ascidians, which are filter-feeding organisms that live attached to rocks or other hard marine surfaces (Buedenbender et al. 2017). Ascidians are known to harbor rich bacterial communities, including members of the phylum Actinobacteria (Schreiber et al. 2016; Chen et al. 2018b). Rare Actinobacteria have also been recovered from these organisms, like the genera *Salinispora* and *Verrucosispora*, identified in the tissues of the colonial ascidian *Eudistoma toealensis* (Steinert et al. 2015), and the genera *Actinoalloteichus*, *Kocuria*, *Micrococcus*, *Micromonospora*, *Mycobacterium*, and *Rhodococcus* isolated from the ascidian *Styela clava* (Chen et al. 2019b).

Efforts to discover ascidian-associated Actinobacteria have led to the isolation of several novel species, like *Gordonia didemni*, obtained from the marine ascidian *Didemnum* sp. (de Menezes et al. 2016), *Streptomyces hyaluromycini*, isolated from the ascidian species *Molgula manhattensis* (Harunari et al. 2016), and *Aeromicrobium halocynthiae*, associated to the marine ascidian *Halocynthia roretzi* (Kim et al. 2010).

Mollusks are invertebrate organisms of the phylum Mollusca that includes the subgroups gastropod (snails, slugs, limpets, and conches), bivalve (mussels, clams, and shellfish), and cephalopod (squid, octopus, and cuttlefish) (Haszprunar and Wanninger 2012; Giribet et al. 2006). Members of this phylum are soft-body organisms that live mostly in marine environments, from coastal waters to the deepest areas, but some groups can also live in freshwater and terrestrial environments (Kershaw 1983). Several studies have reported the presence of Actinobacteria in mollusks: the genera *Dietzia* and *Brevibacterium* were identified in the viscera of the gastropod species *Gibbula umbilicalis*, *Monodonta lineata*, *Nucella lapillus*, and *Patella intermedia* (Pratheepa et al. 2014, 2016); a study on the biodiversity of Actinobacteria in three conical snails, *Conus pulicarius*, *Conus rolani*, and *Conus tribblei*, identified 16 actinobacterial genera, of which strains of the genera *Brevibacterium*, *Gordonia*, *Microbacterium*, and *Streptomyces* were isolated (Peraud et al. 2009). Three novel Actinobacteria species were isolated from mollusks: (1) *Corynebacterium atrinae*, isolated from the gastrointestinal tract of the pen shell *Atrina pectinate*; (2) *Actinomyces haliotis*, obtained from the gut of the abalone *Haliotis discus*; and (3) *Nocardia crassostreae*, recovered from the Pacific oyster *Crassostrea gigas* (Ben-Dov et al. 2009; Friedman et al. 1998; Hyun et al. 2014).

Actinobacteria have also been reported to be part of the microbiome of diverse fish species (Sanchez et al. 2012; Bakke et al. 2015). For example, a culture-independent study showed that Actinobacteria was part of the microbiome of the cod larva *Gadus morhua* and of the juvenile rainbow trout *Oncorhynchus mykiss* (Bakke et al. 2015; Michl et al. 2017). Novel Actinobacteria have been isolated from diverse fish, like the species *Aeromicrobium piscarium*, isolated from the intestinal tract of the spiny head croaker *Collichthys lucidus* (Zhao et al. 2020); *Mycobacterium shottsii*, obtained from the striped bass *Morone saxatilis* (Rhodes et al. 2003); *Mycobacterium stephanolepidis*, retrieved from the marine teleost fish *Stephanolepis cirrhifer* (Fukano et al. 2017); and *Serinibacter salmoneus*, isolated from the intestine of the Japanese whiting *Sillago japonica* (Hamada et al. 2009).

Aside from tunicates, mollusks, and fish, Actinobacteria have also been found associated with sea cucumbers. Sea cucumbers are echinoderm organisms that ingest organic matter from sediments, including bacteria, and have an important ecological role in the recycling of nutrients from the benthic ecosystem (Kim et al. 2017b). Several Actinobacteria associated with sea cucumbers have been reported. Actinobacteria belonging to the genera *Brachybacterium*, *Cellulosimicrobium*, *Glutamicibacter*, *Microbacterium*, *Micrococcus*, *Nocardiopsis*, and *Streptomyces* were isolated from the sea cucumber *Apostichopus japonicus* (Chen et al. 2019a), and a novel actinobacterial species, *Euzebya tangerine*, was isolated from the abdominal epidermis of the sea cucumber *Holothuria edulis* (2010).

1.2.3 Marine Flora: Macroalgae, Seagrass, and Mangrove

Macroalgae, or seaweed, are a group of photosynthetic organisms that are divided according to their pigments—green, red, or brown (Stincone and Brandelli 2020). These organisms host a great diversity of bacteria that may differ according to the season, species, and thallus structure (Valliappan et al. 2014). Several symbiotic relationships, such as the exchange of nutrients, minerals, and secondary metabolites, can be developed between algae and their endophyte and epiphytic bacterial communities (Egan et al. 2013). Macroalgae are known to host a high diversity of Actinobacteria with biotechnological potential (Egan et al. 2013; Uzair et al. 2018). In a study conducted by us, we isolated 90 actinobacterial strains, belonging to the genera *Isoptericola*, *Rhodococcus*, *Nonomuraea*, *Nocardiopsis*, *Microbispora*, *Microbacterium*, and *Streptomyces*, from the macroalgae *Laminaria ochroleuca* collected in a rocky shore in northern Portugal, with many of these strains showing high bioactivities (Girão et al. 2019). The Actinobacteria genera *Agrococcus*, *Arthrobacter*, *Brachybacterium*, *Citricoccus*, *Kocuria*, *Labedella*, *Microbacterium*, *Micrococcus*, *Rhodococcus*, *Salinibacterium*, and *Sanguibacter* were identified in the macroalgae species *Adenocystis utricularis*, *Monostroma hariotii*, *Iridaea cordata*, *Phycodrys antarctica*, *Pyropia endiviifolia*, and *Plocamium cartilagineum* (Leiva et al. 2015). A study on the diversity of cultivable bacteria associated with the surface of three Antarctic macroalgae, *Adenocystis utricularis*, *Iridaea cordata*, and *Monostroma hariotii*, led to the isolation of actinobacterial strains belonging to the Micrococcaceae and Microbacteriaceae families (Alvarado et al. 2018). Novel Actinobacteria have also been isolated from macroalgae, as is the case of the species *Agrococcus jejuensis* (Lee 2008), *Amycolatopsis antarctica* (Wang et al. 2018), and *Phycicoccus jejuensis* (Lee 2006).

Aside from macroalgae, Actinobacteria have also been found in association with other marine flora, such as seagrass. Seagrass is a group of flowering sea plants that colonize tidal and subtidal regions (Valliappan et al. 2014). These plants usually harbor a rich bacterial diversity, in which Actinobacteria are well represented (Wu et al. 2012). Culture-independent techniques have shown that Actinobacteria are part of the microbial community of the seagrass *Zostera marina* (Jensen et al. 2007). The seagrass *Halodule uninervis* was found to host several actinobacterial strains affiliated with the genera *Kocuria*, *Arthrobacter*, *Ornithinimicrobium*, and *Corynebacterium* (Bibi et al. 2018). Rare actinobacterial endophytes belonging to the genera *Saccharomonospora* and *Kocuria* were isolated from root tissues of the seagrass *Cymodocea serrulata* (Jose et al. 2014). Ten Actinobacteria genera (*Actinomycetospora*, *Glycomyces*, *Microbacterium*, *Micromonospora*, *Mycobacterium*, *Nocardiopsis*, *Nonomuraea*, *Saccharomonospora*, *Streptomyces*, and *Verrucosispora*) were isolated from the seagrass *Thalassia hemprichii* (Wu et al. 2012).

Actinobacteria have also been associated with mangroves, which are unique ecosystems composed of salt-tolerant trees adapted to different coastal conditions, located in tropical and subtropical regions (Valliappan et al. 2014). A study characterizing the microbiome of the mangrove plant *Rhizophora stylosa* revealed that Actinobacteria was one of the most abundant phyla in the community (Purahong et al. 2019). Twenty-eight genera of endophytic Actinobacteria were isolated from five mangrove plants collected at the Guangxi Beilun Estuary National Nature Reserve, in China, some of which having been identified as potential new species of the genera *Amnibacterium*, *Marmoricola*, *Mycobacterium*, *Nocardioides*, and *Streptomyces* (Jiang et al. 2018). Several novel endophytic Actinobacteria of mangrove plants have been reported, like the species *Friedmanniella endophytica*, *Jishengella endophytica*, *Marmoricola endophyticus*, *Micromonospora sonneratiae*, *Micromonospora avicenniae*, and *Streptomyces avicenniae*, isolated from the rhizosphere of the mangroves *Kandelia candel*, *Acanthus ilicifolius*, *Thespesia populnea*, *Sonneratia apetala*, and *Avicennia marina* (Xiao et al. 2009; Jiang et al. 2017; Tuo et al. 2016; Li et al. 2013a, b; Xie et al. 2011).

1.3 Bioactive Metabolites Produced by Marine Actinobacteria

The intricacy and multiplicity of actinobacterial secondary metabolism is the outcome of exposure to particular environmental conditions and interaction with other life forms (Barka et al. 2016). As true marine dwellers (Bull et al. 2005), marine Actinobacteria are exposed to high salt concentrations, oligotrophy, different levels of hydrostatic pressure (including extreme ones observed in abyssal habitats), and a soaring assortment of temperatures and light intensity, which forces them to adapt to survive. Such adaptations may translate into molecular diversity and novelty (De Carvalho and Fernandes 2010), and, in this regard, marine Actinobacteria are considered a gold mine of not only new drugs—that are so needed to face urgent problems such as the antibiotic multiresistant bacteria crisis or the increasing incidence of cancer pathologies and other modern-society diseases (Andersson et al. 2020; Busi and Pattnaik 2018)—but also of a wide range of metabolites with biotechnological value able to feed many industries.

1.3.1 Pharmaceuticals

The immense potential of Actinobacteria to produce metabolites with pharmaceutical applications can be illustrated by the fact that of the 242 novel marine microbial compounds described in 2017, more than 60% were produced by strains belonging to this phylum (Carroll et al. 2019). Since the recognition of the bioactive potential and value of marine Actinobacteria, many compounds from their metabolism have been isolated, described, and exploited, including molecules with antibacterial (Le et al. 2016; Liang et al. 2016; Zhang et al. 2016b), antifungal (Pimentel-Elardo et al. 2010; Raju et al. 2010), cytotoxic (Fu et al. 2016; Kim et al. 2017a; Sun et al. 2015), anticancer (Shin et al. 2016; Yang et al. 2013; Ye et al. 2016), antitumor (Gärtner et al. 2011; Hohmann et al. 2009; Lu et al. 2012), anti-inflammatory (Hassan et al. 2015; Kondratyuk et al. 2012; Lee et al. 2016b), antiviral (Kim et al. 2016; Raveh et al. 2013; Strand et al. 2014), antituberculosis (Bürstner et al.

2015; Renner et al. 1999), antimalarial (Bürstner et al. 2015; Cheng et al. 2015; Schulze et al. 2015), immunosuppressive (Asolkar et al. 2009; Oh et al. 2007), antiparasitic (Abdelmohsen et al. 2014b; Yin et al. 2016), antiplasmodial (Maskey et al. 2004; Rakotondraibe et al. 2015), and neuritogenic (Itoh et al. 2003) properties.

Such molecules can play a major role in fighting diseases that affect human health, such as cancer pathologies, for which effective and less aggressive treatments are urgently needed. In the past few years, several anticancer compounds have been isolated from marine-sourced Actinobacteria. Examples include dentigerumycin E, a new cyclic hexapeptide with antiproliferative and antimetastatic activities against several human cancer cells, isolated from a co-culture of a marine *Streptomyces* sp. with a *Bacillus* sp. strain (Shin et al. 2018); streptodepsipeptides P11A and P11B, obtained from a marine *Streptomyces* sp., and with capacity of inhibiting proliferation of glioma cells by blocking glioma cell cycle and downregulating expression of glioma metabolic enzymes (Ye et al. 2017); neaumycin B, a potent polycyclic macrolide inhibitor of glioblastoma, uncovered as a result of the analysis of the metabolome of a marine-derived *Micromonospora* (Kim et al. 2018); and the novel polyketides ansaseomycins A and B, with activity against chronic myelogenous leukemia cells, discovered from a marine *Streptomyces* by a genome mining coupled with a heterologous expression approach (Liu et al. 2019b).

Another worldwide problem that may be addressed by marine actinobacterial secondary metabolism is the emergence of infections caused by antibiotic multiresistant microorganisms. Finding novel antibiotics effective against human pathogenic microorganisms is also in urgent demand, and recent examples of such molecules can be illustrated by the compounds salinaphthoquinones, novel antibacterial compounds isolated from a marine *Salinispora arenicola* strain, recovered from sediments, with activity against *Staphylococcus aureus* and *Enterococcus faecalis* (da Silva et al. 2019); taromycins A and B, two novel peptides displaying potent activity against methicillin-resistant *Staphylococcus aureus* and vancomycin-resistant *Enterococcus faecium* clinical isolates, discovered by awakening and improving the expression of a cryptic biosynthetic gene cluster from a marine *Saccharomonospora* sp. (Reynolds et al. 2018); anthracimycin B, isolated from a deep-sea strain of *Streptomyces cyaneofuscatus*, exhibiting potent antimicrobial activity against Gram-positive bacteria (Rodríguez et al. 2018); and iseolides A–C, three new glycosylated macrolides isolated from a coral-associated *Streptomyces* sp., with potent antifungal activity against the human pathogens *Candida albicans* and *Trichophyton rubrum* (Zhang et al. 2020b).

Several marine actinobacterial bioactive metabolites with potential application in the treatment of other human diseases were also recently uncovered, like the antituberculosis compounds desertomycin G, a macrolide isolated from a marine *Streptomyces althioticus* (Braña et al. 2019), and the cyclodepsipeptide atratumycin retrieved from the metabolism of a deep-sea *Streptomyces atratus* through a genome mining approach (Sun et al. 2019); the antiviral compound streptodepsipeptide SV21, a novel valinomycin analogue derived from a sea cucumber-associated *Streptomyces* sp., with potent antiviral activity against hepatitis C virus (Wibowo et al. 2021); and two new fridamycins H and I isolated from a marine

Actinokineospora spheciospongiae exhibiting inhibitory activity towards *Trypanosoma brucei*, a human parasite and cause of the African trypanosomiasis, or sleeping sickness (Tawfike et al. 2019). Among the various examples of NPs with pharmaceutical relevance produced by marine Actinobacteria, perhaps the main example of success is salinosporamide A, a β-lactone proteasome inhibitor, isolated from a *Salinispora tropica* strain recovered from a marine sediment, that is currently in clinical phase III trials for the treatment of multiple myeloma, solid tumors, and lymphoma (Marx and Burton 2018).

Glancing in more detail the literature from the last 5 years, since 2017, numerous bioactive secondary metabolites have been unearthed from marine-sourced Actinobacteria (Table 1.1). Most of these new biologically active compounds integrate the polyketide antibiotic class, but a chemical multiplicity, from macrolactams to peptides and alkaloids, has been extensively described. Concerning the actinobacterial diversity behind this chemical novelty, more than half of the compounds were retrieved from *Streptomyces* strains, followed by members of the genus *Micromonospora*. Nevertheless, rarer genera such as *Actinokineospora*, *Actinoalloteichus*, *Actinomadura*, *Kocuria*, *Nesterenkonia*, *Nonomuraea*, *Nocardiopsis*, *Pseudonocardia*, *Saccharomonospora*, *Saccharopolyspora*, *Streptomonospora*, *Streptosporangium*, *Salinispora*, and *Verrucosispora* proved to be also a relevant source of bioactive entities worthy of further exploration. These strains were mostly recovered from coastal and deep-sea sediments, but corals, sponges, macroalgae, sea cucumbers, shellfish, tunicates, prawns, seawater, and even a marine mammal were also sources of these microorganisms. Other studies also include mollusks (Lin et al. 2014), polychaetae (Pérez et al. 2016), jellyfish (Hassan et al. 2015), starfish (Shin et al. 2014), and fish (Yin et al. 2016) as source of novel Actinobacteria-derived bioactive compounds. Apart from these compounds showing a defined bioactivity, other novel molecules with still no evident biological activity or under study were also recently discovered, including nocarazepine A, a new diketopiperazine from a *Nocardiopsis* strain isolated from a Cnidaria (Zhou et al. 2017); albisporachelin, a new siderophore (Wu et al. 2018), and saccharopolytide A, a new cyclic tetrapeptide, from a deep-sea *Saccharopolyspora* strain (Xie et al. 2018); two new piperazine-triones (Xu et al. 2019b); and chartrenoline, a novel alkaloid, isolated from marine *Streptomyces* strains (Liu et al. 2019a). These numerous and promising results on marine Actinobacteria prospection reinforce the idea that investing in the exploration of untapped marine ecological niches may be a rewarding route to discover novel pharmaceutically relevant NPs.

1.3.2 Other Bioactive Compounds

Apart from pharmaceuticals, marine-sourced Actinobacteria also encode a panoply of other metabolites with applications in different fields. With a notorious ecological facet, these microorganisms represent a wealthy resource of biotechnological tools for several industries and agriculture. Focusing, for instance, on marine economic

Table 1.1 Novel pharmaceutically relevant bioactive metabolites isolated from marine Actinobacteria since 2017

Source	Compound	Chemical class	Bioactivity	Microorganism	Ref
Sediments	2-Epi-anthracimycin	Polyketide	Cytotoxic	*Streptomyces* sp.	Fukuda et al. (2020)
	Ala-geninthiocin	Thiopeptide	Antibacterial, cytotoxic	*Streptomyces* sp.	Iniyan et al. (2019)
	Anandins A and B	Alkaloid	Cytotoxic	*Streptomyces anandii*	Zhang et al. (2017b)
	Ashimides A and B	Cyclopeptides	Cytotoxic	*Streptomyces* sp.	Shi et al. (2019)
	Atratumycin	Cyclodepsipeptide	Antituberculosis	*Streptomyces atratus*	Sun et al. (2019)
	Aurodox FW05328-1	Macrolactam	Antitumor	*Micromonospora* sp.	Nie et al. (2018)
	Borrelidins C–E	Macrolide	Antibacterial, cytotoxic	*Nocardiopsis* sp.	Kim et al. (2017a)
	Cebulantin	Lanthipeptide	Antibacterial	*Saccharopolyspora cebuensis*	Moon et al. (2019)
	Cytosaminomycin E	Pyrimidine	Cytotoxic	*Streptomyces* sp.	Xu et al. (2019a)
	Dentigerumycin E	Hexapeptide	Anticancer	*Streptomyces* sp.	Shin et al. (2018)
	Divergolides T–W	Macrolide	Cytotoxic	*Streptomyces* sp.	Zhou et al. (2019b)
	Dionemycin	Alkaloid	Antibacterial, cytotoxic	*Streptomyces* sp.	Song et al. (2020)
	Dokdolipids A–C	Fatty acid	Cytotoxic	*Actinoalloteichus hymeniacidonis*	Choi et al. (2019)
	Fluvirucin B6	Macrolactam	Antibacterial	*Nocardiopsis* sp.	Leutou et al. (2018)
	Fradiamines A and E	Siderophore	Antibacterial	*Streptomyces fradiae*	Takehana et al. (2017)
	Grincamycins B–H	Glycoside	Cytotoxic	*Streptomyces lusitanus*	Huang et al. (2012); Zhu et al. (2017)
	Hexaricins	Polyketide	Antioxidant	*Streptosporangium* sp.	Gao et al. (2018)
	Iakyricidins A–D	Piericidin	Antiproliferative	*Streptomyces iakyrus*	Li et al. (2019a)
	Isomethoxyneihumicin	Diketopiperazine	Cytotoxic	*Nocardiopsis alba*	Fukuda et al. (2017)

(continued)

Table 1.1 (continued)

Source	Compound	Chemical class	Bioactivity	Microorganism	Ref
	JBIR-150	Macrolactam	Cytotoxic	*Streptomyces* sp.	Kim et al. (2020a)
	Kendomycins B–D	Polyketide	Antibacterial, cytotoxic	*Verrucosispora* sp.	Zhang et al. (2019c)
	Lactoquinomycin E	Medermycin	Cytotoxic	*Streptomyces* sp.	Zhou et al. (2019a)
	Litoralimycins A and B	Thiopeptide	Cytotoxic	*Streptomonospora* sp.	Khodamoradi et al. (2020)
	Merochlorins E and F Meroindenon	Meroterpenoids	Antibacterial	*Streptomyces* sp.	Ryu et al. (2019)
	Mersaquinone	Tetracene	Antimicrobial	*Streptomyces* sp.	Kim et al. (2020a)
	Microsporanates A–F	Glycoside	Antibacterial	*Micromonospora harpali*	Gui et al. (2017)
	Neoabyssomicins	Polyketide	Antibacterial, antiviral	*Streptomyces koyangensis*	Huang et al. (2018)
	Neo-actinomycins A and B	Chromopeptide	Cytotoxic, antibacterial	*Streptomyces* sp.	Wang et al. (2017a)
	Neothioviridamide	Polythioamide	Cytotoxic	*Streptomyces* sp.	Kawahara et al. (2018)
	Nivelactam B	Macrolactam	Cytotoxic, antifungal	*Streptomyces* sp.	Chen et al. (2018a)
	Nocarterphenyls A–C	*p*-Terphenyls	Cytotoxic	*Nocardiopsis* sp.	Wang et al. (2019a)
	Nocarimidazoles A and B	Alkanoylimidazole	Antimicrobial	*Nocardiopsis* sp.	Karim et al. (2020); Leutou et al. (2015)
	Nocazines F and G	Diketopiperazine	Antibacterial, cytotoxic	*Nocardiopsis* sp.	Sun et al. (2017)
	Pactamides A–F	Macrolactam	Cytotoxic	*Streptomyces pactum*	Saha et al. (2017)
	Paulomycin G	Glycoside	Cytotoxic, antibacterial	*Micromonospora matsumotoense*	Sarmiento-Vizcaíno et al. (2017)
	Phocoenamicins	Spirotetronate	Antibacterial, antituberculosis	*Micromonospora* sp.	Pérez-Bonilla et al. (2018)

	Pyrazolofluostatins A–C	Polyketide	Antioxidant	*Micromonospora rosaria*	Zhang et al. (2017a)
	Rakicidins G–I	Depsipeptide	Antibacterial, cytotoxic	*Micromonospora chalcea*	Chen et al. (2018c)
	Saccharoquinoline	Alkaloid	Cytotoxic	*Saccharomonospora* sp.	Le et al. (2019)
	Salinaphthoquinones A–E	Macrolide	Antibacterial	*Salinispora arenicola*	da Silva et al. (2019)
	Sporalactams	Macrolide	Antituberculosis	*Micromonospora* sp.	Williams et al. 2017)
	Steffimycin E	Tetracyclic quinone	Anti-mycobacterial	*Streptomyces* sp.	Koyama et al. (2020)
	Streptcytosines F–O	Pyrimidine	Cytotoxic	*Streptomyces* sp.	Xu et al. (2019a)
	Streptodepsipeptides P11A and P11B	Peptide	Antitumor	*Streptomyces* sp.	Ye et al. (2017)
	Streptoglutarimides A–J	Glutarimide	Antimicrobial	*Streptomyces* sp.	Zhang et al. (2019a)
	Taromycins A and B	Lipopeptide	Antibacterial	*Saccharomonospora* sp.	Reynolds et al. (2018)
	Terrosamycins A and B	Polycyclic polyether	Antibacterial, anticancer	*Streptomyces* sp.	Sproule et al. (2019)
Corals	Anthracimycin B	Polyketide	Antibacterial	*Streptomyces cyaneofuscatus*	Rodríguez et al. (2018)
	Aranciamycin K	Anthracycline	Cytotoxic	*Streptomyces* sp.	Cong et al. (2019)
	Iseolides A–C	Macrolide	Antifungal	*Streptomyces* sp.	Zhang et al. (2020b)
	Isotirandamycin B	Polyketide	Antibacterial	*Streptomyces* sp.	Cong et al. (2019)
	Lobophorin K	Macrolide	Cytotoxic	*Streptomyces* sp.	Braña et al. (2017a, b)
	Nesteretal A	Polyketide	Anticancer	*Nesterenkonia halobia*	Xie et al. (2019)
	Nocarimidazoles C and D	Alkanoylimidazole	Antimicrobial	*Kocuria* sp.	Karim et al. (2020)
	Pteridicacids C–G	Polyketide	Antibacterial	*Streptomyces* sp.	Nong et al. (2017)

(continued)

Table 1.1 (continued)

Source	Compound	Chemical class	Bioactivity	Microorganism	Ref
Sponges	Actinozine A	Diketopiperazine	Antimicrobial	*Streptomyces* sp.	Shaala et al. (2019)
	Fridamycins H and I	Hydroxyquinones	Antiparasitic	*Actinokineospora spheciospongiae*	Tawfike et al. (2019)
	Heliomycin	Anthraquinone	Anticancer	*Streptomyces* sp.	Abdelfattah et al. (2018)
	IB-96212	Macrolide	Cytotoxic	*Micromonospora* sp.	Fernandez-Chimeno et al. (2000)
	Madurastatins D1 and D2	Siderophore	Antimicrobial	*Actinomadura* sp.	Yan et al. (2019)
	Nocardiopsistins A–C	Polyketide	Antibacterial	*Nocardiosis* sp.	Xu et al. (2018)
	Nocardiotide A	Peptide	Cytotoxic	*Nocardiopsis* sp.	Ibrahim et al. (2018)
	Photopiperazines A–D	Diketopiperazine	Cytotoxic	*Actinomycete*	Kim et al. (2019)
	Rakicidin F	Depsipeptide	Antibacterial	*Streptomyces* sp.	Kitani et al. (2018)
	Tetracinomycin D	Anthraquinone	Anticancer	*Streptomyces* sp.	Abdelfattah et al. (2018)
	Yangpumicins F and G	Enediyne	Antibacterial, cytotoxic	*Micromonospora yangpuensis*	Wang et al. (2019b)
Macroalgae	Desertomycin G	Macrolide	Antituberculosis, anticancer	*Streptomyces althioticus*	Braña et al. (2019)
	Kocumarin	Aromatic acid	Antibacterial	*Kocuria marina*	Uzair et al. (2018)
	Neaumycin B	Macrolide	Anticancer	*Micromonospora* sp.	Kim et al. (2018)
	3-*O*-Methylwailupemycin G Wailupemycins J–L	Polyketide	Cytotoxic, antiviral	*Streptomyces* sp.	Liu et al. (2017)
Seawater	Akazamicin	Polyketide	Cytotoxic	*Nonomuraea* sp.	Yang et al. (2019)
	Arisostatins A and B	Tetrocarcin	Antibacterial, antitumor	*Micromonospora* sp.	Igarashi et al. (2000)
	Branimycins B and C	Macrolide	Antibacterial	*Pseudonocardia carboxydivorans*	Braña et al. (2017a, b)

Marine mammal (*Phocoena phocoena*)	Phocoenamicin	Polyketide	Antibacterial	*Micromonospora* sp.	Ochoa et al. (2018)
Marine plants	Ghanamycins A and B	Lactone	Antimicrobial	*Streptomyces ghanaensis*	Xu et al. (2017)
	Provipeptides A and B	Cyclotetrapeptides	Antibacterial	*Streptomyces* sp.	Betancur et al. (2019)
Ascidians	Caniferolides A–D	Macrolide	Antifungal, antitumor	*Streptomyces caniferus*	Pérez-Victoria et al. (2019)
	Keyicin	Anthracycline	Antibacterial	*Micromonospora* sp.	Adnani et al. (2017a); Adnani et al. (2017b)
	Phallusialides A–E	Alkaloid	Antibacterial	*Micromonospora* sp.	Zhang et al. (2019b)
Sea cucumber	Streptodepsipeptide SV21	Depsipeptide	Antiviral	*Streptomyces cavourensis*	Wibowo et al. (2021)
Shellfish	Streptoseomycin	Macrodilactone	Antibacterial	*Streptomyces seoulensis*	Zhang et al. (2018)
Tunicate	Konamycins A and E Rubromycins CA1 and CA2	Polyketide Polyketide	Antioxidant Antimicrobial	*Streptomyces hyaluromycini*	Harunari et al. (2019)
Prawn	Ansaseomycins A and B	Polyketide	Anticancer	*Streptomyces seoulensis*	Liu et al. (2019b)

activities, biofouling (i.e., the undesirable accumulation of microorganisms, plants, algae, or small animals on surfaces) is a problem that causes extensive damage to structures such as ship hulls, fishnets, or aquaculture facilities. Several industrially relevant and more eco-friendly antifouling molecules have been isolated from marine Actinobacteria. Examples include five novel diketopiperazines with antilarval attachment activity isolated from a deep-sea *Streptomyces* strain (Li et al. 2006); antifouling diketopiperazines able to inhibit diatoms and algae zoospores development obtained from a seaweed-associated *Streptomyces* (Cho et al. 2012); the novel polyhydroxy polyketides nahuoic acids B–E discovered from a marine-derived *Streptomyces* and showing antibiofilm activity against *Shewanella oneidensis* (Nong et al. 2016).

Another meaningful example of the valuable compounds produced by marine Actinobacteria are biosurfactants. Naturally occurring biosurfactants are relevant due to their low toxicity, good biodegradability, and ecological adequacy. Several marine Actinobacteria, namely affiliated with *Streptomyces*, *Nocardiopsis*, and *Brevibacterium* genera, proved to be able to produce such compounds with different applications, for instance, in preventing biofilm formation (Kiran et al. 2010a, b, 2014; Selvin et al. 2016) or even in the cosmetic industry (Das et al. 2013). Marine Actinobacteria are also a notable source of enzymes (Zhao et al. 2016). Such enzymes can act as probiotics in aquaculture (Das et al. 2008) or be used to recycle industrial waste, as they are able to turn over complex biopolymers, namely through the synthesis of keratinases and other industrially relevant enzymes as proteases, lipases, cellulases, and amylases (González et al. 2020; Rajagopal and Kannan 2017). The agriculture sector can also truly benefit from marine actinobacterial secondary metabolism: strains affiliated with the genera *Streptomyces* and *Saccharomonospora* were identified as potential biocide producers through the synthesis of larvicidal, repellent, and ovicidal compounds (Balakrishnan et al. 2017; Karthik et al. 2011); a *Streptomyces* and a *Micromonospora* displaying wheat seed growth-promoting activity were isolated from the rhizosphere of coastal salt marsh plants (Gong et al. 2018); three *Streptomyces* strains isolated from the rhizosphere of *Salicornia bigelovii*, an economically relevant crop that grows in intertidal marine waters, showed not only to promote plant growth and seed production but also to increase the levels of photosynthetic pigments, endogenous auxins, and polyamines (Mathew et al. 2020). The possibilities for bioprospection of marine Actinobacteria are endless. These microorganisms can be exploited in the energy sector as a biofuel source, since they are able to generate economically important byproducts like bioethanol (Muthusamy et al. 2019); can be used as a tool for wastewater treatment, as they synthesize bioflocculants (Awolusi et al. 2020); can act as a bioremediation tool, since they are able to metabolize complex organic compounds and to remove xenobiotics, such as pesticides and heavy metals, from the environment (Alvarez et al. 2017; Kamala et al. 2020); and can also be a source of nontoxic bio-pigments (Chakraborty et al. 2015). It comes clear that many fields can benefit from the chemical repertoire of marine Actinobacteria.

1.4 From the Field to the Laboratory: Best Approaches to Isolate Marine Actinobacteria

Actinobacteria are successful colonizers of various marine ecosystems and exhibit a wide range of morphological and physiological characteristics (Barkal et al. 2016). Members of this phylum require different growth conditions when compared to their terrestrial counterparts (Claverias et al. 2015; Hames-Kocabas and Uzel 2012; Zotchev 2012). A high number of bacteria cannot be cultivated through conventional isolation methods (Overmann and Lepleux 2016). This has been clearly shown by culture-independent studies, and because of the low success of culture-dependent approaches, molecular techniques are being increasingly used to study the diversity and function of microbial communities in environments (Subramani and Sipkema 2019; Jung et al. 2021). These techniques are allowing the discovery of many functional characteristics of Actinobacteria and are contributing to the improvement of isolation strategies, growth conditions, and culture media formulation to recover previously uncultivable Actinobacteria (Kaeberlein et al. 2002; Stewart 2012; Subramani and Sipkema 2019).

The cultivation and isolation of marine Actinobacteria are key steps for bioprospection purposes and for the discovery of novel NP (Fig. 1.2) (Sekurova et al. 2019). To maximize the success in cultivating these microorganisms, it is important to know the environmental factors that can influence their isolation, such as pH, temperature, oxygen, and availability of nutrients, among others (Jiang et al. 2016).

Salt is a very important component for the growth of marine Actinobacteria (Hames-Kocabas and Uzel 2012; Subramani and Aalbersberg 2013; Subramani and Sipkema 2019). Culture media with different concentrations of seawater, artificial seawater, or deionized water supplemented with NaCl have been used for the isolation of marine Actinobacteria (Maldonado et al. 2005; Mincer et al. 2002; Abdelfattah et al. 2016; Undabarrena et al. 2016). Different carbon (chitin, dextrose, glucose, glycerol, maltose, mannitol, soluble starch, and oatmeal) and nitrogen sources (casein, histidine, L-arginine, malt extract, meat extract, peptone, tryptone, and yeast extract) are widely used as components of selective media for the isolation of marine Actinobacteria (Sharma et al. 2021; Shamikh et al. 2020; Axenov-Gribanov et al. 2020; Ribeiro et al. 2020). In addition, extracts of sediments, sponges, or algae are used, alone or as a supplement, to mimic the conditions found in the marine environment (Rego et al. 2019; Zhang et al. 2006a). Oligotrophic media compared to nutrient-rich culture media are generally more effective in isolating rare marine Actinobacteria (Jensen et al. 2005).

Several pre-treatments have been applied to marine samples for selecting slow-growing Actinobacteria, including physical (dry heat, desiccation, freezing, and radiation) (Mincer et al. 2002; Bredholt et al. 2008; Naikpatil and Rathod 2011; Jensen et al. 2005), mechanical (gradient centrifugation and shake with glass beads) (Bredholt et al. 2008; Maldonado et al. 2005), and chemical (chemical compounds and antibiotics) (Bredholt et al. 2008; Ribeiro et al. 2020). Desiccation and heating have been used for selecting sporulating Actinobacteria and inhibiting other Gram-

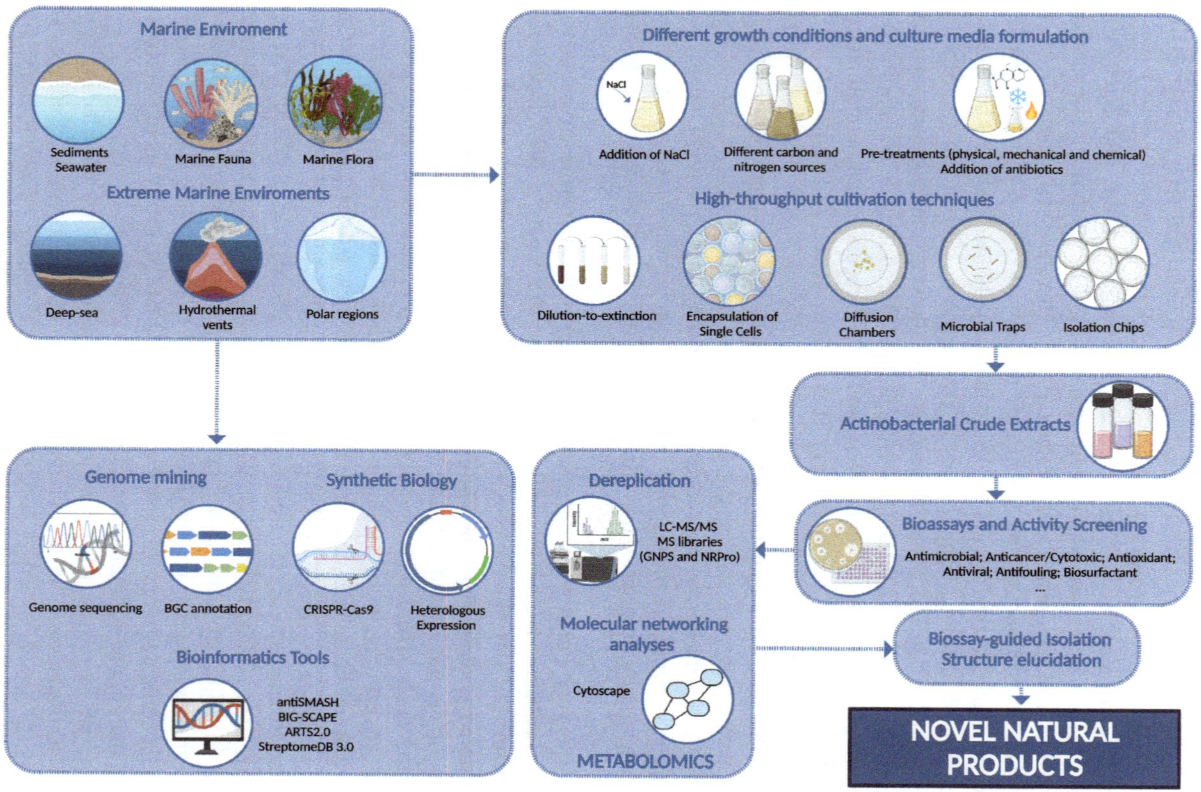

Fig. 1.2 General pipeline for the discovery of novel bioactive compounds from marine Actinobacteria. Created with BioRender.com

positive bacteria (Fenical and Jensen 2006). Treatment with several chemical compounds, such as benzethonium chloride, chloramine-T, phenol, and sodium dodecyl sulfate, has been shown to be very successful in selecting marine Actinobacteria (Bredholt et al. 2008; Hong et al. 2009). Furthermore, the addition of antibacterial and antifungal antibiotics, such as cycloheximide, gentamicin, kanamycin, nalidixic acid, novobiocin, nystatin, penicillin, rifampicin, streptomycin, tunicamycin, and vancomycin, to the culture medium has been also shown very relevant for selecting different genera of marine Actinobacteria (Amin et al. 2020; Bredholt et al. 2008; Claverias et al. 2015; Hong et al. 2009).

Recently, several new and innovative high-throughput cultivation techniques have been employed for the isolation of Actinobacteria that cannot be recovered by classical isolation methods. These techniques include dilution-to-extinction (Connon and Giovannoni 2002), encapsulation of single cells (Zengler et al. 2002), diffusion chambers (Kaeberlein et al. 2002), microbial traps (Gavrish et al. 2008), and isolation chips (Lewis et al. 2010) (Fig. 1.2). By mimicking the conditions found in marine environments and using very diluted microbial concentrations, such strategies also represent a valuable method to promote the growth of rare and slow-growing Actinobacteria (Hames-Kocabas and Uzel 2012; Zotchev 2012).

The dilution-to-extinction method is based on the dilution of natural microbial communities to the point of extinction (Kim et al. 2020b). When in a microtiter plate, this technique allows analyzing several samples, resorting on different culture media and dilution endpoints, and when coupled with oligotrophic culture medium, it allows slow-growing Actinobacteria to grow during long incubation periods without being inhibited by fast-growing bacteria (Connon and Giovannoni 2002; Gontang et al. 2007). Cell encapsulation is an interesting alternative to classic fermentation and seems to be very promising to access uncultivable Actinobacteria and their metabolites (López-García et al. 2014; Mahler et al. 2018). In this methodology, cells from marine samples are mixed with melted agarose and then emulsified with oil to form microcapsules. These structures are then added to a sterile fermentation column, with liquid culture medium, equipped with two sets of filter membranes to prevent free-living cells of contaminating the medium and to retain the encapsulated cells inside (Zengler et al. 2002). Despite the encapsulation separating the cells, the metabolites produced pass through the microcapsules, allowing different bacteria to communicate with each other as in their natural environment (Hames-Kocabas and Uzel 2012). Cell encapsulation has also been shown effective in promoting slow-growing Actinobacteria growth and has been widely used for screening NP (Hug et al. 2018; Toledo et al. 2006).

Another strategy worthy of mention is the use of diffusion chambers, a method designed to simulate real conditions observed in different natural environments (Kaeberlein et al. 2002). For the assembly of these chambers, a membrane (0.03 μm) is glued to a stainless-steel washer, which is then filled with microbial inoculum embedded in agar. After the agar solidifies, a second membrane with the same porosity is glued to the opposite side of the washer. The diffusion chambers can be placed in the original sampling site or in a simulated environment under laboratory conditions. This approach allows the inlet of nutrients in the diffusion chambers and the outlet of secondary metabolites (Gavrish et al. 2008; Kaeberlein et al. 2002; Lewis et al. 2010; Jung et al. 2021). Microbial traps are a similar version of the previously mentioned diffusion chambers, but aim to promote the growth of filamentous Actinobacteria. In this technique, a 0.03 μm membrane is glued to one side of a stainless-steel washer. The chamber is filled with agar and a 0.2 μm membrane is glued to the other side of the washer to allow the passage of thin filaments of Actinobacteria during incubation. After microorganisms' development, the agar is transferred to a petri dish for subculture (Gavrish et al. 2008; Lewis et al. 2010). Similar to microbial traps, isolation chips (iChip) were also inspired on diffusion chambers (Nichols et al. 2010). The iChip consists of several hundred miniature diffusion chambers, each one inoculated with only one environmental cell. These are covered by semipermeable membranes to allow nutrients to diffuse into the cells and inhibit the passage of cells out of the system (Hug et al. 2018). This culture technology has proven to be very important to improve the isolation of bacteria that were once uncultivable (Kaeberlein et al. 2002). Several new drugs have been obtained from Actinobacteria cultivated in iChip platforms (Peoples et al. 2008, 2012; Gavrish et al. 2014).

The application of in situ cultivation methods and the identification of factors that influence the growth of uncultivable bacteria have been proven very important to access a diversity of bacteria hitherto unknown. High-throughput cultivation approaches are not yet widely used for the isolation of marine Actinobacteria; however, it would be a good bet for the discovery of novel species and structurally unique metabolites, since chemical novelty is usually associated with biological novelty.

1.5 New Tools for Bioprospecting Actinobacteria

The history of microbial NP discovery is marked by high and low periods that can be briefly summarized in (1) the "Golden Age," when the discovery rate of microbial bioactive metabolites reached its highest; (2) the desertion of the bioprospection strategy by the pharmaceutical industry, motivated by the promising perspectives of combinatorial chemistry, which caused an abrupt decline in NP discovery; and (3) the re-emergence of interest in NP discovery pipelines, triggered by the genomic era that allowed to uncover valuable information about the full potential of microorganisms as NP producers (Wright 2019). The increasing number of available genome sequences of NP-producing bacteria is changing the way scientists hunt for novel bioactive metabolites. With improved tools and specific software designed for genome mining, a wealth of formerly hidden NP biosynthetic gene clusters (BGCs) have been revealed (Li et al. 2019c). It is truly remarkable that a single *Streptomyces* genome can harbor around 30 NP BGCs, putting Actinobacteria once again under the spotlight for drug drilling. Such genetic potential is not always detected under laboratory conditions, making genome sequencing, synthetic biology, metabolomics, modern mass spectrometry, and in silico tools true tide turners

(Fig. 1.2) (Gruene et al. 2018; Ziemert et al. 2016). For decades, the classical path to discover bioactive metabolites leaned exclusively on chemical/bioactivity-guided screening approaches, but it is now possible to rely on many bioinformatics tools, programs, and databases to gain access to the whole panorama of BGCs in genomic data.

For the identification and analysis of BGCs in both bacterial and fungal genome sequences, antiSMASH is labeled as the most broadly used tool and web server and since 2011 has been assisting researchers to uncover valuable microbial secondary metabolism (Blin et al. 2019). Focusing on large-scale biosynthetic diversity exploration, BIG-SCAPE is a useful tool, as it encodes a computational framework that allows to scale up from single genomes to entire strain collections and microbiome (Navarro-Muñoz et al. 2020). In drug-discovery pipelines, prioritizing BGCs encoding pharmaceuticals with novel modes of action is also a crucial step. In that regard, software as ARTS2.0, which allows automated target directed genome mining in all bacterial taxa as well as in metagenomic data, and the comparison of similar BGCs from different genomes and their putative resistance genes, can make a difference (Mungan et al. 2020). With so many new and different sources of data, organizing all information in databases for an easier access, update, and management is vital. An example, focusing on actinobacterial secondary metabolism, is StreptomeDB 3.0: a public online database that encodes a collection of *Streptomyces* NPs and enables interactive phylogenetic exploration of these microorganisms and their isolated metabolites, as well as the prediction of pharmacokinetic and toxicity profiles (Moumbock et al. 2021).

As mentioned before, the rediscovery of known compounds represents a major tailback in the drug development field. Dereplication plays a major role in addressing this problem, allowing redirecting efforts to unknown entities early in the NP discovery pipeline. Combining increasingly sophisticated analytical techniques, as LC-MS/MS and related, and using MS libraries and databases, it is possible to screen NP potential novelty in terms of structure, molecular formula, and bioactivity in a more accurate, efficient, and faster way (Gaudêncio and Pereira 2015). Examples of such tools include the Global Natural Product Social Molecular Networking (GNPS), a small molecule-focused tandem mass spectrometry data curation and analysis online instrument (Wang et al. 2016), or NRPro, also a mass spectrometry data analysis platform that provides a wide-ranging toolset for both automatic annotation and dereplication of peptidic NPs (Fig. 1.2) (Ricart et al. 2020).

A number of innovative approaches have been recently applied to mine novel NP from the Actinobacteria genetic arsenal, with several inspiring examples of success: four novel 16-demethylrifamycins were discovered by deletion of specific gene clusters in a marine *Micromonospora* sp. strain (Zhou et al. 2019c); CRISPR-Cas9 genome engineering was used to knock out genes encoding for common antibiotic biosynthesis, leading to a shift in the metabolic profile of the producer Actinobacteria that resulted in the uncovering of hidden compounds (Culp et al. 2019); ansaseomycins A and B, two novel compounds with activity against leukemia cells, were discovered in a *Streptomyces* sp. genome by heterologous expression of a cryptic giant type I PKS (polyketide synthase) gene cluster (Liu et al. 2019b); in a similar strategy, the cytotoxic metabolite neothioviridamide was produced by heterologous expression of a *Streptomyces* sp. cryptic RiPP (post-translationally modified peptide) BGC (Kawahara et al. 2018); a new cytotoxic peptide named curacozole was uncovered based on genome mining (Kaweewan et al. 2019); a novel class of small compounds displaying relevant antituberculosis activity were identified by in silico structure-based drug screening (Taira et al. 2017); the use of high temperatures to activate dormant biosynthetic genes led to the discovery of murecholamide, a metabolite from the thermotolerant *Streptomyces* sp. (Saito et al. 2020); spectral networking analysis uncovered noursamycins, chlorinated cyclohexapeptides with antibacterial activity, produced by a *Streptomyces* strain (Mudalungu et al. 2019); combining metabolomic with MS-based molecular networking analysis led to the discovery of lugdunomycin, a novel angucycline-derived polycyclic aromatic polyketide with antimicrobial properties and an unprecedented chemical architecture (Wu et al. 2019).

Several strategies used during the cultivation of the producing actinobacterial strains can also translate into access to novel chemistry: variations on media salinity composition led to the synthesis of lavencidin, a polyene macrolide antibiotic isolated from a *Streptomyces* sp. culture (Yoshioka et al. 2021); the addition of the elicitor N-acetylglucosamine allowed to maximize the chemical diversity of the sponge-derived Actinobacteria *Actinokineospora* sp., granting the discovery of two novel fridamycins active against the parasite *Trypanosoma brucei* (Tawfike et al. 2019); several examples illustrate co-cultivation as an efficient route to the isolation of novel NP from marine-sourced Actinobacteria, which allowed, for instance, the discovery of dentigerumycin E, a new cyclic hexapeptide with antiproliferative and antimetastatic activities against human cancer cells (Shin et al. 2018), of the new anthracycline antibiotic keyicin, effective against *Mycobacterium* sp. (Adnani et al. 2017b), and of several antimalarial compounds (Alhadrami et al. 2021).

As never before, the knowledge and tools to better understand NP biosynthesis, biochemistry, and engineering are available and rapidly upgrading, allowing taking full benefit of the unique properties of these molecules and of their producers.

1.6 Conclusions and Future Perspectives

The ocean represents the largest environment on earth and hosts a remarkable and underexplored actinobacterial diversity. Despite all the inherent challenges, the efforts of the scientific community in the exploration of marine Actinobacteria for the discovery of new biotechnologically relevant NP has been rewarding and a promising source to solve many problems that human society faces. It is a fact that these bacteria colonize several marine habitats and live in association with many organisms, and today, more than ever, the necessary tools to explore their full potential are available. We currently know better where to find these microorganisms, how to isolate them in the laboratory, how to explore their biosynthetic machinery, and how to awake and express cryptic BGCs. Future prospects for the exploration of marine Actinobacteria are bright and with great potential to result in fruitful discoveries to all benefit.

Acknowledgments M. Girão and I. Ribeiro acknowledge the Foundation for Science and Technology (FCT) for the PhD grants SFRH/BD/145646/2019 and SFRH/BD/136357/2018, respectively. M.F. Carvalho wishes to acknowledge CEEC program supported by FCT (CEECIND/02968/2017), Fundo Social Europeu and Programa Operional Potencial Humano. The authors also acknowledge the ACTINODEEPSEA project (PTDC/BIA-MIC/31045/2017), co-financed by COMPETE 2020, Portugal 2020, European Regional Development Fund (ERDF), and FCT, and Strategic Funding UIDB/04423/2020 and UIDP/04423/2020 through national funds provided by FCT and ERDF.

References

Abdelfattah MS, Elmallah MIY, Hawas UW, El-Kassema LTA, Eid MAG (2016) Isolation and characterization of marine-derived actinomycetes with cytotoxic activity from the Red Sea coast. Asian Pac J Trop Biomed 6(8):651–657

Abdelfattah MS, Elmallah MIY, Faraag AHI, Hebishy AMS, Ali NH (2018) Heliomycin and tetracinomycin D: anthraquinone derivatives with histone deacetylase inhibitory activity from marine sponge-associated Streptomyces sp. SP9. 3 Biotech 8(6):1–9

Abdel-Mageed WM, Juhasz B, Lehri B, Alqahtani AS, Nouioui I, Pech-Puch D, Tabudravu JN, Goodfellow M, Rodríguez J, Jaspars M (2020) Whole genome sequence of Dermacoccus abyssi MT1.1 isolated from the Challenger deep of the Mariana Trench reveals phenazine biosynthesis locus and environmental adaptation factors. Mar Drugs 18(3):131

Abdelmohsen UR, Bayer K, Hentschel U (2014a) Diversity, abundance and natural products of marine sponge-associated actinomycetes. Nat Prod Rep 31(3):381–399. https://doi.org/10.1039/c3np70111e

Abdelmohsen UR, Cheng C, Viegelmann C, Zhang T, Grkovic T, Ahmed S, Quinn RJ, Hentschel U, Edrada-Ebel RA (2014b) Dereplication strategies for targeted isolation of new antitrypanosomal actinosporins A and B from a marine sponge associated-Actinokineospora sp. EG49. Mar Drugs 12(3):1220–1244

Adnani N, Braun DR, McDonald BR, Chevrette MG, Currie CR, Bugni TS (2017a) Draft genome sequence of Micromonospora sp. strain WMMB235, a marine Ascidian-associated bacterium. Genome Announc 5(2):e01369–e01316

Adnani N, Chevrette MG, Adibhatla SN, Zhang F, Yu Q, Braun DR, Nelson J, Simpkins SW, McDonald BR, Myers CL (2017b) Coculture of marine invertebrate-associated bacteria and interdisciplinary technologies enable biosynthesis and discovery of a new antibiotic, keyicin. ACS Chem Biol 12(12):3093–3102

Alhadrami HA, Thissera B, Hassan MHA, Behery FA, Ngwa CJ, Hassan HM, Pradel G, Abdelmohsen UR, Rateb ME (2021) Bio-guided isolation of antimalarial metabolites from the coculture of two red sea sponge-derived Actinokineospora and Rhodococcus spp. Mar Drugs 19(2):109

Alvarado P, Huang Y, Wang J, Garrido I, Leiva S (2018) Phylogeny and bioactivity of epiphytic Gram-positive bacteria isolated from three co-occurring antarctic macroalgae. Antonie Van Leeuwenhoek 111(9):1543–1555. https://doi.org/10.1007/s10482-018-1044-6

Alvarez A, Saez JM, Costa JSD, Colin VL, Fuentes MS, Cuozzo SA, Benimeli CS, Polti MA, Amoroso MJ (2017) Actinobacteria: current research and perspectives for bioremediation of pesticides and heavy metals. Chemosphere 166:41–62

Amin DH, Abdallah NA, Abolmaaty A, Tolba S, Wellington EMH (2020) Microbiological and molecular insights on rare Actinobacteria harboring bioactive prospective. Bull Natl Res Cent 44(1):5. https://doi.org/10.1186/s42269-019-0266-8

Andersson DI, Balaban NQ, Baquero F, Courvalin P, Glaser P, Gophna U, Kishony R, Molin S, Tønjum T (2020) Antibiotic resistance: turning evolutionary principles into clinical reality. FEMS Microbiol Rev 44(2):171–188

Asolkar RN, Freel KC, Jensen PR, Fenical W, Kondratyuk TP, Park E-J, Pezzuto JM (2009) Arenamides A–C, cytotoxic NFκB inhibitors from the marine actinomycete Salinispora arenicola. J Nat Prod 72(3):396–402

Awolusi OO, Ademakinwa AN, Ojo A, Erasmus M, Bux F, Agunbiade MO (2020) Marine Actinobacteria bioflocculant: a storehouse of unique biotechnological resources for wastewater treatment and other applications. Appl Sci 10(21):7671

Axenov-Gribanov DV, Kostka DV, Vasilieva UA, Shatilina ZM, Krasnova ME, Pereliaeva EV, Zolotovskaya ED, Morgunova MM, Rusanovskaya OO, Timofeyev MA (2020) Cultivable Actinobacteria first found in baikal endemic algae is a new source of natural products with antibiotic activity. Int J Microbiol 2020:5359816. https://doi.org/10.1155/2020/5359816

Bakke I, Coward E, Andersen T, Vadstein O (2015) Selection in the host structures the microbiota associated with developing cod larvae (Gadus morhua). Environ Microbiol 17(10):3914–3924. https://doi.org/10.1111/1462-2920.12888

Balakrishnan S, Santhanam P, Srinivasan M (2017) Larvicidal potency of marine actinobacteria isolated from mangrove environment against Aedes aegypti and Anopheles stephensi. J Parasit Dis 41(2):387–394

Barka EA, Vatsa P, Sanchez L, Gaveau-Vaillant N, Jacquard C, Klenk H-P, Clément C, Ouhdouch Y, van Wezel GP (2016) Taxonomy, physiology, and natural products of Actinobacteria. Microbiol Mol Biol Rev 80(1):1–43

Barkal LJ, Theberge AB, Guo CJ, Spraker J, Rappert L, Berthier J, Brakke KA, Wang CC, Beebe DJ, Keller NP, Berthier E (2016) Microbial metabolomics in open microscale platforms. Nat Commun 7:10610. https://doi.org/10.1038/ncomms10610

Ben-Dov E, Yosef DZB, Pavlov V, Kushmaro A (2009) Corynebacterium maris sp. nov., a marine bacterium isolated from the mucus of the coral Fungia granulosa. Int J Syst Evol Microbiol 59(10):2458–2463

Betancur LA, Forero AM, Romero-Otero A, Sepúlveda LY, Moreno-Sarmiento NC, Castellanos L, Ramos FA (2019) Cyclic tetrapeptides from the marine strain Streptomyces sp. PNM-161a with activity against rice and yam phytopathogens. J Antibiot 72(10):744–751

Bibi F, Naseer MI, Hassan AM, Yasir M, Al-Ghamdi AAK, Azhar EI (2018) Diversity and antagonistic potential of bacteria isolated from marine grass Halodule uninervis. 3 Biotech 8(1):48. https://doi.org/10.1007/s13205-017-1066-1

Bister B, Bischoff D, Ströbele M, Riedlinger J, Reicke A, Wolter F, Bull AT, Zähner H, Fiedler H-P, Süssmuth RD (2004) Abyssomicin C—A polycyclic antibiotic from a marine Verrucosispora strain as an inhibitor of the p-Aminobenzoic acid/Tetrahydrofolate biosynthesis pathway. Angew Chem Int Ed 43(19):2574–2576

Blin K, Shaw S, Steinke K, Villebro R, Ziemert N, Lee SY, Medema MH, Weber T (2019) antiSMASH 5.0: updates to the secondary metabolite genome mining pipeline. Nucleic Acids Res 47(W1):W81–W87

Braña AF, Sarmiento-Vizcaíno A, Osset M, Pérez-Victoria I, Martín J, De Pedro N, De la Cruz M, Díaz C, Vicente F, Reyes F (2017a) Lobophorin K, a new natural product with cytotoxic activity produced by Streptomyces sp. M-207 associated with the deep-sea coral Lophelia pertusa. Mar Drugs 15(5):144

Braña AF, Sarmiento-Vizcaíno A, Pérez-Victoria I, Otero L, Fernández J, Palacios JJ, Martín J, de la Cruz M, Díaz C, Vicente F (2017b) Branimycins B and C, antibiotics produced by the abyssal actinobacterium Pseudonocardia carboxydivorans M-227. J Nat Prod 80(2):569–573

Braña AF, Sarmiento-Vizcaíno A, Pérez-Victoria I, Martín J, Otero L, Palacios-Gutiérrez JJ, Fernández J, Mohamedi Y, Fontanil T, Salmón M (2019) Desertomycin G, a new antibiotic with activity against mycobacterium tuberculosis and human breast tumor cell lines produced by Streptomyces althioticus MSM3, isolated from the Cantabrian Sea intertidal Macroalgae Ulva sp. Mar drugs 17(2):114

Bredholdt H, Galatenko OA, Engelhardt K, Fjaervik E, Terekhova LP, Zotchev SB (2007) Rare actinomycete bacteria from the shallow water sediments of the Trondheim fjord, Norway: isolation, diversity and biological activity. Environ Microbiol 9(11):2756–2764. https://doi.org/10.1111/j.1462-2920.2007.01387.x

Bredholt H, Fjærvik E, Johnsen G, Zotchev SB (2008) Actinomycetes from sediments in the Trondheim Fjord, Norway: diversity and biological activity. Mar Drugs 6(1):12–24. https://doi.org/10.3390/md6010012

Buedenbender L, Carroll AR, Ekins M, Kurtböke DI (2017) Taxonomic and metabolite diversity of actinomycetes associated with three Australian ascidians. Diversity 9(4):53

Bull AT, Goodfellow M (2019) Dark, rare and inspirational microbial matter in the extremobiosphere: 16 000 m of bioprospecting campaigns. Microbiology 165(12):1252–1264. https://doi.org/10.1099/mic.0.000822

Bull AT, Stach JEM, Ward AC, Goodfellow M (2005) Marine actinobacteria: perspectives, challenges, future directions. Antonie Van Leeuwenhoek 87(1):65–79

Bürstner N, Roggo S, Ostermann N, Blank J, Delmas C, Freuler F, Gerhartz B, Hinniger A, Hoepfner D, Liechty B (2015) Gift from nature: cyclomarin A kills mycobacteria and malaria parasites by distinct modes of action. ChemBioChem 16(17):2433–2436

Busi S, Pattnaik SS (2018) Current status and applications of actinobacteria in the production of anticancerous compounds. In: New and future developments in microbial biotechnology and bioengineering. Elsevier, pp 137–153

Carroll AR, Copp BR, Davis RA, Keyzers RA, Prinsep MR (2019) Marine natural products. Nat Prod Rep 36(1):122–173

Carroll AR, Copp BR, Davis RA, Keyzers RA, Prinsep MR (2020) Marine natural products. Nat Prod Rep 38:362–413

Chakraborty I, Redkar P, Minki M, Sathish Kumar SR, Bhaskara Rao KV (2015) Isolation and characterization of pigment producing marine actinobacteria from mangrove soil and applications of bio-pigments. Der Pharm Lett 7:93–100

Chang X, Liu W, Zhang XH (2011) Spinactinospora alkalitolerans gen. nov., sp. nov., an actinomycete isolated from marine sediment. Int J Syst Evol Microbiol 61(Pt 12):2805–2810. https://doi.org/10.1099/ijs.0.027383-0

Chen P, Zhang L, Guo X, Dai X, Liu L, Xi L, Wang J, Song L, Wang Y, Zhu Y, Huang L, Huang Y (2016) Diversity, biogeography, and biodegradation potential of Actinobacteria in the deep-sea sediments along the Southwest Indian ridge. Front Microbiol 7(1340). https://doi.org/10.3389/fmicb.2016.01340

Chen H, Cai K, Yao R (2018a) A new macrolactam derivative from the marine actinomycete HF-11225. J Antibiot 71(4):477–479

Chen L, Hu J-S, Xu J-L, Shao C-L, Wang G-Y (2018b) Biological and chemical diversity of ascidian-associated microorganisms. Mar Drugs 16(10):362

Chen L, Zhao W, Jiang H-L, Zhou J, Chen X-M, Lian Y-Y, Jiang H, Lin F (2018c) Rakicidins G-I, cyclic depsipeptides from marine Micromonospora chalcea FIM 02-523. Tetrahedron 74(30):4151–4154

Chen L, Du S, Qu W-Y, Guo F-R, Wang G-Y (2019a) Biosynthetic potential of culturable bacteria associated with Apostichopus japonicus. J Appl Microbiol 127(6):1686–1697. https://doi.org/10.1111/jam.14453

Chen L, Wang XN, Fu CM, Wang GY (2019b) Phylogenetic analysis and screening of antimicrobial and antiproliferative activities of culturable bacteria associated with the Ascidian Styela clava from the Yellow Sea, China. Biomed Res Int 2019:7851251. https://doi.org/10.1155/2019/7851251

Cheng KC-C, Cao S, Raveh A, MacArthur R, Dranchak P, Chlipala G, Okoneski MT, Guha R, Eastman RT, Yuan J (2015) Actinoramide A identified as a potent antimalarial from titration-based screening of marine natural product extracts. J Nat Prod 78(10):2411–2422

Cho JY, Kang JY, Hong YK, Baek HH, Shin HW, Kim MS (2012) Isolation and structural determination of the antifouling diketopiperazines from marine-derived Streptomyces praecox 291-11. Biosci Biotechnol Biochem 76(6):1116–1121

Choi B-K, Lee H-S, Kang JS, Shin HJ (2019) Dokdolipids A–C, hydroxylated rhamnolipids from the marine-derived Actinomycete Actinoalloteichus hymeniacidonis. Mar Drugs 17(4):237

Claverias FP, Undabarrena A, Gonzalez M, Seeger M, Camara B (2015) Culturable diversity and antimicrobial activity of Actinobacteria from marine sediments in Valparaiso bay, Chile. Front Microbiol 6(737). https://doi.org/10.3389/fmicb.2015.00737

Cong Z, Huang X, Liu Y, Liu Y, Wang P, Liao S, Yang B, Zhou X, Huang D, Wang J (2019) Cytotoxic anthracycline and antibacterial tirandamycin analogues from a marine-derived Streptomyces sp. SCSIO 41399. J Antibiot 72(1):45–49

Connon SA, Giovannoni SJ (2002) High-throughput methods for culturing microorganisms in very-low-nutrient media yield diverse new marine isolates. Appl Environ Microbiol 68(8):3878–3885

Culp EJ, Yim G, Waglechner N, Wang W, Pawlowski AC, Wright GD (2019) Hidden antibiotics in actinomycetes can be identified by inactivation of gene clusters for common antibiotics. Nat Biotechnol 37(10):1149–1154

da Silva AB, Silveira ER, Wilke DV, Ferreira EG, Costa-Lotufo LV, Torres MCM, Ayala AP, Costa WS, Canuto KM, de Araújo-Nobre AR (2019) Antibacterial Salinaphthoquinones from a strain of the bacterium Salinispora arenicola recovered from the marine sediments of St. Peter and St. Paul Archipelago, Brazil. J Nat Prod 82(7):1831–1838

Das S, Ward LR, Burke C (2008) Prospects of using marine actinobacteria as probiotics in aquaculture. Appl Microbiol Biotechnol 81(3):419–429

Das I, Roy S, Chandni S, Karthik L, Kumar G, Rao KVB (2013) Biosurfactant from marine actinobacteria and its application in cosmetic formulation of toothpaste. Pharm Lett 5(5):1–6

Das L, Deb S, Das SK (2020) Glutamicibacter mishrai sp. nov., isolated from the coral Favia veroni from Andaman Sea. Arch Microbiol 202(4):733–745

Dat TTH, Steinert G, Cuc NTK, Smidt H, Sipkema D (2018) Archaeal and bacterial diversity and community composition from 18 phylogenetically divergent sponge species in Vietnam. PeerJ 6:e4970. https://doi.org/10.7717/peerj.4970

De Carvalho CCCR, Fernandes P (2010) Production of metabolites as bacterial responses to the marine environment. Mar Drugs 8(3):705–727. https://doi.org/10.3390/md8030705

de Menezes CBA, Tonin MF, Silva LJ, de Souza WR, Parma M, de Melo IS, Zucchi TD, Destéfano SAL, Fantinatti-Garboggini F (2015) Marmoricola aquaticus sp. nov., an actinomycete isolated from a marine sponge. Int J Syst Evol Microbiol 65(7):2286–2291

de Menezes CB, Afonso RS, de Souza WR, Parma M, de Melo IS, Zucchi TD, Fantinatti-Garboggini F (2016) Gordonia didemni sp. nov. an actinomycete isolated from the marine ascidium Didemnum sp. Antonie Van Leeuwenhoek 109(2):297–303. https://doi.org/10.1007/s10482-015-0632-y

de Menezes CBA, Afonso RS, de Souza WR, Parma M, de Melo IS, Zucchi TD, Fantinatti-Garboggini F (2017) Williamsia spongiae sp. nov., an actinomycete isolated from the marine sponge Amphimedon viridis. Int J Syst Evol Microbiol 67(5):1260–1265. https://doi.org/10.1099/ijsem.0.001796

de Menezes CBA, Afonso RS, de Souza WR, Parma MM, de Melo IS, Fugita FLS, Moraes LAB, Zucchi TD, Fantinatti-Garboggini F (2019) Williamsia aurantiacus sp. nov. a novel actinobacterium producer of antimicrobial compounds isolated from the marine sponge. Arch Microbiol 201(5):691–698. https://doi.org/10.1007/s00203-019-01633-z

Du ZJ, Miao TT, Lin XZ, Liu QQ, Chen GJ (2013) Flaviflexus huanghaiensis gen. nov., sp. nov., an actinobacterium of the family Actinomycetaceae. Int J Syst Evol Microbiol 63(Pt 5):1863–1867. https://doi.org/10.1099/ijs.0.042044-0

Duncan KR, Haltli B, Gill KA, Kerr RG (2014) Bioprospecting from marine sediments of New Brunswick, Canada: exploring the relationship between total bacterial diversity and Actinobacteria diversity. Mar Drugs 12(2):899–925. https://doi.org/10.3390/md12020899

Egan S, Harder T, Burke C, Steinberg P, Kjelleberg S, Thomas T (2013) The seaweed holobiont: understanding seaweed–bacteria interactions. FEMS Microbiol Rev 37(3):462–476

Fenical W, Jensen PR (2006) Developing a new resource for drug discovery: marine actinomycete bacteria. Nat Chem Biol 2(12):666–673. https://doi.org/10.1038/nchembio841

Fernandez-Chimeno RI, Canedo L, Espliego F, Gravalos D, De La Calle F, Fernandez-Puentes JL, Romero F (2000) IB-96212, a novel cytotoxic macrolide produced by a marine Micromonospora I. Taxonomy, fermentation, isolation and biological activities. J Antibiot 53(5):474–478

Friedman CS, Beaman BL, Chun J, Goodfellow M, Gee A, Hedrick RP (1998) Nocardia crassostreae sp. nov., the causal agent of nocardiosis in Pacific oysters. Int J Syst Evol Microbiol 48(1):237–246

Fu Y, Li Q, Liu K, Xu Y, Wang Y, Jiao N (2012) Oceanitalea nanhaiensis gen. nov., sp. nov., an actinobacterium isolated from seawater. Int J Syst Evol Microbiol 62(Pt 10):2490–2494. https://doi.org/10.1099/ijs.0.038018-0

Fu S, Wang F, Li H, Bao Y, Yang Y, Shen H, Lin B, Zhou G (2016) Secondary metabolites from marine-derived Streptomyces antibioticus strain H74-21. Nat Prod Res 30(21):2460–2467

Fukano H, Wada S, Kurata O, Katayama K, Fujiwara N, Hoshino Y (2017) Mycobacterium stephanolepidis sp. nov., a rapidly growing species related to Mycobacterium chelonae, isolated from marine teleost fish, Stephanolepis cirrhifer. Int J Syst Evol Microbiol 67(8):2811–2817. https://doi.org/10.1099/ijsem.0.002028

Fukuda T, Takahashi M, Nagai K, Harunari E, Imada C, Tomoda H (2017) Isomethoxyneihumicin, a new cytotoxic agent produced by marine Nocardiopsis alba KM6-1. J Antibiot 70(5):590–594

Fukuda T, Nagai K, Kanamoto A, Tomoda H (2020) 2-Epi-anthracimycin, a new cytotoxic agent from the marine-derived actinomycete Streptomyces sp. OPMA00631. J Antibiot 73(8):548–553

Gao C, Guo Z, Lu X, Chen H, Liu L, Yu Z, Chen Y (2018) Hexaricins, pradimicin-like polyketides from a marine sediment-derived Streptosporangium sp. and their antioxidant effects. J Nat Prod 81(9):2069–2074

Gärtner A, Ohlendorf B, Schulz D, Zinecker H, Wiese J, Imhoff JF (2011) Levantilides A and B, 20-membered macrolides from a Micromonospora strain isolated from the Mediterranean deep sea sediment. Mar Drugs 9(1):98–108

Gaudêncio SP, Pereira F (2015) Dereplication: racing to speed up the natural products discovery process. Nat Prod Rep 32(6):779–810

Gavrish E, Bollmann A, Epstein S, Lewis K (2008) A trap for in situ cultivation of filamentous actinobacteria. J Microbiol Methods 72(3):257–262

Gavrish E, Sit CS, Cao S, Kandror O, Spoering A, Peoples A, Ling L, Fetterman A, Hughes D, Bissell A (2014) Lassomycin, a ribosomally synthesized cyclic peptide, kills Mycobacterium tuberculosis by targeting the ATP-dependent protease ClpC1P1P2. Chem Biol 21(4):509–518

Girão M, Ribeiro I, Ribeiro T, Azevedo IC, Pereira F, Urbatzka R, Leão PN, Carvalho MF (2019) Actinobacteria isolated from Laminaria ochroleuca: a source of new bioactive compounds. Front Microbiol 10:683

Giribet G, Okusu A, Lindgren AR, Huff SW, Schrödl M, Nishiguchi MK (2006) Evidence for a clade composed of molluscs with serially repeated structures: monoplacophorans are related to chitons. Proc Natl Acad Sci U S A 103(20):7723–7728. https://doi.org/10.1073/pnas.0602578103

Gong Y, Bai J-L, Yang H-T, Zhang W-D, Xiong Y-W, Ding P, Qin S (2018) Phylogenetic diversity and investigation of plant growth-promoting traits of actinobacteria in coastal salt marsh plant rhizospheres from Jiangsu, China. Syst Appl Microbiol 41(5):516–527

Gontang EA, Fenical W, Jensen PR (2007) Phylogenetic diversity of gram-positive bacteria cultured from marine sediments. Appl Environ Microbiol 73(10):3272–3282. https://doi.org/10.1128/AEM.02811-06

González V, Vargas-Straube MJ, Beys-da-Silva WO, Santi L, Valencia P, Beltrametti F, Cámara B (2020) Enzyme bioprospection of marine-derived Actinobacteria from the Chilean Coast and new insight in the mechanism of keratin degradation in Streptomyces sp. G11C. Mar Drugs 18(11):537

Gozari M, Zaheri A, Jahromi ST, Gozari M, Karimzadeh R (2019) Screening and characterization of marine actinomycetes from the northern Oman Sea sediments for cytotoxic and antimicrobial activity. Int Microbiol 22(4):521–530

Gruene T, Wennmacher JTC, Zaubitzer C, Holstein JJ, Heidler J, Fecteau-Lefebvre A, De Carlo S, Müller E, Goldie KN, Regeni I (2018) Rapid structure determination of microcrystalline molecular compounds using electron diffraction. Angew Chem Int Ed 57(50):16313–16317

Gui C, Zhang S, Zhu X, Ding W, Huang H, Gu Y-C, Duan Y, Ju J (2017) Antimicrobial spirotetronate metabolites from marine-derived Micromonospora harpali SCSIO GJ089. J Nat Prod 80(5):1594–1603

Hamada M, Iino T, Tamura T, Iwami T, Harayama S, Suzuki K (2009) Serinibacter salmoneus gen. nov., sp. nov., an actinobacterium isolated from the intestinal tract of a fish, and emended descriptions of the families Beutenbergiaceae and Bogoriellaceae. Int J Syst Evol Microbiol 59 (Pt 11):2809–2814. https://doi.org/10.1099/ijs.0.011106-0

Hamada M, Tamura T, Shibata C, Yamamura H, Hayakawa M, Suzuki K (2012) Sediminihabitans luteus gen. nov., sp. nov., a new member of the family Cellulomonadaceae isolated from sea sediment. Antonie Van Leeuwenhoek 102(2):325–333. https://doi.org/10.1007/s10482-012-9742-y

Hamada M, Saitou S, Tamura T (2017) Arenivirga flava gen. nov., sp. nov., a new member of the family Microbacteriaceae isolated from a mangrove tidal flat. Int J Syst Evol Microbiol 67(9):3318–3322

Hamada M, Saitou S, Tamura T (2019) Litorihabitans aurantiacus gen. nov., sp. nov., an actinobacterium of the family Beutenbergiaceae isolated from beach sand. Int J Syst Evol Microbiol 69(4):1202–1206

Hames-Kocabas EE, Uzel A (2012) Isolation strategies of marine-derived actinomycetes from sponge and sediment samples. J Microbiol Methods 88(3):342–347. https://doi.org/10.1016/j.mimet.2012.01.010

Han SK, Nedashkovskaya OI, Mikhailov VV, Kim SB, Bae KS (2003) Salinibacterium amurskyense gen. nov., sp. nov., a novel genus of the family Microbacteriaceae from the marine environment. Int J Syst Evol Microbiol 53(Pt 6):2061–2066. https://doi.org/10.1099/ijs.0.02627-0

Harunari E, Hamada M, Shibata C, Tamura T, Komaki H, Imada C, Igarashi Y (2016) Streptomyces hyaluromycini sp. nov., isolated from a tunicate (Molgula manhattensis). J Antibiot (Tokyo) 69(3):159–163. https://doi.org/10.1038/ja.2015.110

Harunari E, Imada C, Igarashi Y (2019) Konamycins A and B and rubromycins CA1 and CA2, aromatic polyketides from the tunicate-derived Streptomyces hyaluromycini MB-PO13T. J Nat Prod 82(6):1609–1615

Hassan HM, Degen D, Jang KH, Ebright RH, Fenical W (2015) Salinamide F, new depsipeptide antibiotic and inhibitor of bacterial RNA polymerase from a marine-derived Streptomyces sp. J Antibiot 68(3):206–209

Haszprunar G, Wanninger A (2012) Molluscs. Curr Biol 22(13):R510–R514

He YQ, Chen RW, Li C, Shi SB, Cui LQ, Long LJ, Tian XP (2020) Actinomarinicola tropica gen. nov. sp. nov., a new marine actinobacterium of the family Iamiaceae, isolated from South China Sea sediment environments. Int J Syst Evol Microbiol 70(6):3852–3858. https://doi.org/10.1099/ijsem.0.004251

Hentschel U, Piel J, Degnan SM, Taylor MW (2012) Genomic insights into the marine sponge microbiome. Nat Rev Microbiol 10(9):641–654. https://doi.org/10.1038/nrmicro2839

Hohmann C, Schneider K, Bruntner C, Irran E, Nicholson G, Bull AT, Jones AL, Brown R, Stach JEM, Goodfellow M (2009) Caboxamycin, a new antibiotic of the benzoxazole family produced by the deep-sea strain Streptomyces sp. NTK 937. J Antibiot 62(2):99–104

Holland LZ (2016) Tunicates. Curr Biol 26(4):R146–R152

Hong K, Gao A-H, Xie Q-Y, Gao HG, Zhuang L, Lin H-P, Yu H-P, Li J, Yao X-S, Goodfellow M (2009) Actinomycetes for marine drug discovery isolated from mangrove soils and plants in China. Mar Drugs 7(1):24–44

Huang H, Yang T, Ren X, Liu J, Song Y, Sun A, Ma J, Wang B, Zhang Y, Huang C (2012) Cytotoxic angucycline class glycosides from the deep sea actinomycete Streptomyces lusitanus SCSIO LR32. J Nat Prod 75(2):202–208

Huang H, Song Y, Li X, Wang X, Ling C, Qin X, Zhou Z, Li Q, Wei X, Ju J (2018) Abyssomicin monomers and dimers from the marine-derived Streptomyces koyangensis SCSIO 5802. J Nat Prod 81(8):1892–1898

Hug JJ, Bader CD, Remškar M, Cirnski K, Müller R (2018) Concepts and methods to access novel antibiotics from actinomycetes. Antibiotics 7(2):44

Hyun D-W, Shin N-R, Kim M-S, Kim PS, Kim JY, Whon TW, Bae J-W (2014) Actinomyces haliotis sp. nov., a bacterium isolated from the gut of an abalone, Haliotis discus hannai. Int J Syst Evol Microbiol 64(2):456–461

Ibrahim AH, Attia EZ, Hajjar D, Anany MA, Desoukey SY, Fouad MA, Kamel MS, Wajant H, Gulder TAM, Abdelmohsen UR (2018) New cytotoxic cyclic peptide from the marine sponge-associated Nocardiopsis sp. UR67. Mar Drugs 16(9):290

Igarashi Y, Takagi K, Kan Y, Kiyonaga Fujii K-IH, Furumai T, Oki T (2000) Arisostatins A and B, new members of tetrocarcin class of antibiotics from Micromonospora sp. TP-A0316 II. Structure determination. J Antibiot 53(3):233–240

Iniyan AM, Sudarman E, Wink J, Kannan RR, Vincent SGP (2019) Ala-geninthiocin, a new broad spectrum thiopeptide antibiotic, produced by a marine Streptomyces sp. ICN19. J Antibiot 72(2):99–105

Itoh T, Kinoshita M, Aoki S, Kobayashi M (2003) Komodoquinone A, a novel neuritogenic anthracycline, from marine Streptomyces sp. KS3. J Nat Prod 66(10):1373–1377

Jensen PR, Dwight R, Fenical W (1991) Distribution of actinomycetes in near-shore tropical marine sediments. Appl Environ Microbiol 57(4):1102–1108

Jensen PR, Gontang E, Mafnas C, Mincer TJ, Fenical W (2005) Culturable marine actinomycete diversity from tropical Pacific Ocean sediments. Environ Microbiol 7(7):1039–1048. https://doi.org/10.1111/j.1462-2920.2005.00785.x

Jensen SI, Kühl M, Priemé A (2007) Different bacterial communities associated with the roots and bulk sediment of the seagrass Zostera marina. FEMS Microbiol Ecol 62(1):108–117. https://doi.org/10.1111/j.1574-6941.2007.00373.x

Jiang Y, Li Q, Chen X, Jiang C (2016) Isolation and cultivation methods of Actinobacteria. In: Actinobacteria-basics and biotechnological applications. InTech

Jiang ZK, Pan Z, Li FN, Li XJ, Liu SW, Tuo L, Jiang MG, Sun CH (2017) Marmoricola endophyticus sp. nov., an endophytic actinobacterium isolated from Thespesia populnea. Int J Syst Evol Microbiol 67(11):4379–4384. https://doi.org/10.1099/ijsem.0.002297

Jiang ZK, Tuo L, Huang DL, Osterman IA, Tyurin AP, Liu SW, Lukyanov DA, Sergiev PV, Dontsova OA, Korshun VA, Li FN, Sun CH (2018) Diversity, novelty, and antimicrobial activity of endophytic Actinobacteria from mangrove plants in beilun estuary national nature Reserve of Guangxi, China. Front Microbiol 9:868. https://doi.org/10.3389/fmicb.2018.00868

Jose PA, Jha B (2017) Intertidal marine sediment harbours Actinobacteria with promising bioactive and biosynthetic potential. Sci Rep 7(1):10041. https://doi.org/10.1038/s41598-017-09672-6

Jose PA, Sivakala KK, Jebakumar SRD (2014) Molecular phylogeny and plant growth promoting traits of endophytic bacteria isolated from roots of seagrass Cymodocea serrulata. IJMS 43 (04):571–579

Jung S-Y, Kim H-S, Song JJ, Lee S-G, Oh T-K, Yoon J-H (2007) Aestuariimicrobium kwangyangense gen. nov., sp. nov., an LL-diaminopimelic acid-containing bacterium isolated from tidal flat sediment. Int J Syst Evol Microbiol 57(9):2114–2118

Jung D, Liu B, He X, Owen JS, Liu L, Yuan Y, Zhang W, He S (2021) Accessing previously uncultured marine microbial resources by a combination of alternative cultivation methods. Microb Biotechnol 14(3):1148–1158. https://doi.org/10.1111/1751-7915.13782

Kaeberlein T, Lewis K, Epstein SS (2002) Isolating "uncultivable" microorganisms in pure culture in a simulated natural environment. Science 296(5570):1127–1129. https://doi.org/10.1126/science.1070633

Kageyama A, Takahashi Y, Yasumoto-Hirose M, Kasai H, Shizuri Y, Omura S (2007) Janibacter corallicola sp. nov., isolated from coral in Palau. J Gen Appl Microbiol 53(3):185–189

Kageyama A, Haga T, Kasai H, Shizuri Y, Omura S, Takahashi Y (2008) Marihabitans asiaticum gen. nov., sp. nov., a meso-diaminopimelic acid-containing member of the family Intrasporangiaceae. Int J Syst Evol Microbiol 58(Pt 10):2429–2432. https://doi.org/10.1099/ijs.0.65210-0

Kai W, Peisheng Y, Rui M, Wenwen J, Zongze S (2017) Diversity of culturable bacteria in deep-sea water from the South Atlantic Ocean. Bioengineered 8(5):572–584

Kamala K, Sivaperumal P, Thilagaraj R, Natarajan E (2020) Bioremediation of Sr2+ ion radionuclide by using marine Streptomyces sp. CuOff24 extracellular polymeric substances. J Chem Technol Biotechnol 95(4):893–903

Kämpfer P, Glaeser SP, Busse H-J, Abdelmohsen UR, Hentschel U (2014) Rubrobacter aplysinae sp. nov., isolated from the marine sponge Aplysina aerophoba. Int J Syst Evol Microbiol 64(3):705–709

Kämpfer P, Glaeser SP, Busse H-J, Abdelmohsen UR, Ahmed S, Hentschel U (2015) Actinokineospora spheciospongiae sp. nov., isolated from the marine sponge Spheciospongia vagabunda. Int J Syst Evol Microbiol 65(3):879–884

Karim MRU, Harunari E, Sharma AR, Oku N, Akasaka K, Urabe D, Sibero MT, Igarashi Y (2020) Nocarimidazoles C and D, antimicrobial alkanoylimidazoles from a coral-derived actinomycete Kocuria sp.: application of 1JC, H coupling constants for the unequivocal determination of substituted imidazoles and stereochemical diversity of anteisoalkyl chains in microbial metabolites. Beilstein J Org Chem 16(1):2719–2727

Karthik L, Gaurav K, Bhaskara Rao KV, Rajakumar G, Abdul Rahuman A (2011) Larvicidal, repellent, and ovicidal activity of marine actinobacteria extracts against Culex tritaeniorhynchus and Culex gelidus. Parasitol Res 108(6):1447–1455

Kawahara T, Izumikawa M, Kozone I, Hashimoto J, Kagaya N, Koiwai H, Komatsu M, Fujie M, Sato N, Ikeda H (2018) Neothioviridamide, a polythioamide compound produced by heterologous expression of a Streptomyces sp. cryptic RiPP biosynthetic gene cluster. J Nat Prod 81(2):264–269

Kaweewan I, Komaki H, Hemmi H, Hoshino K, Hosaka T, Isokawa G, Oyoshi T, Kodani S (2019) Isolation and structure determination of a new cytotoxic peptide, curacozole, from Streptomyces curacoi based on genome mining. J Antibiot 72(1):1–7

Kershaw DR (1983) Phylum Mollusca. In: Animal diversity. Springer, Dordrecht, pp 169–191

Khan ST, Harayama S, Tamura T, Ando K, Takagi M, Kazuo SY (2009) Paraoerskovia marina gen. nov., sp. nov., an actinobacterium isolated from marine sediment. Int J Syst Evol Microbiol 59 (Pt 8):2094–2098. https://doi.org/10.1099/ijs.0.007666-0

Khan ST, Tamura T, Takagi M, Shin-ya K (2010) Streptomyces tateyamensis sp. nov., Streptomyces marinus sp. nov. and Streptomyces haliclonae sp. nov., isolated from the marine sponge Haliclona sp. Int J Syst Evol Microbiol 60(12):2775–2779. https://doi.org/10.1099/ijs.0.019869-0

Khodamoradi S, Stadler M, Wink J, Surup F (2020) Litoralimycins A and B, new cytotoxic thiopeptides from Streptomonospora sp. M2. Mar Drugs 18(6):280

Kim SH, Yang HO, Sohn YC, Kwon HC (2010) Aeromicrobium halocynthiae sp. nov., a taurocholic acid-producing bacterium isolated from the marine ascidian Halocynthia roretzi. Int J Syst Evol Microbiol 60(Pt 12):2793–2798. https://doi.org/10.1099/ijs.0.016618-0

Kim S-H, Ha T-K-Q, Oh WK, Shin J, Oh D-C (2016) Antiviral indolosesquiterpenoid xiamycins C–E from a halophilic actinomycete. J Nat Prod 79(1):51–58

Kim J, Shin D, Kim S-H, Park W, Shin Y, Kim WK, Lee SK, Oh K-B, Shin J, Oh D-C (2017a) Borrelidins C–E: new antibacterial macrolides from a saltern-derived halophilic Nocardiopsis sp. Mar Drugs 15(6):166

Kim TY, Lee JJ, Kim BS, Choi SH (2017b) Whole-body microbiota of sea cucumber (Apostichopus japonicus) from South Korea for improved seafood management. J Microbiol Biotechnol 27(10):1753–1762. https://doi.org/10.4014/jmb.1707.07067

Kim MC, Machado H, Jang KH, Trzoss L, Jensen PR, Fenical W (2018) Integration of genomic data with NMR analysis enables assignment of the full stereostructure of neaumycin B, a potent inhibitor of glioblastoma from a marine-derived Micromonospora. J Am Chem Soc 140(34):10775–10784

Kim MC, Cullum R, Machado H, Smith AJ, Yang I, Rodvold JJ, Fenical W (2019) Photopiperazines A–D, photosensitive interconverting diketopiperazines with significant and selective activity against U87 glioblastoma cells, from a rare, marine-derived Actinomycete of the family Streptomycetaceae. J Nat Prod 82(8):2262–2267

Kim MC, Cullum R, Hebishy A, Mohamed HA, Faraag AHI, Salah NM, Abdelfattah MS, Fenical W (2020a) Mersaquinone, a new tetracene derivative from the marine-derived Streptomyces sp. EG1 exhibiting activity against methicillin-resistant Staphylococcus aureus (MRSA). Antibiotics 9(5):252

Kim S, Park MS, Song J, Kang I, Cho J-C (2020b) High-throughput cultivation based on dilution-to-extinction with catalase supplementation and a case study of cultivating acI bacteria from Lake Soyang. J Microbiol 58(11):893–905

Kiran GS, Anto Thomas T, Selvin J (2010a) Production of a new glycolipid biosurfactant from marine Nocardiopsis lucentensis MSA04 in solid-state cultivation. Colloids Surf B: Biointerfaces 78(1):8–16

Kiran GS, Anto Thomas T, Joseph S, Sabarathnam B, Lipton AP (2010b) Optimization and characterization of a new lipopeptide biosurfactant produced by marine Brevibacterium aureum MSA13 in solid state culture. Bioresour Technol 101(7):2389–2396

Kiran GS, Sabarathnam B, Thajuddin N, Selvin J (2014) Production of glycolipid biosurfactant from sponge-associated marine actinobacterium Brachybacterium paraconglomeratum MSA21. J Surfactant Deterg 17(3):531–542

Kitani S, Ueguchi T, Igarashi Y, Leetanasaksakul K, Thamchaipenet A, Nihira T (2018) Rakicidin F, a new antibacterial cyclic depsipeptide from a marine sponge-derived Streptomyces sp. J Antibiot 71(1):139–141

Kondratyuk TP, Park E-J, Yu R, Van Breemen RB, Asolkar RN, Murphy BT, Fenical W, Pezzuto JM (2012) Novel marine phenazines as potential cancer chemopreventive and anti-inflammatory agents. Mar Drugs 10(2):451–464

Koyama N, Shigeno S, Kanamoto A, Tomoda H (2020) Steffimycin E, a new anti-mycobacterial agent against Mycobacterium avium complex, produced by Streptomyces sp. OPMA02852. J Antibiot 73(8):581–584

Kurahashi M, Fukunaga Y, Sakiyama Y, Harayama S, Yokota A (2010) Euzebya tangerina gen. nov., sp. nov., a deeply branching marine actinobacterium isolated from the sea cucumber Holothuria edulis, and proposal of Euzebyaceae fam. nov., Euzebyales ord. nov. and Nitriliruptoridae subclassis nov. Int J Syst Evol Microbiol 60(Pt 10):2314–2319. https://doi.org/10.1099/ijs.0.016543-0

Lai H, Wei X, Jiang Y, Chen X, Li Q, Jiang Y, Jiang C, Gillerman L (2014) Halopolysporaalba gen. nov., sp. nov., isolated from sediment. Int J Syst Evol Microbiol 64(8):2775–2780

Lai H, Jiang Y, Chen X, Li Q, Jiang C, Jiang Y, Wei X (2017) Haloactinomyces albus gen. nov., sp. nov., isolated from the dead sea. Int J Syst Evol Microbiol 67(5):1163–1168

Lampert Y, Kelman D, Dubinsky Z, Nitzan Y, Hill RT (2006) Diversity of culturable bacteria in the mucus of the Red Sea coral Fungia scutaria. FEMS Microbiol Ecol 58(1):99–108

Lampert Y, Kelman D, Nitzan Y, Dubinsky Z, Behar A, Hill RT (2008) Phylogenetic diversity of bacteria associated with the mucus of Red Sea corals. FEMS Microbiol Ecol 64(2):187–198. https://doi.org/10.1111/j.1574-6941.2008.00458.x

Le TC, Yang I, Yoon YJ, Nam S-J, Fenical W (2016) Ansalactams B–D illustrate further biosynthetic plasticity within the ansamycin pathway. Org Lett 18(9):2256–2259

Le TC, Lee EJ, Lee J, Hong A, Yim C-Y, Yang I, Choi H, Chin J, Cho SJ, Ko J (2019) Saccharoquinoline, a cytotoxic alkaloidal meroterpenoid from marine-derived bacterium Saccharomonospora sp. Mar Drugs 17(2):98

Lee SD (2006) Phycicoccus jejuensis gen. nov., sp. nov., an actinomycete isolated from seaweed. Int J Syst Evol Microbiol 56(Pt 10):2369–2373. https://doi.org/10.1099/ijs.0.64271-0

Lee SD (2008) Agrococcus jejuensis sp. nov., isolated from dried seaweed. Int J Syst Evol Microbiol 58(Pt 10):2297–2300. https://doi.org/10.1099/ijs.0.65731-0

Lee DW, Lee SD (2008) Ponticoccus gilvus gen. nov., sp. nov., a novel member of the family Propionibacteriaceae from seawater. J Microbiol 46(5):508–512

Lee D, Jang JH, Cha S, Seo T (2016a) Thalassiella azotovora gen. nov., sp. nov., a New Member of the Family Kineosporiaceae Isolated from Sea Water in South Korea. Curr Microbiol 73(5):676–683

Lee J, Han C, Lee TG, Chin J, Choi H, Lee W, Paik MJ, Won DH, Jeong G, Ko J (2016b) Marinopyrones A–D, α-pyrones from marine-derived actinomycetes of the family Nocardiopsaceae. Tetrahedron Lett 57(18):1997–2000

Leiva S, Alvarado P, Huang Y, Wang J, Garrido I (2015) Diversity of pigmented Gram-positive bacteria associated with marine macroalgae from Antarctica. FEMS Microbiol Lett 362(24). https://doi.org/10.1093/femsle/fnv206

Leutou AS, Yang I, Kang H, Seo EK, Nam S-J, Fenical W (2015) Nocarimidazoles A and B from a marine-derived actinomycete of the genus Nocardiopsis. J Nat Prod 78(11):2846–2849

Leutou AS, Yang I, Le TC, Hahn D, Lim K-M, Nam S-J, Fenical W (2018) Fluvirucin B6, a new macrolactam isolated from a marine-derived actinomycete of the genus Nocardiopsis. J Antibiot 71(6):609–612

Lewis K, Epstein S, D'onofrio A, Ling LL (2010) Uncultured microorganisms as a source of secondary metabolites. J Antibiot 63(8):468–476

Li X, Dobretsov S, Xu Y, Xiao X, Hung OS, Qian P-y (2006) Antifouling diketopiperazines produced by a deep-sea bacterium, Streptomyces fungicidicus. Biofouling 22(3):187–194

Li H-R, Yu Y, Luo W, Zeng Y-X (2010) Marisediminicola antarctica gen. nov., sp. nov., an actinobacterium isolated from the Antarctic. Int J Syst Evol Microbiol 60(11):2535–2539

Li J, Long L-J, Yang L-L, Xu Y, Wang F-Z, Li Q-X, Zhang S, Li W-J (2012a) Janibacter alkaliphilus sp. nov., isolated from coral Anthogorgia sp. Antonie Van Leeuwenhoek 102(1):157–162

Li J, Yang J, Zhu W-Y, He J, Tian X-P, Xie Q, Zhang S, Li W-J (2012b) Nocardiopsis coralliicola sp. nov., isolated from the gorgonian coral, Menella praelonga. Int J Syst Evol Microbiol 62(7):1653–1658

Li L, Mao YJ, Xie QY, Deng Z, Hong K (2013a) Micromonospora avicenniae sp. nov., isolated from a root of Avicennia marina. Antonie Van Leeuwenhoek 103(5):1089–1096. https://doi.org/10.1007/s10482-013-9888-2

Li L, Tang YL, Wei B, Xie QY, Deng Z, Hong K (2013b) Micromonospora sonneratiae sp. nov., isolated from a root of Sonneratia apetala. Int J Syst Evol Microbiol 63(Pt 7):2383–2388. https://doi.org/10.1099/ijs.0.043570-0

Li K, Liang Z, Chen W, Luo X, Fang W, Liao S, Lin X, Yang B, Wang J, Tang L (2019a) Iakyricidins a–d, antiproliferative piericidin analogues bearing a carbonyl group or cyclic skeleton from Streptomyces iakyrus scsio ns104. J Org Chem 84(19):12626–12631

Li L, Gui Y-H, Xu Q-H, Lin H-W, Lu Y-H (2019b) Spongiactinospora rosea gen. nov., sp. nov., a new member of the family Streptosporangiaceae. Int J Syst Evol Microbiol 69(2):427–433. https://doi.org/10.1099/ijsem.0.003165

Li L, Liu X, Jiang W, Yinhua L (2019c) Recent advances in synthetic biology approaches to optimize production of bioactive natural products in actinobacteria. Front Microbiol 10:2467

Li L, Wang J, Zhou Y-J, Lin H-W, Lu Y-H (2019d) Streptomyces reniochalinae sp. nov. and Streptomyces diacarni sp. nov., from marine sponges. Int J Syst Evol Microbiol 69(1):99–104. https://doi.org/10.1099/ijsem.0.003109

Li L, Xu Q-H, Wang X-T, Lin H-W, Lu Y-H (2019e) Actinomadura craniellae sp. nov., isolated from a marine sponge in the South China Sea. Int J Syst Evol Microbiol 69(4):1207–1212. https://doi.org/10.1099/ijsem.0.003295

Li L, Zhang D, Tang W-Z, Lin H-W, Lu Y-H (2019f) Geodermatophilus marinus sp. nov., isolated from the marine sponge Leucetta chagosensis. Int J Syst Evol Microbiol 69(10):2966–2971

Li L, Zhu H-R, Xu Q-H, Lin H-W, Lu Y-H (2019g) Micromonospora craniellae sp. nov., isolated from a marine sponge, and reclassification of Jishengella endophytica as Micromonospora endophytica comb. nov. Int J Syst Evol Microbiol 69(3):715–720. https://doi.org/10.1099/ijsem.0.003209

Li Q, Zheng Y, Guo A, Chen Y, Zhang S, Li J (2020) Pseudokineococcus galaxeicola sp. nov., isolated from mucus of a stony coral. Int J Syst Evol Microbiol 70(11):5671–5675

Liang Y, Xie X, Chen L, Yan S, Ye X, Anjum K, Huang H, Lian X, Zhang Z (2016) Bioactive polycyclic quinones from marine Streptomyces sp. 182SMLY. Mar Drugs 14(1):10

Lin Z, Koch M, Pond CD, Mabeza G, Seronay RA, Concepcion GP, Barrows LR, Olivera BM, Schmidt EW (2014) Structure and activity of lobophorins from a turrid mollusk-associated Streptomyces sp. J Antibiot 67(1):121–126

Liu H, Chen Z, Zhu G, Wang L, Du Y, Wang Y, Zhu W (2017) Phenolic polyketides from the marine alga-derived Streptomyces sp. OUCMDZ-3434. Tetrahedron 73(36):5451–5455

Liu Y-C, Huang R-M, Bao J, Wu K-Y, Wu H-Y, Gao X-Y, Zhang X-Y (2018) The unexpected diversity of microbial communities associated with black corals revealed by high-throughput Illumina sequencing. FEMS Microbiol Lett 365(15):fny167

Liu C, Yang C, Zeng Y, Shi J, Li L, Li W, Jiao R, Tan R, Ge H (2019a) Chartrenoline, a novel alkaloid isolated from a marine Streptomyces chartreusis NA02069. Chin Chem Lett 30(1):44–46

Liu SH, Wang W, Wang KB, Zhang B, Li W, Shi J, Jiao RH, Tan RX, Ge HM (2019b) Heterologous expression of a cryptic giant type I PKS gene cluster leads to the production of ansaseomycin. Org Lett 21(10):3785–3788

Liu T, Wu S, Zhang R, Wang D, Chen J, Zhao J (2019c) Diversity and antimicrobial potential of Actinobacteria isolated from diverse marine sponges along the Beibu Gulf of the South China Sea. FEMS Microbiol Ecol 95(7). https://doi.org/10.1093/femsec/fiz089

Liu T, Wu S, Zhang R, Wang D, Chen J, Zhao J (2019d) Diversity and antimicrobial potential of Actinobacteria isolated from diverse marine sponges along the Beibu Gulf of the South China Sea. FEMS Microbiol Ecol 95(7):fiz089

López-García MT, Rioseras B, Yagüe P, Álvarez JR, Manteca Á (2014) Cell immobilization of Streptomyces coelicolor: effect on differentiation and actinorhodin production. Int Microbiol 17(2):75

López-Pérez M, Haro-Moreno JM, Iranzo J, Rodriguez-Valera F (2020) Genomes of the "Candidatus Actinomarinales" order: highly streamlined marine epipelagic Actinobacteria. mSystems 5(6). https://doi.org/10.1128/mSystems.01041-20

Lozupone CA, Knight R (2007) Global patterns in bacterial diversity. Proc Natl Acad Sci 104(27):11436–11440

Lu J, Ma Y, Liang J, Xing Y, Xi T, Yuanyuan L (2012) Aureolic acids from a marine-derived Streptomyces sp. WBF16. Microbiol Res 167(10):590–595

Mahler L, Wink K, Julia Beulig R, Scherlach K, Tovar M, Zang E, Martin K, Hertweck C, Belder D, Roth M (2018) Detection of antibiotics synthetized in microfluidic picolitre-droplets by various actinobacteria. Sci Rep 8(1):1–11

Mahmoud HM, Kalendar AA (2016) Coral-associated actinobacteria: diversity, abundance, and biotechnological potentials. Front Microbiol 7:204

Maldonado LA, Fenical W, Jensen PR, Kauffman CA, Mincer TJ, Ward AC, Bull AT, Goodfellow M (2005) Salinispora arenicola gen. nov., sp. nov. and Salinispora tropica sp. nov., obligate marine actinomycetes belonging to the family Micromonosporaceae. Int J Syst Evol Microbiol 55(Pt 5):1759–1766. https://doi.org/10.1099/ijs.0.63625-0

Marx LB, Burton JW (2018) A Total Synthesis of Salinosporamide A. Chem Eur J 24(26):6747–6754

Maskey RP, Helmke E, Kayser O, Fiebig HH, Maier A, Busche A, Laatsch H (2004) Anti-cancer and antibacterial trioxacarcins with high anti-malaria activity from a marine Streptomycete and their absolute stereochemistry. J Antibiot 57(12):771–779

Mathew BT, Torky Y, Amin A, Mourad A-HI, Ayyash MM, El-Keblawy A, Hilal-Alnaqbi A, AbuQamar SF, El-Tarabily KA (2020) Halotolerant marine rhizosphere-competent actinobacteria promote Salicornia bigelovii growth and seed production using seawater irrigation. Front Microbiol 11:552

Meena B, Anburajan L, Vinithkumar NV, Kirubagaran R, Dharani G (2019) Biodiversity and antibacterial potential of cultivable halophilic actinobacteria from the deep sea sediments of active volcanic Barren Island. Microb Pathog 132:129–136. https://doi.org/10.1016/j.micpath.2019.04.043

Michl SC, Ratten JM, Beyer M, Hasler M, LaRoche J, Schulz C (2017) The malleable gut microbiome of juvenile rainbow trout (Oncorhynchus mykiss): diet-dependent shifts of bacterial community structures. PLoS One 12(5):e0177735. https://doi.org/10.1371/journal.pone.0177735

Mincer TJ, Jensen PR, Kauffman CA, Fenical W (2002) Widespread and persistent populations of a major new marine actinomycete taxon in ocean sediments. Appl Environ Microbiol 68(10):5005–5011

Moon K, Xu F, Seyedsayamdost MR (2019) Cebulantin, a cryptic lanthipeptide antibiotic uncovered using bioactivity-coupled HiTES. Angew Chem 131(18):6034–6038

Moumbock AFA, Gao M, Qaseem A, Li J, Kirchner PA, Ndingkokhar B, Bekono BD, Simoben CV, Babiaka SB, Malange YI (2021) StreptomeDB 3.0: an updated compendium of streptomycetes natural products. Nucleic Acids Res 49(D1):D600–D604

Mudalungu CM, von Törne WJ, Voigt K, Rückert C, Schmitz S, Sekurova ON, Zotchev SB, Süssmuth RD (2019) Noursamycins, chlorinated cyclohexapeptides identified from molecular networking of Streptomyces noursei NTR-SR4. J Nat Prod 82(6):1478–1486

Mungan MD, Alanjary M, Blin K, Weber T, Medema MH, Ziemert N (2020) ARTS 2.0: feature updates and expansion of the Antibiotic Resistant Target Seeker for comparative genome mining. Nucleic Acids Res 48(W1):W546–W552

Muthusamy S, Selvan ST, Arunachalam P, Grasian I (2019) Bioconversion and bioethanol production from agro-residues through fermentation process using mangrove-associated actinobacterium Streptomyces olivaceus (MSU3). Biofuels 10(2):167–179

Naikpatil SV, Rathod JL (2011) Selective isolation and antimicrobial activity of rare actinomycetes from mangrove sediment of Karwar. J Ecobiotechnol 3(10):48–53

Najafi A, Moradinasab M, Nabipour I (2018) First record of microbiomes of sponges collected from the Persian Gulf, using tag pyrosequencing. Front Microbiol 9:1500. https://doi.org/10.3389/fmicb.2018.01500

Nathani NM, Mootapally C, Gadhvi IR, Maitreya B, Joshi CG (2020) Marine niche: applications in pharmaceutical sciences, 1st edn. Springer, Singapore

Navarro-Muñoz JC, Selem-Mojica N, Mullowney MW, Kautsar SA, Tryon JH, Parkinson EI, Santos ELCDL, Yeong M, Cruz-Morales P, Abubucker S (2020) A computational framework to explore large-scale biosynthetic diversity. Nat Chem Biol 16(1):60–68

Ng ZY, Fang BZ, Li WJ, Tan GYA (2019) Marinitenerispora sediminis gen. nov., sp. nov., a member of the family Nocardiopsaceae isolated from marine sediment. Int J Syst Evol Microbiol 69(10):3031–3040. https://doi.org/10.1099/ijsem.0.003587

Nichols D, Cahoon N, Trakhtenberg EM, Pham L, Mehta A, Belanger A, Kanigan T, Lewis K, Epstein SS (2010) Use of ichip for high-throughput in situ cultivation of "uncultivable" microbial species. Appl Environ Microbiol 76(8):2445–2450

Nie Y-L, Wu Y-D, Wang C-X, Lin R, Xie Y, Fang D-S, Jiang H, Lian Y-Y (2018) Structure elucidation and antitumour activity of a new macrolactam produced by marine-derived actinomycete Micromonospora sp. FIM05328. Nat Prod Res 32(18):2133–2138

Nimnoi P, Pongsilp N (2020) Marine bacterial communities in the upper gulf of Thailand assessed by Illumina next-generation sequencing platform. BMC Microbiol 20(1):19. https://doi.org/10.1186/s12866-020-1701-6

Nithyanand P, Pandian SK (2009) Phylogenetic characterization of culturable bacterial diversity associated with the mucus and tissue of the coral Acropora digitifera from the Gulf of Mannar. FEMS Microbiol Ecol 69(3):384–394

Nong X-H, Zhang X-Y, Xu X-Y, Wang J, Qi S-H (2016) Nahuoic acids B–E, polyhydroxy polyketides from the marine-derived Streptomyces sp. SCSGAA 0027. J Nat Prod 79(1):141–148

Nong X-H, Wei X-Y, Qi S-H (2017) Pteridic acids C–G spirocyclic polyketides from the marine-derived Streptomyces sp. SCSGAA 0027. J Antibiot 70(11):1047–1052

Nouioui I, Rückert C, Willemse J, van Wezel GP, Klenk HP, Busche T, Kalinowski J, Bredholt H, Zotchev SB (2017) Actinoalloteichus fjordicus sp. nov. isolated from marine sponges: phenotypic, chemotaxonomic and genomic characterisation. Antonie Van Leeuwenhoek 110(12):1705–1717. https://doi.org/10.1007/s10482-017-0920-9

Ochoa JL, Sanchez LM, Koo B-M, Doherty JS, Rajendram M, Huang KC, Gross CA, Linington RG (2018) Marine mammal microbiota yields novel antibiotic with potent activity against Clostridium difficile. ACS Infect Dis 4(1):59–67

Oh D-C, Strangman WK, Kauffman CA, Jensen PR, Fenical W (2007) Thalassospiramides A and B, immunosuppressive peptides from the marine bacterium Thalassospira sp. Org Lett 9(8):1525–1528

Olson JB, Harmody DK, Bej AK, McCarthy PJ (2007) Tsukamurella spongiae sp. nov., a novel actinomycete isolated from a deep-water marine sponge. Int J Syst Evol Microbiol 57(7):1478–1481

Overmann J, Lepleux C (2016) Marine bacteria and archaea: diversity, adaptations, and culturability. In: Stal LJ, Cretoiu MS (eds) The marine microbiome: an untapped source of biodiversity and biotechnological potential. Springer International Publishing, Cham, pp 21–55

Patin NV, Schorn M, Aguinaldo K, Lincecum T, Moore BS, Jensen PR (2017) Effects of actinomycete secondary metabolites on sediment microbial communities. Appl Environ Microbiol 83(4). https://doi.org/10.1128/aem.02676-16

Penesyan A, Kjelleberg S, Egan S (2010) Development of novel drugs from marine surface associated microorganisms. Mar Drugs 8(3):438–459

Peoples AJ, Zhang Q, Millett WP, Rothfeder MT, Pescatore BC, Madden AA, Ling LL, Moore CM (2008) Neocitreamicins I and II, novel antibiotics with activity against methicillin-resistant Staphylococcus aureus and vancomycin-resistant Enterococci. J Antibiot 61(7):457–463

Peoples A, Zhang Q, Moore C, Ling L, Rothfeder M, Lewis K (2012) Macrolactam compounds. Google Patents

Peraud O, Biggs JS, Hughen RW, Light AR, Concepcion GP, Olivera BM, Schmidt EW (2009) Microhabitats within venomous cone snails contain diverse actinobacteria. Appl Environ Microbiol 75(21):6820–6826. https://doi.org/10.1128/AEM.01238-09

Pérez M, Schleissner C, Fernández R, Rodríguez P, Reyes F, Zuñiga P, De La Calle F, Cuevas C (2016) PM100117 and PM100118, new antitumor macrolides produced by a marine Streptomyces caniferus GUA-06-05-006A. J Antibiot 69(5):388–394

Pérez-Bonilla M, Oves-Costales D, De la Cruz M, Kokkini M, Martín J, Vicente F, Genilloud O, Reyes F (2018) Phocoenamicins B and C, new antibacterial spirotetronates isolated from a marine Micromonospora sp. Mar Drugs 16(3):95

Pérez-Victoria I, Oves-Costales D, Lacret R, Martín J, Sánchez-Hidalgo M, Díaz C, Cautain B, Vicente F, Genilloud O, Reyes F (2019) Structure elucidation and biosynthetic gene cluster analysis of caniferolides A–D, new bioactive 36-membered macrolides from the marine-derived Streptomyces caniferus CA-271066. Org Biomol Chem 17(11):2954–2971

Pimentel-Elardo SM, Tiro LP, Grozdanov L, Hentschel U (2008) Saccharopolyspora cebuensis sp. nov., a novel actinomycete isolated from a Philippine sponge (Porifera). Int J Syst Evol Microbiol 58(3):628–632

Pimentel-Elardo SM, Scheuermayer M, Kozytska S, Hentschel U (2009) Streptomyces axinellae sp. nov., isolated from the Mediterranean sponge Axinella polypoides (Porifera). Int J Syst Evol Microbiol 59(6):1433–1437. https://doi.org/10.1099/ijs.0.007856-0

Pimentel-Elardo SM, Kozytska S, Bugni TS, Ireland CM, Moll H, Hentschel U (2010) Anti-parasitic compounds from Streptomyces sp. strains isolated from Mediterranean sponges. Mar Drugs 8(2):373–380

Pratheepa VK, Silva M, Vasconcelos V (2014) Viscera-associated bacterial diversity among intertidal gastropods from Northern-Atlantic Coast of Portugal. Curr Microbiol 68(2):140–148. https://doi.org/10.1007/s00284-013-0450-2

Pratheepa V, Alex A, Silva M, Vasconcelos V (2016) Bacterial diversity and tetrodotoxin analysis in the viscera of the gastropods from Portuguese coast. Toxicon 119:186–193. https://doi.org/10.1016/j.toxicon.2016.06.003

Purahong W, Sadubsarn D, Tanunchai B, Wahdan SFM, Sansupa C, Noll M, Wu YT, Buscot F (2019) First insights into the microbiome of a mangrove tree reveal significant differences in taxonomic and functional composition among plant and soil compartments. Microorganisms 7(12). https://doi.org/10.3390/microorganisms7120585

Rajagopal G, Kannan S (2017) Systematic characterization of potential cellulolytic marine actinobacteria Actinoalloteichus sp. MHA15. Biotechnol Rep 13:30–36

Raju R, Piggott AM, Conte MM, Capon RJ (2010) Heronamides A–C, new polyketide macrolactams from an Australian marine-derived Streptomyces sp. A biosynthetic case for synchronized tandem electrocyclization. Org Biomol Chem 8(20):4682–4689

Rakotondraibe LH, Rasolomampianina R, Park H-Y, Li J, Slebodnik C, Brodie PJ, Blasiak LC, Hill R, TenDyke K, Shen Y (2015) Antiproliferative and antiplasmodial compounds from selected Streptomyces species. Bioorg Med Chem Lett 25(23):5646–5649

Ramaprasad EVV, Mahidhara G, Sasikala C, Ramana CV (2018) Rhodococcus electrodiphilus sp. nov., a marine electro active actinobacterium isolated from coral reef. Int J Syst Evol Microbiol 68(8):2644–2649

Raveh A, Delekta PC, Dobry CJ, Peng W, Schultz PJ, Blakely PK, Tai AW, Matainaho T, Irani DN, Sherman DH (2013) Discovery of potent broad spectrum antivirals derived from marine actinobacteria. PLoS One 8(12):e82318

Rego A, Raio F, Martins TP, Ribeiro H, Sousa AGG, Séneca J, Baptista MS, Lee CK, Cary SC, Ramos V, Carvalho MF, Leão PN, Magalhães C (2019) Actinobacteria and cyanobacteria diversity in terrestrial antarctic microenvironments evaluated by culture-dependent and independent methods. Front Microbiol 10(1018). https://doi.org/10.3389/fmicb.2019.01018

Renner MK, Shen Y-C, Cheng X-C, Jensen PR, Frankmoelle W, Kauffman CA, Fenical W, Lobkovsky E, Clardy J (1999) Cyclomarins A−C, new antiinflammatory cyclic peptides produced by a marine bacterium (Streptomyces sp.). J Am Chem Soc 121(49):11273–11276

Reynolds KA, Luhavaya H, Li J, Dahesh S, Nizet V, Yamanaka K, Moore BS (2018) Isolation and structure elucidation of lipopeptide antibiotic taromycin B from the activated taromycin biosynthetic gene cluster. J Antibiot 71(2):333–338

Rhodes MW, Kator H, Kotob S, van Berkum P, Kaattari I, Vogelbein W, Quinn F, Floyd MM, Butler WR, Ottinger CA (2003) Mycobacterium shottsii sp. nov., a slowly growing species isolated from Chesapeake Bay striped bass (Morone saxatilis). Int J Syst Evol Microbiol 53 (Pt 2):421–424. https://doi.org/10.1099/ijs.0.02299-0

Ribeiro I, Girão M, Alexandrino DAM, Ribeiro T, Santos C, Pereira F, Mucha AP, Urbatzka R, Leão PN, Carvalho MF (2020) Diversity and bioactive potential of actinobacteria isolated from a coastal marine sediment in northern portugal. Microorganisms 8(11). https://doi.org/10.3390/microorganisms8111691

Ricart E, Pupin M, Müller M, Lisacek F (2020) Automatic annotation and dereplication of tandem mass spectra of peptidic natural products. Anal Chem 92(24):15862–15871

Rodríguez V, Martín J, Sarmiento-Vizcaíno A, De la Cruz M, García LA, Blanco G, Reyes F (2018) Anthracimycin B, a potent antibiotic against gram-positive bacteria isolated from cultures of the deep-sea actinomycete Streptomyces cyaneofuscatus M-169. Mar Drugs 16(11):406

Ryu M-J, Hwang S, Kim S, Yang I, Oh D-C, Nam S-J, Fenical W (2019) Meroindenon and merochlorins E and F, antibacterial meroterpenoids from a marine-derived sediment bacterium of the genus Streptomyces. Org Lett 21(15):5779–5783

Saha S, Zhang W, Zhang G, Zhu Y, Chen Y, Liu W, Yuan C, Zhang Q, Zhang H, Zhang L (2017) Activation and characterization of a cryptic gene cluster reveals a cyclization cascade for polycyclic tetramate macrolactams. Chem Sci 8(2):1607–1612

Saito S, Kato W, Ikeda H, Katsuyama Y, Ohnishi Y, Imoto M (2020) Discovery of "heat shock metabolites" produced by thermotolerant actinomycetes in high-temperature culture. J Antibiot 73(4):203–210

Sanchez LM, Wong WR, Riener RM, Schulze CJ, Linington RG (2012) Examining the fish microbiome: vertebrate-derived bacteria as an environmental niche for the discovery of unique marine natural products. PLoS One 7(5):e35398. https://doi.org/10.1371/journal.pone.0035398

Sarmiento Vizcaíno A, Iglesias VG, Braña AJF, Ramírez AM, Fernández JLA, Díaz LAG, Blanco MGB (2015) Myceligenerans cantabricum sp. nov., a barotolerant actinobacterium isolated from a deep cold-water coral. Int J Syst Evol Microbiol 65(4)

Sarmiento-Vizcaíno A, Braña AF, Pérez-Victoria I, Martín J, De Pedro N, De la Cruz M, Díaz C, Vicente F, Acuña JL, Reyes F (2017) Paulomycin g, a new natural product with cytotoxic activity against tumor cell lines produced by deep-sea sediment derived Micromonospora matsumotoense M-412 from the Avilés canyon in the Cantabrian Sea. Mar Drugs 15(9):271

Sayed AM, Hassan MHA, Alhadrami HA, Hassan HM, Goodfellow M, Rateb ME (2020) Extreme environments: microbiology leading to specialized metabolites. J Appl Microbiol 128(3):630–657. https://doi.org/10.1111/jam.14386

Schmitt S, Tsai P, Bell J, Fromont J, Ilan M, Lindquist N, Perez T, Rodrigo A, Schupp PJ, Vacelet J, Webster N, Hentschel U, Taylor MW (2012) Assessing the complex sponge microbiota: core, variable and species-specific bacterial communities in marine sponges. ISME J 6(3):564–576. https://doi.org/10.1038/ismej.2011.116

Schreiber L, Kjeldsen KU, Funch P, Jensen J, Obst M, López-Legentil S, Schramm A (2016) Endozoicomonas are specific, facultative symbionts of sea squirts. Front Microbiol 7(1042). https://doi.org/10.3389/fmicb.2016.01042

Schulze CJ, Navarro G, Ebert D, DeRisi J, Linington RG (2015) Salinipostins A–K, long-chain bicyclic phosphotriesters as a potent and selective antimalarial chemotype. J Org Chem 80(3):1312–1320

Sekurova ON, Schneider O, Zotchev SB (2019) Novel bioactive natural products from bacteria via bioprospecting, genome mining and metabolic engineering. Microb Biotechnol 12(5):828–844. https://doi.org/10.1111/1751-7915.13398

Selvin J, Sathiyanarayanan G, Lipton AN, Al-Dhabi NA, Arasu MV, Kiran GS (2016) Ketide synthase (KS) domain prediction and analysis of iterative type II PKS gene in marine sponge associated actinobacteria producing biosurfactants and antimicrobial agents. Front Microbiol 7: 63

Shaala LA, Youssef DTA, Badr JM, Harakeh SM, Genta-Jouve G (2019) Bioactive diketopiperazines and nucleoside derivatives from a sponge-derived Streptomyces species. Mar Drugs 17(10):584

Shamikh YI, El Shamy AA, Gaber Y, Abdelmohsen UR, Madkour HA, Horn H, Hassan HM, Elmaidomy AH, Alkhalifah DHM, Hozzein WN (2020) Actinomycetes from the red sea sponge Coscinoderma mathewsi: isolation, diversity, and potential for bioactive compounds discovery. Microorganisms 8(5):783

Sharma AK, Kikani BA, Singh SP (2021) Diversity and phylogeny of actinomycetes of Arabian sea along the Gujarat Coast. Geomicrobiol J 38(4):347–364. https://doi.org/10.1080/01490451.2020.1860165

Sheikh M, Rathore DS, Gohel S, Singh SP (2019) Cultivation and characteristics of the Marine Actinobacteria from the Sea water of Alang, Bhavnagar

Shi J, Zeng YJ, Zhang B, Shao FL, Chen YC, Xu X, Sun Y, Xu Q, Tan RX, Ge HM (2019) Comparative genome mining and heterologous expression of an orphan NRPS gene cluster direct the production of ashimides. Chem Sci 10(10):3042–3048

Shin HJ, Lee H-S, Lee JS, Shin J, Lee MA, Lee H-S, Lee Y-J, Yun J, Kang JS (2014) Violapyrones H and I, new cytotoxic compounds isolated from Streptomyces sp. associated with the marine starfish Acanthaster planci. Mar Drugs 12(6):3283–3291

Shin YS, Cha HY, Lee B-S, Kang SU, Hwang HS, Kwon HC, Kim C-H, Choi EC (2016) Anti-cancer effect of luminacin, a marine microbial extract, in head and neck squamous cell carcinoma progression via autophagic cell death. Cancer Res Treat 48(2):738

Shin D, Byun WS, Moon K, Kwon Y, Bae M, Um S, Lee SK, Oh D-C (2018) Coculture of marine Streptomyces sp. with Bacillus sp. produces a new piperazic acid-bearing cyclic peptide. Front Chem 6:498

Silva FS, Souza DT, Zucchi TD, Pansa CC, de Figueiredo Vasconcellos RL, Crevelin EJ, de Moraes LA, Melo IS (2016) Streptomyces atlanticus sp. nov., a novel actinomycete isolated from marine sponge Aplysina fulva (Pallas, 1766). Antonie Van Leeuwenhoek 109(11):1467–1474. https://doi.org/10.1007/s10482-016-0748-8

Sivasankar P, Rekadwad B, Poongodi S, Sivakumar K, Parli BV, Anil Kumar N (2018) Bioinformatics delimitation of the psychrophilic and psychrotolerant actinobacteria isolated from the Polar Frontal waters of the Southern Ocean. Data Brief 18:576–584

Soest V, Rob WM, Boury-Esnault N, Vacelet J, Dohrmann M, Erpenbeck D, De Voogd NJ, Santodomingo N, Vanhoorne B, Kelly M, Hooper JNA (2012) Global diversity of sponges (porifera). PLoS One 7(4):e35105. https://doi.org/10.1371/journal.pone.0035105

Solano G, Rojas-Jimenez K, Jaspars M, Tamayo-Castillo G (2009) Study of the diversity of culturable actinomycetes in the North Pacific and Caribbean coasts of Costa Rica. Antonie Van Leeuwenhoek 96(1):71–78. https://doi.org/10.1007/s10482-009-9337-4

Song Y, Yang J, Yu J, Li J, Yuan J, Wong N-K, Ju J (2020) Chlorinated bis-indole alkaloids from deep-sea derived Streptomyces sp. SCSIO 11791 with antibacterial and cytotoxic activities. J Antibiot 73(8):542–547

Souza DT, da Silva FSP, da Silva LJ, Crevelin EJ, Moraes LAB, Zucchi TD, Melo IS (2017) Saccharopolyspora spongiae sp. nov., a novel actinomycete isolated from the marine sponge Scopalina ruetzleri (Wiedenmayer, 1977). Int J Syst Evol Microbiol 67(6):2019–2025

Sproule A, Correa H, Decken A, Haltli B, Berrué F, Overy DP, Kerr RG (2019) Terrosamycins A and B, bioactive polyether ionophores from Streptomyces sp. RKND004 from Prince Edward Island sediment. Mar Drugs 17(6):347

Steinert G, Taylor MW, Schupp PJ (2015) Diversity of actinobacteria associated with the marine Ascidian Eudistoma toealensis. Mar Biotechnol (NY) 17(4):377–385. https://doi.org/10.1007/s10126-015-9622-3

Stewart EJ (2012) Growing unculturable bacteria. J Bacteriol 194(16):4151–4160. https://doi.org/10.1128/jb.00345-12

Stincone P, Brandelli A (2020) Marine bacteria as source of antimicrobial compounds. Crit Rev Biotechnol 40(3):306–319. https://doi.org/10.1080/07388551.2019.1710457

Strand M, Carlsson M, Uvell H, Islam K, Edlund K, Cullman I, Altermark B, Mei Y-F, Elofsson M, Willassen N-P (2014) Isolation and characterization of anti-adenoviral secondary metabolites from marine actinobacteria. Mar Drugs 12(2):799–821

Subramani R, Aalbersberg W (2013) Culturable rare Actinomycetes: diversity, isolation and marine natural product discovery. Appl Microbiol Biotechnol 97(21):9291–9321. https://doi.org/10.1007/s00253-013-5229-7

Subramani R, Sipkema D (2019) Marine rare Actinomycetes: a promising source of structurally diverse and unique novel natural products. Mar Drugs 17(5):249. https://doi.org/10.3390/md17050249

Sun W, Dai S, Jiang S, Wang G, Liu G, Wu H, Li X (2010) Culture-dependent and culture-independent diversity of Actinobacteria associated with the marine sponge Hymeniacidon perleve from the South China Sea. Antonie Van Leeuwenhoek 98(1):65–75

Sun M, Zhang X, Hao H, Li W, Lu C (2015) Nocarbenzoxazoles A–G, benzoxazoles produced by halophilic Nocardiopsis lucentensis DSM 44048. J Nat Prod 78(8):2123–2127

Sun M, Chen X, Li W, Lu C, Shen Y (2017) New diketopiperazine derivatives with cytotoxicity from Nocardiopsis sp. YIM M13066. J Antibiot 70(6):795–797

Sun C, Yang Z, Zhang C, Liu Z, He J, Liu Q, Zhang T, Jianhua J, Ma J (2019) Genome mining of Streptomyces atratus SCSIO ZH16: discovery of atratumycin and identification of its biosynthetic gene cluster. Org Lett 21(5):1453–1457

Supong K, Suriyachadkun C, Pittayakhajonwut P, Suwanborirux K, Thawai C (2013a) Micromonospora spongicola sp. nov., an actinomycete isolated from a marine sponge in the Gulf of Thailand. J Antibiot 66(9):505–509

Supong K, Suriyachadkun C, Suwanborirux K, Pittayakhajonwut P, Thawai C (2013b) Verrucosispora andamanensis sp. nov., isolated from a marine sponge. Int J Syst Evol Microbiol 63(11):3970–3974

Taira J, Morita K, Kawashima S, Umei T, Baba H, Maruoka T, Komatsu H, Sakamoto H, Sacchettini JC, Aoki S (2017) Identification of a novel class of small compounds with anti-tuberculosis activity by in silico structure-based drug screening. J Antibiot 70(11):1057–1064

Takehana Y, Umekita M, Hatano M, Kato C, Sawa R, Igarashi M (2017) Fradiamine A, a new siderophore from the deep-sea actinomycete Streptomyces fradiae MM456M-mF7. J Antibiot 70(5):611–615

Tawfike A, Attia EZ, Desoukey SY, Hajjar D, Makki AA, Schupp PJ, Edrada-Ebel RA, Abdelmohsen UR (2019) New bioactive metabolites from the elicited marine sponge-derived bacterium Actinokineospora spheciospongiae sp. nov. AMB Express 9(1):1–9

Thawai C, Rungjindamai N, Klanbut K, Tanasupawat S (2017) Nocardia xestospongiae sp. nov., isolated from a marine sponge in the Andaman Sea. Int J Syst Evol Microbiol 67(5):1451–1456

Tian XP, Tang SK, Dong JD, Zhang YQ, Xu LH, Zhang S, Li WJ (2009a) Marinactinospora thermotolerans gen. nov., sp. nov., a marine actinomycete isolated from a sediment in the northern South China Sea. Int J Syst Evol Microbiol 59(Pt 5):948–952. https://doi.org/10.1099/ijs.0.005231-0

Tian XP, Zhi XY, Qiu YQ, Zhang YQ, Tang SK, Xu LH, Zhang S, Li WJ (2009b) Sciscionella marina gen. nov., sp. nov., a marine actinomycete isolated from a sediment in the northern South China Sea. Int J Syst Evol Microbiol 59(Pt 2):222–228. https://doi.org/10.1099/ijs.0.001982-0

Toledo G, Green W, Gonzalez RA, Christoffersen L, Podar M, Chang HW, Hemscheidt T, Trapido-Rosenthal HG, Short JM, Bidigare RR (2006) High throughput cultivation for isolation of novel marine microorganisms. Oceanography 19(2):120–124

Tuo L, Pan Z, Li FN, Lou I, Guo M, Lee SM, Chen L, Hu L, Sun CH (2016) Friedmanniella endophytica sp. nov., an endophytic actinobacterium isolated from bark of Kandelia candel. Int J Syst Evol Microbiol 66(8):3057–3062. https://doi.org/10.1099/ijsem.0.001146

Ue H, Matsuo Y, Kasai H, Yokota A (2011) Miniimonas arenae gen. nov., sp. nov., an actinobacterium isolated from sea sand. Int J Syst Evol Microbiol 61(1):123–127

Undabarrena A, Beltrametti F, Claverías FP, González M, Moore ERB, Seeger M, Cámara B (2016) Exploring the diversity and antimicrobial potential of marine actinobacteria from the comau fjord in Northern Patagonia, Chile. Front Microbiol 7:1135

Uzair B, Menaa F, Khan BA, Mohammad FV, Ahmad VU, Djeribi R, Menaa B (2018) Isolation, purification, structural elucidation and antimicrobial activities of kocumarin, a novel antibiotic isolated from actinobacterium Kocuria marina CMG S2 associated with the brown seaweed Pelvetia canaliculata. Microbiol Res 206:186–197

Valliappan K, Sun W, Li Z (2014) Marine actinobacteria associated with marine organisms and their potentials in producing pharmaceutical natural products. Appl Microbiol Biotechnol 98(17):7365–7377

Vester JK, Glaring MA, Stougaard P (2015) Improved cultivation and metagenomics as new tools for bioprospecting in cold environments. Extremophiles 19(1):17–29

Wang M, Carver JJ, Phelan VV, Sanchez LM, Garg N, Peng Y, Nguyen DD, Watrous J, Kapono CA, Luzzatto-Knaan T (2016) Sharing and community curation of mass spectrometry data with Global Natural Products Social Molecular Networking. Nat Biotechnol 34(8):828–837

Wang Q, Zhang Y, Wang M, Tan Y, Hu X, He H, Xiao C, You X, Wang Y, Gan M (2017a) Neo-actinomycins A and B, natural actinomycins bearing the 5H-oxazolo[4,5-b]phenoxazine chromophore, from the marine-derived Streptomyces sp. IMB094. Sci Rep 7(1):1–8

Wang S-B, Wang J-X, Liu X-H, Fan Y-P, Jiang R, Liu M-H, Liu X-Z, Wu D-M (2017b) Diversity and abundance of bacteria in the surface seawater of the Changjiang Estuary and its adjacent areas. Evol Ecol Res 18(5):515–530

Wang J, Leiva S, Huang J, Huang Y (2018) Amycolatopsis antarctica sp. nov., isolated from the surface of an Antarctic brown macroalga. Int J Syst Evol Microbiol 68(7):2348–2356. https://doi.org/10.1099/ijsem.0.002844

Wang D, Wang Y, Ouyang Y, Fu P, Zhu W (2019a) Cytotoxic p-Terphenyls from a marine-derived Nocardiopsis species. J Nat Prod 82(12):3504–3508

Wang Z, Wen Z, Liu L, Zhu X, Shen B, Yan X, Duan Y, Huang Y (2019b) Yangpumicins F and G, enediyne congeners from Micromonospora yangpuensis DSM 45577. J Nat Prod 82(9):2483–2488

Wang M, Ma Y, Feng C, Cai L, Li W (2020) Diversity of pelagic and benthic bacterial assemblages in the Western Pacific Ocean. Front Microbiol 11:1730. https://doi.org/10.3389/fmicb.2020.01730

Webster NS, Taylor MW (2012) Marine sponges and their microbial symbionts: love and other relationships. Environ Microbiol 14(2):335–346. https://doi.org/10.1111/j.1462-2920.2011.02460.x

Wibowo JT, Kellermann MY, Köck M, Putra MY, Murniasih T, Mohr KI, Wink J, Praditya DF, Steinmann E, Schupp PJ (2021) Anti-infective and antiviral activity of valinomycin and its analogues from a sea cucumber-associated bacterium, Streptomyces sp. SV 21. Mar Drugs 19(2):81

Williams DE, Dalisay DS, Chen J, Polishchuck EA, Patrick BO, Narula G, Ko M, Av-Gay Y, Li H, Magarvey N (2017) Aminorifamycins and sporalactams produced in culture by a Micromonospora sp. isolated from a Northeastern-Pacific marine sediment are potent antibiotics. Org Lett 19(4):766–769

Wright GD (2019) Unlocking the potential of natural products in drug discovery. Microb Biotechnol 12(1):55–57

Wu H, Chen W, Wang G, Dai S, Zhou D, Zhao H, Guo Y, Ouyang Y, Li X (2012) Culture-dependent diversity of Actinobacteria associated with seagrass (Thalassia hemprichii). Afr J Microbiol Res 6:87–94

Wu J-F, Li J, You Z-Q, Zhang S (2014) Prauserella coralliicola sp. nov., isolated from the coral Galaxea fascicularis. Int J Syst Evol Microbiol 64(10):3341–3345

Wu Q, Deering RW, Zhang G, Wang B, Li X, Sun J, Chen J, Zhang H, Rowley DC, Wang H (2018) Albisporachelin, a new hydroxamate type siderophore from the deep ocean sediment-derived actinomycete Amycolatopsis albispora WP1T. Mar Drugs 16(6):199

Wu C, van der Heul HU, Melnik AV, Lübben J, Dorrestein PC, Minnaard AJ, Choi YH, van Wezel GP (2019) Lugdunomycin, an angucycline-derived molecule with unprecedented chemical architecture. Angew Chem Int Ed 58(9):2809–2814

Xiao J, Wang Y, Luo Y, Xie SJ, Ruan JS, Xu J (2009) Streptomyces avicenniae sp. nov., a novel actinomycete isolated from the rhizosphere of the mangrove plant Avicennia mariana. Int J Syst Evol Microbiol 59(Pt 10):2624–2628. https://doi.org/10.1099/ijs.0.009357-0

Xiao C, Huang H, Ye J, Wu X, Zhu J, Zhan B, Bao S (2011) Ornithinibacter aureus gen. nov., sp. nov., a novel member of the family Intrasporangiaceae. Int J Syst Evol Microbiol 61(3):659–664

Xie QY, Wang C, Wang R, Qu Z, Lin HP, Goodfellow M, Hong K (2011) Jishengella endophytica gen. nov., sp. nov., a new member of the family Micromonosporaceae. Int J Syst Evol Microbiol 61(Pt 5):1153–1159. https://doi.org/10.1099/ijs.0.025288-0

Xie C-L, Niu S, Xia J-M, Peng K, Zhang G-Y, Yang X-W (2018) Saccharopolytide A, a new cyclic tetrapeptide with rare 4-hydroxy-proline moieties from the deep-sea derived actinomycete Saccharopolyspora cebuensis MCCC 1A09850. Nat Prod Res 32(14):1627–1631

Xie C-L, Chen R, Yang S, Xia J-M, Zhang G-Y, Chen C-H, Zhang Y, Yang X-W (2019) Nesteretal A, a novel class of cage-like polyketide from marine-derived actinomycete Nesterenkonia halobia. Org Lett 21(20):8174–8177

Xu J-H, Gu K-B, Zhang D-J, Li Y-G, Tian L (2017) Ghanamycins A and B, two novel γ-butyrolactones from marine-derived Streptomyces ghanaensis TXC6-16. J Antibiot 70(6):733–736

Xu D, Nepal KK, Chen J, Harmody D, Zhu H, McCarthy PJ, Wright AE, Wang G (2018) Nocardiopsistins AC: new angucyclines with anti-MRSA activity isolated from a marine sponge-derived Nocardiopsis sp. HB-J378. Synth Syst Biotechnol 3(4):246–251

Xu C-D, Zhang H-J, Ma Z-J (2019a) Pyrimidine nucleosides from Streptomyces sp. SSA28. J Nat Prod 82(9):2509–2516

Xu X, Han J, Lin R, Polyak SW, Song F (2019b) Two new piperazine-triones from a marine-derived Streptomycetes sp. strain SMS636. Mar Drugs 17(3):186

Yan J-X, Chevrette MG, Braun DR, Harper MK, Currie CR, Bugni TS (2019) Madurastatin D1 and D2, oxazoline containing siderophores isolated from an Actinomadura sp. Org Lett 21(16):6275–6279

Yang N, Song F (2018) Bioprospecting of novel and bioactive compounds from marine actinomycetes isolated from South China Sea sediments. Curr Microbiol 75(2):142–149

Yang X-W, Peng K, Liu Z, Zhang G-Y, Li J, Wang N, Steinmetz A, Liu Y (2013) Strepsesquitriol, a rearranged zizaane-type sesquiterpenoid from the deep-sea-derived actinomycete Streptomyces sp. SCSIO 10355. J Nat Prod 76(12):2360–2363

Yang T, Yamada K, Zhou T, Harunari E, Igarashi Y, Terahara T, Kobayashi T, Imada C (2019) Akazamicin, a cytotoxic aromatic polyketide from marine-derived Nonomuraea sp. J Antibiot 72(4):202–209

Ye X, Anjum K, Song T, Wang W, Yu S, Huang H, Lian X-Y, Zhang Z (2016) A new curvularin glycoside and its cytotoxic and antibacterial analogues from marine actinomycete Pseudonocardia sp. HS7. Nat Prod Res 30(10):1156–1161

Ye X, Anjum K, Song T, Wang W, Liang Y, Chen M, Huang H, Lian X-Y, Zhang Z (2017) Antiproliferative cyclodepsipeptides from the marine actinomycete Streptomyces sp. P11-23B downregulating the tumor metabolic enzymes of glycolysis, glutaminolysis, and lipogenesis. Phytochemistry 135:151–159

Yi H, Schumann P, Sohn K, Chun J (2004) Serinicoccus marinus gen. nov., sp. nov., a novel actinomycete with L-ornithine and L-serine in the peptidoglycan. Int J Syst Evol Microbiol 54 (Pt 5):1585–1589. https://doi.org/10.1099/ijs.0.03036-0

Yi H, Schumann P, Chun J (2007) Demequina aestuarii gen. nov., sp. nov., a novel actinomycete of the suborder Micrococcineae, and reclassification of Cellulomonas fermentans Bagnara et al. 1985 as Actinotalea fermentans gen. nov., comb. nov. Int J Syst Evol Microbiol 57(1):151–156

Yin F, Sun P, Tang B, Gong H, Ke Q, Li A (2016) Anti-parasitic effects of Leptomycin B isolated from Streptomyces sp. CJK17 on marine fish ciliate Cryptocaryon irritans. Vet Parasitol 217: 89–94

Yoshioka T, Igarashi Y, Namba T, Ueda S, Pait IGU, Nihira T, Kitani S (2021) Lavencidin, a polyene macrolide antibiotic from Streptomyces lavendulae FRI-5. J Antibiot 74:359–362

Zengler K, Toledo G, Rappé M, Elkins J, Mathur EJ, Short JM, Keller M (2002) Cultivating the uncultured. Proc Natl Acad Sci 99(24):15681–15686

Zhang H, Lee YK, Zhang W, Lee HK (2006a) Culturable actinobacteria from the marine sponge Hymeniacidon perleve: isolation and phylogenetic diversity by 16S rRNA gene-RFLP analysis. Antonie Van Leeuwenhoek 90(2):159–169. https://doi.org/10.1007/s10482-006-9070-1

Zhang H, Zheng W, Huang J, Luo H, Jin Y, Zhang W, Liu Z, Huang Y (2006b) Actinoalloteichus hymeniacidonis sp. nov., an actinomycete isolated from the marine sponge Hymeniacidon perleve. Int J Syst Evol Microbiol 56(10):2309–2312. https://doi.org/10.1099/ijs.0.64217-0

Zhang L, Xi L, Ruan J, Huang Y (2012) Micromonospora yangpuensis sp. nov., isolated from a sponge. Int J Syst Evol Microbiol 62(2):272–278

Zhang X-Y, He F, Wang G-H, Bao J, Xu X-Y, Qi S-H (2013) Diversity and antibacterial activity of culturable actinobacteria isolated from five species of the South China Sea gorgonian corals. World J Microbiol Biotechnol 29(6):1107–1116

Zhang DF, Wang HF, Xiong ZJ, Tian XP, Liu L, Zhang XM, Jiang Z, Zhang S, Li WJ (2014) Mariniluteicoccus flavus gen. nov., sp. nov., a new member of the family Propionibacteriaceae, isolated from a deep-sea sediment. Int J Syst Evol Microbiol 64(Pt 3):1051–1056. https://doi.org/10.1099/ijs.0.058404-0

Zhang G, Wang S, Wang L (2016a) Sediminivirga luteola gen. nov., sp. nov., a member of the family Brevibacteriaceae, isolated from marine sediment. Int J Syst Evol Microbiol 66(3):1494–1498

Zhang Y, Adnani N, Braun DR, Ellis GA, Barns KJ, Parker-Nance S, Guzei IA, Bugni TS (2016b) Micromonohalimanes A and B: antibacterial halimane-type diterpenoids from a marine Micromonospora species. J Nat Prod 79(11):2968–2972

Zhang W, Yang C, Huang C, Zhang L, Zhang H, Zhang Q, Yuan C-S, Zhu Y, Zhang C (2017a) Pyrazolofluostatins A–C, pyrazole-fused benzo [a] fluorenes from South China Sea-derived Micromonospora rosaria SCSIO N160. Org Lett 19(3):592–595

Zhang Y-M, Liu B-L, Zheng X-H, Huang X-J, Li H-Y, Zhang Y, Zhang T-T, Sun D-Y, Lin B-R, Zhou G-X (2017b) Anandins A and B, two rare steroidal alkaloids from a marine Streptomyces anandii H41-59. Mar Drugs 15(11):355

Zhang B, Wang KB, Wang W, Bi SF, Mei YN, Deng XZ, Jiao RH, Tan RX, Ge HM (2018) Discovery, biosynthesis, and heterologous production of streptoseomycin, an anti-microaerophilic bacteria macrodilactone. Org Lett 20(10):2967–2971

Zhang D, Yi W, Ge H, Zhang Z, Wu B (2019a) Bioactive streptoglutarimides A–J from the marine-derived Streptomyces sp. ZZ741. J Nat Prod 82(10):2800–2808

Zhang F, Braun DR, Chanana S, Rajski SR, Bugni TS (2019b) Phallusialides A–E, pyrrole-derived alkaloids discovered from a marine-derived Micromonospora sp. bacterium using MS-based metabolomics approaches. J Nat Prod 82(12):3432–3439

Zhang S, Xie Q, Sun C, Tian X-P, Gui C, Qin X, Zhang H, Ju J (2019c) Cytotoxic kendomycins containing the carbacylic ansa scaffold from the marine-derived Verrucosispora sp. SCSIO 07399. J Nat Prod 82(12):3366–3371

Zhang Q, Ye J, Yu L, Lin D, Tang K (2020a) Serinicoccus hydrothermalis sp. nov., isolated from shallow-sea hydrothermal systems off Kueishantao Island. Int J Syst Evol Microbiol 70(5):3139–3144

Zhang Z, Zhou T, Harunari E, Oku N, Igarashi Y (2020b) Iseolides A–C, antifungal macrolides from a coral-derived actinomycete of the genus Streptomyces. J Antibiot 73(8):534–541

Zhao XQ, Xu XN, Chen LY (2016) Production of enzymes from marine actinobacteria. Adv Food Nutr Res 78:137–151

Zhao LH, Xing X, Liu YY, Sha S, Song C, Sun YH, Xianyu DW, Zhu Q (2020) Aeromicrobium piscarium sp. nov., isolated from the intestine of Collichthys lucidus. Int J Syst Evol Microbiol 70(10):5280–5286. https://doi.org/10.1099/ijsem.0.004407

Zhou H, Yang X, Li F, Yi X, Yu L, Gao C, Huang R (2017) A new diketopiperazine of Nocardiopsis alba isolated from Anthogorgia caerulea. Chem Nat Compd 53(2):338–340

Zhou B, Huang Y, Zhang H-J, Li J-Q, Ding W-J (2019a) Nitricquinomycins A-C, uncommon naphthopyrrolediones from the Streptomyces sp. ZS-A45. Tetrahedron 75(30):3958–3961

Zhou L-M, Kong F-D, Xie Q-Y, Ma Q-Y, Hu Z, Zhao Y-X, Luo D-Q (2019b) Divergolides T–W with apoptosis-inducing activity from the mangrove-derived Actinomycete Streptomyces sp. KFD18. Mar Drugs 17(4):219

Zhou Q, Luo GC, Zhang H, Tang GL (2019c) Discovery of 16-demethylrifamycins by removing the predominant polyketide biosynthesis pathway in Micromonospora sp. strain TP-A0468. Appl Environ Microbiol 85(4)

Zhou Y, Pei S, Xie F, Gu L, Zhang G (2020) Saccharopolyspora coralli sp. nov. a novel actinobacterium isolated from the stony coral Porites. Int J Syst Evol Microbiol 70(5):3241–3246

Zhu X, Duan Y, Cui Z, Wang Z, Li Z, Zhang Y, Ju J, Huang H (2017) Cytotoxic rearranged angucycline glycosides from deep sea-derived Streptomyces lusitanus SCSIO LR32. J Antibiot 70(7):819–822

Ziemert N, Alanjary M, Weber T (2016) The evolution of genome mining in microbes–a review. Nat Prod Rep 33(8):988–1005

Zotchev SB (2012) Marine actinomycetes as an emerging resource for the drug development pipelines. J Biotechnol 158(4):168–175. https://doi.org/10.1016/j.jbiotec.2011.06.002

Exploration of Insects and Mollusks for New Secondary Metabolites from Actinobacteria

2

Chandra Risdian, Nasim Safaei, Michael Steinert, and Joachim Wink

Abstract

Actinobacteria are Gram-positive bacteria, which have been known for their roles in various fields in human life, especially in medicine. They constitute a highly diverse phylum with an unrivalled metabolic versatility. Finding novel antibiotics is highly important since drug-resistant infections are a serious threat to people's health. Since Actinobacteria, especially those from the genus *Streptomyces*, have been known as prolific producers of antibiotics, isolating rare or novel actinobacterial species from uncommon habitats like the ones associated with insects and mollusks might increase the chance to discover novel active compounds. Numerous actinobacterial strains have been isolated from insects (ants, termites, bees, wasps, beetles, and bugs) and mollusks (nudibranchs, snails, and clams) with different genera and bioactive properties. Moreover, rare Actinobacteria like *Actinocorallia*, *Verrucosispora*, *Sphaerisporangium*, *Actinocatenispora*,

Chandra Risdian and Nasim Safaei contributed equally to this work.

C. Risdian
Microbial Strain Collection (MISG), Helmholtz Centre for Infection Research (HZI), Braunschweig, Germany

Research Unit for Clean Technology, National Research and Innovation Agency Republic of Indonesia (BRIN), Bandung, Indonesia
e-mail: Chandra.Risdian@helmholtz-hzi.de

N. Safaei · J. Wink (✉)
Microbial Strain Collection (MISG), Helmholtz Centre for Infection Research (HZI), Braunschweig, Germany
e-mail: Nasim.Safaei@helmholtz-hzi.de; Joachim.Wink@helmholtz-hzi.de

M. Steinert
Institut für Mikrobiologie, Technische Universität Braunschweig, Braunschweig, Germany
e-mail: m.steinert@tu-bs.de

R. V. Rai, J. A. Bai (eds.), *Natural Products from Actinomycetes*,
https://doi.org/10.1007/978-981-16-6132-7_2

Agromyces, and *Saccharothrix* were isolated from these invertebrates. Various active compounds such as furamycins, anthracyclines, lobophorins, antimycins, julichromes, nocapyrones, pulicatins, and saccharothrixmicines examplify that insects and mollusks represent valuable sources for interesting secondary metabolites.

Keywords

Actinobacteria · Drug discovery · Insects · Mollusks

2.1 Introduction

The phylum Actinobacteria is one of the largest taxonomic groups among Gram-positive bacteria. They are characterized by high guanine and cytosine (G + C) content in their genome, and some genera form mycelium structures and reproduce by sporulation. Actinobacteria are generally well studied because of their importance in biotechnology, medicine, and agriculture. One of the most studied genera of Actinobacteria is *Streptomyces*, which is known to produce various secondary metabolites important for human medicine, including anti-infective, antitumor, and immunosuppressant agents (Barka et al. 2016; Seipke et al. 2012b).

One of the problems in the health issue nowadays is the rapid development and spread of resistant bacteria that needs to be resolved. Finding new antibiotics is therefore essential to overcome this threat. Actinobacteria are still promising sources for novel antibiotic compounds when the investigation of drug discovery is focused on new Actinobacteria species or genera from underexplored habitats (Müller and Wink 2014).

Actinobacteria can be discovered in both terrestrial and aquatic (including marine) habitats. The majority of them are free-living bacteria; however, some species are tightly associated with higher organisms like invertebrates, where they contribute to the acquisition of nutrients. They secrete extracellular enzymes such as cellulases, chitinases, lipases, amylases, and proteases that hydrolyze a wide range of polysaccharides and insoluble complex organic polymers. In addition to providing nutritional resources, symbiotic Actinobacteria also protect their hosts against pathogens, parasitoids, or predators by the production of secondary metabolites. Since many Actinobacteria genera and species are still waiting for their isolation and Actinobacteria are producers of highly complex metabolites, they still represent a worthwhile source for novel compounds. Especially the isolation of rare Actinobacteria and strains from uncommon habitats will foster the chance to find new antifungal, antiviral, antitumor, immunosuppressive, antiparasitic, and antibacterial agents (Barka et al. 2016; Seipke et al. 2012b; Cambronero-Heinrichs et al. 2019). Taking this into account, we here provide an overview of Actinobacteria isolated from underexplored habitats such as insects and mollusks.

2.2 Secondary Metabolites from Actinobacteria

Actinobacteria can produce various secondary metabolites with diverse structure and function. They have generated around two-thirds of all clinically used natural product antibiotics. Some of the genera known to produce antimicrobial compounds are *Streptomyces*, *Micromonospora*, *Kitasatospora*, *Amycolatopsis*, *Actinoplanes*, *Saccharopolyspora*, *Nocardiopsis*, *Pseudonocardia*, and *Nocardia*. In 1940, the first antibiotic, actinomycin, was isolated from *Streptomyces antibioticus*. After that time, there have been many antibiotics discovered from different Actinobacteria species (Barka et al. 2016; Risdian et al. 2019).

Streptomycin and gentamicin are aminoglycoside antibiotics produced by *Streptomyces griseus* (Waksman et al. 1946) and *Micromonospora purpurea* (Weinstein et al. 1963), respectively. They inhibit the protein synthesis of bacteria by blocking the 30S ribosome subunit (de Lima Procópio et al. 2012). Erythromycin was firstly isolated from *Streptomyces erythraeus* (recent name: *Saccharopolyspora erythraea*) (McGuire et al. 1952). It is a macrolide antibiotic and its mechanism of action is by targeting the 50S ribosomal subunit resulting in inhibition of the protein synthesis of bacteria (Jelić and Antolović 2016). Vancomycin and ristocetin were isolated from *Amycolatopsis orientalis* and *Nocardia lurida*, respectively. Both compounds belong to glycopeptide antibiotics that inhibit bacterial cell wall biosynthesis by binding to its peptidoglycan structure (Zhang et al. 2018; Nielsen et al. 1982).

Amphotericin B, an antifungal compound, is produced by *Streptomyces nodosus* and belongs to polyene macrolide. This type of molecule can disrupt eukaryotic cell membranes due to its high affinity to ergosterol which is the primary sterol in fungal cell membranes (Caffrey et al. 2001). Nystatin-like *Pseudonocardia* polyene (NPP) has a similar structure and mechanism of antifungal activity like amphotericin. The difference is that it contains a disaccharide moiety and that it is isolated from *Pseudonocardia autotrophica* (Kim et al. 2018). Transvalensin is produced by the pathogenic actinobacterium *Nocardia transvalensis*. It is an antifungal agent that has a thiazolidine zinc complex in its structure (Hoshino et al. 2004). A 14-membered macrolide, rustmicin, is an antifungal agent produced by *Micromonospora narashinoensis* 980-MC_1 (Takatsu et al. 1985). Its activity is by targeting inositol phosphoceramide synthase, causing ceramide accumulation and inhibiting sphingolipid biosynthesis (Mandala et al. 1998). *Amycolatopsis* sp. M39 was found to synthesize a glycosylated macrolactam, macrotermycin, which is active against bacteria and fungi. However, the mechanism of action is still not clear (Beemelmanns et al. 2017).

Taken together, since the discovery of streptomycin, the first antibiotic isolated from Actinobacteria and many other bioactive compounds discovered to date, these microbes saved millions of lives. However, due to the intensity of research in this field, there is a problem of frequent rediscovery of the same chemical compounds. By new approaches such as "genome mining" and the exploration of uncommon habitats, a renaissance in the field of Actinobacteria seems to be in sight.

2.3 Insect-Microbe Interactions

Insects in the phylum Arthropoda have characteristics such as three pairs of jointed legs, three-segmented bodies, a couple of antennae, and exoskeletons made of chitin. Some of them also may have one or two pairs of wings and the others not. There are six largest orders according to the number of identified species. They are Coleoptera (e.g., beetles), Lepidoptera (e.g., butterflies), Hymenoptera (e.g., bees), Diptera (e.g., flies), Hemiptera (e.g., true bugs), and Orthoptera (e.g., grasshoppers) (https://www.nps.gov/teachers/classrooms/insect-orders.htm).

As the most plentiful animal class on earth, insects have developed symbiotic associations embracing microorganisms as their partners in daily life. Many microbial symbionts are transmitted vertically from one generation to the next. They are essential for the reproduction or survival of the host. Symbiotic microorganisms help their host's metabolism and physiological capability by providing them essential vitamins and other nutrients or aiding to digest complex nutritional compounds. They can also help their corresponding insects or the host's food stocks from some unpleasant organisms like parasites, pathogens, or parasitoids. The bacteria associated with insects also benefit by residing in a suitable environment characterized by minimum competitors and plentiful nutrients provided by their host (Kaltenpoth et al. 2006; Salem et al. 2013; Mogouong et al. 2020).

Members of the phylum Actinobacteria have been associated with many different insects due to their metabolic capacity to utilize various nutritional resources and capability in producing secondary metabolites with antibiotic activity. Therefore, they can be very supportive or even essential for the survival or the environmental adaptation of their host (Hulcr et al. 2011; Salem et al. 2013).

2.4 Actinobacteria Isolated from Ants

Mutualistic and antagonistic interactions between fungus-growing ants, Actinobacteria, fungal cultivar, and fungal pathogens are highly interesting and intensively studied (Fig. 2.1). Fungus-growing *Allomerus* ants are plant symbionts that utilize a sooty mold (order Chaetothyriales) on their host plants to build traps for their insect prey (Heil and McKey 2003; Ruiz-González et al. 2011; Dejean et al. 2005). The interaction between the ant and the fungus is suggested to be the result of fungiculture (Defossez et al. 2009). The fungiculture that *Allomerus* ants apply is by grooming and weeding, which means that the ants cultivate the sooty mold that they need and remove the other fungal contaminants (Ruiz-González et al. 2011).

Allomerus ants were reported to be associated with Actinobacteria that can produce antifungal compounds. It is suggested that the microbes aid them by inhibiting the non-cultivar fungi in the plants that the ants are associated with. From the cuticles of workers of two *Allomerus* species (*A. decemarticulatus* and *A. octoarticulatus*) obtained from 19 trees, Seipke et al. (2012a) isolated seven actinobacterial strains including six strains of the genus *Streptomyces* and one strain from the genus *Amycolatopsis* by using lysogeny broth (LB-Lennox) agar. Some of

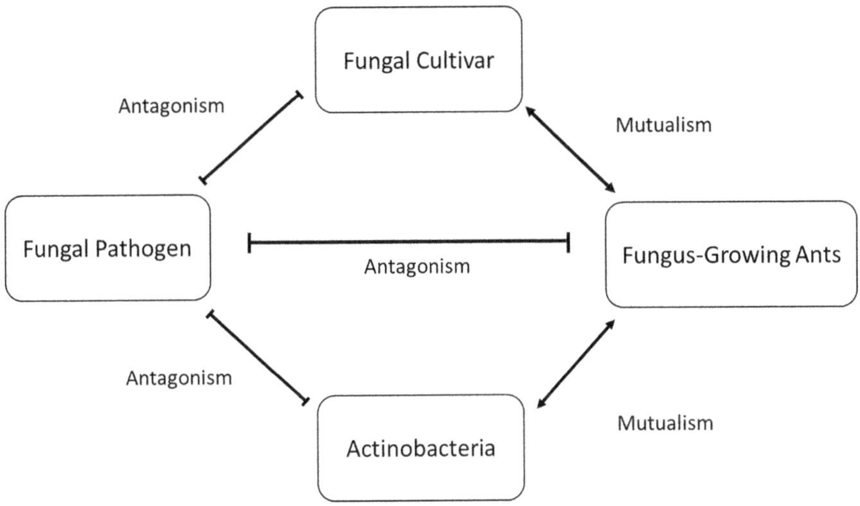

Fig. 2.1 Mutualistic and antagonistic interactions between fungus-growing ants, Actinobacteria, fungal cultivar, and fungal pathogen

the isolated strains had a similarity of 100% to the known species, such as *Amycolatopsis australiensis* and *Streptomyces misionensis*. The trees belong to the species *Hirtella physophora* and *Cordia nodosa*, which were sampled from French Guiana.

The fungus-growing ants (tribe Attini) cultivate basidiomycetous fungi for their nutrition. They are associated with filamentous Actinobacteria that can produce antifungal compounds for inhibiting microfungal parasites from the genus *Escovopsis* (Currie et al. 1999; Cafaro and Currie 2005) (Fig. 2.1). Some parts of the cuticle of the fungus-growing ants are covered with a powdery, whitish-gray layer made from filamentous Actinobacteria of the genus *Streptomyces*. It is suggested that all 22 species of attine ants from 8 genera (*Myrmicocrypta*, *Apterostigma*, *Mycocepurus*, *Cyphomyrmex*, *Sericomyrmex*, *Trachymyrmex*, and leaf-cutting genera like *Acromyrmex* and *Atta*) were associated with *Streptomyces* (Currie et al. 1999). Furthermore, three genera of fungus-growing ants (tribe Attini), i.e., *Acromyrmex*, *Trachymyrmex*, and *Apterostigma*, were also reported to contain Actinobacteria from the genus *Pseudonocardia* on cuticles of workers. The isolation of actinobacterial strains was conducted using chitin-agar plates containing nystatin (10,000 units/mL) as an antifungal agent (Cafaro and Currie 2005).

Another study conducted by Wang et al. (2020) revealed hundreds of actinobacterial isolates from three ant species, i.e., *Camponotus japonicus*, *Lasius fuliginosus*, and *Lasius flavus*. The isolation of Actinobacteria was conducted by using six different media, i.e., humic acid-vitamin (HV) agar, Gause's synthetic (GS) agar no. 1, chitin agar (CA), tap water-yeast extract (TWYE) agar, sodium succinate-asparagine (SSA) agar, and xylan-arginine (XA) agar. Six families of isolated Actinobacteria comprising Streptomycetaceae, Micromonosporaceae,

Promicromonosporaceae, Nocardiaceae, Streptosporangiaceae, and Thermomonosporaceae were identified. The majority of the isolates belonged to the *Streptomyces*, followed by *Micromonospora* (Micromonosporaceae) and *Nocardia* (Nocardiaceae). The other isolated Actinobacteria, i.e., *Promicromonospora*, *Nonomuraea*, *Verrucosispora*, *Phytohabitans*, *Streptosporangium*, *Microbispora*, and *Actinocorallia*, were isolated in low numbers. Two antifungal compounds (furamycins I (**1**) and II (**2**)) and a novel polyene amide compound were isolated from one of the isolated *Streptomyces* strains.

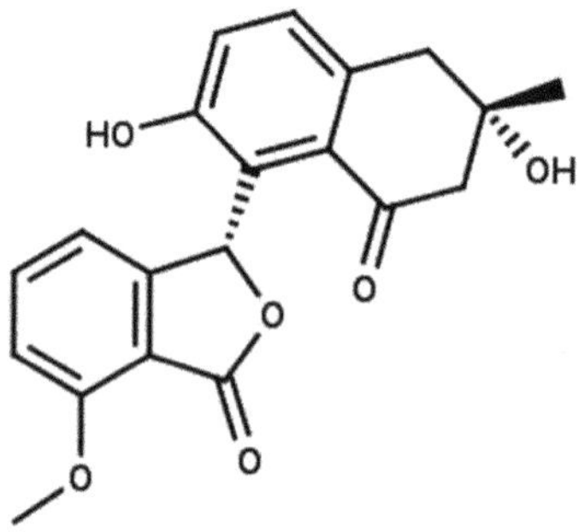

Furamycin I (**1**)

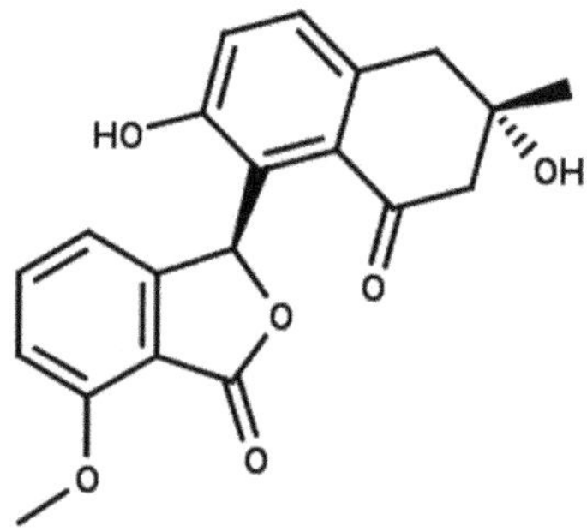

Furamycin II (**2**)

2.5 Actinobacteria Isolated from Termites

The presence of the cellulolytic Actinobacteria in the gut of termites was suggested to have a beneficial role for the insect to obtain their nutrition. Isolation of the cellulolytic actinobacterial strains from termites was conducted by using an enrichment medium containing Avicel in the first step, followed by an isolation agar medium comprising filter paper Whatman No. 1. Four different termites, i.e., *Macrotermes*, *Armitermes*, *Odontotermes*, and *Microcerotermes* spp., are known to be associated with cellulolytic Actinobacteria from the genera *Streptomyces* and *Micromonospora* in their hindguts (Pasti and Belli 1985).

In another study, using chitin and microcrystalline media as selective low-nutrient media, actinobacterial strains from the genera *Streptomyces*, *Actinomadura*, and *Micromonospora* were isolated from fungus-growing termites (genera *Macrotermes*, *Microtermes*, and *Odontotermes*). The actinobacterial isolates were found in the abdomen, head, and exoskeleton of the termites, with a high percentage of the bacteria residing in the abdomen. Additionally, the comb of the termites was also determined to be a habitat for Actinobacteria species (Visser et al. 2012).

Benndorf et al. (2018) revealed 97 Actinobacteria isolates covering ten families from the gut, exoskeleton, and comb of *Macrotermes natalensis* by employing chitin and microcrystalline medium. *Streptomyces* was the predominant genus of the isolated Actinobacteria, and seven strains from four genera (*Streptomyces*, *Nocardia*, *Actinomadura*, and *Sphaerisporangium*) were putatively novel species. Three new compounds (*cyclo*(N-Me-L-3,5-dichlorotyrosine-DhB), rubrominin A (**3**), and rubrominin B (**4**)) were isolated from the extract of one of the isolated strains, *Actinomadura* sp. RB29. However, bioactivities of the novel isolated compounds are not yet identified.

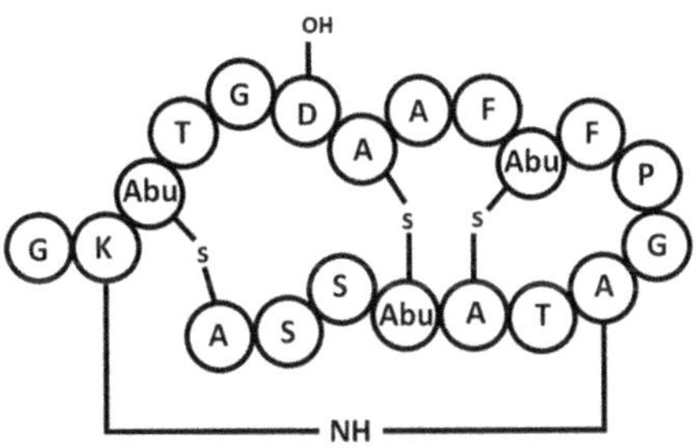

Rubrominin A (**3**)

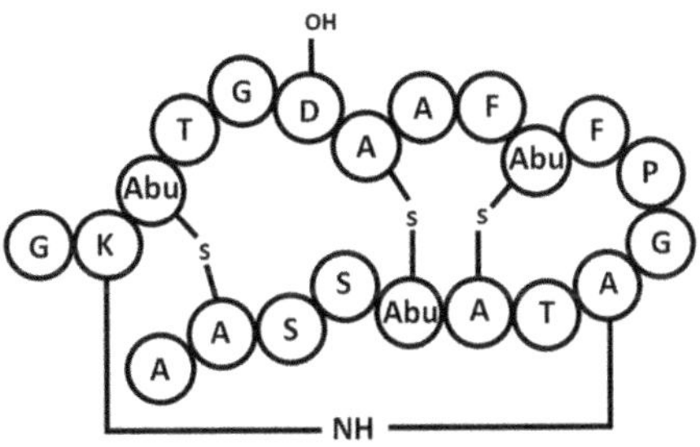

Rubrominin B (**4**)

2.6 Actinobacteria Isolated from Bees

Bees host a highly specific and surprisingly limited microbiome, including Actinobacteria. Mohr and Tebbe (2006) studied bacterial diversity in the gut of three bees, i.e., the honey bee *Apis mellifera* ssp. *carnica*, the bumblebee *Bombus terrestris*, and the red mason bee *Osmia bicornis*. The investigation was performed using a culture-independent method based on DNA extracted from the gut of adults and larvae of the insects. They found that Actinobacteria, such as *Propionibacterium*, *Rhodococcus*, and *Corynebacterium*, had a small portion (7%) of the total groups of bacteria found in the gut of the three bee species.

Cambronero-Heinrichs et al. (2019) reported culturable Actinobacteria from the stingless bee *Tetragonisca angustula*. Using oatmeal agar and chitin agar, they found the majority of Actinobacteria was isolated from the worker bee and pollen, which includes *Streptomyces*, *Nocardiopsis*, *Gordonia*, *Actinocatenispora*, *Actinomadura*, and *Micromonospora*. The study revealed that actinobacterial filamentous structures were detected on the outer surface of adult bees, such as thorax setae, jaws, and pollen combs of the posterior leg, using a scanning electron microscope. Some of the isolated actinobacterial strains showed activity against Gram-positive bacteria and fungi. The majority of them were *Streptomyces* and only one strain of *Actinomadura*. The antimicrobial activity of the actinobacterial strains suggested that they might protect bees from pathogenic Gram-positive bacteria and fungi.

A study by Rodríguez-Hernández et al. (2019) reported that seven Actinobacteria (six strains of *Streptomyces* and one strain of *Micromonospora*) were isolated from the eusocial stingless bee species *Melipona scutellaris* by using chitin media. Six lobophorin (**5**) compounds and eleven anthracycline (**6**) molecules were isolated from *Streptomyces* sp. ICBG1323 and *Micromonospora* sp. ICBG1321, respectively. These compounds are known to possess antibacterial activity against *Paenibacillus larvae*.

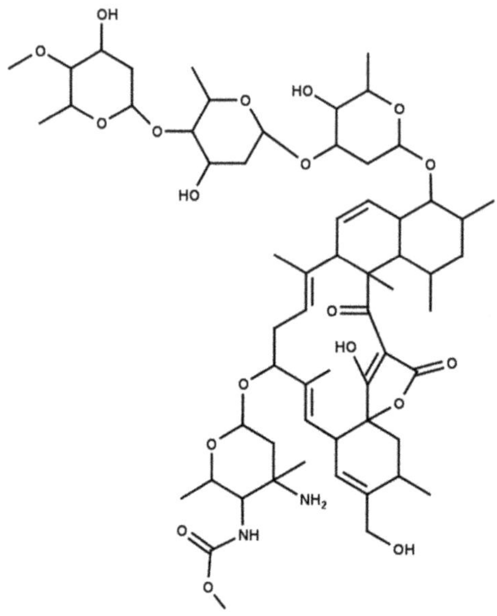

Lobophorin A (**5**)

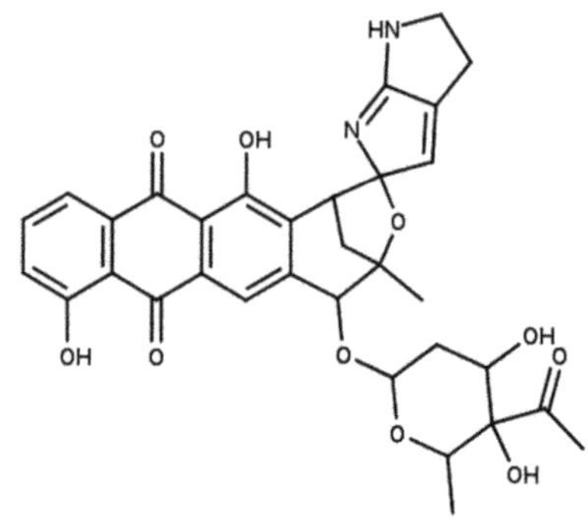

Kosinostatin (**6**)

2.7 Actinobacteria Isolated from Wasps

Ecology-driven natural product discovery approaches including the chemical analyses of symbiotic microorganisms demonstrated that *Streptomyces* strains are also associated with *Sirex noctilio*, an invasive wood-feeding wasp. The Actinobacteria were isolated from the body of the adult wasps after separating the mycangia by using a chitin medium. *Streptomyces* strains isolated from the wasps were suggested to have the capability to digest cellulose after studying their biochemical and genomic properties. Therefore, the symbiotic Actinobacteria may have a role in assisting *S. noctilio* in acquiring their nutrient (Adams et al. 2011).

Digger wasps, such as the European beewolf (*Philanthus triangulum*, Hymenoptera, Crabronidae) and from the genus *Trachypus*, were reported to associate with endosymbiotic *Streptomyces*. By employing transmission and scanning electron microscopy, *Streptomyces* species were detected in the antennal glands of the female wasps from which the actinobacterial symbionts are secreted into the brood cell. However, the isolation and cultivation of the bacteria with the standard technique was not successful. The identification was, therefore, conducted by extracting DNA from the whole beewolf antennae and analyzing the 16S rRNA gene sequence. As *Streptomyces* are known to have the ability to produce antifungal compounds, the secretion of the bacteria to the wasp's offspring may protect them from the fungal infection (Kaltenpoth et al. 2006, 2010).

Poulsen et al. (2011) reported more than 200 isolates of *Streptomyces* from 33 mud daubers (25 of *Sceliphron caementarium* and 8 *Chalybion californicum*) by employing chitin medium. Chemical analysis of the extract from 15 *Streptomyces* strains of the isolated Actinobacteria revealed 11 various secondary metabolites. One of them was sceliphrolactam (**7**), a new macrocyclic lactam that has antifungal activity, and the others are streptazoline, streptazon B, daunomycin, bafilomycins (two compounds), antimycins (four compounds), and mycangimycin.

Streptomyces sp. M54 was found to be associated with the eusocial wasp *Polybia plebeja* (Matarrita-Carranza et al. 2017, 2021). The bacterium was isolated from immature insect *Polybia plebeja* using chitin agar (Matarrita-Carranza et al. 2017). It was reported that the extract of the strain was active against *Hirsutella citriformis*, *Staphylococcus aureus*, and *Candida albicans*. *Hirsutella citriformis* is known as a natural fungal enemy of the insect, suggesting that *Streptomyces* sp. M54 can be beneficial for its host. The genome analysis of *Streptomyces* sp. M54 revealed 30 biosynthetic gene clusters for secondary metabolites. Some macrotetrolides or macrocyclic ionophore compounds such as nonactin, monactin, dinactin, trinactin, and tetranactin were isolated from the strain which were active against *Staphylococcus aureus* (Matarrita-Carranza et al. 2021).

2.8 Actinobacteria Isolated from Beetles

The majority of insect defensive symbioses characterized so far involve Actinobacteria. This is not surprising as Actinobacteria, especially *Streptomyces*, are well-known producers of bioactive secondary metabolites. Ambrosia beetles cultivate fungi for their food in their nest located in the xylem of the dead or dying trees. Two genera of ambrosia beetles, *Xyleborus* and *Xyleborinus*, cultivate fungi mostly of the Ascomycota genera *Ambrosiella* (Ceratocystidaceae: Microascales) and *Raffaelea* (Ophiostomataceae: Ophiostomatales). They are real fungus farmers and obligatorily depend on their fungal symbiont for their nutrition. Other fungal symbionts of these beetles like *Nectria*, *Penicillium*, and *Aspergillus* are also found in low numbers in their nests. They act as competitors, parasites, or pathogens of the insect. Besides fungal symbionts, Actinobacteria are associated with ambrosia beetles. *Streptomyces* strains were isolated by using chitin agar from two species of fungus-farming ambrosia beetles, i.e., *Xyleborinus saxesenii* and *Xyleborus affinis*. Although the actinobacterial strains were found in adults and larvae of the insects, their number was less than that in the gallery. It is suggested that

Streptomyces may protect the ambrosia beetles by inhibiting the parasitic fungi, probably by producing cycloheximide (**8**) (Grubbs et al. 2020).

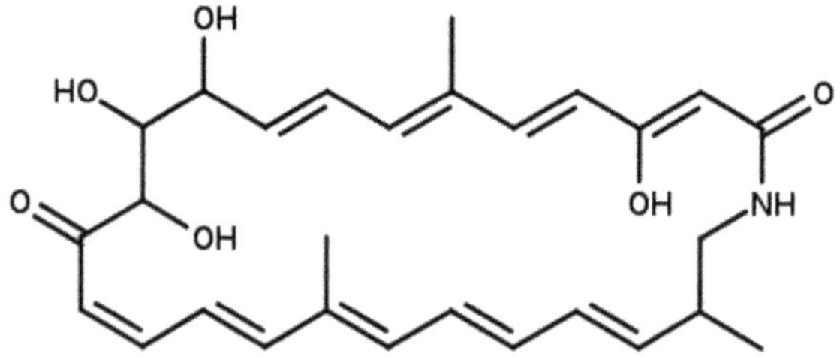

Sceliphrolactam (7)

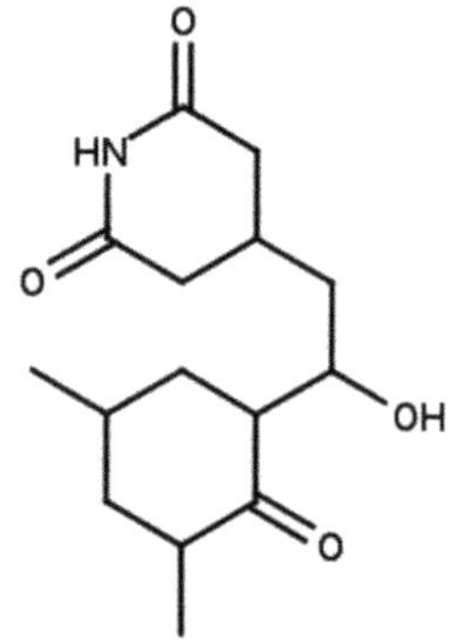

Cycloheximide (**8**)

Dendroctonus, a genus of bark beetles, is also known to be associated with fungi which are important for the nutrition of their larvae. Besides fungi, *Dendroctonus* is also associated with Actinobacteria, especially from the genus *Streptomyces*. A study conducted by Hulcr et al. (2011) with ten *Dendroctonus* species suggested that *Streptomyces* could be found in the whole crushed body, body surface, gallery frass, and larval chamber by using chitin agar. Especially for *Dendroctonus frontalis*, *Streptomyces* also could be isolated from hindgut and mycangia. Most *Streptomyces* strains were found in gallery frass and larval chamber, followed by whole crushed beetles, and beetle surface.

Another species from Actinobacteria, *Promicromonospora pachnodae* (recent name: *Xylanimicrobium pachnodae*), was isolated from the hindgut of larvae of the scarab beetle *Pachnoda marginata* employing carboxymethylcellulose or xylan containing agar under aerobic and anaerobic conditions (Cazemier et al. 2003). It is suggested that the bacteria help the beetle larvae by producing xylanases and endoglucanases for lignocellulose degradation, which is the nutrition of the larvae.

2.9 Actinobacteria Isolated from Bugs

Actinobacteria are known as important contributors to the process of plant biomass decomposition. Therefore, it is not surprising that Actinobacteria have also been identified as widespread symbionts of eukaryotes, helping herbivores to gain access to plant biomass. The cotton stainer (*Dysdercus fasciatus*) and the red firebug (*Pyrrhocoris apterus*), e.g., both from Pyrrhocoridae family, harbor the actinobacterial symbionts *Coriobacterium glomerans* and *Gordonibacter* sp. (Salem et al. 2013). These nutritional symbionts were detected by analyzing DNA samples

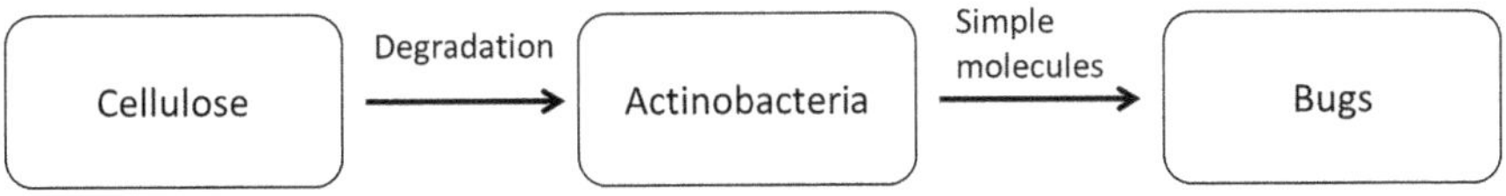

Fig. 2.2 Role of Actinobacteria as nutrition providers for bugs

from the M3 region of their midgut, and it was suggested that the bacteria degrade complex plant material like cellulose and provide vitamins (Fig. 2.2).

2.10 Mollusca-Microbe Interactions

The second-largest phylum of invertebrates after Arthropoda (including insects) is Mollusca (Brusca and Brusca 2003). The phylum Mollusca represents an enormous diversity of species with eight distinct classes (Ruppert and Barnes 1994). Gastropoda (e.g., nudibranchs, snails, slugs), Bivalvia (e.g., mussels, clams, oysters), Cephalopoda (e.g., octopuses, squids, cuttlefish), Monoplacophora (e.g., *Neopilina*), Scaphopoda, Polyplacophora (e.g., chiton), Solenogastres (e.g., *Neomenia*), and Caudofoveata (e.g., *Chaetoderma*, *Limifossor*) (Parkhaev 2017; Pati et al. 2015).

Except for the two sister taxa of Gastropoda and Bivalvia (known as Pleistomollusca), other classes are exclusively marine habitants. Bivalvia is able to colonize freshwater and marine habitats, whereas gastropods live in a wide range of ecological niches from the highest alpine regions to the deepest sea vents (Kocot et al. 2011).

Mollusks can be utilized in the food industry, homeopathic remedies, ornaments, lime production, and as pollution indicators, or some species can clean the environment (Pati et al. 2015). During the past few decades, there was an enormous interest in the discovery of natural products derived from organisms especially marine organism-associated bacteria (Karuppiah et al. 2016). For many cultures around the world, mollusks have a long background in producing useful medicinal products (Benkendorff 2010). In China, Cephalopoda (especially *Sepia* sp.), including squid and cuttlefish's egg, meat, bone, and ink, are used in a wide range of traditional medicine (Hu et al. 1999). In South Africa, Polyplacophora (some species of chiton) are used to cause vaginal spasm and prevent bedwetting in children (Herbert et al. 2003).

Ziconotide, isolated from the venom of predatory marine cone snail (*Conus magus*), was the first marine drug approved for clinical use as a treatment for chronic pain (Prommer 2006). Later, the compounds dolastatin 10 (brentuximab vedotin), kahalalide F, and ES-285 were isolated from *Dolabella auricularia* (sea slug), *Elysia rufescens* (sea slug), and *Mactromeris polynyma* (bivalve), respectively, and passed through clinical trials (Luypaert et al. 2020). It has been hypothesized that the original producers of bioactive compounds are symbiotic microorganisms. Especially in marine habitats, symbiotic microbes produce large amounts of metabolites to protect marine animals from predators and competitors (Ji et al. 2020). Bewley et al. (1996) separated the sponge cell populations by centrifugation and analyzed the obtained fractions. The result showed that it is possible to locate the cytotoxic macrolide swinholide A and antifungal cyclic peptide in the heterotrophic unicellular bacteria and in the filamentous bacteria, respectively, that were both isolated from the sponge *Theonella swinhoei*.

An example compound from mollusks is a pentapeptide (dolastatin 10) which was isolated from the sea slug *Dorabella auricularia* and later found out to be a secondary metabolite from cyanobacterium *Symploca* sp. (Luesch et al. 2001). The julimycins or julichromes, known as aromatic polyketides, were first isolated from terrestrial *Streptomyces shiodaensis* in the 1960s and later identified from marine gastropod mollusk-associated *Streptomyces sampsonii* SCSIO 054 in 2020 (Dong et al. 2020). Metagenomic approaches helped to delineate the contribution of either host, microbial associate, or both in metabolite production by following putative genes encoding the targeting metabolites (Ji et al. 2020).

Currently, next-generation sequencing (NGS) and metagenomic approaches allow us to gain insight into microbial communities (Mioduchowska et al. 2020). Most of the studies have focused on the gut microbiome of land, freshwater, and marine snails and slugs. The gut bacteria are involved in many physiological processes such as digestion, reproduction, and immunity (Hu et al. 2018). More studies focused on finding bacterial sources to produce novel enzymes with biotechnological potential (Joynson et al. 2017).

Several studies on the microbiome of mollusks revealed that the abundance of the phylum Actinobacteria is less than 10% among the microbial community. A study was conducted to compare bacterial communities in freshwater mussel species (*Cyclonaias asperata*, *Fusconaia cerina*, *Lampsilis ornata*, and *Obovaria unicolor*) and the overlying water of the river. The results revealed that Firmicutes and Planctomycetes were the most abundant phyla in the gut of mussel species and the phylum Actinobacteria represented 5.4% of the population. However, overlying water contained a high proportion of the phyla Actinobacteria and Verrucomicrobia (Weingarten et al. 2019). Another culture-independent study on an eastern oyster, *Crassostrea virginica*, revealed that the contribution of the phylum Actinobacteria is less than 10% (King et al. 2012). Studies on the microbiome of the freshwater snails *Potamopyrgus antipodarum*, *Radix auricularia*, and *Pomacea canaliculata* showed that the phylum Actinobacteria is not the dominant phylum (Hu et al. 2018; Takacs-Vesbach et al. 2016; Li et al. 2019). A recent study on the bacterial community composition of the freshwater snail *Physa acuta* was in line with other studies and revealed that the phylum Actinobacteria made up just 2% of the total population (Safaei et al. 2021).

To better understand the role of mollusk-associated Actinobacteria, culture-dependent approaches in parallel have been proven useful. *Streptomyces* sp. SCRC-A20, isolated from marine mollusk in Aburastubo, Kanagawa Prefecture, Japan, was reported to produce aburatubolactam A (**9**). The compound has cytotoxic and antimicrobial activity as well as superoxide generation inhibition activity, which probably benefits to protect the host from its antagonists and oxidative stress (Bae et al. 1996; Henderson and Phillips 2008). This example demonstrates that although Actinobacteria seem not to be the prominent phylum in mollusks, they still may have important functions in their hosts.

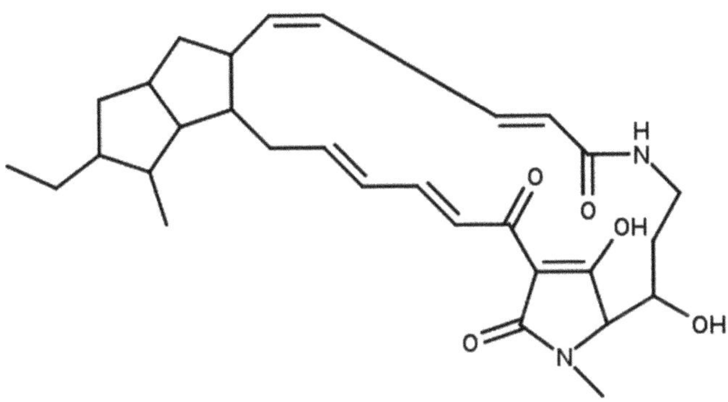

Aburatubolactam A (**9**)

2.11 Actinobacteria Isolated from Nudibranch

Some Actinobacteria were isolated from nudibranch, and many of them had promising properties like antimicrobial and antitumor activity. *Streptomyces* sp. NPC 8 was isolated from Indonesian nudibranch (*Chromodoris* sp.) using starch nitrate agar. The strain was reported to produce an extract, which could inhibit the growth of *Enterobacter* sp., *Proteus* sp., and *Staphylococcus* sp. (Riyanti et al. 2009). Abdelrahman et al. (2021) isolated some actinobacterial strains from three Red Sea nudibranch species (*Ceratosoma trilobatum*, *Chromodoris quadricolor*, and *Jorunna funebris*). The isolation of the Actinobacteria was performed using various isolation media such as marine agar, R2A agar, actinomycete isolation agar (AIA), starch casein agar, ISP2 agar, and M1 agar. The isolated actinobacterial strains belonged to the genera *Streptomyces*, *Nocardiopsis*, *Rhodococcus*, and *Kocuria*. Generally, the isolated strains from the genera *Nocardiopsis* and *Streptomyces* were reported to have promising bioactivity. The extract of *Nocardiopsis* sp. 13BY, for example, has strong antimicrobial and antitumor activities.

2.12 Actinobacteria Isolated from Snails

Peraud et al. investigated the Actinobacteria community in three cone snails: *Conus pulicarius*, *C. rolani*, and *C. tribblei*. Isolation of Actinobacteria was conducted by using marine agar 2216, ISP2, and R2A agar. Fifty-five species belonging to 16 different genera with four common genera (*Brevibacterium*, *Gordonia*,

Microbacterium, and *Streptomyces*) were isolated. Some isolates showed promising bioactivity in a neurological assay (Peraud et al. 2009). *Streptomyces* sp. CP32, one of the strains isolated from *C. pulicarius*, was revealed to produce neuroactive thiazoline metabolites pulicatins (**10**) A–E (Lin et al. 2010).

Nocardiopsis alba CR167, a strain associated with *C. rolani*, could produce nocapyrones (**11**) H–Q and nocapyrones A–C. Previously, nocapyrones A–D were isolated from *Nocardiopsis* sp. HB383 associated with the marine sponge *Halichondria panicea* (Schneemann et al. 2010). Nocapyrones B and H are known to be active against nearly all neuronal cell types (Lin et al. 2013b). *Streptomyces* sp. CN48 and *Streptomyces* sp. CT3a were isolated from *Chicoreus nobilis* and *Conus tribblei*, respectively. Nobilamides (**12**) A–H, *N*-acetyl-L-phenylalanyl-L-leucinamide, and two known compounds (A-3302-A and A-3302-B) were isolated from *Streptomyces* sp. CN48 and *Streptomyces* sp. CT3a. Nobilamide B and A-3302-A are regarded as the long-acting antagonists of transient receptor potential vanilloid-1 (TRPV1) which is known as the primary mediator of pain and inflammation (Lin et al. 2011). *Streptomyces* sp. 1053U.I.1a.1b was isolated from *Lienardia totopotens* and was reported to synthetize peptide-polyketide compounds totopotensamides A (**13**) and B (**14**) (Lin et al. 2012).

Streptomyces sampsonii SCSIO 054, derived from *Batillaria zonalis* using ISP2 agar medium containing 3.0% sea salt as the isolation medium, was confirmed to produce new compounds julichromes (julichromes Q_{11} and Q_{12}), five known julichromes (julichromes Q_{10}, $Q_{3.5}$, $Q_{3.3}$, Q_{6}, $Q_{6.6}$), and four known anthraquinones (chrysophanol, 4-acetylchrysophanol, islandicin, huanglongmycin A). Julichrome Q_{12} showed antibacterial activity against *Micrococcus luteus* and *Bacillus subtilis*. Whereas julichromes Q_{11} (**15**), Q_{10}, Q_{6}, and $Q_{6.6}$ showed inhibitory activities against *Staphylococcus aureus* and *Staphylococcus simulans* AKA1 (Dong et al. 2020). A novel species of *Streptomyces* with the potential of producing an angucycline-like aromatic polyketide compound (emycin A) was isolated from the freshwater snail *Physa acuta* employing 5336 agar (Safaei et al. 2021). Thirteen cellulolytic Actinobacteria from five genera (*Streptomyces*, *Cellulosimicrobium*, *Agromyces*, *Microbacterium*, and *Nocardiopsis*) were isolated from the giant land snail *Achatina fulica* using solid minimal media (MM) containing carboxymethylcellulose (CMC) (Pinheiro et al. 2015). *Gordonia* sp. 647 W.R.1a.05 was isolated from *Conus circumcisus*. The strain was found to produce circumcin A (**16**) and 11 known analogs which were identified to have various biological activities such as neuroactivity, anti-coccidial, antiviral, and antioxidant activities (Lin et al. 2013a). These examples from investigations of the recent years suggest that snails may harbor highly interesting yet undiscovered Actinobacteria.

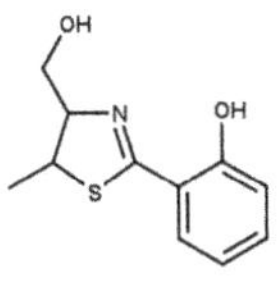

Pulicatin A (**10**)

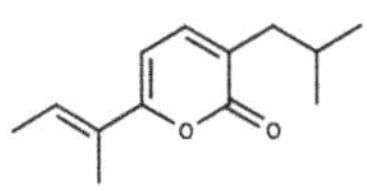

Nocapyrone H (**11**)

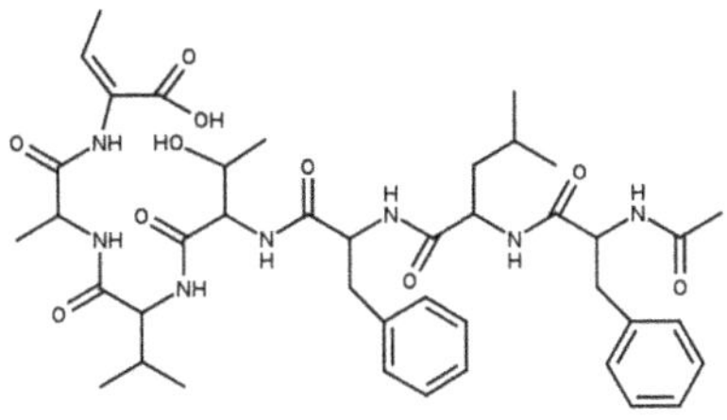

Nobilamide A (**12**)

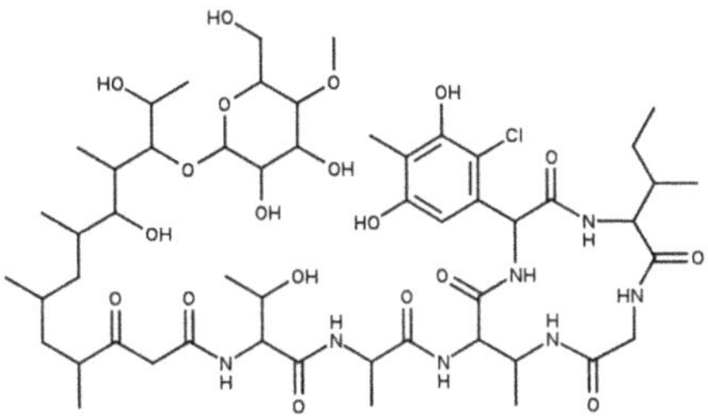

Totopotensamide A (**13**)

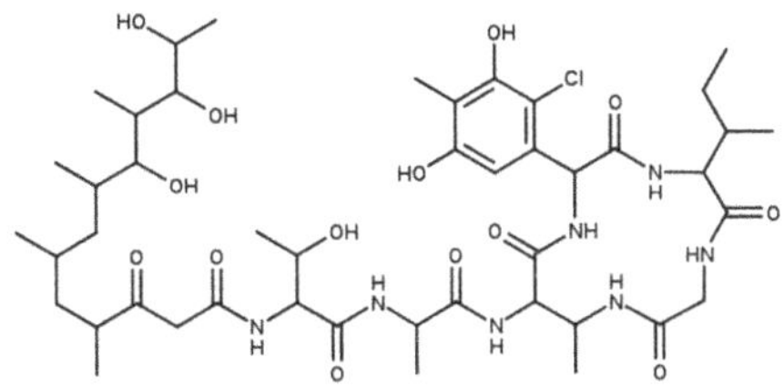

Totopotensamide B (**14**)

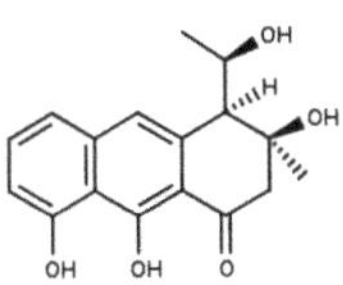

Julichrome Q_{11} (**15**)

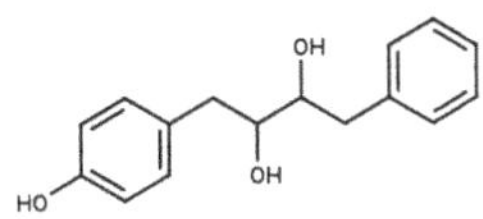

Circumcin A (**16**)

2.13 Actinobacteria Isolated from Clams

Mollusk-associated bacteria were isolated from *Anadara broughtonii* (a species of ark clam) using seawater medium (SWM), marine 2216 agar, and tryptic soya agar (Romanenko et al. 2008). One of the isolated strains is *Saccharothrix espanaensis* An 113, which produced novel angucyclines, saccharothrixmicines (**17**) A and B (**18**), and three diketopiperazines (**19–21**). The identified diketopiperazines were cyclo-(D-leucyl-trans-4-hydroxy-L-proline), cyclo-(L-leucyl-trans-4-hydroxy-L-proline), and cyclo-(L-phenylalanyl-cis-4-hydroxy-D-proline) that showed antimicrobial activity against *Vibrio alginolyticus* and *Vibrio parahaemolyticus* (Kalinovskaya et al. 2010). The saccharothrixmicines showed antimicrobial activity against *Candida albicans* and *Xanthomonas* sp. (Kharel et al. 2012; Kalinovskaya et al. 2010). El-Shatoury et al. isolated 63 actinomycetes, which belong to ten genera and two unidentified isolates from the marine shellfish *Donax trunculus anatinus* using starch casein agar (SCA) added with natural seawater. The majority of their metabolic extracts showed antimicrobial activities. Extract from isolates *Streptomyces* 23-2B showed high antitumor activity against Ehrlich's ascites carcinoma (El-Shatoury et al. 2009). Since clams interact intensively with the water column, encounter rates with microbes are usually high. Therefore, the chance to discover more interesting new Actinobacteria in this habitat seems also to be high.

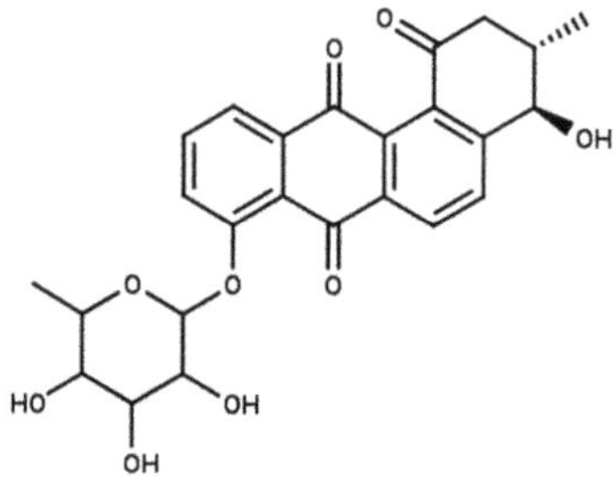

Saccharothrixmicine A (**17**)

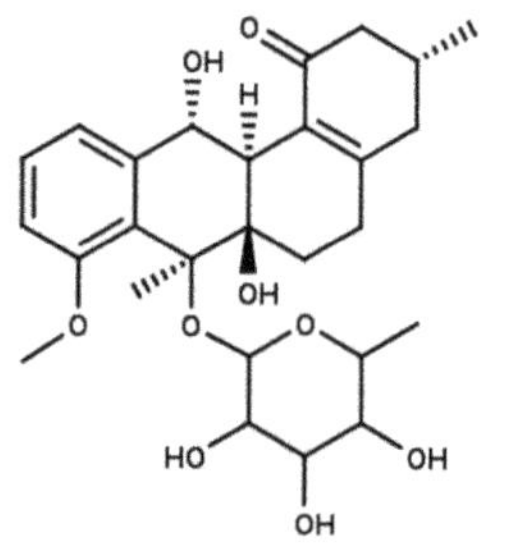

Saccharothrixmicine B (**18**)

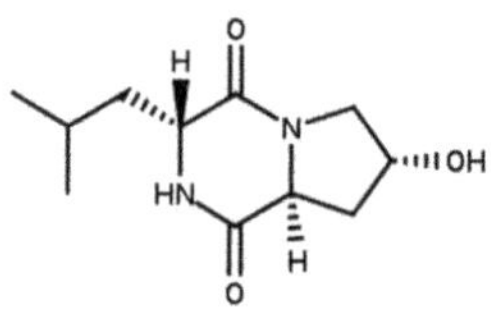

cyclo-(D-leucyl-*trans*-4-hydroxy-L-proline) (**19**)

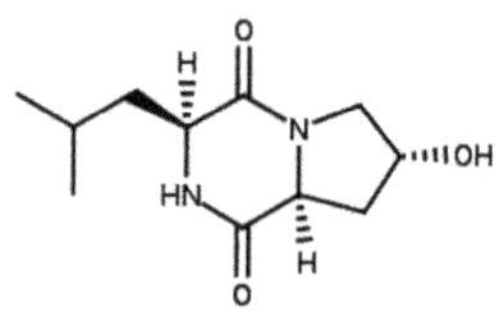

cyclo- (L-leucyl-*trans*-4-hydroxy-L-proline) (**20**)

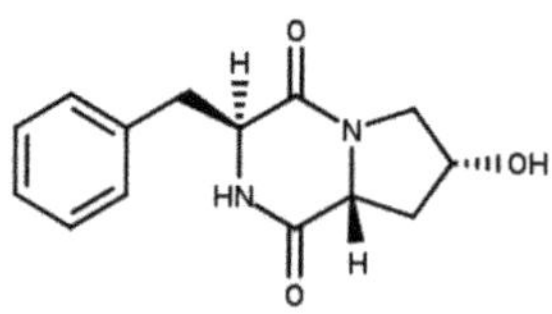

cyclo-(L- phenylalanyl-*cis*-4-hydroxy-D-proline) (**21**)

2.14 Conclusion and Future Perspectives

Insects and mollusks engage in a remarkable diversity of symbiotic associations involving actinobacterial partners (Table 2.1). Many of these partnerships benefit the host by nutritional supplementation or the degradation of complex dietary compounds. Additionally, an increasing number of defensive associations are being described where the symbionts protect their respective hosts or the host's food resources from pathogens and parasites. In the recent years, this area of research has largely been stimulated by the search for new classes of natural products. However, it should also be a pivotal aim to understand the biological functions of the interactions between Actinobacteria and their host organisms since they can guide the search for new compounds in unusual and underexplored environments. With the finding of some rare actinobacterial strains and active compounds produced

Table 2.1 Some of the Actinobacteria isolated from insects and mollusks

Host	Actinobacteria	Isolated compound (bioactivity)	References
Ants	*Actinocorallia* *Amycolatopsis* *Microbispora* *Micromonospora* *Nocardia* *Nonomuraea* *Phytohabitans* *Promicromonospora* *Pseudonocardia* *Streptomyces* *Streptosporangium* *Verrucosispora*	Furamycins (antifungal agents)	Cafaro and Currie (2005) Currie et al. (1999) Seipke et al. (2012a) Wang et al. (2020)
Termites	*Actinomadura* *Micromonospora* *Nocardia* *Sphaerisporangium* *Streptomyces*	Dichlorinated diketopiperazine derivative (unknown bioactivity) Rubrominins (unknown bioactivity)	Benndorf et al. (2018) Pasti and Belli (1985) Visser et al. (2012)
Bees	*Actinocatenispora* *Actinomadura* *Corynebacterium* *Gordonia* *Micromonospora* *Nocardiopsis* *Propionibacterium* *Rhodococcus* *Streptomyces*	Anthracyclines (antibacterial agents) Lobophorins (antibacterial agents)	Cambronero-Heinrichs et al. (2019) Mohr and Tebbe (2006) Rodríguez-Hernández et al. (2019)
Wasps	*Streptomyces*	Antimycins (antifungal agents) Bafilomycins (antifungal and cytotoxic agents) Daunomycin (cytotoxic antibiotic agent) Mycangimycin (antifungal agent) Nonactin (antimicrobial agent) Sceliphrolactam (antifungal agent) Streptazoline (antimicrobial agent) Streptazon B (antimicrobial agent)	Adams et al. (2011) Kaltenpoth et al. (2006) Kaltenpoth et al. (2010) Matarrita-Carranza et al. (2021) Poulsen et al. (2011)
Beetles	*Promicromonospora* (*Xylanimicrobium*) *Streptomyces*	Cycloheximide (antifungal agent)	Cazemier et al. (2003) Grubbs et al. (2020) Hulcr et al. (2011)
Bugs	*Coriobacterium* *Gordonibacter*	–	Salem et al. (2013)
Nudibranch	*Kocuria* *Nocardiopsis* *Rhodococcus* *Streptomyces*	–	Abdelrahman et al. (2021) Riyanti et al. (2009)

(continued)

Table 2.1 (continued)

Host	Actinobacteria	Isolated compound (bioactivity)	References
Snails	*Agromyces* *Brevibacterium* *Cellulosimicrobium* *Gordonia* *Microbacterium* *Nocardiopsis* *Streptomyces*	Anthraquinones (unknown bioactivity) Circumcin A (unknown bioactivity) Julichromes (antibacterial agent) Nobilamides (analgesic agents) Nocapyrones (neuroactive agent) Pulicatins (neuroactive agent) Totopotensamides (unknown bioactivity)	Dong et al. (2020) Lin et al. (2010) Lin et al. (2011) Lin et al. (2012) Lin et al. (2013a) Lin et al. (2013b) Peraud et al. (2009) Pinheiro et al. (2015) Schneemann et al. (2010)
Clams	*Saccharothrix* *Streptomyces*	Diketopiperazines (antimicrobial agent) Saccharothrixmicines (antimicrobial agent)	El-Shatoury et al. (2009) Kalinovskaya et al. (2010) Romanenko et al. (2008)

by Actinobacteria associated with insects and mollusks, it is encouraged to explore other kinds of insects and mollusks to solve the problem in drug discovery.

Acknowledgments M.S. and J.W. were supported by a grant of the Deutsche Forschungsgemeinschaft (DFG STE 838/11-1; DFG WI 4609/2-1).

References

Abdelrahman SM, Patin NV, Hanora A, Aboseidah A, Desoky S, Desoky SG, Stewart FJ, Lopanik NB (2021) The natural product biosynthetic potential of Red Sea nudibranch microbiomes. PeerJ 9:1–26. https://doi.org/10.7717/peerj.10525

Adams AS, Jordan MS, Adams SM, Suen G, Goodwin LA, Davenport KW, Currie CR, Raffa KF (2011) Cellulose-degrading bacteria associated with the invasive woodwasp *Sirex noctilio*. ISME J 5(8):1323–1331. https://doi.org/10.1038/ismej.2011.14

Bae M, Yamada K, Ijuin Y, Tsuji T, Yazawa K, Uemura D (1996) Aburatubolactam A, a novel inhibitor of superoxide anion generation from a marine microorganism. Heterocycl Commun 2(4):315–318. https://doi.org/10.1515/HC.1996.2.4.315

Barka EA, Vatsa P, Sanchez L, Gaveau-Vaillant N, Jacquard C, Klenk H-P, Clément C, Ouhdouch Y, van Wezel GP (2016) Taxonomy, physiology, and natural products of Actinobacteria. Microbiol Mol Biol Rev 80(1):1–43. https://doi.org/10.1128/MMBR.00019-15

Beemelmanns C, Ramadhar TR, Kim KH, Klassen JL, Cao S, Wyche TP, Hou Y, Poulsen M, Bugni TS, Currie CR, Clardy J (2017) Macrotermycins A-D, glycosylated macrolactams from a termite-associated *Amycolatopsis* sp. M39. Org Lett 19(5):1000–1003. https://doi.org/10.1021/acs.orglett.6b03831

Benkendorff K (2010) Molluscan biological and chemical diversity: secondary metabolites and medicinal resources produced by marine molluscs. Biol Rev 85(4):757–775

Benndorf R, Guo H, Sommerwerk E, Weigel C, Garcia-Altares M, Martin K, Hu H, Küfner M, de Beer ZW, Poulsen M, Beemelmanns C (2018) Natural products from Actinobacteria associated with fungus-growing termites. Antibiotics 7(3):1–25. https://doi.org/10.3390/antibiotics7030083

Bewley CA, Holland ND, Faulkner DJ (1996) Two classes of metabolites from *Theonella swinhoei* are localized in distinct populations of bacterial symbionts. Experientia 52(7):716–722. https://doi.org/10.1007/BF01925581

Brusca RC, Brusca GJ (2003) Invertebrates. Basingstoke

Cafaro MJ, Currie CR (2005) Phylogenetic analysis of mutualistic filamentous bacteria associated with fungus-growing ants. Can J Microbiol 51(6):441–446. https://doi.org/10.1139/w05-023

Caffrey P, Lynch S, Flood E, Finnan S, Oliynyk M (2001) Amphotericin biosynthesis in *Streptomyces nodosus*: deductions from analysis of polyketide synthase and late genes. Chem Biol 8(7):713–723. https://doi.org/10.1016/S1074-5521(01)00046-1

Cambronero-Heinrichs JC, Matarrita-Carranza B, Murillo-Cruz C, Araya-Valverde E, Chavarría M, Pinto-Tomás AA (2019) Phylogenetic analyses of antibiotic-producing *Streptomyces* sp. isolates obtained from the stingless-bee *Tetragonisca angustula* (Apidae: Meliponini). Microbiology (United Kingdom) 165(3):292–301. https://doi.org/10.1099/mic.0.000754

Cazemier AE, Verdoes JC, Reubsaet FAG, Hackstein JHP, van der Drift C, Op den Camp HJM (2003) *Promicromonospora pachnodae* sp. nov., a member of the (hemi)cellulolytic hindgut flora of larvae of the scarab beetle *Pachnoda marginata*. Antonie Van Leeuwenhoek 83(2): 135–148. https://doi.org/10.1023/A:1023325817663

Currie CR, Scottt JA, Summerbell RC, Malloch D (1999) Fungus-growing ants use antibiotic-producing bacteria to control garden parasites. Nature 398(6729):701–704. https://doi.org/10.1038/19519

de Lima Procópio RE, da Silva IR, Martins MK, de Azevedo JL, de Araújo JM (2012) Antibiotics produced by *Streptomyces*. Braz J Infect Dis 16(5):466–471. https://doi.org/10.1016/j.bjid.2012.08.014

Defossez E, Selosse M-A, Dubois M-P, Mondolot L, Faccio A, Djieto-Lordon C, McKey D, Blatrix R (2009) Ant-plants and fungi: a new threeway symbiosis. New Phytol 182(4):942–949. https://doi.org/10.1111/j.1469-8137.2009.02793.x

Dejean A, Solano PJ, Ayroles J, Corbara B, Orivel J (2005) Arboreal ants build traps to capture prey. Nature 434(7036):973–973. https://doi.org/10.1038/434973a

Dong Y, Ding W, Sun C, Ji X, Ling C, Zhou Z, Chen Z, Chen X, Ju J (2020) Julichrome Monomers from marine gastropod mollusk-associated *Streptomyces* and stereochemical revision of julichromes Q3 · 5 and Q3 · 3. Chem Biodivers 17(4):e2000057

El-Shatoury SA, El-Shenawy NS, Abd El-Salam IM (2009) Antimicrobial, antitumor and in vivo cytotoxicity of actinomycetes inhabiting marine shellfish. World J Microbiol Biotechnol 25(9): 1547–1555

Grubbs KJ, Surup F, Biedermann PHW, McDonald BR, Klassen JL, Carlson CM, Clardy J, Currie CR (2020) Cycloheximide-producing *Streptomyces* associated with *Xyleborinus saxesenii* and *Xyleborus affinis* fungus-farming ambrosia beetles. Front Microbiol 11:1–12. https://doi.org/10.3389/fmicb.2020.562140

Heil M, McKey D (2003) Protective ant-plant interactions as model systems in ecological and evolutionary research. Annu Rev Ecol Evol Syst 34(1):425–553. https://doi.org/10.1146/annurev.ecolsys.34.011802.132410

Henderson JA, Phillips AJ (2008) Total synthesis of aburatubolactam A. Angew Chem Int Ed Engl 47(44):8499–8501. https://doi.org/10.1002/anie.200803593

Herbert DG, Hamer ML, Mander M, Mkhize N, Prins F (2003) Invertebrate animals as a component of the traditional medicine trade in KwaZulu-Natal, South Africa. Afr Invertebr 44(2):1–8

Hoshino Y, Mukai A, Yazawa K, Uno J, Ishikawa J, Ando A, Fukai T, Mikami Y (2004) Transvalencin A, a thiazolidine zinc complex antibiotic produced by a clinical isolate of *Nocardia transvalensis*: I. Taxonomy, fermentation, isolation and biological activities. J Antibiot (Tokyo) 57(12):797–802. https://doi.org/10.7164/antibiotics.57.797

Hu SY, Kong YC, But PP (1999) An enumeration of Chinese materia medica. Chinese University Press

Hu Z, Chen X, Chang J, Yu J, Tong Q, Li S, Niu H (2018) Compositional and predicted functional analysis of the gut microbiota of *Radix auricularia* (Linnaeus) via high-throughput Illumina sequencing. PeerJ 6:e5537

Hulcr J, Adams AS, Raffa K, Hofstetter RW, Klepzig KD, Currie CR (2011) Presence and diversity of *Streptomyces* in *Dendroctonus* and sympatric bark beetle galleries across North America. Microb Ecol 61(4):759–768. https://doi.org/10.1007/s00248-010-9797-0

Jelić D, Antolović R (2016) From erythromycin to azithromycin and new potential ribosome-binding antimicrobials. Antibiotics 5(3):29. https://doi.org/10.3390/antibiotics5030029

Ji X, Dong Y, Ling C, Zhou Z, Li Q, Ju J (2020) Elucidation of the tailoring steps in julichrome biosynthesis by marine gastropod mollusk-associated *Streptomyces sampsonii* SCSIO 054. Org Lett 22(17):6927–6931

Joynson R, Pritchard L, Osemwekha E, Ferry N (2017) Metagenomic analysis of the gut microbiome of the common black slug *Arion ater* in search of novel lignocellulose degrading enzymes. Front Microbiol 8:2181

Kalinovskaya NI, Kalinovsky AI, Romanenko LA, Dmitrenok PS, Kuznetsova TA (2010) New angucyclines and antimicrobial diketopiperazines from the marine mollusk-derived actinomycete *Saccharothrix espanaensis* An 113. Nat Prod Commun 5(4):597–602

Kaltenpoth M, Goettler W, Dale C, Stubblefield JW, Herzner G, Roeser-Mueller K, Strohm E (2006) "*Candidatus* Streptomyces philanthi", an endosymbiotic streptomycete in the antennae of *Philanthus* digger wasps. Int J Syst Evol Microbiol 56(6):1403–1411. https://doi.org/10.1099/ijs.0.64117-0

Kaltenpoth M, Schmitt T, Polidori C, Koedam D, Strohm E (2010) Symbiotic streptomycetes in antennal glands of the South American digger wasp genus *Trachypus* (Hymenoptera, Crabronidae). Physiol Entomol 35(2):196–200. https://doi.org/10.1111/j.1365-3032.2010.00729.x

Karuppiah V, Sun W, Li Z (2016) Natural products of Actinobacteria derived from marine organisms. In: Studies in natural products chemistry, vol 48. Elsevier, pp 417–446

Kharel MK, Pahari P, Shepherd MD, Tibrewal N, Nybo SE, Shaaban KA, Rohr J (2012) Angucyclines: biosynthesis, mode-of-action, new natural products, and synthesis. Nat Prod Rep 29(2):264–325

Kim HJ, Han CY, Park JS, Oh SH, Kang SH, Choi SS, Kim JM, Kwak JH, Kim ES (2018) Nystatin-like *Pseudonocardia* polyene B1, a novel disaccharide-containing antifungal heptaene antibiotic. Sci Rep 8(1):1–8. https://doi.org/10.1038/s41598-018-31801-y

King GM, Judd C, Kuske CR, Smith C (2012) Analysis of stomach and gut microbiomes of the eastern oyster (*Crassostrea virginica*) from coastal Louisiana, USA. PLoS One 7(12):e51475

Kocot KM, Cannon JT, Todt C, Citarella MR, Kohn AB, Meyer A, Santos SR, Schander C, Moroz LL, Lieb B, Halanych KM (2011) Phylogenomics reveals deep molluscan relationships. Nature 477(7365):452–456. https://doi.org/10.1038/nature10382

Li LH, Lv S, Lu Y, Bi DQ, Guo YH, Wu JT, Yue ZY, Mao GY, Guo ZX, Zhang Y, Tang YF (2019) Spatial structure of the microbiome in the gut of *Pomacea canaliculata*. BMC Microbiol 19(1): 1–9

Lin Z, Antemano RR, Hughen RW, Tianero MD, Peraud O, Haygood MG, Concepcion GP, Olivera BM, Light A, Schmidt EW (2010) Pulicatins A-E, neuroactive thiazoline metabolites from cone snail-associated bacteria. J Nat Prod 73(11):1922–1926

Lin Z, Reilly CA, Antemano R, Hughen RW, Marett L, Concepcion GP, Haygood MG, Olivera BM, Light A, Schmidt EW (2011) Nobilamides A-H, long-acting transient receptor potential vanilloid-1 (TRPV1) antagonists from mollusk-associated bacteria. J Med Chem 54(11): 3746–3755

Lin Z, Flores M, Forteza I, Henriksen NM, Concepcion GP, Rosenberg G, Haygood MG, Olivera BM, Light AR, Cheatham TE III, Schmidt EW (2012) Totopotensamides, polyketide-cyclic peptide hybrids from a mollusk-associated bacterium *Streptomyces* sp. J Nat Prod 75(4): 644–649

Lin Z, Marett L, Hughen RW, Flores M, Forteza I, Ammon MA, Concepcion GP, Espino S, Olivera BM, Rosenberg G, Haygood MG, Light AR, Schmidt EW (2013a) Neuroactive diol and acyloin metabolites from cone snail-associated bacteria. Bioorg Med Chem Lett 23(17):4867–4869. https://doi.org/10.1016/j.bmcl.2013.06.088

Lin Z, Torres JP, Ammon MA, Marett L, Teichert RW, Reilly CA, Kwan JC, Hughen RW, Flores M, Tianero MD, Peraud O (2013b) A bacterial source for mollusk pyrone polyketides. Chem Biol 20(1):73–81. https://doi.org/10.1016/j.chembiol.2012.10.019

Luesch H, Moore RE, Paul VJ, Mooberry SL, Corbett TH (2001) Isolation of dolastatin 10 from the marine cyanobacterium *Symploca* species VP642 and total stereochemistry and biological evaluation of its analogue symplostatin 1. J Nat Prod 64(7):907–910

Luypaert T, Hagan JG, McCarthy ML, Poti M (2020) Status of marine biodiversity in the anthropocene. In: Jungblut S, Liebich V, Bode-Dalby M (eds) YOUMARES 9—the oceans: our research, our future. Springer, Cham. https://doi.org/10.1007/978-3-030-20389-4_4

Mandala SM, Thornton RA, Milligan J, Rosenbach M, Garcia-Calvo M, Bull HG, Harris G, Abruzzo GK, Flattery AM, Gill CJ, Bartizal K, Dreikorn S, Kurtz MB (1998) Rustmicin, a potent antifungal agent, inhibits sphingolipid synthesis at inositol phosphoceramide synthase. J Biol Chem 273(24):14942–14949. https://doi.org/10.1074/jbc.273.24.14942

Matarrita-Carranza B, Moreira-Soto RD, Murillo-Cruz C, Mora M, Currie CR, Pinto-Tomas AA (2017) Evidence for widespread associations between neotropical hymenopteran insects and Actinobacteria. Front Microbiol 8:1–17. https://doi.org/10.3389/fmicb.2017.02016

Matarrita-Carranza B, Murillo-Cruz C, Avendaño R, Ríos MI, Chavarría M, Gómez-Calvo ML, Tamayo-Castillo G, Araya JJ, Pinto-Tomás AA (2021) *Streptomyces* sp. M54: an actinobacteria associated with a neotropical social wasp with high potential for antibiotic production. Antonie Van Leeuwenhoek 114:379–398. https://doi.org/10.1007/s10482-021-01520-y

McGuire JM, Bunch RL, Anderson RC, Boaz HE, Flynn EH, Powell HM, Smith JW (1952) Ilotycin, a new antibiotic. Antibiot Chemother (Northfield) 2(6):281–283

Mioduchowska M, Zając K, Bartoszek K, Madanecki P, Kur J, Zając T (2020) *16S rRNA* gene-based metagenomic analysis of the gut microbial community associated with the DUI species *Unio crassus* (Bivalvia: Unionidae). J Zool Syst Evol Res 58(2):615–623

Mogouong J, Constant P, Lavallée R, Guertin C (2020) Gut microbiome of the emerald ash borer, *Agrilus planipennis* Fairmaire, and its relationship with insect population density. FEMS Microbiol Ecol 96(8):1–12. https://doi.org/10.1093/femsec/fiaa141

Mohr KI, Tebbe CC (2006) Diversity and phylotype consistency of bacteria in the guts of three bee species (*Apoidea*) at an oilseed rape field. Environ Microbiol 8(2):258–272. https://doi.org/10.1111/j.1462-2920.2005.00893.x

Müller R, Wink J (2014) Future potential for anti-infectives from bacteria—how to exploit biodiversity and genomic potential. Int J Med Microbiol 304(1):3–13. https://doi.org/10.1016/j.ijmm.2013.09.004

Nielsen RV, Hyldig-Nielsen F, Jacobsen K (1982) Biological properties of ristocetin-Ψ-aglycone. J Antibiot (Tokyo) 35(11):1561–1564. https://doi.org/10.7164/antibiotics.35.1561

Parkhaev PY (2017) Origin and the early evolution of the phylum Mollusca. Paleontol J 51(6): 663–686

Pasti MB, Belli ML (1985) Cellulolytic activity of actinomycetes isolated from termites (Termitidae) gut. FEMS Microbiol Lett 26(1):107–112. https://doi.org/10.1111/j.1574-6968.1985.tb01574.x

Pati P, Sahu BK, Panigrahy RC (2015) Marine molluscs as a potential drug cabinet: an overview. Indian J Geo-Mar Sci 44(7):961–970

Peraud O, Biggs JS, Hughen RW, Light AR, Concepcion GP, Olivera BM, Schmidt EW (2009) Microhabitats within venomous cone snails contain diverse actinobacteria. Appl Environ Microbiol 75(21):6820–6826. https://doi.org/10.1128/AEM.01238-09

Pinheiro GL, Correa R, Soares R, Cardoso A, Chaia C, Clementino MM, Garcia E, De Souza W, Frasés S (2015) Isolation of aerobic cultivable cellulolytic bacteria from different regions of the gastrointestinal tract of giant land snail *Achatina fulica*. Front Microbiol 6:860

Poulsen M, Oh DC, Clardy J, Currie CR (2011) Chemical analyses of wasp-associated *Streptomyces* bacteria reveal a prolific potential for natural products discovery. PLoS One 6(2):e16763. https://doi.org/10.1371/journal.pone.0016763

Prommer E (2006) Ziconotide: a new option for refractory pain. Drugs Today 42(6):369–378

Risdian C, Mozef T, Wink J (2019) Biosynthesis of polyketides in *Streptomyces*. Microorganisms 7(5):124

Riyanti R, Widada J, Radjasa OK (2009) Isolation and screening of antimicrobial producing-actinomycetes symbionts in nudibranch. Indones J Biotechnol 14:1132–1138. https://doi.org/10.22146/ijbiotech.7807

Rodríguez-Hernández D, Melo WGP, Menegatti C, Lourenzon VB, do Nascimento FS, Pupo MT (2019) Actinobacteria associated with stingless bees biosynthesize bioactive polyketides against bacterial pathogens. New J Chem 43(25):10109–10117. https://doi.org/10.1039/C9NJ01619H

Romanenko LA, Uchino M, Kalinovskaya NI, Mikhailov VV (2008) Isolation, phylogenetic analysis and screening of marine mollusc-associated bacteria for antimicrobial, hemolytic and surface activities. Microbiol Res 163(6):633–644. https://doi.org/10.1016/j.micres.2006.10.001

Ruiz-González MX, Malé PJG, Leroy C, Dejean A, Gryta H, Jargeat P, Quilichini A, Orivel J (2011) Specific, non-nutritional association between an ascomycete fungus and *Allomerus* plant-ants. Biol Lett 7(3):475–479. https://doi.org/10.1098/rsbl.2010.0920

Ruppert EE, Barnes RD (1994) Invertebrate zoology. Saunders College Publishing, Fort Worth

Safaei N, Mast Y, Steinert M, Huber K, Bunk B, Wink J (2021) Angucycline-like aromatic polyketide from a novel *Streptomyces* species reveals freshwater snail *Physa acuta* as underexplored reservoir for antibiotic-producing actinomycetes. Antibiotics 10(1):22

Salem H, Kreutzer E, Sudakaran S, Kaltenpoth M (2013) Actinobacteria as essential symbionts in firebugs and cotton stainers (Hemiptera, Pyrrhocoridae). Environ Microbiol 15(7):1956–1968. https://doi.org/10.1111/1462-2920.12001

Schneemann I, Ohlendorf B, Zinecker H, Nagel K, Wiese J, Imhoff JF (2010) Nocapyrones A-D, γ-pyrones from a *Nocardiopsis* strain isolated from the marine sponge *Halichondria panicea*. J Nat Prod 73(8):1444–1447

Seipke RF, Barke J, Ruiz-Gonzalez MX, Orivel J, Yu DW, Hutchings MI (2012a) Fungus-growing *Allomerus* ants are associated with antibiotic-producing actinobacteria. Antonie Van Leeuwenhoek 101(2):443–447. https://doi.org/10.1007/s10482-011-9621-y

Seipke RF, Kaltenpoth M, Hutchings MI (2012b) *Streptomyces* as symbionts: an emerging and widespread theme? FEMS Microbiol Rev 36(4):862–876. https://doi.org/10.1111/j.1574-6976.2011.00313.x

Takacs-Vesbach C, King K, Van Horn D, Larkin K, Neiman M (2016) Distinct bacterial microbiomes in sexual and asexual *Potamopyrgus antipodarum*, a New Zealand freshwater snail. PLoS One 11(8):e0161050

Takatsu T, Nakayama H, Shimazu A, Furihata K, Ikeda K, Furihata K, Seto H, Otake N (1985) Rustmicin, a new macrolide antibiotic active against wheat stem rust fungus. J Antibiot (Tokyo) 38(12):1806–1809

Visser AA, Nobre T, Currie CR, Aanen DK, Poulsen M (2012) Exploring the potential for actinobacteria as defensive symbionts in fungus-growing termites. Microb Ecol 63(4): 975–985. https://doi.org/10.1007/s00248-011-9987-4

Waksman SA, Reilly HC, Johnstone DB (1946) Isolation of streptomycin-producing strains of *Streptomyces griseus*. J Bacteriol 52(3):393–397. https://jb.asm.org/content/52/3/393.long

Wang Z, Yu Z, Zhao J, Zhuang X, Cao P, Guo X, Liu C, Xiang W (2020) Community composition, antifungal activity and chemical analyses of ant-derived actinobacteria. Front Microbiol 11:1–12. https://doi.org/10.3389/fmicb.2020.00201

Weingarten EA, Atkinson CL, Jackson CR (2019) The gut microbiome of freshwater Unionidae mussels is determined by host species and is selectively retained from filtered seston. PLoS One 14(11):1–17

Weinstein MJ, Luedemann GM, Oden EM, Wagman GH, Rosselet JP, Marquez JA, Coniglio CT, Charney W, Herzog HL, Black J (1963) Gentamicin, a new antibiotic complex from *Micromonospora*. J Med Chem 6(4):463–464. https://doi.org/10.1021/jm00340a034

Zhang J, Guo C, Chen W, De Lajudie P, Zhang Z, Shang Y, Wang ET (2018) *Mesorhizobium wenxiniae* sp. nov., isolated from chickpea (*Cicer arietinum* L.) in China. Int J Syst Evol Microbiol 68(6):1930–1936. https://doi.org/10.1099/ijsem.0.002770

Polar Actinobacteria: A Source of Biosynthetic Diversity

3

Adriana Rego, Maria de Fátima Carvalho, Pedro Leão, and Catarina Magalhães

Abstract

Actinobacteria are widely known for the production of secondary metabolites, many of which have found application in the clinic. Despite being one of the most extensively explored bacterial phyla, Actinobacteria from polar regions are still overlooked. In this chapter, the taxonomic diversity of Actinobacteria isolated from the poles, their associated secondary metabolites, and the diversity of biosynthetic genes that encode for polyketides and non-ribosomal peptides will be covered.

Keywords

Natural products (NPs) · Biosynthetic gene clusters (BGCs) · Polyketides (PKs) · Non-ribosomal peptides (NRPs) · Polar Actinobacteria · Extreme environments · Antarctica · Arctic · Bioprospection · High-throughput sequencing (HTS)

A. Rego · M. de Fátima Carvalho
Interdisciplinary Centre of Marine and Environmental Research (CIIMAR/CIMAR) University of Porto, Matosinhos, Portugal

ICBAS Institute of Biomedical Sciences Abel Salazar, University of Porto, Porto, Portugal
e-mail: adriana.rego@ciimar.up.pt; mcarvalho@ciimar.up.pt

P. Leão
Interdisciplinary Centre of Marine and Environmental Research (CIIMAR/CIMAR) University of Porto, Matosinhos, Portugal
e-mail: pleao@ciimar.up.pt

C. Magalhães (✉)
Interdisciplinary Centre of Marine and Environmental Research (CIIMAR/CIMAR) University of Porto, Matosinhos, Portugal

Department of Biology, Faculty of Sciences, University of Porto, Porto, Portugal
e-mail: cmagalhaes@ciimar.up.pt

R. V. Rai, J. A. Bai (eds.), *Natural Products from Actinomycetes*,
https://doi.org/10.1007/978-981-16-6132-7_3

3.1 Introduction

Actinobacteria are regarded as the most prolific bacterial phylum in terms of the diversity of natural products (NPs). They are responsible for the synthesis of a plethora of compounds that have found clinical application, including two-thirds of currently administered naturally derived antibiotics (Barka et al. 2016). Around 70% of all the described microbial secondary metabolites are produced by Actinobacteria (Subramani and Aalbersberg 2012), with a wide range of biological activities reported, including antibacterial, antifungal, cytotoxic, anti-inflammatory, and antiviral (Manivasagan et al. 2014).

Inhabiting wide-ranging environments, aquatic (including marine environments) and terrestrial (Barka et al. 2016), and distributed from the tropics to the poles (Rego et al. 2019), Actinobacteria have been considered an inexhaustible source of new metabolites until the beginning of the twenty-first century (Genilloud 2017). The over-exploration of soil-derived Actinobacteria led to increasing re-isolation of known species and rediscovery of known compounds (Dhaneesha et al. 2017). This has triggered the prospection of Actinobacteria from overlooked ecosystems, including those in marine and extreme environments, such as the deep-sea (Tian et al. 2013) and polar regions (Prasad et al. 2014; Buchwał and Szczuciński 2015; Liao et al. 2019; Silva et al. 2020).

In parallel, the development of modern molecular and sequencing technologies, most notably high-throughput sequencing (HTS) technologies, has uncovered the remarkable biosynthetic richness of bacteria, including Actinobacteria, from extreme environments (Benaud et al. 2019; Borsetto et al. 2019).

This chapter will cover the current knowledge on Actinobacteria dwelling polar regions, including their taxonomic and secondary metabolite diversity, with a particular focus on their biosynthetic genes and the associated potential for encoding the production of novel molecules.

3.1.1 Actinobacteria from the Poles

Polar habitats, Arctic and Antarctica, occupy roughly similar areas on our planet but in opposite regions (around 14,500,000 km^2). These englobe some of the most inhospitable ecosystems on earth, usually characterized by extreme temperatures accompanied by strong winds, UV radiation, and drought. Such extreme environmental constraints have been shown to impact the distribution of bacterial communities and to shape their adaptational strategies. Entry into dormancy states (Goordial et al. 2017), colonization of microenvironments (Walker and Pace 2007), and production of secondary metabolites (Tian et al. 2017) are some of the strategies adopted by members of polar bacterial communities to survive extreme conditions.

In the poles, Actinobacteria have been reported as the dominant bacterial phylum across several ecosystems, including cold arid soils (Pointing et al. 2009; Van Goethem et al. 2016; Goordial et al. 2017; Rego et al. 2019; Araujo et al. 2020), and have been isolated from diverse locations such as pack ice (Brinkmeyer et al.

2003), permafrost soil (Hansen et al. 2007), and saline lakes (Lawson et al. 2000). Although the molecular basis behind Actinobacteria dominance in polar areas is unclear, metabolic activity at subzero temperatures (Soina et al. 2004), the formation of spores (Mohammadipanah and Wink 2016), and cyst-like resting forms for non-sporulating species (Soina et al. 2004) are some characteristics of these organisms that allow for survival in such extreme conditions.

Actinobacteria are among the most extensively explored bacterial phyla, particularly for the production of bioactive compounds (Subramani and Aalbersberg 2012; Subramani and Sipkema 2019; Girão et al. 2019). However, polar Actinobacteria are relatively poorly explored. This represents opportunities for the discovery of novel biological and chemical diversity. In fact, new Actinobacteria phylotypes have been described recently from Arctic and Antarctica (Table 3.1), including members of the well-known *Streptomyces* genus (Le Roes-Hill et al. 2009; Kamjam et al. 2019) as well as rare Actinobacteria genera (Li et al. 2010; Liao et al. 2019). This new diversity, together with the extreme stimuli that bacteria are subjected to in such environments, makes polar Actinobacteria an untapped source for biotechnological applications. Despite this promising potential, only a few NPs have been isolated from polar Actinobacteria (Table 3.2).

3.1.2 Compounds Isolated from Polar Actinobacteria

Around 70% of all the described microbial NPs are produced by Actinobacteria (Subramani and Aalbersberg 2012), of which only 3% have been isolated from polar bacteria (Lebar et al. 2007). This occurs in part due to the difficulty in cultivating polar bacteria in the laboratory (Svenson 2013) and to the inaccessibility to these remote environments.

The presence of a higher percentage of phylogenetically divergent microorganisms with unique adaptations to their polar habitats (Zhao et al. 2008) is expected to be translated in the synthesis of singular NPs. The extreme conditions under which they develop render polar bacteria a particularly interesting target for biotechnological applications. For example, the production of cold-adapted enzymes (Jeon et al. 2009; Yu et al. 2011; Lu et al. 2010), anti-freezing molecules (Muñoz et al. 2017), and fatty acids, particularly polyunsaturated fatty acids (PUFAs) (Yoshida et al. 2016), are commonly reported in polar bacteria (Spijkerman et al. 2012).

The potential of polar Actinobacteria for the production of bioactive compounds has been surveyed mostly through bioassay screening methods, including for the production of antimicrobial (Lavin et al. 2016; O'Brien et al. 2004; Giudice et al. 2007; Lee et al. 2012; Tomova et al. 2015; Buchwał and Szczuciński 2015; Tedesco et al. 2016) and cytotoxic compounds (Gao et al. 2012; Moon et al. 2014; Dhaneesha et al. 2017; Shen et al. 2019, 2020).

Novel NPs recovered from polar Actinobacteria strains include compounds with distinct bioactivities such as antiviral, antibiotic, and cytotoxic (Table 3.2). In 2005, an angucyclinone antibiotic, frigocyclinone (**1**) (Bruntner et al. 2005), and the first

Table 3.1 New Actinobacteria taxa isolated from polar environments

Microorganism	Isolation source	Reference
Friedmanniella antarctica	Sandstone sample containing a cryptoendolithic microbial community from Linnaeus Terrace, McMurdo Dry Valleys, Antarctica	Schumann et al. (1997)
Nocardioides aquaticus	Water sample of the hypersaline Ekho Lake, Antarctica	Lawson et al. (2000)
Friedmanniella lacustris	Water sample of the hypersaline Ekho Lake, Antarctica	Lawson et al. (2000)
Modestobacter multiseptatus	Soil samples from Linnaeus Terrace, McMurdo Dry Valleys, Antarctica	Mevs et al. (2000)
Arthrobacter flavus	Cyanobacterial mat sample from a pond in McMurdo Dry Valleys, Antarctica	Reddy et al. (2000)
Micrococcus antarcticus	Chinese Great-Wall station in Antarctica	Liu et al. (2000)
Nesterenkonia lacusekhoensis	Hypersaline Ekho Lake (Vestfold Hills), East Antarctica	Collins et al. (2002)
Arthrobacter roseus	Cyanobacterial mat sample from a pond located in McMurdo, Antarctica	Reddy et al. (2002)
Leifsonia rubra	Cyanobacterial mat sample from a pond in Wright Valley, McMurdo, Antarctica	Reddy et al. (2003a)
Leifsonia aurea	Cyanobacterial mat sample from a pond in Wright Valley, McMurdo, Antarctica	Reddy et al. (2003a)
Kocuria polaris	Cyanobacterial mat sample from a pond located in McMurdo Dry Valley, Antarctica	Reddy et al. (2003b)
Rhodoglobus vestalii	Antarctic Dry Valley Lake	Sheridan et al. (2003)
Arthrobacter gangotriensis	Penguin rookery soil sample in Antarctica	Gupta et al. (2004)
Arthrobacter kerguelensis	Penguin rookery soil sample in Antarctica	Gupta et al. (2004)
Pseudonocardia antarctica	Moraine sample from the McMurdo Dry Valleys, Antarctica	Prabahar et al. (2004)
Micromonospora endolithica	Sandstone rocks of Linnaeus Terrace, Antarctica	Hirsch et al. (2004)
Arthrobacter ardleyensis	Antarctic Ardley Island lake sediment	Chen et al. (2005)
Sanguibacter antarcticus	Sea sand on King George Island, Antarctica	Hong et al. (2008)
Arthrobacter psychrochitiniphilus	Adelie penguin guano from Antarctica	Wang et al. (2009)
Leifsonia antarctica	Spade core sediment sample from the Antarctic Ocean	Pindi et al. (2009)
Demequina lutea	Permafrost soil collected from the Adventdalen Valley, Spitsbergen, Northern Norway	Finster et al. (2009)
Streptomyces hypolithicus	Base of a translucent quartz rock in Miers Valley, eastern Antarctica	Le Roes-Hill et al. (2009)

(continued)

Table 3.1 (continued)

Microorganism	Isolation source	Reference
Marisediminicola antarctica	Sandy intertidal sediment sample collected from the coastal area off the Chinese Antarctic Zhongshan Station, East Antarctica	Li et al. (2010)
Arthrobacter antarcticus	Sediment samples collected from a depth of 400 m near the Larsemann Hills area, Antarctica	Pindi et al. (2010)
Arthrobacter livingstonensis	Moss-covered soil from Livingston Island, Antarctica	Ganzert et al. (2011a)
Arthrobacter cryotolerans	Moss-covered soil from Livingston Island, Antarctica	Ganzert et al. (2011a)
Streptomyces fildesensis	Soil sample collected from the Fildes Peninsula, King George Island, Antarctica	Li et al. (2011)
Leifsonia psychrotolerans	Moss-covered soil from Livingston Island, Antarctica	Ganzert et al. (2011b)
Barrientosiimonas humi	Soil collected from Barrientos Island in the Antarctic	Lee et al. (2013)
Nocardiopsis fildesensis	Soil sample collected from the Fildes Peninsula, King George Island, West Antarctica	Xu et al. (2014)
Nocardioides antarcticus	Marine sediment of Ardley Cove, King George Island, Antarctica	Deng et al. (2015)
Raineyella antarctica	Rhizosphere of moss *Leptobryum* sp. collected at the shore of Lake Zub in Antarctica	Pikuta et al. (2016)
Sanguibacter gelidistatuariae	Naturally formed ice sculpture on the shore of Lake Podprudnoye in Antarctica	Pikuta et al. (2017)
Rhodococcus psychrotolerans	Rhizosphere of *Deschampsia antarctica* collected at King George Island, Antarctic Peninsula	Silva et al. (2018)
Amycolatopsis antarctica	Surface of an Antarctic brown macroalgae, collected from the Punta Rodriguez site of the King George Island, Antarctica	Wang et al. (2018)
Nakamurella antarctica	Tundra soil sampled near the Antarctic Peninsula, South Shetland Islands	Da et al. (2019)
Streptomyces septentrionalis	Frozen soil sample which was collected from the Arctic region	Kamjam et al. (2019)
Streptomyces polaris	Frozen soil sample which was collected from the Arctic region	Kamjam et al. (2019)
Glaciihabitans arcticus	Soil sampled at the Arctic region in Cambridge Bay, NU, Canada	Dahal and Kim (2019)
Tessaracoccus antarcticus	Soil sample of Fildes Peninsula, Antarctica	Zhou et al. (2020)
Pseudarthrobacter psychrotolerans	Antarctic soil collected from the Cape Burk area	Shin et al. (2020)
Marisediminicola senii	Glacier fed sediment sample collected from the Queen Maud Land, near India's Maitri station in Antarctica	Jani et al. (2021)

bridged angucyclinone, gephyromycin (**2**) (Bringmann et al. 2005), together with other known antibiotics, were isolated from two *Streptomyces griseus* strains, isolated from an Antarctica soil sample. In 2007, a new sulfur-containing natural

Table 3.2 NPs isolated from Arctic and Antarctic Actinobacteria

Species	Compounds	Reported bioactivity	Isolation source	Reference
Streptomyces griseus strain NTK 97	Frigocyclinone (**1**)	Antibiotic	Terrestrial sample from Antarctica	Bruntner et al. (2005)
Streptomyces griseus strain NTK 14	Gephyromycin (**2**)	Glutaminergic activity (agonist) towards neuronal cells	Terrestrial sample from Antarctica	Bringmann et al. (2005)
Microbispora aerata	Microbiaeratin (**3**)	Antiproliferative and cytotoxic	Antarctic Livingston Island	Ivanova et al. (2007)
Nocardiopsis sp.	Cyclo-(L-Pro-L-Met) (**4**)	Antiangiogenic agent	Arctic seaweed	Shin et al. (2010)
Nocardia dassonvillei	*N*-(2-hydroxyphenyl)-2-phenazinamine (NHP) (**5**)	Anticancer and antifungal	Sediment sample from the Arctic Ocean	Gao et al. (2012)
Streptomyces nitrosporeus	Nitrosporeusines A (**6**) and B (**7**)	Antiviral	Sediments from Arctic Chukchi Sea	Yang et al. (2013)
Streptomyces sp.	Arcticoside (**8**) and C-1027 chromophore-V (**9**)	Antifungal and cytotoxic	Surface sediment taken at the East Siberian continental margin	Moon et al. (2014)
Nocardiopsis sp. SCSIO KS107	Germicidin H (**10**) and 4-hydroxymucidone (**11**)	Not described	Seashore sediment from Antarctica	Zhang et al. (2016)
Streptomyces sp. OUCMDZ-4348	Cyclamenols B (**12**), C (**13**), and D (**14**)	Selective inhibition against the gastric carcinoma cell line N87 (cyclamenol B)	Sand sample from Antarctica	Shen et al. (2019)
Streptomyces sp. OUCMDZ-4348	Cyclamenols E (**15**) and F (**16**)	Moderate inhibitory activity against the gastric carcinoma cell line N87 (cyclamenol E)	Sand sample from Antarctica	Shen et al. (2020)

alkaloid named microbiaeratin (**3**) (Ivanova et al. 2007) was isolated from the culture filtrate of *Microbispora aerata*, which has been recovered from penguin excrements collected on the Antarctic Livingston Island (Fig. 3.1).

A rare bioactive diketopiperazine, cyclo-(L-Pro-L-Met) (**4**), was also recovered from a marine *Nocardiopsis* sp. 03N67, isolated from a seaweed collected on an Arctic expedition (Shin et al. 2010). In 2012, Gao et al. (2012) found a new secondary metabolite, *N*-(2-hydroxyphenyl)-2-phenazinamine (NHP) (**5**), and six

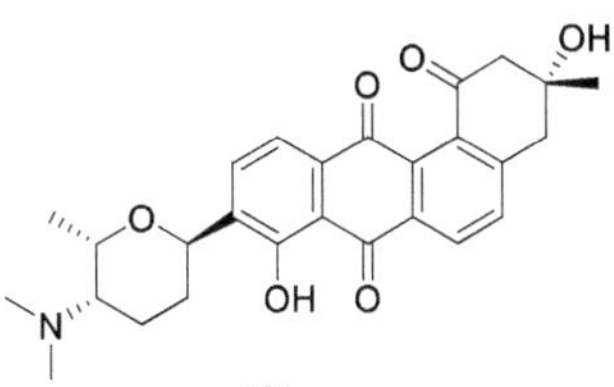

(1)

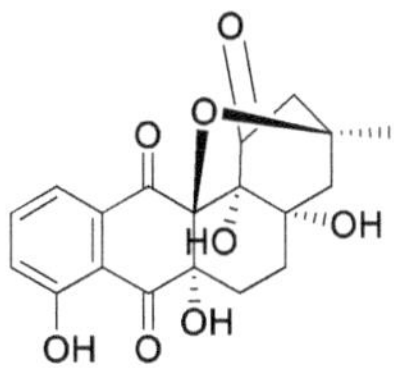

(2)

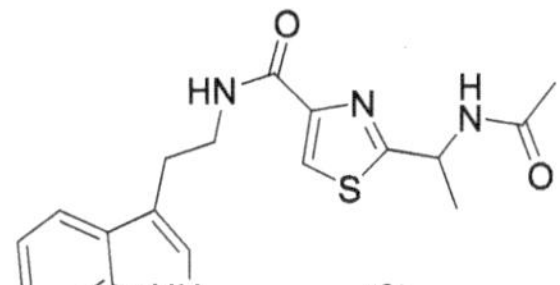

(3)

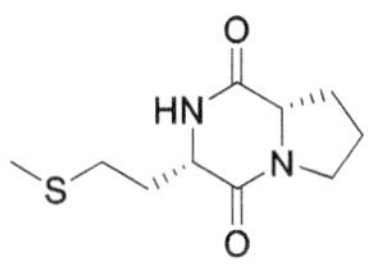

(4)

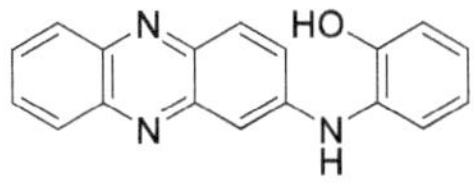

(5)

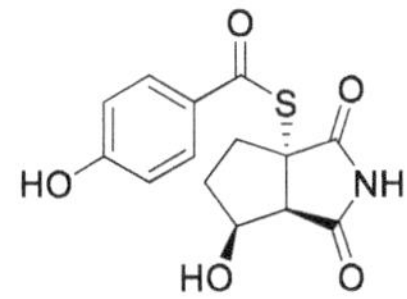

(6)

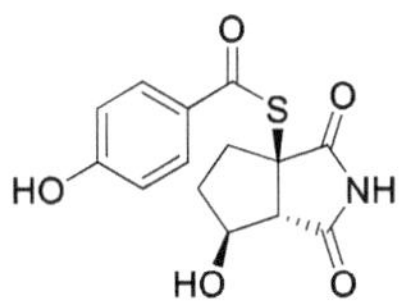

(7)

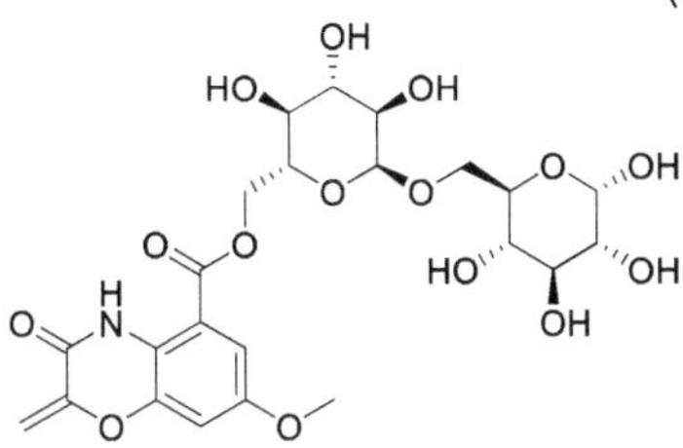

(8)

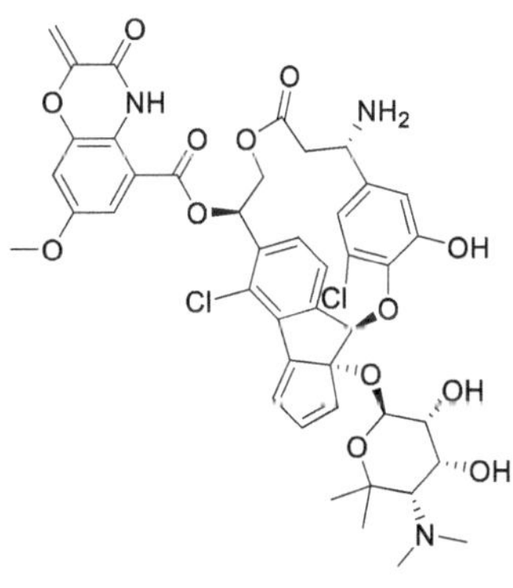

(9)

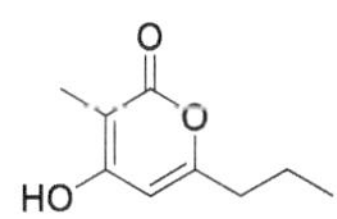

(10)

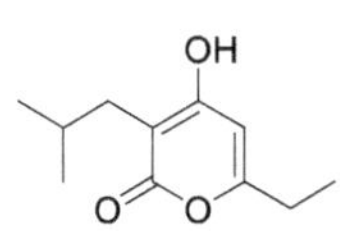

(11)

Fig. 3.1 Chemical structures of secondary metabolites derived from polar Actinobacteria

known antibiotics from a *Nocardia* strain isolated from a sediment sample from the Arctic Ocean.

Streptomyces nitrosporeus, isolated from Arctic Chukchi Sea sediments, was found to produce two unprecedented thioester-bearing alkaloids, nitrosporeusines A (**6**) and B (**7**) (Yang et al. 2013). Two new benzoxazine compounds, arcticoside (**8**) and C-1027 chromophore-V (**9**) (Moon et al. 2014), were isolated from an Arctic marine *Streptomyces* strain obtained from a sediment sample. The exploration of a *Nocardiopsis* strain isolated from an Antarctic seashore sediment sample led to the discovery of two new α-pyrones, germicidin H (**10**) and 4-hydroxymucidone (**11**), together with the known compound 7-hydroxymucidone (Zhang et al. 2016).

More recently, new polycyclic macrolactams (cyclamenols B–D) (**12–14**) (Shen et al. 2019) and bicyclic macrolactams (cyclamenols E–F) (**15–16**) (Shen et al. 2020) were isolated from an Antarctic *Streptomyces* sp. OUCMDZ-4348.

The majority of the described NPs (Table 3.2) have been isolated from *Streptomyces* species (Fig. 3.2) associated to marine sediment samples. This is not surprising, since the genus *Streptomyces* encloses the largest number of marine Actinobacteria species (Yang et al. 2013) and is reported as an extremely fruitful producer of NPs. Further, rare actinobacterial genera are known to be more recalcitrant to isolation and cultivation in laboratory, and selective isolation methods are usually required (Pulschen et al. 2017; Rego et al. 2019).

On previously described studies (Table 3.2), Actinobacteria strains were isolated through traditional cultivation methods such as by serial dilution technique (Shin et al. 2010; Gao et al. 2012). It is widely known that probably less than 1% of the microorganisms can be isolated through traditional culturing techniques (Kennedy et al. 2010), and in the polar areas, this percentage is likely even smaller (Lambrechts

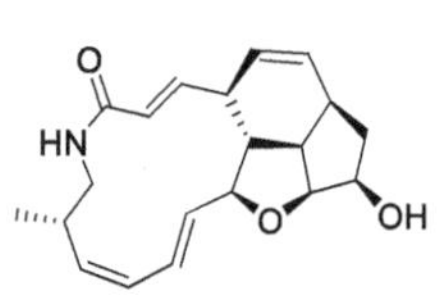

(12)

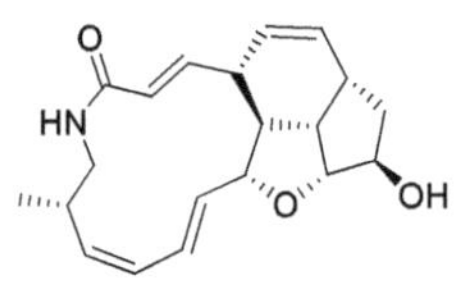

(13)

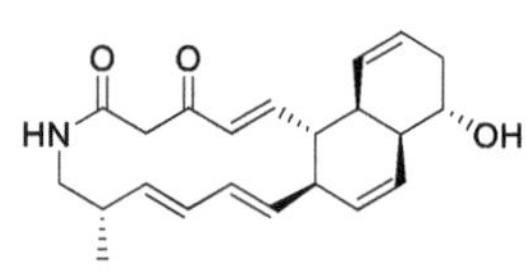

(14)

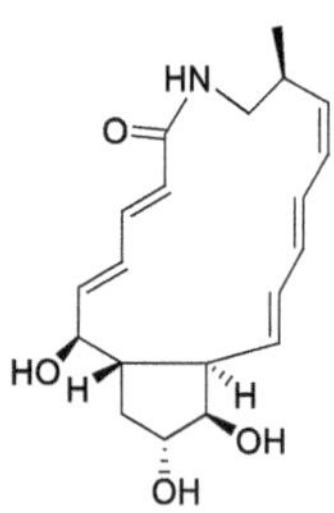

(15)

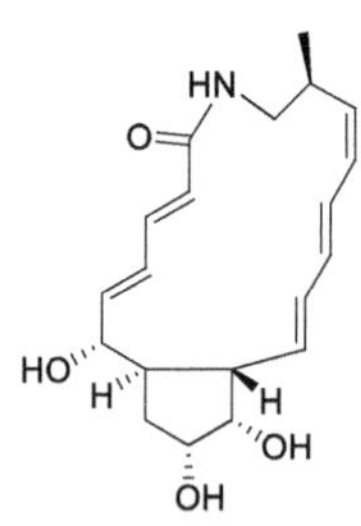

(16)

Fig. 3.1 (continued)

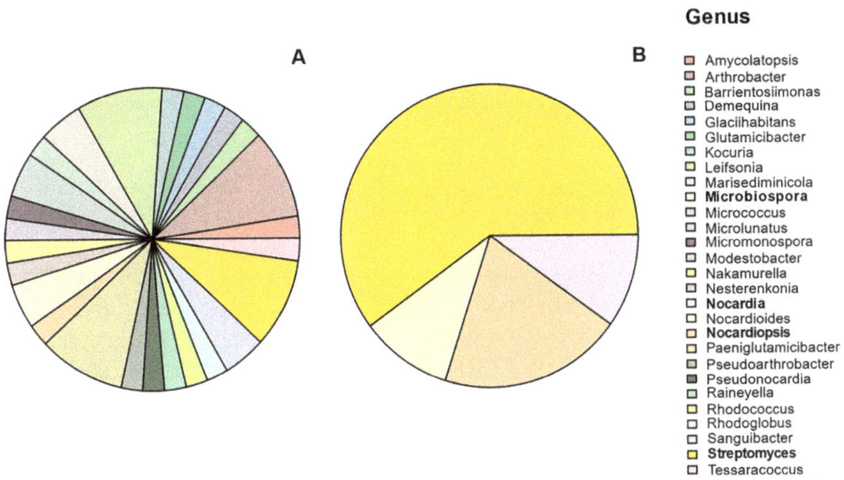

Fig. 3.2 Pie chart with distribution of new Actinobacteria genera isolated from polar environments (**a**) and the polar Actinobacteria (genus) from which new compounds were isolated (**b**)

et al. 2019). This would leave the majority of the bacteria and their biosynthetic potential out of reach for conventional culture-based NP discovery approaches. With the emerging massive-parallel sequencing technologies and the advent of (meta)-genome mining studies, a great diversity of unknown and low-abundance phylotypes and a large amount of silent or cryptic gene clusters potentially encoding the production of new bioactive metabolites are being reported. This highlights both the bottleneck that culture-dependent approaches represent in discovery strategies and the promise of heterologous expression approaches to reveal the cryptic diversity of bacterial NPs (Huo et al. 2019).

3.2 Diversity of Biosynthetic Gene Clusters in Polar Actinobacteria

One of the main challenges faced by the traditional NP research is the ability to isolate and grow the microorganisms in laboratory. The advance of new sequencing technologies in the beginning of the twenty-first century (Maclean et al. 2009) dictated a new golden era of NP research. With the increasing number of DNA sequences deposited in public databases, not only genome mining studies but also homology-based PCR screening of biosynthetic genes started to be used. To screen the biosynthetic potential, several studies have taken advance of PCR amplification and HTS technologies to survey the diversity of polyketide synthases (PKS) and non-ribosomal peptide synthetases (NRPS)—enzymes coding for two of the most important families of NPs, polyketides (PKs) and non-ribosomal peptides (NRPs) (Wang et al. 2014). In fact, these modern molecular and sequencing techniques have revealed an untapped potential of polar bacteria, and Actinobacteria in particular, for the production of novel NPs (Benaud et al. 2019; Rego et al. 2020; Borsetto et al. 2019). This has revealed that known NPs isolated from polar Actinobacteria represent just the tip of the iceberg regarding the vast diversity of NPs encoded in polar Actinobacteria-derived BGCs.

3.2.1 PCR-Based Approach

One of the first reports of biosynthetic diversity analysis on polar regions dates from 2008, in which Zhao and co-authors (2008) studied the phylogenetic diversity of type I PKS and NRPS genes in two Antarctica sediment samples. The authors used primer pairs for ketosynthase (KS) (Moffitt and Neilan 2001) and adenylation (AD) (Neilan et al. 1999) domains that were initially designed for Cyanobacteria. In this study, only one PKS KS domain was identified, with 52% identity to *Streptomyces avermitilis*, and none of the NRPS AD domains were affiliated with the Actinobacteria group, suggesting that the assessed Actinobacteria were not rich PKS or NRPS producers.

In 2015, Amos et al. (2015) developed a new PCR-based assay for the discovery of biosynthetic gene clusters (BGCs), including the design of new primer pairs targeting the rare microbial biosphere, including rare Actinobacteria. The primers were tested on Antarctic soil samples from Mars Oasis and recovered sequences with high similarities to NRPS sequences found in a wide range of Actinobacteria, including the rare genera *Thermomonospora* and *Saccharothrix* and the genus *Streptomyces*. These sequences were phylogenetically divergent from representative sequences, which suggested that these belonged to a yet undiscovered NRPS pathway. PCR-based sequence approaches allowed also the identification of nine halogenase genes in 60 Arctic marine Actinobacteria using in-house designed and previously reported PCR primers (Liao et al. 2016). The halogenases were predicted to be involved in halogenation of indole groups, antitumor agent ansamitocin-like substrates, or unknown peptide-like compounds. The authors considered that this PCR-based screening approach could facilitate the bioprospection of halometabolite-producing Actinobacteria from the Arctic (Liao et al. 2016).

Streptomyces artemisiae MCCB 248 was isolated from the Arctic fjord Kongsfjorden (Dhaneesha et al. 2017) and its screening, by PCR, using degenerate primers previously designed for *Streptomyces* (Izumikawa et al. 2003; Ayuso-Sacido and Genilloud 2005), revealed 82% similarity to known biosynthetic genes of *Streptomyces*, indicating the likely production of a novel secondary metabolite.

More recently, the traditional PCR-based approach was adapted to HTS technologies to survey the biosynthetic potential of the bacterial communities directly from environmental DNA (Benaud et al. 2019; Borsetto et al. 2019; Rego et al. 2020). Benaud et al. (2019) employed an amplicon sequencing strategy targeting PKS and NRPS genes, using PacBio RS II, to survey over 200 soils across 12 East Antarctic and also Arctic sites from Alexandra Fiord and Svalbard regions. The assessed soils were comprised of high relative abundances of Actinobacteria, and the majority of the recovered NRPS AD domains were taxonomically assigned also to Actinobacteria. Most of the recovered KS and AD domain sequences were identified as encoding for potentially novel metabolites, given the low shared nucleotide sequence homology (<70%) to known BGCs, highlighting the potential of Actinobacteria from the assessed samples. Borsetto et al. (2019) used amplicon sequencing (Illumina MiSeq) to explore the diversity of NRPS and PKS genes from 13 soils across the world, including Antarctica soils (Mars Oasis). The resulting data highlighted a rich endemic diversity of these harsh environments with a high percentage of domain reads unmatched to known sequences, suggesting that these genes encode enzymes for new PKs and NRPs.

In a study developed by our team (Rego et al. 2020), an amplicon strategy (using Illumina MiSeq) was employed to survey the diversity of PKS and NRPS genes across maritime Antarctica, Maxwell Bay. Recovered KS domain sequences detected in high abundance across all samples were taxonomically assigned to rare Actinobacteria genera (*Actinoalloteichus*, *Kitasatospora*, *Mycobacterium*, and *Rhodococcus*) and to the two most NP-rich genera, *Streptomyces* and *Micromonospora*. In an attempt to determine the relatedness of the recovered sequences to known BGCs, sequences were matched against the MIBiG database (Medema et al. 2015). For KS domain, in total, over 40% of the matches were associated taxonomically to *Streptomyces* species. The OTU assigned with the highest identity (90%) to MIBiG database corresponded to a BGC associated to the production of spinosad (Waldron et al. 2001), a nonantibiotic macrocyclic lactone with insecticide activity, isolated from the rare actinobacterial genus *Saccharopolyspora*. For AD domain, the top match obtained had 96% identity to the BGC associated with the production of aurantimycin (Rego et al. 2020), which is a gram-positive specific antibiotic, initially isolated from *Streptomyces aurantiacus* IMET 43917 (Gräfe et al. 1995; Zhao et al. 2016). All these findings underline the potential of Antarctica Actinobacteria for the production of new bioactive compounds.

3.2.2 Genomic and Metagenomic Approaches

The development of genome sequencing methods and mining studies has revealed the unexplored potential hidden in cultivated bacterial genomes, even for well-explored bacterial genera, such as *Streptomyces* (Bentley et al. 2002; Ikeda et al. 2003). Genome mining studies have revealed BGCs to represent over 10% of actinobacterial genomes (Udwary et al. 2007; Nett et al. 2009), with a huge proportion of cryptic and silent BGCs and the majority of the actinobacterial BGCs detected not linked to any known product (Letzel et al. 2017), thus likely to encode novel NPs.

Only a few studies have been conducted on genomes of polar Actinobacteria strains with the goal of exploring their biosynthetic gene diversity. Recently, a rare actinobacterial species, *Marisediminicola antarctica* ZS314T, and the only type strain of the novel actinobacterial genus *Marisediminicola*, were obtained from intertidal sediments in East Antarctica (Liao et al. 2019). Its genome was sequenced and five BGCs were identified, including a gene cluster responsible for the biosynthesis of a C50 carotenoid, four additional BGCs of an unknown oligosaccharide, salinixanthin, alkylresorcinol derivatives, and NRPS or amino acid-derived compounds. The authors suggested the isolated strain is probably endemic to extreme environments and were able to identify a new xanthorhodopsin gene together with the analysis of the salinixanthin BGC. These findings have increased the knowledge on the biosynthetic potential of this rare actinobacterial species.

Besides the analysis of single genomes from isolated Actinobacteria, recently culture-dependent approaches and metagenomics revealed high relative abundance of Actinobacteria in sediment cores from Antarctic and sub-Arctic locations (Millán-Aguiñaga et al. 2019). In this study, a molecular network analysis of metabolic extracts of selected isolated strains revealed polar Actinobacteria as a rich source of previously unreported chemistry.

3.3 Conclusions and Future Perspectives

Conventional NP isolation programs have massively underestimated the biosynthetic diversity of bacteria. Recent studies have emphasized the untapped biosynthetic potential of polar Actinobacteria, given the phylogenetic divergence of PKS and NRPS conserved domains and the low sequence homology to BGC sequences of known NPs. The presence of a higher ratio of uncultured bacteria in polar regions implies a promising future for the bioprospection of new microbial biosynthetic diversity. Much of the current knowledge on polar Actinobacteria and their ability to produce bioactive compounds is based on the chemical exploration of laboratory cultures or the analysis of whole genomes of cultured strains. However, our improved ability to characterize the natural complex microbial communities using HTS technologies, with the application of sophisticated bioinformatic tools, allowing for the reconstruction of metagenome-assembled genomes (MAGs), will in the near future illuminate the true potential of the Actinobacteria diversity and enable the discovery of new biosynthetic functions with promising biotechnological applications. In the future, the use of heterologous expression of BGCs from environmental DNA as a means to functionally characterize the biosynthetic pathways holds the most promise to uncover new NPs without the necessity of bacterial isolation and cultivation.

Acknowledgments The authors acknowledge funding from the Portuguese Science and Technology Foundation (FCT) through grants to CM (PTDC/CTA-AMB/30997/2017; PTDC/CTA-AMB/4946/2020), a PhD scholarship to AR (SFRH/BD/140567/2018), and through the project and the Strategic Funding UIDB/04423/2020 and UIDP/04423/2020 through national funds provided by FCT and European Regional Development Fund (ERDF) in the framework of the program PT2020. The authors also acknowledge funding from the structured program of R&D&I ATLANTIDA—Platform for the monitoring of the North Atlantic Ocean and tools for the sustainable exploitation of the marine resources (reference NORTE-01-0145-FEDER-000040), supported by the North Portugal Regional Operational Programme (NORTE2020), through the ERDF.

References

Amos GCA, Borsetto C, Laskaris P, Krsek M, Berry AE, Newsham KK, Calvo-Bado L, Pearce DA, Vallin C, Wellington EMH (2015) Designing and implementing an assay for the detection of rare and divergent NRPS and PKS clones in European, Antarctic and Cuban soils. PLoS One 10(9):1–15. https://doi.org/10.1371/journal.pone.0138327

Araujo R, Gupta VVSR, Reith F, Bissett A, Mele P, Franco CMM (2020) Biogeography and emerging significance of actinobacteria in Australia and Northern Antarctica soils. Soil Biol Biochem 146:107805. https://doi.org/10.1016/j.soilbio.2020.107805

Ayuso-Sacido A, Genilloud O (2005) New PCR primers for the screening of NRPS and PKS-I Systems in actinomycetes: detection and distribution of these biosynthetic gene sequences in major taxonomic groups. Microb Ecol 49(1):10–24. https://doi.org/10.1007/s00248-004-0249-6

Barka EA, Vatsa P, Sanchez L, Gaveau-Vaillant N, Jacquard C, Klenk H-P, Clément C, Ouhdouch Y, Van Wezel P (2016) Taxonomy, physiology, and natural products of Actinobacteria. Microbiol Mol Biol Rev 80(1):1–43. https://doi.org/10.1128/MMBR.00019-15.Address

Benaud N, Zhang E, van Dorst J, Brown MV, Kalaitzis JA, Neilan BA, Ferrari BC (2019) Harnessing long-read amplicon sequencing to uncover NRPS and Type I PKS gene sequence diversity in polar desert soils. FEMS Microbiol Ecol 95(4):fiz031. https://doi.org/10.1093/femsec/fiz031

Bentley SD, Chater KF, Cerdeño-Tárraga AM, Challis GL, Thomson NR, James KD, Harris DE et al (2002) Complete genome sequence of the model Actinomycete Streptomyces coelicolor A3 (2). Nature 417(6885):141–147. https://doi.org/10.1038/417141a

Borsetto C, Amos GCA, Da Rocha UN, Mitchell AL, Finn RD, Laidi RF, Vallin C, Pearce DA, Newsham KK, Wellington EMH (2019) Microbial community drivers of PK/NRP gene diversity in selected global soils. Microbiome 7(78):1–11

Bringmann G, Lang G, Maksimenka K, Hamm A, Gulder TAM, Dieter A, Bull AT et al (2005) Gephyromycin, the First bridged angucyclinone, from Streptomyces Griseus strain NTK 14. Phytochemistry 66(11 Spec. Iss.):1366–1373. https://doi.org/10.1016/j.phytochem.2005.04.010

Brinkmeyer R, Knittel K, Jürgens J, Weyland H, Amann R, Helmke E (2003) Diversity and structure of bacterial communities in Arctic versus Antarctic Pack Ice. Appl Environ Microbiol 69(11):6610–6619. https://doi.org/10.1128/AEM.69.11.6610-6619.2003

Bruntner C, Binder T, Pathom-Aree W, Goodfellow M, Bull AT, Potterat O, Puder C et al (2005) Frigocyclinone, a novel angucyclinone antibiotic produced by a Streptomyces Griseus strain from Antarctica. J Antibiot 58(5):346–349. https://doi.org/10.1038/ja.2005.43

Buchwał A, Szczuciński W (2015) Isolation, identification and screening of Actinobacteria in volcanic soil of deception island (the Antarctic) for antimicrobial metabolites. Pol Polar Res 36(1):51–65. https://doi.org/10.1515/popore-2015-0001

Chen M, Xiao X, Wang P, Zeng X, Wang F (2005) Arthrobacter ardleyensis sp. nov., isolated from Antarctic lake sediment and deep-sea sediment. Arch Microbiol 183(4):301–305. https://doi.org/10.1007/s00203-005-0772-y

Collins MD, Lawson PA, Labrenz M, Tindall BJ, Weiss N, Hirsch P (2002) Nesterenkonia lacusekhoensis sp. nov., isolated from hypersaline Ekho Lake, East Antarctica, and emended description of the Genus Nesterenkonia. Int J Syst Evol Microbiol 52(4):1145–1150. https://doi.org/10.1099/ijs.0.02118-0

Da X, Zhao Y, Zheng R, Wang L, Chang X, Zhang Y, Yang J, Peng F (2019) Nakamurella Antarctica sp. nov., isolated from Antarctica South Shetland Islands soil. Int J Syst Evol Microbiol 69(12):3710–3715. https://doi.org/10.1099/ijsem.0.003507

Dahal RH, Kim J (2019) Glaciihabitans arcticus sp. nov., a psychrotolerant bacterium isolated from Arctic soil. Int J Syst Evol Microbiol 69(8):2492–2497. https://doi.org/10.1099/ijsem.0.003520

Deng S, Chang X, Zhang Y, Ren L, Jiang F, Zhihao Q, Peng F (2015) Nocardioides antarcticus sp. nov, isolated from marine sediment. Int J Syst Evol Microbiol 65(8):2615–2621. https://doi.org/10.1099/ijs.0.000309

Dhaneesha M, Benjamin Naman C, Krishnan KP, Sinha RK, Jayesh P, Valsamma J, Bright Singh IS, Gerwick WH, Sajeevan TP (2017) Streptomyces artemisiae MCCB 248 isolated from Arctic Fjord sediments has unique PKS and NRPS biosynthetic genes and produces potential new anticancer natural products. 3 Biotech 7(1):1–10. https://doi.org/10.1007/s13205-017-0610-3

Finster KW, Herbert RA, Kjeldsen KU, Schumann P, Lomstein BA (2009) Demequina lutea sp. nov., isolated from a high Arctic permafrost soil. Int J Syst Evol Microbiol 59(4): 649–653. https://doi.org/10.1099/ijs.0.004929-0

Ganzert L, Bajerski F, Mangelsdorf K, Lipski A, Wagner D (2011a) Arthrobacter livingstonensis sp. nov. and arthrobacter cryotolerans sp. nov., salt-tolerant and psychrotolerant species from Antarctic soil. Int J Syst Evol Microbiol 61(4):979–984. https://doi.org/10.1099/ijs.0.021022-0

Ganzert L, Bajerski F, Mangelsdorf K, Lipski A, Wagner D (2011b) Leifsonia psychrotolerans sp. nov., a psychrotolerant species of the family microbacteriaceae from Livingston Island, Antarctica. Int J Syst Evol Microbiol 61(8):1938–1943. https://doi.org/10.1099/ijs.0.021956-0

Gao X, Lu Y, Xing Y, Ma Y, Lu J, Bao W, Wang Y, Xi T (2012) A novel anticancer and antifungus phenazine derivative from a marine actinomycete BM-17. Microbiol Res 167(10):616–622. https://doi.org/10.1016/j.micres.2012.02.008

Genilloud O (2017) Actinomycetes: still a source of novel antibiotics. Nat Prod Rep 34(10): 1203–1232. https://doi.org/10.1039/c7np00026j

Girão M, Ribeiro I, Ribeiro T, Azevedo IC, Pereira F, Urbatzka R, Leão PN, Carvalho MF (2019) Actinobacteria isolated from Laminaria ochroleuca: a source of new bioactive compounds. Front Microbiol 10:1–13. https://doi.org/10.3389/fmicb.2019.00683

Giudice AL, Bruni V, Michaud L (2007) Characterization of Antarctic psychrotrophic bacteria with antibacterial activities against terrestrial microorganisms. J Basic Microbiol 47(6):496–505. https://doi.org/10.1002/jobm.200700227

Goordial J, Davila A, Greer CW, Cannam R, DiRuggiero J, McKay CP, Whyte LG (2017) Comparative activity and functional ecology of permafrost soils and lithic niches in a hyper-arid polar desert. Environ Microbiol 19(2):443–458. https://doi.org/10.1002/elsc.201200179

Gräfe U, Schlegel R, Ritzau M, Ihn W, Dornberger K, Stengel C, Fleck WF, Gutsche W, Härtl A, Paulus EF (1995) Aurantimycins, new depsipeptide antibiotics from Streptomyces Aurantiacus IMET 43917. Production, isolation, structure elucidation, and biological activity. J Antibiot 48(2):119–125. http://www.ncbi.nlm.nih.gov/pubmed/7706121

Gupta P, Reddy GSN, Delille D, Shivaji S (2004) Arthrobacter gangotriensis sp. nov. and arthrobacter kerguelensis sp. nov. from Antarctica. Int J Syst Evol Microbiol 54(6): 2375–2378. https://doi.org/10.1099/ijs.0.63110-0

Hansen AA, Herbert RA, Mikkelsen K, Jensen LL, Kristoffersen T, Tiedje JM, Lomstein BA, Finster KW (2007) Viability, diversity and composition of the bacterial community in a high arctic permafrost soil from Spitsbergen, Northern Norway. Environ Microbiol 9(11): 2870–2884. https://doi.org/10.1111/j.1462-2920.2007.01403.x

Hirsch P, Mevs U, Kroppenstedt RM, Schumann P, Stackebrandt E (2004) Cryptoendolithic actinomycetes from Antarctic sandstone rock samples: micromonospora endolithica sp. nov. and two isolates related to Micromonospora coerulea Jensen 1932. Syst Appl Microbiol 27(2): 166–174. https://doi.org/10.1078/072320204322881781

Hong SG, Lee YK, Kim JH, Chun J, Lee HK (2008) Sanguibacter antarcticus sp. nov., isolated from Antarctic sea sand. Int J Syst Evol Microbiol 58(1):50–52. https://doi.org/10.1099/ijs.0.65031-0

Huo L, Hug JJ, Chengzhang F, Bian X, Zhang Y, Müller R (2019) Heterologous expression of bacterial natural product biosynthetic pathways. Nat Prod Rep 36(10):1412–1436. https://doi.org/10.1039/c8np00091c

Ikeda H, Ishikawa J, Hanamoto A, Shinose M, Kikuchi H, Shiba T, Sakaki Y, Hattori M, Omura S (2003) Complete genome sequence and comparative analysis of the industrial microorganism streptomyces avermitilis. Nat Biotechnol 21(5):526–531. https://doi.org/10.1038/nbt820

Ivanova V, Kolarova M, Aleksieva K, Gräfe U, Dahse H-M, Laatsch H (2007) Microbiaeratin, a new natural indole alkaloid from a microbispora aerata strain, isolated from Livingston Island, Antarctica. Prep Biochem Biotechnol 37(2):161–168. https://doi.org/10.1080/10826060701199122

Izumikawa M, Murata M, Tachibana K, Ebizuka Y, Fujii I (2003) Cloning of modular type I polyketide synthase genes from salinomycin producing strain of streptomyces albus. Bioorg Med Chem 11(16):3401–3405. https://doi.org/10.1016/S0968-0896(03)00337-7

Jani K, Kajale S, Shetye M, Palkar S, Sharma A (2021) Marisediminicola senii sp. nov. isolated from Queen Maud Land, Antarctica. Int J Syst Evol Microbiol 71. https://doi.org/10.1099/ijsem.0.004641

Jeon JH, Kim JT, Kang SG, Lee JH, Kim SJ (2009) Characterization and its potential application of two esterases derived from the arctic sediment metagenome. Mar Biotechnol 11(3):307–316. https://doi.org/10.1007/s10126-008-9145-2

Kamjam M, Nopnakorn P, Zhang L, Peng F, Deng Z, Hong K (2019) Streptomyces polaris sp. nov. and streptomyces septentrionalis sp. nov., isolated from frozen soil. Antonie Van Leeuwenhoek 112(3):375–387. https://doi.org/10.1007/s10482-018-1166-x

Kennedy J, Flemer B, Jackson SA, Lejon DPH, Morrissey JP, O'Gara F, Dobson ADW (2010) Marine metagenomics: new tools for the study and exploitation of marine microbial metabolism. Mar Drugs 8(3):608–628. https://doi.org/10.3390/md8030608

Lambrechts S, Willems A, Tahon G (2019) Uncovering the uncultivated majority in Antarctic soils: toward a synergistic approach. Front Microbiol 10(242):1–19. https://doi.org/10.3389/fmicb.2019.00242

Lavin PL, Yong ST, Wong CMVL, De Stefano M (2016) Isolation and characterization of Antarctic psychrotroph streptomyces sp. strain INACH3013. Antarct Sci 28(6):433–442. https://doi.org/10.1017/S0954102016000250

Lawson PA, David Collins M, Peter S, Tindall BJ, Hirsch P, Labrenz M (2000) New LL-diaminopimelic acid-containing actinomycetes from hypersaline, heliothermal and meromictic Antarctic Ekho lake: nocardioides aquaticus sp. nov. and friedmanniella lacustris sp. nov. Syst Appl Microbiol 23(2):219–229. https://doi.org/10.1016/S0723-2020(00)80008-9

Le Roes-Hill M, Rohland J, Meyers PR, Cowan DA, Burton SG (2009) Streptomyces hypolithicus sp. nov., isolated from an Antarctic hypolith community. Int J Syst Evol Microbiol 59(8):2032–2035. https://doi.org/10.1099/ijs.0.007971-0

Lebar MD, Heimbegner JL, Baker BJ (2007) Cold-water marine natural products. Nat Prod Rep 24(4):774–797. https://doi.org/10.1039/b516240h

Lee L-H, Cheah Y-K, Sidik SM, Mutalib N-SA, Tang Y-L, Lin H-P, Hong K (2012) Molecular characterization of Antarctic actinobacteria and screening for antimicrobial metabolite production. World J Microbiol Biotechnol 28(5):2125–2137. https://doi.org/10.1007/s11274-012-1018-1

Lee LH, Cheah YK, Sidik SM, Xie QY, Tang YL, Lin HP, Mutalib NSA, Hong K (2013) Barrientosiimonas humi gen. nov., sp. nov., an actinobacterium of the family dermacoccaceae. Int J Syst Evol Microbiol 63(1):241–248. https://doi.org/10.1099/ijs.0.038232-0

Letzel A-C, Li J, Amos G, Millán-Aguiñaga N, Ginigini J, Abdelmohsen UR, Gaudêncio SP, Ziemert N, Moore BS, Jensen PR (2017) Genomic insights into specialized metabolism in the marine actinomycete salinispora. Environ Microbiol 19(9):3660–3673. https://doi.org/10.1111/1462-2920.13867

Li HR, Yu Y, Luo W, Zeng YX (2010) Marisediminicola antarctica gen. nov., sp. nov., an actinobacterium isolated from the Antarctic. Int J Syst Evol Microbiol 60(11):2535–2539. https://doi.org/10.1099/ijs.0.018754-0

Li J, Tian XP, Zhu TJ, Yang LL, Li WJ (2011) Streptomyces fildesensis sp. nov., a novel streptomycete isolated from Antarctic soil. Antonie Van Leeuwenhoek 100(4):537–543. https://doi.org/10.1007/s10482-011-9609-7

Liao L, Chen R, Jiang M, Tian X, Liu H, Yu Y, Fan C, Chen B (2016) Bioprospecting potential of halogenases from arctic marine actinomycetes. BMC Microbiol 16(1):1–9. https://doi.org/10.1186/s12866-016-0662-2

Liao L, Su S, Zhao B, Fan C, Zhang J, Li H, Chen B (2019) Biosynthetic potential of a novel Antarctic actinobacterium marisediminicola Antarctica ZS314T revealed by genomic data mining and pigment characterization. Mar Drugs 17(7):1–15. https://doi.org/10.3390/md17070388

Liu H, Xu Y, Ma Y, Zhou P (2000) Characterization of micrococcus antarcticus sp. nov., a psychrophilic bacterium from Antarctica. Int J Syst Evol Microbiol 50(2):715–719. https://doi.org/10.1099/00207713-50-2-715

Lu M, Wang S, Fang Y, Li H, Liu S, Liu H (2010) Cloning, expression, purification, and characterization of cold-adapted α-amylase from pseudoalteromonas arctica GS230. Protein J 29(8):591–597. https://doi.org/10.1007/s10930-010-9290-0

Maclean D, Jones JDG, Studholme DJ (2009) Application of 'next-generation' sequencing technologies to microbial genetics. Nat Rev Microbiol 7(2):96–97. https://doi.org/10.1038/nrmicro2088

Manivasagan P, Venkatesan J, Sivakumar K, Kim SK (2014) Pharmaceutically active secondary metabolites of marine actinobacteria. Microbiol Res 169(4):262–278. https://doi.org/10.1016/j.micres.2013.07.014

Medema MH, Kottmann R, Yilmaz P, Cummings M, Biggins JB, Blin K, De Bruijn I et al (2015) Minimum information about a biosynthetic gene cluster. Nat Chem Biol 11(9):625–631. https://doi.org/10.1038/nchembio.1890

Mevs U, Stackebrandt E, Schumann P, Gallikowski CA, Hirsch P (2000) Modestobacter multiseptatus gen. nov., sp. nov., a budding actinomycete from soils of the Asgard range (Transantarctic mountains). Int J Syst Evol Microbiol 50(1):337–346. https://doi.org/10.1099/00207713-50-1-337

Millán-Aguiñaga N, Soldatou S, Brozio S, Munnoch JT, Howe J, Hoskisson PA, Duncan KR (2019) Awakening ancient polar Actinobacteria: diversity evolution and specialized metabolite potential. Microbiology 165(11):1169–1180. https://doi.org/10.1099/mic.0.000845

Moffitt MC, Neilan BA (2001) On the presence of peptide synthetase and polyketide synthase genes in the cyanobacterial genus nodularia. FEMS Microbiol Lett 196(2):207–214. https://doi.org/10.1016/S0378-1097(01)00070-2

Mohammadipanah F, Wink J (2016) Actinobacteria from arid and desert habitats: diversity and biological activity. Front Microbiol 6(1541):1–10. https://doi.org/10.3389/fmicb.2015.01541

Moon K, Ahn CH, Shin Y, Won TH, Ko K, Lee SK, Oh KB, Shin J, Nam SI, Oh DC (2014) New benzoxazine secondary metabolites from an arctic actinomycete. Mar Drugs 12(5):2526–5238. https://doi.org/10.3390/md12052526

Muñoz PA, Márquez SL, González-Nilo FD, Márquez-Miranda V, Blamey JM (2017) Structure and application of antifreeze proteins from Antarctic bacteria. Microb Cell Fact 16(1):1–13. https://doi.org/10.1186/s12934-017-0737-2

Neilan BA, Dittmann E, Rouhiainen L, Bass RA, Schaub V, Sivonen K, Börner T (1999) Nonribosomal peptide synthesis and toxigenicity of cyanobacteria. J Bacteriol 181(13):4089–4097

Nett M, Ikeda H, Moore BS (2009) Genomic basis for natural product biosynthetic diversity in the actinomycetes. Nat Prod Rep 26(11):1362–1384. https://doi.org/10.1039/b817069j.Genomic

O'Brien A, Sharp R, Russell NJ, Roller S (2004) Antarctic bacteria inhibit growth of food-borne microorganisms at low temperatures. FEMS Microbiol Ecol 48(2):157–167. https://doi.org/10.1016/j.femsec.2004.01.001

Pikuta EV, Menes RJ, Bruce AM, Lyu Z, Patel NB, Liu Y, Hoover RB, Busse HJ, Lawson PA, Whitman WB (2016) Raineyella antarctica gen. nov., sp. nov., a psychrotolerant, D-amino-acid-utilizing anaerobe isolated from two geographic locations of the southern hemisphere. Int J Syst Evol Microbiol 66(12):5529–5536. https://doi.org/10.1099/ijsem.0.001552

Pikuta EV, Lyu Z, Williams MD, Patel NB, Liu Y, Hoover RB, Busse HJ, Lawson PA, Whitman WB (2017) Sanguibacter gelidistatuariae sp. nov., a novel psychrotolerant anaerobe from an ice sculpture in Antarctica, and emendation of descriptions of the family sanguibacteraceae, the Genus Sanguibacter and Species S. Antarcticus, S. Inulinus, S. Kedieii, S. Ma. Int J Syst Evol Microbiol 67(5):1442–1450. https://doi.org/10.1099/ijsem.0.001838

Pindi PK, Hara Kishore K, Reddy GSN, Shivaji S (2009) Description of Leifsonia kafniensis sp. nov. and Leifsonia Antarctica sp. nov. Int J Syst Evol Microbiol 59(6):1348–1352. https://doi.org/10.1099/ijs.0.006643-0

Pindi PK, Manorama R, Begum Z, Shivaji S (2010) Arthrobacter antarcticus sp. nov., isolated from an Antarctic marine sediment. Int J Syst Evol Microbiol 60(10):2263–2266. https://doi.org/10.1099/ijs.0.012989-0

Pointing SB, Chan Y, Lacap DC, Lau MCY, Jurgens JA, Farrell RL (2009) Highly specialized microbial diversity in hyper-arid polar desert. Proc Natl Acad Sci U S A 106(47):19964–19969. https://doi.org/10.1073/pnas.0908274106

Prabahar V, Dube S, Reddy GSN, Shivaji S (2004) Pseudonocardia Antarctica sp. nov. an actinomycetes from McMurdo Dry Valleys, Antarctica. Syst Appl Microbiol 27(1):66–71. https://doi.org/10.1078/0723-2020-00249

Prasad S, Manasa P, Buddhi S, Tirunagari P, Begum Z, Rajan S, Shivaji S (2014) Diversity and bioprospective potential (cold-active enzymes) of cultivable marine bacteria from the Subarctic Glacial Fjord, Kongsfjorden. Curr Microbiol 68(2):233–238. https://doi.org/10.1007/s00284-013-0467-6

Pulschen AA, Bendia AG, Fricker AD, Pellizari VH, Galante D, Rodrigues F (2017) Isolation of uncultured bacteria from antarctica using long incubation periods and low nutritional media. Front Microbiol 8:1–12. https://doi.org/10.3389/fmicb.2017.01346

Reddy GSN, Aggarwal RK, Matsumoto GI, Shivaji S (2000) Arthrobacter flavus sp. nov., a psychrophilic bacterium isolated from a pond in McMurdo Dry Valley, Antarctica. Int J Syst Evol Microbiol 50(4):1553–1561. https://doi.org/10.1099/00207713-50-4-1553

Reddy GSN, Prakash JSS, Matsumoto GI, Stackebrandt E (2002) Arthrobacter roseus sp. nov. a psychrophilic bacterium isolated from an Antarctic cyanobacterial mat sample. Int J Syst Evol Microbiol 52(3):1017–1021. https://doi.org/10.1099/ijs.0.02131-0

Reddy GSN, Prakash JSS, Srinivas R, Matsumoto GI, Shivaji S (2003a) Leifsonia rubra sp. nov. and leifsonia aurea sp. nov., psychrophiles from a Pond in Antarctica. Int J Syst Evol Microbiol 53(4):977–984. https://doi.org/10.1099/ijs.0.02396-0

Reddy GSN, Prakash JSS, Prabahar V, Matsumoto GI, Stackebrandt E, Shivaji S (2003b) Kocuria polaris sp. nov., an orange-pigmented psychrophilic bacterium isolated from an Antarctic cyanobacterial mat sample. Int J Syst Evol Microbiol 53(1):183–187. https://doi.org/10.1099/ijs.0.02336-0

Rego A, Raio F, Martins TP, Ribeiro H, Sousa AGG, Séneca J, Baptista MS et al (2019) Actinobacteria and cyanobacteria diversity in terrestrial Antarctic microenvironments evaluated by culture-dependent and independent methods. Front Microbiol 10. https://doi.org/10.3389/fmicb.2019.01018

Rego A, Sousa AGG, Santos JP, Pascoal F, Canário J, Leão PN, Magalhães C (2020) Diversity of bacterial biosynthetic genes in maritime Antarctica. Microorganisms 8(2):1–22. https://doi.org/10.3390/microorganisms8020279

Schumann P, Prauser H, Iuiiney FA, Stackebrandt E, Hirsch P (1997) Friedmanniella Antarctica gen. nov., sp. noc., an LL-diaminopimelic acid-containing actinomycete from Antarctic sandstone. Int J Syst Bacteriol 47(2):278–283

Shen J, Fan Y, Zhu G, Chen H, Zhu W, Peng F (2019) Polycyclic macrolactams generated via intramolecular Diels-Alder reactions from an Antarctic streptomyces species. Org Lett 21(12):4816–4820. https://doi.org/10.1021/acs.orglett.9b01710

Shen J, Wang J, Chen H, Wang Y, Zhu W, Peng F (2020) Cyclamenols E and F, two diastereoisomeric bicyclic macrolactams with a cyclopentane moiety from an Antarctic: streptomyces species. Org Chem Front 7(2):310–317. https://doi.org/10.1039/c9qo01215j

Sheridan PP, Loveland-Curtze J, Miteva VI, Brenchley JE (2003) Rhodoglobus vestalii gen. nov., sp. nov., a novel psychrophilic organism isolated from an Antarctic Dry Valley Lake. Int J Syst Evol Microbiol 53(4):985–994. https://doi.org/10.1099/ijs.0.02415-0

Shin HJ, Mojid Mondol MA, Yu TK, Lee HS, Lee YJ, Jung HJ, Kim JH, Kwon HJ (2010) An angiogenesis inhibitor isolated from a marine-derived actinomycete, Nocardiopsis Sp. 03N67. Phytochem Lett 3(4):194–197. https://doi.org/10.1016/j.phytol.2010.07.005

Shin Y, Lee BH, Lee KE, Park W (2020) Pseudarthrobacter psychrotolerans sp. nov., a cold-adapted bacterium isolated from Antarctic soil. Int J Syst Evol Microbiol 70(12). https://doi.org/10.1099/ijsem.0.004505

Silva LJ, Souza DT, Genuario DB, Hoyos HAV, Santos SN, Rosa LH, Zucchi TD, Melo IS (2018) Rhodococcus psychrotolerans sp. nov., isolated from rhizosphere of Deschampsia Antarctica. Antonie Van Leeuwenhoek 111(4):629–636. https://doi.org/10.1007/s10482-017-0983-7

Silva LJ, Crevelin EJ, Souza DT, Lacerda-Júnior GV, de Oliveira VM, Ruiz ALTG, Rosa LH, Moraes LAB, Melo IS (2020) Actinobacteria from Antarctica as a source for anticancer discovery. Sci Rep 10(1):1–15. https://doi.org/10.1038/s41598-020-69786-2

Soina VS, Mulyukin AL, Demkina EV, Vorobyova EA, El-Registan GI (2004) The structure of resting bacterial populations in soil and subsoil permafrost. Astrobiology 4(3):345–358. https://doi.org/10.1089/ast.2004.4.345

Spijkerman E, Wacker A, Weithoff G, Leya T (2012) Elemental and fatty acid composition of snow algae in arctic habitats. Front Microbiol 3:1–15. https://doi.org/10.3389/fmicb.2012.00380

Subramani R, Aalbersberg W (2012) Marine actinomycetes: an ongoing source of novel bioactive metabolites. Microbiol Res 167(10):571–580. https://doi.org/10.1016/j.micres.2012.06.005

Subramani R, Sipkema D (2019) Marine rare actinomycetes: a promising source of structurally diverse and unique novel natural products. Mar Drugs 17. https://doi.org/10.3390/md17050249

Svenson J (2013) MabCent: Arctic marine bioprospecting in Norway. Phytochem Rev 12(3): 567–578. https://doi.org/10.1007/s11101-012-9239-3

Tedesco P, Maida I, Esposito FP, Tortorella E, Subko K, Ezeofor CC, Zhang Y et al (2016) Antimicrobial activity of monoramnholipids produced by bacterial strains isolated from the Ross Sea (Antarctica). Mar Drugs 14(5). https://doi.org/10.3390/md14050083

Tian XP, Long LJ, Li SM, Zhang J, Xu Y, He J, Li J et al (2013) Pseudonocardia antitumoralis sp. nov., a deoxynyboquinone-producing actinomycete isolated from a deep-sea sediment. Int J Syst Evol Microbiol 63(Part 3):893–899. https://doi.org/10.1099/ijs.0.037135-0

Tian Y, Li YL, Zhao FC (2017) Secondary metabolites from polar organisms. Mar Drugs 15(3). https://doi.org/10.3390/md15030028

Tomova I, Stoilova-Disheva M, Lazarkevich I, Vasileva-Tonkova E (2015) Antimicrobial activity and resistance to heavy metals and antibiotics of heterotrophic bacteria isolated from sediment and soil samples collected from two Antarctic Islands. Front Life Sci 8(4):348–357. https://doi.org/10.1080/21553769.2015.1044130

Udwary DW, Zeigler L, Asolkar RN, Singan V, Lapidus A, Fenical W, Jensen PR, Moore BS (2007) Genome sequencing reveals complex secondary metabolome in the marine Actinomycete Salinispora Tropica. Proc Natl Acad Sci U S A 104(25):10376–10381. https://doi.org/10.1073/pnas.0700962104

Van Goethem MW, Makhalanyane TP, Valverde A, Cary SC, Cowan DA (2016) Characterization of bacterial communities in lithobionts and soil niches from Victoria Valley, Antarctica. FEMS Microbiol Ecol 92(4):1–22. https://doi.org/10.1093/femsec/fiw051

Waldron C, Matsushima P, Rosteck PR, Broughton MC, Turner J, Madduri K, Crawford KP, Merlo DJ, Baltz RH (2001) Cloning and analysis of the spinosad biosynthetic gene cluster of Saccharopolyspora spinosa. Chem Biol 8(5):487–499. https://doi.org/10.1016/s1074-5521(01)00029-1

Walker JJ, Pace NR (2007) Endolithic microbial ecosystems. Annu Rev Microbiol 61:331–347. https://doi.org/10.1146/annurev.micro.61.080706.093302

Wang F, Gai Y, Chen M, Xiao X (2009) Arthrobacter psychrochitiniphilus sp. nov., a psychrotrophic bacterium isolated from Antarctica. Int J Syst Evol Microbiol 59(11): 2759–2762. https://doi.org/10.1099/ijs.0.008912-0

Wang H, Fewer DP, Holm L, Rouhiainen L, Sivonen K (2014) Atlas of nonribosomal peptide and polyketide biosynthetic pathways reveals common occurrence of nonmodular enzymes. Proc Natl Acad Sci U S A 111(25):9259–9264. https://doi.org/10.1073/pnas.1401734111

Wang J, Leiva S, Huang J, Huang Y (2018) Amycolatopsis antarctica sp. nov., isolated from the surface of an Antarctic brown macroalga. Int J Syst Evol Microbiol 68(7):2348–2356. https://doi.org/10.1099/ijsem.0.002844

Xu S, Yan L, Zhang X, Wang C, Feng G, Li J (2014) Nocardiopsis fildesensis sp. nov., an actinomycete isolated from soil. Int J Syst Evol Microbiol 64(Part 1):174–179. https://doi.org/10.1099/ijs.0.053595-0

Yang A, Si L, Shi Z, Tian L, Liu D, Zhou D, Proksch P, Lin W (2013) Nitrosporeusines A and B, unprecedented thioester-bearing alkaloids from the arctic streptomyces nitrosporeus. Org Lett 15(20):5366–5369. https://doi.org/10.1021/ol4026809

Yoshida K, Hashimoto M, Hori R, Adachi T, Okuyama H, Orikasa Y, Nagamine T, Shimizu S, Ueno A, Morita N (2016) Bacterial long-chain polyunsaturated fatty acids: their biosynthetic genes, functions, and practical use. Mar Drugs 14(5). https://doi.org/10.3390/md14050094

Yu EY, Kwon MA, Lee M, Oh JY, Choi JE, Lee JY, Song BK, Hahm DH, Song JK (2011) Isolation and characterization of cold-active family VIII esterases from an Arctic soil metagenome. Appl Microbiol Biotechnol 90(2):573–581. https://doi.org/10.1007/s00253-011-3132-7

Zhang H, Saurav K, Yu Z, Mándi A, Kurtán T, Li J, Tian X, Zhang Q, Zhang W, Zhang C (2016) α-Pyrones with diverse hydroxy substitutions from three marine-derived nocardiopsis strains. J Nat Prod 79(6):1610–1618. https://doi.org/10.1021/acs.jnatprod.6b00175

Zhao J, Yang N, Zeng R (2008) Phylogenetic analysis of type I polyketide synthase and nonribosomal peptide synthetase genes in Antarctic sediment. Extremophiles 12(1):97–105. https://doi.org/10.1007/s00792-007-0107-9

Zhao H, Wang L, Wan D, Qi J, Gong R, Deng Z, Chen W (2016) Characterization of the aurantimycin biosynthetic gene cluster and enhancing its production by manipulating two pathway-specific activators in streptomyces Aurantiacus JA 4570. Microb Cell Fact 15(1):1. https://doi.org/10.1186/s12934-016-0559-7

Zhou LY, Zhang JY, Chen XY, Du ZJ, Mu DS (2020) Tessaracoccus antarcticus sp. nov., a rhodopsin-containing bacterium from an Antarctic environment and emended description of the genus Tessaracoccus. Int J Syst Evol Microbiol 70(3):1555–1561. https://doi.org/10.1099/ijsem.0.003930

Actinobacteria from Arid Environments and Their Biotechnological Applications

4

Feiyang Xie, Pharada Rangseekaew, and Wasu Pathom-aree

Abstract

Actinobacteria are a large group of beneficial Gram-positive bacteria that play an important role in sustainable biotechnology. Desert, as a representative of arid environment, is known to harbour diverse Actinobacteria with biotechnological potential. These cultivable Actinobacteria are adapted with a special survival mechanism to combat various stresses in the desert biomes. Both the emergence of multidrug-resistant (MDR) diseases and the sharply decreasing trend in agricultural productivity caused by climate change are indicators of an emergency quest for exploration of these desert Actinobacteria for the discovery of novel bioactive metabolites, especially those with the antibacterial, anticancer, antifungal, anti-inflammatory, anti-MDR and antitumour activities. In this chapter, we provide an overview of the diversity of novel Actinobacteria from desert environments including their bioactive compounds reported during the period from 2010 until 2021.

F. Xie
PhD's Degree Program in Applied Microbiology, Faculty of Science, Chiang Mai University, Under the CMU Presidential Scholarship, Chiang Mai, Thailand

Department of Biology, Faculty of Science, Chiang Mai University, Chiang Mai, Thailand
e-mail: feiyang_x@cmu.ac.th

P. Rangseekaew
Department of Biology, Faculty of Science, Chiang Mai University, Chiang Mai, Thailand
e-mail: Pharada_ra@cmu.ac.th

W. Pathom-aree (✉)
Department of Biology, Faculty of Science, Chiang Mai University, Chiang Mai, Thailand

Research Center in Microbial Diversity and Sustainable Utilization, Faculty of Science, Chiang Mai University, Chiang Mai, Thailand
e-mail: wasu.p@cmu.ac.th

R. V. Rai, J. A. Bai (eds.), *Natural Products from Actinomycetes*,
https://doi.org/10.1007/978-981-16-6132-7_4

Keywords

Actinobacteria · Desert · Arid environment · Isolation strategy · Bioactive compounds · Secondary metabolites · Plant growth promotion · Biocatalyst · Multidrug-resistant pathogens

4.1 Introduction

4.1.1 Desert Environment

Arid environments which lack vegetation and animal represent more than one-third of the Earth's surface area. These specific habitats are characterized by deficit moisture, low annual precipitation and organic matter and potential evapotranspiration (Goudie 2013). Arid environments can be divided into arid, extremely arid, hyper-arid and extreme hyper-arid habitats, based on the ratio of mean annual rainfall to mean annual evaporation. Most arid regions are deserts which considered as the largest continental ecosystems occupy about 30% of an Earth's surface (Neilson et al. 2012). Currently, there are 35 deserts on the planet with half of them located in Asia. The deserts can be classified based on their proximity to the oceans or continentality, different levels of latitudes and the rate of precipitation. For example, the Atacama and Namib Deserts are coastal deserts near the ocean, while the Gobi Desert in Mongolia and the Indian Himalayas Desert are cold deserts with extremely low temperatures (Goudie 2013; Thomas 2011). Vostok Desert in Antarctica is a cold polar desert with an annual precipitation of less than 50 mm (Goudie 2013).

Microbial resources in desert achieve great attention among microbiologists since water availability is the primary influence factor for all life forms. Microorganisms can survive under low water activity environments through expending more energy to cease cell activity but remain viable (Connon et al. 2007). Low moisture and water, limited fog and water, extreme temperature and low organic carbon source, such arid biomes significantly limit the growth of microorganism. However, some microorganisms are adapted to live under these harsh abiotic conditions. Microorganisms that can grow optimally under these extreme conditions are named extremophiles, whereas extremotolerant used to describe organisms that tolerate such conditions (MacElroy 1974; Sayed et al. 2020). Both chemical (pH, organic carbon, oxidants) and physical conditions (temperature, water activity, radiation) of the desert can be used as selective parameters for special microorganisms called extremotrophs (Bull 2011). Microorganisms that thrive in arid regions adjust their access to nutrient required for their growth. As a result, the microbial community composition is highly reflected in biomes' environmental conditions, especially those endemic to arid habitats.

4.1.2 Actinobacteria from Desert

Actinobacteria, one of the largest phyla of the Bacteria domain, are a group of Gram-positive bacteria with a high ratio of guanine plus cytosine (G + C) in their DNA content. The main morphology of Actinobacteria is either filamentous or non-filamentous. Actinobacteria are normally present in the upper layer of soil with high population density (10^6–10^9 cells per gram of soil) (Goodfellow and Williams 1983). Vegetative Actinobacteria can be found in dry soil as its favour with low humidity environments (Barka et al. 2016). Currently, the phylum of Actinobacteria is divided into 6 classes (Actinobacteria, Acidimicrobiia, Coriobacteriia, Nitriliruptoria, Rubrobacteria and Thermoleophilia), 22 orders, 54 families, 250 genera and at least 3000 species (Amaresan et al. 2020). They are generally abundant in soils rich in organic matter with a slightly alkaline condition (Barka et al. 2016). However, it is becoming apparent that they are also abundant in several extreme environments including arid or desert habitats (Okoro et al. 2009).

In the past decades, innumerable actinobacterial taxa are continued to be recovered from desert biomes. Among the Actinobacteria, members of the orders Acidimicrobiales, Actinomycetales, Geodermatophilales, Micrococcales, Micromonosporales, Nakamurellales and Pseudonocardiales were found abundantly in 34 deserts worldwide, belonging to 115 new species, 44 genera and 19 families, respectively (Table 4.1). Acidotolerant, alkaliphilic, alkalitolerant, halotolerant, haloalkalitolerant, psychrotolerant, thermotolerant and xerophilous Actinobacteria are found among desert Actinobacteria (Lubsanova et al. 2014). For example, *Geodermatophilus arenarius* and *G. siccatus* are xerophilic Actinobacteria isolated from the Sahara Desert. Another alkalitolerant actinobacterium, *Citricoccus alkalitolerans*, was discovered from eastern desert of Egypt (Montero-Calasanz et al. 2012; Harwani 2013; Li et al. 2005). A moderate thermophilic xerotolerant *Streptomyces fumigatiscleroticus* was isolated from Mongolian desert soil (Lubsanova et al. 2014). *Streptomyces aburaviensis*, a halotolerant alkaliphilic strain, was recovered from a saline Kutch Desert in India (Thumar et al. 2010).

The highest numbers of new actinobacterial species are from members of *Streptomyces* (14 species), followed by the genus *Geodermatophilus* (9 species). Unlike the prolific bioactive compound producing *Streptomyces*, most members of *Geodermatophilus* were unique in their adaptive response to the desert environment. For example, *G. pulveris* and *G. sabuli* were gamma-radiation actinomycetes, whereas *G. tzadiensis* is a UV radiation-resistant actinobacterium. Other new species were isolated from desert plants, rhizosphere soil and rock surface, respectively. Actinobacteria recovered from desert plants include *Aeromicrobium endophyticum*, *Labedella phragmitis*, *L. populi*, *Microbacterium karelineae*, *M. halophytorum*, *Nesterenkonia populi*, *Nocardiopsis* sp., *Prauserella endophytica* and *Promicromonospora panici*. Species such as *Kribbella deserti*, *Nakamurella deserti*, *Nesterenkonia rhizosphaerae*, *Nocardioides deserti* and *Streptomyces netropsis* were reported from rhizosphere soils. *Modestobacter multiseptatus* and *Pseudonocardia nigra* were discovered from desert rock surface.

Table 4.1 Diversity and distribution of desert Actinobacteria during the past decades

Family	Genus	Species	Location	References
Actinopolysporaceae	*Actinopolyspora*	*Actinopolyspora mzabensis*	Sahara Desert, Algeria	Meklat et al. (2013)
Bogoriellaceae	*Georgenia*	*Georgenia alba*	Arid area, Saudi Arabia	Li et al. (2019d)
		Georgenia deserti		Hozzein et al. (2018)
Cellulomonadaceae	*Cellulomonas*	*Cellulomonas telluris*	Badain Jaran Desert, China	Shi et al. (2020)
Dietziaceae	*Dietzia*	*Dietzia kunjamensis*	Kunjam Pass, Indian Himalayas	Mayilraj et al. (2006b)
	Williamsia	*Williamsia* sp.	Antarctic Desert	Guerrero et al. (2014)
Geodermatophilaceae	*Blastococcus*	*Blastococcus atacamensis*	Atacama Desert, Chile	Castro et al. (2018a)
		Blastococcus deserti	Gurbantunggut Desert, China	Yang et al. (2019)
	Geodermatophilus	*Geodermatophilus africanus*	Sahara Desert, Chad	del Carmen Montero-Calasanz et al. (2013b)
		Geodermatophilus arenarius		Montero-Calasanz et al. (2012)
		Geodermatophilus chilensis	Atacama Desert, Chile	Castro et al. (2018b)
		Geodermatophilus pulveris	Sahara Desert, Libya	Hezbri et al. (2016)
		Geodermatophilus sabuli	Sahara Desert, Tunisia	Hezbri et al. (2015)
		Geodermatophilus saharensis	Sahara Desert, Chad	Montero-Calasanz et al. (2013a)
		Geodermatophilus siccatus		del Carmen Montero-Calasanz et al. (2013c)
		Geodermatophilus telluris		Montero-Calasanz et al. (2013b)
		Geodermatophilus tzadiensis		del Carmen Montero-Calasanz et al. (2013a)

	Modestobacter	*Modestobacter altitudinis*	Atacama Desert, Chile	Golińska et al. (2020b)
		Modestobacter caceresii		Busarakam et al. (2016b)
		Modestobacter excelsi		Golińska et al. (2020a)
Ilumatobacteraceae	*Desertimonas*	*Desertimonas flava*	Gurbantunggut Desert, China	Asem et al. (2018)
Jiangellaceae	*Jiangella*	*Jiangella asiatica*	Karakum Desert, Turkmenistan	Saygin et al. (2020c)
		Jiangella aurantiaca		
		Jiangella ureilytica		
		Jiangella gansuensis	Gansu Desert, China	Song et al. (2005)
Intrasporangiaceae	*Janibacter*	*Janibacter* sp.	Arid and saline sediment, Tunisia	Khessairi et al. (2014)
	Ornithinicoccus	*Ornithinicoccus halotolerans*	Xinjiang Desert, China	Zhang et al. (2016)
	Labedella	*Labedella phragmitis*	Taklamakan Desert, China	Li et al. (2019c)
		Labedella populi		
	Microbacterium	*Microbacterium album*	Arid area, Saudi Arabia	Yang et al. (2018b)
		Microbacterium deserti		
		Microbacterium halophytorum	Gurbantunggut Desert, China	Li et al. (2018a)
		Microbacterium suaedae		Zhu et al. (2019)
		Microbacterium karelineae	Taklamakan Desert, China	Zhu et al. (2021)
	Mycetocola	*Mycetocola manganoxydans*		Luo et al. (2012)

(continued)

Table 4.1 (continued)

Family	Genus	Species	Location	References
	Planctomonas	*Planctomonas deserti*		Liu et al. (2019a)
Micrococcaceae	*Arthrobacter*	*Arthrobacter deserti*	Turpan Desert, China	Hu et al. (2016)
		Arthrobacter mobilis	Cholistan Desert, Pakistan	Ye et al. (2020)
	Nesterenkonia	*Nesterenkonia populi*	Taklamakan Desert, China	Liu et al. (2015b)
		Nesterenkonia rhizosphaerae	Fukang Desert, China	Wang et al. (2014)
Micromonosporaceae	*Actinoplanes*	*Actinoplanes deserti*	Turpan Desert, China	Habib et al. (2018)
	Micromonospora	*Micromonospora acroterricola*	Atacama Desert, Chile	Carro et al. (2019b)
		Micromonospora arida		Carro et al. (2019a)
		Micromonospora inaquosa		
		Micromonospora deserti	Karakum Desert, Turkmenistan	Saygin et al. (2020a)
Motilibacteraceae	*Motilibacter*	*Motilibacter aurantiacus*	Cholistan Desert, Pakistan	Liu et al. (2020d)
		Motilibacter deserti		
Nakamurellaceae	*Nakamurella*	*Nakamurella deserti*	Taklamakan Desert, China	Liu et al. (2019b)
Nocardiaceae	*Nocardia*	*Nocardia* sp.	Xinjiang Desert, China	Zhang et al. (2020)
Nocardioidaceae	*Aeromicrobium*	*Aeromicrobium endophyticum*	Taklamakan Desert, China	Li et al. (2019a)
		Aeromicrobium halotolerans	Turpan Desert, China	Yan et al. (2016)

	Kribbella	*Kribbella deserti*	Desert soil of Hangjin Banner, Ordos, Inner Mongolia, Northern China	Sun et al. (2017)
		Kribbella turkmenica	Karakum Desert, Turkmenistan	Saygin et al. (2019b)
	Nocardioides	*Nocardioides deserti*	Taklamakan Desert, China	Tuo et al. (2015)
		Nocardioides thalensis	Thal Desert, Pakistan	Khan et al. (2017)
		Nocardioides vastitatis	Taklamakan Desert, China	Liu et al. (2020c)
	Tenggerimyces	*Tenggerimyces mesophilus*	Tengger Desert, China	Sun et al. (2015)
	Nocardiopsis	*Nocardiopsis alkaliphila*	Eastern Desert, Egypt	Hozzein et al. (2004)
		Nocardiopsis benisuefensis	Beni-Suef Governorate, Egypt	Hozzein et al. (2014)
		Nocardiopsis sp.	Taklamakan Desert, China	Chen et al. (2015)
Promicromonosporaceae	*Promicromonospora*	*Promicromonospora panici*	Sahara Desert, Tunisia	Guesmi et al. (2021)
Propionibacteriaceae	*Auraticoccus*	*Auraticoccus cholistanensis*	Cholistan Desert, Pakistan	Cheema et al. (2020)
	Desertihabitans	*Desertihabitans aurantiacus*	Badain Jaran Desert, China	Sun et al. (2019)
		Desertihabitans brevis	Taklamakan Desert, China	Liu et al. (2020b)

(continued)

Table 4.1 (continued)

Family	Genus	Species	Location	References
Pseudonocardiaceae	*Actinophytocola*	*Actinophytocola algeriensis*	Sahara Desert, Algeria	Bouznada et al. (2016b)
		Actinophytocola gilvus	Tengger Desert, China	Sun et al. (2014)
	Amycolatopsis	*Amycolatopsis deserti*	Arid Australian soils	Busarakam et al. (2016a)
		Amycolatopsis granulosa		Zucchi et al. (2012a)
		Amycolatopsis ruanii	Atacama Desert, Chile	
		Amycolatopsis thermalba		
		Amycolatopsis thermophila	Arid Australian soils	Zucchi et al. (2012b)
		Amycolatopsis viridis		
		Amycolatopsis vastitatis	Atacama Desert, Chile	Idris et al. (2018)
	Lechevalieria	*Lechevalieria atacamensis*		Okoro et al. (2010)
		Lechevalieria deserti		
		Lechevalieria roselyniae		
	Lentzea	*Lentzea chajnantorensis*		Idris et al. (2017b)
		Lentzea isolaginshaensis	Isolaginsha Desert, China	Wang et al. (2019)
	Prauserella	*Prauserella endophytica*	Taklamakan Desert, China	Liu et al. (2015a)

		Prauserella isguenensis	Béni-Isguen (Mzab), Algeria	Saker et al. (2015)
		Prauserella shujinwangii	Xinjiang Desert, China	Liu et al. (2014)
	Pseudonocardia	*Pseudonocardia nigra*	Atacama Desert, Chile	Trujillo et al. (2017)
	Saccharopolyspora	*Saccharopolyspora deserti*	Arid area, Saudi Arabia	Yang et al. (2018a)
	Saccharothrix	*Saccharothrix algeriensis*	Sahara Desert, Algeria	Zitouni et al. (2004)
		Saccharothrix ghardaiensis		Bouznada et al. (2017)
		Saccharothrix hoggarensis		Boubetra et al. (2013b)
		Saccharothrix isguenensis		Bouznada et al. (2016a)
		Saccharothrix tamanrassetensis		Boubetra et al. (2015)
		Saccharothrix deserti	Xinjiang Desert, China	Liu et al. 2020a
		Saccharothrix tharensis	Thar Desert, India	Ibeyaima et al. (2018)
	Yuhushiella	*Yuhushiella deserti*	Xinjiang Desert, China	Mao et al. (2011)
		Yuhushiella sp.	Thar Desert, India	Ibeyaima et al. (2016)

(continued)

Table 4.1 (continued)

Family	Genus	Species	Location	References
Streptomycetaceae	*Streptomyces*	*Streptomyces aburaviensis*	Kutch Desert, India	Thumar et al. (2010)
		Streptomyces aridus	Atacama Desert, Chile	Idris et al. (2017a)
		Streptomyces asenjonii		Goodfellow et al. (2017)
		Streptomyces altiplanensis		Cortés-Albayay et al. (2019)
		Streptomyces cahuitamycinicus	Karakum Desert, Turkmenistan	Saygin et al. (2020b)
		Streptomyces desertarenae	Gurbantunggut Desert, China	Li et al. (2019b)
		Streptomyces deserti	Atacama Desert, Chile	Santhanam et al. (2012)
		Streptomyces dengpaensis	Tibet Autonomous Desert, China	Li et al. (2018b)
		Streptomyces bullii	Atacama Desert, Chile	Santhanam et al. (2013)
		Streptomyces fukangensis	Fukang Desert, China	Zhang et al. (2013)
		Streptomyces fragilis	Arid area, Saudi Arabia	Nithya et al. (2017)
		Streptomyces leeuwenhoekii	Atacama Desert, Chile	Castro et al. (2018c)
		Streptomyces netropsis	Baja California Desert, Mexico	Abdelmoteleb and González-Mendoza (2020)
		Streptomyces sannurensis	Eastern Desert of Egypt	Hozzein et al. (2011)
		Streptomyces taklimakanensis	Xinjiang Desert, China	Yuan et al. (2020)

Streptosporangiaceae	*Desertiactinospora*	*Desertiactinospora gelatinilytica*	Karakum Desert, Turkmenistan	Saygin et al. (2019a)
	Nonomuraea	*Nonomuraea deserti*		Saygin et al. (2020d)
		Nonomuraea diastatica		
		Nonomuraea longispora		
		Nonomuraea mesophila		
		Nonomuraea terrae	Eastern Anatolia, Turkey	Ay (2020)
	Streptosporangium	*Streptosporangium algeriense*	Sahara Desert, Algeria	Boubetra et al. (2016)
		Streptosporangium becharense	Béni-Abbès, Béchar, Algeria	Chaouch et al. (2016)
Thermomonosporaceae	*Actinomadura*	*Actinomadura deserti*	Xinjiang Desert, China	Cao et al. (2018)

4.2 Isolation Strategies for Desert Actinobacteria

The isolation techniques to recover Actinobacteria in arid biomes are important because of their various preferences for growth conditions (Goodfellow and Fiedler 2010; Kurtböke 2012). Several approaches have been recommended to isolate specific actinobacterial taxa based on their biological properties (Balagurunathan et al. 2020; Goodfellow 2010; Tiwari and Gupta 2013). Consequently, the strategy of isolating taxonomically diverse Actinobacteria from the desert can be summarized in Fig. 4.1. This strategy involves the collection of environmental samples, pre-treatment of samples, inoculation with selective isolation media supplemented with antibiotics and incubation under appropriate conditions followed by dereplication and recognition of target actinobacterial taxa. These procedures have been proved by studies on acidophilic (Golińska et al. 2013), endophytic (Wang et al. 2020), halophilic (Saker et al. 2015) and motile (Ye et al. 2020) Actinobacteria and other genera, namely *Amycolatopsis* (Idris et al. 2018), *Desertiactinospora* (Saygin et al. 2019a), *Labedella* (Li et al. 2019b), *Micromonospora* (Carro et al. 2019b) and *Motilibacter* (Liu et al. 2020d).

4.2.1 Pre-treatment of Environmental Samples

The physical or chemical pre-treatment of environmental samples can effectively enhance the selection of Actinobacteria. The effectiveness of pre-treatment techniques depends on the separation of actinobacterial propagules from particulate substrates followed by the recovery of actinobacteria within dispensed soil aggregates (Goodfellow and Fiedler 2010; Williams et al. 1984). The physical

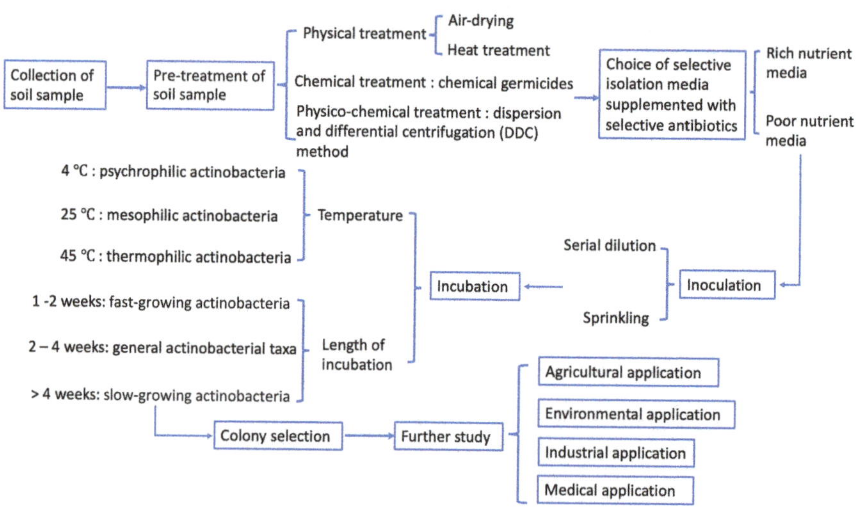

Fig. 4.1 General isolation strategies of recovering novel diverse Actinobacteria from desert

treatments include an air-drying method for isolating spore-forming taxa (Sembiring et al. 2000), and heat (≥100 °C) samples for an hour help inhibit the growth of unwanted bacteria (Nonomura and Ohara 1969). Both methods increase the ratio of Actinobacteria to bacteria on selective isolation plates, thereby facilitating the Actinobacteria recovery. For example, the spore-forming genera *Actinomadura*, *Microtetraspora* and *Pseudonocardia* have been reported from air-dried soil samples (Bulina et al. 1997; Terekhova 2003). Several novel species from the genera *Microbispora*, *Streptomyces* and *Thermomonospora* were isolated from heated soil samples (Hayakawa et al. 1996; Manfio et al. 2003; Nonomura and Ohara 1969). For chemical pre-treatments, samples are treated with germicides, such as benzethonium chloride, chlorhexidine gluconate and phenol, to target specific taxa, especially members of the family Streptosporangiaceae (Goodfellow 2010; Hayakawa et al. 1988, 1991a, 1995; Yamamura et al. 2003). These treatments eliminate vegetative cells of aerobic, endospore-forming bacilli and pseudomonads, enhancing the recovery of Actinobacteria. For example, members of the *Microbispora* genus was discovered after the treatment with chlorhexidine gluconate and phenol (Hayakawa et al. 1991b). A physico-chemical treatment named dispersion and differential centrifugation (DDC) procedure also preferentially increases the yield of actinobacterial propagules from environmental samples, as exemplified from the isolation of *Streptomyces* species followed DDC method (Atalan et al. 2000; Hopkins et al. 1991; Maldonado et al. 2005).

4.2.2 Selective Media and Antibiotics

Selective isolation media are designed to encourage the growth of desired microorganisms with differential nutritional components while suppressing the growth of unwanted microbes. Rich nutrient media such as trypticase soy agar (TSA) have been recommended for the isolation of many actinobacterial genera. For example, the genera *Jiangella*, *Nocardioides*, *Rhodococcus* and *Streptomyces* were reported from dilution plates on TSA (Li et al. 2019a; Mayilraj et al. 2006a; Saygin et al. 2020c; Tuo et al. 2015). However, these rich media failed in the recovery of most Actinobacteria from arid environments. Media with limited nutrient ingredients are more effective to yield the highest numbers of Actinobacteria on isolation plates. Studies results suggested that poor nutrient media were more favourable for the growth of desert Actinobacteria (Covington et al. 2018; Passari et al. 2018). For example, Reasoner's 2A (R2A) agar as a low nutrient medium was successfully used for the isolation of *Aeromicrobium*, *Desertihabitans*, *Geodermatophilus*, *Labedella*, *Microbacterium*, *Micromonospora* and *Saccharopolyspora* from deserts worldwide (Carro et al. 2019a; Hezbri et al. 2016; Li et al. 2019a, b; Liu et al. 2020b; Yang et al. 2018b).

Numerous microorganisms are desert inhabitants in deserts in particular fungi and other fast-growing bacteria hence address the need for using selective antibiotics to eliminate unwanted microbes. Antifungal antibiotics such as cycloheximide and nystatin and antibacterial agents (nalidixic acid and vancomycin) are commonly used (Gregory and Lacey 1962; Hayakawa et al. 1996; Williams and Davies 1965). Bacitracin, colistin, penicillin G and polymyxin B are mainly used to isolate slow-growing Actinobacteria through controlling the growth of fast-growing bacteria (Nonomura and Ohara 1969; Williams and Davies 1965). Likewise, selective isolation media supplemented with antibiotics are widely used to obtain specific Actinobacteria taxa (Goodfellow 2010). For example, *Lentzea chajnantorensis* and *Modestobacter excelsi* were recovered from isolation plates supplemented with cycloheximide and nystatin (Golińska et al. 2020a, b; Idris et al. 2017b), while *Streptosporangium becharense* was isolated from chitin-vitamin agar with polymyxin (Chaouch et al. 2016).

4.2.3 Growth Conditions of Desert Actinobacteria

It is well known that incubation conditions are contributing to the growth of Actinobacteria under particular temperature and length of incubation. Inoculated plates incubated at 4 °C, 25 °C and 45 °C were found suitable for the recovery of psychrophilic, mesophilic and thermophilic taxa, respectively. Psychrophilic species, such as *Kocuria aegyptia* and *Actinoalloteichus spitiensis*, were isolated from cold deserts of the Indian Himalayas (Li et al. 2006; Singla et al. 2005). The genera *Labedella* and *Motilibacter* represent mesophiles (Li et al. 2019b; Liu et al. 2020d), whereas the genera *Amycolatopsis* and *Geodermatophilus* are thermophiles (Busarakam et al. 2016a; Harwani 2013). The incubation time for 1–2 weeks is required by fast-growing Actinobacteria, such as *Microbacterium album* and *Nocardioides deserti* (Tuo et al. 2015; Yang et al. 2018b). Other actinobacterial taxa grow well after 2–4 weeks, such as *Micromonospora* and *Modestobacter* (Busarakam et al. 2016b; Carro et al. 2019b), whereas slow-growing strains may need at least up to 4 weeks such as *Labedella* and *Motilibacter* (Li et al. 2019b; Liu et al. 2020d).

4.2.4 Selection of Actinobacterial Colonies

Diverse cultural and morphological characteristics of Actinobacteria colonies can be observed on the isolation plates. In general, the production of substrate and aerial mycelium and soluble pigments are major characteristics which can used to assign Actinobacteria into group. Actinobacteria can also be assigned to specific taxa based on diagnostic morphological features, such as the presence of spore vesicles or spore chain morphology. For example, *Streptomyces* can produce filamentous leathery colonies with aerial spore mass, while *Amycolatopsis* strains form branched filamentous colonies covered by powdery white sterile aerial hyphae (Atalan et al. 2000; Busarakam et al. 2016a; Tan et al. 2006). It is not usually possible to distinguish between species of the same actinobacterial genus on selective isolation plates. Hence, putative colonies need to be sub-cultured on suitable growth media and preserved for further study as they may be members of novel taxa with biotechnological potential.

4.3 Biotechnological Applications

4.3.1 Environmental and Industrial Applications

Actinobacteria have a unique ability to synthesize enzymes which help their growth in the desert environments (Quinn et al. 2020). For example, an endophytic Actinobacteria, *Janibacter* sp. FAS23, has the ability of pentachlorophenol (PCP) degradation, which can be used to treat PCP-contaminated environments (Khessairi et al. 2014). Two new oxidative enzymes, type I Baeyer-Villiger monooxygenases (BVMOs), were discovered from *Streptomyces leeuwenhoekii* C34, isolated from the Atacama Desert. These enzymes are known as industrial potential biocatalysts. The new Atacama BVMOs displayed NADPH-dependent BV oxidation activity on ketones, sulphoxidation activity on sulphides. This enzyme also showed high apparent melting temperature (45 °C) and water-miscible cosolvent tolerance (Gran-Scheuch et al. 2018). A thermostable α-amylase was discovered from *Streptomyces fragilis* DA7-7 with industrial potential. Another thermostable enzyme named Am-fluorinase was discovered from *Actinopolyspora mzabensis* with potential biocatalytic application, whereas *Cellulomonas telluris* was reported as an endoglucanase-producing actinobacterium (Busarakam et al. 2016a; Shi et al. 2020; Sooklal et al. 2020).

4.3.2 Agricultural Applications

Rapid global population increase is responsible for a continuous rising in food demand. However, climate change has caused biotic and abiotic stresses to plants. Such harsh environmental conditions are significantly affecting agricultural productivity (Raza et al. 2019). Abiotic stresses such as drought, cold, heat, salinity and other physical-chemical factors can induce more risk on crop yield and food security. Desert Actinobacteria can play an important role in promoting plant growth for sustainable agriculture through controlling and minimizing biotic or abiotic stresses. For example, several strains reported from the Atacama Desert have displayed potential plant growth-promoting properties, which indicated that Actinobacteria from arid environments have great agriculture potential (Gaete et al. 2020). Two phytohormones, indole acetic acid (IAA) and gibberellins, well-known plant growth promoters, were obtained from *Streptomyces netropsis* A-ICA. *S. netropsis* A-ICA can also be used as a biocontrol agent since it showed antifungal ability against plant pathogens (Abdelmoteleb and González-Mendoza 2020).

4.3.3 Health Science Applications

The indiscriminate use of antibiotics has led to a serious consequence of antibiotic-resistant pathogens and threaten medical practices back to the situation of the pre-antibiotic era (Talbot et al. 2006). The emergence and spread of multidrug-resistant (MDR) pathogens, such as the ESKAPE bacteria (***E**nterococcus faecium*, ***S**taphylococcus aureus*, ***K**lebsiella pneumoniae*, ***A**cinetobacter baumannii*, ***P**seudomonas aeruginosa* and ***E**nterobacter* species), are responsible for high mortality rates of infections across the world (Genilloud 2014; Wright 2012). Therefore, there is an urgent need for the discovery of new bioactive compounds to combat these MDR pathogens. In this aspect, Actinobacteria from the desert environment have shown promising results in terms of novel secondary metabolisms with antibacterial, anticancer, antifungal, anti-inflammatory, anti-MDR and antitumour bioactivity as concluded in Table 4.2. The majority of these new compounds show antibacterial ability against a panel of Gram-positive and/or Gram-negative bacteria. For example, *Streptomyces asenjonii* KNN 42.f was evaluated against *Staphylococcus aureus*, *Bacillus subtilis*, *Escherichia coli*, *Enterococcus faecalis* and *Mycobacterium smegmatis* (Abdelkader et al. 2018). *Saccharothrix* sp. SA198 was strongly active against *B. subtilis* and *E. faecalis* and also fungi such as *Ascochyta fabae*, *Aspergillus carbonarius*, *Fusarium culmorum*, *Fusarium equiseti*, *Mucor ramannianus* and *Penicillium expansum* (Boubetra et al. 2013a). Anti-MDR species, such as *Micromonospora arida* and *M. inaquosa*, showed high activity against a panel of bacterial and fungal strains including multidrug-resistant *Klebsiella pneumoniae* ATCC 700603, while *Saccharothrix algeriensis* produced new dithiolopyrrolones to overcome MDR bacteria such as *Staphylococcus aureus* (Carro et al. 2019a; Strub et al. 2008). Antitumour producing strains *M. chalcea* LB4 and LB41 were active against HepG2 cells, and *S. leeuwenhoekii* C38 showed moderate activity against adenocarcinoma and breast carcinoma cells (Carro et al. 2019a; Nachtigall et al. 2011). It is well accepted that the most potent bioactive compound producing strains belong to the genus *Streptomyces*, such as *Streptomyces* sp. DA3-7 inhibited the growth of *Cryptococcus neoformans*, *Enterococcus faecalis*, *Escherichia coli* and *Salmonella typhimurium*, and strong cytotoxicity against MCF-7 human breast adenocarcinoma cells (Nithya et al. 2018). Potent antimicrobial activity against methicillin-resistant *Staphylococcus aureus* and vancomycin-resistant *Enterococcus* species was found in *Streptomyces* sp. SAS09, SAS13 and SAS15 (Masand et al. 2018). In addition, *Yuhushiella* sp. TD-032 has the inhibitory activity against Gram-positive and Gram-negative bacteria, especially clinical isolate of *Klebsiella pneumoniae* (Ibeyaima et al. 2016).

4.4 Conclusion and Future Perspectives

The increasing threats of agricultural, environmental, industrial and medical issues are raising the demand for novel bioactive compounds. In this chapter, we provide an overview of the diversity of members of the phylum Actinobacteria recovered from

Table 4.2 Bioactive compounds of desert Actinobacteria

Organisms	Bioactive compound	Bioactivity	References
Jiangella gansuensis YIM 002^T	Jiangrines A–E	Anti-inflammatory activity	Han et al. (2014)
	Pyrrolezanthine		
	Jiangrine F	No inflammatory activity	
	Pristinamycin	Anti-staphylococcal infections	Jiao et al. (2017)
Lentzea chajnantorensis H45^T	Lentzeosides A–F	Anti-HIV-1 integrase activity	Wichner et al. (2017)
Micromonospora arida LB32^T	ND	Anti-multidrug-resistant bacteria and antifungal activity	Carro et al. (2019a)
Micromonospora inaquosa LB39^T	ND		
Micromonospora chalcea LB4 and LB41	ND	Antitumour activity	
Nocardia sp. XJ31	Nocardimicins J–R	Viral polymerase inhibitor	Zhang et al. (2020)
	Brasiliquinone E	Anti-bacillus Calmette-Guérin (BCG) activity	
Nocardiopsis sp. 38-7L-1	ND	Anti-*Pseudomonas aeruginosa*	Chen et al. (2015)
Saccharothrix algeriensis SA 233^T	Five dithiolopyrrolone antibiotics	Anti-multi-antibiotic-resistant bacteria, anticancer and antiallergic activities	Strub et al. (2008); Zitouni et al. (2004)
Saccharothrix sp. SA198	Compounds A4 and A5	Antifungal and antibacterial activities	Boubetra et al. (2013a)
Streptomyces aburaviensis Kut-8	ND	Antimicrobial and cytotoxic activities	Ramirez-Rodriguez et al. (2018); Thumar et al. (2010)
Streptomyces asenjonii KNN 35.1b^T	ND	Antimicrobial activity	Goodfellow et al. (2017)
Streptomyces asenjonii KNN 42.f	Asenjonamides A–C	Antibacterial activity	Abdelkader et al. (2018)
Streptomyces leeuwenhoekii C34^T	Chaxalactins A–C	Antibacterial activity	Rateb et al. (2011b)
	Chaxamycins A–C	Anti-intrinsic ATPase activity of heat shock protein 90 (Hsp90)	Rateb et al. (2011a)
	Chaxamycin D	Anti-methicillin-resistant *Staphylococcus aureus* (MRSA)	
	Leepeptin	ND	Gomez-Escribano et al. (2019)

(continued)

Table 4.2 (continued)

Organisms	Bioactive compound	Bioactivity	References
Streptomyces leeuwenhoekii C38	Atacamycins A–C	Antibacterial, antitumour and cytotoxic activities	Nachtigall et al. (2011)
Streptomyces leeuwenhoekii C58	Chaxapeptin	Anti-human lung cancer cell line A549	Elsayed et al. (2015)
Streptomyces leeuwenhoekii C79	Four new specialized metabolites (structures = ND)	Antibacterial activity	Fiedler et al., unpublished results; Idris (2016)
Streptomyces sp. 8P21H-1	Acetyl-griseoviridin	No antibacterial activity	Wang et al. (2020)
	Desulphurizing griseoviridin	Antibacterial activity	
	Griseoviridin		
Streptomyces sp. BS30	Two compounds (ND)	Anti-*Aspergillus niger* 2CA936 and anti-yeast activities	Souagui et al. (2017)
Streptomyces sp. D25	ND	Antioxidant and anti-biofilm-forming bacteria	Radhakrishnan et al. (2016)
Streptomyces sp. DA3-7	Pyridine alkaloid: pyridine-2,5-diacetamide	Antimicrobial activity	Nithya et al. (2018)
Streptomyces sp. DB634	Abenquines A–D	Antibacterial, anti-dermatophytic fungal and anti-phosphodiesterase type 4b activities	Schulz et al. (2011)
Streptomyces sp. SAS02	New anthracyclines	Anticancer activity	Masand et al. (2018)
Streptomyces sp. SAS09	New anthracycline glycoside	Antibacterial activity	
	New macrolide		
Streptomyces sp. SAS13	New anthracycline glycoside		
	New macrolide		
Streptomyces sp. SAS15	New macrolide		
Yuhushiella sp. TD-032	ND	Antimicrobial activity	Ibeyaima et al. (2016)

ND not determined

desert, a representative of arid environment, and their biotechnological applications based on their biological activities. To obtain taxonomically diverse Actinobacteria, effective isolation techniques are needed which involve pre-treatment of desert sample, selective isolation media, antibiotics and optimized growth conditions. The exploration of Actinobacteria from desert showed promising in terms of biotechnological potential in particular compounds with antibacterial, anticancer, antifungal, anti-inflammatory, anti-MDR and antitumour activities. It is evident that Actinobacteria are common in desert environments as 108 new species in 44 genera and 19 families were isolated from 34 deserts worldwide during 2010–2021. At least 73 novel bioactive compounds were found from 26 strains. These data strongly support the view that desert Actinobacteria are a promising source of novel bioactive compounds which merit exploitation. Research on desert Actinobacteria is expected to continuously increase due to their potential application in various fields. Rapid accumulating data from whole genome sequencing coupled with the current advancements in "omics" tools such as comparative genomics and genome mining will undoubtedly provide new opportunity for biotechnological explorations from these emerging diverse desert Actinobacteria.

Acknowledgements This research work is partially supported by Chiang Mai University. FX is supported by the CMU Presidential Scholarship for her PhD study. PR is grateful to the Graduate School, Chiang Mai University, for the TA/RA scholarship for 2019–2021.

References

Abdelkader MSA, Philippon T, Asenjo JA, Bull AT, Goodfellow M, Ebel R, Jaspars M, Rateb ME (2018) Asenjonamides A–C, antibacterial metabolites isolated from *Streptomyces asenjonii* strain KNN 42.f from an extreme-hyper arid Atacama desert soil. J Antibiot 71(4):425–431

Abdelmoteleb A, González-Mendoza D (2020) A novel streptomyces rhizobacteria from desert soil with diverse anti-fungal properties. Rhizosphere 16:100243. https://doi.org/10.1016/j.rhisph.2020.100243

Amaresan N, Kumar MS, Annapurna K, Kumar K, Sankaranarayanan A (eds) (2020) Beneficial microbes in agro-ecology: bacteria and fungi. Academic, Cambridge

Asem MD, Shi L, Jiao J-Y, Wang D, Han M-X, Dong L, Liu F, Salam N, Li W-J (2018) *Desertimonas flava* gen. nov., sp. nov. isolated from a desert soil, and proposal of *Ilumatobacteraceae* fam. nov. Int J Syst Evol Microbiol 68(11):3593–3599

Atalan E, Manfio GP, Ward AC, Kroppenstedt RM, Goodfellow M (2000) Biosystematic studies on novel streptomycetes from soil. Antonie Van Leeuwenhoek 77(4):337–353

Ay H (2020) *Nonomuraea terrae* sp. nov., isolated from arid soil. Arch Microbiol 202:2197–2205

Balagurunathan R, Radhakrishnan M, Shanmugasundaram T, Gopikrishnan V, Jerrine J (2020) Sample collection, isolation, and diversity of *Actinobacteria*. In: Balagurunathan R, Radhakrishnan M, Shanmugasundaram T, Gopikrishnan V, Jerrine J (eds) Protocols in actinobacterial research. Springer, New York, pp 1–24

Barka EA, Vatsa P, Sanchez L, Gaveau-Vaillant N, Jacquard C, Klenk H-P, Clément C, Ouhdouch Y, van Wezel GP (2016) Taxonomy, physiology, and natural products of *Actinobacteria*. Microbiol Mol Biol Rev 80(1):1–43

Boubetra D, Sabaou N, Zitouni A, Bijani C, Lebrihi A, Mathieu F (2013a) Taxonomy and chemical characterization of new antibiotics produced by *Saccharothrix* SA198 isolated from a Saharan soil. Microbiol Res 168(4):223–230

Boubetra D, Zitouni A, Bouras N, Mathieu F, Lebrihi A, Schumann P, Spröer C, Klenk H-P, Sabaou N (2013b) *Saccharothrix hoggarensis* sp. nov., an actinomycete isolated from Saharan soil. Int J Syst Evol Microbiol 63(2):549–553

Boubetra D, Zitouni A, Bouras N, Schumann P, Spröer C, Klenk H-P, Sabaou N (2015) *Saccharothrix tamanrassetensis* sp. nov., an actinomycete isolated from Saharan soil. Int J Syst Evol Microbiol 65(4):1316–1320

Boubetra D, Bouras N, Zitouni A, Schumann P, Spröer C, Sabaou N, Klenk H-P (2016) *Streptosporangium algeriense* sp. nov., an actinobacterium isolated from desert soil. Int J Syst Evol Microbiol 66(2):1034–1038

Bouznada K, Bouras N, Mokrane S, Chaouch FC, Zitouni A, Pötter G, Spröer C, Klenk H-P, Sabaou N (2016a) *Saccharothrix isguenensis* sp. nov., an actinobacterium isolated from desert soil. Int J Syst Evol Microbiol 66(11):4785–4790

Bouznada K, Bouras N, Schumann P, Spröer C, Sabaou N, Klenk H-P (2016b) *Actinophytocola algeriensis* sp. nov., an actinobacterium isolated from Saharan soil. Int J Syst Evol Microbiol 66(7):2760–2765

Bouznada K, Bouras N, Mokrane S, Chaouch FC, Zitouni A, Pötter G, Spröer C, Klenk H-P, Sabaou N (2017) *Saccharothrix ghardaiensis* sp. nov., an actinobacterium isolated from Saharan soil. Antonie Van Leeuwenhoek 110(3):399–405

Bulina TI, Alferova IV, Terekhova LP (1997) A new method for the isolation of actinomycetes with the use of microwave irradiation of soil samples. Mikrobiologiya 66(2):278–282

Bull AT (2011) *Actinobacteria* of the extremobiosphere. In: Horikoshi K, Antranikian G, Bull AT, Robb FT, Stetter KO (eds) Extremophiles handbook. Springer, Tokyo, pp 1203–1240

Busarakam K, Brown R, Bull AT, Tan GYA, Zucchi TD, da Silva LJ, de Souza WR, Goodfellow M (2016a) Classification of thermophilic actinobacteria isolated from arid desert soils, including the description of *Amycolatopsis deserti* sp. nov. Antonie Van Leeuwenhoek 109(2):319–334

Busarakam K, Bull AT, Trujillo ME, Riescro R, Sangal V, van Wezel GP, Goodfellow M (2016b) *Modestobacter caceresii* sp. nov., novel actinobacteria with an insight into their adaptive mechanisms for survival in extreme hyper-arid Atacama Desert soils. Syst Appl Microbiol 39(4):243–251

Cao C, Xu T, Liu J, Cai X, Sun Y, Qin S, Jiang J, Huang Y (2018) *Actinomadura deserti* sp. nov., isolated from desert soil. Int J Syst Evol Microbiol 68(9):2930–2935

Carro L, Castro JF, Razmilic V, Nouioui I, Pan C, Igual JM, Jaspars M, Goodfellow M, Bull AT, Asenjo JA (2019a) Uncovering the potential of novel micromonosporae isolated from an extreme hyper-arid Atacama Desert soil. Sci Rep 9(1):1–16

Carro L, Golinska P, Nouioui I, Bull AT, Igual JM, Andrews BA, Klenk H-P, Goodfellow M (2019b) *Micromonospora acroterricola* sp. nov., a novel actinobacterium isolated from a high altitude Atacama Desert soil. Int J Syst Evol Microbiol 69(11):3426–3436

Castro JF, Nouioui I, Sangal V, Choi S, Yang S-J, Kim B-Y, Trujillo ME, Riesco R, Montero-Calasanz MDC, Rahmani TPD (2018a) *Blastococcus atacamensis* sp. nov., a novel strain adapted to life in the Yungay core region of the Atacama Desert. Int J Syst Evol Microbiol 68(9):2712–2721

Castro JF, Nouioui I, Sangal V, Trujillo ME, de Carmen Montero-Calasanz M, Rahmani T, Bull AT, Asenjo JA, Andrews BA, Goodfellow M (2018b) *Geodermatophilus chilensis* sp. nov., from soil of the Yungay core-region of the Atacama Desert, Chile. Syst Appl Microbiol 41(5): 427–436

Castro JF, Razmilic V, Gomez-Escribano JP, Andrews B, Asenjo J, Bibb M (2018c) The 'gifted' actinomycete *Streptomyces leeuwenhoekii*. Antonie Van Leeuwenhoek 111(8):1433–1448

Chaouch FC, Bouras N, Mokrane S, Zitouni A, Schumann P, Spröer C, Sabaou N, Klenk H-P (2016) *Streptosporangium becharense* sp. nov., an actinobacterium isolated from desert soil. Int J Syst Evol Microbiol 66(7):2484–2490

Cheema MT, Ye J-J, Li F-N, Lu Q-P, Abbas M, Sajid I, Huang D-L, Liu S-W, Sun C-H (2020) *Auraticoccus cholistanensis* sp. nov., an actinomycete isolated from soil of the Cholistan Desert, and emended description of the genus *Auraticoccus*. Int J Syst Evol Microbiol 70(5):3179–3185

Chen C, Liu J-M, Jiang Z-K, Li X-J, Liu S-W, Tuo L, Dai S-J, Yang G-L, Wang F-F, Huang D-L, Jiang M-G, Sun C (2015) Endophytic actinomycetes from psammophytes of taklamakan desert against *Pseudomonas aeruginosa* and study on bioactive product of strain 38-7L-1. Chin J Antibiot 40:81–87, 115

Connon SA, Lester ED, Shafaat HS, Obenhuber DC, Ponce A (2007) Bacterial diversity in hyperarid Atacama desert soils. J Geophys Res Biogeosci 112(4):1–9

Cortés-Albayay C, Dorador C, Schumann P, Schniete JK, Herron P, Andrews B, Asenjo J, Nouioui I (2019) *Streptomyces altiplanensis* sp. nov., an alkalitolerant species isolated from Chilean Altiplano soil, and emended description of *Streptomyces chryseus* (Krasil'nikov et al. 1965) Pridham 1970. Int J Syst Evol Microbiol 69(8):2498–2505

Covington BC, Spraggins JM, Ynigez-Gutierrez AE, Hylton ZB, Bachmann BO (2018) Response of secondary metabolism of hypogean actinobacterial genera to chemical and biological stimuli. Appl Environ Microbiol 84(19):e01125–e01118. https://doi.org/10.1128/AEM.01125-18

del Carmen Montero-Calasanz M, Göker M, Broughton WJ, Cattaneo A, Favet J, Pötter G, Rohde M, Spröer C, Schumann P, Klenk H-P (2013a) *Geodermatophilus tzadiensis* sp. nov., a UV radiation-resistant bacterium isolated from sand of the Saharan desert. Syst Appl Microbiol 36(3):177–182

del Carmen Montero-Calasanz M, Göker M, Pötter G, Rohde M, Spröer C, Schumann P, Gorbushina AA, Klenk H-P (2013b) *Geodermatophilus africanus* sp. nov., a halotolerant actinomycete isolated from Saharan desert sand. Antonie Van Leeuwenhoek 104(2):207–216

del Carmen Montero-Calasanz M, Göker M, Rohde M, Schumann P, Pötter G, Spröer C, Gorbushina AA, Klenk H-P (2013c) *Geodermatophilus siccatus* sp. nov., isolated from arid sand of the Saharan desert in Chad. Antonie Van Leeuwenhoek 103(3):449–456

Elsayed SS, Trusch F, Deng H, Raab A, Prokes I, Busarakam K, Asenjo JA, Andrews BA, Van West P, Bull AT, Goodfellow M, Yi Y, Ebel R, Jaspars M, Rateb ME (2015) Chaxapeptin, a lasso peptide from extremotolerant *Streptomyces leeuwenhoekii* strain C58 from the hyperarid Atacama Desert. J Org Chem 80(20):10252–10260

Gaete A, Mandakovic D, González M (2020) Isolation and identification of soil bacteria from extreme environments of Chile and their plant beneficial characteristics. Microorganisms 8(8): 1213–1225

Genilloud O (2014) The re-emerging role of microbial natural products in antibiotic discovery. Antonie Van Leeuwenhoek 106(1):173–188

Golińska P, Wang D, Goodfellow M (2013) *Nocardia aciditolerans* sp. nov., isolated from a spruce forest soil. Antonie Van Leeuwenhoek 103(5):1079–1088

Golińska P, del Carmen Montero-Calasanz M, Świecimska M, Yaramis A, Igual JM, Bull AT, Goodfellow M (2020a) *Modestobacter excelsi* sp. nov., a novel actinobacterium isolated from a high altitude Atacama Desert soil. Syst Appl Microbiol 43:126051. https://doi.org/10.1016/j.syapm.2019.126051

Golińska P, Świecimska M, del Carmen Montero-Calasanz M, Yaramis A, Igual JM, Bull AT, Goodfellow M (2020b) *Modestobacter altitudinis* sp. nov., a novel actinobacterium isolated from Atacama Desert soil. Int J Syst Evol Microbiol 70(5):3513–3527

Gomez-Escribano JP, Castro JF, Razmilic V, Jarmusch SA, Saalbach G, Ebel R, Jaspars M, Andrews B, Asenjo JA, Bibb MJ (2019) Heterologous expression of a cryptic gene cluster from *Streptomyces leeuwenhoekii* C34^T yields a novel lasso peptide, leepeptin. Appl Environ Microbiol. https://doi.org/10.1128/AEM.01752-19

Goodfellow M (2010) Selective isolation of Actinobacteria. In: Baltz RH, Demain AL, Davies JE, Bull AT, Junker B, Katz L, Lynd LR, Masurekar P, Reeves CD, Zhao H (eds) Manual of industrial microbiology and biotechnology, 3rd edn. ASM Press, Washington, DC, pp 13–27

Goodfellow M, Fiedler HP (2010) A guide to successful bioprospecting: informed by actinobacterial systematics. Antonie Van Leeuwenhoek 98(2):119–142

Goodfellow M, Williams ST (1983) Ecology of actinomycetes. Annu Rev Microbiol 37(1): 189–216

Goodfellow M, Busarakam K, Idris H, Labeda DP, Nouioui I, Brown R, Kim BY, del Carmen Montero-Calasanz M, Andrews BA, Bull AT (2017) *Streptomyces asenjonii* sp. nov., isolated from hyper-arid Atacama Desert soils and emended description of *Streptomyces viridosporus* Pridham et al. 1958. Antonie Van Leeuwenhoek 110(9):1133–1148

Goudie AS (ed) (2013) Arid and semi-arid geomorphology. CUP, New York

Gran-Scheuch A, Trajkovic M, Parra L, Fraaije MW (2018) Mining the genome of *Streptomyces leeuwenhoekii*: two new Type I Baeyer–Villiger Monooxygenases from Atacama Desert. Front Microbiol 9:1609. https://doi.org/10.3389/fmicb.2018.01609

Gregory PH, Lacey ME (1962) Isolation of thermophilic actinomycetes. Nature 195:95. https://doi.org/10.1038/195096a0

Guerrero LD, Makhalanyane TP, Aislabie JM, Cowan DA (2014) Draft genome sequence of *Williamsia* sp. strain D3, isolated from the Darwin Mountains, Antarctica. Genome Announc 2(1):e01230–e01213. https://doi.org/10.1128/genomeA.01230-13

Guesmi S, Nouioui I, Pujic P, Dubost A, Najjari A, Ghedira K, Igual JM, Cherif A, Klenk H, Sghaier H (2021) Draft genome sequence of *Promicromonospora panici* sp. nov., a novel ionizing-radiation-resistant actinobacterium isolated from roots of the desert plant *Panicum turgidum*. Extremophiles 25(1):25–38

Habib N, Khan IU, Chu X, Xiao M, Li S, Fang B-Z, Zhi X-Y, Li W-J (2018) *Actinoplanes deserti* sp. nov., isolated from a desert soil sample. Antonie Van Leeuwenhoek 111(12):2303–2310

Han L, Gao C, Jiang Y, Guan P, Liu J, Li L, Xu L, Huang X (2014) Jiangrines A–F and jiangolide from an actinobacterium, *Jiangella gansuensis*. J Nat Prod 77(12):2605–2610

Harwani D (2013) Biodiversity of rare thermophilic actinomycetes in the great Indian Thar desert: an overview. Indo Am J Pharm Res 3:934–939

Hayakawa M, Ishizawa K, Nonomura H (1988) Distribution of rare actinomycetes in Japanese soils. J Ferment Technol 66(4):367–373

Hayakawa M, Kajiura T, Nonomura H (1991a) New methods for the highly selective isolation of *Streptosporangium* and *Dactylosporangium* from soil. J Ferment Bioeng 72(5):327–333

Hayakawa M, Sadakata T, Kajiura T, Nonomura H (1991b) New methods for the highly selective isolation of *Micromonospora* and *Microbispora* from soil. J Ferment Bioeng 72(5):320–326

Hayakawa M, Momose Y, Kajiura T, Yamazaki T, Tamura T, Hatano K, Nonomura H (1995) A selective isolation method for *Actinomadura viridis* in soil. J Ferment Bioeng 79(3):287–289

Hayakawa M, Takeuchi T, Yamazaki T (1996) Combined use of trimethoprim with nalidixic acid for the selective isolation and enumeration of actinomycetes from soil. Actinomycetologica 10(2):80–90

Hezbri K, Ghodhbane-Gtari F, del Carmen Montero-Calasanz M, Sghaier H, Rohde M, Schumann P, Klenk H-P, Gtari M (2015) *Geodermatophilus sabuli* sp. nov., a γ-radiation-resistant actinobacterium isolated from desert limestone. Int J Syst Evol Microbiol 65(10): 3365–3372

Hezbri K, Ghodhbane-Gtari F, del Carmen Montero-Calasanz M, Nouioui I, Rohde M, Spröer C, Schumann P, Klenk H-P, Gtari M (2016) *Geodermatophilus pulveris* sp. nov., a gamma-radiation-resistant actinobacterium isolated from the Sahara desert. Int J Syst Evol Microbiol 66(10):3828–3834

Hopkins DW, Macnaughton SJ, O'Donnell AG (1991) A dispersion and differential centrifugation technique for representatively sampling microorganisms from soil. Soil Biol Biochem 23(3): 217–225

Hozzein WN, Li W-J, Ali MLA, Hammouda O, Mousa AS, Xu L-H, Jiang C-L (2004) *Nocardiopsis alkaliphila* sp. nov., a novel alkaliphilic actinomycete isolated from desert soil in Egypt. Int J Syst Evol Microbiol 54(1):247–252

Hozzein WN, Ali MLA, Hammouda O, Mousa AS, Goodfellow M (2011) *Streptomyces sannurensis* sp. nov., a new alkaliphilic member of the genus *Streptomyces* isolated from Wadi Sannur in Egypt. Afr J Microbiol Res 5(11):1329–1334

Hozzein WN, Ali MLA, Hammouda O, Mousa AS (2014) Characterization of a new alkaliphilic Nocardiopsis strain from the desert of Egypt. J Pure Appl Microbiol 8:3285–3290

Hozzein WN, Yang Z-W, Alharbi SA, Alsakkaf WAA, Asem MD, Xiao M, Salam N, Li W-J (2018) *Georgenia deserti* sp. nov., a halotolerant actinobacterium isolated from a desert sample. Int J Syst Evol Microbiol 68(4):1135–1139

Hu Q-W, Chu X, Xiao M, Li C-T, Yan Z-F, Hozzein WN, Kim C-J, Zhi X-Y, Li W-J (2016) *Arthrobacter deserti* sp. nov., isolated from a desert soil sample. Int J Syst Evol Microbiol 66(5): 2035–2040

Ibeyaima A, Rana J, Dwivedi A, Gupta S, Sharma SK, Saini N, Sarethy IP (2016) Characterization of *Yuhushiella* sp. TD-032 from the Thar Desert and its antimicrobial activity. J Adv Pharm Technol Res 7(2):32–36

Ibeyaima A, Singh AK, Lal R, Gupta S, Goodfellow M, Sarethy IP (2018) *Saccharothrix tharensis* sp. nov., an actinobacterium isolated from the Thar Desert, India. Antonie Van Leeuwenhoek 111(11):2141–2147

Idris H (2016) Actinobacterial diversity in Atacama Desert habitats as a road map to biodiscovery. Dissertation, Newcastle University, UK

Idris H, Labeda DP, Nouioui I, Castro JF, del Carmen Montero-Calasanz M, Bull AT, Asenjo JA, Goodfellow M (2017a) *Streptomyces aridus* sp. nov., isolated from a high altitude Atacama Desert soil and emended description of *Streptomyces noboritoensis* Isono et al. 1957. Antonie Van Leeuwenhoek 110(5):705–717

Idris H, Nouioui I, Asenjo JA, Bull AT, Goodfellow M (2017b) *Lentzea chajnantorensis* sp. nov., an actinobacterium from a very high altitude Cerro Chajnantor gravel soil in northern Chile. Antonie Van Leeuwenhoek 110(6):795–802

Idris H, Nouioui I, Pathom-Aree W, Castro JF, Bull AT, Andrews BA, Asenjo JA, Goodfellow M (2018) *Amycolatopsis vastitatis* sp. nov., an isolate from a high altitude subsurface soil on Cerro Chajnantor, northern Chile. Antonie Van Leeuwenhoek 111(9):1523–1533

Jiao JY, Carro L, Liu L, Gao XY, Zhang XT, Hozzein WN, Lapidus A, Huntemann M, Reddy TBK, Varghese N, Hadjithomas M, Ivanova NN, Göker M, Pillay M, Eisen JA, Woyke T, Klenk H-P, Kyrpides NC, Li WJ (2017) Complete genome sequence of *Jiangella gansuensis* strain YIM 002^{T} (DSM 44835^{T}), the type species of the genus *Jiangella* and source of new antibiotic compounds. Stand Genomic Sci 12(1):1–8

Khan IU, Hussain F, Habib N, Xiao M, Ahmed I, Amin A, Zhi X-Y, Li W-J (2017) *Nocardioides thalensis* sp. nov., isolated from a desert. Int J Syst Evol Microbiol 67(8):2848–2852

Khessairi A, Fhoula I, Jaouani A, Turki Y, Cherif A, Boudabous A, Hassen A, Ouzari H (2014) Pentachlorophenol degradation by *Janibacter* sp., a new actinobacterium isolated from saline sediment of arid land. Biomed Res Int 2014:296472. https://doi.org/10.1155/2014/296472

Kurtböke DI (2012) Biodiscovery from rare actinomycetes: an eco-taxonomical perspective. Appl Microbiol Biotechnol 93(5):1843–1852

Li W-J, Chen H-H, Zhang Y-Q, Kim C-J, Park D-J, Lee J-C, Xu L-H, Jiang C-L (2005) *Citricoccus alkalitolerans* sp. nov., a novel actinobacterium isolated from a desert soil in Egypt. Int J Syst Evol Microbiol 55(1):87–90

Li W-J, Zhang Y-Q, Schumann P, Chen H-H, Hozzein WN, Tian X-P, Xu L-H, Jiang C-L (2006) *Kocuria aegyptia* sp. nov., a novel actinobacterium isolated from a saline, alkaline desert soil in Egypt. Int J Syst Evol Microbiol 56(4):733–737

Li Y-R, Zhu Z-N, Li Y-Q, Xiao M, Han M-X, Wadaan MAM, Hozzein WN, An D-D, Li W-J (2018a) *Microbacterium halophytorum* sp. nov., a novel endophytic actinobacterium isolated from halophytes. Int J Syst Evol Microbiol 68(12):3928–3934

Li Y, Li Y, Wang Lw, Bao J (2018b) *Streptomyces dengpaensis* sp. nov., an actinomycete isolated from desert soil. Int J Syst Evol Microbiol 68(10):3322–3326

Li F-N, Liao S-L, Liu S-W, Jin T, Sun C-H (2019a) *Aeromicrobium endophyticum* sp. nov., an endophytic actinobacterium isolated from reed (*Phragmites australis*). J Microbiol 57(9): 725–731

Li L-Y, Yang Z-W, Asem MD, Fang B-Z, Salam N, Alkhalifah DHM, Hozzein WN, Nie G-X, Li W-J (2019b) *Streptomyces desertarenae* sp. nov., a novel actinobacterium isolated from a desert sample. Antonie Van Leeuwenhoek 112(3):367–374

Li F-N, Lu Q, Liao S-L, Jin T, Li W, Sun C-H (2019c) *Labedella phragmitis* sp. nov. and *Labedella populi* sp. nov., two endophytic actinobacteria isolated from plants in the Taklamakan Desert and emended description of the genus Labedella. Syst Appl Microbiol 42:126004. https://doi.org/10.1016/j.syapm.2019.126004

Li L-Y, Yang Z-W, Asem MD, Salam N, Xiao M, Alkhalifah DHM, Hozzein WN, Nie G-X, Li W-J (2019d) *Georgenia alba* sp. nov., a novel halotolerant actinobacterium isolated from a desert sand sample. Antonie Van Leeuwenhoek 112(2):203–209

Liu M, Zhang L, Ren B, Yang N, Yu X, Wang J, Ding L, Liu X, Liu Z, Goodfellow M (2014) *Prauserella shujinwangii* sp. nov., from a desert environment. Int J Syst Evol Microbiol 64(11): 3833–3837

Liu J-M, Habden X, Guo L, Tuo L, Jiang Z-K, Liu S-W, Liu X-F, Chen L, Li R-F, Zhang Y-Q (2015a) *Prauserella endophytica* sp. nov., an endophytic actinobacterium isolated from *Tamarix taklamakanensis*. Antonie Van Leeuwenhoek 107(6):1401–1409

Liu J-M, Tuo L, Habden X, Guo L, Jiang Z-K, Liu X-F, Chen L, Zhang Y-Q, Sun C-H (2015b) *Nesterenkonia populi* sp. nov., an actinobacterium isolated from *Populus euphratica*. Int J Syst Evol Microbiol 65(5):1474–1479

Liu S-W, Li F-N, Zheng H-Y, Qi X, Huang D-L, Xie Y-Y, Sun C-H (2019a) *Planctomonas deserti* gen. nov., sp. nov., a new member of the family *Microbacteriaceae* isolated from soil of the Taklamakan desert. Int J Syst Evol Microbiol 69(3):616–624

Liu S, Li F-N, Qi X, Xie Y-Y, Sun C-H (2019b) *Nakamurella deserti* sp. nov., isolated from rhizosphere soil of *Reaumuria* in the Taklamakan desert. Int J Syst Evol Microbiol 69(1): 214–219

Liu J, Sun Y, Liu J, Wu Y, Cao C, Li R, Jiang J (2020a) *Saccharothrix deserti* sp. nov., an actinomycete isolated from desert soil. Int J Syst Evol Microbiol 70(3):1882–1887

Liu S-W, Li F-N, Liu H-Y, Yu L-Y, Sun C-H (2020b) *Desertihabitans brevis* sp. nov., an actinobacterium isolated from sand of the Taklamakan desert, and emended description of the genus *Desertihabitans*. Int J Syst Evol Microbiol 70(2):1166–1171

Liu S-W, Xue C-M, Li F-N, Sun C-H (2020c) *Nocardioides vastitatis* sp. nov., isolated from Taklamakan desert soil. Int J Syst Evol Microbiol 70(1):77–82

Liu S-W, Ye J-J, Lu Q-P, Cheema MT, Abbas M, Huang D-L, Sajid I, Sun C-H (2020d) *Motilibacter deserti* sp. nov. and *Motilibacter aurantiacus* sp. nov., two novel actinobacteria isolated from soil of Cholistan Desert and emended description of the genus *Motilibacter*. Syst Appl Microbiol 43(6):126150. https://doi.org/10.1016/j.syapm.2020.126150

Lubsanova DA, Zenova GM, Kozhevin PA, Manucharova NA, Shvarov AP (2014) Filamentous *Actinobacteria* of the saline soils of arid territories. Moscow Univ Soil Sci Bull 69(2):88–92

Luo X, Wang J, Zeng X-C, Wang Y, Zhou L, Nie Y, Dai J, Fang C (2012) *Mycetocola manganoxydans* sp. nov., an actinobacterium isolated from the Taklamakan desert. Int J Syst Evol Microbiol 62(12):2967–2970

MacElroy RD (1974) Some comments on the evolution of extremophiles. Biosystems 6(1):74–75

Maldonado LA, Stach JEM, Pathom-aree W, Ward AC, Bull AT, Goodfellow M (2005) Diversity of cultivable actinobacteria in geographically widespread marine sediments. Antonie Van Leeuwenhoek 87(1):11–18

Manfio GP, Atalan E, Zakrzewska-Czerwinska J, Mordarski M, Rodríguez C, Collins MD, Goodfellow M (2003) Classification of novel soil streptomycetes as *Streptomyces aureus* sp. nov., *Streptomyces laceyi* sp. nov. and *Streptomyces sanglieri* sp. nov. Antonie Van Leeuwenhoek 83(3):245–255

Mao J, Wang J, Dai H-Q, Zhang Z-D, Tang Q-Y, Ren B, Yang N, Goodfellow M, Zhang L-X, Liu Z-H (2011) *Yuhushiella deserti* gen. nov., sp. nov., a new member of the suborder *Pseudonocardineae*. Int J Syst Evol Microbiol 61(3):621–630

Masand M, Sivakala KK, Menghani E, Thinesh T, Anandham R, Sharma G, Sivakumar N, Jebakumar SRD, Jose PA (2018) Biosynthetic potential of bioactive *Streptomycetes* isolated from arid region of the Thar desert, Rajasthan (India). Front Microbiol 9:687. https://doi.org/10.3389/fmicb.2018.00687

Mayilraj S, Krishnamurthi S, Saha P, Saini HS (2006a) *Rhodococcus kroppenstedtii* sp. nov., a novel actinobacterium isolated from a cold desert of the Himalayas, India. Int J Syst Evol Microbiol 56(5):979–982

Mayilraj S, Suresh K, Kroppenstedt RM, Saini HS (2006b) *Dietzia kunjamensis* sp. nov., isolated from the Indian Himalayas. Int J Syst Evol Microbiol 56(7):1667–1671

Meklat A, Bouras N, Zitouni A, Mathieu F, Lebrihi A, Schumann P, Spröer C, Klenk H-P, Sabaou N (2013) *Actinopolyspora mzabensis* sp. nov., a halophilic actinomycete isolated from an Algerian Saharan soil. Int J Syst Evol Microbiol 63(10):3787–3792

Montero-Calasanz MC, Göker M, Pötter G, Rohde M, Spröer C, Schumann P, Gorbushina AA, Klenk H-P (2012) *Geodermatophilus arenarius* sp. nov., a xerophilic actinomycete isolated from Saharan desert sand in Chad. Extremophiles 16(6):903–909

Montero-Calasanz MC, Göker M, Pötter G, Rohde M, Spröer C, Schumann P, Gorbushina AA, Klenk H-P (2013a) *Geodermatophilus saharensis* sp. nov., isolated from sand of the Saharan desert in Chad. Arch Microbiol 195(3):153–159

Montero-Calasanz MC, Göker M, Pötter G, Rohde M, Spröer C, Schumann P, Klenk H-P, Gorbushina AA (2013b) *Geodermatophilus telluris* sp. nov., an actinomycete isolated from Saharan desert sand. Int J Syst Evol Microbiol 63(6):2254–2259

Nachtigall J, Kulik A, Helaly S, Bull AT, Goodfellow M, Asenjo JA, Maier A, Wiese J, Imhoff JF, Süssmuth RD, Fiedler HP (2011) Atacamycins A-C, 22-membered antitumor macrolactones produced by *Streptomyces* sp. C38. J Antibiot 64(12):775–780

Neilson JW, Quade J, Ortiz M, Nelson WM, Legatzki A, Tian F, LaComb M, Betancourt JL, Wing RA, Soderlund CA, Maier RM (2012) Life at the hyperarid margin: novel bacterial diversity in arid soils of the Atacama Desert, Chile. Extremophiles 16(3):553–566

Nithya K, Muthukumar C, Kadaikunnan S, Alharbi NS, Khaled JM, Dhanasekaran D (2017) Purification, characterization, and statistical optimization of a thermostable α-amylase from desert actinobacterium *Streptomyces fragilis* DA7-7. Biotechnology 7(5):1–13

Nithya K, Muthukumar C, Biswas B, Alharbi NS, Kadaikunnan S, Khaled JM, Dhanasekaran D (2018) Desert actinobacteria as a source of bioactive compounds production with a special emphases on Pyridine-2,5-diacetamide a new pyridine alkaloid produced by *Streptomyces* sp. DA3-7. Microbiol Res 207:116–133

Nonomura H, Ohara Y (1969) Distribution of actinomycetes in soil. VI. A culture method effective for both preferential isolation and enumeration of *Microbispora* and *Streptosporangium* strains in soil. J Ferment Technol 47:463–469

Okoro CK, Brown R, Jones AL, Andrews BA, Asenjo JA, Goodfellow M, Bull AT (2009) Diversity of culturable actinomycetes in hyper-arid soils of the Atacama Desert, Chile. Antonie Van Leeuwenhoek 95(2):121–133

Okoro CK, Bull AT, Mutreja A, Rong X, Huang Y, Goodfellow M (2010) Lechevalieria atacamensis sp. nov., *Lechevalieria deserti* sp. nov. and *Lechevalieria roselyniae* sp. nov., isolated from hyperarid soils. Int J Syst Evol Microbiol 60(2):296–300

Passari AK, Mishra VK, Gupta VK, Singh BP (2018) Methods used for the recovery of culturable endophytic *Actinobacteria*: an overview. In: Singh BP, Passari AK, Gupta VK (eds) New and future developments in microbial biotechnology and bioengineering, vol 43. Elsevier, Amsterdam, pp 1–11

Quinn GA, Banat AM, Abdelhameed AM, Banat IM (2020) *Streptomyces* from traditional medicine: sources of new innovations in antibiotic discovery. J Med Microbiol 69(8):1040–1048

Radhakrishnan M, Gopikrishnan V, Vijayalakshmi G, Kumar V (2016) In *vitro* antioxidant activity and antimicrobial activity against biofilm forming bacteria by the pigment from Desert soil *Streptomyces* sp D25. J Appl Pharm Sci 6:148–150

Ramirez-Rodriguez L, Stepanian-Martinez B, Morales-Gonzalez M, Diaz L (2018) Optimization of the cytotoxic activity of three *Streptomyces* strains isolated from Guaviare River sediments (Colombia, South America). Biomed Res Int 2018:2839356. https://doi.org/10.1155/2018/2839356

Rateb ME, Houssen WE, Arnold M, Abdelrahman MH, Deng H, Harrison WTA, Okoro CK, Asenjo JA, Andrews BA, Ferguson G, Bull AT, Goodfellow M, Ebel R, Jaspars M (2011a) Chaxamycins A-D, bioactive ansamycins from a hyper-arid desert *Streptomyces* sp. J Nat Prod 74(6):1491–1499

Rateb ME, Houssen WE, Harrison WTA, Deng H, Okoro CK, Asenjo JA, Andrews BA, Bull AT, Goodfellow M, Ebel R, Jaspars M (2011b) Diverse metabolic profiles of a *Streptomyces* strain isolated from a hyper-arid environment. J Nat Prod 74:1965–1971

Raza A, Razzaq A, Mehmood SS, Zou X, Zhang X, Lv Y, Xu J (2019) Impact of climate change on crops adaptation and strategies to tackle its outcome: a review. Plants 8(2):34. https://doi.org/10.3390/plants8020034

Saker R, Bouras N, Meklat A, Zitouni A, Schumann P, Spröer C, Sabaou N, Klenk H-P (2015) *Prauserella isguenensis* sp. nov., a halophilic actinomycete isolated from desert soil. Int J Syst Evol Microbiol 65(5):1598–1603

Santhanam R, Okoro CK, Rong X, Huang Y, Bull AT, Andrews BA, Asenjo JA, Weon H, Goodfellow M (2012) *Streptomyces deserti* sp. nov., isolated from hyper-arid Atacama Desert soil. Antonie Van Leeuwenhoek 101(3):575–581

Santhanam R, Rong X, Huang Y, Andrews BA, Asenjo JA, Goodfellow M (2013) *Streptomyces bullii* sp. nov., isolated from a hyper-arid Atacama Desert soil. Antonie Van Leeuwenhoek 103(2):367–373

Sayed AM, Hassan MHA, Alhadrami HA, Hassan HM, Goodfellow M, Rateb ME (2020) Extreme environments: microbiology leading to specialized metabolites. J Appl Microbiol 128(3): 630–657

Saygin H, Ay H, Guven K, Cetin D, Sahin N (2019a) *Desertiactinospora gelatinilytica* gen. nov., sp. nov., a new member of the family *Streptosporangiaceae* isolated from the Karakum Desert. Antonie Van Leeuwenhoek 112(3):409–423

Saygin H, Ay H, Guven K, Sahin N (2019b) *Kribbella turkmenica* sp. nov., isolated from the Karakum Desert. Int J Syst Evol Microbiol 69(8):2533–2540

Saygin H, Ay H, Guven K, Cetin D, Sahin N (2020a) *Micromonospora deserti* sp. nov., isolated from the Karakum Desert. Int J Syst Evol Microbiol 70(1):282–291

Saygin H, Ay H, Guven K, Cetin D, Sahin N (2020b) *Streptomyces cahuitamycinicus* sp. nov., isolated from desert soil and reclassification of *Streptomyces galilaeus* as a later heterotypic synonym of *Streptomyces bobili*. Int J Syst Evol Microbiol 70(4):2750–2759

Saygin H, Ay H, Guven K, Sahin N (2020c) Genome-based classification of three novel actinobacteria from the Karakum Desert: *Jiangella asiatica* sp. nov., *Jiangella aurantiaca* sp. nov. and *Jiangella ureilytica* sp. nov. Int J Syst Evol Microbiol 70(3):1993–2002

Saygin H, Nouioui I, Ay H, Guven K, Cetin D, Klenk H-P, Goodfellow M, Sahin N (2020d) Polyphasic classification of *Nonomuraea* strains isolated from the Karakum Desert and description of *Nonomuraea deserti* sp. nov., *Nonomuraea diastatica* sp. nov., *Nonomuraea longispora* sp. nov. and *Nonomuraea mesophila* sp. nov. Int J Syst Evol Microbiol 70(1):636–647

Schulz D, Beese P, Ohlendorf B, Erhard A, Zinecker H, Dorador C, Imhoff JF (2011) Abenquines A–D: aminoquinone derivatives produced by *Streptomyces* sp. strain DB634. J Antibiot 64(12): 763–768

Sembiring L, Ward AC, Goodfellow M (2000) Selective isolation and characterisation of members of the *Streptomyces violaceusniger* clade associated with the roots of *Paraserianthes falcataria*. Antonie Van Leeuwenhoek 78(3):353–366

Shi Y-L, Sun Y, Ruan Z-Y, Su J, Yu L-Y, Zhang Y-Q (2020) *Cellulomonas telluris* sp. nov., an endoglucanase-producing actinobacterium isolated from Badain Jaran desert sand. Int J Syst Evol Microbiol 70(1):631–635

Singla AK, Mayilraj S, Kudo T, Krishnamurthi S, Prasad GS, Vohra RM (2005) *Actinoalloteichus spitiensis* sp. nov., a novel actinobacterium isolated from a cold desert of the Indian Himalayas. Int J Syst Evol Microbiol 55(6):2561–2564

Song L, Li W-J, Wang Q-L, Chen G-Z, Zhang Y-S, Xu L-H (2005) *Jiangella gansuensis* gen. nov., sp. nov., a novel actinomycete from a desert soil in north-west China. Int J Syst Evol Microbiol 55(2):881–884

Sooklal SA, De Koning C, Brady D, Rumbold K (2020) Identification and characterisation of a fluorinase from *Actinopolyspora mzabensis*. Protein Expr Purif 166:105508. https://doi.org/10.1016/j.pep.2019.105508

Souagui Y, Grosdemange-Billiard C, Tritsch D, Kecha M (2017) Antifungal molecules produced by a new salt-tolerant and alkaliphilic *Streptomyces* sp. BS30 isolated from an arid soil. Proc Natl Acad Sci India Sect B Biol Sci 87(2):527–535

Strub C, Brandam C, Meyer X, Lebrihi A (2008) Investigations of *Saccharothrix algeriensis* growth on synthetic media. J Biosci Bioeng 106(2):148–153

Sun H-M, Zhang T, Yu L-Y, Lu X-X, Mou X-Z, Zhang Y-Q (2014) *Actinophytocola gilvus* sp. nov., isolated from desert soil crusts, and emended description of the genus *Actinophytocola* Indananda et al. 2010. Int J Syst Evol Microbiol 64(9):3120–3125

Sun H-M, Zhang T, Wei Y-Z, Liu H-Y, Yu L-Y, Zhang Y-Q (2015) *Tenggerimyces mesophilus* gen. nov., sp. nov., a member of the family *Nocardioidaceae*. Int J Syst Evol Microbiol 65(10):3359–3364

Sun J-Q, Xu L, Guo Y, Li W-L, Shao Z-Q, Yang Y-L, Wu X-L (2017) *Kribbella deserti* sp. nov., isolated from rhizosphere soil of *Ammopiptanthus mongolicus*. Int J Syst Evol Microbiol 67(3):692–696

Sun Y, Wang H, Zhang T, Liu W-H, Liu H-Y, Yu L-Y, Zhang Y-Q (2019) *Desertihabitans aurantiacus* gen. nov., sp. nov., a novel member of the family *Propionibacteriaceae*. Int J Syst Evol Microbiol 69(8):2486–2491

Talbot GH, Bradley J, Edwards JE Jr, Gilbert D, Scheld M, Bartlett JG (2006) Bad bugs need drugs: an update on the development pipeline from the antimicrobial availability task force of the infectious diseases society of America. Clin Infect Dis 42(5):657–668

Tan GYA, Ward AC, Goodfellow M (2006) Exploration of *Amycolatopsis* diversity in soil using genus-specific primers and novel selective media. Syst Appl Microbiol 29(7):557–569

Terekhova L (2003) Isolation of actinomycetes with the use of microwaves and electric pulses. In: Kurböke I (ed) Selective isolation of rare actinomycetes. Nambour, Queensland, pp 82–101

Thomas DSG (2011) Arid environments: their nature and extent. In: Thomas DSG (ed) Arid Zone Geomorphology: process, form and change in drylands. Wiley, Blackwell, pp 1–16

Thumar JT, Dhulia K, Singh SP (2010) Isolation and partial purification of an antimicrobial agent from halotolerant alkaliphilic *Streptomyces aburaviensis* strain Kut-8. World J Microbiol Biotechnol 26(11):2081–2087

Tiwari K, Gupta RK (2013) Diversity and isolation of rare actinomycetes: an overview. Crit Rev Microbiol 39(3):256–294

Trujillo ME, Idris H, Riesco R, Nouioui I, Igual JM, Bull AT, Goodfellow M (2017) *Pseudonocardia nigra* sp. nov., isolated from Atacama Desert rock. Int J Syst Evol Microbiol 67(8):2980–2985

Tuo L, Dong Y-P, Habden X, Liu J-M, Guo L, Liu X-F, Chen L, Jiang Z-K, Liu S-W, Zhang Y-B (2015) *Nocardioides deserti* sp. nov., an actinobacterium isolated from desert soil. Int J Syst Evol Microbiol 65(5):1604–1610

Wang H-F, Zhang Y-G, Chen J-Y, Hozzein WN, Li L, Wadaan MAM, Zhang Y-M, Li W-J (2014) *Nesterenkonia rhizosphaerae* sp. nov., an alkaliphilic actinobacterium isolated from rhizosphere soil in a saline-alkaline desert. Int J Syst Evol Microbiol 64(12):4021–4026

Wang L, Li Y, Li Y (2019) *Lentzea isolaginshaensis* sp. nov., an actinomycete isolated from desert soil. Antonie Van Leeuwenhoek 112(4):633–639

Wang T, Li F, Lu Q, Wu G, Jiang Z, Liu S, Habden X, Razumova EA, Osterman IA, Sergiev PV (2020) Studies on diversity, novelty, antimicrobial activity, and new antibiotics of cultivable endophytic actinobacteria isolated from psammophytes collected in Taklamakan Desert. J Pharm Anal. https://doi.org/10.1016/j.jpha.2020.06.004

Wichner D, Idris H, Houssen WE, McEwan AR, Bull AT, Asenjo JA, Goodfellow M, Jaspars M, Ebel R, Rateb ME (2017) Isolation and anti-HIV-1 integrase activity of lentzeosides A–F from extremotolerant *lentzea* sp. H45, a strain isolated from a high-altitude Atacama Desert soil. J Antibiot 70(4):448–453

Williams ST, Davies FL (1965) Use of antibiotics for selective isolation and enumeration of actinomycetes in soil. Microbiology 38(2):251–261

Williams ST, Goodfellow M, Vickers JC (1984) New microbes from old habitats? In: Kelly DP, Carr G (eds) The microbe II: prokaryotes and eukaryotes. Cambridge University Press, Cambridge, pp 219–256

Wright GD (2012) Antibiotics: a new hope. Chem Biol 19(1):3–10

Yamamura H, Hayakawa M, Iimura Y (2003) Application of sucrose-gradient centrifugation for selective isolation of *Nocardia* spp. from soil. J Appl Microbiol 95(4):677–685

Yan Z-F, Lin P, Chu X, Kook M, Li C-T, Yi T-H (2016) *Aeromicrobium halotolerans* sp. nov., isolated from desert soil sample. Arch Microbiol 198(5):423–427

Yang Z-W, Salam N, Asem MD, Fang B-Z, Lan L, Xiao M, Wadaan MAM, Hozzein WN, Li W-J (2018a) *Saccharopolyspora deserti* sp. nov., a novel halotolerant actinobacterium isolated from a desert. Int J Syst Evol Microbiol 68(3):860–864

Yang Z-W, Salam N, Mohany M, Chinnathambi A, Alharbi SA, Xiao M, Hozzein WN, Li W-J (2018b) *Microbacterium album* sp. nov. and *Microbacterium deserti* sp. nov., two halotolerant actinobacteria isolated from desert soil. Int J Syst Evol Microbiol 68(1):217–222

Yang Z-W, Asem MD, Li X, Li L-Y, Salam N, Alkhalifah DHM, Hozzein WN, Nie G-X, Li W-J (2019) *Blastococcus deserti* sp. nov., isolated from a desert sample. Arch Microbiol 201(2): 193–198

Ye J-J, Liu S-W, Lu Q-P, Cheema MT, Abbas M, Sajid I, Huang D-L, Sun C-H (2020) *Arthrobacter mobilis* sp. nov., a novel actinobacterium isolated from Cholistan desert soil. Int J Syst Evol Microbiol 70(10):5445–5452

Yuan L-L, Zhang L-L, Luo X-X, Xia Z-F, Sun B-B, Zeng H (2020) *Streptomyces taklimakanensis* sp. nov., an actinomycete isolated from the Taklimakan desert. Antonie Van Leeuwenhoek 113(7):1023–1031

Zhang Y-G, Wang H-F, Liu Q, Hozzein WN, Wadaan MAM, Cheng J, Chen Y-J, Zhang Y-M, Li W-J (2013) *Streptomyces fukangensis* sp. nov., a novel alkaliphilic actinomycete isolated from a saline-alkaline soil. Antonie Van Leeuwenhoek 104(6):1227–1233

Zhang Y-G, Wang H-F, Yang L-L, Guo J-W, Xiao M, Huang M-J, Hozzein WN, Li W-J (2016) *Ornithinicoccus halotolerans* sp. nov., and emended description of the genus *Ornithinicoccus*. Int J Syst Evol Microbiol 66(4):1894–1899

Zhang L, Zhang J, Ren B, Lu W, Hou C, Wang J, Ma X, Ma R, Liu M, Liu Z (2020) Characterization of anti-BCG benz[α]anthraquinones and new siderophores from a Xinjiang desert–isolated rare actinomycete *Nocardia* sp. XJ31. Appl Microbiol Biotechnol 104(19):8267–8278

Zhu Z-N, Li Y-R, Li Y-Q, Xiao M, Han M-X, Wadaan MAM, Hozzein WN, An D-D, Li W-J (2019) *Microbacterium suaedae* sp. nov., isolated from Suaeda aralocaspica. Int J Syst Evol Microbiol 69(2):411–416

Zhu Q, Yang C, Luo X, Zhang L, Xia Z (2021) *Microbacterium karelineae* sp. nov. isolated from a halophyte plant in the Taklamakan desert. Int J Syst Evol Microbiol. https://doi.org/10.1099/ijsem.0.004629

Zitouni A, Lamari L, Boudjella H, Badji B, Sabaou N, Gaouar A, Mathieu F, Lebrihi A, Labeda DP (2004) *Saccharothrix algeriensis* sp. nov., isolated from Saharan soil. Int J Syst Evol Microbiol 54(4):1377–1381

Zucchi TD, Tan GYA, Bonda ANV, Frank S, Kshetrimayum JD, Goodfellow M (2012a) *Amycolatopsis granulosa* sp. nov., *Amycolatopsis ruanii* sp. nov. and *Amycolatopsis thermalba* sp. nov., thermophilic actinomycetes isolated from arid soils. Int J Syst Evol Microbiol 62(6): 1245–1251

Zucchi TD, Tan GYA, Goodfellow M (2012b) *Amycolatopsis thermophila* sp. nov. and *Amycolatopsis viridis* sp. nov., thermophilic actinomycetes isolated from arid soil. Int J Syst Evol Microbiol 62(1):168–172

5 Endophytic Actinobacteria in Biosynthesis of Bioactive Metabolites and Their Application in Improving Crop Yield and Sustainable Agriculture

Yadollah Bahrami, Yaser Delbari, Kimia Rafiei Buzhani, Elham Kakaei, Yaser Mohassel, Sasan Bouk, and Christopher M. M. Franco

Abstract

Agrochemicals are broadly applied into agriculture to control pests and phytopathogens. The excessive use of agrochemical poses a huge concern for the future of agriculture. Plants are associated with numerous microbes and possess a dynamic microbiome. Endophytic Actinobacteria (EA) are a prolific source of bioactive secretion and exudation products and offer renewed attention towards attaining sustainable agriculture. Endophytic Actinobacteria can be used as a 'biofertilizer' as they have the capability to increase plant growth, crop yields and plant-resistant pathogenicity by rendering available the insoluble and

Y. Bahrami
Medical Biology Research Center, Faculty of Medicine, Kermanshah University of Medical Sciences, Kermanshah, Iran

Department of Medical Biotechnology, School of Medicine, College of Medicine and Public Health, Flinders University, Adelaide, SA, Australia

Department of Medical Biotechnology, Faculty of Medicine, Kermanshah University of Medical Sciences, Kermanshah, Iran

Y. Delbari · K. R. Buzhani · E. Kakaei · S. Bouk
Department of Medical Biotechnology, Faculty of Medicine, Kermanshah University of Medical Sciences, Kermanshah, Iran

Y. Mohassel
Department of Clinical Biochemistry, Faculty of Medicine, Kermanshah University of Medical Sciences, Kermanshah, Iran

C. M. M. Franco (✉)
Department of Medical Biotechnology, School of Medicine, College of Medicine and Public Health, Flinders University, Adelaide, SA, Australia
e-mail: chris.franco@flinders.edu.au

R. V. Rai, J. A. Bai (eds.), *Natural Products from Actinomycetes*,
https://doi.org/10.1007/978-981-16-6132-7_5

complex essential macronutrients to the plant. They synthese phytohormones, siderophores, hydrogen cyanide and ammonia for the host plants.

Our results cover over 85 new plant growth-promoting Actinobacteria (PGPB) species, that have been isolated from different plants, having various biological activities. This chapter describes the beneficial and positive impact of EA on the host plant growth, performance, health and resilience to pathogens and stresses and crop productivities. The knowledge obtained from this chapter can help us to understand the pivotal application of EA in advancing agricultural sustainability.

This chapter also describes the structure of over 30 compounds associated with biocontrol activity, with the majority being alkaloids, polyketides, flavonoids, peptides and terpenoids. They are produced by different actinobacterial genera, with *Streptomyces* being the predominant contributor generating 25 compounds exclusively.

Our study reveals that EA, by producing a multitude of bioactive compounds, play a significant role in plant growth and crop yields and have the potential to improve sustainability in the agriculture industry. This chapter addresses the harnessing of EA as biofertilizers, a term used to include microbial inoculants that are biocontrol and plant growth-promoting agents and that have a role in the alleviation of abiotic and biotic stresses.

Keywords

Endophytic Actinobacteria · Plant growth promotion · Chemical diversity · Secondary metabolites · Phytohormone · Siderophore · Phytopathogens · Biocontrol · Biofertilizer · Crop yields

5.1 Introduction

Over the past century, the world's population has grown from about 1.8 billion to more than 8 billion and is predicted to reach 9.7 billion by 2050 (Elferink and Schierhorn 2016). This, along with the increase in standard of living of most people worldwide, has led to a significant increase in the demand for food globally. This need must be met mainly through either increasing agricultural lands or enhancing the efficiency of agricultural systems, in which the latter has to account for environmental risks and climate change. In addition, due to biological stresses such as plant pathogens and abiotic stresses, namely salinity, drought and depletion of rich soils, plant growth and crop yields are currently at risk (Yadav et al. 2020). Controlling these risks is of particular importance; therefore, designing and utilizing appropriate strategies are required. At present, pesticides and agrochemicals are widely used to control plant pests and enhance crop yields. However, using them will probably have more disadvantages in the long run than the aforementioned advantages. These disadvantages are the emergence of pesticide-resistant strains and the difficulty in managing them; destruction of soil texture; salinization and pollution of groundwater; the destructive effect on biodiversity resulting in the poor quality of agricultural products (Kumar and Kumar 2019).

To overcome the above-mentioned issues, various strategies are being used to increase food production. Using biofertilizers or bioinoculants is one of these approaches. Numerous studies reported the plant-associated microbiota and their mutual and beneficial interactions (Tian et al. 2020). Biofertilizers are living microorganisms that have beneficial effects on plant growth and health (Mącik et al. 2020). These microorganisms can improve plant yields and thus crop productions through various approaches such as increasing the absorption of minerals, namely nitrogen, phosphorus and potassium, production of secondary metabolites and controlling plant pathogens, especially fungi (Kour et al. 2020). These microorganisms may also influence plant metabolism and hence alter the secretion profile of root exudates and metabolites, which in turn alters the microbiome and colonization of selected beneficial microorganisms (Araujo et al. 2019). Despite the lack of toxicity and eco-friendly attributes of biofertilizers, there are concerns about the consistency of their efficacy, lack of detailed knowledge on the complex relationship between microorganisms, their cost-effectiveness and their rapid function, which need to be considered and addressed (Sharma et al. 2018). Biocontrol activity is considered as a viable alternative to the existing chemical methods to overcome plant pathogens (Vurukonda et al. 2018). Biocontrol is a method in which a pathogenic microorganism is controlled by a beneficial one. This method uses properties such as the production of antibiotics, competition for resources or the production of enzymes by microorganisms which may reside in the rhizosphere, phyllosphere, endosphere or various plant surfaces (Gomes et al. 2018). Microorganisms that directly stimulate plant growth by producing secondary metabolites (SMs), or indirectly control pathogens, have been isolated from plant microbiomes. These bacteria which can colonize the rhizosphere, tissue surfaces and endosphere are very diverse and can promote plant growth through the acquisition of soil nutrients, tolerating stress and resistance to pathogens (dos Santos and Olivares 2021). Although the competitive and cooperative interaction of plants and endophytic Actinobacteria still remains unclear, and to what extent these interactions are beneficial or deleterious depends on many factors. The plant-associated microbiota offer a special opportunity to improve the agricultural industry (Arif et al. 2020).

Actinobacteria are a major part of the microbiomes of plants, indicating the establishment of the formation of a strong interaction with the host plants (Kõiv et al. 2015). Actinobacteria are ubiquitous Gram-positive bacteria with high GC content in their genome, differentiated from other orders of bacteria by forming branched filaments and many other distinctive morphological structures (Barka et al. 2016). Actinobacteria produce a broad spectrum of bioactive secondary metabolites which positively affect the host growth, which in turn makes the microorganism capable of antagonizing plant pathogens (Vurukonda et al. 2018). Clearly, endophytic Actinobacteria which colonize internal tissues of plants will produce metabolites which switch on plant genes, and should contain novel biosynthetic pathways for secondary metabolites (Zhao et al. 2018). These bacteria can increase the plant biomass and crop yields by growth promotion of lateral roots, hight and weight of plants (Pillay and Nowak 1997). Endophytic Actinobacteria can directly engage with plant pathogens through the production of lytic enzymes such as chitinases and cellulases or synthesizing components like hydrogen cyanide (HCN) (Daulagala 2021). They can also indirectly affect plant immunity and protect them against diseases by competition for resources and production of siderophores, organic compounds, enzymes and phytohormones (Ansari et al. 2020). The beneficial effect of EA on the ability of plants to cope with drought, temperature, heavy metals, salinity, nitrogen deficiency and pathogens results in the enhancement of crop yields. The purpose of this chapter is to reveal and harness the potential of EA to improve plant growth and fitness of plants and hence to increase the plant productivity and crop yields. In addition, it also addresses the harnessing of EA as biostimulants and sustainable biocontrol agents for agriculture and chemical-free organic farming in the near future, as well as the alleviation of abiotic and biotic stresses.

This study aims to present the latest knowledge on the chemical structure of bioactive compounds, siderophores, enzymes and other plant growth-promoting agents produced by EA to manage the daily stresses of plant host and hence to facilitate the growth of plant host and crop yields. This chapter also aims to highlight the importance of plant-associated EA and their application in the agriculture industry in the near future. This review has covered the resources available in citation systems and databases reported in PubMed, Scopus, ScienceDirect and Web of Knowledge since 2000.

5.2 Endophytic Microbes

Endophytes are microorganisms, typically bacteria and fungi, that reside within plants asymptomatically and play pivotal roles in growth development, resilience to pathogens, stress tolerance and crop yields of plants. Endophytic microbes colonizing plant tissues either are transmitted vertically from generation to generation via seed or can be horizontally obtained from the surrounding soil or atmosphere (Frank et al. 2017). Endophytic microbes include diverse phyla, namely Actinobacteria, Acidobacteria, Deinococcus–Thermus, Bacteroidetes, Firmicutes, Proteobacteria and Verrucomicrobia, with the most dominant and studied endophytic bacteria being Proteobacteria, Firmicutes and Actinobacteria, respectively (Rana et al. 2020a). *Streptomyces* is the most prevalent genus reported among endophytic Actinobacteria (Kaewkla and Franco 2013). Actinobacteria are one of the major groups of endophytes which, in addition to their therapeutic uses (Delbari et al. 2020), have a broad potential and value in food and agricultural applications. The EA can stimulate plant growth through both direct and indirect mechanisms, namely nutrient solubilization, production of phytohormones, production of enzymes including 1-aminocyclopropane-1-carboxylic acid (ACC) deaminase, siderophores, induction of pathogen resistance and aiding resistance to biotic and abiotic stresses like those imposed by high salinity and drought (Conn et al. 2008).

5.3 Endophytic Actinobacteria for Sustainable Agriculture Improvement

Plant root exudates play a crucial role in selecting and augmenting the types of bacteria by the plants. The nature, concentration and chemical structure of the exudate components and the corresponding capability of microbes determine the type of interaction between microbes and the plant as epiphytic/endophytic/rhizospheric. Many reports have documented the beneficial relationship between the EA and host plants showing that EA can influence plant growth either directly or indirectly. Direct modes comprise iron (Fe) chelation, phosphate (P) solubilization, 1-aminocyclopropane-1-carboxylic acid (ACC) deaminase activation and phytohormone production, while indirect mechanisms consist of antibiotic and secondary metabolite production, pathogen suppression by outcompeting them for macro- and micronutrients, siderophore production, nanoparticle synthesis, secretion of lytic enzymes and establishment of the plant systemic resistance (Tewari et al. 2019). Endophytic Actinobacteria can produce a variety of secondary metabolites and different enzymes to reduce or prevent the deleterious effects of plant pathogens, counteract various stresses in their host as well as stimulate plant growth and provide competitive advantages to their host (Zhao et al. 2018). Endophytic microorganisms are essential for plant growth and health, and during plant–pathogen interactions, endophytic microbial communities of plants undergo profound changes, which indicates the close interaction between them (Kõiv et al. 2015). Therefore, endophytes cooperate with plants through various mechanisms and exert their biocontrol effects, including the production of phytohormones and siderophores, solubilizing insoluble soil phosphate and making it available to plants, production of various enzymes and antagonistic effects against various pathogens, especially fungi. To deliver and introduce bioinoculants to the plant, many approaches including seed pelleting, seed dressing, seed priming (seed treatment), soil drenching (amendment), foliar spray and root dipping are widely used (Yashavantha Rao et al. 2020).

5.4 Biological Control

Plants are generally exposed to many bacterial and fungal pathogens which adversely influence the crop growth and productivity worldwide. The unsystematic and indiscriminate use of chemical agents for handling plant diseases may lead to the emergence of resistant strains and elimination of beneficial microorganisms in addition to the adverse effects on the products. Recently, EA have received much attention as an alternative to agrochemicals (Franco 2021). Thereby, biocontrol of disease is an alternative and safe way to manage plant diseases and minimize the negative effects of the long-term applications of agrochemicals due to its eco-friendly nature, cost-efficient and long-lasting effects. Biocontrol microbes can restrain phytopathogens directly by generating antagonistic compounds, competition for iron, detoxification or degradation of virulence factors or indirectly by induction of systemic resistance (ISR) in plants against certain diseases; signal interference; competition for nutrients and niches; and interference with activity, survival, germination and sporulation of the pathogen (Verma et al. 2017). Seed treatment with endophytic *Streptomyces* sp. strain EN27 and *Micromonospora* sp. strain EN43 provided high levels of the resistance by *Arabidopsis thaliana* to fungal infections of *Erwinia carotovora* and *F. oxysporum* (Conn et al. 2008).

Glomerella cingulata is a pathogenic fungus causing anthracnose disease in strawberry plant, leading to $35 million financial loss to farmers in Japan annually, and is responsible for a 50% reduction in crop production (Marian et al. 2020). Therefore, Marian and colleagues investigated the effects of *Streptomyces* isolated from this plant on the fungus in question to identify its latent potential (Marian et al. 2020). Overall, they isolated 217 endophytic Actinobacteria and applied them against the *G. cingulata* pathogen, of which *MBFA-172* and *MBFA-227* strains resulted in 96.0% and 90.0% reduction in symptoms of anthracnose disease, respectively. However, for all strains, this value was 48.9% on average. Further in vivo experiments also confirmed the antifungal effects of these two strains, but in vitro studies using this strain against the fungus did not show a significant inhibitory effect, indicating that the strain does not produce an antifungal compound, but by using indirect methods such as mineral absorption, it helps the plant to resist anthracnose disease.

Sclerotinia is another fungal disease caused by *Sclerotinia sclerotiorum*. The pathogen is capable of infecting more than 400 species of plants and causing significant economic losses to the agricultural industry (Liu et al. 2019a). They isolated and identified 1574 actinobacterial colonies belonging to 70 species of 14 different genera from soybean. In addition to demographic changes of endophytes in healthy and *S. sclerotiorum*-infected plants, the team investigated the antifungal effects of nine metabolites identified from strains DAAG3-11, DGS1-1, DGS3-15 and DDPA2-14. These strains showed inhibition against *S. sclerotiorum* from 54.1 to 87.6%. Xu et al. (2019) reported *Streptomyces hygroscopicus* OsiSh-2 from rice that exhibits potential antifungal activities against the pathogen *Magnaporthe oryzae* which can be used as potential biocontrol agent against rice blast disease. The results also showed that spraying the 5% concentration of filtered extract reduced the disease index from 56.7% to 41.3% and from 47.0% to 36.6% in routine culture, respectively. A bioactive metabolite 1-vinyl-β-carboline-3-carboxylic acid generated by endophytic Actinobacteria *Microbispora* sp. *LGMB259* reported from the host plant *Vochysia divergens* exhibited antagonistic activity towards *Kocuria* B-1106 and *Micrococcus* NRRL B2618 (Savi et al. 2015).

Coombs et al. (2004) documented that endophytic *Microbispora* and *Nocardioides* spp. strains possess antagonistic activity against phytopathogen *Gaeumannomyces graminis* var. *tritici* (Ggt) in wheat and *R. solani*, which can be utilized as biocontrol agents in the field (Barnett et al. 2019). The results of these studies approve harnessing of endophytic Actinobacteria in agriculture to biocontrol the pathogens, which can lead to reduced use of harmful chemicals. Many reports are available that EA are also capable of producing a wide range of extracellular enzymes, namely chitinases, β-1,3-glucanases, amylases, lipases and cellulases, that act as antifungal agents (El-Tarabily 2003; Passari et al. 2015).

Defence-related enzymes inhibit the growth of fungi by attacking the cell wall and contribute to the biocontrol of plant diseases. The inhibitory effects of cell wall-degrading lytic enzymes, namely cellulases and glucanases, isolated from EA against *Pythium* spp. and *Pythium* disease are also reported (El-Tarabily 2006). The presence of chitinase gene in the collected isolates was also documented (El-Tarabily 2003; Passari et al. 2015). On the other hand, the fungal cell wall contains chitin, cellulose, a variety of proteins and lipids which endophytes can destroy by producing hydrolysing enzymes such as chitinase and lipase (Goudjal et al. 2018). Passari et al. (2016) isolated 15 endophytic Actinobacteria from *Solanum lycopersicum*, in which eight of the isolates exhibited antagonistic and PGP activities. They stated the production of cellulase observed among seven of the eight isolates indicated their growth inhibitory activity against fungal pathogens. They also pointed that *Streptomyces* sp. strain DBT210 was the most efficient strain in the production of fungal cell wall-degrading chitinolytic enzyme (Passari et al. 2016). Twenty-one lipase-producers and 35 cellulase-producer strains were isolated from *Lycium ruthenicum* Murr, many of which belong to the order Actinobacteria (Liu et al. 2019c). Zhang et al. (2014) explored a new prenylated indole derivative from *Streptomyces* sp. Neau-D50 with antagonistic activity towards the plant pathogenic fungi *Colletotrichum orbiculare*, *Phytophthora capsici*, *Corynespora cassiicola* and *F. oxysporum*.

Quecine et al. (2008) isolated 25 *Streptomyces* strains from soybean and citrus and evaluated the chitinase production activity of these strains against *Phytophthora parasitica*, *Guignardia citricarpa*, *Rhizoctonia solani*, *Colletotrichum sublineolum*, *Pythium* sp. and *Fusarium oxysporum* and contended that most strains have chitinolytic activity. Although the biological control agents are valuable alternatives to classical pesticides, they still make up a small proportion of global crop protection markets (Raymaekers et al. 2020).

Endophytic Actinobacteria are also able to produce secondary metabolites with biocontrol properties. These compounds have the ability to combat plant pathogens with antagonistic activity. Secondary metabolites with biocontrol potential isolated from endophytic Actinobacteria are presented in Table 5.1. Over 30 bioactive compounds with biocontrol property have been reported; however, the structures of only 27 are assigned. The structure of these 27 bioactive compounds produced by EA is illustrated in Fig. 5.1. These compounds possess different chemical structures varying from small peptides to polyketides, alkaloids and flavonoids. *Streptomyces* are the major contributors, generating over 83% of compounds, exclusively.

Table 5.1 Secondary metabolites and enzymes from endophytic Actinobacteria with biocontrol activity

Metabolite name (chemical structure)	Chemical class	Strain	Host plant	Ref.
14-Hydroxyisochainin (**1**)	Polyene	*Streptomyces* sp.	*Arabidopsis thaliana*	Ayuso-Sacido and Genilloud (2005)
7-*O*-β-D-Glucosyl-genistein (**2**)	Isoflavone	*Mycobacterium* sp. LGMB471	*Vochysia divergens*	Savi et al. (2019)
7-*O*-β-D-Glucosyl-daidzein (**3**)	Isoflavone	*Mycobacterium* sp. LGMB471	*Vochysia divergens*	Savi et al. (2019)
7,30-di-(g,g-dimethylallyloxy)-5-hydroxy-40 methoxyflavone (**4**)	Flavonoid	*Streptomyces* sp. MA-12	*Myoporum bontioides*	Ding et al. (2013)
Antimycin A18 (**5**)	Peptide	*Streptomyces albidoflavus* I07A-01824	*Bruguiera gymnorrhiza*	Yan et al. (2010)
Borrelidins J and K (**6–7**)	Polyketide	*Streptomyces* sp. NA06554	*Aster tataricus*	Zhang et al. (2018)
Cedarmycins A and B (**8–9**)	Terpenoid	*Streptomyces* sp. TP-A0456	*Cryptomeria japonica*	Igarashi (2004)
Chitinase	Hydrolase	*Streptomyces hygroscopicus*	*Peanut*	Quecine et al. (2008)
Chitinase	Hydrolase	*Streptomyces* sp. SNL2	Tomato	Quecine et al. (2008)
Cyclo-(L-Ile-L-Glu-L-Ile-L-Glu) (**10**)	Peptide	*Streptomyces* sp. 447	Medicinal plants	Abdalla (2017)
Cyclo-(L-Leu-L-Glu-L-Leu-L-Glu) (**11**)	Peptide	*Streptomyces* sp. 447	Medicinal plants	Abdalla (2017)
Cyclo-(L-Val-L-Glu-L-Val-L-Glu) (**12**)	Peptide	*Streptomyces* sp. 447	Medicinal plants	Abdalla (2017)
Fistupyrone (**13**)	NA	*Streptomyces* sp. TP-A0569	Onion *Allium fistulosum*	Igarashi et al. (2000)
Fungichromin (**14**)	Polyene	*Streptomyces* sp. WP-1	*Pinus dabeshanensis*	Peng et al. (2020)
Lokisin (**15**)	Lipopeptide	*Pseudomonas koreensis*	*Nicotiana tabacum*	Gu et al. (2020)
Naphthomycins L, M and N (**15–17**)	Polyketide	*Streptomyces* sp. CS	NA	Yang et al. (2012)
Organophosphate (**18**)	Organophosphorus	*Streptomyces albus* CAI-21	Sorghum	Gopalakrishnan et al. (2020)
Phenol, 2,4-bis(1,1-dimethylethyl)	Phenol	*Kutzneria* sp. TSII	Mangrove	Devi et al. (2021)

S-Adenosyl-*N*-acetylhomocysteine (**19**)	Peptide	*Micromonospora* sp. PC1052	*P. candollei* var. *candollei*	Boonsnongcheep et al. (2017)
Strepimidazoles A, B, C, D, E, F and G (**20–26**)	Alkaloid	*Streptomyces* sp. PKU-EA00015	*Salvia miltiorrhiza*	Sun et al. (2020)

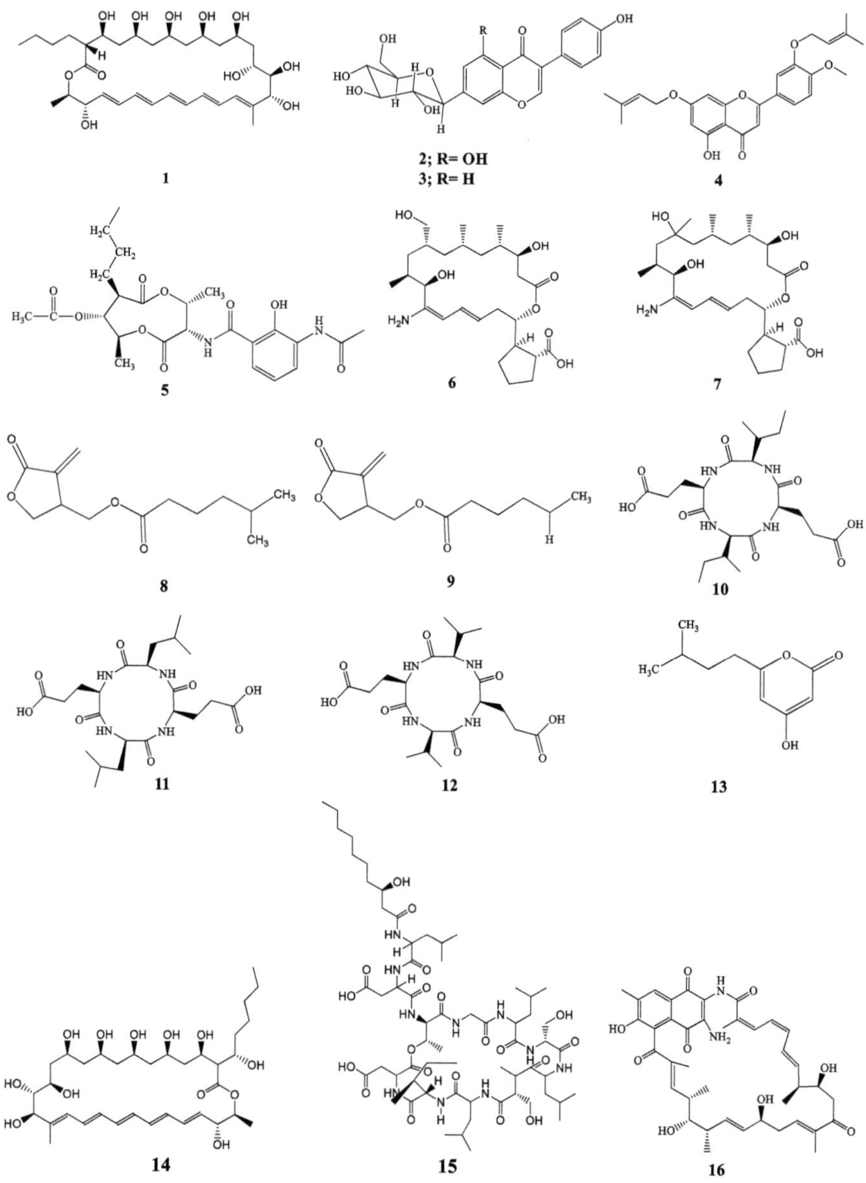

Fig. 5.1 The chemical structures of bioactive secondary metabolites produced by endophytic Actinobacteria having biocontrol activity

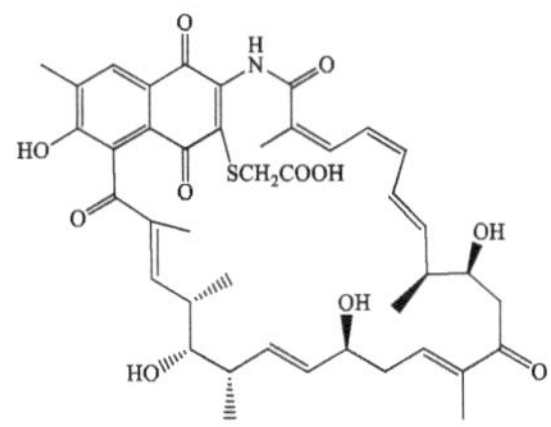

17

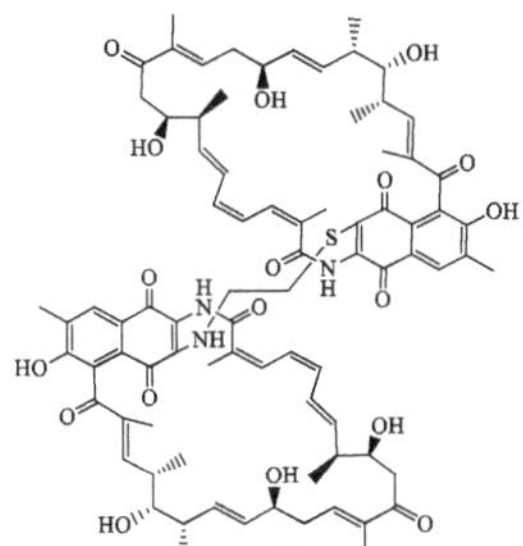

18

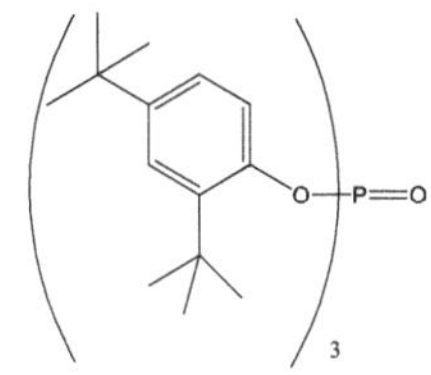

19

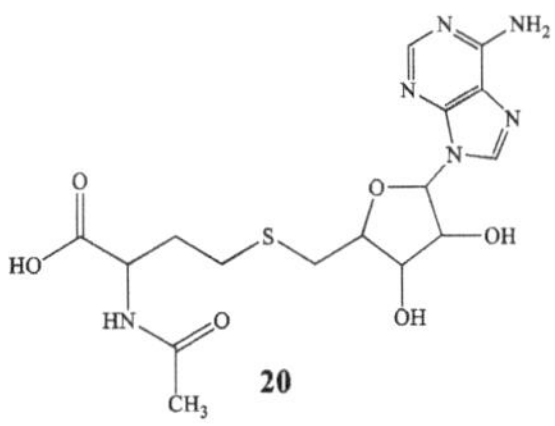

20

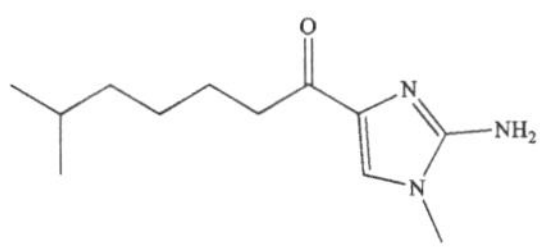

21

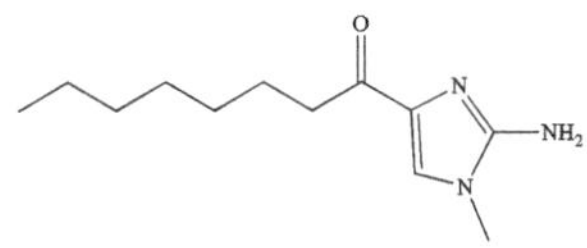

22

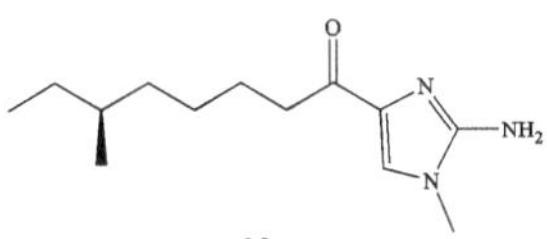

23

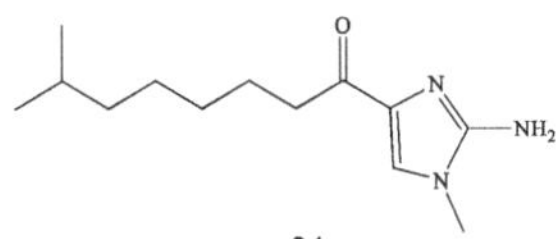

24

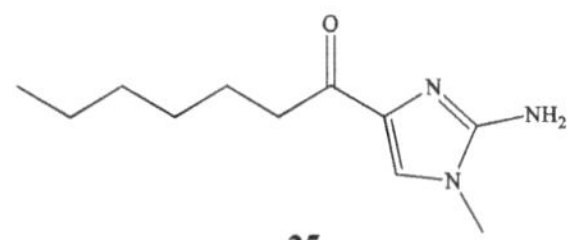

25

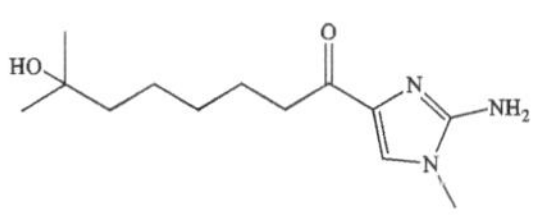

26

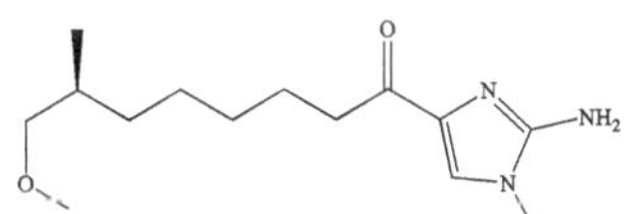

27

Fig. 5.1 (continued)

5.5 Abiotic Stress: Reduction, Alleviation and Control

Plants constantly deal with living organisms and variable environmental factors surrounding them for survival. Additionally, owing to climate change and global warming, plants are facing an escalated number of abiotic stresses, which adversely influence their growth and yield. Abiotic stress is growth inhibitory to the plant, varying from oxidative stress, UV rays and high salinity to drought and heavy metal toxicity. There are many reports that highlight the adverse effects of abiotic stress can be mitigated by employing EA. Salinity stress, mainly due to irrigation practices, can disrupt plants' ionic and osmotic equilibria, causing damage ranging from stunted growth to death (Zhu 2001). Plants mainly try to overcome these stresses by regulating growth and producing different enzymes encoded by stress resistance-related genes. Meanwhile, some bacteria including endophytes are able to withstand salinity and drought and help their host plant in ways such as spore production and production of various enzymes and hormones. Alishahi and co-workers isolated 316 species of endophytes from roots of the halophyte *Suaeda* sp. plant, of which 15.6% belonged to the order Actinobacteria (Alishahi et al. 2020). They reported that only *Zhihengliuella halotolerans* and *Brachybacterium* sp. are able to grow in semi-solid N-free (NFb) medium enriched with 6% NaCl. Evaluation of the nitrogen fixation ability of these strains in saline soil conditions, in TSB and NFb environments, showed that they could tolerate salt concentrations up to 20%. Also, nitrogen fixation ability of these strains in salinity effectively increases root growth and germination up to 70% in in vivo studies. A halophyte plant, *Limonium sinense*, was tested for its endophytic community composition (Qin et al. 2018). Of the 29 identified endophytic strains, the majority belonged to the order Actinobacteria and the genus *Glutamicibacter*. Of these 15 strains were able to withstand salinity stress and ensure plant growth by producing 1-aminocyclopropane-1-carboxylic acid deaminase (ACC deaminase). Many reports have emphasized PGP microbes that produce ACC deaminase augment plant growth by diminishing the ethylene level, eliciting salt tolerance and lowering drought stress in plants. In addition, the randomly selected strains KLBMP 5180, KLBMP 5252 and KLBMP 5285 confirmed the results of in vivo studies in the presence of 85 mM NaCl in the role of endophytes in salinity tolerance (Zhang et al. 2019). This study, which investigated the endophytic community composition of plants *Distichlis spicata* and *Pluchea absinthioides* collected from very hot Atacama deserts as well as *Gaultheria mucronata* and *Hieracium pilosella* from very cold Patagonia areas, indicated that up to 23% of the known endophytic populations in both ecosystems belonged to the Actinobacteria order.

Heavy metals such as cadmium pollute the environment and have destructive effects on it. One available method, bioremediation, is to use bacteria to remove the contaminant. Endophytic Actinobacteria are able to collect heavy metals by producing factors such as auxin, ACC deaminase and siderophores (Siripan et al. 2018). Heavy metals such as Cr and Cd stimulate oxidative stress in metabolic pathways of plants, leading to synthesis of free radicals. The abscisic acid (ABA) level is elevated in the presence of Cr and Cd, which corroborate the role of ABA in establishing tolerance towards heavy metals in plants (Singh et al. 2020).

5.6 Plant Growth Promotion

Plant and endophyte can develop a variety of relationships with each other to produce metabolites including (a) endophytic bacteria-mediated induction of secondary metabolite production in host plant, (b) production and manipulation of host-specific compounds by endophytes and (c) the modulation of bioactive products from the endophyte and vice versa (Ludwig-Müller 2015). Endophytic Actinobacteria, in addition to helping plants cope with biotic stress, have been shown to enhance plant growth by producing numerous secondary metabolites (SMs) and enzymes as well as promoting their resistance to various stresses

(Orozco-Mosqueda et al. 2020). These SM producers are able to (1) induce the plant growth by stimulating the production of plant phytohormones, (2) elevate nitrogen fixation, (3) enhance the accessibility of crucial nutrients; solubilize inorganic phosphate and/or other nutrients and (4) tolerate, resist or inhibit deleterious microorganisms. Plant growth-promoting bacteria (PGPB) can solubilize the mineral element in the soil, stimulate the plant resistance to pathogens and facilitate the colonization of rhizobia in the roots. As a result, this leads to better acquisition of nutrients and therefore boosts plant growth and crop yields.

5.6.1 Siderophore Production

Iron (Fe) as one of the most critical micronutrients for crop production acts a crucial role in several enzymatic activities. Owing to low availability of iron in many soil environments, Actinobacteria under low iron concentrations are able to produce siderophores. Siderophores are water-soluble, small chelating compounds with a high ferric iron affinity and low molecular weight ranging from 200 to 2000 Da. Approximately 500 different types of siderophores have been reported, of which about 270 have been structurally characterized (Rana et al. 2020b). Two soluble forms of Fe are consumed by microbes and plants, Fe^{2+} and Fe^{3+}, which are insoluble in neutral pH. Siderophores are a Fe^{3+}-adsorbing molecules which can chelate ambient (soil) iron when the plant is deficient in this vital mineral and provides the iron for plant growth rather than the pathogens, and thus depriving them of this essential element (Albelda-Berenguer et al. 2019). This molecule promotes plant growth in different conditions with the help of iron-dependent enzymes in photosynthesis, chlorophyll production and NO_2 and SO_4 reduction (Verma et al. 2011). Since iron is largely present in the soil but difficult to access without the help of a molecule that can absorb it, endophyte-produced siderophores offer the plant the ability to resist environmental pathogens (Passari et al. 2015). The siderophore desferrioxamine E produced by streptomycetes is capable of sequestering a panel of ions, namely Mn, Co, Li, Cu, Zn, Cd, Ni, Al and Mg (Nakouti and Hobbs 2013). It is documented that *Actinomadura madurae*, *Nocardia asteroides* and *Streptomyces griseus* generate hydroxamate and salicylate types of siderophores that assist in defending plants from pathogens (Kannahi and Senbagam 2014). On the other hand, siderophores are capable of protecting the plant against heavy metal stress by activating the expression of antioxidant enzymes in addition to absorbing other heavy metals, such as Al, Pb, Cr and Cd (Złoch et al. 2016), and lessen the toxicity of the heavy metal on the plant.

5.6.2 Nitrogen Fixation

Nitrogen as the main mineral nutrient of the plant is the major constituent of enzyme protein, nucleotides, energy production and chlorophyll production. The plants obtain nitrogen from the soil in the form of ammonia, nitrite or nitrate. Endophytes are able to convert the molecular form of nitrogen (gaseous form (N_2)), which is the naturally occurring form, to the form of ammonium (NH_4^+) and nitrate (NO_3^-), the ready-to-use forms of nitrogen (Prasad et al. 2020; Trujillo et al. 2014).

The N_2-fixing *Frankia*, isolated from the nodules of actinorhizal plants, stimulate the growth and abiotic resistance for several plants, including *Alnus rubra*, *Hippophae tibetana*, *Elaeagnus angustifolia* and *Casuarina glauca* (Tian et al. 2020; Van Nguyen and Pawlowski 2017).

5.6.2.1 Interaction with *Rhizobium*

It is evident that the entry and colonization of endophytes is a complex phenomenon and multiple factors such as signalling molecules, proteins and/or the secretory products of both plants and microbes are involved to shape it (Kumar et al. 2020). Plants are actively recruiting their microorganisms from surrounding microbial reservoirs including the soil/rhizosphere, the phyllosphere, the spermosphere (the exterior of germinated seed), the anthosphere (the external environment of flowers) and the carposphere (the external fruit environment) (Compant et al. 2019).

Root microbiota are generally horizontally transferred. Root exudates are rich in organic substances and play a crucial role in the colonization of endophytic bacteria within the host tissues. It has been proposed that host- and microbe-derived molecules have a key role in the development of symbiotic relationship. Although EA have been mainly investigated for antagonistic properties, they can promote the growth of beneficial *Rhizobium* bacteria by establishing a beneficial relationship. Through a beneficial connection, they can increase the growth of *Rhizobium* bacteria. *Rhizobium* bacteria, found mainly in the nodules, are major contributors to plant fertility (Matiru and Dakora 2004). Le and co-workers have investigated the effects of endophytic Actinobacteria isolated from several medicinal plants on few species of rhizobia (Le et al. 2016). Among the extracted endophytic strains, *Streptomyces LuP3*, *LuP30*, *LuP47B* and *EN23* strains increased the growth of *Rhizobium* strains in vitro. These results were confirmed for *in planta* studies as the selected *Streptomyces* strains, when used in combination with *Sinorhizobium meliloti RRI 128* strain, significantly increased nodulation and root dry weight. One reason for these results was the contribution of endophytes in facilitating nutrient uptake, especially their enhancement of nitrogen fixation ability of the rhizobial partner. The dry weight of plants was also increased by 48%, indicating a positive effect of endophytic strain co-inoculation with rhizobia. The simultaneous co-inoculation model of endophytic strains with other commercial strains can be considered as an appropriate option for increasing crop production and utilization in industrial agriculture.

5.6.3 Phosphate Solubilization

Phosphorus is one of the minerals that is essential to improve the quality and increase the production of agricultural products, though in many soils it is locked up or immobilized. Therefore, sustainable approaches are required to make it available since food demand is growing rapidly (Boubekri et al. 2021). In addition to positive effects on plant metabolism, this mineral can also play a key role in essential plant functions such as cell division (Mitra et al. 2020). Phosphorus can be made readily available to the host with the help of symbiotic endophytes. Today, large quantities of phosphate fertilizers are used in agriculture, while only a small fraction can be taken up by plants, and the rest undergoes a *precipitation* process and forms a complex with metal ions in soil (Soumare et al. 2020). Many endophytic bacteria, by producing various organic acids like tartrate, lactate, succinate, gluconate, ketogluconate, acetate, citrate, oxalate and glycolate, are able to solubilize insoluble phosphorus in the rhizosphere and facilitate its uptake by the plant; therefore reducing the large quantities of chemical fertilizer released to the environment, resulting in significant economic benefits (Mesta et al. 2018). Actinobacteria are a promising group of microorganisms that have been shown to act as effective P solubilizers and chemical fertilizer alternatives. Endophytic Actinobacteria were cultured on Pikovskaya's agar to evaluate their phosphate solubility and PGP activity (Kaur et al. 2013). Out of 321 isolates, five strains showed phosphate solubilizing activity. In another study, several EA were isolated from *Solanum lycopersicum*, and from these isolates, *Streptomyces* sp. DBT204 was the most promising strain to be developed as a biofertilizer due to its PGP traits, ammonia production and phosphate solubilization (Passari et al. 2016). For the first time in New Zealand, *Streptomyces* sp. UKCW/B and *Nocardia* sp. TP1BA1B isolated from *Pseudowintera colorata* (horopito) showed phosphate solubilization trait on TCP agar (Purushotham et al. 2018). Phosphate solubilization, along with other properties of EA, makes them as significant contributors for biofertilizers.

5.6.4 ACC Deaminase Activation

Ethylene is a stress-stimulated phytohormone which can prevent plant growth. Some EA can drop the level of ethylene in the plant by breaking down the plant-synthesized ethylene precursor 1-aminocyclopropane-1-carboxylic acid (ACC). This enzyme catalyses the cleavage of 1-aminocyclopropane-1-carboxylic acid (ACC) into NH_3 and α-ketobutyrate. ACC deaminase modulates stress as it results in the reduction of the ethylene and ACC levels within the plant cell, while the latter acts as a precursor molecule for the production of ethylene. Low ethylene C_2H_4 levels increase plant biomass as well as expression of defence-related proteins, but with the emergence of stress, the increased level of ethylene leads to the growth inhibition and even plant death (Robison et al. 2001). Plants and endophytes support low levels of ethylene in the plant by maintaining low levels of ACC deaminase (ACCD). But in response to abiotic stresses, the expression of ACC oxidase and consequently the ethylene production are elevated; thereby, this increment will result in the increase of ACCD in order to balance the ethylene levels and tolerate the stress (Glick 2014). As a result of this process, despite environmental stress, endophytes will continue to accommodate plant growth by producing ACC deaminase. ACC deaminase regulates the ethylene level after such a burst, keeping ethylene levels below the threshold where growth is impeded. In a study of *Glycine max* endophytic Actinobacteria, a *Streptomyces* sp. NEAU-S7GS2 isolated from the roots was found to be effective against *Sclerotinia sclerotiorum*, a plant pathogen (Liu et al. 2019b). It has been reported that ACCD enzyme is involved in increasing PGP activity. In addition, soil contamination has been shown to inhibit plant growth by inducing ethylene production, while this does not occur in plants that coexist with ACCD-producing bacteria (Arshad et al. 2007). In a similar study, *Streptomyces venezuelae* ATCC 10712 was selected to evaluate its PGP activity when it carries a mutation in *acdS*, the ACCD gene (Yoolong et al. 2019). Comparison of salinity tolerance of *Oryza sativa* showed a far greater tolerance when it coexisted with the bacterium carrying the mutant, producing more ACCD in comparison with the wild-type bacterium. For the first time, Govindasamy et al. isolated 179 EA from cactus (*Opuntia ficus-indica*), all of which showed ACC deaminase activity following PGP analysis (Govindasamy et al. 2021). Another study investigated two salt-tolerant endophytes, namely *Kocuria rhizophila* KF875448 (14 ASP), an actinobacterium, and *Cronobacter sakazakii* KM042090 (OF115), an enterobacterium, for PGP traits. Their results determined ACC deaminase activity for both strains was a great tool for enhancing plant growth and stress resistance (Afridi et al. 2019).

5.6.5 Phytohormone Production

Phytohormones, as one of the most crucial secondary metabolites, plant growth-promoting (PGP) substances and bio-controllers, play major roles in the growth of plants. Gibberellic acid, auxins, indole pyruvic acid, abscisic acid (ABA), jasmonic acid and cytokines are examples of metabolites synthesized by endophytic Actinobacteria, as plant phytohormones, to support the plants to combat pathogens, tolerate plant stresses and also promote their host growth (Swarnalakshmi et al. 2016). Auxin, cytokinin, gibberellins, ethylene and abscisic acid are five well-recognized plant growth regulators (PGRs). The most dynamic and active auxin in plants is indole-3-acetic acid (IAA). Amino acid tryptophan is the prime precursor for the biosynthesis of IAA. Indole-3-acetic acid, a phytohormone as the most vital auxin, is a plant growth-promoting hormone secreted by endophytes (Dochhil et al. 2013; Shutsrirung et al. 2013; Passari et al. 2016). This phytohormone not only stimulates plant cell division by controlled degradation of the cell wall and increasing root mass and the availability of soil nutrients but also, in cooperation with the ACCD enzyme, increases plant tolerance to environmental stress by decreasing the levels of ethylene hormone in the plant (Glick 2014). In general, IAA has effects on the cell division in plants, initiation of seed and tuber germination, nodule formation, development of xylem and root, photosynthesis, and mediates tolerance and resistance by plants towards many stressful conditions. Pteridic acids A and B isolated from *Streptomyces hygroscopicus* are reported to possess plant growth-promoting activity (Igarashi et al. 2002). This phytohormone is also able to promote plant growth by stimulating the production of various siderophores (Lacava and De Sousa 2016). Lin and Xu (2013) by examining the streptomycetes of *Taxus chinensis* and

Artemisia annua reported the endophytic strain *Streptomyces* sp. En-1, which was able to produce IAA. It could also colonize *Arabidopsis thaliana*, and they concluded that the production of IAA was a key trait.

5.7 Crop Yields

In general, EA can be used as biofertilizers or biocides (Shivlata and Satyanarayana 2017) due to their biosynthetic potential. Today, the use of biofertilizers and biocontrol agents is an indispensable approach to maintain soil and environmental quality that is ecologically sound as well as improving crop yields as a result of low or no toxic effects. EA with plant growth-promoting, antipathogenic and disease-resistance properties elicit the host to produce more crops with better quality produce (Goudjal et al. 2016). It has been documented the use of bacterial endophytes in the agricultural field may induce early flowering, seed setting and seed germination (Pahari et al. 2017). Seed transmission of endophytic Actinobacteria can enhance productivity, as well as protection against plant diseases (Suwitchayanon et al. 2018). Their application also reduces the use of pesticides and chemical fertilizers in agriculture and thus decreases environmental issues (Qin et al. 2017). Borah and Thakur (2020) have identified 46 endophytic Actinobacteria associated with *Camellia* sp. belonging to 9 genera. Subsequently, they reported that endophytes can affect plant growth by investigating the ability of these strains to produce plant growth-promoting factors such as siderophore and chitinase and their ability to solubilize environmental phosphorus. In addition, for in vivo PGP studies, the three strains *SA1*, *T1LA3* and *SA14* that exhibited the highest PGP activity were cultured along with three commercial strains of tea plant, and the results were compared to the control group. The results showed a significant increase in growth components of tea plants, indicating the positive effects of endophytic strain co-inoculation with commercial strains in the agricultural industry. *Vigna unguiculata* seed priming with *Serratia* sp. *ZoB14* was shown to have an increased growth in bacteria-treated plants compared to the control plants (Sabu et al. 2017). Table 5.2 summarizes the most recent studies on endophytic Actinobacteria to evaluate their plant growth-promoting effects.

Our data indicated that over 85 new plant growth-promoting bacteria (PGPB) species have been reported from different plants during the past two decades with different biological activities. Our data also revealed that the majority of plant growth-promoting bacteria (PGPB) are *Streptomyces*. To sum up, endophytic Actinobacteria are rich and very valuable resources for agricultural use; therefore, they can be used as an alternative to the conventional chemical fertilizers that cause salinity and other environmental stresses. Having in-depth knowledge on the interaction between the EA and the host plants is vital for the bioprospecting potential applications of them in agricultural, medicinal, pharmaceutical and cosmetics industries.

Table 5.2 Recently identified Actinobacteria and their PGP activities and bioactive products along with their host plants since 2000

Species	Host plant	PGP activities and secondary metabolites	Ref.
Actinoplanes campanulatus, Micromonospora chalcea, Streptomyces spiralis	*Cucumis sativus*	In vivo PGP activity, antagonistic activity	El-Tarabily et al. (2010)
Glutamicibacter	*Limonium sinense*	Salinity tolerance, ACCD and IAA production, P solubilizing	Qin et al. (2018)
Microbacterium	*Salix caprea*	Heavy metal resistance, ACCD, IAA and siderophore production	Kuffner et al. (2010)
Microbacterium	*Pteris vittata*	Heavy metal resistance, IAA and siderophore production, P solubilization	Xu et al. (2016)
Microbacterium	*Suaeda*	Salinity tolerance	Alishahi et al. (2020)
Microbispora rosea, Streptomyces olivochromogenes	Cabbage	Antagonistic activity	Lee et al. (2008)
Micromonospora	*Anastatica hierochuntica*	Keratinolytic activity	El-Bondkly and El-Gendy (2010)
Rhodococcus	*Lotus corniculatus, Oenothera biennis*	ACCD, IAA and siderophore production, P solubilizing, phytoremediation	Pawlik et al. (2017)
Streptomyces, Micromonospora	Herbaceous plants	Antagonistic activity, chitinase activity	Kim et al. (2012)
Streptomyces aureofaciens CMUAc130	*Zingiber officinale, Alpinia galanga*	Antagonistic activity	Taechowisan and Lumyong (2003)
Streptomyces aureofaciens CMUAc130	Herbaceous plants	Chitinase activity	Taechowisan et al. (2003a)
Streptomyces caeruleatus GIMN4	Leguminous plants	Antagonistic activity	Mingma et al. (2014)
Streptomyces chartreusis	Sugarcane	In vivo PGP activity, nitrogen fixation potential	Wang et al. (2017)
Streptomyces galbus	*Rhododendron*	Antagonistic activity	Shimizu (2007)
Streptomyces hygroscopicus OsiSh-2	*Oryza sativa*	In vivo PGP activity, antagonistic activity, chitinase activity	Xu et al. (2019)
Streptomyces lydicus DSM 40002	*Oryza sativa*	In vivo PGP activity, antagonistic activity, IAA and siderophore production, P solubilizing	Mingma and Duangmal (2018)

(continued)

Table 5.2 (continued)

Species	Host plant	PGP activities and secondary metabolites	Ref.
Streptomyces rochei ERY1	*Eryngium foetidum* L.	Antagonistic activity	Suwitchayanon et al. (2018)
Streptomyces	*Jatropha curcas*	In vivo PGP activity, ACCD activity, IAA and siderophore production, P solubilizing, nitrogen fixation potential	Qin et al. (2015)
Streptomyces	*Oryza sativa*	Antagonistic activity	Kampapongsa and Kaewkla (2016)
Streptomyces	*Glycyrrhiza* sp.	Antagonistic activity	Zhao et al. (2016)
Streptomyces	Herbaceous plants	Antagonistic activity, chitinase activity, gibberellic acid, IAA and siderophore production, P solubilizing	Singh and Gaur (2016)
Streptomyces	*Lycium ruthenicum* Murr	P solubilizing, cellulase and lipase activity, nitrogen fixation potential	Liu et al. (2019c)
Streptomyces	Herbaceous plants	Antagonistic activity	Chen et al. (2019)
Streptomyces	*Thymus roseus*	Antagonistic activity, siderophore production, chitinase activity	Musa et al. (2020)
Streptomyces	Herbaceous plants	Antagonistic activity	Taechowisan et al. (2003b)
Streptomyces	*Musa acuminata*	Antagonistic activity	Cao et al. (2004)
Streptomyces	Tomato plants	Antagonistic activity, siderophore production	Tan et al. (2006)
Streptomyces	*Azadirachta indica*	In vivo PGP activity, antagonistic activity, IAA and siderophore production, P solubilizing	Verma et al. (2011)
Streptomyces	Tomato plants	Antagonistic activity, ACCD and siderophore production	Tan et al. (2011)
Streptomyces	*Zea mays* L.	Antagonistic activity, chitinase activity	Costa et al. (2012)
Streptomyces	*Aerva tomentosa*	In vivo PGP activity, salinity and drought tolerance, production of IAA	Yandigeri et al. (2012)
Streptomyces	*Centella asiatica*	In vivo PGP activity, IAA	Dochhil et al. (2013)

(continued)

Table 5.2 (continued)

Species	Host plant	PGP activities and secondary metabolites	Ref.
Streptomyces	Herbaceous plants	IAA production	Goudjal et al. (2013)
Streptomyces	*Limonium sinense*	ACCD, IAA and siderophore production, in vivo PGP activity, nitrogen fixation potential	Qin et al. (2014)
Streptomyces	*Alisma orientale*	In vivo PGP activity, antagonistic activity	Li et al. (2014)
Streptomyces	Herbaceous plants	In vivo PGP activity, antagonistic activity, IAA and siderophore production, P solubilizing	Zamoum et al. (2015)
Streptomyces	*Solanum lycopersicum*	Antagonistic activity, IAA, cellulase and ACCD activity, P solubilizing	Passari et al. (2016)
Streptomyces	*Betula pendula* L., *Alnus glutinosa*	Siderophore production, heavy metal resistance	Złoch et al. (2016)
Streptomyces	Herbaceous plants	Antagonistic activity, IAA and siderophore production, chitinase activity	Liu et al. (2016)
Streptomyces	Sugarcane	In vivo PGP activity, nitrogen fixation potential, ACCD, IAA and siderophore production, P solubilizing	Kruasuwan and Thamchaipenet (2016)
Streptomyces	Pearl millet	Antagonistic activity	Jogaiah et al. (2016)
Streptomyces	Sorghum	In vivo PGP activity, antagonistic activity	Patel et al. (2018)
Streptomyces BPSAC34 and *Leifsonia xyli* BPSAC24	Herbaceous plants	Antagonistic activity, IAA, siderophore production, HCN production, chitinase activity	Passari et al. (2015)
Streptomyces En-1	*Taxus chinensis*, *Artemisia annua*	Production of IAA	Lin and Xu (2013)
Streptomyces GMKU	*Oryza sativa*	Siderophore production	Rungin et al. (2012)
Streptomyces GMKU 336	*Oryza sativa*	In vivo PGP activity, salinity tolerance, ACCD production	Jaemsaeng et al. (2018)
Streptomyces HAAG3-15	Cucumber	Antagonistic activity	Cao et al. (2020)

(continued)

Table 5.2 (continued)

Species	Host plant	PGP activities and secondary metabolites	Ref.
Streptomyces KLBMP	*Limonium sinense*	In vivo PGP activity, salinity tolerance, ACCD and IAA production	Qin et al. (2017)
Streptomyces MBCN152-1	Cabbage	Antagonistic activity	Hassan et al. (2017)
Streptomyces MBCu-56	*Cucumis sativus*, *Cucurbita moschata*	Antagonistic activity	Shimizu et al. (2009)
Streptomyces PRY2RB2, *Streptomyces* UKCW/B and *Nocardia* TP1BA1B	*Pseudowintera colorata*	In vivo PGP activity, antagonistic activity, siderophore production, P solubilizing	Purushotham et al. (2018)
Streptomyces SCAU5283 and SCAU5215	*Glycyrrhiza inflata*	Antagonistic activity, production of IAA and siderophore, P solubilizing, salinity tolerance, chitinase activity	Zhao et al. (2018)
Streptomyces S96	*Musa* sp.	Antagonistic activity	Cao et al. (2005)
Streptomyces pluricoloresce	Tomato plants	Antagonistic activity, IAA and siderophore production, P solubilizing,	Fialho de Oliveira et al. (2010)
Streptomyces	*Pinus thunbergii*	Antagonistic activity, chitinase, pectinase and cellulase activity	Kataoka and Futai (2011)
Streptomyces, *Micromonospora*, *Actinopolyspora*, *Saccharopolyspora*	*Aloe vera*, *Mentha arvensis*, *Ocimum sanctum*	Antagonistic activity, IAA and siderophore production, P solubilizing	Gangwar et al. (2014)
Streptomyces, *Nocardia*, *Nocardiopsis*, *Spirillospora*, *Microbispora*, *Micromonospora*	*Citrus reticulata*	In vivo PGP activity, IAA production	Shutsrirung et al. (2013)
Streptomyces, *Actinobacteria*, *Nonomuraea*, *Actinomadura*, *Pseudonocardia*, *Nocardia*	*Aquilaria crassna*	Production of IAA and siderophore, nitrogen fixation potential, protease activity	Nimnoi et al. (2010)
Streptomyces sp. HOU1	Vegetable plants	In vivo PGP activity, antagonistic activity	Kunasakdakul et al. (2012)

5.8 Conclusion and Future Perspectives

The ongoing development of sustainable practices while maintaining the improvement of crop yields and plant productivity is a major challenge for the agriculture industries. Plant-associated microbiomes and their bioproducts will play a significant role in achieving the required plant performance and productivity. Actinobacteria being the repository of natural bioactive compounds have shown that they play a pivotal role in many aspects of this beneficial relationship with their host plants. Secondary metabolites synthesized by EA play a crucial role in future biocontrol approaches of plant pathogens. EA are compatible alternatives to the agrochemicals and can also be effectively employed as biofertilizers, cost-effective bioinoculants for sustainable agriculture and chemical-free production in the near future.

Endophytic Actinobacteria, by regulating their metabolites (either elevating auxins, cytokines, gibberellins and antibiotics or reducing the ethylene levels), are able to increase the growth of host plants.

Endophytic Actinobacteria are proficient macronutrient solubilizers. In addition to siderophores, EA also assist plant growth by making other minerals such as nitrogen (Wang et al. 2017), phosphorus (Mingma and Duangmal 2018), potassium and zinc available since they form a mycelial network outside the plant. Further, studies need to be focused more on mineral solubilizing EA, as they are crucial for plant growth and productivities. Enzymes produced and excreted by endophytic Actinobacteria require to be explored more as they are valuable and useful at scientific and industrial levels.

There is a need to augment the current delivery systems for inoculants to improve host plant colonization as well as consider the timing of endophyte addition and the duration of colonization in the life cycle of the plant. Applying the EA as biofertilizer and biocontrol inoculants on a large scale in the field relies on proper formulations; thus, the development of appropriate formulations to deliver them to the field is future research.

Our current knowledge on the key processes elaborated in plant adaptations to abiotic stress conditions is still very limited. Therefore, there is a need to decipher and understand the mechanisms of plant resilience/tolerance/adaptation and mitigation tactics to abiotic stresses.

Even though considerable progress has been achieved in recent decades, implementation and application of EA to improve crop yields as biofertilizers still requires further research. Despite considerable research done on the positive impact of EA on their host plants, there is still limited information available on their mechanism of action which is required to optimize the inoculant efficacy and engineer the microbiome for improved functionalities.

The results of this study reveal that plant-associated endophytic Actinobacteria are more effective as biofertilizers to counter the negative effects of biotic and abiotic stressors but also offer the possibility of discovering bioactive products with potential for biocontrol activity. The effective application of endophytic Actinobacteria in the form of bioinoculants and commercial biofertilizers for all crops accompanied by a decline in the usage of agrochemicals promises a role for green technology in achieving the goal of sustainable agriculture.

Author Contributions Y.B. designed and planned the work. Y.D., K.R., E.K., Y.M. and Y.B., writing original draft preparation; Y.B. and C.F., writing review, revising and editing; S.B. and K. R., drawing chemical structures.

Conflicts of Interest The authors declare no conflict of interest.

References

Abdalla MA (2017) Three new cyclotetrapeptides isolated from *Streptomyces* sp. 447. Nat Prod Res 31(9):1014–1021. https://doi.org/10.1080/14786419.2016.1263849

Afridi MS, Amna S, Mahmood T, Salam A, Mukhtar T, Mehmood S, Ali J, Khatoon Z, Bibi M, Javed MT, Sultan T, Chaudhary HJ (2019) Induction of tolerance to salinity in wheat genotypes by plant growth promoting endophytes: involvement of ACC deaminase and antioxidant enzymes. Plant Physiol Biochem 139:569–577. https://doi.org/10.1016/j.plaphy.2019.03.041

Albelda-Berenguer M, Monachon M, Joseph E, Gadd GM, Sariaslani S (eds) (2019) Siderophores: from natural roles to potential applications. Adv Appl Microbiol 106:193–225

Alishahi F, Alikhani HA, Khoshkholgh-Sima NA, Etesami H (2020) Mining the roots of various species of the halophyte *Suaeda* for halotolerant nitrogen-fixing endophytic bacteria with the potential for promoting plant growth. Int Microbiol 23(3):415–427. https://doi.org/10.1007/s10123-019-00115-y

Ansari WA, Krishna R, Zeyad MT, Singh S, Yadav A (2020) Endophytic actinomycetes-mediated modulation of defense and systemic resistance confers host plant fitness under biotic stress conditions. In: Singh RP, Manchanda G, Maurya IK, Wei Y (eds) Microbial versatility in varied environments: microbes in sensitive environments. Springer, Singapore, pp 167–180

Araujo R, Dunlap C, Barnett S, Franco CMM (2019) Decoding wheat endosphere–rhizosphere microbiomes in *Rhizoctonia solani* infested soils challenged by *Streptomyces* biocontrol agents. Front Plant Sci 10:1–13. https://doi.org/10.3389/fpls.2019.01038

Arif I, Batool M, Schenk PM (2020) Plant microbiome engineering: expected benefits for improved crop growth and resilience. Trends Biotechnol 38(12):1385–1396

Arshad M, Saleem M, Hussain S (2007) Perspectives of bacterial ACC deaminase in phytoremediation. Trends Biotechnol 25(8):356–362. https://doi.org/10.1016/j.tibtech.2007.05.005

Ayuso-Sacido A, Genilloud O (2005) New PCR primers for the screening of NRPS and PKS-I systems in actinomycetes: detection and distribution of these biosynthetic gene sequences in major taxonomic groups. Microb Ecol 49(1):10–24

Barka EA, Vatsa P, Sanchez L, Gaveau-Vaillant N, Jacquard C, Klenk H-P, Clément C, Ouhdouch Y, van Wezel GP (2016) Taxonomy, physiology, and natural products of actinobacteria. Microbiol Mol Biol Rev 80(1):1–43

Barnett SJ, Ballard RA, Franco CMM (2019) Field assessment of microbial inoculants to control Rhizoctonia root rot on wheat. Biol Control 132:152–160. https://doi.org/10.1016/j.biocontrol.2019.02.019

Boonsnongcheep P, Nakashima T, Takahashi Y, Prathanturarug S (2017) Diversity of endophytic actinomycetes isolated from roots and root nodules of *Pueraria candollei* Grah. ex Benth. and the analyses of their secondary metabolites. Chiang Mai J Sci 44:1–14

Borah A, Thakur D (2020) Phylogenetic and functional characterization of culturable endophytic actinobacteria associated with *Camellia* spp. for growth promotion in commercial tea cultivars. Front Microbiol 11:318. https://doi.org/10.3389/fmicb.2020.00318

Boubekri K, Soumare A, Mardad I, Lyamlouli K, Hafidi M, Ouhdouch Y, Kouisni L (2021) The Screening of potassium-and phosphate-solubilizing actinobacteria and the assessment of their ability to promote wheat growth parameters. Microorganisms 9(3):470

Cao L, Qiu Z, Dai X, Tan H, Lin Y, Zhou S (2004) Isolation of endophytic actinomycetes from roots and leaves of banana (*Musa acuminata*) plants and their activities against *Fusarium oxysporum* f. sp. cubense. World J Microbiol Biotechnol 20(5):501–504. https://doi.org/10.1023/B:WIBI.0000040406.30495.48

Cao L, Qiu Z, You J, Tan H, Zhou S (2005) Isolation and characterization of endophytic streptomycete antagonists of *Fusarium* wilt pathogen from surface-sterilized banana roots. FEMS Microbiol Lett 247(2):147–152. https://doi.org/10.1016/j.femsle.2005.05.006

Cao P, Li C, Wang H, Yu Z, Xu X, Wang X, Zhao J, Xiang W (2020) Community structures and antifungal activity of root-associated endophytic actinobacteria in healthy and diseased cucumber plants and *Streptomyces* sp. HAAG3-15 as a promising biocontrol agent. Microorganisms 8(2):236. https://doi.org/10.3390/microorganisms8020236

Chen P, Zhang C, Xiong Y, Xing K, Qin S (2019) Community composition and metabolic potential of endophytic actinobacteria from coastal salt marsh plants in Jiangsu, China. Front Microbiol 10:1063

Compant S, Samad A, Faist H, Sessitsch A (2019) A review on the plant microbiome: ecology, functions, and emerging trends in microbial application. J Adv Res 19:29–37. https://doi.org/10.1016/j.jare.2019.03.004

Conn VM, Walker AR, Franco CMM (2008) Endophytic actinobacteria induce defense pathways in *Arabidopsis thaliana*. Mol Plant-Microbe Interact 21(2):208–218. https://doi.org/10.1094/mpmi-21-2-0208

Coombs JT, Michelsen PP, Franco CMM (2004) Evaluation of endophytic actinobacteria as antagonists of *Gaeumannomyces graminis* var. *tritici* in wheat. Biol Control 29(3):359–366. https://doi.org/10.1016/j.biocontrol.2003.08.001

Costa LE d O, de Queiroz MV, Borges ACC, de Moraes CA, de Araújo EF (2012) Isolation and characterization of endophytic bacteria isolated from the leaves of the common bean *(Phaseolus vulgaris)*. Braz J Microbiol 43(4):1562–1575. https://doi.org/10.1590/S1517-83822012000400041

Daulagala PWHKP (2021) Chitinolytic endophytic bacteria as biocontrol agents for phytopathogenic fungi and nematode pests: a review. Asian J Res Bot 5(3):14–24

Delbari Y, Mohassel Y, Bahrami Y, Kakaie E, Mostafaie A (2020) A review on isolation and identification of endophytic actinobacteria, their chemical structure, bioactive compounds, and potential medical-pharmaceutical applications. J Mazandaran Univ Med Sci 30(186):195–217

Devi TS, Vijay K, Vidhyavathi RM, Kumar P, Giovarthanan M, Kavitha T (2021) Antifungal activity and molecular docking of phenol, 2,4-bis(1,1-dimethylethyl) produced by plant growth-promoting actinobacterium kutzneria sp. Strain tsii from mangrove sediments. Arch Microbiol 203(14). https://doi.org/10.1007/s00203-021-02397-1

Ding W-J, Zhang S-Q, Wang J-H, Lin Y-X, Liang Q-X, Zhao W-J, Li C-Y (2013) A new di-O-prenylated flavone from an actinomycete *Streptomyces* sp. MA-12. J Asian Nat Prod Res 15(2): 209–214

Dochhil H, Dkhar MS, Barman DINA (2013) Seed germination enhancing activity of endophytic *Streptomyces* isolated from indigenous ethno-medicinal plant *Centella asiatica*. Int J Pharm Bio Sci 4(1):B256–B262

dos Santos LF, Olivares FL (2021) Plant microbiome structure and benefits for sustainable agriculture. Curr Plant Biol 26:100198

El-Bondkly AM, El-Gendy MMA (2010) Keratinolytic activity from new recombinant fusant AYA2000, derived from endophytic *Micromonospora* strains. Can J Microbiol 56(9):748–760

Elferink M, Schierhorn F (2016) Global demand for food is rising. Can we meet it? Harv Bus Rev. April 07, 2016. https://store.hbr.org/search.php?search_query=Florian%20Schierhorn&section=product

El-Tarabily KA (2003) An endophytic chitinase-producing isolate of *Actinoplanes missouriensis*, with potential for biological control of root rot of lupin caused by *Plectosporium tabacinum*. Aust J Bot 51(3):257–266. https://doi.org/10.1071/BT02107

El-Tarabily KA (2006) Rhizosphere-competent isolates of streptomycete and non-streptomycete actinomycetes capable of producing cell-wall-degrading enzymes to control *Pythium aphanidermatum* damping-off disease of cucumber. Botany 84(2):211–222

El-Tarabily KA, Hardy GESJ, Sivasithamparam K (2010) Performance of three endophytic actinomycetes in relation to plant growth promotion and biological control of *Pythium aphanidermatum*, a pathogen of cucumber under commercial field production conditions in the United Arab Emirates. Eur J Plant Pathol 128(4):527–539

Fialho de Oliveira M, da Silva MG, Van Der Sand ST (2010) Anti-phytopathogen potential of endophytic actinobacteria isolated from tomato plants (*Lycopersicon esculentum*) in southern Brazil, and characterization of *Streptomyces* sp. R18(6), a potential biocontrol agent. Res Microbiol 161(7):565–572. https://doi.org/10.1016/j.resmic.2010.05.008

Franco C (2021) Inoculation effects in the rhizosphere: diversity and function. In: Vadakattu G, Sharma A (eds) Rhizosphere biology: interactions between microbes and plants. Springer, pp 339–356

Frank AC, Guzmán JPS, Shay JE (2017) Transmission of bacterial endophytes. Microorganisms 5(4):70. https://doi.org/10.3390/microorganisms5040070

Gangwar M, Dogra S, Gupta UP, Kharwar RN (2014) Diversity and biopotential of endophytic actinomycetes from three medicinal plants in India. Afr J Microbiol Res 8(2):184–191

Glick BR (2014) Bacteria with ACC deaminase can promote plant growth and help to feed the world. Microbiol Res 169(1):30–39. https://doi.org/10.1016/j.micres.2013.09.009

Gomes E d B, Dias LRL, de Cassia MR (2018) Actinomycetes bioactive compounds: biological control of fungi and phytopathogenic insect. Afr J Biotechnol 17(17):552–559

Gopalakrishnan S, Sharma R, Srinivas V, Naresh N, Mishra SP, Ankati S, Pratyusha S, Govindaraj M, Gonzalez SV, Nervik S, Simic N (2020) Identification and characterization of a *Streptomyces* albus strain and its secondary metabolite organophosphate against charcoal rot of sorghum. Plants 9(12):1727. https://doi.org/10.3390/plants9121727

Goudjal Y, Toumatia O, Sabaou N, Barakate M, Mathieu F, Zitouni A (2013) Endophytic actinomycetes from spontaneous plants of Algerian Sahara: indole-3-acetic acid production and tomato plants growth promoting activity. World J Microbiol Biotechnol 29(10):1821–1829. https://doi.org/10.1007/s11274-013-1344-y

Goudjal Y, Zamoum M, Sabaou N, Mathieu F, Zitouni A (2016) Potential of endophytic *Streptomyces* spp. for biocontrol of Fusarium root rot disease and growth promotion of tomato seedlings. Biocontrol Sci Technol 26(12):1691–1705. https://doi.org/10.1080/09583157.2016.1234584

Goudjal Y, Zamoum M, Sabaou N, Zitouni A (2018) Chapter 7—Endophytic actinobacteria from native plants of Algerian Sahara: potential agents for biocontrol and promotion of plant growth. In: Singh BP, Gupta VK, Passari AK (eds) New and future developments in microbial biotechnology and bioengineering. Elsevier, pp 109–124

Govindasamy V, George P, Ve Ramesh S, Sureshkumar P, Rane J, Minhas P (2021) Characterization of root-endophytic actinobacteria from cactus (*Opuntia ficus-indica*) for plant growth promoting traits. Res Square. https://doi.org/10.21203/rs.3.rs-482515/v1

Gu Y, Ma Y-N, Wang J, Xia Z, Wei H-L (2020) Genomic insights into a plant growth-promoting pseudomonas koreensis strain with cyclic lipopeptide-mediated antifungal activity. MicrobiologyOpen 9:e1092. https://doi.org/10.1002/mbo3.1092

Hassan N, Nakasuji S, Elsharkawy MM, Naznin HA, Kubota M, Ketta H, Shimizu M (2017) Biocontrol potential of an endophytic *Streptomyces* sp. Strain MBCN152-1 against *Alternaria brassicicola* on Cabbage Plug seedlings. Microbes Environ 32(2):133–141. https://doi.org/10.1264/jsme2.ME17014

Igarashi Y (2004) Screening of novel bioactive compounds from plant-associated actinomycetes. Actinomycetologica 18(2):63–66

Igarashi Y, Ogawa M, Sato Y, Saito N, Yoshida R, Kunoh H, Onaka H, Furumai T (2000) Fistupyrone, a novel inhibitor of the infection of Chinese cabbage by *Alternaria brassicicola*, from *Streptomyces* sp. TP-A0569. J Antibiot 53(10):1117–1122

Igarashi Y, Iida T, Sasaki T, Saito N, Yoshida R, Furumai T (2002) Isolation of actinomycetes from live plants and evaluation of antiphytopathogenic activity of their metabolites. Actinomycetologica 16(1):9–13

Jaemsaeng R, Jantasuriyarat C, Thamchaipenet A (2018) Molecular interaction of 1-aminocyclopropane-1-carboxylate deaminase (ACCD)-producing endophytic *Streptomyces* sp. GMKU 336 towards salt-stress resistance of *Oryza sativa* L. cv. KDML105. Sci Rep 8(1): 1950. https://doi.org/10.1038/s41598-018-19799-9

Jogaiah S, Kurjogi M, Govind SR, Huntrike SS, Basappa VA, Tran L-SP (2016) Isolation and evaluation of proteolytic actinomycete isolates as novel inducers of pearl millet downy mildew disease protection. Sci Rep 6:30789. https://doi.org/10.1038/srep30789

Kaewkla O, Franco CMM (2013) Rational approaches to improving the isolation of endophytic actinobacteria from Australian native trees. Microb Ecol 65(2):384–393. https://doi.org/10.1007/s00248-012-0113-z

Kampapongsa D, Kaewkla O (2016) Biodiversity of endophytic actinobacteria from jasmine rice (*Oryza sativa* L. KDML 105) grown in Roi-Et Province, Thailand and their antimicrobial activity against rice pathogens. Ann Microbiol 66(2):587–595. https://doi.org/10.1007/s13213-015-1140-z

Kannahi M, Senbagam N (2014) Studies on siderophore production by microbial isolates obtained from rhizosphere soil and its antibacterial activity. J Chem Pharm Res 6:1142–1145

Kataoka R, Futai K (2011) Endophytic actinomycetes from *Pinus thunbergii* and their antifungal activity against *Cylindrocladium* sp. Arch Phytopathol Plant Protect 44(19):1852–1861. https://doi.org/10.1080/03235400903308990

Kaur T, Sharma D, Kaur A, Manhas RK (2013) Antagonistic and plant growth promoting activities of endophytic and soil actinomycetes. Arch Phytopathol Plant Protect 46(14):1756–1768. https://doi.org/10.1080/03235408.2013.777169

Kim T-U, Cho S-H, Han J-H, Shin YM, Lee HB, Kim SB (2012) Diversity and physiological properties of root endophytic actinobacteria in native herbaceous plants of Korea. J Microbiol 50(1):50–57. https://doi.org/10.1007/s12275-012-1417-x

Kõiv V, Roosaare M, Vedler E, Kivistik PA, Toppi K, Schryer DW, Remm M, Tenson T, Mäe A (2015) Microbial population dynamics in response to *Pectobacterium atrosepticum* infection in potato tubers. Sci Rep 5(1):1–18

Kour D, Rana KL, Yadav AN, Yadav N, Kumar M, Kumar V, Vyas P, Dhaliwal HS, Saxena AK (2020) Microbial biofertilizers: bioresources and eco-friendly technologies for agricultural and environmental sustainability. Biocatal Agric Biotechnol 23:101487

Kruasuwan W, Thamchaipenet A (2016) Diversity of culturable plant growth-promoting bacterial endophytes associated with sugarcane roots and their effect of growth by co-inoculation of diazotrophs and actinomycetes. J Plant Growth Regul 35(4):1074–1087. https://doi.org/10.1007/s00344-016-9604-3

Kuffner M, De Maria S, Puschenreiter M, Fallmann K, Wieshammer G, Gorfer M, Strauss J, Rivelli AR, Sessitsch A (2010) Culturable bacteria from Zn- and Cd-accumulating *Salix caprea* with differential effects on plant growth and heavy metal availability. J Appl Microbiol 108(4): 1471–1484. https://doi.org/10.1111/j.1365-2672.2010.04670.x

Kumar V, Kumar P (2019) Pesticides in agriculture and environment: impacts on human health. In: Contaminants in agriculture and environment: health risks and remediation, vol 1. pp 76–95

Kumar A, Droby S, Singh VK, Singh SK, White JF (2020) Chapter 1—Entry, colonization, and distribution of endophytic microorganisms in plants. In: Kumar A, Radhakrishnan EK (eds) Microbial endophytes. Woodhead Publishing, pp 1–33

Kunasakdakul K, Suwitchayanon P, Phuakjaiphaoe C (2012) Antifungal pathogen activities and growth promotion of endophytic actinomycetes on brassica seedling. Chiang Mai Univ J Nat Sci 11(1 Special Issue):7–12

Lacava PT, de Sousa CP (2016) Role of endophytic actinomycetes in crop protection: plant growth promotion and biological control. In: Subramaniam G, Arumugam S, Rajendran V (eds) Plant growth promoting Actinobacteria: a new avenue for enhancing the productivity and soil fertility of grain legumes. Springer, Singapore, pp 147–160

Le XH, Franco CMM, Ballard RA, Drew EA (2016) Isolation and characterisation of endophytic actinobacteria and their effect on the early growth and nodulation of lucerne (*Medicago sativa* L.). Plant Soil 405(1–2):13–24. https://doi.org/10.1007/s11104-015-2652-9

Lee SO, Choi GJ, Choi YH, Jong KS, Park DJ, Kim CJ, Kim J-C (2008) Isolation and characterization of endophytic actinomycetes from Chinese cabbage roots as antagonists to *Plasmodiophora brassicae*. J Microbiol Biotechnol 18(11):1741–1746. https://doi.org/10.4014/jmb.0800.108

Li X, Huang P, Wang Q, Xiao L, Liu M, Bolla K, Zhang B, Zheng L, Gan B, Liu X, Zhang L, Zhang X (2014) Staurosporine from the endophytic *Streptomyces* sp. strain CNS-42 acts as a potential biocontrol agent and growth elicitor in cucumber. Antonie Van Leeuwenhoek 106(3): 515–525. https://doi.org/10.1007/s10482-014-0220-6

Lin L, Xu X (2013) Indole-3-acetic acid production by endophytic *Streptomyces* sp. En-1 isolated from medicinal plants. Curr Microbiol 67(2):209–217. https://doi.org/10.1007/s00284-013-0348-z

Liu X, Dou G, Ma Y (2016) Potential of endophytes from medicinal plants for biocontrol and plant growth promotion. J Gen Plant Pathol 82(3):165–173. https://doi.org/10.1007/s10327-016-0648-9

Liu C, Zhuang X, Yu Z, Wang Z, Wang Y, Guo X, Xiang W, Huang S (2019a) Community structures and antifungal activity of root-associated endophytic actinobacteria of healthy and diseased soybean. Microorganisms 7(8):243

Liu D, Yan R, Fu Y, Wang X, Zhang J, Xiang W (2019b) Antifungal, plant growth-promoting, and genomic properties of an endophytic actinobacterium *Streptomyces* sp. NEAU-S7GS2. Front Microbiol 10:2077. https://doi.org/10.3389/fmicb.2019.02077

Liu Y-H, Wei Y-Y, Mohamad OAA, Salam N, Zhang Y-G, Guo J-W, Li L, Egamberdieva D, Li W-J (2019c) Diversity, community distribution and growth promotion activities of endophytes associated with halophyte *Lycium ruthenicum* Murr. 3 Biotech 9(4):144. https://doi.org/10.1007/s13205-019-1678-8

Ludwig-Müller J (2015) Plants and endophytes: equal partners in secondary metabolite production? Biotechnol Lett 37(7):1325–1334. https://doi.org/10.1007/s10529-015-1814-4

Mącik M, Gryta A, Frąc M (2020) Biofertilizers in agriculture: an overview on concepts, strategies and effects on soil microorganisms. Adv Agron 162:31–87

Marian M, Ohno T, Suzuki H, Kitamura H, Kuroda K, Shimizu M (2020) A novel strain of endophytic *Streptomyces* for the biocontrol of strawberry anthracnose caused by *Glomerella cingulata*. Microbiol Res 234:126428

Matiru VN, Dakora FD (2004) Potential use of rhizobial bacteria as promoters of plant growth for increased yield in landraces of African cereal crops. Afr J Biotechnol 3(1):1–7

Mesta SC, Onkarappa R, Meghana SH, Manu HM, Kavana S, Talib M (2018) Phosphate solubilizing endophytic actinomycetes from mangrove plants *Rhizophora mucronata* and *Sonneratia caseolaris* and its effect on Seedling vigour. Res J Pharm Technol 11(3):1172–1178

Mingma R, Duangmal K (2018) Characterization, antifungal activity and plant growth promoting potential of endophytic actinomycetes isolated from rice (*Oryza sativa* L.). Chiang Mai J Sci 45(7):2652–2665

Mingma R, Pathom-aree W, Trakulnaleamsai S, Thamchaipenet A, Duangmal K (2014) Isolation of rhizospheric and roots endophytic actinomycetes from Leguminosae plant and their activities to inhibit soybean pathogen, *Xanthomonas campestris* pv. glycine. World J Microbiol Biotechnol 30(1):271–280. https://doi.org/10.1007/s11274-013-1451-9

Mitra D, Anđelković S, Panneerselvam P, Senapati A, Vasić T, Ganeshamurthy AN, Chauhan M, Uniyal N, Mahakur B, Radha TK (2020) Phosphate-solubilizing microbes and biocontrol agent for plant nutrition and protection: current perspective. Commun Soil Sci Plant Anal 51(5): 645–657. https://doi.org/10.1080/00103624.2020.1729379

Musa Z, Ma J, Egamberdieva D, Mohamad OAA, Abaydulla G, Liu Y, Li W-J, Li L (2020) Diversity and antimicrobial potential of cultivable endophytic actinobacteria associated with the medicinal plant *Thymus roseus*. Front Microbiol 11:191

Nakouti I, Hobbs G (2013) A new approach to studying ion uptake by actinomycetes. J Basic Microbiol 53(11):913–916. https://doi.org/10.1002/jobm.201200407

Nimnoi P, Pongsilp N, Lumyong S (2010) Endophytic actinomycetes isolated from *Aquilaria crassna* Pierre ex Lec and screening of plant growth promoters production. World J Microbiol Biotechnol 26(2):193–203. https://doi.org/10.1007/s11274-009-0159-3

Orozco-Mosqueda M d C, Glick BR, Santoyo G (2020) ACC deaminase in plant growth-promoting bacteria (PGPB): an efficient mechanism to counter salt stress in crops. Microbiol Res 235: 126439. https://doi.org/10.1016/j.micres.2020.126439

Pahari A, Pradhan A, Nayak SK, Mishra BB (2017) Bacterial siderophore as a plant growth promoter. In: Patra JK, Vishnuprasad CN, Das G (eds) Microbial biotechnology: volume 1. Applications in agriculture and environment. Springer SingaporeBook Section, Singapore, pp 163–180. 978-981-10-6847-8

Passari AK, Mishra VK, Gupta VK, Yadav MK, Saikia R, Singh BP (2015) In vitro and in vivo plant growth promoting activities and DNA fingerprinting of antagonistic endophytic actinomycetes associates with medicinal plants. PLoS One 10(9):e0139468. https://doi.org/10.1371/journal.pone.0139468

Passari AK, Chandra P, Zothanpuia, Mishra VK, Leo VV, Gupta VK, Kumar B, Singh BP (2016) Detection of biosynthetic gene and phytohormone production by endophytic actinobacteria associated with *Solanum lycopersicum* and their plant-growth-promoting effect. Res Microbiol 167(8):692–705. https://doi.org/10.1016/j.resmic.2016.07.001

Patel JK, Madaan S, Archana G (2018) Antibiotic producing endophytic *Streptomyces* spp. colonize above-ground plant parts and promote shoot growth in multiple healthy and pathogen-challenged cereal crops. Microbiol Res 215:36–45. https://doi.org/10.1016/j.micres.2018.06.003

Pawlik M, Cania B, Thijs S, Vangronsveld J, Piotrowska-Seget Z (2017) Hydrocarbon degradation potential and plant growth-promoting activity of culturable endophytic bacteria of *Lotus corniculatus* and *Oenothera biennis* from a long-term polluted site. Environ Sci Pollut Res 24(24):19640–19652. https://doi.org/10.1007/s11356-017-9496-1

Peng C, An D, Ding W-X, Zhu Y-X, Ye L, Li J (2020) Fungichromin production by *Streptomyces* sp. WP-1, an endophyte from *Pinus dabeshanensis*, and its antifungal activity against *Fusarium oxysporum*. Appl Microbiol Biotechnol 104(24):10437–10449. https://doi.org/10.1007/s00253-020-10996-z

Pillay VK, Nowak J (1997) Inoculum density, temperature, and genotype effects on in vitro growth promotion and epiphytic and endophytic colonization of tomato (*Lycopersicon esculentum* L.) seedlings inoculated with a pseudomonad bacterium. Can J Microbiol 43(4):354–361

Prasad M, Srinivasan R, Chaudhary M, Mahawer SK, Jat LK (2020) Chapter 3—Endophytic bacteria: role in sustainable agriculture. In: Kumar A, Singh VK (eds) Microbial endophytes. Woodhead Publishing, pp 37–60

Purushotham N, Jones E, Monk J, Ridgway H (2018) Community structure of endophytic actinobacteria in a New Zealand native medicinal plant *Pseudowintera colorata* (Horopito) and their influence on plant growth. Microb Ecol 76(3):729–740. https://doi.org/10.1007/s00248-018-1153-9

Qin S, Zhang Y-J, Yuan B, Xu P-Y, Xing K, Wang J, Jiang J-H (2014) Isolation of ACC deaminase-producing habitat-adapted symbiotic bacteria associated with halophyte *Limonium sinense* (Girard) Kuntze and evaluating their plant growth-promoting activity under salt stress. Plant Soil 374(1–2):753–766. https://doi.org/10.1007/s11104-013-1918-3

Qin S, Miao Q, Feng W-W, Wang Y, Zhu X, Xing K, Jiang J-H (2015) Biodiversity and plant growth promoting traits of culturable endophytic actinobacteria associated with *Jatropha curcas*

L. growing in Panxi dry-hot valley soil. Appl Soil Ecol 93:47–55. https://doi.org/10.1016/j.apsoil.2015.04.004

Qin S, Feng W-W, Wang T-T, Ding P, Xing K, Jiang J-H (2017) Plant growth-promoting effect and genomic analysis of the beneficial endophyte *Streptomyces* sp. KLBMP 5084 isolated from halophyte *Limonium sinense*. Plant Soil 416(1):117–132

Qin S, Feng W-W, Zhang Y-J, Wang T-T, Xiong Y-W, Xing K, Cann I (2018) Diversity of bacterial microbiota of coastal halophyte *Limonium sinense* and amelioration of salinity stress damage by symbiotic plant growth-promoting actinobacterium *Glutamicibacter halophytocola* KLBMP 5180. Appl Environ Microbiol 84(19):e01533–e01518. https://doi.org/10.1128/AEM.01533-18

Quecine M, Araújo W, Marcon J, Gai C, Azevedo J, Pizzirani-Kleiner AA (2008) Chitinolytic activity of endophytic *Streptomyces* and potential for biocontrol. Lett Appl Microbiol 47:486–491. https://doi.org/10.1111/j.1472-765X.2008.02428.x

Rana KL, Kour D, Kaur T, Devi R, Yadav AN, Yadav N, Dhaliwal HS, Saxena AK (2020a) Endophytic microbes: biodiversity, plant growth-promoting mechanisms and potential applications for agricultural sustainability. Antonie Van Leeuwenhoek 113(8):1075–1107. https://doi.org/10.1007/s10482-020-01429-y

Rana KL, Kour D, Yadav AN, Yadav N, Saxena AK (2020b) Chapter 16—Agriculturally important microbial biofilms: biodiversity, ecological significances, and biotechnological applications. In: Yadav MK, Singh BP (eds) New and future developments in microbial biotechnology and bioengineering: microbial biofilms. Elsevier, pp 221–265

Raymaekers K, Ponet L, Holtappels D, Berckmans B, Cammue BPA (2020) Screening for novel biocontrol agents applicable in plant disease management—a review. Biol Control 144:104–240. https://doi.org/10.1016/j.biocontrol.2020.104240

Robison MM, Griffith M, Pauls KP, Glick BR (2001) Dual role for ethylene in susceptibility of tomato to *Verticillium* wilt. J Phytopathol 149(7–8):385–388. https://doi.org/10.1111/j.1439-0434.2001.tb03867.x

Rungin S, Indananda C, Suttiviriya P, Kruasuwan W, Jaemsaeng R, Thamchaipenet A (2012) Plant growth enhancing effects by a siderophore-producing endophytic streptomycete isolated from a Thai jasmine rice plant (*Oryza sativa* L. cv. KDML105). Antonie Van Leeuwenhoek 102(3): 463–472. https://doi.org/10.1007/s10482-012-9778-z

Sabu R, Aswani R, Jishma P, Jasim B, Mathew J, Radhakrishnan EK (2017) Plant growth promoting endophytic *Serratia* sp. ZoB14 protecting ginger from fungal pathogens. Proc Natl Acad Sci India Sect B Biol Sci 89(1):213–220

Savi DC, Shaaban KA, Vargas N, Ponomareva LV, Possiede YM, Thorson JS, Glienke C, Rohr J (2015) *Microbispora* sp. LGMB259 endophytic actinomycete isolated from Vochysia divergens (*Pantanal, Brazil*) producing β-carbolines and indoles with biological activity. Curr Microbiol 70(3):345–354. https://doi.org/10.1007/s00284-014-0724-3

Savi DC, Shaaban KA, Gos FMW, Thorson JS, Glienke C, Rohr J (2019) Secondary metabolites produced by Microbacterium sp. LGMB471 with antifungal activity against the phytopathogen *Phyllosticta citricarpa*. Folia Microbiol 64(3):453–460

Sharma S, Rana VS, Kumari M, Mishra P (2018) Biofertilizers: boon for fruit production. Pharmacogn Phytochem 7(5):3244–3247

Shimizu M (2007) Cultivation of disease-resistant, tissue-cultured seedlings using endophytic actinomycetes. J Gen Plant Pathol 73(6):426–427. https://doi.org/10.1007/s10327-007-0051-7

Shimizu M, Yazawa S, Ushijima Y (2009) A promising strain of endophytic *Streptomyces* sp. for biological control of cucumber anthracnose. J Gen Plant Pathol 75(1):27–36. https://doi.org/10.1007/s10327-008-0138-9

Shivlata L, Satyanarayana T (2017) Actinobacteria in agricultural and environmental sustainability. In: Singh JS, Seneviratne G (eds) Agro-environmental sustainability: vol 1: managing crop health. Springer International Publishing, Cham, pp 173–218

Shutsrirung A, Chromkaew Y, Pathom-Aree W, Choonluchanon S, Boonkerd N (2013) Diversity of endophytic actinomycetes in mandarin grown in northern Thailand, their phytohormone production potential and plant growth promoting activity. Soil Sci Plant Nutr 59(3):322–330. https://doi.org/10.1080/00380768.2013.776935

Singh SP, Gaur R (2016) Evaluation of antagonistic and plant growth promoting activities of chitinolytic endophytic actinomycetes associated with medicinal plants against sclerotium rolfsii in chickpea. J Appl Microbiol 121:506–518. https://doi.org/10.1111/jam.13176

Singh S, Kumar V, Dhanjal DS, Sidhu GK, Datta S, Kumar S, Singh J (2020) Chapter 5—Endophytic microbes in abiotic stress management. In: Kumar A, Singh VK (eds) Microbial endophytes. Woodhead Publishing, pp 91–123

Siripan O, Thamchaipenet A, Surat W (2018) Enhancement of the efficiency of Cd phytoextraction using bacterial endophytes isolated from *Chromolaena odorata*, a Cd hyperaccumulator. Int J Phytoremediation 20(11):1096–1105

Soumare A, Boubekri K, Lyamlouli K, Hafidi M, Ouhdouch Y, Kouisni L (2020) From isolation of phosphate solubilizing microbes to their formulation and use as biofertilizers: status and needs. Front Bioeng Biotechnol 7:425. https://doi.org/10.3389/fbioe.2019.00425

Sun X, Wang G, Xiao H, Jiang J, Xiao D, Xing B, Li A, Zhang Y, Sun K, Xu Y, Guo L, Yang D, Ma M (2020) Strepimidazoles a-g from the plant endophytic *Streptomyces* sp. PKU-EA00015 with inhibitory activities against a plant pathogenic fungus. J Nat Prod 83(7):2246–2254. https://doi.org/10.1021/acs.jnatprod.0c00362

Suwitchayanon P, Chaipon S, Chaichom S, Kunasakdakul K (2018) Potentials of *Streptomyces* rochei ERY1 as an endophytic actinobacterium inhibiting damping-off pathogenic fungi and growth promoting of cabbage seedling. Chiang Mai J Sci 45(2):692–700

Swarnalakshmi K, Senthilkumar M, Ramakrishnan B (2016) Endophytic actinobacteria: nitrogen fixation, phytohormone production, and antibiosis. In: Subramaniam G, Arumugam S, Rajendran V (eds) Plant growth promoting Actinobacteria: a new avenue for enhancing the productivity and soil fertility of grain legumes. Springer, Singapore, pp 123–145

Taechowisan T, Lumyong S (2003) Activity of endophytic actinomycetes from roots of *Zingiber officinale* and *Alpinia galanga* against phytopathogenic fungi. Ann Microbiol 53(3):291–298

Taechowisan T, Peberdy JF, Lumyong S (2003a) Chitinase production by endophytic *Streptomyces aureofaciens* CMUAc130 and its antagonism against phytopathogenic fungi. Ann Microbiol 53(4):447–461

Taechowisan T, Peberdy JF, Lumyong S (2003b) Isolation of endophytic actinomycetes from selected plants and their antifungal activity. World J Microbiol Biotechnol 19(4):381–385. https://doi.org/10.1023/A:1023901107182

Tan HM, Cao LX, He ZF, Su G, Lin B, Zhou SN (2006) Isolation of endophytic actinomycetes from different cultivars of tomato and their activities against *Ralstonia solanacearum* in vitro. World J Microbiol Biotechnol 22(12):1275–1280. https://doi.org/10.1007/s11274-006-9172-y

Tan H, Zhou S, Deng Z, He M, Cao L (2011) Ribosomal-sequence-directed selection for endophytic streptomycete strains antagonistic to *Ralstonia solanacearum* to control tomato bacterial wilt. Biol Control 59(2):245–254. https://doi.org/10.1016/j.biocontrol.2011.07.018

Tewari S, Shrivas VL, Hariprasad P, Sharma S (2019) Harnessing endophytes as biocontrol agents. In: Ansari RA, Mahmood I (eds) Plant health under biotic stress: vol 2: microbial interactions. Springer, Singapore, pp 189–218

Tian L, Lin X, Tian J, Ji L, Chen Y, Tran L-SP, Tian C (2020) Research advances of beneficial microbiota associated with crop plants. Int J Mol Sci 21(5):1792

Trujillo ME, Bacigalupe R, Pujic P, Igarashi Y, Benito P, Riesco R, Médigue C, Normand P (2014) Genome features of the endophytic actinobacterium *Micromonospora lupini* strain Lupac 08: on the process of adaptation to an endophytic life style? PLoS One 9(9):e108522. https://doi.org/10.1371/journal.pone.0108522

Van Nguyen T, Pawlowski K (2017) *Frankia* and actinorhizal plants: symbiotic nitrogen fixation. In: Mehnaz S (ed) Rhizotrophs: plant growth promotion to bioremediation. Springer, Singapore, pp 237–261

Verma VC, Singh SK, Prakash S (2011) Bio-control and plant growth promotion potential of siderophore producing endophytic *Streptomyces* from *Azadirachta indica* A. Juss. J Basic Microbiol 51(5):550–556. https://doi.org/10.1002/jobm.201000155

Verma P, Yadav AN, Kumar V, Singh DP, Saxena AK (2017) Beneficial plant-microbes interactions: biodiversity of microbes from diverse extreme environments and its impact for crop improvement. In: Singh DP, Singh HB, Prabha R (eds) Plant-microbe interactions in agro-ecological perspectives: volume 2: microbial interactions and agro-ecological impacts. Springer, Singapore, pp 543–580

Vurukonda SSKP, Giovanardi D, Stefani E (2018) Plant growth promoting and biocontrol activity of *Streptomyces* spp. as endophytes. Int J Mol Sci 19(4):952

Wang Z, Solanki MK, Pang F, Singh RK, Yang L-T, Li Y-R, Li H-B, Zhu K, Xing Y-X (2017) Identification and efficiency of a nitrogen-fixing endophytic actinobacterial Strain from Sugarcane. Sugar Tech 19(5):492–500. https://doi.org/10.1007/s12355-016-0498-y

Xu J-Y, Han Y-H, Chen Y, Zhu L-J, Ma LQ (2016) Arsenic transformation and plant growth promotion characteristics of As-resistant endophytic bacteria from As-hyperaccumulator *Pteris vittata*. Chemosphere 144:1233–1240. https://doi.org/10.1016/j.chemosphere.2015.09.102

Xu T, Cao L, Zeng J, Franco CMM, Yang Y, Hu X, Liu Y, Wang X, Gao Y, Bu Z, Shi L, Zhou G, Zhou Q, Liu X, Zhu Y (2019) The antifungal action mode of the rice endophyte *Streptomyces hygroscopicus* OsiSh-2 as a potential biocontrol agent against the rice blast pathogen. Pestic Biochem Physiol 160:58–69. https://doi.org/10.1016/j.pestbp.2019.06.015

Yadav AN, Kaur T, Kour D, Rana KL, Yadav N, Rastegari AA, Kumar M, Paul D, Sachan SG, Saxena AK (2020) Chapter 16—Saline microbiome: biodiversity, ecological significance, and potential role in amelioration of salt stress. In: Rastegari AA, Yadav AN, Yadav N (eds) New and future developments in microbial biotechnology and bioengineering. Elsevier, pp 283–309

Yan L-L, Han N-N, Zhang Y-Q, Yu L-Y, Chen J, Wei Y-Z, Li Q-P, Tao L, Zheng G-H, Yang S-E (2010) Antimycin A 18 produced by an endophytic *Streptomyces albidoflavus* isolated from a mangrove plant. J Antibiot 63(5):259

Yandigeri MS, Meena KK, Singh D, Malviya N, Singh DP, Solanki MK, Yadav AK, Arora DK (2012) Drought-tolerant endophytic actinobacteria promote growth of wheat (*Triticum aestivum*) under water stress conditions. Plant Growth Regul 68(3):411–420. https://doi.org/10.1007/s10725-012-9730-2

Yang Y-H, Fu X-L, Li L-Q, Zeng Y, Li C-Y, He Y-N, Zhao P-J (2012) Naphthomycins L–N, Ansamycin antibiotics from *Streptomyces* sp. CS. J Nat Prod 75(7):1409–1413. https://doi.org/10.1021/np300109s

Yashavantha Rao HC, Mohana NC, Satish S (2020) Chapter 13—Biocommercial aspects of microbial endophytes for sustainable agriculture. In: Kumar A, Radhakrishnan EK (eds) Microbial endophytes. Woodhead Publishing, pp 323–347

Yoolong S, Kruasuwan W, Phạm HTT, Jaemsaeng R, Jantasuriyarat C, Thamchaipenet A (2019) Modulation of salt tolerance in Thai jasmine rice (*Oryza sativa* L. cv. KDML105) by *Streptomyces venezuelae* ATCC 10712 expressing ACC deaminase. Sci Rep 9(1):1275. https://doi.org/10.1038/s41598-018-37987-5

Zamoum M, Goudjal Y, Sabaou N, Barakate M, Mathieu F, Zitouni A (2015) Biocontrol capacities and plant growth-promoting traits of endophytic actinobacteria isolated from native plants of Algerian Sahara. J Plant Dis Protect 122(5–6):215–233

Zhang J, Wang J-D, Liu C-X, Yuan J-H, Wang X-J, Xiang W-S (2014) A new prenylated indole derivative from endophytic actinobacteria *Streptomyces* sp. neau-D50. Nat Prod Res 28:431–437. https://doi.org/10.1080/14786419.2013.871546

Zhang L, Shi J, Liu CL, Xiang L, Ma SY, Li W, Jiao RH, Tan RX, Ge HM (2018) New borrelidin derivatives from an endophytic *Streptomyces* sp. Tetrahedron Lett 59(51):4517–4520. https://doi.org/10.1016/j.tetlet.2018.11.023

Zhang Q, Acuña JJ, Inostroza NG, Mora ML, Radic S, Sadowsky MJ, Jorquera MA (2019) Endophytic bacterial communities associated with roots and leaves of plants growing in Chilean extreme environments. Sci Rep 9(1):4950. https://doi.org/10.1038/s41598-019-41160-x

Zhao K, Zhao C, Liao P, Zhang Q, Li Y, Liu M, Ao X, Gu Y, Liao D, Xu K, Yu X, Xiang Q, Huang C, Chen Q, Zhang L, Zhang X, Penttinen P (2016) Isolation and antimicrobial activities of actinobacteria closely associated with liquorice plants *glycyrrhiza glabra* L. and *glycyrrhiza inflate* BAT. in Xinjiang, China. Microbiology (United Kingdom) 162(7):1135–1146. https://doi.org/10.1099/mic.0.000301

Zhao K, Li J, Zhang X, Chen Q, Liu M, Ao X, Gu Y, Liao D, Xu K, Ma M (2018) Actinobacteria associated with *Glycyrrhiza inflata* Bat. are diverse and have plant growth promoting and antimicrobial activity. Sci Rep 8(1):1–13

Zhu J-K (2001) Plant salt tolerance. Trends Plant Sci 6(2):66–71. https://doi.org/10.1016/S1360-1385(00)01838-0

Złoch M, Thiem D, Gadzała-Kopciuch R, Hrynkiewicz K (2016) Synthesis of siderophores by plant-associated metallotolerant bacteria under exposure to Cd(2.). Chemosphere 156:312–325. https://doi.org/10.1016/j.chemosphere.2016.04.130

An Overview on Natural Product from Endophytic Actinomycetes

6

Govindan Nadar Rajivgandhi, R. T. V. Vimala, G. Ramachandran, Chelliah Chenthis Kanisha, N. Manoharan, and Wen-Jun Li

Abstract

In recent years, all the chemical-mediated compounds, drugs, and other compounds are inactivated due to the development of resistant effect. For this reason, there is an urgent need to prevent these kinds of infections with natural sources. In this regard, endophytic actinomycete is an excellent reservoir for production of novel chemical metabolites with excellent byproduct in biomedical, pharmaceutical, and industrial applications. The endophytic actinomycetes are grown in inside of the plant, and it also utilize the plant hormones, carbon,

G. N. Rajivgandhi
State Key Laboratory of Biocontrol, Guangdong Provincial Key Laboratory of Plant Resources and Southern Marine Science and Engineering Guangdong Laboratory (Zhuhai), School of Life Sciences, Sun Yat-Sen University, Guangzhou, People's Republic of China

Department of Marine Science, Bharathidasan University, Tiruchirappalli, Tamil Nadu, India

R. T. V. Vimala · N. Manoharan
Department of Biotechnology, Bharathidasan University, Tiruchirappalli, Tamilnadu, India

G. Ramachandran
Department of Marine Science, Bharathidasan University, Tiruchirappalli, Tamil Nadu, India

C. C. Kanisha
Department of Nanotechnology, Noorul Islam Centre for Higher Education, Kumaracoil, Tamil Nadu, India

W.-J. Li (✉)
State Key Laboratory of Biocontrol, Guangdong Provincial Key Laboratory of Plant Resources and Southern Marine Science and Engineering Guangdong Laboratory (Zhuhai), School of Life Sciences, Sun Yat-Sen University, Guangzhou, People's Republic of China

State Key Laboratory of Desert and Oasis Ecology, Xinjiang Institute of Ecology and Geograph, Chinese Academy of Sciences, Urumqi, People's Republic of China
e-mail: liwenjun3@mail.sysu.edu.cn

R. V. Rai, J. A. Bai (eds.), *Natural Products from Actinomycetes*,
https://doi.org/10.1007/978-981-16-6132-7_6

nitrogen and various other nutrients. These factors are important advantages for endophtyic actinomycetes. The difficult environment of endophytic actinomycetes creates enhanced role in all the sectors including biomedical, biopharmaceuticals, and bioindustries. Altogether, based on the advantages, the current study is focused in the field of endophytic actinomycetes and their product role in all biomedical and industrial sectors.

Keywords

Actinomycetes · Endophytic nature · Marine environment · Secondary metabolites · Industrial application · Enzyme production

6.1 Introduction

Microorganisms are important sources for various antibiotic productions involving antimicrobial, antiviral, anticancer, and anti-parasitic activity. To date, about 25,0000 bioactive compounds can be reported by different microorganisms, and a few of them can be frequently used in agriculture, food, pharmacology, and biomedical applications (Weib et al. 2018). Among these, 10,000 compounds can be synthesized by actinomycetes with high biological activities. Almost 80% of the actinomycetes-derived compounds can be used in many applications (Ramachandran et al. 2019). Actinomycetes are the group of filamentous gram-positive bacteria within the order of *Actinomycetales*. It is rich in G + C content and distributed widely in the different terrestrial and marine environments (Kim et al. 2020). Among the bacteria, actinomycetes exhibit a structurally diverse and broad spectrum of biologically active compounds.

Actinomycetes producing antibiotics are present in different ecological habitats (Sujarit et al. 2020), as well as in fresh water and other marine environments. It is commonly referred to as the storehouse of antibiotics which are mainly used in various industrial and biomedical applications. The significance of endophytic actinomycetes is listed out in Fig. 6.1.

Among the actinomycetes genus, *Streptomyces* is the supreme antibiotic producer, which synthesized 7600 active compounds and is currently used in the pharmaceutical industry and biomedical applications (Leetanasaksakul and Thamchaipenet 2018). In recent years, many potent therapeutic lead molecules have been isolated and characterized from *Streptomyces* species, and they have immense potential as they are well-known producers of various lead molecules with novel biological properties (Sujarit et al. 2020). The majority of the *Streptomyces* compounds have potential biological activities and are used in various agricultural, food, pharmaceutical, and medicinal industries. But, some of them are used as herbicide and growth-promoting agents.

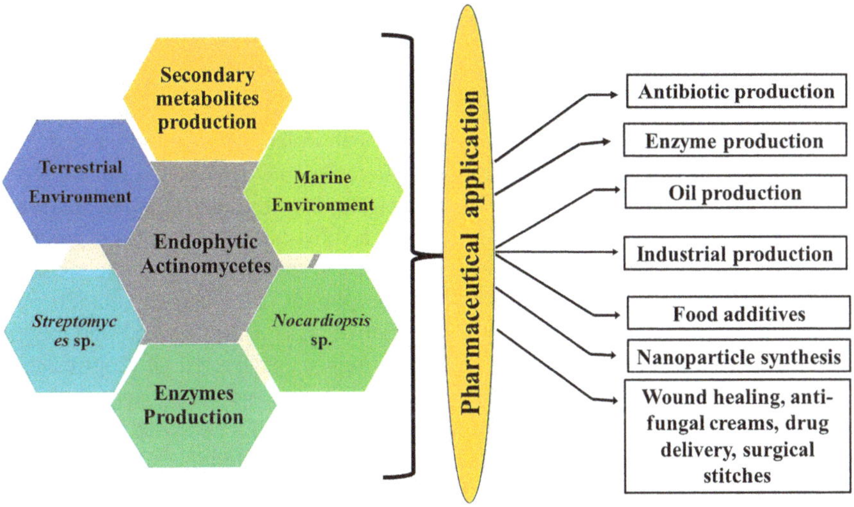

Fig. 6.1 Significance of endophytic actinomycetes

6.2 Endophytic Actinomycetes

The microbes which survive inside of the plant and animal tissues without causing any infections are referred to as endophytic microbes. They can be isolated from the living areas of crops, medicinal plants, woody plants, ferns, and club mosses. Among the endophytic microbes, actinomycetes are very effective in synthesizing novel bioactive metabolites (Peng et al. 2015). It increases the growth-promoting substances of the plant (Ramachandran et al. 2018).

Among the endophytic actinomycetes, *Streptomyces* is the most frequent genus isolated in the form of endophytic actinomycetes with excellent bioactivities (Nimnoia et al. 2010). Notably, alnumycin, munumbicins A to D, and coronamycins were derived from various *Streptomyces* species for different biological activities including antimicrobial, anticancer, immunosuppressive, larvicidal, and antiviral agents (Benhadj et al. 2019). In addition, the two important anticancer drugs such as anthraquinones and lupinacidins A and B were synthesized by the genus *Streptomyces*. Further, non-*Streptomyces* sp. of *Micromonospora*, *Microbispora*, *Nocardia*, *Pseudonocardia*, and *Streptosporangium* was also considered as an isolated endophytic actinomycetes (Supong et al. 2016). Recently, many secondary metabolites can be reported from non-*Streptomyces* with excellent bioactivities against various infections and diseases. However, endophytic actinomycetes are considered a promising source in search of novel antimicrobial compounds which act as alternatives to combat various infection-causing pathogens.

6.3 Endophytic Actinomycetes from Various Sources

Endophytic actinomycetes can play an important role in the field such as agricultural, medicinal, and biopharmaceutical applications (Barbosa et al. 2020). Recently, more endophytic actinomycetes have been reported from plant sources (Ramachandran et al. 2019). Plant-derived endophytic actinomycetes may represent an unexplored reservoir of new species and an important source for identifying the unique unknown compounds. The previous study has reported a different kind of natural products such as alkaloids, benzopyranones, benzoquinones, flavonoids, phenols, steroids, terpenoids, tetralones, xanthones, and other compounds (Kuncharoen et al. 2019).

Recently, Ramachandran et al. (2019) reported that plant-mediated endophytic actinomycetes exhibit excellent bioactivities against multidrug-resistant uropathogens (Bunbamrung et al. 2020). In addition, the endophytic actinomycetes can be produced in the plant phytochemical compounds with increased bioactivities due to the plant nutrients and different growth hormones (Djinni et al. 2013). Plant leaf and roots are commonly used for the treatment of various infectious diseases like AIDS, malaria, typhoid, jaundice, inflammation, scalds, dengue fever, liver diseases, and stomach-related diseases (Tanvir et al. 2016).

Endophytic microbes and their secondary metabolite productions are questionable due to the underexplored plant region (Betancur et al. 2020). For example, the phytochemical components and secondary metabolites of *P. cuspidatum* reported that it exhibits excellent antimicrobial, antiviral, antioxidant, antibiofilm, larvicidal, and antitumor activities (Valipour et al. 2018). Previously, some of the endophytic compounds derived from endophytic actinomycetes have been reported. Examples are emodin, piceid, polydatin, and resveratrol (Georgousaki et al. 2020). Moreover, Wonglom et al. (2019) reported that plant-derived endophytic actinomycetes are involved in the production of secondary metabolites such as siderophore, hormones, and phospholipid solubilization. In addition, it was reported that the endophytic actinomycetes genus, *Streptomyces*, helped to promote the gene expression in rice under salt stress conditions (Ramachandran et al. 2019).

6.4 Marine Endophytic Actinomycetes

The discoveries of novel chemical metabolites from terrestrial sources are decreased frequently due to the detection of existing compounds. So, the researcher highly focused on marine microbes from diverse environments for their ability to synthesize novel bioactive metabolites (Rajivgandhi et al. 2018). Compared with terrestrial rainforest, marine sources exhibit secondary metabolites. Moreover, the emergence of multidrug-resistant pathogen causing infectious diseases causes a major threat to human health. Therefore, there is an urgent need for novel antibiotics with potential chemical constituents to fight against spontaneously evolved pathogens (Supong et al. 2016). In addition, 70% of the earth is surrounded by oceans and life originated from the sea hypothetically (Al-Ansari et al. 2020). Ocean nature is extremely varied when compared with terrestrial ones, which produce some undefined chemical constituents. To the best of our knowledge, less than 1% of the actinomycetes from marine sources have been discovered, reported, and documented until now. So, most of the researchers are currently concentrated in various marine sources including bacteria, fungi, sponges, and endophytes for novel antimicrobial metabolite detection (Ramachandran et al. 2018).

To combat these multidrug-resistant pathogens to currently available antibiotics, there is an emerging need for periodic replacement of existing antibiotics through screening of unexplored habitats (Ramachandran et al. 2019). Therefore, researchers are very interested to discover the secondary metabolites from unexplored marine sources (Sathish Kumar and Kokati Venkata 2012). To date, approximately 25,000 bioactive compounds have been reported from marine microorganisms. Among which, 12,500 compounds are synthesized by actinomycetes indicating 50% of all bioactive compounds discovered. Among actinomycetes, 10,000 compounds are produced by genus *Streptomyces*. Therefore, *Streptomyces* species are primary antibiotic-producing organisms that can be exploited by the pharmaceutical industry (Valli et al. 2012). It can tolerate high temperatures, pH, NaCl, and various other environmental conditions. In addition, few novel compounds from *Streptomyces* have been reported from extreme, cold-adaptive, and high salt conditions.

Currently, more than 60% of actinomycetes synthesized antibiotics can be performed against various infections, particularly the genus *Streptomyces* compound (Al-Dhabi et al. 2020). Previously, only 11 genera are reported by various researchers in the year of 1970, followed by 100 genera in 2005 and 220 genera in 2010 (Rajivgandhi et al. 2018). However, the physiological, chemical, and structural features of rare actinomycete isolation and identifications can be confirmed by selective isolation media only. To date, 15,000 actinomycete secondary metabolites are reported, of which 11,000 are synthesized from *Streptomyces* species and 4000 are synthesized from non-*Streptomyces* species (Leetanasaksakul and Thamchaipenet 2018). All the non-*Streptomyces* species derived compounds have the potential ability to inhibit multidrug-resistant bacteria, fungi, and virus than *Streptomyces*. Non-*Streptomyces* and reported compounds were presented in Fig. 6.2.

6.5 Biomedical Applications of Marine Endophytic Actinomycetes

The extensive use of antibiotics and inadequate and unauthorized antibiotics usage are the main reason for bacteria to resist the antimicrobial drugs. The use of extensive antibiotics may affect the bacterial cell wall, increase of drug efflux pumps, enzymatic modification, quorum sensing, biofilm formation, target site modification, and acquisition of alternative metabolic pathways. Therefore, researchers have started to work on marine organisms as they exhibit numerous natural products with a unique biological activities (Valli et al. 2012). Some of the unique antimicrobial compounds from actinomycetes act against various multidrug-resistant bacteria and fungi (Khebizi et al. 2018). The chemical diversity of the marine environment is entirely

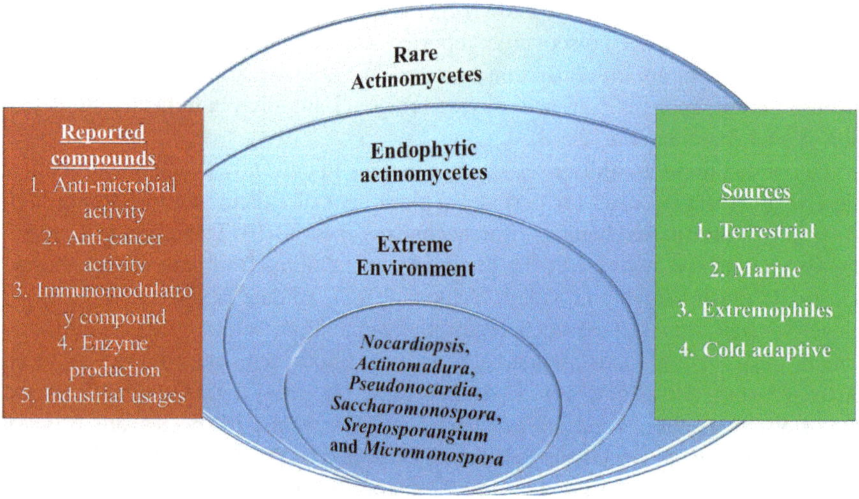

Fig. 6.2 Non-*Streptomyces* and their important secondary metabolites

different from other environments, and it produced some important molecules such as peptides, polypeptides, alkaloids, sterols, terpenoids, lactones, halogenated compounds, nucleosides, etc.

Further, researchers are specifically interested to work on endophytic actinomycetes of marine sources for various secondary metabolites. Based on the previous study, to date almost 200 endophytic actinomycetes can be isolated and identified from various marine sources, in seaweed and seagrasses (Supong et al. 2016; Barbosa et al. 2020). All the marine endophytic actinomycetes can be entirely different than the previously reported terrestrial strains. They have also produced some interesting bioactive compounds that may be used in biomedical applications. Recently, the research group of Wen-Jun Li reported more number of marine endophytic actinomycetes such as *Micromonospora*, *Saccharomonospora*, *Mycobacterium*, *Actinomycetospora*, *Nonomuraea*, *Verrucosispora*, *Nocardiopsis*, *Microbacterium*, and *Glycomyces* with excellent biomedical applications (Ramachandran et al. 2018).

Recently, Ramachandran et al. (2019) reported that the endophytic actinomycete of the mangrove plant, *Avicennia marina*, exhibited excellent antimicrobial activity against multidrug-resistant gram-negative bacteria. Similarly, seagrass-associated endophytic actinomycetes exhibited higher antimicrobial and anticancer activities against multidrug-resistant bacteria through internal cell damages (Ravikumar et al. 2012). Recently, EL-Gendy et al. (2018) reported that the marine endophytic *Streptomyces* species had more antibacterial and anticancer activity against methicillin-resistant *Staphylococcus aureus* and colon cancer HCT-16, liver cancer HepG-2, and lung cancer A549 cell lines. The endophytic actinomycetes, *Nocardiopsis* GRG 1 (KT235640) strain, isolated from seaweed and its crude extract had an excellent antibiofilm activity (Rajivgandhi et al. 2018). Previously,

Ramachandran et al. (2018) and Sathish Kumar and Kokati Venkata (2012) reported that the endophytic actinomycete *Nocardiopsis* sp. derived from mangrove plant of *Avicennia marina* exhibited an excellent antimicrobial activity against multidrug-resistant gram-negative and gram-positive bacteria at the very lowest concentration of crude extract (EL-Gendy et al. 2018). Therefore, endophytic actinomycetes are producing valuable unexplored products than actinomycetes. Herewith, some of the industrially important actinomycetes and their application have been listed below.

6.6 Industrially Important Actinomycetes and Their Application

Synthetic biology is a field of science that involves redesigning organisms for useful purposes by engineering them to have new abilities. Therefore, genetic engineering methods are being used to increase the secondary metabolite production using actinomycetes in industries. Moreover, gene deletion and gene addition processes of recombinant DNA are very effective for secondary metabolite production. However, the novel compound can be detected by novel pathways due to the heterologous expression resulted from genetic engineering methods (Mukhtar et al. 2017).

Most of the commercially available drugs can be obtained at a large scale by industrial process of fermentation, chemical synthesis, or semisynthetic processes (Benhadj et al. 2019). Most of the industries are using *Streptomyces* as a sole source for drug production (Matsuo et al. 2020). *Streptomyces* can synthesize some other complex structures of the compounds including polysaccharides, alcohols, amino acids, and aromatic compounds. Moreover, it can produce some important enzymes such as amylase, chitinase, cellulase, glucanase, and protease (Hozzein et al. 2019). For example, penicillin producing microbes such as *Penicillin chrysogenum* and riboflavin, *Ashbya gossypii* was reported effectively against various infections (Okazawa et al. 2019). Previous reports of actinomycetes and their secondary metabolites against various applications are listed in Table 6.1. Similarly, the novel bioactive compounds from various endophytic actinomycetes and their biomedical applications are presented in Table 6.2. In addition, Table 6.3 represents the industrial enzymes produced by actinomycetes useful for different applications.

6.7 Secondary Metabolites of Endophytic Actinomycetes

Recently, researchers are interested to identify the new classes of bioactive compounds to inhibit the growth of drug-resistant pathogens (Kim et al. 2020). Interestingly, endophytic environments provide the ecological niches for bacteria, fungi, and actinomycetes. Particularly, endophytic actinomycetes were reported to produce new classes of chemical compounds against various infections (Al-Dhabi et al. 2020). These rare endophytic actinomycetes produced the compound with potential antibiotic nature (Djinni et al. 2013). To date, 3500 new compounds have been produced from 100 endophytic actinomycetes. Recently, Al-Ansari et al.

Table 6.1 *Streptomyces* and their application in various fields

S. no.	Actinomycetes	Antibiotic name	Sources	Applications	References
1	*Streptomyces* sp. CA-129531	Trichostatin derivatives	Soil	Tyrosinase inhibitors	Georgousaki et al. (2020)
2	*Streptomyces levis*	Ethyl 2-propylphenyl ester	Soil	Anti-carcinogenic activity	Valipour et al. (2018)
3	*Streptomyces* sp. M4	Salvianolic acid B	Soil	Antifungal activity	Rajesh and Manhas (2020)
4	*Streptomyces* sp. PNM-9	Methyl-*N*-(2′-phenylethyl)-butanamide and 3-methyl-*N*-(2′-phenylethyl)-butanamide	Soil	Anti-phytopathogenic	Betancur et al. (2020)
5	*Streptomyces puniceus* strain AS13	(1-(2,4-dihydroxy-6-methylphenyl)-ethanone)	Soil	Antitumor	Hussain et al. (2018)
6	*Streptomyces angustmyceticus* NR8-2	β-1,3-glucanase	Soil	Cell wall degradation in fungi	Wonglom et al. (2019)
7	*Streptomyces* sp. Al-Dhabi-100	1-(2,6-Dimethyl-4-propoxyphenyl)propan-1-one (7%) and ethyl 2-propylphenyl ester	Soil	Antibacterial, antitubercular, and antioxidant	Al-Dhabi et al. (2020)
8	*Streptomyces avermitilis*	Phthoxazolin A	Soil	Plant pathogenicity	Suroto et al. (2017)
9	*Streptomyces* sp. strain HG29	Oligomycins A and E	Soil	Phytopathogenic fungi	Khebizi et al. (2018)
10	*Streptomyces antibioticus* strain	Quercetin 3-O-glucoside	Soil	Antimicrobial	Sholkamy et al. (2020)

Table 6.2 Various marine endophytic actinomycetes and their secondary metabolites against various biomedical applications

Endophytic actinomycete	Sources	Antibiotic name	Application	References
Nocardia caishijiensis and *Pseudonocardia carboxydivorans*	Plant leaf	7-Octadecenamide, 9, 12 octadecandienamide	Antioxidant and antimicrobial	Tanvir et al. (2016)
Micromonospora musae sp.	Plant root	Diphosphatidylglycerol, phosphatidylethanolamine	Antimicrobial	Kuncharoen et al. (2019)
Streptomyces coeruleorubidus GRG 4 (KY457708)	Algae	Bis(2-ethylhexyl) phthalate	Antimicrobial and anticancer	Rajivgandhi et al. (2018)
Streptomyces sp. TBRC7642	Plant root	Abyssomicins Y–Z, methyl aeruginoate, desferri-ferrithocin-4-hydroxyphenethylester, streptomethiocins A–B, furaquinocin I, and streptolactone	Antibacterial, antitubercular, antimalarial, and cytotoxic	Bunbamrung et al. (2020)
Streptomyces sp. RS-9 and *Streptomyces* sp. MS-26	Invertebrate	1-Methyl-2-acetyl-6-acetoxy-7-methoxy-4-(dimethylamino)	Antimicrobial and anticancer	El-gendy et al. (2018)
Nocardiopsis sp.	Algae	Hexahydro-tetrahydroxy-6-methyl-1,12-dioxo-, [4S-(4-α,4a,al)]–(2-naphthacene)	Antimicrobial	Rajivgandhi et al. (2018)
Streptomyces sundarbansensis WR1L1S8	Algae	[2-hydroxy-5-((6-hydroxy-4-oxo-4Hpyran-2-yl)methyl)-2-propylchroman-4-one]	Antimicrobial and anticancer	Djinni et al. (2013)
Streptomyces sp. Al-Dhabi-100	Algae	1-(2,6-Dimethyl-4-propoxy phenyl)propan-1-one (7%) and ethyl 2-propylphenyl ester	Antibacterial, antitubercular, and antioxidant	Al-Dhabi et al. (2020)
Nocardiopsis **sp. GRG 1 (KT235640)**	Algae	Pyrrolo[1,2-a]pyrazine-1,4-dione, hexahydro-3-(2-methylpropyl)	Antibiofilm and anti-quorum sensing	Rajivgandhi et al. (2018)
Micromonospora lupini sp.	Algae	Anthraquinones	Antitumor	Igarashi et al. (2007)
Streptomyces sp. *BCC72023*	Rice	Efomycins M, oxohygrolidin, polyethers, abierixin, and 9-O-methylabierixi	Antimicrobial	Supong et al. (2016)
Streptomyces sp. GMKU 336	Rice	1-Aminocyclopropane-1-carboxylate deaminase	Antimicrobial	Jaemsaeng et al. (2018)
Nocardiopsis **sp. GRG 2 (KT 235641)**	Algae	1,4-Diaza-2,5-dioxo-3-isobutyl bicyclo[4.3.0] nonane	Antimicrobial and anti-ESBLs	Rajivgandhi et al. (2018)

Table 6.3 Industrial enzymes produced by actinomycetes useful for different applications

Enzyme	Producing strain	Industrial Applications	References
Cellulase	*Streptomyces ruber*	Detergent	Kar and Ray (2008)
	Thermobifida halotolerans	Paper and pulp	Yang and Liu (2008)
Protease	*Streptomyces pactum*	Pharmaceutical	Wietzorrek and Bibb (1997)
Keratinase	*Actinomadura keratinilytica*	Leather	Habbeche et al. (2014)
Amylase	*Streptomyces erumpens*	Detergent	El-Sersy et al. (2010)
	Thermobifida fusca	Paper and pulp	Zhang et al. (2011)
Xylanase	*Streptomyces* sp.	Paper and pulp	Priya et al. (2012)
Lipase	*Streptomyces exfoliates*	Paper and pulp	Aly et al. (2012)
	Nocardiopsis alba	Detergent	Gandhimathi et al. (2009)
Chitinase	*Streptomyces thermoviolaceus*	Textile	Bhattacharya et al. (2007)
Pectinase	*Streptomyces lydicus*	Beverage Textile	Jacob et al. (2008)

(2020) reported that the crude extract of marine *Streptomyces* sp. AS11 exhibited higher antimicrobial activity against *S. aureus*. Different parameters such as pH, temperature, NaCl, carbon sources, and nitrogen sources of the culture medium significantly played a significant role in secondary metabolite production (Ramachandran et al. 2018). Further, it has shown excellent anticancer activity against HeLa cell line at IC_{50} concentration of 0.250 mg/mL.

Seaweed-derived endophytic actinomycetes *Streptomyces coeruleorubidus* GRG 4 (KY457708) exhibited antimicrobial activity and anticancer activity at 75 μg/mL and 100 μg/mL, respectively. The researcher reported that the actinomycetes can utilize the nutrient from seaweed and enhance the bioactivity against multidrug-resistant bacteria and A549 lung cancer cells. In 2020, the secondary metabolites of *Streptomyces* sp. produced novel compounds against various bacteria with potential antioxidant activity. GCMS result of synthesized secondary metabolites exhibited two different novel compounds 1-(2,6-dimethyl-4-propoxyphenyl)propan-1-one (7%) and ethyl 2-propylphenyl ester (11.9%). Both the compounds have excellent antibacterial activity against *E. faecalis*, *B. subtilis*, *S. aureus*, *S. epidermidis*, and *K. pneumoniae* at MIC of 62.5 mg/mL, 31.25 mg/mL, 125 mg/mL, 200 mg/mL, and 125 mg/mL, respectively (Al-Dhabi et al. 2020). The result has been supported by the result of Rajivgandhi et al. (2018). In which, the marine endophytic actinomycete compound was derived from the mangrove plant of *Rhizophora mucronata.* It has an excellent antibiofilm activity against biofilm-producing *Pseudomonas aeruginosa* and *Klebsiella pneumoniae*. Similarly, the result was supported by Ramachandran et al. (2019). Marine algae *Caulerpa racemosa*-derived endophytic actinomycetes exhibited high inhibitory activity against multidrug-resistant urinary tract pathogens such as *Escherichia coli*, *Pseudomonas aeruginosa*, *Proteus mirabilis*, *Klebsiella pneumoniae*, and *Enterobacter* sp.

Interestingly, the supportive evidence of Rajivgandhi et al. (2018) reported that the algae-derived endophytic actinomycetes *Nocardiopsis* sp. exhibited a novel compound with an excellent antibiofilm activity against biofilm-producing *P. mirabilis* and *E. coli*. The novel compound of pyrrolo[1,2-a]pyrazine-1,4-dione, hexahydro-3-(2-methylpropyl) was reported from the endophytic actinomycetes Nocardiopsis sp. GRG 1 (KT235640). It has shown excellent antibacterial activity against ESBL-producing gram-negative bacteria. The rare endophytic actinomycetes strain *Nocardiopsis* sp. GRG 2 (KT 235641) produced an undefined novel compound 1,4-diaza-2,5-dioxo-3-isobutyl bicycle nonane. It has excellent anti-ESBL properties against multidrug-resistant ESBL-producing pathogens (Rajivgandhi et al. 2018). The supportive evidence of Djinni et al. (2013) and EL-Gendy et al. (2018) reported that the marine endophytic actinomycetes are an excellent bioactive secondary metabolite producer. The reported compounds are effectively inhibited in the methicillin-resistant *Staphylococcus aureus* and various gram-negative bacteria. The rate of inhibition is very high due to the various unexplored environment such as temperature, pH, NaCl, stress, carbon, and nitrogen sources. Bioactive metabolites derived from actinomycetes for pharmaceutical industrial applications are listed in Table 6.4.

6.8 Conclusion and Future Prospectives

Marine actinomycetes require special environmental conditions for growth compared to terrestrial actinomycetes. Moreover, the structure and functions of the endophytic actinomycetes are still unclear. High-throughput molecular techniques and metagenomics study should help to understand the exact role. Sometimes, the endophytic actinomycetes are unculturable due to the unknown details of taxonomy, physiology, and other environmental parameters including cultivation temperature, pH, stress tolerance, NaCl, oxygen, and other nutritional requirement.

In contrast, the study has focused on the endophytic actinomycetes and their secondary metabolite productions for various biomedical applications. At present, there are very few reports and research going on the study of endophytic actinomycetes against infections. Under immense stress, unpredictable environmental conditions, high salt conditions, and additional nutrients, the endophytic actinomycetes can play a significant role in diversity and secondary metabolite production. In the future, this new kind of actinomycetes and their bioactive compounds are alternative drug choices for various infections. Therefore, extensive research on endophytic actinomycetes from various unexplored sources is required for establishing their significance in diversity and secondary metabolite production.

Table 6.4 Bioactive metabolites derived from actinomycetes for pharmaceutical industrial application

Biological activity	Metabolite	Commercial product	Producing organism
Antibiotic	Penicillins	Penicillins G and V, ampicillin, methicillin,	*Penicillium* sp. *Aspergillus* sp.
	Cephalosporins	Cefoxitin Cefaclor Cefotaxime	*Acremonium* sp. *Emericellopsis* sp. *Durans Streptomyces clavuligerus*
	Thienamycin	Imipenem	*Streptomyces cattleya*
	Erythromycin	Azithromycin	*Saccharopolyspora erythraea*
	Vancomycin	Vancocin	*Streptomyces orientalis*
	Streptomycin		*Streptomyces griseus*
	Gentamycin		*Micromonospora purpurea*
	Tetracyclines	Tigecycline	*Streptomyces* sp. *Dactylosporangium* sp. *Actinomadura brunnea*
	Daptomycin	Cubicin	*Streptomyces roseosporus*
Antitumor	Bleomycin	–	*Streptomyces verticillus*
	Doxorubicin	–	*Streptomyces peucetius*
	Daunorubicin	–	*Streptomyces peucetius* and other *Streptomyces* sp.
Antifungal	Amphotericin B	Caspofungin acetate	*Streptomyces nodosus*
	Pneumocandin B_0	–	*Glarea lozoyensis*
	Validamycin	–	*Streptomyces malaysiesis*
	Lomofugin	–	*Streptomyces lomodesis*
	Antimycin	–	*Streptomyces lucius*
Antibacterial	Rifamycin	–	*Micromonospora rifamycinica*
	Roseoflavin	–	*Streptomyces davawesis*

Acknowledgement All the authors gratefully acknowledge the National Natural Science Foundation of China (Project Approval Numbers: 41950410573, 91951205 and 31670009) and Postdoctoral Science Foundation of China (Project Approval Number: 2019M663213) for financial support for this work.

References

Al-Ansari M, Kalaiyarasi M, Almalki MA, Vijayaraghavan P (2020) Optimization of medium components for the production of antimicrobial and anticancer secondary metabolites from *Streptomyces* sp. AS11 isolated from the marine environment. J King Saud Univ Sci 32:1993–1998

Al-Dhabi NA, Esmail GA, Ghilan AKM, ValanArasu M, Duraipandiyan V (2020) Metabolite profiling of *Streptomyces* sp. Al-Dhabi-100 isolated from the marine environment in Saudi

Arabia with anti-bacterial, anti-tubercular and anti-oxidant potentials. J King Saud Univ Sci 32: 1628–1633

Aly MM, Tork S, Al-Garni SM, Nawar L (2012) Production of lipase from genetically improved Streptomyces exfoliates LP10 isolated from oilcontaminated soil. Afr J Microbiol Res 6:1125–1137

Barbosa F, Pinto E, Kijjo A, Pinto M, Sousa E (2020) Targeting antimicrobial drug resistance with marine natural products. Int J Antimicrob Agent 56:106005. https://doi.org/10.1016/j.ijantimicag

Benhadj M, Gacemi-Kirane DJ, Menasria T, Guebla K, Ahmane Z (2019) Screening of rare actinomycetes isolated from natural wetland ecosystem (Fetzara Lake, northeastern Algeria) for hydrolytic enzymes and antimicrobial activities. J King Saud Univ Sci 31:706–712

Betancur LA, Forero AM, Vinchira-Villarraga DM, Cárdenas JD, Romero-Otero A, Chagas FO, Pupo MT, Castellanos L, Ramos FA (2020) NMR-based metabolic profiling to follow the production of anti-phytopathogenic compounds in the culture of the marine strain *Streptomyces* sp. PNM-9. Microbiol Res 239:126507

Bhattacharya D, Nagpure A, Gupta RK (2007) Bacterial chitinases: properties and potential. Crit Rev Biotechnol 27:21–28

Bunbamrung N, Intaraudom C, Dramae A, Thawai C, Tadtonge S, Auncharoen P, Pittayakhajonwut P (2020) Antibacterial, antitubercular, antimalarial and cytotoxic substances from the endophytic Streptomyces sp. TBRC7642. Phytochemistry 172:112275

Djinni I, Defant AM, Kecha K, Mancini I (2013) Metabolite profile of marine-derived endophytic *Streptomyces sundarbansensis* WR1L1S8 by liquid chromatography–mass spectrometry and evaluation of culture conditions on antibacterial activity and mycelial growth. J Appl Microbiol 116:39–50

EL-Gendy MMAA, Mohamed ZK, Hekal NZ, Ali FM, Yousef AEM (2018) Production of bioactive metabolites from different marine endophytic Streptomyces species and testing them against methicillin-resistant Staphylococcus aureus (MRSA) and cancer cell lines. J Biotechnol Comput Biol Bionanotechnol 99:13–35

El-Sersy NA, Abd-Elnaby H, Abou-Elela GM, Ibrahim HAH, El-Toukhy NMK (2010) Optimization, economization and characterization of cellulase produced by marine Streptomyces ruber. Afr J Biotechnol 9:6355–6364

Gandhimathi R, Seghal Kiran G, Hema TA, Selvin J, Raviji TR, Shanmughapriya S (2009) Production and characterization of lipopeptide biosurfactant by a sponge-associated marine actinomycetes Nocardiopsis alba MSA10. Bioprocess Biosyst Eng 32:825–835

Georgousaki K, Tsafantakis N, Gumeni S, Gonzalez I, Mackenzie TA, Reyes F, Lambert C, Trougakos IP, Genilloud O, Fokialakis N (2020) Screening for tyrosinase inhibitors from actinomycetes; identification of trichostatin derivatives from *Streptomyces* sp. CA-129531 and scale up production in bioreactor. Bioorg Med Chem Lett 30:126952

Habbeche A, Saoudi B, Jaouadi B, Haberra S, Kerouaz B, Boudelaa M (2014) Purification and biochemical characterization of a detergent-stable keratinase from a newly thermophilic actinomycete Actinomadura keratinilytica strain Cpt29 isolated from poultry compost. J Biosci Bioeng 117:413–421

Hozzein WN, Abuelsoud W, Wadaan MAM, Shuikand AM, Selim S, Jaouni SA, Elgawad HA (2019) Exploring the potential of actinomycetes in improving soil fertility and grain quality of economically important cereals. Sci Total Environ 651:2787–2798

Hussain A, Rather MA, Dar MS, Dangroo NA, Aga MA, Qayum A, Shah AM, Ahmad Z, Dar MJ, Hassan QP (2018) *Streptomyces puniceus* strain AS13., Production, characterization and evaluation of bioactive metabolites: a new face of dinactin as an antitumor antibiotic, Microbiol Res 207:196–202

Igarashi Y, Trujillo ME, Martinez-Molin E, Yanase S, Miyanaga S, Obata T, Sakurai H, Saiki I, Fujitad T, Furumaia T (2007) Antitumor anthraquinones from an endophytic actinomycete *Micromonospora lupini* sp. nov. Bioorg Med Chem Lett 17:3702–3705

Jacob N, Asha Poorna C, Prema P (2008) Purification and partial characterization of polygalacturonase from Streptomyces lydicus. Bioresour Technol 99:6697–6701

Jaemsaeng R, Jantasuriyarat C, Thamchaipenet A (2018) Positive role of 1 aminocyclopropane-1-carboxylate deaminase-producing endophytic *Streptomyces* sp. GMKU 336 on flooding resistance of mung bean. Agric Nat Resour 52:330–336

Kar S, Datta T, Chandra Ray R (2008) Optimization of thermostable α-amylase production by Streptomyces erumpens MTCC 7317 in solid-state fermentation using cassava fibrous residue. Braz Arch Biol Technol 53:301–309

Khebizi N, Boudjella H, Bijani C, Bouras N, Klenk HP, Pont F, Mathieu F, Sabaou N (2018) Oligomycins A and E, major bioactive secondary metabolites produced by *Streptomyces* sp. strain HG29 isolated from a Saharan soil. J Mycol Med 28:150–160

Kim JH, Choi JY, Park DH, Park DJ, Parka MG, Kimb SY, Jub YJ, Kima JY, Wanga M, Kimb CJ, Jea YH (2020) Isolation and characterization of the insect growth regulatory substances from actinomycetes. Comp Biochem Physiol Part C 228:108651

Kuncharoen N, Kudo T, Yuki M, Okuma M, Booncharoen A, Tanasupawat S (2019) *Micromonospora musae* sp. nov., an endophytic actinomycete isolated from roots of *Musa* species. Syst Appl Microb 42:126020

Leetanasaksakul K, Thamchaipenet A (2018) Potential anti-biofilm producing marine actinomycetes isolated from sea sediments in Thailand. Agric Nat Res 52:228–233

Matsuo H, Nakanishi J, Noguchi Y, Kitagawa K, Shigemura K, Sunazuka T, Takahashi O, Omura S, Nakashima T (2020) Nanaomycin K, a new epithelial mesenchymal transition inhibitor produced by the actinomycete "*Streptomyces rosa* subsp. notoensis" OS-3966. J Biosci Bioeng 129:291–295

Mukhtar S, Zaheer A, Aiysha D, Malik KA, Mehnaz S (2017) Actinomycetes: a source of industrially important enzymes. J Proteomics Bioinform 10:316–319

Nimnoia P, Pongsilp N, Lumyonga S (2010) Genetic diversity and community of endophytic actinomycetes within the roots of Aquilaria crassna Pierre ex Lec assessed by Actinomycetes-specific PCR and PCR-DGGE of 16S rRNA gene. Biochem Syst Ecol 38:595–601

Okazawa A, Yamanishi K, Katsuyama N, Kitazawa S, Ogawa T, Ohta D (2019) Identification of novel cytochrome P450 monooxygenases from actinomycetes capable of intermolecular oxidative CeC coupling reactions. J Biosci Bioeng 129:23–30

Peng Q, Zhi-Xiang F, Jie-Wei T, Zu-Chao L, Lei W, Zhi-Gang Z, Yi-Wen C, Yong-Qiang T (2015) Diversity, bioactivities, and metabolic potentials of endophytic actinomycetes isolated from traditional medicinal plants in Sichuan. China. Chin J Nat Med 13:0942–0953

Priya BS, Stalin T, Selvam Kuppusamy K (2012) Efficient utilization of xylanase and lipase producing thermophilic marine actinomycetes (Streptomyces albus and Streptomyces hygroscopicus) in the production of ecofriendly alternative energy from waste. Afr J Biotechnol 11:14320–14325

Rajesh MS, Manhas K (2020) Purification and characterization of salvianolic acid B from *Streptomyces* sp. M4 possessing antifungal activity against fungal phytopathogens. Microbiol Res 237: 126478

Rajivgandhi G, Ramachandran G, Maruthupandy M, Saravanakumar S, Manoharan N, Viji R (2018) Antibacterial effect of endophytic actinomycetes from marine algae against multi drug resistant gram negative bacteria. Exam Mar Biol Oceanogr 1:1–8

Ramachandran G, Rajivgandhi G, Maruthupandy M, Manoharan N (2018) Isolation and identification of antibacterial compound from marine endophytic actinomycetes against multi drug resistant bacteria. Ann Microbiol Immunol 1:1003

Ramachandran G, Rajivgandhi G, Maruthupandy M, Manoharan N (2019) Extraction and partial purification of secondary metabolites from endophytic actinomycetes of marine green algae *Caulerpa racemosa* against multi drug resistant uropathogens. Biocatal Agric Biotechnol 17: 750–757

Ravikumar S, Gnanadesigan M, Saravanan A, Monisha N, Brindha V, Muthumari S (2012) Antagonistic properties of seagrass associated Streptomyces sp. RAUACT-1: a source for anthraquinone rich compound. Asian Pac J Trop Med 5(11):887–890

Sathish Kumar SR, Kokati Venkata BH (2012) In-vitro antimicrobial activity of marine actinobacteria against multidrug resistance *Staphylococcus aureus*. Asian Pac J Trop Biomed 2(10):787–792

Sholkamy EN, Ahmed MS, Yasser MM, Mostafa AA (2020) Antimicrobial quercetin 3-O-glucoside derivative isolated from *Streptomyces antibioticus* strain ess_amA8. J King Saud Univ Sci 32:1838–1844

Sujarit K, Pathom-aree W, Mori M, Dobashi K, Shiomi K, Lumyong S (2020) *Streptomyces palmae* CMU-AB204T, an antifungal producing-actinomycete, as a potential biocontrol agent to protect palm oil producing trees from basal stem rot disease fungus, *Ganoderma boninense*. Biol Control 148:104307

Supong K, Thawai C, Choowong W, Kittiwongwattana C, Thanaboripat D, Laosinwattana C, Koohakan P, Parinthawong N, Pittayakhajonwut P (2016) Antimicrobial compounds from endophytic *Streptomyces* sp. BCC72023 isolated from rice (Oryza sativa L.). Res Microb 167:290–298

Suroto DA, Kitani S, Miyamoto KT, Sakihama Y, Arai M, Ikeda H, Nihira T (2017) Activation of cryptic phthoxazolin A production in *Streptomyces avermitilis* by the disruption of autoregulator-receptor homologue AvaR3. J Biosci Bioeng 124:611–617

Tanvir R, Sajid I, Hasnain S, Kulik A, Gron S (2016) Rare actinomycetes *Nocardia caishijiensis* and *Pseudonocardia carboxydivorans* as endophytes, their bioactivity and metabolites evaluation. Microbiol Res 185:22–35

Valipour B, Mohammadi SM, Abedelahi A, Maragheh BFA, Naderali E, Dehnad A, Charoudeh HN (2018) Culture filtrate ether extracted metabolites from Streptomyces Levis ABRIINW111 increased apoptosis and reduced proliferation in acute lymphoblastic leukemia. Biomed Pharm 108:216–223

Valli S, Suvathi Sugasini S, Aysha OS, Nirmala P, Vinoth Kumar P, Reena A (2012) Antimicrobial potential of Actinomycetes species isolated from marine environment. Asian Pac J Trop Biomed 2(6):469–473

Weib J, Hea L, Niu G (2018) Regulation of antibiotic biosynthesis in actinomycetes: perspectives and challenges. Synth Syst Biotechnol 3:229–235

Wietzorek A, Bibb M (1997) A novel family of proteins that regulate antibiotic production in Streptomyces appears to contain an OmpR-like DNA-binding fold. Mol Microbiol 25:1181–1184

Wonglom P, Suwannarach N, Lumyong S, Ito S, Matsui K, Sunpapao A (2019) *Streptomyces angustmyceticus* NR8-2 as a potential microorganism for the biological control of leaf spots of *Brassica rapa* subsp. *pekinensis* caused by *Colletotrichum* sp. and *Curvularia lunata*. Biol Control 138:104046

Yang CH, Liu WH (2008) Purification and properties of an acetylxylan esterase from Thermobifida fusca. Enzym Microb Technol 42:181–186

Zhang F, Chen J-J, Ren W-Z, Nie G-X, Ming H, Tang S-K, Li W-J (2011) Cloning, expression and characterization of an alkaline thermostable GH9 endoglucanase from Thermobifida halotolerans YIM 90462. Bioresour Technol:10143–10146

Regulation of Secondary Metabolites Through Signaling Molecules in *Streptomyces*

7

Kenji Arakawa and Toshihiro Suzuki

Abstract

The bacterial genus *Streptomyces* is well characterized by an ability to produce a wide variety of secondary metabolites including antibiotics, anticancer agents, antiviral drugs, herbicides, pesticides, insecticides, and immunosuppressants. Their biosynthesis is strictly controlled by small diffusible signaling molecules that constitute a signaling molecule/receptor regulatory systems. These signaling molecule/receptor regulatory systems are extensively studied in various *Streptomyces* species to understand the regulatory function for secondary metabolite production. This chapter focuses on the *Streptomyces* signaling molecules in view of structural classification, distribution, and biosynthesis. In addition, this chapter reviews the signaling molecule-dependent regulatory systems including the receptor, the target activator, and the pseudo-receptor that are responsible for controlling secondary metabolite production in *Streptomyces*. Application for activation of silent secondary metabolites by manipulation of regulatory systems is also briefly described.

Keywords

Streptomyces · Signaling molecule · Regulatory system · Secondary metabolites · Biosynthesis

K. Arakawa (✉)
Unit of Biotechnology, Graduate School of Integrated Sciences for Life, Hiroshima University, Hiroshima, Japan
e-mail: karakawa@hiroshima-u.ac.jp

T. Suzuki
Department of Fermentation Science, Faculty of Applied Biosciences, Tokyo University of Agriculture, Tokyo, Japan
e-mail: ts206188@nodai.ac.jp

R. V. Rai, J. A. Bai (eds.), *Natural Products from Actinomycetes*,
https://doi.org/10.1007/978-981-16-6132-7_7

7.1 Introduction

The filamentous bacterial genus *Streptomyces* is well characterized by an ability to produce a wide variety of secondary metabolites including antibiotics, anticancer agents, antiviral drugs, herbicides, pesticides, insecticides, and immunosuppressants. Their production is strictly controlled by small diffusible signaling molecules that constitute a signaling molecule/receptor regulatory systems (Bibb 2005; Takano 2006; Horinouchi and Beppu 2007; Niu et al. 2016; Arakawa 2018; Kong et al. 2019; Martin and Liras 2020) (Fig. 7.1). The most extensively studied signaling molecule/receptor regulatory system is A-factor/ArpA in streptomycin producer *Streptomyces griseus* (Ohnishi et al. 1999). At early stage fermentation, a specific receptor protein ArpA binds to the promoter region of its target gene *adpA* and represses its transcription. When A-factor reaches the critical concentration (around nM order), an A-factor/ArpA complex dissociates from the *adpA* promoter region, leading to the *adpA* transcription. AdpA acts as a transcriptional activator (homologous to an AraC/XylS family transcriptional activator) and binds to its target genes (AdpA regulons) to activate both secondary metabolite production and morphological development in *S. griseus*. These signaling molecule/receptor regulatory systems are extensively studied in various *Streptomyces* species to understand the regulatory function for secondary metabolite production. Genes involved in the regulatory systems are listed in Table 7.1 (details are described in Sect. 7.3).

In this chapter, we describe the *Streptomyces* signaling molecules and their dependent regulatory systems to induce secondary metabolite production.

7.2 Structural Diversity of the *Streptomyces* Signaling Molecules

The low-molecular-weight signaling molecules induce secondary metabolite production at nano-molar concentrations in *Streptomyces*. The *Streptomyces* signaling molecules hitherto discovered are classified into three groups; (1) γ-butyrolactone-

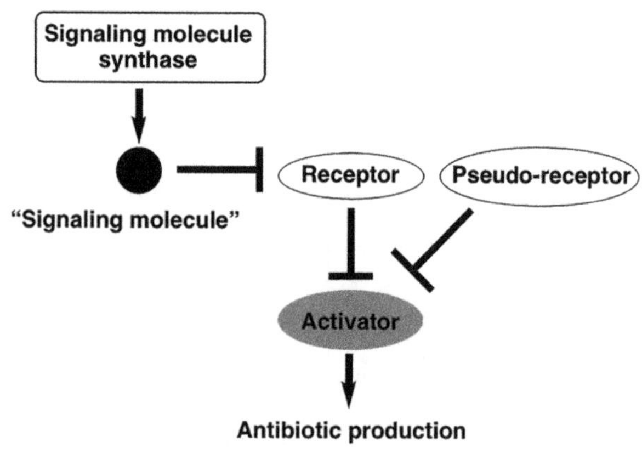

Fig. 7.1 Schematic model of the signaling molecule/receptor system for antibiotic production in *Streptomyces*

Table 7.1 Gene sets in the regulatory systems in various *Streptomyces* species

Streptomyces strains	Signaling molecule biosynthesis gene	Signaling molecule (s)	Signaling molecule receptor gene	Pseudo-receptor gene	Target activator gene(s) of the signaling molecule/receptor system	Secondary metabolite(s)
S. griseus	*afsA*	A-factor	*arpA*	SGR6382	*adpA*	Streptomycin, grixazone
S. rochei	*srrX*	SRB1, SRB2	*srrA*	*srrB*	*srrY*	Lankacidin, lankamycin
S. fradiae	N.D.	N.D.	*tylP*	*tylQ*	*tylS*	Tylosin
S. coelicolor	*mmfL*	MMFs	*mmfR*			Methylenomycin
S. coelicolor	*scbA*	SCB1-3 (SCB4-8)	*scbR*	*scbR2*	*kasO*	Coelimycin P-1
S. virginiae	*barX*	Virginia butanolides	*barA*	*barB*		Virginiamycin
S. venezuelae	*jadW1*	SVB1	*jadR3*	*jadR2*	*jadR1*	Jadomycin
S. ansochromogenes	*sabA*	SAB1-3	*sabR1*	*sabR2*	*cprC*	Nikkomycin
S. lavendulae	*farX*	IM-2	*farA*	*farR2*		Showdomycin
S. avermitilis	*aco*	Avenolide	*avaR1*	*avaR2*	*aveR*	Avermectin
S. ambofaciens	N.D	N.D.	*alpZ*	*alpW*	*alpV*	Kinamycin

N.D. not determined

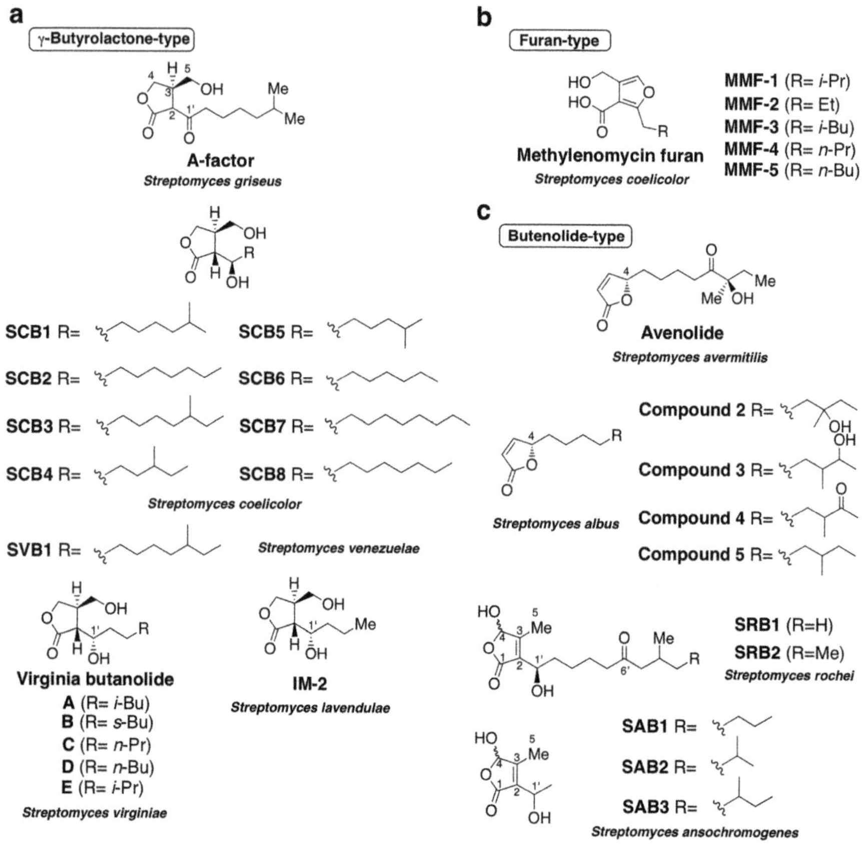

Fig. 7.2 Examples of *Streptomyces* signaling molecules. (**a**) γ-butyrolactone-type, (**b**) furan-type, (**c**) butenolide-type, and two other molecules, (**d**) PI factor and (**e**) *N*-methylphenylalanyl-dehydrobutyrine diketopiperazine. Me methyl, *i*-Bu *iso*-butyl, *s*-Bu *sec*-butyl, *n*-Pr *normal*-propyl, *n*-Bu *normal*-butyl, *i*-Pr *iso*-propyl, Et ethyl

type, (2) furan-type, and (3) butenolide-type (Fig. 7.2). The γ-butyrolactone-type molecules are the most studied family of signaling molecules, and they contain the 2,3-disubstituted γ-butyrolactone skeletons (Fig. 7.2a). They are further divided into three subclasses based on their oxidation degree and their stereochemistry at C-1′. The first subclass, the 1′-keto type, contains A-factor in *S. griseus* (Khokhlov et al. 1967; Hara and Beppu 1982). The second subclass, the (1′*R*)-hydroxyl type, contains IM-2 in *Streptomyces lavendulae* (Sato et al. 1989), SCB1-8 in *Streptomyces coelicolor* (Takano et al. 2005; Hsiao et al. 2009; Sidda et al. 2016), and SVB1 in *Streptomyces venezuelae* (Zou et al. 2014). The third subclass, the (1′*S*)-hydroxyl type, contains virginiae butanolides in *Streptomyces virginiae* (Kondo et al. 1989). About 60% of *Streptomyces* species use γ-butyrolactone-type signaling molecules to induce antibiotic production (Hara and Beppu 1982; Yamada 1995). Hitherto identified γ-butyrolactone-type are 19 molecules in 8 strains (Kong et al. 2019).

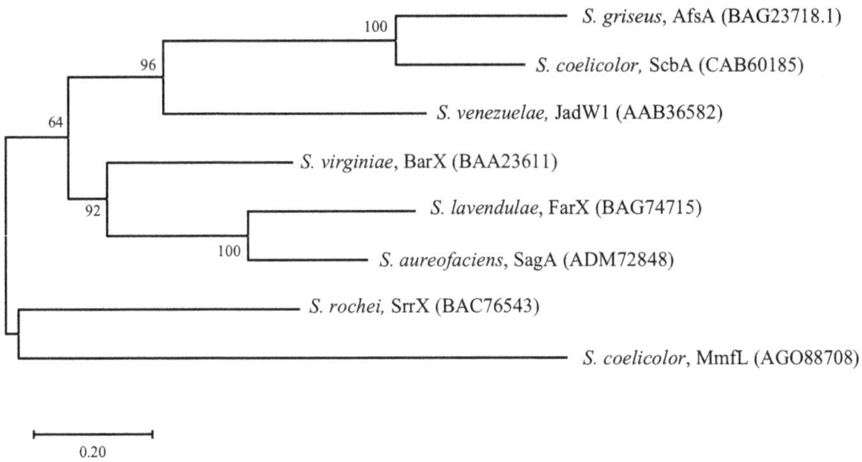

Fig. 7.3 Phylogenetic tree of the AfsA-homolog proteins in several *Streptomyces* species using the maximum likelihood method. Phylogenetic tree was conducted with MEGA-X version 10.2.0 software. Numbers given in parentheses are accession numbers in the GenBank. Bootstrap values, expressed as a percentage of 1,000 replications, are given at branching points

Additional types of signaling molecules, furan-type and butenolide-type, have been isolated from several *Streptomyces* species. Five furan-type signaling molecules named methylenomycin furans 1–5 (Fig. 7.2b) were identified as inducers for methylenomycin production in *S. coelicolor* (Corre et al. 2008). Butenolide-type molecules were further classified into two subclasses, 4-monosubstituted and 2,3-disubstituted 4-hydroxy butenolides. A 4-monosubstituted butenolide named avenolide (Fig. 7.2c) induces avermectin production in *Streptomyces avermitilis* at 4 nM concentration (Kitani et al. 2011). Another four avenolide-like butenolides were isolated from *Streptomyces albus* J1074 (Ahmed et al. 2017; Nguyen et al. 2018). Two 2,3-disubstituted 4-hydroxy butenolide molecules named SRB1 and SRB2 induce lankacidin and lankamycin production in *Streptomyces rochei* at 40–50 nM concentration (Arakawa et al. 2012). Additional three SRB-like butenolides SAB1-3 were identified as nikkomycin inducers in *Streptomyces ansochromogenes* (Wang et al. 2018). Regarding the signaling molecule biosynthesis genes, the γ-butyrolactone-type, furan-type, and butenolide (SRB)-type could be synthesized by AfsA-homolog, whereas butenolide (avenolide)-type could be synthesized by two genes, an acyl-CoA oxidase gene (i) and a cytochrome P450 hydroxylase gene (Kitani et al. 2011). Phylogenetic tree of the AfsA-homolog proteins was shown in Fig. 7.3. Recently, biosynthetic intermediates of signaling molecules have been reported. The *S. rochei* mutant of a P450 monooxygenase gene *srrO* accumulated 6′-deoxo-SRBs as new signaling molecule derivatives, whose inducing activities were 1/100-folds when compared with natural SRBs (Teshima et al. 2020).

Distribution of three types of signaling molecules could be estimated by several independent research groups using disruptant of signaling molecule biosynthesis gene as an indicator strain (Hara and Beppu 1982; Yamada 1995; Thao et al. 2017). Based on the reports for distribution of the γ-butyrolactone-type molecules (Hara and Beppu 1982; Yamada 1995), the furan- and butenolide-type molecules could share the remaining 40% of signaling molecules. Nihira and co-workers showed the avenolide activity in 24% of 51 actinomycetes (Thao et al. 2017). Taking into account the presence of novel-type signaling molecules and signaling molecule-free strains, these findings indicate the unique distribution of furan-type and SRB-like butenolide molecules in *Streptomyces*.

These three types of signaling molecules (Fig. 7.2a–c) harbor five-membered heterocycle rings with four carbons and one oxygen; however, two other signaling molecules have quite different structures. One is PI factor (Fig. 7.2d) isolated from *Streptomyces natalensis* for pimaricin production (Recio et al. 2004), while the other is *N*-methylphenylalanyl-dehydrobutyrine diketopiperazine (Fig. 7.2e) from *Streptomyces globisporus* for landomycin E production (Matselyukh et al. 2015). These signaling molecules hitherto identified were discovered by restoration of antibiotic production through supplementation of culture fractions in the antibiotic-deficient strains. Hence, unprecedented signaling molecules may be discovered through a combination of large-scale fermentation and comprehensive screening of culture fractions using bioassay.

7.3 The Signaling Molecule-Dependent Regulatory Pathway for Secondary Metabolite Production

The signaling molecule-dependent regulatory pathways in *Streptomyces* typically contain the following gene sets: signaling molecule biosynthesis gene, receptor gene, target gene of receptor, and pseudo-receptor gene (Fig. 7.1). The well-characterized regulatory cascades for secondary metabolite production are shown in Table 7.1, for streptomycin and grixazone production in *S. griseus* (Hara and Beppu 1982; Onaka et al. 1995; Ohnishi et al. 1999, 2005), coelimycin P1 in *S. coelicolor* (Takano et al. 2001, 2005; Hsiao et al. 2009; Gottelt et al. 2010; Li et al. 2015), jadomycin in *S. venezuelae* (Yang et al. 1995; Wang and Vining 2003; Wang et al. 2009; Xu et al. 2010; Zou et al. 2014), showdomycin in *S. lavendulae* (Sato et al. 1989; Kitani et al. 2001), virginiamycin in *S. virginiae* (Kondo et al. 1989; Kinoshita et al. 1997; Kawauchi et al. 2000), avermectins in *S. avermitilis* (Kitani et al. 2011; Wang et al. 2014a, b; Zhu et al. 2016), tylosin in *Streptomyces fradiae* (Bate et al. 1999, 2002; Stratigopoulos and Cundliffe 2002), kinamycins in *Streptomyces ambofaciens* (Aigle et al. 2005; Bunet et al. 2008, 2011), lankacidin and lankamycin in *S. rochei* (Arakawa et al. 2007, 2012; Yamamoto et al. 2008; Suzuki et al. 2010; Arakawa 2014, 2018; Misaki et al. 2020), and nikkomycin in *S. ansochromogenes* (Wang et al. 2018). Understanding of the whole regulatory pathway including signaling molecule/receptor system, downstream activator gene (s), and its (their) direct target(s) for secondary metabolite production will be a great value for and rational activation of silent secondary metabolite production, together with titer improvement in *Streptomyces*.

7.4 The Receptors of Regulatory Signaling Molecules

The signaling molecule receptors constitute a major class of TetR-type regulators, which harbor a conserved DNA-binding helix-turn-helix motif at the N-terminus and a ligand-binding pockets at the C-terminus (Yu et al. 2010; Xu and Yang 2019). Their phylogenetic trees were shown in Fig. 7.4. The signaling molecule receptor genes usually locate adjacent to the signaling molecule synthase genes (Biarnes-Carrera et al. 2015); hence, the signaling molecule/receptor systems are able to predict in *Streptomyces* species (Niu et al. 2016). Many signaling molecule receptor genes were characterized by their mutational analysis; however, the comprehensive analysis of the signaling molecules/receptor system was quite limited due to the difficulty for isolation of signaling molecules (extremely low concentration around 10^{-9} M in culture), for example, A-factor/ArpA in *S. griseus* (Ohnishi et al. 1999, 2005), SCBs/ScbR in *S. coelicolor* (Takano et al. 2001), SVBs/JadR3 in *S. venezuelae* (Zou et al. 2014), virginiae butanolides/BarA in *S. virginiae* (Okamoto et al. 1995; Kinoshita et al. 1997), IM-2/FarA in *S. lavendulae* (Kitani et al. 2001), avenolide/AvaR1 in *S. avermitilis* (Kitani et al. 2011), SRBs/SrrA in *S. rochei* (Yamamoto et al. 2008), SABs/SabR1 in *S. ansochromogenes* (Wang et al. 2018), and MMFs/MmfR (Zhou et al. 2021) (Table 7.1 and Fig. 7.2).

Other representative examples of signaling receptors, whose ligands have not been identified, are as follows: TylP for tylosin production in *S. fradiae* (Bate et al.

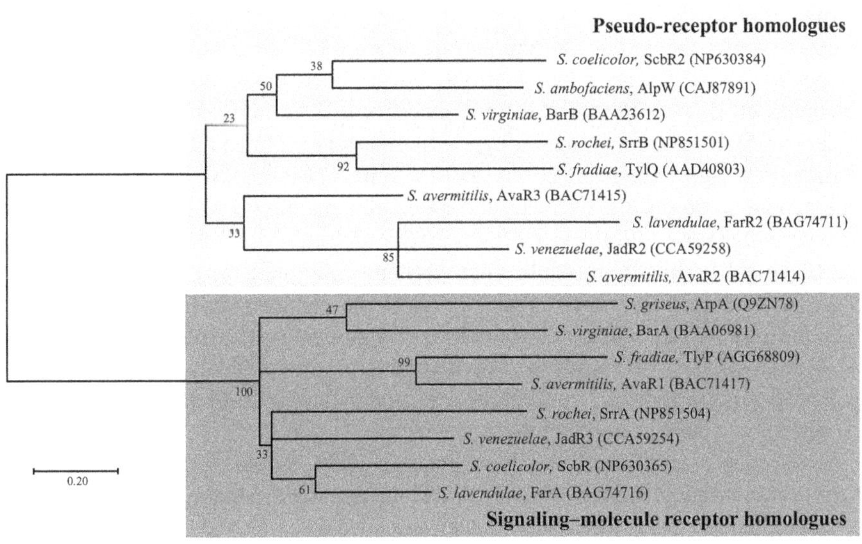

Fig. 7.4 Phylogenetic tree of the signaling molecule receptor proteins and pseudo-receptor proteins in several *Streptomyces* species using the maximum likelihood method. Phylogenetic tree was conducted with MEGA-X version 10.2.0 software. Numbers given in parentheses are accession numbers in the GenBank. Bootstrap values, expressed as a percentage of 1,000 replications, are given at branching points

1999), AlpZ for kinamycin in *S. ambofaciens* (Bunet et al. 2008; Vicente et al. 2020), SpbR for pristinamycin in *Streptomyces pristinaespiralis* (Handel et al. 2020), SagR for auricin in *Streptomyces aureofaciens* (Mingyar et al. 2015), and BulR1 for tacrolimus (FK506) in *Streptomyces tsukubaensis* (Salehi-Najafabadi et al. 2014).

7.5 The Pseudo-receptor Proteins: Their Function and Interactive Ligands

The pseudo-receptors are widely found in *Streptomyces* species (Table 7.1), and many of them act as negative regulators for antibiotic production, for example, TylQ for tylosin production in *Streptomyces fradiae* (Stratigopoulos and Cundliffe 2002), BarB for virginiamycin in *Streptomyces virginiae* (Matsuno et al. 2004), ScbR2 for coelimycin P-1 in *Streptomyces coelicolor* (Gottelt et al. 2010), AlpW for orange pigment kinamycin in *Streptomyces ambofaciens* (Bunet et al. 2008), AvaR2 for avermectin in *Streptomyces avermitilis* (Zhu et al. 2016), and SrrB for lankacidin and lankamycin in *S. rochei* (Arakawa et al. 2007; Misaki et al. 2020). Their phylogenetic trees were shown in Fig. 7.4. Their apparent differences are the location to the signaling molecule synthase genes. Although the signaling molecule receptor genes usually locate adjacent to the signaling molecule synthase genes, the pseudo-receptor genes have no relationship with the synthase genes (Biarnes-Carrera et al. 2015).

In general, the pseudo-receptors are insensitive to endogenous signaling molecules. BarB has no binding affinity to virginiae butanolides in *S. virginiae* (Matsuno et al. 2004). ScbR2 from *S. coelicolor* does not bind to the signaling molecules SCBs but to two antibiotics, actinorhodin (ACT) and undecylprodigiosin (RED) (Fig. 7.5) (Xu et al. 2010). JadR2 from *S. venezuelae* binds not only to endogenous antibiotics jadomycin and chloramphenicol (Fig. 7.5) as a ligand but does to exogenous antibiotics ACT and RED (Xu et al. 2010). Thus, ScbR2 and JadR2 are able to bind to multiple antibiotics and coordinate their biosynthesis (Zou et al. 2014). In *S. avermitilis*; AvaR2 binds to the endogenous signaling molecule avenolide, but not to endogenous antibiotics, oligomycin and avermectin (Zhu et al. 2016). In *S. rochei*, SrrB binds to the endogenous signaling molecules SRBs, but not to endogenous polyketide antibiotics lankacidin and lankamycin, together with exogenous antibiotics including aromatic polyketide chlortetracycline, aminoglycoside antibiotic kanamycin, and β-lactam antibiotic ampicillin even at 1 mM concentration (Misaki et al. 2020). Thus, the ligand-binding properties of the pseudo-receptors are variable in *Streptomyces* species.

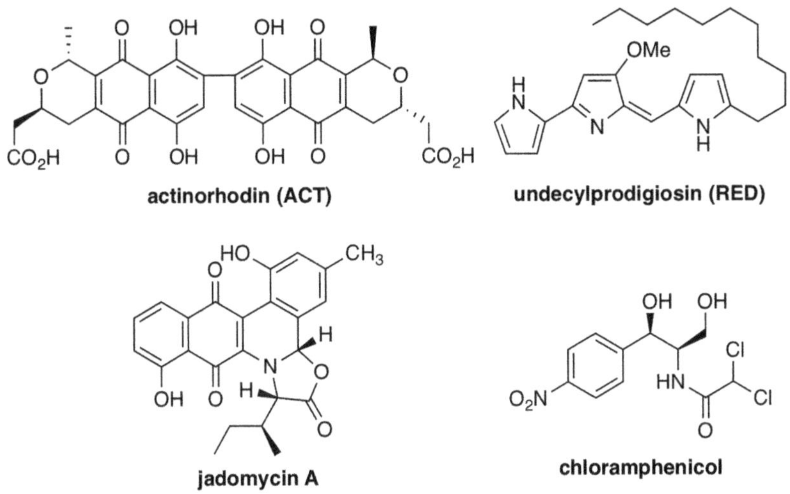

Fig. 7.5 Antibiotics function as regulators for the pseudo-receptors. *Me* methyl

7.6 Signaling Molecules/Antibiotics Act as Intra-/Interspecies Signals

Some signaling molecules/antibiotics act as intraspecies signals among actinomycetes. SCB3 in *S. coelicolor* acts as a signaling molecule for production of jadomycin in *S. venezuelae*, since their signaling molecules (SCB3 in *S. coelicolor* and SVB1 in *S. venezuelae*) have same chemical structures (Zou et al. 2014). Jadomycin, produced by *S. venezuelae*, triggers RED production by binding to pseudo-receptor ScbR2 in *S. coelicolor* (Wang et al. 2014a, b). Addition of avermectin led to induce cryptic secondary metabolites in *S. albus* J1074, while butenolide-type signaling molecules in *S. albus* act as intraspecies signals to induce avermectin production in *S. avermitilis* (Nguyen et al. 2018). Thus, exogenous signaling molecules and antibiotics are effective tools for induction of secondary metabolite production in *Streptomyces*.

Remarkably, some streptomycetes have antibiotic-producing ability of *Aspergillus* species (Netzker et al. 2015). Coculture of *Aspergillus nidulans* with 58 culture collection strains led to activate the fungal biosynthetic gene cluster for archetypal polyketide orsellinic acid and its derivative, lecanoric acid (Schroeckh et al. 2009). In a similar manipulation, *Streptomyces rapamycinicus* interacted with the human pathogenic fungus *Aspergillus fumigatus* which led to activate the silent polyketide synthase coding for the prenylated polyphenol (fumicycline A) (König et al. 2013). Extensive investigation of physical interspecies interaction between actinomycetes and other microbes, including *Aspergillus*, *Bacillus*, *Penicillium*, and others together with symbionts, will lead to accumulation of novel natural (in some cases "hybrid") metabolites.

7.7 The *Streptomyces* Antibiotic Regulatory Proteins (SARP): One of the Main Targets of Signaling Molecule/Receptor System

The signaling molecule/receptor regulatory system generally has a target activator gene that activates the downstream gene(s) for antibiotic production in *Streptomyces* species. In *S. griseus*, an *adpA* gene, an *araC/xylS*-family activator gene, is a direct target of ArpA. The gene product of *adpA* then activates many genes (AdpA regulon) responsible for secondary metabolite production and morphological development (Ohnishi et al. 2005). Accumulated data indicates that *Streptomyces* antibiotic regulatory protein (SARP)-type regulators are one of main targets of signaling molecule/receptor regulatory system (Bibb 2005). SARPs are widely distributed among actinomycetes such as *Streptomyces*, *Micromonospora*, *Nocardia*, and *Amycolatopsis* (Krause et al. 2020) and play an important role in transcriptional activation of these biosynthetic genes (Bibb 2005). SARPs are characterized by the presence of an OmpR-like DNA-binding domain at the N-terminus (Wietzorrek and Bibb 1997) that binds to the conserved consensus sequence of the target DNA (Fernandez-Moreno et al. 1991; Madduri and Hutchinson 1995). Representative SARPs for secondary metabolite production are the following: ActII-ORF4 for actinorhodin production in *S. coelicolor* (Takano et al. 1992; Arias et al. 1999), RedD for undecylprodigiosin in *S. coelicolor* (Takano et al. 1992), DnrI for daunorubicin in *S. peucetius* (Madduri and Hutchinson 1995; Ajithkumar et al. 2013), CcaR for cephamycin in *S. clavuligerus* (Pérez-Llarena et al. 1997), FdmR1 for fredericamycin in *S. griseus* ATTCC 49344 (Chen et al. 2008), TylS for tylosin in *S. fradiae* (Bate et al. 2002; Cundliffe 2008), NanR1 and NanR2 for nanchangmycin in *S. nangchangensis* (Yu et al. 2012), and SrrY and SrrZ for lankamycin and lankacidin in *S. rochei* (Suzuki et al. 2010). Phylogenetic tree of SARP-type activators is shown in Fig. 7.6.

SARPs specifically bind to a heptameric direct repeat sequence (5′-TCGAGXX-3′) at the promoter region of their targets, among which a 4-bp spacers are located (Arias et al. 1999; Sheldon et al. 2002; Liu et al. 2013). Molecular studies of AfsR protein in *S. coelicolor* revealed that a 8 bp distance from the −10 region of the promoter (Tanaka et al. 2007) is important for tripartite complex of the *afsS* promoter (a target of SARP)/AfsR (SARP)/RNA polymerase. The DNA-binding sequences including heptameric direct repeats and 4-bp spacers at the promoter of target DNA (s) have been determined in several *Streptomyces* species (Fig. 7.7). The characterized DR sequences for secondary metabolite production are the following: SrrY binding at upstream of *srrZ* for lankamycin production in *S. rochei* (Suzuki et al. 2010), ActII-ORF4 binding at upstream of both *actI-ORF1* and *actI-ORFA* for actinorhodin in *S. coelicolor* (Arias et al. 1999), DnrI binding at upstream of *dnrG* and *dspE* for daunorubicin in *S. peucetius* (Tang et al. 1996), FdmR1 binding at upstream of *fdmR1*, *fdmD*, and *fdmV* for fredericamycin in *S. griseus* ATCC 49344 (Chen et al. 2008), Aur1PR4 binding at upstream of *sal3* for auricin in *S. aureofaciens* (Rehakova et al. 2013), GriR binding at upstream of *griJ* and *griC*

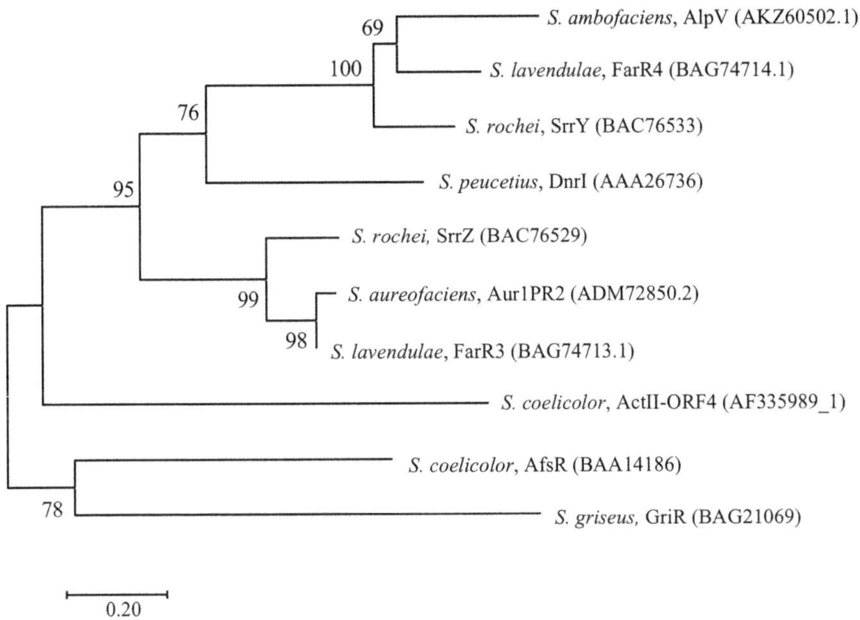

Fig. 7.6 Phylogenetic tree of the *Streptomyces* antibiotic regulatory proteins (SARPs) in several *Streptomyces* species using the maximum likelihood method. Phylogenetic tree was conducted with MEGA-X version 10.2.0 software. Numbers given in parentheses are accession numbers in the GenBank. Bootstrap values, expressed as a percentage of 1,000 replications, are given at branching points

SARP	Target gene	
SrrY	*srrZ*	GCGCCGCTTCCGCTTGGAGTGGGGCTCGAACGGCGCTGGAGATACCGGAGCCGGACGGGAGAACGCGGCTACGGTGCAGA**C**G
ActII-ORF4	*actI*-ORF1	GCATCGAGGGGTCCCGTATCGGCCTTCGAGCCTCCTTCGAGCCACGGGGCCGACGATGACG**A**CGACCACCGGACGAACGCAT
ActII-ORF4	*actVI*-ORFA	CTGGCAGCGCCCACTCGAACCCGGCTCGAGGGAACTTCGAGGCCACCGGGCGACCGTGGCCGT**A**TGACCATCACCGCGCTTC
DnrI	*dnrG*	TCGAGCCGCGCTTGACGACCCGGCTTCCAGAGGGGGTCGAGCGGCGCCGCAGAGCCTCGGAG**T**CCGGCGTGACCGCACCTGT
DnrI	*dpsE*	TCGACCCCCTCTGGAAGCCGGGTCGTCAAGCGCGGCTCGAGCGCGAGATGTGAGGGTGGGCCC**T**CCGTCGAAGGGAGGACCA
FdmR1	*fdmR1*	GCTCACCGTCCTTGTCGAGTACCGCTCAAGCAGAAGTCAAGTAATCGCCCGCACGCTCAC**G**AAGCGCATCAGGCCGTAGGGA
FdmR1	*fdmD*	GCGGAAATCTGACGTCGAGCTGGCCTGGACCGGCCCTCAAGCCGCCCGGGCCAGCATCAC**G**TCCCGACAGCCCGCCGGCCGC
FdmR1	*fdmV*	CCGGGCCGGAACCATCCAGCGCCCCTCCACGCGGTTTCGAGCGGTGCGGGAAACGCTGGA**C**GCCGACGGCAAGCCAAGGTGA
Aur1PR4	*sal3*	CGCGCGGTACGGCGTCGAGCGGCTTTCCAGCGGAGCTGGACGCGCCGCCCCGAGGGTGGGC**G**CATCGCCCCCGTATGGCGTG
GriR	*griJ*	CGCAATAGCCCGCCGGTGCGACGGAGAGACGGCCCCGATATCCCCCTCCTGGAAGCTCGTCGC**A**GTGCTCGGGAGGCCGGGC
GriR	*griC*	CGGACACGTTTCACGACACCCGGGCGATACGTGGCCGATAGGCCCCGCCCGCACGATGGGTGCC**G**TACCGCCCTTGCCCCGT
PapR2	*papR1*	TGCCGTGACGCCGTGTCAGGCGGGCGGTGCGCAGGTGTGGTTGCGTACGGCGGTCAGGCAGCCGTCGTTGCAGGTGGGACGG

Fig. 7.7 Alignment of the promoter region containing SARP-binding heptameric direct repeat and 4-bp spacer sequences determined in several *Streptomyces* species. The possible −10 regions are shaded in gray. The boldfaces indicate the transcription start sites (TSS), and the underlines indicate the heptameric direct repeat sequences. TSS of *papR1* is unknown

for grixazone in *S. griseus* (Higashi et al. 2007), and PapR2 binding at upstream of *papR1* for pristinamycin in *S. pristinaespiralis* (Mast et al. 2015).

7.8 Natural Product Discovery Through Manipulation of Regulatory System

One of the important aspects in understanding the *Streptomyces* regulatory system is an application for systematic natural products discovery. Manipulation of regulatory genes often causes activation of "silent" secondary metabolites (Arakawa 2018; Zerikly and Challis 2009; Rutledge and Challis 2015; Olano et al. 2008). In *S. ambofaciens*, mutation of the pseudo-receptor gene *alpW* caused kinamycin production (Bunet et al. 2011). A mutant of the repressor gene *ksbC* accumulated β-carboline compound kitasetaline in *Kitasatospora setae* (Aroonsri et al. 2012). In *S. rochei*, the triple mutations on two biosynthetic gene clusters for lankacidin and lankamycin and *srrB* led to accumulate azoxyalkene compound KA57-A (Kunitake et al. 2015). Overexpression of SARP gene is also a powerful strategy for activation of secondary metabolite production, since they only require simple genetic manipulation, introducing one SARP gene expression construct into the strains of interests. Overexpression of an SARP-7 gene in *Streptomyces* sp. MSC090213JE08 led to accumulate amide-containing polyene ishigamide, whose biosynthetic gene cluster located adjacent to the SARP-7 gene (Du et al. 2016). In *S. fungicidicus*, the enduracidin production increased fourfold when a SARP gene located in the enduracidin biosynthesis gene cluster was overexpressed in vivo (Chen et al. 2019). In *Streptomyces* sp. KO-7888, the recombinant of a SARP gene, *speR*, produced two novel lipopeptides, sarpeptins A and B (Koomsiri et al. 2019).

7.9 Conclusion and Future Perspective

Here we described the signaling molecule/receptor system for secondary metabolite production in *Streptomyces*. *Streptomyces* species have more than 30 secondary metabolite biosynthetic gene clusters in their genome; however, around 80–90% of them are weakly expressed or not at all. Genome mining is extensively studied by many researchers in natural product chemistry together with chemical and synthetic biology. Deciphering the complex regulatory cascade together with the structural diversity of *Streptomyces* signaling molecules may lead to activate "silent" secondary metabolite gene clusters.

References

Ahmed Y, Rebets Y, Tokovenko B, Brotz E, Luzhetskyy A (2017) Identification of butenolide regulatory system controlling secondary metabolism in *Streptomyces albus* J1074. Sci Rep 7: 9784

Aigle B, Pang X, Decaris B, Leblond P (2005) Involvement of AlpV, a new member of the *Streptomyces* antibiotic regulatory protein family, in regulation of the duplicated type II polyketide synthase *alp* gene cluster in *Streptomyces ambofaciens*. J Bacteriol 187:2491–2500

Ajithkumar V, Karuppasamy K, Prasad R (2013) Regulation of daunorubicin biosynthesis in *Streptomyces peucetius*—feed forward and feedback transcriptional control. J Basic Microbiol 2013(53):636–644

Arakawa K (2014) Genetic and biochemical analysis of the antibiotic biosynthetic gene clusters on the *Streptomyces* linear plasmid. Biosci Biotechnol Biochem 78:183–189

Arakawa K (2018) Manipulation of metabolic pathways controlled by signaling molecules, inducers of antibiotic production, for genome mining in *Streptomyces* spp. Antonie Van Leeuwenhoek 111:743–751

Arakawa K, Mochizuki S, Yamada K, Noma T, Kinashi H (2007) γ-Butyrolactone autoregulator-receptor system involved in lankacidin and lankamycin production and morphological differentiation in *Streptomyces rochei*. Microbiology 153:1817–1827

Arakawa K, Tsuda N, Taniguchi A, Kinashi H (2012) The butenolide signaling molecules SRB1 and SRB2 induce lankacidin and lankamycin production in *Streptomyces rochei*. ChemBioChem 13:1447–1457

Arias P, Fernandez-Moreno MA, Malpartida F (1999) Characterization of the pathway-specific positive transcriptional regulator for actinorhodin biosynthesis in *Streptomyces coelicolor* A3 (2) as a DNA-binding protein. J Bacteriol 181:6958–6968

Aroonsri A, Kitani S, Hashimoto J, Kosone I, Izumikawa M, Komatsu M, Fujita N, Takahashi Y, Shin-ya K, Ikeda H, Nihira T (2012) Pleiotropic control of secondary metabolism and morphological development by KsbC, a butyrolactone autoregulator receptor homologue in *Kitasatospora setae*. Appl Environ Microbiol 78:8015–8024

Bate N, Butler AR, Gandecha AR, Cundliffe E (1999) Multiple regulatory genes in the tylosin biosynthetic cluster of *Streptomyces fradiae*. Chem Biol 6:617–624

Bate N, Stratigopoulos G, Cundliffe E (2002) Differential roles of two SARP-encoding regulatory genes during tylosin biosynthesis. Mol Microbiol 43:449–458

Biarnes-Carrera M, Breitling R, Takano E (2015) Butyrolactone signalling circuits for synthetic biology. Curr Opin Chem Biol 28:91–98

Bibb MJ (2005) Regulation of secondary metabolism in streptomycetes. Curr Opin Microbiol 8: 208–215

Bunet R, Mendes MV, Rouhier N, Pang X, Hotel L, Leblond P, Aigle B (2008) Regulation of the synthesis of the angucyclinone antibiotic alpomycin in *Streptomyces ambofaciens* by the autoregulator receptor AlpZ and its specific ligand. J Bacteriol 190:3293–3305

Bunet R, Song L, Mendes MV, Corre C, Hotel L, Rouhier N, Framboisier X, Leblond P, Challis GL, Aigle B (2011) Characterization and manipulation of the pathway-specific late regulator AlpW reveals *Streptomyces ambofaciens* as a new producer of kinamycins. J Bacteriol 193: 1142–1153

Chen Y, Wendt-Pienkowski E, Shen B (2008) Identification and utility of FdmR1 as a *Streptomyces* antibiotic regulatory protein activator for fredericamycin production in *Streptomyces griseus* ATCC 49344 and heterologous hosts. J Bacteriol 190:5587–5596

Chen YW, Liu XC, Lv FX, Li P (2019) Characterization of three regulatory genes involved in enduracidin biosynthesis and improvement of enduracidin production in *Streptomyces fungicidicus*. J Appl Microbiol 127:1698–1705

Corre C, Song L, O'Rourke S, Chater KF, Challis GL (2008) 2-Alkyl-4-hydroxymethylfuran-3-carboxylic acids, antibiotic production inducers discovered by *Streptomyces coelicolor* genome mining. Proc Natl Acad Sci U S A 105:17510–17515

Cundliffe E (2008) Control of tylosin biosynthesis in *Streptomyces fradiae*. J Microbiol Biotechnol 18:1485–1491

Du D, Katsuyama Y, Onaka H, Fujie M, Satoh N, Shin-Ya K, Ohnishi Y (2016) Production of a novel amide-containing polyene by activating a cryptic biosynthetic gene cluster in *Streptomyces* sp. MSC090213JE08. Chembiochem 17:1464–1471

Fernandez-Moreno MA, Caballero JL, Hopwood DA, Malpartida F (1991) The act cluster contains regulatory and antibiotic export genes, direct targets for translational control by the *bldA* tRNA gene of *Streptomyces*. Cell 66:769–780

Gottelt M, Kol S, Gomez-Escribano JP, Bibb M, Takano E (2010) Deletion of a regulatory gene within the *cpk* gene cluster reveals novel antibacterial activity in *Streptomyces coelicolor* A3(2). Microbiology 156:2343–2353

Handel F, Kulik A, Mast Y (2020) Investigation of the autoregulator-receptor system in the pristinamycin producer *Streptomyces pristinaespiralis*. Front Microbiol 11:580990

Hara O, Beppu T (1982) Mutants blocked in streptomycin production in *Streptomyces griseus*—the role of A-factor. J Antibiot 35:349–358

Higashi T, Iwasaki Y, Ohnishi Y, Horinouchi S (2007) A-Factor and phosphate depletion signals are transmitted to the grixazone biosynthesis genes via the pathway-specific transcriptional activator GriR. J Bacteriol 189:3515–3524

Horinouchi S, Beppu T (2007) Hormonal control by A-factor of morphological development and secondary metabolism in Streptomyces. Prod Jpn Acad Ser B Phys Biol Sci 83:277–295

Hsiao NH, Nakayama S, Merlo ME, de Vries M, Bunet R, Kitani S, Nihira T, Takano E (2009) Analysis of two additional signaling molecules in *Streptomyces coelicolor* and the development of a butyrolactone-specific reporter system. Chem Biol 16:951–960

Kawauchi R, Akashi T, Kamitani Y, Sy A, Wangchaisoonthorn U, Nihira T, Yamada Y (2000) Identification of an AfsA homologue (BarX) from *Streptomyces virginiae* as a pleiotropic regulator controlling autoregulator biosynthesis, virginiamycin biosynthesis and virginiamycin M1 resistance. Mol Microbiol 36:302–313

Khokhlov AS, Tovarova II, Borisova LN, Pliner SA, Shevchenko LN, Kornitskaia EI, Ivkina NS, Rapoport IA (1967) The A-factor, responsible for streptomycin biosynthesis by mutant strains of *Actinomyces streptomycini*. Dokl Akad Nauk SSSR 177:232–235

Kinoshita H, Ipposhi H, Okamoto S, Nakano H, Nihira T, Yamada Y (1997) Butyrolactone autoregulator receptor protein (BarA) as a transcriptional regulator in *Streptomyces virginiae*. J Bacteriol 179:6986–6993

Kitani S, Yamada Y, Nihira T (2001) Gene replacement analysis of the butyrolactone autoregulator receptor (FarA) reveals that FarA acts as a Novel regulator in secondary metabolism of *Streptomyces lavendulae* FRI-5. J Bacteriol 183:4357–4363

Kitani S, Miyamoto KT, Takamatsu S, Herawati E, Iguchi H, Nishitomi K, Uchida M, Nagamitsu T, Omura S, Ikeda H, Nihira T (2011) Avenolide, a *Streptomyces* hormone controlling antibiotic production in *Streptomyces avermitilis*. Proc Natl Acad Sci U S A 108: 16410–16415

Kondo K, Higuchi Y, Sakuda S, Nihira T, Yamada Y (1989) New virginiae butanolides from *Streptomyces virginiae*. J Antibiot 42:1873–1876

Kong D, Wang X, Nie J, Niu G (2019) Regulation of antibiotic production by signaling molecules in *Streptomyces*. Front Microbiol 10:2927

König CC, Scherlach K, Schroeckh V, Horn F, Nietzsche S, Brakhage AA, Hertweck C (2013) Bacterium induces cryptic meroterpenoid pathway in the pathogenic fungus *Aspergillus fumigatus*. ChemBioChem 14:938–942

Koomsiri W, Inahashi Y, Leetanasaksakul K, Shiomi K, Takahashi YK, O Mura S, Samborskyy M, Leadlay PF, Wattana-Amorn P, Thamchaipenet A, Nakashima T (2019) Sarpeptins A and B, lipopeptides produced by *Streptomyces* sp. KO-7888 overexpressing a specific SARP regulator. J Nat Prod 82:2144–2151

Krause J, Handayani I, Blin K, Kulik A, Mast Y (2020) Disclosing the potential of the SARP-type regulator PapR2 for the activation of antibiotic gene clusters in *Streptomycetes*. Front Microbiol 11:225

Kunitake H, Hiramatsu T, Kinashi H, Arakawa K (2015) Isolation and biosynthesis of an azoxyalkene compound produced by a multiple gene disruptant of *Streptomyces rochei*. ChemBioChem 16:2237–2243

Li X, Wang J, Li S, Ji J, Wang W, Yang K (2015) ScbR- and ScbR2-mediated signal transduction networks coordinate complex physiological responses in *Streptomyces coelicolor*. Sci Rep 5: 14831

Liu G, Chater KF, Chandra G, Niu G, Tan H (2013) Molecular regulation of antibiotic biosynthesis in *Streptomyces*. Mol Biol Rev 77:112–143

Madduri K, Hutchinson CR (1995) Functional characterization and transcriptional analysis of a gene cluster governing early and late steps in daunorubicin biosynthesis in *Streptomyces peucetius*. J Bacteriol 177:3879–3884

Martin J, Liras P (2020) The balance metabolism safety net: integration of stress signals by interacting transcriptional factors in *Streptomyces* and related actinobacteria. Front Microbiol 10:3120

Mast Y, Guezguez J, Handel F, Schinko E (2015) A complex signaling cascade governs pristinamycin biosynthesis in *Streptomyces pristinaespiralis*. Appl Environ Microbiol 81: 6621–6636

Matselyukh B, Mohammadipanah F, Laatsch H, Rohr J, Efremenkova O, Khilya V (2015) N-methylphenylalanyl-dehydrobutyrine diketopiperazine, anA-factor mimic that restores antibiotic biosynthesis and morphogenesis in *Streptomyces globisporus* 1912-B2 and *Streptomyces griseus* 1439. J Antibiot 68:9–14

Matsuno K, Yamada Y, Lee CK, Nihira T (2004) Identification by gene deletion analysis of *barB* as a negative regulator controlling an early process of virginiamycin biosynthesis in *Streptomyces virginiae*. Arch Microbiol 181:52–59

Mingyar E, Feckova L, Novakova R, Bekeova C, Kormanec J (2015) A gamma-butyrolactone autoregulator-receptor system involved in the regulation of auricin production in *Streptomyces aureofaciens* CCM 3239. Appl Microbiol Biotechnol 99:309–325

Misaki Y, Yamamoto S, Suzuki T, Iwakuni M, Sasaki H, Takahashi Y, Inada K, Kinashi H, Arakawa K (2020) SrrB, a pseudo-receptor protein, acts as a negative regulator for lankacidin and lankamycin production in *Streptomyces rochei*. Front Microbiol 11:1089

Netzker T, Fischer J, Weber J, Mattern DJ, König CC, Valiante V, Schroeckh V, Brakhage AA (2015) Microbial communication leading to the activation of silent fungal secondary metabolite gene clusters. Front Microbiol 6:299

Nguyen TB, Kitani S, Shimma S, Nihira T (2018) Butenolides from *Streptomyces albus* J1074 act as external signals to stimulate avermectin production in *Streptomyces avermitilis*. Appl Environ Microbiol 84:e2791–e2717

Niu G, Chater KF, Tian Y, Zhang J, Tan H (2016) Specialized metabolites regulating antibiotic biosynthesis in *Streptomyces* spp. FEMS Microbiol Rev 40:554–573

Ohnishi Y, Kameyama S, Onaka H, Horinouchi S (1999) The A-factor regulatory cascade leading to streptomycin production in *Streptomyces griseus*: identification of a target gene of the A-factor receptor. Mol Microbiol 34:102–111

Ohnishi Y, Yamazaki H, Kato JY, Tomono A, Horinouchi S (2005) AdpA, a central transcriptional regulator in the A-factor regulatory cascade that leads to morphological development and secondary metabolism in *Streptomyces griseus*. Biosci Biotechnol Biochem 69:431–439

Okamoto S, Nakamura K, Nihira T, Yamada Y (1995) Virginiae butanolide binding protein from *Streptomyces virginiae*. Evidence that VbrA is not the virginiae butanolide binding protein and reidentification of the true binding protein. J Biol Chem 270:12319–12326

Olano C, Lombó F, Méndez C, Salas JA (2008) Improving production of bioactive secondary metabolites in actinomycetes by metabolic engineering. Metab Eng 10:281–292

Onaka H, Ando N, Nihira T, Yamada Y, Beppu T, Horinouchi S (1995) Cloning and characterization of the A-factor receptor gene from *Streptomyces griseus*. J Bacteriol 177:6083–6092

Pérez-Llarena FJ, Liras P, Rodríguez-García A, Martín JF (1997) A regulatory gene (*ccaR*) required for cephamycin and clavulanic acid production in *Streptomyces clavuligerus*: amplification results in overproduction of both beta-lactam compounds. J Bacteriol 179:2053–2059

Recio E, Colinas A, Rumbero A, Aparicio JF, Martín JF (2004) PI factor, a novel type quorum-sensing inducer elicits pimaricin production in *Streptomyces natalensis*. J Biol Chem 279: 41586–41593

Rehakova A, Novakova R, Feckova L, Mingyar E, Kormanec J (2013) A gene determining a new member of the SARP family contributes to transcription of genes for the synthesis of the angucycline polyketide auricin in *Streptomyces aureofaciens* CCM 3239. FEMS Microbiol Lett 346:45–55

Rutledge PJ, Challis GL (2015) Discovery of microbial natural products by activation of silent biosynthetic gene clusters. Nat Rev Microbiol 13:509–523

Salehi-Najafabadi Z, Barreiro C, Rodríguez-García A, Cruz A, López GE, Martín JF (2014) The gamma-butyrolactone receptors BulR1 and BulR2 of *Streptomyces tsukubaensis*: tacrolimus (FK506) and butyrolactone synthetases production control. Appl Microbiol Biotechnol 98: 4919–4936

Sato K, Nihira T, Sakuda S, Yanagimoto M, Yamada Y (1989) Isolation and structure of a new butyrolactone autoregulator from *Streptomyces* sp. FRI-5. J Ferment Bioeng 68:170–173

Schroeckh V, Scherlach K, Nützmann H-W, Shelest E, Schmidt-Heck W, Schuemann J, Martin K, Hertweck C, Brakhage AA (2009) Intimate bacterial-fungal interaction triggers biosynthesis of archetypal polyketides in Aspergillus nidulans. Proc Natl Acad Sci U S A 106:14558–14563

Sheldon PJ, Busarow SB, Hutchinson CR (2002) Mapping the DNA-binding domain and target sequences of the *Streptomyces peucetius* daunorubicin biosynthesis regulatory protein, DnrI. Mol Microbiol 44:449–460

Sidda JD, Poon V, Song L, Wang W, Yang K, Corre C (2016) Overproduction and identification of butyrolactones SCB1–8 in the antibiotic production superhost *Streptomyces* M1152. Org Biomol Chem 14:6390–6393

Stratigopoulos G, Cundliffe E (2002) Expression analysis of the tylosin-biosynthetic gene cluster: pivotal regulatory role of the *tylQ* product. Chem Biol 9:71–78

Suzuki T, Mochizuki S, Yamamoto S, Arakawa K, Kinashi H (2010) Regulation of lankamycin biosynthesis in *Streptomyces rochei* by two SARP genes, *srrY* and *srrZ*. Biosci Biotechnol Biochem 74:819–827

Takano E (2006) γ-Butyrolactones: *Streptomyces* signaling molecules regulating antibiotic production and differentiation. Curr Opin Microbiol 9:287–294

Takano E, Gramajo HC, Strauch E, Andres N, White J, Bibb MJ (1992) Transcriptional regulation of the *redD* transcriptional activator gene accounts for growth-phase-dependent production of the antibiotic undecylprodigiosin in *Streptomyces coelicolor* A3(2). Mol Microbiol 6:2797–2804

Takano E, Chakaraburtty R, Nihira T, Yamada Y, Bibb MJ (2001) A complex role for the γ-butyrolactone SCB1 in regulating antibiotic production in *Streptomyces coelicolor* A3(2). Mol Microbiol 41:1015–1028

Takano E, Kinoshita H, Mersinias V, Bucca G, Hotchkiss G, Nihira T, Smith CP, Bibb M, Wohlleben W, Chater K (2005) A bacterial hormone (the SCB1) directly controls the expression of a pathway-specific regulatory gene in the cryptic type I polyketide biosynthetic gene cluster of *Streptomyces coelicolor*. Mol Microbiol 56:465–479

Tanaka A, Takano Y, Ohnishi Y, Horinouchi S (2007) AfsR recruits RNA polymerase to the *afsS* promoter: a model for transcriptional activation by SARPs. J Mol Biol 369:322–333

Tang L, Grimm A, Zhang YX, Hutchinson CR (1996) Purification and characterization of the DNA-binding protein DnrI, a transcriptional factor of daunorubicin biosynthesis in *Streptomyces peucetius*. Mol Microbiol 22:801–813

Teshima A, Hadae N, Tsuda N, Arakawa K (2020) Functional analysis of P450 monooxygenase SrrO in the biosynthesis of butenolide-type signaling molecules in *Streptomyces rochei*. Biomolecules 10:1237

Thao NB, Kitani S, Nitta H, Tomioka T, Nihira T (2017) Discovering potential *Streptomyces* hormone producers by using disruptants of essential biosynthetic genes as indicator strains. J Antibiot 70:1004–1008

Vicente CM, Girardet JM, Hôtel L, Aigle B (2020) Molecular dynamics to elucidate the DNA-Binding activity of AlpZ, a member of the gamma-butyrolactone receptor family in *Streptomyces ambofaciens*. Front Microbiol 11:1255

Wang L, Vining LC (2003) Control of growth, secondary metabolism and sporulation in *Streptomyces venezuelae* ISP5230 by *jadW1*, a member of the *afsA* family of gamma-butyrolactone regulatory genes. Microbiology 149:1991–2004

Wang L, Tian X, Wang J, Yang H, Fan K, Xu G, Yang K, Tan H (2009) Autoregulation of antibiotic biosynthesis by binding of the end product to an atypical response regulator. Proc Natl Acad Sci U S A 106:8617–8622

Wang JB, Zhang F, Pu JY, Zhao J, Zhao QF, Tang GL (2014a) Characterization of AvaR1, an autoregulator receptor that negatively controls avermectins production in a high avermectin-producing strain. Biotechnol Lett 36:813–819

Wang W, Ji J, Wang J, Li S, Pan G, Fan K, Yang K (2014b) Angucyclines as signals modulate the behaviors of *Streptomyces coelicolor*. Proc Natl Acad Sci U S A 111:5688–5693

Wang W, Zhang J, Liu X, Li D, Li Y, Tian Y, Tan H (2018) Identification of a butenolide signaling system that regulates nikkomycin biosynthesis in *Streptomyces*. J Biol Chem 293:20029–20040

Wietzorrek A, Bibb MJ (1997) A novel family of proteins that regulates antibiotic production in streptomycetes appears to contain an OmpR-like DNA-binding fold. Mol Microbiol 25:1181–1184

Xu G, Yang S (2019) Regulatory and evolutionary roles of pseudo γ-butyrolactone receptors in antibiotic biosynthesis and resistance. Appl Microbiol Biotechnol 103:9373–9378

Xu G, Wang J, Wang L, Tian X, Yang H, Fan K, Yang K, Tan H (2010) "Pseudo" γ-butyrolactone receptors respond to antibiotic signals to coordinate antibiotic biosynthesis. J Biol Chem 285: 27440–27448

Yamada Y (1995) Butyrolactone autoregulators, inducers of secondary metabolites, in *Streptomyces*. Actinomycetol 9:57–65

Yamamoto S, He Y, Arakawa K, Kinashi H (2008) γ-Butyrolactone-dependent expression of the SARP gene *srrY* plays a central role in the regulatory cascade leading to lankacidin and lankamycin production in *Streptomyces rochei*. J Bacteriol 190:1308–1316

Yang K, Han L, Vining LC (1995) Regulation of jadomycin B production in *Streptomyces venezuelae* ISP5230: involvement of a repressor gene, *jadR2*. J Bacteriol 177:6111–6117

Yu Z, Reichheld SE, Savchenko A, Parkinson J, Davidson AR (2010) A comprehensive analysis of structural and sequence conservation in the TetR family transcriptional regulators. J Mol Biol 400:847–864

Yu Q, Du A, Liu T, Deng Z, He X (2012) The biosynthesis of the polyether antibiotic nanchangmycin is controlled by two pathway-specific transcriptional activators. Arch Microbiol 194:415–426

Zerikly M, Challis GL (2009) Strategies for the discovery of new natural products by genome mining. ChemBioChem 10:625–633

Zhou S, Bhukya H, Malet N, Harrison PJ, Rea D, Belousoff MJ, Venugopal H, Sydor PK, Styles KM, Song L, Cryle MJ, Alkhalaf LM, Fülöp V, Challis GL, Corre C (2021) Molecular basis for control of antibiotic production by a bacterial hormone. Nature 590:463–467

Zhu J, Sun D, Liu W, Chen Z, Li J, Wen Y (2016) AvaR2, a pseudo γ-butyrolactone receptor homologue from *Streptomyces avermitilis*, is a pleiotropic repressor of avermectin and avenolide biosynthesis and cell growth. Mol Microbiol 102:562–578

Zou Z, Du D, Zhang Y, Zhang J, Niu G, Tan H (2014) A γ-butyrolactone-sensing activator/repressor, JadR3, controls a regulatory mini-network for jadomycin biosynthesis. Mol Microbiol 94:490–505

New Strategies to Activate Secondary Metabolism in *Streptomyces* 8

Paula Yagüe, Nathaly Gonzalez-Quiñonez, Gemma Fernández-García, Sergio Alonso-Fernández, and Angel Manteca

Abstract

Streptomycetes are important industrial bacteria from which two thirds of clinically relevant secondary metabolites, especially antibiotics, were discovered. Antibiotic resistance became one of the major challenges in clinic, and new antibiotics are urgently required. Natural microbial strains are still the best source to find new scaffolds and families of antibiotics. Next-generation sequencing analyses revealed that streptomycetes harbour many cryptic secondary metabolite pathways, i.e. pathways not expressed in the laboratory. A typical *Streptomyces* strain produces around 4 secondary metabolites in the lab, but it encodes 20–50 secondary metabolite pathways. To activate *Streptomyces* secondary metabolite cryptic pathways is one of the main challenges in industry. Streptomycetes have complex developmental cycles, which include mycelium differentiation and sporulation. Secondary metabolism is conditioned by differentiation but differentiation in liquid cultures still not fully understood, due to the fact that most *Streptomyces* do not sporulate in these conditions. Here, we review the strategies to activate cryptic pathways in streptomycetes.

Keywords

Streptomyces · Screening · Antibiotics · Secondary metabolism · Differentiation · Liquid cultures · Bioreactors

---

Paula Yagüe and Nathaly Gonzalez-Quiñonez contributed equally to this work.

P. Yagüe · N. Gonzalez-Quiñonez · G. Fernández-García · S. Alonso-Fernández · A. Manteca (✉)
Área de Microbiología, Departamento de Biología Funcional, IUOPA, ISPA, Facultad de Medicina, Universidad de Oviedo, Oviedo, Spain
e-mail: yaguepaula@uniovi.es; mantecaangel@uniovi.es

R. V. Rai, J. A. Bai (eds.), *Natural Products from Actinomycetes*,
https://doi.org/10.1007/978-981-16-6132-7_8

8.1 *Streptomycetes* Are Complex Mycelial Sporulating Bacteria and the Main Source of Bioactive Compounds

8.1.1 *Streptomyces* Development

The classical *Streptomyces* developmental cycle defined by Waksman (1959) and (Wildermuth (1970) focused on sporulation: a vegetative mycelium (substrate mycelium) differentiates to a new mycelium that starts to grow into the air (aerial mycelium) that suffers a programmed cell death (PCD), and the remaining viable hyphae differentiate into unigenomic spores. Most streptomycetes do not sporulate in liquid cultures, but secondary metabolite production is usually performed in liquid (flasks or bioreactors). More recently, a young, compartmentalised mycelium (MI) was discovered after spore germination (Manteca et al. 2005). MI hyphae suffer a programmed cell death (Manteca et al. 2006), and the remaining viable segments differentiate to a multinucleated mycelium (MII) which corresponds to the substrate and aerial mycelium. Sporulation is absent in liquid cultures, but the MI and MII also exist in these culture conditions. Interestingly, MII is the antibiotic producer mycelium in solid and liquid cultures (Manteca et al. 2008). Transcriptomic and proteomic experiments corroborated that MI is the vegetative mycelium, while MII is the antibiotic producer mycelium (Yague et al. 2014; Rioseras et al. 2018).

8.1.2 *Streptomyces* Differentiation and Industrial Fermentations

The basic knowledge about differentiation in liquid cultures has been generated recently. MII was demonstrated to be the antibiotic-producing mycelium in solid sporulating cultures and liquid not-sporulating cultures (Manteca et al. 2008). Until then, optimisation of industrial fermentation was performed empirically for each strain and compound (Manteca et al. 2008). Differentiation was demonstrated to be one of the keys to optimise *Streptomyces* secondary metabolite production in bioreactors (Rioseras et al. 2014). Recently, some genetic determinants controlling pellet and clump formation were discovered. The *mat* gene cluster (van Dissel et al. 2015) and the *cslA*, *glxA* and *dtpA* genes modulate mycelial aggregation (Chaplin et al. 2015; Petrus et al. 2016; Petrus and Claessen 2014) and might be used to modulate pellet and clump formation in industrial fermentations.

8.1.3 Novel Drug Discovery Approaches

Two thirds of the secondary metabolites used in clinic, especially antibiotics, were discovered in *Streptomyces* strains. Clavulanic acid (*S. clavuligerus*) (Reading and Cole 1977), neomycin (*S. marinensis*) (Waksman and Lechevalier 1949), chloramphenicol (S. *venezuelae*) (Shapiro and Vining 1983), the insecticide avermectin (*S. avermitilis*) (Ikeda et al. 1999), the immunosuppressant tacrolimus

(*S. tsukubaensis*) (Karlsson et al. 1991), kanamycin (*S. kanamyceticus*) (Kharel et al. 2004) and potent antitumoral platenolides (*S. platensis*) (Sakai et al. 2004) are examples of the vast amount of bioactive compounds produced by members of the *Streptomyces* genus. During the past 30 years only three new classes of antibiotics, mutilins, lipopeptides and oxazolidinones, have been introduced in clinic (Fischbach and Walsh 2009). The last new antibiotic scaffold discovered corresponded to platensimycin, produced by *S. platensis*, identified in 2006 (Takahashi and Omura 2003). Unfortunately, further work is necessary to improve platensimycin synthesis and stability before using this antibiotic in clinic (reviewed in (Lu and You 2010)). Simultaneously, microbial resistance to antibiotics increased dramatically. New antibiotics are urgently required in clinic to fight against microbial resistance.

Next-generation sequencing analyses revealed that a typical *Streptomyces* strain produces around 4 secondary metabolites in the lab, but its genome encodes 20–50 secondary metabolite pathways (cryptic pathways not expressed under usual laboratory conditions) (Genilloud 2014; Manteca and Yague 2018).

8.1.4 Screening for New Bioactive Compounds from Natural Microorganisms

High-throughput screening (HTS) from natural microorganisms needs the correct compound library and the correct target assay(s). Natural compound libraries, especially those coming from streptomycetes, were the most successful to discover antimicrobials (Fischbach and Walsh 2009). New antimicrobials can be found activating cryptic pathways, which indeed are the majority of pathways (Genilloud 2014; Manteca and Yague 2018). New strains obtained from niches that have not been well explored or new synthetic compound libraries can also be screened (Manteca and Yague 2018). Changing the target of the assay used to test the antimicrobial activity can also contribute to the discovery of new antimicrobials (Yague et al. 2020).

8.2 New Strategies for High-Throughput Screening from Natural Microorganisms

HTS for bioactive compounds stands on three pillars: first, search of the microbial strains to be tested, mainly actinomycetes, but also fungi and other microorganisms; second, culture and prepare the cellular extracts of these strains, trying to activate cryptic pathway; and third, the screening assay to identify bioactive compounds from the cellular extracts. In this chapter, we will focus in actinomycetes and in the state of the art of the new strategies to access to new strains (Sect. 8.2.1) and to activate secondary metabolite production (Sect. 8.2.2).

8.2.1 New Strategies for Actinomycete Strain Isolation

8.2.1.1 Exploring New Niches

One the most common strategies to look for new secondary metabolites is the isolation of new streptomycetes from new environments. During the last decade, marine environments have been the most productive niches to find new species of *Streptomyces (*Lam 2006*)*. New *Streptomyces* strains have also been isolated from other exotic environments such as the Himalaya (Bhattacharjee et al. 2012), islands (Charousova et al. 2017), Atacama Desert (Idris et al. 2017), Antarctica (Sanyal et al. 2018) and extremophile habitats (Bonilla et al. 2018; Ben Abdallah et al. 2018; Bull et al. 2018).

8.2.1.2 Symbiotic Relationships

Streptomycetes were classically considered as free-living soil bacteria. However, they have symbiotic relationships with plants, fungi and animals (Seipke et al. 2012), and these symbiotic interactions condition secondary metabolism (Seipke et al. 2012). One of the first discovered *Streptomyces* symbiotic interactions was with plant roots (Schrey and Tarkka 2008). Streptomycetes are present in the rhizosphere, providing a natural defence against fungi (Schrey and Tarkka 2008). *Streptomyces* also interacts with insects, providing protection against pathogens (Kroiss et al. 2010; Engl et al. 2018). New bioactive compounds, such as natalamycin A, geldanamycin (Kim et al. 2014), dentigerumycin (Wyche et al. 2017), termisoflavone A (Kang et al. 2016) or microtermolides A and B (Laureti et al. 2011), were isolated from *Streptomyces* strains isolated from termites. The antifungals mycangimycin (Qian et al. 2020) and frontalamides A and B (Onaka 2017; Marmann et al. 2014) were discovered from *Streptomyces* strains isolated from beetles. Novel bioactive compounds have been isolated from new *Streptomyces* strains isolated from ants (Piel 2004; Yu et al. 2017; Perez et al. 2011; Wang et al. 2014).

8.2.1.3 Non-culturable Microorganisms

Non-culturable microorganisms, i.e. microbes that cannot be cultured under usual laboratory conditions, are a vast majority (Li et al. 2009). Genome sequencing revealed the big pharmacological potential of uncultured bacteria, and novel cultivation methods, such as isolation chip (iChip), are very promising to look for novel secondary metabolites (Kealey et al. 2017). The combination of iChip, with the screening of unexplored niches can contribute to the discovery of new bioactive compounds (Kealey et al. 2017).

8.2.2 Old and New Strategies to Activate and Explore Secondary Metabolism in Actinomycetes

Actinomycete strains need to be screened for the production of novel secondary metabolites. With this aim, strains need to be cultured to prepare extracts containing

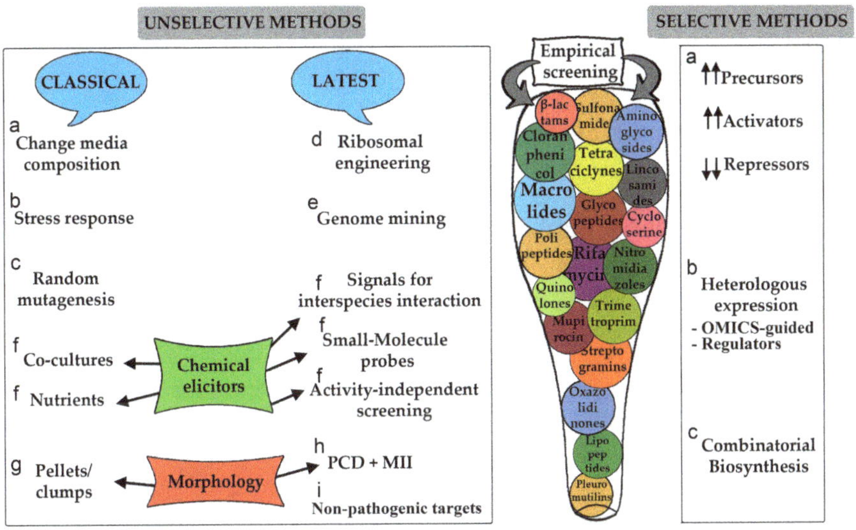

Fig. 8.1 Outline of the different strategies to activate *Streptomyces* secondary metabolism reviewed in this chapter

the compounds to be screened. In 2013, Arryn Craney et al. (Craney et al. 2013) classified the methods used to enhance secondary metabolism in unselective and selective methods. Unselective methods are used to the screening for new activities, whereas selective methods are used to improve the production of known compounds (Craney et al. 2013). This classification is not exclusive, and, in occasions, unselective methods can be used to improve the selective production of a known compound, or selective methods can be used to unselectively improve production. However this classification allows to nicely organise the most important methods used for secondary metabolism activation (Fig. 8.1).

8.2.2.1 Selective Methods to Improve the Production of Known Compound

8.2.2.1.1 Engineering Precursor Biosynthesis and Specific Regulators

Each secondary metabolite biosynthetic pathway has several precursors that sometimes limit the overall production. A strategy widely used to improve secondary metabolite production enhances the biosynthetic pathways of these precursors or even, adds them to the cultures.

Secondary metabolism is highly regulated, and each specific pathway has positive and negative regulators. Engineering the expression of these regulators can improve the final yield of specific bioactive compounds (Yi et al. 2018; Razmilic et al. 2018). Omic methodologies (transcriptomics, proteomics and metabolomics) have been successfully used to identify rate-limiting biosynthetic steps, as well as precursor biosynthetic pathways, needed for secondary metabolite biosynthesis (Tan and Liu 2017).

8.2.2.1.2 Heterologous Expression, Omics-Guided Secondary Metabolite Synthesis

Heterologous expression is very useful to express a unique gene, for instance, recombinant proteins or small biochemical biosynthetic pathways (Ece et al. 2017). However, in the case of *Streptomyces*, heterologous expression usually works only among *Streptomyces* strains (Novakova et al. 2018; Kawahara et al. 2018). A recent work achieved good results optimising heterologous biosynthesis, in *Streptomyces*, of spinosad (Tan et al. 2017). In this work, the authors found biosynthetic rate-limiting steps, by transcriptomics, proteomics and metabolomics.

Heterologous expression of pleiotropic regulators is achieving good results in activating cryptic secondary metabolite pathways. For instance, the widely conserved *Streptomyces coelicolor* pleiotropic regulator *Afs*Q activates the production of new compounds in several streptomycetes (Daniel-Ivad et al. 2017).

8.2.2.1.3 Combinatorial Biosynthesis

Combinatorial biosynthesis consists in altering the structure of secondary metabolites in vivo. For this purpose, biosynthetic pathways are combined by genetic engineering into one or several microorganisms that make sequential modifications into the specific compounds, producing new and altered structures (Floss 2006). Combinatorial biosynthesis has been largely developed over the last 20 years, and it has been enhanced, thanks to genome mining and synthetic biology (Olano et al. 2010; González et al. 2016; Baltz 2018).

8.2.2.2 Unselective Methods

8.2.2.2.1 Modifying the Culture Medium

Streptomyces strains grow well in rich complex culture media. However, antibiotic production often needs some specific nutrients. Sometimes defined medium can be better for determined compounds. Optimisation of nutritional requirements has been classically used to improve secondary metabolite production (Vastrad and Neelagund 2014; Alduina et al. 2018; Jakeman et al. 2006; Saudagar and Singhal 2007).

8.2.2.2.2 Inducing Stress

Stress is one of the triggers of secondary metabolism in microbes, and different stresses have been used to improve secondary metabolite biosynthesis in *Streptomyces* fermentations. Heat and ethanol, salt, hydrostatic pressure, acid and feed limitation have been successfully used to improve *Streptomyces* secondary metabolism production (reviewed in (Yoon and Nodwell 2014)).

8.2.2.2.3 Random Mutagenesis

Classical random mutagenesis (*N*-methyl-*N*′-nitro-*N*-nitrosoguanidine, MNNG; ethyl methanesulfonate, EMS; nitrous acid, NA. ultraviolet light, X-rays) is routinely used to generate overproducing *Streptomyces* strains (Baltz 2016; Khaliq et al. 2009; Korbekandi et al. 2010). Transposons have been also successfully used for *Streptomyces* random mutagenesis. Transposition creates single mutations that can be easily mapped, facilitating the identification of the mutated genes (Fedashchin et al. 2015; Xu et al. 2017).

8.2.2.2.4 Ribosomal Engineering

Some antibiotic resistances against antibiotics which target ribosomal proteins trigger secondary metabolism, including some cryptic pathways (Craney et al. 2013). Ribosomal protein gene mutagenesis was successfully used to enhance secondary metabolism, for example, in *Streptomyces coelicolor (*Wang et al. 2008*)*. Why this strategy works is unknown. It has been proven that in *S. coelicolor*, some positive secondary metabolite regulators are activated (Craney et al. 2013). At least, 22 antibiotics from 20 *Streptomyces* strains were induced by ribosome engineering (reviewed in (Zhu et al. 2019)). Interestingly, this methodology, was also reported to work in other actinomycetes such as *Planobispora rosea* or *Amycolatopsis orientalis*, as well as in other bacteria such as *Escherichia coli* or *Bacillus* (Zhu et al. 2019).

8.2.2.2.5 Genome Mining

In the postgenomic era, the combination of efficient genome sequencing and efficient bioinformatic analyses allowed to identify actinomycete secondary metabolism gene clusters. Interestingly, most of these clusters are silent and are not expressed in the lab (Genilloud 2014). During the last 15 years, the number of sequenced *Streptomyces* genomes increased exponentially. More than 1500 *Streptomyces* genomes are accessible in the RefSeq database (reviewed in (Lee et al. 2020)). In addition, several bioinformatic tools (BAGEL, ClustScan, CLUSEAN, NP.searcher, PRISM, antiSMASH) have been developed to identify potential secondary metabolism clusters, including cryptic pathways (Lee et al. 2020). Despite the enormous potential of genome mining, the main limitation of this technique is that secondary pathway finding depends on DNA homologies. Consequently, secondary metabolites encoded by putative new biosynthetic pathways without homologies with those from other secondary metabolites could not be found by this way. The other limitation of genome mining is that cryptic pathways need to be activated in the lab, many times introducing strong promoters (Laureti et al. 2011) or by heterologous expression (Qian et al. 2020).

8.2.2.2.6 Co-cultures and Elicitors

Elicitors are small diffusible molecules modulating *Streptomyces* differentiation and secondary metabolism. Elicitors can be used as signals for improving secondary metabolite production (Onaka 2017).

Elicitors activating *Streptomyces* secondary metabolism can be produced by other microorganisms in co-cultures (Onaka 2017; Marmann et al. 2014). Co-cultures usually include species that have symbiotic relationships with *Streptomyces (*Piel 2004*;* Yu et al. 2017*)* or pathogenic partners (Perez et al. 2011; Wang et al. 2014; Sung et al. 2017).

A complex mix of fungal cell walls and filtered fungi cultures affect positively natamycin production (Wang et al. 2013). GlcNac and phosphate can activate differentiation and antibiotic production in *S. coelicolor* through the activation of *actII*-ORF4/*redZ* genes in some conditions (Liu et al. 2013).

Chemical elicitors combined with activity-independent screening were successfully used to identify several cryptic biosynthetic genes that could not be identified by usual activity-guided methods (Pimentel-Elardo et al. 2015).

8.2.2.2.7 Strategies to Modulate Secondary Metabolism Based on Macroscopic Morphology (Pellet and Clump Formation)

Morphology in liquid cultures (pellets and clumps) correlates with the production of secondary metabolites (Manteca et al. 2008). The *mat* gene cluster (van Dissel et al. 2015) and the *cslA*, *glxA* and *dtpA* genes (Chaplin et al. 2015; Petrus et al. 2016; Petrus and Claessen 2014) control mycelium aggregation. This knowledge has the potential to control morphology in industrial fermentations.

Some streptomycetes are able to sporulate in liquid cultures (Geneviève Girard et al. 2013). Some strains, which normally do not sporulate in flasks, are able to sporulate in bioreactors (Rioseras et al. 2014). Sporulation stops secondary metabolite, hurting antibiotic production (Rioseras et al. 2014).

Streptomyces morphology in liquid cultures has been monitored by laser diffraction (Ronnest et al. 2012) and flow (van Veluw et al. 2012; Petrus et al. 2014). Mathematical models have been developed to predict the behaviour of *Streptomyces* pellets/clumps in liquid cultures (Celler et al. 2012; Nieminen et al. 2013).

Oxygen transfer rate (OTR) and oxygen uptake rate (OUR) are crucial for development and production in bioreactors (Rioseras et al. 2014). OTR and OUR, together with pH, viscosity, agitation, dissolved oxygen levels and surface tension, directly affect mycelium morphology and directly affect *Streptomyces* morphology (Ha et al. 2017).

8.2.2.2.8 Strategies to Modulate Secondary Metabolism Based on Mycelium Differentiation and Programmed Cell Death

Programmed cell death is a key event in *Streptomyces* development, triggering differentiation of the antibiotic production (Manteca et al. 2008). PCD was quantified measuring the fluorescence of vital stains (Yague et al. 2010). PCD and MII differentiation can be promoted modifying the culture conditions. This approach was successfully used to induce the production of apigenin (Marin et al. 2017) and microbial mTGase (Lucia Treppiccione et al. 2017).

8.2.2.2.9 Strategies to Modulate Secondary Metabolism Based on the Use of Non-pathogenic Targets to Test the Biological Activity

Antimicrobial screening usually tests the effects of natural or synthetic molecules against pathogens. McAuley et al. (McAuley et al. 2019) changed this paradigm, testing the effect of synthetic compounds against the sporulation of the non-pathogenic bacterium *Streptomyces venezuelae*. By means of this methodology (McAuley et al. 2019), they discovered a novel DNA-targeting antibiotic. The use of non-pathogenic microbes and bacteria differentiation as antimicrobial assay targets is an important advance that will contribute to find novel antimicrobials in high-throughput screening campaigns (Yague et al. 2020). New assays to test bioactive compounds might contribute to find novel antimicrobials.

8.3 Conclusions

Antimicrobial resistances are growing at a faster rate than our capacity to discover new antimicrobials. *Streptomycetes* still is the most promising source of natural bioactive compounds. New multidisciplinary strategies to avoid rediscovering the same compounds, and to access to the vast number of cryptic pathways, are necessary. Successful the discovery of new scaffolds/families of antibiotics from natural strains depends on the existence of these compounds in nature. Cryptic pathways are very promising, but, at present, there is no evidence indicating that secondary metabolites encoded by cryptic pathways include families of antibiotics different to those already discovered. The next years will be critical to know if the new strategies to activate secondary metabolism described in this chapter will allow the discovery of new antibiotics that can contribute to face antimicrobial resistances.

Acknowledgments We thank the "Ministerio de Ciencia, Innovación Universidades/Agencia Estatal de Investigación/Fondo Europeo de Desarrollo Regional" (RTI2018-093978-B-I00), the "Consejería de Empleo, Industria y Turismo del Principado de Asturias" (FC-GRUPIN-IDI/2018/000120) and the European Research Council (ERC PoCStrepCryptPath—835857) for financial support. Sergio Alonso was funded by a Severo Ochoa fellowship (FICYT, Consejería de Educación y Ciencia, Asturias, Spain).

References

Alduina R, Sosio M, Donadio S (2018) Complex regulatory networks governing production of the glycopeptide A40926. Antibiotics (Basel) 7(2):30

Baltz RH (2016) Genetic manipulation of secondary metabolite biosynthesis for improved production in Streptomyces and other actinomycetes. J Ind Microbiol Biotechnol 43(2–3):343–370

Baltz RH (2018) Synthetic biology, genome mining, and combinatorial biosynthesis of NRPS-derived antibiotics: a perspective. J Ind Microbiol Biotechnol 45(7):635–649

Ben Abdallah M, Karray F, Kallel N, Armougom F, Mhiri N, Quemeneur M et al (2018) Abundance and diversity of prokaryotes in ephemeral hypersaline lake Chott El Jerid using Illumina Miseq sequencing, DGGE and qPCR assays. Extremophiles 22(5):811–823

Bhattacharjee K, Banerjee S, Joshi SR (2012) Diversity of Streptomyces spp. in Eastern Himalayan region—computational RNomics approach to phylogeny. Bioinformation 8(12):548–554

Bonilla JO, Kurth DG, Cid FD, Ulacco JH, Gil RA, Villegas LB (2018) Prokaryotic and eukaryotic community structure affected by the presence of an acid mine drainage from an abandoned gold mine. Extremophiles 22(5):699–711

Bull AT, Idris H, Sanderson R, Asenjo J, Andrews B, Goodfellow M (2018) High altitude, hyper-arid soils of the Central-Andes harbor mega-diverse communities of actinobacteria. Extremophiles 22(1):47–57

Celler KPC, van Loosdrecht MC, van Wezel GP (2012) Structured morphological modeling as a framework for rational strain design of Streptomyces species. Antonie Van Leeuwenhoek 102 (3):409–423

Chaplin AK, Petrus ML, Mangiameli G, Hough MA, Svistunenko DA, Nicholls P et al (2015) GlxA is a new structural member of the radical copper oxidase family and is required for glycan deposition at hyphal tips and morphogenesis of Streptomyces lividans. Biochem J 469 (3):433–444

Charousova I, Medo J, Halenarova E, Javorekova S (2017) Antimicrobial and enzymatic activity of actinomycetes isolated from soils of coastal islands. J Adv Pharm Technol Res 8(2):46–51

Craney A, Ahmed S, Nodwell J (2013) Towards a new science of secondary metabolism. J Antibiot (Tokyo) 66(7):387–400

Daniel-Ivad M, Hameed N, Tan S, Dhanjal R, Socko D, Pak P et al (2017) An engineered allele of afsQ1 facilitates the discovery and investigation of cryptic natural products. ACS Chem Biol 12 (3):628–634

Ece S, Lambertz C, Fischer R, Commandeur U (2017) Heterologous expression of a Streptomyces cyaneus laccase for biomass modification applications. AMB Express 7(1):86

Engl T, Kroiss J, Kai M, Nechitaylo TY, Svatos A, Kaltenpoth M (2018) Evolutionary stability of antibiotic protection in a defensive symbiosis. Proc Natl Acad Sci U S A 115(9):E2020–E20E9

Fedashchin A, Cernota WH, Gonzalez MC, Leach BI, Kwan N, Wesley RK et al (2015) Random transposon mutagenesis of the Saccharopolyspora erythraea genome reveals additional genes influencing erythromycin biosynthesis. FEMS Microbiol Lett 362(22):fnv180

Fischbach MA, Walsh CT (2009) Antibiotics for emerging pathogens. Science 325 (5944):1089–1093

Floss HG (2006) Combinatorial biosynthesis—potential and problems. J Biotechnol 124 (1):242–257

Geneviève Girard BAT, Sangal V, Mascini N, Hoskisson PA, Goodfellow M, van Wezel GP (2013) A novel taxonomic marker that discriminates between morphologically complex actinomycetes. Open Biol 3(10):130073

Genilloud O (2014) The re-emerging role of microbial natural products in antibiotic discovery. Antonie Van Leeuwenhoek 106(1):173–188

González A, Rodríguez M, Braña AF, Méndez C, Salas JA, Olano C (2016) New insights into paulomycin biosynthesis pathway in Streptomyces albus J1074 and generation of novel derivatives by combinatorial biosynthesis. Microb Cell Fact 15:56

Ha S, Lee KJ, Lee SI, Gwak HJ, Lee JH, Kim TW et al (2017) Optimization of Herbicidin A production in submerged culture of Streptomyces scopuliridis M40. J Microbiol Biotechnol 27 (5):947–955

Idris H, Labeda DP, Nouioui I, Castro JF, Del Carmen Montero-Calasanz M, Bull AT et al (2017) Streptomyces aridus sp. nov., isolated from a high altitude Atacama Desert soil and emended description of Streptomyces noboritoensis Isono et al. 1957. Antonie Van Leeuwenhoek 110 (5):705–717

Ikeda H, Nonomiya T, Usami M, Ohta T, Omura S (1999) Organization of the biosynthetic gene cluster for the polyketide anthelmintic macrolide avermectin in Streptomyces avermitilis. Proc Natl Acad Sci U S A 96(17):9509–9514

Jakeman DL, Graham CL, Young W, Vining LC (2006) Culture conditions improving the production of jadomycin B. J Ind Microbiol Biotechnol 33(9):767–772

Kang HR, Lee D, Benndorf R, Jung WH, Beemelmanns C, Kang KS et al (2016) Termisoflavones A-C, isoflavonoid glycosides from termite-associated Streptomyces sp. RB1. J Nat Prod 79 (12):3072–3078

Karlsson H, Truedsson L, Nassberger L (1991) The immunosuppressive agent FK506 inhibits in vitro expression of membrane-bound and soluble interleukin-2 receptors on resting but not on activated human lymphocytes. Immunol Lett 30(1):129–132

Kawahara T, Izumikawa M, Kozone I, Hashimoto J, Kagaya N, Koiwai H et al (2018) Neothioviridamide, a polythioamide compound produced by heterologous expression of a Streptomyces sp. cryptic RiPP biosynthetic gene cluster. J Nat Prod 81(2):264–269

Kealey C, Creaven CA, Murphy CD, Brady CB (2017) New approaches to antibiotic discovery. Biotechnol Lett 39(6):805–817

Khaliq S, Akhtar K, Afzal Ghauri M, Iqbal R, Mukhtar Khalid A, Muddassar M (2009) Change in colony morphology and kinetics of tylosin production after UV and gamma irradiation mutagenesis of Streptomyces fradiae NRRL-2702. Microbiol Res 164(4):469–477

Kharel MK, Subba B, Basnet DB, Woo JS, Lee HC, Liou K et al (2004) A gene cluster for biosynthesis of kanamycin from Streptomyces kanamyceticus: comparison with gentamicin biosynthetic gene cluster. Arch Biochem Biophys 429(2):204–214

Kim KH, Ramadhar TR, Beemelmanns C, Cao S, Poulsen M, Currie CR et al (2014) Natalamycin A, an Ansamycin from a termite-associated Streptomyces sp. Chem Sci 5 (11):4333–4338

Korbekandi H, Darkhal P, Hojati Z, Abedi D, Hamedi J, Pourhosein M (2010) Overproduction of clavulanic acid by UV mutagenesis of Streptomyces clavuligerus. Iran J Pharm Res 9 (2):177–181

Kroiss J, Kaltenpoth M, Schneider B, Schwinger MG, Hertweck C, Maddula RK et al (2010) Symbiotic Streptomycetes provide antibiotic combination prophylaxis for wasp offspring. Nat Chem Biol 6(4):261–263

Lam KS (2006) Discovery of novel metabolites from marine actinomycetes. Curr Opin Microbiol 9 (3):245–251

Laureti L, Song L, Huang S, Corre C, Leblond P, Challis GL et al (2011) Identification of a bioactive 51-membered macrolide complex by activation of a silent polyketide synthase in Streptomyces ambofaciens. Proc Natl Acad Sci U S A 108(15):6258–6263

Lee N, Hwang S, Kim J, Cho S, Palsson B, Cho BK (2020) Mini review: genome mining approaches for the identification of secondary metabolite biosynthetic gene clusters in Strepto myces. Comput Struct Biotechnol J 18:1548–1556

Li X, Guo J, Dai S, Ouyang Y, Wu H, Sun W et al (2009) Exploring and exploiting microbial diversity through metagenomics for natural product drug discovery. Curr Top Med Chem 9 (16):1525–1535

Liu G, Chater KF, Chandra G, Niu G, Tan H (2013) Molecular regulation of antibiotic biosynthesis in Streptomyces. Microbiol Mol Biol Rev 77(1):112–143

Lu X, You Q (2010) Recent advances on platensimycin: a potential antimicrobial agent. Curr Med Chem 17(12):1139–1155

Lucia Treppiccione AO, Luongo D, Maurano F, Manteca Á, Lombó F, Rossi M (2017) Development of gluten with immunomodulatory properties using mTG-active food grade supernatants from Streptomyces mobaraensis cultures. J Funct Foods 34:390–397

Manteca A, Yague P (2018) Streptomyces differentiation in liquid cultures as a trigger of secondary metabolism. Antibiotics (Basel) 7(2):41

Manteca A, Fernandez M, Sanchez J (2005) Mycelium development in Streptomyces antibioticus ATCC11891 occurs in an orderly pattern which determines multiphase growth curves. BMC Microbiol 5:51

Manteca A, Fernandez M, Sanchez J (2006) Cytological and biochemical evidence for an early cell dismantling event in surface cultures of Streptomyces antibioticus. Res Microbiol 157 (2):143–152

Manteca A, Alvarez R, Salazar N, Yague P, Sanchez J (2008) Mycelium differentiation and antibiotic production in submerged cultures of Streptomyces coelicolor. Appl Environ Microbiol 74(12):3877–3886

Marin L, Gutierrez-Del-Rio I, Yague P, Manteca A, Villar CJ, Lombo F (2017) De novo biosynthesis of apigenin, luteolin, and eriodictyol in the actinomycete Streptomyces albus and production improvement by feeding and spore conditioning. Front Microbiol 8:921

Marmann A, Aly AH, Lin W, Wang B, Proksch P (2014) Co-cultivation—a powerful emerging tool for enhancing the chemical diversity of microorganisms. Mar Drugs 12(2):1043–1065

McAuley S, Huynh A, Howells A, Walpole C, Maxwell A, Nodwell JR (2019) Discovery of a novel DNA gyrase-targeting antibiotic through the chemical perturbation of Streptomyces venezuelae sporulation. Cell Chem Biol 26(9):1274–82.e4

Nieminen L, Webb S, Smith MC, Hoskisson PA (2013) A flexible mathematical model platform for studying branching networks: experimentally validated using the model actinomycete, Streptomyces coelicolor. PLoS One 8(2):e54316

Novakova R, Nunez LE, Homerova D, Knirschova R, Feckova L, Rezuchova B et al (2018) Increased heterologous production of the antitumoral polyketide mithramycin A by engineered Streptomyces lividans TK24 strains. Appl Microbiol Biotechnol 102(2):857–869

Olano C, Mendez C, Salas JA (2010) Post-PKS tailoring steps in natural product-producing actinomycetes from the perspective of combinatorial biosynthesis. Nat Prod Rep 27(4):571–616

Onaka H (2017) Novel antibiotic screening methods to awaken silent or cryptic secondary metabolic pathways in actinomycetes. J Antibiot (Tokyo) 70(8):865–870

Perez J, Munoz-Dorado J, Brana AF, Shimkets LJ, Sevillano L, Santamaria RI (2011) Myxococcus xanthus induces actinorhodin overproduction and aerial mycelium formation by Streptomyces coelicolor. Microb Biotechnol 4(2):175–183

Petrus ML, Claessen D (2014) Pivotal roles for Streptomyces cell surface polymers in morphological differentiation, attachment and mycelial architecture. Antonie Van Leeuwenhoek 106 (1):127–139

Petrus ML, van Veluw GJ, Wosten HA, Claessen D (2014) Sorting of Streptomyces cell pellets using a complex object parametric analyzer and sorter. J Vis Exp 84:e51178

Petrus ML, Vijgenboom E, Chaplin AK, Worrall JA, van Wezel GP, Claessen D (2016) The DyP-type peroxidase DtpA is a Tat-substrate required for GlxA maturation and morphogenesis in Streptomyces. Open Biol 6(1):150149

Piel J (2004) Metabolites from symbiotic bacteria. Nat Prod Rep 21(4):519–538

Pimentel-Elardo SM, Sorensen D, Ho L, Ziko M, Bueler SA, Lu S et al (2015) Activity-independent discovery of secondary metabolites using chemical elicitation and cheminformatic inference. ACS Chem Biol 10(11):2616–2623

Qian Z, Bruhn T, D'Agostino PM, Herrmann A, Haslbeck M, Antal N et al (2020) Discovery of the Streptoketides by direct cloning and rapid heterologous expression of a cryptic PKS II gene cluster from Streptomyces sp. Tu 6314. J Org Chem 85(2):664–673

Razmilic V, Castro JF, Andrews B, Asenjo JA (2018) Analysis of metabolic networks of Streptomyces leeuwenhoekii C34 by means of a genome scale model: prediction of modifications that enhance the production of specialized metabolites. Biotechnol Bioeng 115(7):1815–1828

Reading C, Cole M (1977) Clavulanic acid: a beta-lactamase-inhibiting beta-lactam from Streptomyces clavuligerus. Antimicrob Agents Chemother 11(5):852–857

Rioseras B, Lopez-Garcia MT, Yague P, Sanchez J, Manteca A (2014) Mycelium differentiation and development of Streptomyces coelicolor in lab-scale bioreactors: programmed cell death, differentiation, and lysis are closely linked to undecylprodigiosin and actinorhodin production. Bioresour Technol 151:191–198

Rioseras B, Shliaha PV, Gorshkov V, Yague P, Lopez-Garcia MT, Gonzalez-Quinonez N et al (2018) Quantitative proteome and phosphoproteome analyses of streptomyces coelicolor reveal proteins and phosphoproteins modulating differentiation and secondary metabolism. Mol Cell Proteomics 17(8):1591–1611

Ronnest NP, Stocks SM, Lantz AE, Gernaey KV (2012) Comparison of laser diffraction and image analysis for measurement of Streptomyces coelicolor cell clumps and pellets. Biotechnol Lett 34 (8):1465–1473

Sakai T, Sameshima T, Matsufuji M, Kawamura N, Dobashi K, Mizui Y (2004) Pladienolides, new substances from culture of Streptomyces platensis Mer-11107. I. Taxonomy, fermentation, isolation and screening. J Antibiot (Tokyo) 57(3):173–179

Sanyal A, Antony R, Samui G, Thamban M (2018) Microbial communities and their potential for degradation of dissolved organic carbon in cryoconite hole environments of Himalaya and Antarctica. Microbiol Res 208:32–42

Saudagar PS, Singhal RS (2007) Optimization of nutritional requirements and feeding strategies for clavulanic acid production by Streptomyces clavuligerus. Bioresour Technol 98(10):2010–2017

Schrey SD, Tarkka MT (2008) Friends and foes: streptomycetes as modulators of plant disease and symbiosis. Antonie Van Leeuwenhoek 94(1):11–19

Seipke RF, Kaltenpoth M, Hutchings MI (2012) Streptomyces as symbionts: an emerging and widespread theme? FEMS Microbiol Rev 36(4):862–876

Shapiro S, Vining LC (1983) Nitrogen metabolism and chloramphenicol production in Streptomyces venezuelae. Can J Microbiol 29(12):1706–1714

Sung AA, Gromek SM, Balunas MJ (2017) Upregulation and identification of antibiotic activity of a marine-derived Streptomyces sp. via co-cultures with human pathogens. Mar Drugs 15(8):250

Takahashi Y, Omura S (2003) Isolation of new actinomycete strains for the screening of new bioactive compounds. J Gen Appl Microbiol 49(3):141–154

Tan GY, Liu T (2017) Rational synthetic pathway refactoring of natural products biosynthesis in actinobacteria. Metab Eng 39:228–236

Tan GY, Deng K, Liu X, Tao H, Chang Y, Chen J et al (2017) Heterologous biosynthesis of Spinosad: an omics-guided large polyketide synthase gene cluster reconstitution in Streptomyces. ACS Synth Biol 6(6):995–1005

van Dissel D, Claessen D, Roth M, van Wezel GP (2015) A novel locus for mycelial aggregation forms a gateway to improved Streptomyces cell factories. Microb Cell Fact 14:44

van Veluw GJ, Petrus ML, Gubbens J, de Graaf R, de Jong IP, van Wezel GP et al (2012) Analysis of two distinct mycelial populations in liquid-grown Streptomyces cultures using a flow cytometry-based proteomics approach. Appl Microbiol Biotechnol 96(5):1301–1312

Vastrad BM, Neelagund SE (2014) Optimization of medium composition for the production of neomycin by Streptomyces fradiae NCIM 2418 in solid state fermentation. Biotechnol Res Int 2014:674286

Waksman SA (1959) Strain specificity and production of antibiotic substances. X. Characterization and classification of species within the Streptomyces Griseus Group. Proc Natl Acad Sci U S A 45(7):1043–1047

Waksman SA, Lechevalier HA (1949) Neomycin, a new antibiotic active against streptomycin-resistant bacteria, including tuberculosis organisms. Science 109(2830):305–307

Wang G, Hosaka T, Ochi K (2008) Dramatic activation of antibiotic production in Streptomyces coelicolor by cumulative drug resistance mutations. Appl Environ Microbiol 74(9):2834–2840

Wang D, Yuan J, Gu S, Shi Q (2013) Influence of fungal elicitors on biosynthesis of natamycin by Streptomyces natalensis HW-2. Appl Microbiol Biotechnol 97(12):5527–5534

Wang Y, Wang L, Zhuang Y, Kong F, Zhang C, Zhu W (2014) Phenolic polyketides from the co-cultivation of marine-derived Penicillium sp. WC-29-5 and Streptomyces fradiae 007. Mar Drugs 12(4):2079–2088

Wildermuth H (1970) Development and organization of the aerial mycelium in Streptomyces coelicolor. J Gen Microbiol 60(1):43–50

Wyche TP, Ruzzini AC, Beemelmanns C, Kim KH, Klassen JL, Cao S et al (2017) Linear peptides are the major products of a biosynthetic pathway that encodes for cyclic depsipeptides. Org Lett 19(7):1772–1775

Xu Z, Wang Y, Chater KF, Ou HY, Xu HH, Deng Z et al (2017) Large-scale transposition mutagenesis of Streptomyces coelicolor identifies hundreds of genes influencing antibiotic biosynthesis. Appl Environ Microbiol 83(6):e02889–e02816

Yague P, Manteca A, Simon A, Diaz-Garcia ME, Sanchez J (2010) New method for monitoring programmed cell death and differentiation in submerged Streptomyces cultures. Appl Environ Microbiol 76(10):3401–3404

Yague P, Rodriguez-Garcia A, Lopez-Garcia MT, Rioseras B, Martin JF, Sanchez J et al (2014) Transcriptomic analysis of liquid non-sporulating Streptomyces coelicolor cultures demonstrates the existence of a complex differentiation comparable to that occurring in solid sporulating cultures. PLoS One 9(1):e86296

Yague P, Gonzalez-Quinonez N, Manteca A (2020) Nonpathogenic bacteria as targets in antimicrobial high-throughput screening. Trends Microbiol 28(1):1–2

Yi JS, Kim M, Kim EJ, Kim BG (2018) Production of pikromycin using branched chain amino acid catabolism in Streptomyces venezuelae ATCC 15439. J Ind Microbiol Biotechnol 45 (5):293–303

Yoon V, Nodwell JR (2014) Activating secondary metabolism with stress and chemicals. J Ind Microbiol Biotechnol 41(2):415–424

Yu J, Liu Q, Chen C, Qi X (2017) Antifungal activity change of Streptomyces rimosus MY02 mediated by confront culture with other microorganism. J Basic Microbiol 57(3):276–282

Zhu S, Duan Y, Huang Y (2019) The application of ribosome engineering to natural product discovery and yield improvement in Streptomyces. Antibiotics (Basel) 8(3):133

Novel Agroactive Secondary Metabolites from Actinomycetes in the Past Two Decades with Focus on Screening Strategies and Discovery

9

Kaimei Wang, Shaoyong Ke, Wei Fang, Zhaoyuan Wu, and Yani Zhang

Abstract

Actinomycetes produce plenty of bioactive secondary metabolites. Since the golden age of antibiotics, actinomycetes have been the kernel of discovery of pharmaceutical and agrochemical entities. Many important pesticides originated directly from secondary metabolites of actinomycetes or based on the lead compounds from actinomycetes have been developed and shared rather high portion of the pesticide market. The discovery of new leads for the developments of new agrochemicals, especially eco-friendly agrochemicals, becomes more and more difficult now; however, the structure diversity and extensive activity of secondary metabolites from actinomycetes attract great interest of the researchers and multi-national agrochemical companies. In this review, we summarized the techniques applied in the discovery of agroactive secondary metabolites from actinomycetes, with focus on the screening strategies applied and some promising agroactive secondary metabolites.

Keywords

Actinomycetes · Secondary metabolites · Agroactivity · Screening · Chemical screening · Genome mining · Synthetic biology · Metabolic engineering

K. Wang (✉) · S. Ke (✉) · W. Fang · Z. Wu · Y. Zhang
Hubei Biopesticide Engineering Research Centre, Hubei Academy of Agricultural Sciences, Wuhan, China

National Biopesticide Engineering Research Centre, Wuhan, China
e-mail: kaimei.wang@nberc.com; shaoyong.ke@nberc.com; wei.fang@nberc.com; zhaoyuan.wu@nberc.com; yani.zhang@nberc.com

R. V. Rai, J. A. Bai (eds.), *Natural Products from Actinomycetes*,
https://doi.org/10.1007/978-981-16-6132-7_9

9.1 Introduction

Agriculture, especially crop production, has been threatened by insect pests, diseases, and weeds for a long time. Synthetic agrochemicals had secured the crop production and sustainable food supply for the increasing population in the world. With the frequent applications of agrochemicals in the pest control, it caused the development of resistance of pests, the unbalance of natural enemies and other environmental issues, and the residue of agrochemicals in food products and environment. With more concerns of the negative effects of agrochemicals on the environment and food safety, the development of eco-friendly, safe to human, and highly effective pesticides have been gaining much attention.

A lot of such pesticides had been developed or under development (Umetsu and Shirai 2020), while natural products played important roles in such trend of development of novel pesticides. As reviewed by Sparks, the sales of pesticides directly derived from natural products accounted about 21% of global total pesticide sales in 2014, and the share of pesticide in which the natural product model existed might rise up to 53% (Sparks et al. 2017). Actinomycetes are an inexhaustible source of bioactive natural products (Takahashi and Nakashima 2018). Actinomycetes were found to produce diverse secondary metabolites which showed insecticidal, acaricidal, bactericidal, fungicidal, and herbicidal or plant growth regulating activities. The agroactive secondary metabolites from actinomycetes had been developed as agrochemicals for pest control, from which avermectins, spinosyns, milbemycins, validamycin, and some others were among the most important. The agroactive secondary metabolites were found in diverse actinomycetes isolated from samples collected from different habitats. The discovery of novel agroactive secondary metabolites was accelerated by activity screening, analytical chemistry, and molecular biological techniques. In spite of the difficulties in the discovery of novel agroactive secondary metabolites from actinomycetes, we still see the ascending figures of agroactive secondary metabolites from actinomycetes.

In the past two decades, more than 450 novel agroactive secondary metabolites from actinomycetes were reported. Compared with the bioactive numbers from actinomycetes, the number of agroactive metabolites from actinomycetes was rather low. It might be caused by the low ratio of actinomycete secondary metabolites screened for agroactivities. Anyway, the success rate of agroactive secondary metabolites as the leads for the development of new agrochemicals was much higher than the randomly synthesized chemicals. We have studied the literatures (publications, patents, and other source of literature) for the novel agroactive secondary metabolites from actinomycetes in the past two decades. We summarized in this review the discovery of agroactive secondary metabolites from actinomycetes, with focus on new techniques and strategies applied and some promising agroactive secondary metabolites (high potential for the direct development of pesticides or as leads for the development of new pesticides). We also analyzed the impacts of sample sources and taxonomic diversity on the discovery of novel agroactive secondary metabolites from actinomycetes. We hope that it would be helpful for accelerating the pesticide lead discovery from actinomycetes and the development of pesticides with agroactive secondary metabolites from actinomycetes as leads.

9.2 Exploring Actinomycetes Producing Agroactive Secondary Metabolites from New Habitats

The discovery of agroactive secondary metabolites focused on the terrestrial actinomycetes, mainly the actinomycetes from soil. With the increasing difficulties for discovering novel metabolites, researchers, especially the natural product chemists, turned their eyes to the untapped habitats, such as sea, antarctic, desert, and endophytes. Actinomycetes under such specific habitats have adapted to the extreme conditions or special physiological environment. Such adaption might relate to the production of specific secondary metabolites. It might increase the opportunities of discovery of novel agroactive secondary metabolites.

9.2.1 Actinomycetes from Marine Environment

Marine environment, as the largest area on our planet, has been seldom tapped for the exploitation of actinobacteria for a long time. With the achievement in the sampling techniques, researchers have the capability now to access the samples from different marine environments. Samples included soil at the coast, sediments at the intertidal zone, plants at intertidal mudflat, floating plants, mangroves and related soil, marine animals, and sediments of deep sea. Mangrove ecosystems are among the most favorite types for sampling, as they can be easily accessed. The sediments and the mangrove plants from mangrove ecosystems were collected for the isolation of actinomycetes which might produce agroactive secondary metabolites. *Streptomyces flavotricini* Y12-26, an endophytic actinomycete from mangrove plant *Acanthus ilicifolius*, produced a new analogue of bafilomycin-bafilomycin K, which showed antifungal activity against *Pyricularia oryzae*, and *Colletotrichum lagenarium* (Zhang et al. 2011a, b). *Streptomyces cheonanensis* VUK-A isolated from sediment of the Coringa mangrove ecosystem produced 2-methyl butyl propyl phthalate and diethyl phthalate which showed moderate antibacterial activity against *Xanthomonas malvacearum* and *Xanthomonas campestris* and strong antifungal activity against *Fusarium solani*, with MIC of 16 mg/L. Two diketopiperazines were produced by *Pseudonocardia endophytica* VUK-10 isolated from the sediment of mangrove ecosystem (Mangamuri et al. 2016a) and showed strong to moderate activity to *X. malvacearum* and *F. solani*. P. *endophytica* VUK-10 also produced *N*-(4-aminocyclooctyl)-3,5-dinitrobenzamide which showed moderate antibacterial and antifungal activity against X. *malvacearum* and F. *solani* (Mangamuri et al. 2016b). *Streptomyces* sp. AMC from the rhizosphere soil of *Rhizophora mangle* in a well-preserved mangrove forest produced bafilomycins B1 and B2 (Crevelin et al. 2013). Some actinomycetes isolated from the sediment of deep sea also produced novel agroactive secondary metabolites. *Streptomyces somaliensis* SCSIO ZH66 from sediment at a depth of 3536 m of the South China Sea produced somamycins A–D which were antifungal against *Fusarium oxysporum* MHKW and *Alternaria brassicae* BCHB with the MIC values of 1.6 and 3.1 μg/mL, respectively (Hou et al. 2020). Pan et al. isolated bafilomycin B1 and C1 from *Streptomyces cavourensis* NA4 which was isolated from sediment collected at the depth of 1464 m from the South China Sea (Pan et al. 2015). *Streptomyces* sp. NTK-937 isolated from Atlantic Ocean deep-sea sediment at the southern edge of the Saharan debris flow near the Canary Islands at a depth of 3814 m produced caboxamycin which showed strong antibacterial activity against *Xanthomonas* sp. and *Ralstonia solanacearum* (Hohmann et al. 2009). Many other actinomycetes were isolated from the sediments at seashore, intertidal zone, and coast. An endophytic *Streptomyces ghanaensis* TXC6-16 from *Tamarix chinensis* grown in the intertidal zone produced ghanamycins A and B which showed antibacterial activity against *Erwinia* sp. and *Pseudomonas syringae* and weak activity against *Colletotrichum* sp. and *F. oxysporum* (Xu et al. 2017). Iseolides A–C produced by *Streptomyces* sp. DC4-5 from stony coral *Dentrophyllia* showed potent antifungal activity against *Glomerella cingulata* and weak antibacterial activity against *R. solanacearum* (Zhang et al. 2020a, c). Endophytic *Streptomyces* sp. PNM-161a produced provipeptides A and B which showed antibacterial activity against *Burkholderia* and antifungal activity against *Colletotrichum* (Betancur et al. 2019).

9.2.2 Actinomycetes from Extreme Terrestrial Environment

The actinomycetes from extreme terrestrial environment were prolific producers of agroactive secondary metabolites. Such extreme terrestrial environments included the desert, high altitude area (might be extremely low temperature area), salty and alkaline soil, aquatic environment, or contaminated environment. An unidentified *Streptomyces* strain from the soil collected from the Atacama Desert in Chile produced abenquines which inhibited the growth of bloom-forming cyanobacteria (Schulz et al. 2011; Nain-Perez et al. 2016, 2017). Atrovimycin from *Streptomyces atrovirens* LQ13 isolated from saline-alkali desert soil strongly inhibited the growth of *F. oxysporum* f. sp. *cucumerinum* with IC_{50} of 6.99 μM (Liu et al. 2019). *Streptomyces* sp. TN256 isolated from Saharan soil sample produced several antifungal metabolites against *Fusarium* sp. (Smaoui et al. 2012). Antimycin A17, a new macrodiolide from *Streptomyces* sp. GAAS7310 which was isolated from a soil sample collected in Tianshan mountain, Xinjiang, inhibited the growth of *Colletotrichum nigrum*, *Cladosporium* sp., and *Alternaria solani*, (Chen et al. 2005).

Several diketopiperazines from *Streptomyces cavourensis* TN638 isolated from soil sample collected at an industrial waste area showed antibacterial activity against *Agrobacterium tumefaciens* and antifungal activity against *Rhizoctonia* and *Fusarium* (Kaaniche et al. 2020). Pyrocoll produced by a novel alkaliphilic *Streptomyces* sp. AK 409 which was isolated from a steel waste tip soil sample showed strong growth inhibition to *Botrytis cinerea* (Dieter et al. 2003).

9.2.3 Terrestrial Plant- or Animal-Associated Actinomycetes

The co-evolution of endophytic actinomycetes with host plants resulted mutual dependency between host plants and endophytic actinomycetes. The endophytic actinomycetes might produce agroactive secondary metabolites to enhance the host for their resistance to insects and phytopathogens. More than 60 novel agroactive secondary metabolites were found from the cultures of endophytic actinomycetes isolated from terrestrial plants. The number of agroactive secondary metabolites from terrestrial endophytic actinomycetes was more than 10% of the agroactive secondary metabolites from actinomycetes. Coronamycin was isolated from an endophytic *Streptomyces* sp. MSU-2110 from the stem of *Monstera* sp. and showed antioomycete activity against *Pythium* and *Phytophthora* (Ezra et al. 2004). Endophytic *Streptomyces* sp. BS-1 from *Stemona sessilifolia* produced endostemonines A–J which showed strong lethal activity to *Aphis gossypii* with LC_{50} range of 3.55–32 mg/L (Zhao et al. 2020). *Streptomyces ambofaciens* F3, an endophyte of root of *Platycladus orientalis*, produced ambobactin which showed antioomycete activity against *Pythium ultimum* and *Phytophthora cinnamomi* (Wei et al. 2015). Dehydroxyaquayamycin B produced by endophytic *Streptomyces blastomycetica* strain F4-20 showed growth inhibition to *Valsa*, *Colletotrichum*, and *Fusarium* (Yan et al. 2017). A-79197-2 and A-79197-3 produced by *Dactylosporangium aurantiacum* SANK 61299, an endophyte from the leaf of *Cucubalus* sp., inhibited the seed germination of *Brassica rapa* (Kizuka et al. 2002). *Streptomyces* sp. NRRRL3062, an endophytic actinomycetes from snakevine (*Kennedia nigriscans*), produced munumbicins A–D, E-4, and E-5 showing strong antioomycete activity against P. *ultimum* (Castillo et al. 2002, 2006). Endophytic *Streptomyces* sp. PKU-EA00015 isolated from *Salvia miltiorrhiza* Bunge produced strepimidazoles A–G which showed different inhibition to the growth of *Verticillium dahliae* (Sun et al. 2020).

Rhizosphere, as a special ecological niche, attracts many microbes by root exudates from host plants. Many actinomycetes were isolated from the rhizosphere of crops and other plants and produced many novel agroactive secondary metabolites. *Streptomyces griseocarneus* from rhizosphere soil of *Heveabra siliensis* produced 1-methoxypyrrole-2-carboxamide which showed antibacterial activity against *X. campestris* pv. *oryzae* (Wattanasuepsin et al. 2017). *Amycolatopsis* sp. YIM PH20520 strain was isolated from the rhizosphere soil of *Panax notoginseng* and produced *echinosporin* and 7-deoxyechinosporin which also showed good antifungal activity against root rot fungal pathogens of *P. notoginseng* (Xu et al. 2019). 6-Hydroxyphenazine-1-carboxamide and methyl 6-carbamoylphenazine-1-carboxylate, produced by *Streptomyces diastaticus* subsp. *ardesiacus* strain YIM PH20246 isolated from rhizosphere soil of *P. notoginseng*, showed antifungal activity against four species of root rot fungal pathogens of *P. notoginseng* (Hu et al. 2019). *S. diastaticus* subsp. *ardesiacus* YIM PH20095 was also isolated from rhizosphere soil of *P. notoginseng* and produced streptophenazines A–C which inhibited the growth of root rot fungal pathogens of *P.notoginseng*(Chen et al. 2019). *Streptomyces* sp. 39PL from rhizosphere soil of maize produced luminacin I which was phytotoxic to *Lemna minor* (Martinez et al. 2020). *Actinokineospora bangkokensis* strain 44EHW from rhizosphere soil produced thailandins A and B which showed broad antifungal activity against fungi causing anthracnose disease (Intra et al. 2016). The actinomycetes in rhizosphere soil might benefit the host plants through the production of antimicrobial metabolites. Targeted isolation of actinomycetes from rhizosphere soil might increase the opportunity to discover new or novel agroactive secondary metabolites.

Endophenazines A–D were found in the cultures of several *Streptomyces anulatus* strains which were isolated from the intestine of woodlice, beetles, and millipedes. Endophenazine D inhibited the growth of *Botrytis cinerea* and *Lemna minor* (Gebhardt et al. 2002). *Streptomyces antibioticus* strain ess_amA8 isolated from the insect *Tapinoma simrothi* produced a quercetin 3-*O*-glucoside derivative which showed antifungal activity against *Fusarium moniliforme* (Sholkamy et al. 2020).

9.3 Phylogenetic Analysis of the Agroactive Secondary Metabolites-Producing Actinomycetes Strains

As we analyzed the phylogenetic source of the producing actinomycetes of novel agroactive secondary metabolites discovered in the past two decades, we found that more than 85% of novel agroactive secondary metabolites-producing actinomycetes were of the genus *Streptomyces* and most of the streptomycete producers were unidentified. Of the agroactive secondary metabolites-producing actinomycetes, nearly 15% were rare actinomycetes or non-streptomycetes. The ratio of streptomycete producer to whole actinomycete producers was much higher than that of bioactive secondary metabolites from *Streptomyces* to the whole bioactive metabolites from actinomycetes as demonstrated in Berdy's review (Berdy 2012). It might be caused by that many bioactive secondary metabolites from non-*Streptomyces* actinomycetes were not evaluated for their agroactivity due to the interests of the researchers or the purposes of the studies.

Of the rare actinomycetes which produce novel secondary metabolites, they had a very wide range of taxonomic distribution, like *Actinokineospora*, *Actinomadura*, *Allostreptomyces*, *Amycolatopsis*, *Dactylosporangium*, *Kutzneria*, *Lechevalieria*, *Microbacterium*, *Microbispora*, *Nocardia*, *Nocardiopsis*, *Pseudonocardia*, *Pseudosporangium*, *Saccharopolyspora*, *Streptoverticillium*, etc.

Based on the collection of literatures we accessed, the average number of novel agroactive secondary metabolites per rare actinomycete strain producing agroactive secondary metabolites were 3.4, which was about one times higher than the average number of novel agroactive secondary metabolites per streptomycete strain producing novel agroactive secondary metabolites. There might be two possibilities: one is that rare actinomycetes have higher potential for producing novel secondary metabolites; the other is that rare actinomycetes are less tapped and give higher opportunities for the researches to discover novel agroactive secondary metabolites from them. Nevertheless, it is necessary for researchers to keep eyes on rare actinomycetes for lead discovery for the development of new agrochemicals for crop protection.

9.4 New Techniques for Accelerating the Process of Discovery of Novel Agroactive Secondary Metabolites

For the discovery of agroactive secondary metabolites from actinomycetes, many strategies had been applied to diversify the secondary metabolites. Such strategies included the selective isolation of actinomycetes, especially rare actinomycetes, the screening for agroactivity, and diversifying the secondary metabolites through traditional cultivation or genetic operation. Agroactivity-based screening resulted high rate of rediscovery of known compounds because it did not distinguish the compounds based only on activity. It must be combined with other screening techniques for the discovery of novel agroactive secondary metabolites. In this section, we will focus on the chemical screening, gene screening, genome mining, and targeted biosynthesis through synthetic biology.

9.4.1 Chemical Screening

The chemical screening for new agroactive secondary metabolites depends on the progress in analytical techniques and related instrument. In the past two decades, the most popularly used method was liquid chromatography coupled with mass spectrometer. Regarding the mass spectrometer used for chemical screening, the types of MS include single quadrupole MS, tandem quadrupole MS, ion trap MS, time of flight MS, etc. With LC/MS, the researchers could easily get the molecular weight and UV spectra of the compounds produced by actinomycetes. The patterns of the fragmentation of the compounds could also be specific for a compound or compounds with special functional group. In the information of MW, UV of detected compounds were matched with the commercial database, like Chapman and Hall Dictionary of Natural Products, AntiBase, or self-constructed database for dereplication. Evans et al. developed a mass spectrometric method to detect actinomycete secondary metabolites containing phosphonate and phosphinate group (Evans et al. 2013). Phosphonates preferentially fragment to give the m/z63 ion. Based on the fragmentation patterns of phosphonates and phosphorylated compounds, it could distinguish phosphonates of interest from highly abundant phosphorylated compounds and phosphate from the cell extracts and culture media. Ju et al. screened the putative phosphonate producing actinomycetes obtained from gene screening with ^{31}P NMR spectroscopy and found that at least 45 strains produced phosphonates (Ju et al. 2015). The wide application of LC/MS in natural product research make chemical screening with LC/MS a conventional method of dereplication in the discovery of new agroactive metabolites from actinomycetes. There were several reports newly published which applied ^{1}H NMR metabolomic approach to study the profile of secondary metabolites (Saadouli et al. 2020;

Betancur et al. 2020), but the information of ^{1}H NMR spectra were not enough to determine the structures of the compounds.

9.4.2 Gene Screening

With understanding of the biosynthesis pathways of secondary metabolites in actinomycetes, the genes responsible for the key enzymes in the biosynthesis of secondary metabolites would be used as the target of gene screening for the presence of a specific type of secondary metabolites. For some studies, polyketide synthases (PKSs) or non-ribosomal peptide synthetase (NRPS) genes were screened with PCR, but it only provided the presence of polyketides or non-ribosomal peptides in actinomycetes, without further information of the secondary metabolites. As the biosynthesis pathways of phosphonates started with phosphoenolpyruvate mutase (PepM), Ju et al. designed PCR primers based on the DNA fragments of *pepM* to screen a collection of 10,000 actinomycetes for the production of phosphonic acid natural products (Ju et al. 2015). They really found 278 strains were *pepM+*. Interestingly, all the known phosphonate producing actinomycetes, like the producer of phosalacine, were among the *pepM+* strains. It indicated that screening for *pepM* accelerated the finding of new phosphonates.

9.4.3 Genome Mining

The potential of production of secondary metabolites by actinomycetes was much higher than what we see with analytical instruments, as a lot of genes responsible for the biosynthesis of secondary metabolites were silent, or the titer of the secondary metabolites was too low to be detected by analytical instruments. The cost for DNA sequencing dropped dramatically in the past two decades; it is possible to sequence the genomic DNA for understanding biosynthetic potential of actinomycetes. The genome mining of 10,000 actinomycetes strains identified 78 discrete biosynthetic gene clusters of phosphonates. The genome mining also revealed new producers of plumbemycin and phosphonothrixin which showed agroactivity (Ju et al. 2015). Based on the analysis of biosynthetic gene clusters of phosphonates, Ju et al. selected three *Streptomyces* sp. strains for the isolation of novel phosphonates which resulted in the discovery of argolaphos A and B, phosphono-cystoximates, and valinophos. It is possible to use this strategy for the discovery of other actinomycete natural products. Genome mining could be combined with one strain-many compounds (OSMAC) strategy for discovering the new compounds deduced from genome mining.

9.5 Traditional Methods for Discovering New Agroactive Secondary Metabolites from Actinomycetes

Classical mutagenesis and bioconversion are still used for studying the potential of microorganisms to produce novel or new compounds. Many studies demonstrated the success of classical methods for finding novel agroactive secondary metabolites from actinomycetes.

9.5.1 Mutagenesis

The traditional mutagenesis for discovering new agroactive secondary metabolites from actinomycetes include physical mutagenesis and chemical mutagenesis. A mutant of *Streptomyces bingchenggensis* X-4, which was obtained through the treatment with UV and *N*-methyl-*N*′-nitro-*N*-nitrosoguanidine, produced milbemycin β15 and seco-milbemycins E and F which showed acaricidal and nematicidal activity but weaker than commercial milbemycins (Zhang et al. 2011a, b). Li and coworkers found 23, 24-didehydro-13α-hydroxy milbemycin A3 and 24,30-didehydro-13α-hydroxy milbemycin A3 and two other doramectin analogues from a mutant of *Streptomyces avermitilis* NEAU1069 obtained in the same way above (Li et al. 2014, 2015); the first two milbemycin analogues showed more potent insecticidal activity against *Brevicoryne brassicae* than the mixture of milbemycins A3/A4. The new nemadectin analogues from this mutant showed weaker acaricidal and insecticidal activity. Mutant Y-3 of *Streptomyces microflavus* NEAU3 obtained in the same way above produced two new congeners of nemadectin (Yang et al. 2013), of which the new analogue of nemadectin with C-27 of carbonyl instead of C-27 of methylene in nemadectin showed much higher acaricidal and nematicidal activity than nemadectin and milbemycins A3/A4. The classical mutagenesis might be a way to find the new analogues of known secondary metabolites from actinomycetes, but it will be very difficult to select right mutant producing new analogues because it is time-consuming.

9.5.2 Bioconversion

Shake flask fermentation of *Saccharopolyspora erythraea* ATCC 11635 by adding the mixture of tenvermectins A and B led to the isolation of 4″-*O*-glucosyl tenvermectin A and 4″-*O*-glucosyl tenvermectin B, both showing nematocidal and acaricidal activity against *Bursaphelenchus xylophilus* and *Tetranychus cinnabarinus*, respectively (Wan et al. 2017). Biotransformation of nemadectin α with *Mucor ramannianus* yielded a nemadectin analogue with a phosphate group at C-23, while transformation with *Streptomyces griseus* yielded two analogues of nemadectin α with C-29 as hydroxymethyl group and carboxyl group, respectively. Biotransformation of nemadectin β with *S. griseus* yielded an analogue with C-28 as hydroxymethyl group (Fang et al. 2001).

9.6 The Designed Biosynthesis of New Agroactive Secondary Metabolites in Post-omics Era

With the rapid development of sequencing technology, it becomes much easier to sequence the genome of agroactive actinomycetes. Through the genome analysis tools and genetic engineering techniques, the biosynthesis pathways of agroactive secondary metabolites from actinomycetes can be elucidated quickly. Genes and related regulatory genes responsible for the biosynthesis could be easily cloned and expressed homogeneously or heterogeneously. A lot of efforts had been put forward to study the biosynthesis of those commercialized insecticides from actinomycetes, like avermectins, milbemycins, spinosyns, polyoxins, and nikkomycins. Researchers had used gene replacement, knockdown, and inactivation and introduced heterogeneous genes to diversify the analogues of the above compounds for the discovery of novel and potent derivatives. Many new derivatives or hybrid metabolites were discovered.

9.6.1 Gene Replacement

With understanding of the functions of PKS genes in the biosynthesis of polyketides, it is possible to design the biosynthesis of targeted compounds through gene replacement. The replacement of *ave*DH2-KR2 in avermectin biosynthetic gene cluster in *S. avermitilis* NA-108 with milDH2-ER2-KR2 from milbemycin producing *S. bingchenggensis* led to the high yield of ivermectin B1a, while replacement of *ave*LAT-ACP encoding the loading module of avermectin PKS in *S. avermitilis* NA-108 with the *mil*LAT-ACP encoding the loading module of milbemycin PKS in *S. bingchenggensis* led to the production of two new congeners of ivermectin-25-ethyl and 25-methyl ivermectin. The mixture of these two ivermectin derivatives showed stronger nematicidal activity than ivermectin and milbemycin A3/A4 and stronger insecticidal activity against second instar larvae of *Mythimna separata* than milbemycin A3/A4 (Zhang et al. 2015). The replacement of *aveA1* in *S. avermitilis* G8-17 with *mil*A1 from *Streptomyces hygroscopicus* resulted in the production of novel avermectin and milbemycin, such as tenvermectins A and B (Huang et al. 2015); tenvermectins C and D (Wan et al. 2017); $\Delta^{2,3}$-temverectom and 13α-hydroxy milbemycin β_{11} (Zhang et al. 2020b); 13α-hydroxy-4-ethyl milbemycin β_3, 13α-hydroxy-25-ethyl milbemycin β_3, and 13α-hydroxy milbemycin β_3 (Pan et al. 2015); and 13α-*O*-α-L-oleandrosyl milbemycin β_3, 13α-*O*-α-L-oleandrosyl-25-ethyl milbemycin β_3, and 13α-*O*-α-L-oleandrosyl-25-isopropyl milbemycin β_3 (Pan et al. 2016a, b). All of these new analogues of avermectins or milbemycins still showed strong acaricidal activity against adult mites of *Tetranychus* sp. (Huang et al. 2015; Wan et al. 2017; Zhang et al. 2020b). The replacement of *AveA3* in *S. avermitilis* TM24 with milbemycin biosynthetic gene *milA3* resulted in the production of tennovel 16-membered macrolactones which showed acaricidal activity against *T. cinnabarinus* and nematicidal activity against *B. xylophilus* (Feng et al. 2019). Kim et al. constructed the mutants through the replacement of AveA1 and

AveA3 (or module 7 in AveA3) of in the avermectin high-producing strain *S. avermitilis* SA-01 with MilA1 and MilA3 (or module 7 in MilA3) of the milbemycin PKS which increased the production of C5-*O*-methylated congener milbemycins B2, B3, and G as major products (Kim et al. 2017). The replacement of the PKS genes or modules resulted in the production of analogues with different modified group.

9.6.2 Gene Overexpression

The overexpression of specific genes in the biosynthetic gene cluster of secondary metabolites not only results in the higher production of target compounds but also induces the production of novel metabolites. Herbicidins were a group of secondary metabolites from *Streptomyces* sp. with herbicidal activity. The overexpression of *hcdR2* in *Streptomyces mobaraensis* US-43 resulted in the production of herbicidins O–S (Shi et al. 2019).

9.6.3 Gene Deletion or Disruption

The disruption of milF gene in *Streptomyces bingchenggensis* resulted in the mutant BCJ60 which produced 5-oxomilbemycins A3/A4 (Wang et al. 2014). This mutant also produced 27-methoxylmilbemycin α31 and 27-oxomilbemycin α31 (Li et al. 2017a) and 4,25-diethyl-4,25-demethyl-milbemycin β3 and 27-formaldehyde-milbemycin β14 (Li et al. 2017b). All these four new analogues of milbemycin showed potent acaricidal activity against *T. cinnabarinus* and *B. xylophilus* as milbemycins A3/A4 mixture. The disruption of neoC which encodes 2-deoxystreptamine (2-DOS) synthases which is crucial for the biosynthesis of neomycin in *Streptomyces fradiae* resulted in the production of new derivatives of neomycin once feeding the mutant with different synthetic 2-DOS derivatives. The disruption of busO gene in *Saccharopolyspora pogona* which produced C-21 butenyl spinosyns resulted in the accumulation of 17-hydroxy butenyl-spinosyn and 17-(3″-*O*-methylglucosyl)-butenyl-spinosyn (Hahn et al. 2006). Most of such neomycin derivatives still showed antibacterial activity against *Pseudomonas solanacearum*, *Erwinia carotovora*, *Xanthomonas oryzae*, and *X. campestris* (Shi et al. 2011). The deletion of first PKS module of the geldanamycin biosynthesis gene in *Streptomyces* sp. LZ35 resulted in the production of cuevaenes C–E by the mutant, while cuevaene C showed antifungal activity against *Fusarium verticillioides* and *Rhizoctonia solani* (Deng et al. 2014). The disruption of sanL in *Streptomyces ansochromogenes* with homogeneous recombination resulted in the production of new analogues of nikkomycins-nikkomycins Px and Pz once the mutant was fed with nicotinic acid (Feng et al. 2014). The disrupted and modified gene clusters of nikkomycins with *nik*B or *nik*B and *nik*Q deletion were introduced into the polyoxin non-producing mutants of *Streptomyces aureochromogenes* generated through gene deletion. The introduction of the disrupted nikkomycin gene clusters resulted in the production of polyoxin N and nikkomycin-polyoxin hybrids-nikkoxins B–D (Zhai et al. 2012). Polyoxin N and nikkoxin D showed more potent antifungal activity against phytopathogenic fungi than nikkomycins and polyoxin A and B. The introduction of modified nikkomycin gene cluster with *nik*O and *nik*Q disruption into a polyoxin C producing mutant with *pol*O-P disrupted resulted in the production of nikkoxins E–G (Qi et al. 2015), and the introduction of an extra copy of *nikS* gene enhanced the production of nikkoxins E–G. Nikkoxins E–G showed more potent antifungal activity against several phytopathogenic fungi.

9.6.4 Expression of Introduced Heterogeneous Genes

The introduction of genes required for the biosynthesis of the dipeptidyl moiety of polyoxin from *Streptomyces cacaoi* into a *S. ansochromogenes* mutant producing only nikkomycin X resulted in the production of several new hybrid antibiotics-polynik A and polyoxin N, and both compounds showed stronger inhibitory activity to phytopathogenic fungi than polyoxin B (Li et al. 2011). The heterogeneous expression of polyoxin biosynthetic gene cluster from *S. cacaoi* in the *sanN* inactivated mutant of *S. ansochromogenes* resulted in the isolation of two novel polyoxins-polyoxins O and P. Polyoxin P showed similar antifungal activity as polyoxin N (Li et al. 2012).

9.7 Novel Agroactive Secondary Metabolites from Actinomycetes in the Past Two Decades

In the past two decades, hundreds of novel agroactive secondary metabolites were discovered from actinomycetes through classical screening or genetic manipulation. Because of the limit of the context in this article, we do not intend to cover all the novel or new agroactive secondary metabolites discovered. We just present some examples of the novel agroactive secondary metabolites naturally produced with promising insecticidal, fungicidal, herbicidal, or plant growth regulating activities or with special structures which might be interesting to chemists for structure optimization or derivation.

9.7.1 Novel Insecticidal and Acaricidal Metabolites

Lewer et al. discovered a new series of very potent insecticidal compounds structurally related to spinosyn A from *Saccharopolyspora pogona*, and this series of compounds contained a but-1-enyl group at C-21 instead of ethyl group in the classical spinosyns. Some butenyl-spinosyns showed higher or comparable activity against tobacco budworm as spinosyn A, and higher activity against cotton aphid than spinosyn A, and comparable activity against whitefly as spinosyns A and D (Lewer et al. 2009). A new 24-membered macrolide, 5′-*epi*-SPA-6952A, was isolated from *Streptomyces diastatochromogenes* showing strong insecticidal activity against *Pieris rapae* under field condition, and the control effects of 5′-*epi*-SPA-6952A on *Pieris rapae* were 89.47% and 82.16% at the dosage of 90 g/hm^2 and 45 g/hm^2 after 48 h (Cao et al. 2019; Zhang et al. 2019). Antimycin A_9 was isolated from the culture of *Streptomyces* sp. K01-0031 and showed strong insecticidal against *Artemia salina* (Shiomi et al. 2005).

9.7.2 Novel Antifungal or Bactericidal Metabolites

In this section, it summarized the antifungal or bactericidal secondary metabolites from actinomycetes which were active against bacterial of fungal plant pathogens (as shown in Table 9.1). Chloroxaloterpins A and B were isolated from *Streptomyces* sp. SN194 and showed remarkable inhibitory activity against the spore germination and moderate effects on the mycelial growth of *B. cinerea* (Bi and Yu 2016). Two novel 32-membered macrolide compounds, named novonestmycins A and B, were isolated from the soil strain *Streptomyces phytohabitans* HBERC-20821, which showed strong activities against the phytopathogenic fungi *Corynespora cassiicola*, *R. solani* and *Septoria nodorum*, with MIC values of 0.78, 0.39, and 0.78 μg/mL, respectively (Wan et al. 2015). Munumbicins A–D, E4, and E5 were isolated from an endophytic *Streptomyces* sp. NRRL 3052 in the snakevine (*Kennedia nigriscans*) (Castillo et al. 2002, 2006), and all these six newly described compounds showed strong activity against *Pythium ultimum* with MIC of 0.2–5 μg/mL and

Table 9.1 Some examples of antifungal or antibacterial secondary metabolites from actinomycetes

Actinomycetes	Active metabolites	Targets	References
Streptomyces sp. SN194	Chloroxaloterpins A and B	*B. cinerea*	Bi and Yu (2016)
Streptomyces phytohabitans HBERC-20821	Novonestmycins A and B	*Corynespora cassiicola*, *R. solani*, and *Septoria nodorum*	Wan et al (2015)
Streptomyces sp. NRRL 3052	Munumbicins A–D, E4, and E5	*Phytophthora cinnamomi*, *Pythium ultimum*, and other phytopathogenic fungi *Pseudomonas syringae*	Castillo et al. (2002, 2006)
32,33-Didehydroroflamycoin	*Streptomyces durmitorensis* MS405 *Streptomyces rectiviolaceus* DY46	*Alternaria alternata*, *Colletotrichum acutatum* and *Penicillium expansum*, *Botrytis cinerea*	Milisavljevic et al. (2015) Kim et al. (2019)
Cyclothiazomycins B1 and B2	*Streptomyces* sp. A307	*Botrytis cinerea*, *Septoria nodorum*	Hashimoto et al. (2006)
Echinosporin and 7-deoxyechinosporin	*Amycolatopsis* sp. YIM PH20520	Root rot pathogens of *P. notoginseng*	Xu et al. (2019)

Phytophthora cinnamomi with MIC of 1.5–15.6 μg/mL. Munumbicins A–D showed moderate to strong antibacterial activity against *P. syringae* with MIC of 0.2–15.6 μg/mL and moderate to strong antifungal activity against several species of phytopathogenic fungi. 32,33-Didehydroroflamycoin was found from *Streptomyces durmitorensis* MS405 (Milisavljevic et al. 2015) and *Streptomyces rectiviolaceus* DY46 (Kim et al. 2019), and this pentane macrolide showed broad spectra of antifungal activity against phytopathogenic fungi (Kim et al. 2019; Milisavljevic et al. 2015). 32,33-Didehydroroflamycoin inhibited the fruit necrosis of apple caused by *Alternaria alternata*, *Colletotrichum acutatum*, and *Penicillium expansum* (Milisavljevic et al. 2015) and reduced the disease development of gray mold on tomato fruit (Kim et al. 2019). Cyclothiazomycins B1 and B2 were found in the culture of *Streptomyces* sp. A307. Cyclothiazomycin B1 showed strong antifungal activity against *B. cinerea* and *S. nodorum* with IC_{50} of 0.2 and 4.1 nmol/mL (Hashimoto et al. 2006). Echinosporin and 7-deoxyechinosporin from the culture of *Amycolatopsis* sp. YIM PH20520 showed rather strong antifungal activity against the root rot pathogen of *P. notoginseng* with MIC between 32–128 μg/mL (Xu et al. 2019).

9.7.3 Novel Herbicidal, Phytotoxic, or Plant Growth Regulating Metabolites

Shi et al. had reviewed the herbicidal, phytotoxic, or plant growth regulating metabolites from actinomycetes (Shi et al. 2020). Here we just selected several typical compounds discovered in the past two decades. Ribofuranosyl triazolone was isolated from the culture of *Actinomadura madurae* and showed broad spectrum of postemergence herbicidal activity against broadleaf weeds and grasses, such as *Helianthus annuus*, *Ipomoea hederacea*, *Echinochloa crus-galli*, *Avena fatua*, *Abutilon theophrasti*, *Setaria faberi*, and *Alopecurus myosuroides*, in a greenhouse pot assay. Ribofuranosyl triazolone inhibited adenylosuccinate synthetase after the phosphorylation in plants (Schmitzer et al. 2000). A terphenyl glycoside, terfestatin A, was produced by *Streptomyces* sp. F40 (Yamazoe et al. 2004) and inhibited the expression of auxin-inducible genes in *Arabidopsis thaliana* and antagonized the response in roots by auxin. Two novel derivatives of terfestatin A, terfestatins B and C, were produced by *Streptomyces* sp. RM-5-8, but there were no reports on the herbicidal activity of them. *Streptomyces* sp. strain-329 produced two herbicidal metabolites, 329-C1 and 329-C3, which showed stronger herbicidal activity against many weeds than cycloheximide (Bo et al. 2019). A-87774-1–3 from *Streptomyces* sp. SANK 61805 showed strong herbicidal activity against wild mustard, velvetleaf, and redroot pigweed at 50 ppm under glasshouse condition, and A-87774-2 strongly killed many species of weeds at 200 g/ha and 125 g/ha through stem and leaf treatment in pot assay. A-87774-2 and A-87774-3 showed excellent herbicidal effects in soil treatment in pot assay and paddy field treatment (Shigematsu et al. 2010). *N*-Phenylpropanamide and *N*-(naphthalene-1-yl)-propanamide were from *Streptomyces* sp. KA1-3 strongly inhibited the seed germination of *Cyperus rotundus* but promoted the growth of *Sesamum indicum* (Priyadharsini et al. 2013a, b). Diethyl 7-hydroxytrideca-2,5,8,11-tetraenedioate from *Streptomyces* sp. DDBH019 showed stronger herbicidal activity against *E. crus-galli*, *Amaranthus spinosus*, and *C. rotundus* than commercial 2,4-D (Priyadharsini et al. 2017).

9.7.4 Novel Nematicidal Metabolites

A novel macrocyclic lactone and a new congener of nemadectin from *S. microflavus* neau3 showed comparable nematicidal activity against *Caenorhabditis elegans* as mixture of milbemycins A3/A4 (Xiang et al. 2010; Wang et al. 2014). Yoon et al. isolated 3-benzyl-1,4-diaza-2,5-dioxobicyclo[4.3.0]nonane from *S. cacaoi* GY525 which showed moderate nematicidal activity against eggs and J2 of root-knot nematodes (Yoon et al. 2012). Antibiotic 1100-50, a nucleoside antibiotic from *Streptomyces lavendulae* SANK 64297, showed nematicidal activity against *Meloidogyne hapla*, and 24-h pretreatment of *M. hapla* with antibiotic 1100-50 inhibited the root-knot formation in cucumber pot assay, even at 0.05 ppm (Takatsu et al. 2003). It seemed that the nematicidal activity of antibiotic 1100-50 was caused by the inhibition of RNA synthesis.

9.7.5 Novel Antiphytoviral Metabolites

There were few reports of the discovery of secondary metabolites from actinomycetes which were active against plant virus. ε-Poly-L-lysine from *Streptomyces ahygroscopicus* strain STZ showed strong protective and curative activity against tobacco mosaic virus through half-leaf method with *Nicotiana glutinosa*. With the treatment of 500–1000 μg/mL of ε-poly-L-lysine, it showed protective and curative activity against tobacco mosaic virus with control effects of 71.4 and 84.2% and 59.3 and 68.7%, respectively, in half-leaf assay (Chen et al. 2019).

9.8 The Development of Agrochemicals with Leads from Actinomycete and Their Application in Crop Protection

Several agrochemicals were developed with agroactive secondary metabolites from actinomycetes as leads. Spinetoram, a chemically modified mixture of spinosyn J/L produced by *Saccharopolyspora spinosa*, was developed by Dow AgroSciences LLC (currently Corteva Agriscience) (Galm and Sparks 2016). The artificial neural network modeling and traditional quantitative structure activity relationship (QSAR) approaches on efficacy improvement of spinosyns resulted in the design of spinetoram—with rhamnose-3′-*O*-ethylation and hydrogenation of the 5,6-double bond. The mixture showed broader-spectrum and more potent activities. The scientists at Dow discovered rhamnosyl-3′-desmethyl spinosyn J (and L) through high-throughput screening in a long-term strain improvement of S. *spinosa.* A single point mutation with the *spnK* gene encoding the corresponding 3′-*O*-methyltransferase resulted in the accumulation of spinosyns J and L. Spinetoram was semi-synthesized through rhamnose-3′-*O*-ethylating in spinosyn L and 3′-*O*-ethylation and 5,6 double-bond reduction of spinosyn J. Spinetoram exhibited increased efficacy, broader spectrum, and improved residual activity than spinosad. It also had similar mammalian and environmental toxicological properties. The compound was launched in 2007 and used for the control of many important crop pests like codling mother, leaf miners, diamondback moth, thrips, cutworms, and others. For its excellent performance, it won the Presidential Green Chemistry award in the USA in 2008. Another example was the fungicide-fenpicoxamid which was developed with UK-2A from an unidentified *Streptomyces* strains lead. UK-2A was found from *Streptomyces* sp. 517-02 and showed antifungal activities against various filamentous fungi (Ueki et al. 1996; Hanafi et al. 1996; Shibata et al. 1998). UK-2A showed a novel mode of action which involved inhibition of mitochondrial complex III via binding to the Qi ubiquinone binding site (Ueki and Taniguchi 1997; Machida et al. 1999). But the highly antifungal activity in vitro of UK-2A could not be transformed to the control effect in greenhouse and field trial against key fungal diseases. Semi-synthetic modification of UK-2A focusing on the modification of the picolinamide ring, the macrocycle isobutyl ester position, and the macrocycle benzyl position was carried out at Dow AgroSciences (currently Corteva Agriscience) (Meyer et al. 2005; Owen et al. 2018, 2019a, b). Fenpicoxamid, an analogue of UK-2A through the modification of picolinamide OH group with isobutyryloxymethyl ether provided strong control of *Zymoseptoria tritici* that caused wheat leaf blotch which were resistant to current cereal fungicides. The mean control effects of fenpicoxamid on wheat leaf blotch was 82.1% in multiple European field trials (Owen et al. 2017). Fenpicoxamid was converted by *Z. tritici* to UK-2A which was more effective against *Z. tritici.* Fenpicoxamid was registered in the European Union in 2018 and approved for sale in France in late 2020 (CortevaAgrisciences 2020), but its launch in the UK was delayed till 2021. It is estimated the peak sale of fenpicoxamid will be $350m USD annually. In spite of limited number of agrochemicals developed with actinomycete secondary metabolites as leads in the first two decades in the new century, the diverse agroactive secondary metabolites from actinomycetes still have high potential in the development of new AI of natural product-based agrochemicals.

9.9 Conclusion

Although no novel agroactive secondary metabolites discovered from actinomycetes in the past two decades had been developed as agrochemicals or used as lead of commercial agrochemicals, the discovery of novel agroactive secondary metabolites from actinomycetes is still a hot area for natural product chemists or agrochemists. Actinomycetes from untapped or rarely tapped environment, like marine sediment, mangrove, desert, and plant- or insect-associated niche, have higher rate of production of novel agroactive secondary metabolites. Rare actinomycetes have the higher potential to produce novel agroactive secondary metabolites. In this review, we do not cover the culturing methods to diversify the secondary metabolites produced by actinomycetes, as it is well understood by researchers in this area. In most cases, the discovery of bioactive secondary metabolites was activity-targeted, especially for the development of pharmaceuticals and agrochemicals. The activity-guided isolation of compounds led to the discovery of novel active compounds. In this review, we do not cover the progress in bioassay methods for discovery of novel agroactive secondary metabolites. As we know, the more the sensitivity of the assays, the more chance to find new metabolites. Highly sensitive assays must be combined with chemical screening to dereplicate the known agroactive metabolite-producing actinomycetes to accelerate the discovery process. The understanding of biosynthesis pathways of secondary metabolites makes it possible to use gene screening with a specific primer for identifying the putative actinomycetes producing the compound with a special functional group. We could understand the biosynthetic potential of actinomycetes by genome mining, in spite that some metabolites might be silent. Such silent metabolites might be produced through genetic manipulation or OSMAC strategy. Genetic manipulation of actinomycetes can also result in the production of hybrid metabolites and novel analogues of known agroactive metabolites. In the post-omics era, the combination of all the above strategies, including microbiology, chemistry, and molecular biology, will accelerate the discovery of novel agroactive secondary metabolites and support the development of new agrochemicals with agroactive secondary metabolites as leads.

Acknowledgements The work of the authors is funded by grants of National Key R&D Program of China (2017YFD0201205) and Hubei Innovation Centre of Agricultural Science and Technology (2019-620-000-001-27).

References

Berdy J (2012) Thoughts and facts about antibiotics: where we are now and where we are heading. J Antibiot (Tokyo) 65(8):385–395. https://doi.org/10.1038/ja.2012.27

Betancur LA, Forero AM, Romero-Otero A, Sepulveda LY, Moreno-Sarmiento NC, Castellanos L, Ramos FA (2019) Cyclic tetrapeptides from the marine strain Streptomyces sp. PNM-161a with activity against rice and yam phytopathogens. J Antibiot (Tokyo) 72(10):744–751. https://doi.org/10.1038/s41429-019-0201-0

Betancur LA, Forero AM, Vinchira-Villarraga DM, Cardenas JD, Romero-Otero A, Chagas FO, Pupo MT, Castellanos L, Ramos FA (2020) NMR-based metabolic profiling to follow the production of anti-phytopathogenic compounds in the culture of the marine strain *Streptomyces* sp. PNM-9. Microbiol Res 239:126507. https://doi.org/10.1016/j.micres.2020.126507

Bi Y, Yu Z (2016) Diterpenoids from *Streptomyces* sp. SN194 and their antifungal activity against *Botrytis cinerea*. J Agric Food Chem 64(45):8525–8529. https://doi.org/10.1021/acs.jafc.6b03645

Bo AB, Kim JD, Kim YS, Sin HT, Kim HJ, Khaitov B, Ko YK, Park KW, Choi JS (2019) Isolation, identification and characterization of *Streptomyces* metabolites as a potential bioherbicide. PLoS One 14(9):e0222933. https://doi.org/10.1371/journal.pone.0222933

Cao H, Liu W, Chang Y, Chen H, Zhang Y, Xu W, Tao L (2019) 5′-Epi-SPA-6952A, a new insecticidal 24-membered macrolide produced by *Streptomyces diastatochromogenes* SSPRC-11339. Nat Prod Res 33(5):659–664. https://doi.org/10.1080/14786419.2017.1405401

Castillo UF, Strobel GA, Ford EJ, Hess WM, Porter H, Jensen J, Albert H, Robison R, Condron MAM, Teplow DB, Stevens D, Yaver D (2002) Munumbicins, wide-spectrum antibiotics produced by *Streptomyces* NRRL 30562, endophytic on *Kennedia nigriscans*. Microbiology 148:2675–2685

Castillo UF, Strobel GA, Mullenberg K, Condron MM, Teplow DB, Folgiano V, Gallo M, Ferracane R, Mannina L, Viel S, Codde M, Robison R, Porter H, Jensen J (2006) Munumbicins E-4 and E-5: novel broad-spectrum antibiotics from *Streptomyces* NRRL 3052. FEMS Microbiol Lett 255(2):296–300. https://doi.org/10.1111/j.1574-6968.2005.00080.x

Chen G, Lin B, Lin Y, Xie F, Lu W, Fong W-F (2005) A new fungicide produced by a *Streptomyces* sp. gaas7310. J Antibiot 58(8):519–522

Chen X, Hu LF, Huang XS, Zhao LX, Miao CP, Chen YW, Xu L-H, Han L, Li YQ (2019) Isolation and characterization of new phenazine metabolites with antifungal activity against root-rot pathogens of *Panax notoginseng* from *Streptomyces*. J Agric Food Chem 67(41):11403–11407. https://doi.org/10.1021/acs.jafc.9b04191

CortevaAgrisciences LLC (2020) Corteva Agriscience Launches Inatreq™ Active in Cereals in Europe. https://www.corteva.com/resources/media-center/corteva-launches-inatreq-active-in-cereals-in-europe.html. Accessed 03-20-2020

Crevelin EJ, Canova SP, Melo IS, Zucchi TD, da Silva RE, Moraes LA (2013) Isolation and characterization of phytotoxic compounds produced by *Streptomyces* sp. AMC 23 from red mangrove (*Rhizophora mangle*). Appl Biochem Biotechnol 171(7):1602–1616. https://doi.org/10.1007/s12010-013-0418-5

Deng JJ, Lu CH, Li YY, Li SR, Shen YM (2014) Cuevaenes C-E: three new triene carboxylic derivatives from *Streptomyces* sp. LZ35DeltagdmAI. Beilstein J Org Chem 10:858–862. https://doi.org/10.3762/bjoc.10.82

Dieter A, Hamm A, Fiedler HP, Goodfellow M, Mueller WEG, Brun R, Bei W, Bringmann G (2003) Pyrocoll, an antibiotic, antiparasitic and antitumor compound produced by a novel alkaliphilic *Streptomyces* strain. J Antibiot 56(7):639–646

Evans BS, Zhao C, Gao J, Evans CM, Ju KS, Doroghazi JR, van der Donk WA, Kelleher NL, Metcalf WW (2013) Discovery of the antibiotic phosacetamycin via a new mass spectrometry-based method for phosphonic acid detection. ACS Chem Biol 8(5):908–913. https://doi.org/10.1021/cb400102t

Ezra D, Castillo UF, Strobel GA, Hess WM, Porter H, Jensen JB, Condron MAM, Teplow DB, Sears J, Maranta M, Hunter M, Weber B, Yaver D (2004) Coronamycins, peptide antibiotics produced by a verticillate *Streptomyces* sp. (MSU-2110) endophytic on *Monstera* sp. Microbiology (Reading) 150(Pt 4):785–793. https://doi.org/10.1099/mic.0.26645-0

Fang K, Schlingmann G, Enos A, Carter GT (2001) New biotransformation products of nemadectins. J Antibiot 54(10):805–809

Feng C, Ling H, Du D, Zhang J, Niu G, Tan H (2014) Novel nikkomycin analogues generated by mutasynthesis in *Streptomyces ansochromogenes*. Microb Cell Fact 13(59). http://www.microbialcellfactories.com/content/13/1/59

Feng Y, Yu Z, Zhang S, Xue Z, Huang J, Zhang H, Wan X, Chen A, Wang J (2019) Isolation and characterization of new 16-membered macrolides from the aveA3 gene replacement mutant strain *Streptomyces avermitilis* TM24 with acaricidal and nematicidal activities. J Agric Food Chem 67(17):4782–4792. https://doi.org/10.1021/acs.jafc.9b00079

Galm U, Sparks TC (2016) Natural product derived insecticides: discovery and development of spinetoram. J Ind Microbiol Biotechnol 43:185–193

Gebhardt K, Schimana J, Krastel P, Dettner K, Rheinheimer J, Zeeck A, Fiedler H-P (2002) Endophenazines A-D, new phenazine antibiotics from the arthropod associated endosymbiont *Streptomyces anulatus* I. Taxonomy, fermentation, isolation and biological activities. J Antibiot 55(9):534–541

Hahn DR, Gustafson G, Waldron C, Bullard B, Jackson JD, Mitchell J (2006) Butenyl-spinosyns, a natural example of genetic engineering of antibiotic biosynthetic genes. J Ind Microbiol Biotechnol 33:94–104. https://doi.org/10.1007/s10295-005-0016-9

Hanafi M, Shibata K, Ueki M, Taniguchi M (1996) UK-2A, B, C and D, novel antifungal antibiotics from Streptomyces sp. 517-02. II. Structural elucidation. J Antibiot 49:1226–1231

Hashimoto M, Murakami T, Funahashi K, Tokunaga T, Nihei K, Okuno T, Kimura T, Naoki H, Himeno H (2006) An RNA polymerase inhibitor, cyclothiazomycin B1, and its isomer. Bioorg Med Chem 14(24):8259–8270. https://doi.org/10.1016/j.bmc.2006.09.006

Hohmann C, Schneider K, Bruntner C, Irran E, Nicholson G, Bull AT, Jones AL, Brown R, Stach JEM, Goodfellow M, Beil W, Kraemer M, Imhoft JF, Suessmuth RD, Fiedler H-P (2009) Caboxamycin, a new antibiotic of the benzoxazole family produced by the deep-sea strain *Streptomyces* sp. NTK 937. J Antibiot 62(2):99–104

Hou L, Liu Z, Yu D, Li H, Ju J, Li W (2020) Targeted isolation of new polycyclic tetramate macrolactams from the deep sea-derived *Streptomyces somaliensis* SCSIO ZH66. Bioorg Chem 101:103954. https://doi.org/10.1016/j.bioorg.2020.103954

Hu L, Chen X, Han L, Zhao L, Miao C, Huang X, Chen Y, Li P, Li Y (2019) Two new phenazine metabolites with antimicrobial activities from soil-derived *Streptomyces* species. J Antibiot (Tokyo) 72(7):574–577. https://doi.org/10.1038/s41429-019-0163-2

Huang J, Chen AL, Zhang H, Yu Z, Li MH, Li N, Lin J-T, Bai H, Wang J-D, Zheng YG (2015) Gene replacement for the generation of designed novel avermectin derivatives with enhanced acaricidal and nematicidal activities. Appl Environ Microbiol 81(16):5326–5334. https://doi.org/10.1128/AEM.01025-15

Intra B, Bechthold A, Euanorasetr J, Paululat T, Panbangred W (2016) Thailandins A and B, new polyene macrolactone compounds isolated from *Actinokineospora bangkokensis* strain 44EHWT, possessing antifungal activity against anthracnose fungi and pathogenic yeasts. J Agric Food Chem 64(25):5171–5179. https://doi.org/10.1021/acs.jafc.6b01119

Ju KS, Gao J, Doroghazi JR, Wang KK, Thibodeaux CJ, Li S, Metzger E, Fudala J, Su J, Zhang JK, Lee J, Cioni JP, Evans BS, Hirota R, Labeda DP, van der Donk W, Metcalf WW (2015) Discovery of phosphonic acid natural products by mining the genomes of 10,000 actinomycetes. Proc Natl Acad Sci U S A 112(39):12175–12180. https://doi.org/10.1073/pnas.1500873112

Kaaniche F, Hamed A, Elleuch L, Chakchouk-Mtibaa A, Smaoui S, Karray-Rebai I, Koubaa I, Arcile G, Allouche N, Mellouli L (2020) Purification and characterization of seven bioactive compounds from the newly isolated *Streptomyces cavourensis* TN638 strain via solid-state fermentation. Microb Pathog 142:104106. https://doi.org/10.1016/j.micpath.2020.104106

Kim MS, Cho WJ, Song MC, Park SW, Kim K, Kim E, Lee N, Nam S-J, Oh K-H, Yoon YJ (2017) Engineered biosynthesis of milbemycins in the avermectin high-producing strain *Streptomyces avermitilis*. Microb Cell Fact 16(1):9. https://doi.org/10.1186/s12934-017-0626-8

Kim JD, Park MY, Jeon BJ, Kim BS (2019) Disease control efficacy of 32,33-didehydroroflamycoin produced by *Streptomyces rectiviolaceus* strain DY46 against gray mold of tomato fruit. Sci Rep 9(1):13533. https://doi.org/10.1038/s41598-019-49779-6

Kizuka M, Enokita R, Shibata K, Okamoto Y, Inoue Y, Okazaki T (2002) Studies on Actinomycetes isolated from plant leaves—new plant growth inhibitors A-79197-2 and -3 from *Dactylosporangium aurantiacum* SANK 61299. Actinomycetologica 16:14–16

Lewer P, Hahn DR, Karr LL, Duebelbeis DO, Gilbert JR, Crouse GD, Worden T, Sparks TC, Edwards PMR, Graupner PR (2009) Discovery of the butenyl-spinosyn insecticides: novel macrolides from the new bacterial strain *Saccharopolyspora pogona*. Bioorg Med Chem 17(12):4185–4196. https://doi.org/10.1016/j.bmc.2009.02.035

Li J, Li L, Tian Y, Niu G, Tan H (2011) Hybrid antibiotics with the nikkomycin nucleoside and polyoxin peptidyl moieties. Metab Eng 13(3):336–344. https://doi.org/10.1016/j.ymben.2011.01.002

Li J, Li L, Feng C, Chen Y, Tan H (2012) Novel polyoxins generated by heterologously expressing polyoxin biosynthetic gene cluster in the sanN inactivated mutant of *Streptomyces ansochromogenes*. Microb Cell Fact 11:135

Li JS, Wang JD, Yang LY, Zhang J, Zhang H, Wang XJ, Xiang WS (2014) New doramectin analogs from mutant *Streptomyces avermitilis* NEAU1069-3. J Antibiot (Tokyo) 67(2):187–189. https://doi.org/10.1038/ja.2013.97

Li LJ, Zhou D, Chen AL, Huang J, Zhang H, Wang JD, Xiang WS (2015) Two new alpha-class milbemycin metabolites from mutant *Streptomyces avermitilis* NEAU1069-3. J Antibiot (Tokyo) 68(5):354–356. https://doi.org/10.1038/ja.2014.159

Li JS, Du MN, Zhang H, Zhang J, Zhang SY, Wang HY, Chen AL, Wang JD, Xiang WS (2017a) New milbemycin metabolites from the genetically engineered strain *Streptomyces bingchenggensis* BCJ60. Nat Prod Res 31(7):780–784. https://doi.org/10.1080/14786419.2016.1244194

Li JS, Zhang H, Zhang SY, Wang HY, Zhang J, Chen AL, Wang JD, Xiang WS (2017b) New macrocyclic lactones with acaricidal and nematocidal activities from a genetically engineered strain *Streptomyces bingchenggensis* BCJ60. J Asian Nat Prod Res 19(4):339–346. https://doi.org/10.1080/10286020.2016.1211641

Liu Q, Liu Z, Sun C, Shao M, Ma J, Wei X, Zhang T, Li W, Ju J (2019) Discovery and biosynthesis of atrovimycin, an antitubercular and antifungal cyclodepsipeptide featuring vicinal-dihydroxylated cinnamic acyl chain. Org Lett 21(8):2634–2638. https://doi.org/10.1021/acs.orglett.9b00618

Machida K, Takimoto H, Miyoshi H, Taniguchi M (1999) UK-2A, B, C and D, novel antifungal antibiotics from *Streptomyces* sp. 517-02. V. Inhibition mechanism of bovine heart mitochondrial cytochrome bc1 by the novel antibiotic UK-2A. J Antibiot 52:748–753

Mangamuri U, Muvva V, Poda S, Naragani K, Munaganti RK, Chitturi B, Yenamandra V (2016a) Bioactive metabolites produced by *Streptomyces cheonanensis* VUK-A from *Coringa* mangrove sediments: isolation, structure elucidation and bioactivity. 3 Biotech 6(1):63. https://doi.org/10.1007/s13205-016-0398-6

Mangamuri UK, Vijayalakshmi M, Poda S, Manavathi B, Chitturi B, Yenamandra V (2016b) Isolation and biological evaluation of N-(4-aminocyclooctyl)-3,5-dinitrobenzamide, a new semisynthetic derivative from the Mangrove-associated actinomycete *Pseudonocardia endophytica* VUK-10. 3 Biotech 6(2):158. https://doi.org/10.1007/s13205-016-0472-0

Martinez AFC, Mello FMP, Zucchi TD, Melo IS, Moraes LAB (2020) Tandem mass spectrometry methods to accelerate the identification of phytotoxic metabolites produced by *Streptomyces* sp. 39 PL. Nat Prod Res 34(2):210–216. https://doi.org/10.1080/14786419.2018.1525713

Meyer KG, Rogers RB, Yao C, Niyaz NM, Adamski Butz JL, Nader BS (2005) Derivatives of UK-2A. US Patent 6861390, 1 Mar 2005

Milisavljevic M, Zivkovic S, Pekmezovic M, Stankovic N, Vojnovic S, Vasiljevic B, Senerovic L (2015) Control of human and plant fungal pathogens using pentaene macrolide 32,33-didehydroroflamycoin. J Appl Microbiol 118(6):1426–1434. https://doi.org/10.1111/jam.12811

Nain-Perez A, Barbosa LCA, Maltha CRA, Forlani G (2016) First total synthesis and phytotoxic activity of *Streptomyces* sp. metabolites abenquines. Tetrahedron Lett 57(16):1811–1814. https://doi.org/10.1016/j.tetlet.2016.03.038

Nain-Perez A, Barbosa LCA, Maltha CRA, Forlani G (2017) Natural abenquines and their synthetic analogues exert algicidal activity against bloom-forming cyanobacteria. J Nat Prod 80(4):813–818. https://doi.org/10.1021/acs.jnatprod.6b00629

Owen WJ, Yao C, Myung K, Kemmitt G, Leader A, Meyer KG, Bowling A, Slanec T, Kramer VJ (2017) Biological characterization of fenpicoxamid, a new fungicide with utility in cereals and other crops. Pest Manag Sci 73:2005–2016. https://doi.org/10.1002/ps.4588

Owen WJ, Meyer KG, Slanec TJ, Wang NX, Meyer ST, Niyaz NM, Rogers RB, Bravo-Altamirano-K, Herrick JL, Yao C (2018) Synthesis and biological activity of analogs of the antifungal antibiotic UK-2A. I. Impact of picolinamide ring replacement. Pest Manag Sci 75:413–426. https://doi.org/10.1002/ps.5130

Owen WJ, Meyer KG, Meyer ST, Li F, Slanec TJ, Wang NX, Yao C (2019a) Synthesis and biological activity of analogs of the antifungal antibiotic UK-2A. II. Impact of modifications to the macrocycle benzyl position. Pest Manag Sci 75:1831–1840. https://doi.org/10.1002/ps.5329

Owen WJ, Meyer KG, Slanec TJ, Meyer ST, Wang NX, Fitzpatrick GM, Niyaz NN, Nugent J, Ricks MJ, Rogers RB, Yao C (2019b) Synthesis and biological activity of analogs of the antifungal antibiotic UK-2A. III. Impact of modifications to the macrocycle isobutyryl ester position. Pest Manag Sci 76:277–286. https://doi.org/10.1002/ps.5511

Pan HQ, Yu SY, Song CF, Wang N, Hua HM, Hu JC, Wang SJ (2015) Identification and characterization of the antifungal substances of a novel *Streptomyces cavourensis* NA4. J Microbiol Biotechnol 25(3):353–357. https://doi.org/10.4014/jmb.1407.07025

Pan J-J, Wan X, Zhang H, Chen Z, Huang J, Yang B, Chen A, Wang J-D (2016a) Three new milbemycins from a genetically engineered strain *Streptomyces avermitilis* MHJ1011. J Antibiot (Tokyo) 69(2):104–107. https://doi.org/10.1038/ja.2015.90

Pan JJ, Wan X, Zhang SY, Huang J, Zhang H, Chen AL, Wang J-D (2016b) Three new 16-membered macrolide compounds from a genetically engineered strain *Streptomyces avermitilis* MHJ1011. Bioorg Med Chem Lett 26(14):3376–3379. https://doi.org/10.1016/j.bmcl.2016.04.033

Priyadharsini P, Dhanasekaran D, Kanimozhi B (2013a) Isolation and Structural Characterization of N-(naphthalene-1-yl)propanamide, a herbicidal compound from *Streptomyces* sp. KA1-3. In: Velu RK (ed) Microbiological research in agroecosystem management. Springer, New Delhi, pp 187–195

Priyadharsini P, Dhanasekaran D, Kanimozhi B (2013b) Isolation, structural identification and herbicidal activity of N-phenylpropanamide from *Streptomyces* sp. KA1-3. Arch Phytopathol Plant Prot 46(3):364–373. https://doi.org/10.1080/03235408.2012.758418

Priyadharsini P, Dhanasekaran D, Gopinath PM, Ramanathan K, Shanthi V, Chandraleka S, Biswas B (2017) Spectroscopic identification and molecular modeling of diethyl 7-hydroxytrideca-2,5,8,11-tetraenedioate: a herbicidal compound from *Streptomyces* sp. Arab J Sci Eng 42(6):2217–2227. https://doi.org/10.1007/s13369-016-2401-2

Qi J, Liu J, Wan D, Cai YS, Wang Y, Li S, Wu P, Feng X, Qiu G, Yang S, Chen W, Deng Z (2015) Metabolic engineering of an industrial polyoxin producer for the targeted overproduction of designer nucleoside antibiotics. Biotechnol Bioeng 112(9):1865–1871. https://doi.org/10.1002/bit.25594

Saadouli I, Zendah El Euch I, Trabelsi E, Mosbah A, Redissi A, Ferjani R, Fhoula I, Cherif A, Sabatier J-M, Sewald N, Ouzari HI (2020) Isolation, characterization and chemical synthesis of large spectrum antimicrobial cyclic dipeptide (l-leu-l-pro) from *Streptomyces misionensis* V16R3Y1 bacteria extracts. A novel ^{1}H NMR metabolomic approach. Antibiotics (Basel) 9(5):270. https://doi.org/10.3390/antibiotics9050270

Schmitzer PR, Graupner PR, Chapin EL, Fields SC, Gilbert JR, Gray JA, Peacock CL, Gerwick BC (2000) Ribofuranosyl triazolone: a natural product herbicide with activity on adenylosuccinate synthetase following phosphorylation. J Nat Prod 63(6):777–781. https://doi.org/10.1021/np990590i

Schulz D, Beese P, Ohlendorf B, Erhard A, Zinecker H, Dorador C, Imhoff JF (2011) Abenquines A-D: aminoquinone derivatives produced by *Streptomyces* sp. strain DB634. J Antibiot (Tokyo) 64(12):763–768. https://doi.org/10.1038/ja.2011.87

Shi G, Zhang X, Wu L, Xie J, Tao K, Hou T (2011) Mutational biosynthesis of neomycin analogs by a mutant of neomycin-producing *Streptomyces fradiae*. Folia Microbiol (Praha) 56(6):555–561. https://doi.org/10.1007/s12223-011-0082-5

Shi Y, Gu R, Li Y, Wang X, Ren W, Li X, Wang L, Xie Y, Hong B (2019) Exploring novel herbicidin analogues by transcriptional regulator overexpression and MS/MS molecular networking. Microb Cell Fact 18(1):175. https://doi.org/10.1186/s12934-019-1225-7

Shi L, Wu Z, Zhang Y, Zhang Z, Fang W, Wang Y, Wan Z, Wang K, Ke S (2020) Herbicidal secondary metabolites from actinomycetes: structure diversity, modes of action, and their roles in the development of herbicides. J Agric Food Chem 68(1):17–32

Shibata K, Hanafi M, Fujii J, Sakanaka O, Iinuma K, Ueki M, Taniguchi M (1998) UK-2A, B, C and D, novel antifungal antibiotics from Streptomyces sp. 517-02 III. Absolute configuration of an antifungal antibiotic, UK-2A, and consideration of its conformation. J Antibiot 51:1113–1116

Shigematsu Y, Sugie Y, Kizuka M (2010) A-87774 compound or salt thereof, method for producing the compound or salt thereof, and agrochemical containing the compound or salt thereof as active ingredient. EU Patent No. EP2468879B1

Shiomi K, Hatae K, Hatano H, Matsumoto A, Takahashi Y, Jiang C, Tomoda H, Kobayashi S, Tanaka H, Omura S (2005) A new antibiotic, antimycin A9, produced by *Streptomyces* sp. K01-0031. J Antibiot 58(1):74–78

Sholkamy EN, Ahmed MS, Yasser MM, Mostafa AA (2020) Antimicrobial quercetin 3-*O*-glucoside derivative isolated from *Streptomyces antibioticus* strain ess_amA8. J King Saud Univ Sci 32:1838–1844

Smaoui S, Mathieu F, Elleuch L, Coppel Y, Merlina G, Karray-Rebai I, Mellouli L (2012) Taxonomy, purification and chemical characterization of four bioactive compounds from new *Streptomyces* sp. TN256 strain. World J Microbiol Biotechnol 28(3):793–804. https://doi.org/10.1007/s11274-011-0872-6

Sparks TC, Hahn DR, Garizi NV (2017) Natural products, their derivatives, mimics and synthetic equivalents: role in agrochemical discovery. Pest Manag Sci 73(4):700–715. https://doi.org/10.1002/ps.4458

Sun X, Wang G, Xiao H, Jiang J, Xiao D, Xing B, Li A, Zhang Y, Sun K, Xu Y, Guo L, Yang D, Ma M (2020) Strepimidazoles A-G from the plant endophytic *Streptomyces* sp. PKU-EA00015 with inhibitory activities against a plant pathogenic fungus. J Nat Prod 83(7):2246–2254. https://doi.org/10.1021/acs.jnatprod.0c00362

Takahashi Y, Nakashima T (2018) Actinomycetes, an inexhaustible source of naturally occurring antibiotics. Antibiotics (Basel) 7:45. https://doi.org/10.3390/antibiotics7020045

Takatsu T, Horiuchi N, Ishikawa M, Wanibuchi K, Moriguchi T, Takahashi S (2003) 1100-50, a novel nematocide from *Streptomyces lavendulae* SANK 64297. J Antibiot 56(3):306–309

Ueki M, Taniguchi M (1997) The mode of action of UK-2A and UK-3A, novel antifungal antibiotics from Streptomyces sp. 517-02. J Antibiot 50:1052–1057

Ueki M, Abe K, Hanafi M, Shibata K, Tanaka T, Taniguchi M (1996) UK-2A, B, C and D, novel antifungal antibiotics from Streptomyces sp. 517-02. I. Fermentation, isolation, and biological properties. J Antibiot 49:639–643

Umetsu N, Shirai Y (2020) Development of novel pesticides in the 21st century. J Pest Sci 45(2):54–74. https://doi.org/10.1584/jpestics.D20-201

Wan Z, Fang W, Shi L, Wang K, Zhang Y, Zhang Z, Wu Z, Yang Z, Gu Y (2015) Novonestmycins A and B, two new 32-membered bioactive macrolides from *Streptomyces phytohabitans* HBERC-20821. J Antibiot (Tokyo) 68:185–190. https://doi.org/10.1038/ja.2014.123

Wan X, Zhang SY, Zhang H, Zhai J, Huang J, Chen AL, Wang JD (2017) New tenvermectin analogs obtained by microbial conversion with *Saccharopolyspora erythraea*. J Antibiot (Tokyo) 70(2):190–192. https://doi.org/10.1038/ja.2016.91

Wang HY, Zhang J, Zhang YJ, Zhang B, Liu CX, He HR, Wang XJ, Xiang WS (2014) Combined application of plasma mutagenesis and gene engineering leads to 5-oxomilbemycins A3/A4 as main components from *Streptomyces bingchenggensis*. Appl Microbiol Biotechnol 98(23):9703–9712. https://doi.org/10.1007/s00253-014-5970-6

Wattanasuepsin W, Intra B, Euanorasetr J, Watanabe Y, Mingma R, Fukasawa W, Mori M, Maatsumoto A, Shiomi K, Panbangred W (2017) 1-Methoxypyrrole-2-carboxamide—a new pyrrole compound isolated from *Streptomyces griseocarneus* SWW368. J Gen Appl Microbiol 63(4):207–211. https://doi.org/10.2323/jgam.2016.10.006

Wei S, Zhang W, Ji Z (2015) Structure and antibacterial activity of ambobactin, a new telomycin-like cyclic depsipeptide antibiotic produced by *Streptomyces ambofaciens* F3. Molecules 20(9):16278–16289. https://doi.org/10.3390/molecules200916278

Xiang W-S, Wang J-D, Wang M, Wang X-J (2010) New nemadectin congener from *Streptomyces microflavus* neau3: fermentation, isolation, structure elucidation and biological activities. J Antibiot 63:171–175

Xu JH, Gu KB, Zhang DJ, Li YG, Tian L (2017) Ghanamycins A and B, two novel gamma-butyrolactones from marine-derived *Streptomyces ghanaensis* TXC6-16. J Antibiot (Tokyo) 70(6):733–736. https://doi.org/10.1038/ja.2017.37

Xu X, Han L, Zhao L, Chen X, Miao C, Hu L, Huang X, Chen Y, Li Y (2019) Echinosporin antibiotics isolated from *Amycolatopsis* strain and their antifungal activity against root-rot pathogens of the *Panax notoginseng*. Folia Microbiol (Praha) 64(2):171–175. https://doi.org/10.1007/s12223-018-0642-z

Yamazoe A, Hayashi K, Kuboki A, Ohira S, Nozaki H (2004) The isolation, structural determination, and total synthesis of terfestatin A, a novel auxin signaling inhibitor from *Streptomyces* sp. Tetrahedron Lett 45:4. https://doi.org/10.1016/j.tetlet.2004.09.055

Yan H, Li Y, Zhang XY, Zhou WY, Feng TJ (2017) A new cytotoxic and anti-fungal C-glycosylated benz[alpha]anthraquinone from the broth of endophytic *Streptomyces* blastomycetica strain F4-20. J Antibiot (Tokyo) 70(3):301–303. https://doi.org/10.1038/ja.2016.126

Yang LY, Wang JD, Zhang J, Xue CY, Zhang H, Wang XJ, Xiang WS (2013) New nemadectin congeners with acaricidal and nematocidal activity from *Streptomyces microflavus* neau3 Y-3. Bioorg Med Chem Lett 23(20):5710–5713. https://doi.org/10.1016/j.bmcl.2013.08.002

Yoon GY, Lee YS, Lee SY, Park RD, Hyun HN, Nam Y, Kim KY (2012) Effects on Meloidogyne incognita of chitinase, glucanase and a secondary metabolite from *Streptomyces cacaoi* GY525. Nematology 14(2):175–184

Zhai L, Lin S, Qu D, Hong X, Bai L, Chen W, Deng Z (2012) Engineering of an industrial polyoxin producer for the rational production of hybrid peptidyl nucleoside antibiotics. Metab Eng 14(4):388–393. https://doi.org/10.1016/j.ymben.2012.03.006

Zhang BX, Zhang H, Wang XJ, Wang JD, Liu CX, Xiang WS (2011a) New milbemycins from mutant *Streptomyces bingchenggensis* X-4. J Antibiot (Tokyo) 64(11):753–756. https://doi.org/10.1038/ja.2011.75

Zhang DJ, Wei G, Wang Y, Si CC, Tian L, Tao LM, Li YG (2011b) Bafilomycin K, a new antifungal macrolide from *Streptomyces flavotricini* Y12-26. J Antibiot (Tokyo) 64(5):391–393. https://doi.org/10.1038/ja.2011.12

Zhang J, Yan YJ, An J, Huang SX, Wang XJ, Xiang WS (2015) Designed biosynthesis of 25-methyl and 25-ethyl ivermectin with enhanced insecticidal activity by domain swap of avermectin polyketide synthase. Microb Cell Fact 14:152. https://doi.org/10.1186/s12934-015-0337-y

Zhang Y, Fan Y, Gao J, Xu W, Xu Z, Liu Y, Li Z, Tao L (2019) A new 24-membered macrolide shows insecticidal activity against *Pieris rapae* potentially through induction of programmed cell death. Food Agric Immunol 30(1):727–742. https://doi.org/10.1080/09540105.2019.1626808

Zhang D, Lu Y, Chen H, Wu C, Zhang H, Chen L, Chen X (2020a) Antifungal peptides produced by actinomycetes and their biological activities against plant diseases. J Antibiot (Tokyo) 73: 265–282. https://doi.org/10.1038/s41429-020-0287-4

Zhang H, Feng Y, Zhang S, Huang J, Xue Z, Wang J (2020b) Two new sixteen-membered macrolides from the genetically engineered strain *Streptomyces avermitilis* MHJ1011. Nat Prod Res 34(20):2959–2963

Zhang Z, Zhou T, Harunari E, Oku N, Igarashi Y (2020c) Iseolides A-C, antifungal macrolides from a coral-derived actinomycete of the genus *Streptomyces*. J Antibiot (Tokyo) 73(8):534–541. https://doi.org/10.1038/s41429-020-0304-7

Zhao H, Yang A, Zhang N, Li S, Yuan T, Ding N, Zhang S, Bao S, Wang C, Zhang Y, Wang X, Hu L (2020) Insecticidal endostemonines A-J produced by endophytic *Streptomyces* from *Stemona sessilifolia*. J Agric Food Chem 68(6):1588–1595. https://doi.org/10.1021/acs.jafc.9b06755

Quorum Sensing and Quorum Quenching Metabolites in Actinomycetes

10

Jamuna A. Bai and Ravishankar V. Rai

Abstract

Quorum sensing or cell-to-cell communication in bacteria regulates gene expression and phenotype production based on population density. Basically, there are four classes of signaling molecules, which are also known as autoinducers. Gram-negative bacteria synthesize the acyl-homoserine lactones (AHLs) or autoinducer type 1, whereas Gram-positive bacteria produce small peptides or autoinducer type 4 signaling molecules. Similarly, autoinducer type 2 is used by bacteria for interspecies signaling and autoinducer type 3 is synthesized by gut bacteria and has a role in host-pathogen interaction. However, the actinomycetes comprised of *Streptomyces* and non-*Streptomyces* genus, unlike other Gram-positive bacteria produce γ-butyrolactones (GBLs) as signal molecules. These molecules structurally resemble AHLs and regulate morphological development and secondary metabolite production in actinomycetes. The first γ butyrolactone characterized and studied was the A-factor. In recent years, several other γ-butyrolactones have been identified in different species of *Streptomyces* and non-*Streptomyces*. Apart from GBLs, certain actinomycetes also utilize peptides and antibiotics as signal molecules. In this chapter, we describe the quorum sensing signal mechanism in actinomycetes and the various types of signal molecules produced by *Streptomyces* and non-*Streptomyces* genus. The regulatory role of the signal molecules has also been discussed. The advantages and limitation of conventional and new techniques used for the detection of QS signals and receptors in actinomycetes have been listed. The quorum quenching enzymes and QS inhibitory compounds produced by actinomycetes are also mentioned.

J. A. Bai
Department of Microbiology, School of Life Sciences, JSS AHER, Mysore, Karnataka, India

R. V. Rai (✉)
Department of Studies in Microbiology, University of Mysore, Mysore, Karnataka, India

R. V. Rai, J. A. Bai (eds.), *Natural Products from Actinomycetes*,
https://doi.org/10.1007/978-981-16-6132-7_10

Keywords

Actinomycetes · Quorum sensing · γ-Butyrolactone · *Streptomyces* · Antibiotics

10.1 Introduction

Bacteria can regulate gene expression based on their population or cell numbers by producing signaling molecules also known as autoinducers. This mechanism of bacterial signaling for a cell-to-cell communication is referred to as quorum sensing (QS) (Nealson 1977; Miller and Bassler 2001). Certain bacteria use quorum sensing to regulate many phenotypes including sporulation, competence factor production, nitrogen fixation, and antibiotic synthesis (Rutherford and Bassler 2012). Pathogenic bacteria utilize the QS mechanism to synthesize virulence factors and for biofilm formation (De Kievit and Iglewski 2000). Eukaryotes such as fungi also utilize QS signaling molecules such as farnesol (Albuquerque and Casadevall 2012). There are many instances where the QS signal molecules or their effectors produced by one species are used by other species and higher organisms, and this is known as intercellular/interspecies and interkingdom signaling (Williams 2007). The QS mechanism involves the production, secretion, and reception of signal molecules. Some of the bacteria produce enzymes capable of degrading the signaling molecules and this is termed as quorum quenching activity (Grandclément et al. 2016). Generally, in the Gram-negative bacteria, the cognate receptors are regulatory proteins and on recognizing signal molecules, bind to these and undergo a conformational change. Further, the signal receptor dimer binds to the promoter of the target gene and results in its expression (Papenfort and Bassler 2016). In the Gram-positive bacteria, autoinducing peptides are the signaling molecules and the signal reception and QS mechanism use two-component histidine sensor kinase system (Sturme et al. 2002). The signaling molecules are classified into four groups: (1) autoinducer 1 or acyl-homoserine lactones (AHLs) produced by Gram-negative bacteria, (2) autoinducer 2 also known as an universal signaling molecule, (3) autoinducer 3 produced by certain gut pathogens such as *Escherichia coli*, and (4) autoinducer 4 or autoinducing polypeptides produced by Gram-positive bacteria (Pereira et al. 2013). Though Gram-positive bacteria use peptides as QS signals, there are certain *Actinobacteria* which can synthesize and use γ-butyrolactones as signaling molecules (Takano 2006).

The study on quorum sensing systems in bacteria has gained importance as it provides a better understanding of the regulatory mechanisms, expression of various phenotypes, and metabolite production in bacteria, which could lead to their potential applications in medicine, agriculture, and biotechnology (Choudhary and Schmidt-Dannert 2010).

Actinomycetes are the Gram-positive bacteria belonging to the *Actinobacteria* phylum and are the major source of diverse and novel secondary metabolites of therapeutic and industrial importance. The bioactive discovered from these bacteria include antibiotics, chemotherapeutics, immunosuppressants, fungicides, and herbicides (Weber et al. 2003). Though the first QS signal molecule identified was the γ-butyrolactones from *Streptomyces*, very few QS systems have been characterized so far from actinomycetes (Horinouchi and Beppu 1992; Takano et al. 2000). The present chapter focuses on quorum sensing molecules and the phenotypes regulated by the QS system in various *Streptomyces* and non-*Streptomyces* actinomycetes.

10.2 QS Signals in Actinomycetes

In the 1960s, A. S. Khokhlov discovered a diffusible self-regulatory molecule in *Streptomyces griseus*. The compound named as A-factor was identified as 13-carbon (C_{13}) γ-butyrolactone. Studies showed that the compound concurrently induced both sporulation and production of antibiotic streptomycin in *S. griseus* (Khokhlov et al. 1967). The molecular mechanism in the A-factor regulatory network was elucidated. The specific receptor and transcriptional activators modulated the synthesis of secondary metabolite and differentiation of morphology (Ando et al. 1997). Subsequent studies showed that the γ-butyrolactone (GBL) quorum sensing system is unique to *Actinobacteria*. The GBL system has a resemblance to the AHL-based QS system of the Gram-negative bacteria. The signaling molecule GBL is structurally similar to the AHL, and its signal receptor also acts as a response regulator (Choi et al. 2003). Unlike the AHLs which are involved in the regulation of various phenotypes, the γ-butyrolactones regulate mainly antibiotic synthesis and morphological differentiation (Takano et al. 2000). In the AHL QS system, the signal synthases are AHL synthase, whereas in the GBL system of *S. griseus*, AfsA, an enzyme is involved in A-factor biosynthesis (Nishida et al. 2007).

GBL signal production has been deciphered in 60% of known *Streptomyces* species (Yamada 1999). *S. griseus*, *S. virginiae*, *S. coelicolor*, *S. lavendulae*, and *S. rochei* have been studied in detail for GBL signals and their receptors. Around 20 GBL-based signals have been elucidated from actinomycetes (Takano et al. 2000). The GBLs have a characteristic core structure of 2 (1′ oxo or 1′hydroxyalkyl)-3-hydroxymethylbutyrolactone with slight variations in each species as shown in Fig. 10.1 (Willey and Gaskell 2011). Based on the length, branching, and presence of hydroxyl group in alkyl side chain, the GBLs are classified into three groups: A-factor type (1′-keto-type), virginiae butanolide (VB) type ((1′S)-hydroxy type), and IM-2 type ((1′R)-hydroxy type) (Arakawa et al. 2012). Some of the *Streptomyces* species synthesizing these GBLs are as follows: (a) A-factor type is synthesized by *S. griseus*, (b) VB-A-E are produced by *S. virginiae*, (c) IM-2 is observed in *S. lavendulae*, and (d) SCB1 has been isolated from *S. coelicolor* (Martin and Liras 2012). Non-*Streptomyces* species are also known to synthesize the GBL-like autoregulators. These types of signaling molecules are especially used by *Micromonospora*, *Actinoplanes*, and *Amycolatopsis* (Choi et al. 2003).

Certain actinomycetes use molecules other than GBLs for signaling and regulation of antibiotic production (Yamada 1999). For example, *S. violaceoruber* and

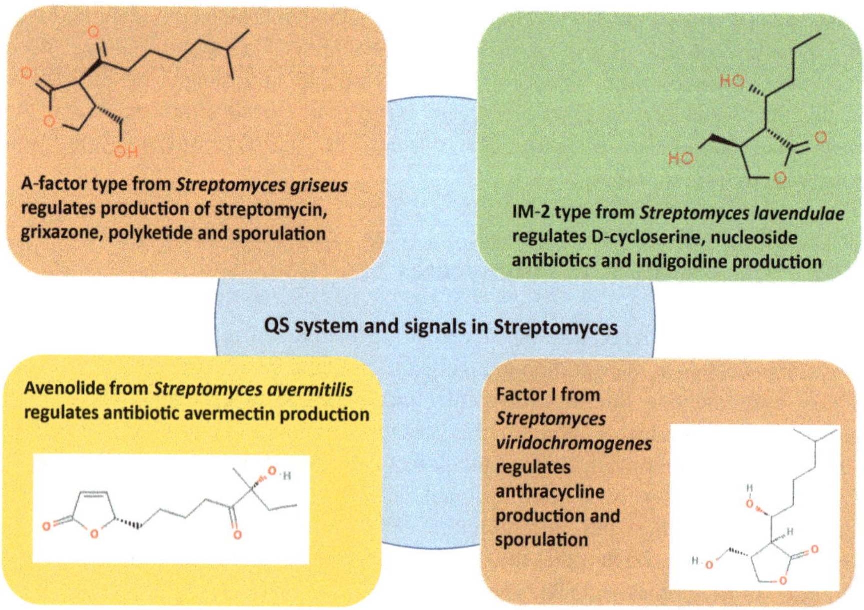

Fig. 10.1 γ-Butyrolactone signals and the GBL regulated phenotypes in *Streptomyces* sp.

S. coelicolor synthesize methylenomycin furan, *S. avermitilis* produces avenolide, and *S. natalensis* synthesizes small molecules such as PI factor (Martin and Liras 2012).

In the GBL QS system, the signaling molecules are synthesized in the early exponential growth phase. The GBLs diffuse through the cell membrane and accumulate based on the cell population (Horinouchi and Beppu 2007). GBL concentration can be around ~100 nM as the cells reach the mid-exponential phase (Tomono et al. 2005a). On reaching a critical threshold concentration, the quorum sensing mechanism is switched on resulting in physiological and morphological differentiation (Yang et al. 2015). The GBLs are recognized by TetR-like receptor proteins. The receptor signals are bound to promoters and repress transcription of the target genes. Only when threshold concentration of signal is found, the receptor dissociates from the promoter region and forms complex with the signal resulting in the expression of a derepressed target gene (Nishida et al. 2007). This mechanism is observed in ArpA, the A-factor receptor in *S. griseus*, BarA in *S. virginiae* that binds virginiae butalonides, JadR3 in *S. venezuelae* which senses SVB1, FarA in *S. lavendulae* that binds to IM-2, and AvaR1 in *S. avermitilis* which detects avenolides (Onaka and Horinouchi 1997; Kinoshita et al. 1997; Kitani et al. 1999, 2011; Zou et al. 2014). The antibiotic synthesis involves a signaling cascade wherein the receptors have a regulatory role. The receptors also control the expression of various other regulatory genes (Kitani et al. 2001). In the last step of the signal relay, the pathway-specific regulator directly activates biosynthetic genes for antibiotic synthesis (Bignell et al. 2007; Suzuki et al. 2010; Mast et al. 2015).

It has been observed that in actinomycetes, a large number of diverse GBL receptors are involved in regulatory circuits. For example, in *S. venezuelae*, the GBL receptor is JadR3 protein which is also the positive regulator for the production of jadomycin. The JadR3 protein regulates antibiotic production in two ways. Firstly, it represses GBL synthase jadW transcription and negatively controls GBL production. Secondly, it positively regulates by controlling transcription of jadomycin cluster activator protein, JadR1 (Zou et al. 2014). Similarly, the ScbR and FarA, the receptor proteins in *S. coelicolor* and *S. lavendulae*, respectively, act as both a repressor and an activator (Takano et al. 2001; Kurniawan et al. 2014). In *S. avermitilis*, the GBL receptor AvaR1 represses the transcription of *aco* gene which is the central regulator for the production of the antiparasitic avermectin. However, avermectin production is induced when AvaR1 binds to the signal avenolide (Wang et al. 2014a). However, in *S. tsukubaensis*, the GBL receptor, BulR1, is a positive regulator and induces the production of tacrolimus, an immunosuppressant (Salehi-Najafabadi et al. 2014). In *S. coelicolor* A3(2), a GBL pseudo-receptor known as CprB is present. It has been successfully crystallized and its structure elucidated (Bhukya et al. 2014). However, it has been difficult to crystallize GBL receptors in other organisms.

In *Streptomyces* species, apart from GBL receptors playing a regulatory role, certain auxiliary proteins have also been discovered which can modulate the production of antibiotics. One such example, is the protein *Streptomyces* antibiotic regulatory protein (SARP) FarR4 present in *S. lavendulae* FRI-5, which is specific for antibiotics virginiamycins M and S. This protein negatively affects the GBL signal IM-2 production (Kurniawan et al. 2014). In *S. virginiae*, until the strain has acquired resistance to virginiamycin antibiotics, there is no induction of antibiotic production. A GBL homologue pseudo-receptor protein BarB is involved in this regulation. It is a repressor protein and prevents transcription of the SARP gene *vmsR* and thus represses induction of virginiamycin M and S production gene (Matsuno et al. 2004). Similar regulation of pristinamycin antibiotic production by the GBL system has been observed in *S. pristinaespiralis* (Mast and Wohlleben 2014; Biarnes-Carrera et al. 2015).

10.3 QS in *Streptomyces* Species

Streptomycetes are soil bacteria which have also been isolated from marine sediments, and these grow as mycelia and are known to sporulate. These bacteria produce a diverse range of antibiotics of therapeutic value and also as biocontrol agents. The antibiotics are produced from complex biosynthetic pathways and their synthesis is growth phase-dependent (Choi et al. 2003). In liquid culture, the antibiotics are produced in stationary phase, whereas in surface fermentation, they are produced during aerial hyphae formation (Takano 2006). In the *Streptomyces* genus, the γ-butyrolactones QS system induces secondary metabolism and controls

Table 10.1 Quorum sensing signaling systems and their regulated phenotypes in *Streptomyces* species

Streptomyces species	GBL synthase/ receptor	GBL signal	GBL regulated phenotypes	Reference
S. acidiscabies	SabA/ SabR	Not identified	Not identified	Healy et al. (2009)
S. ambofaciens	AlpA/ AlpZ	Not identified	Not identified	Pang et al. (2004)
Streptomyces albidoflavus	–	A-factor	Induces formation of aerial mycelium and sporulation	Kiss et al. (2008)
S. aureofaciens	SagA/ SagR	Not identified	Not identified	Mingyar et al. (2015)
S. avermitilis	*aco*/ AvaR1	Avenolide	Production of avermectin antibiotic	Kitani et al. (2011)
S. chattanoogensis	*ScgA-ScgX/ ScgR*	Not identified	Induction of pimaricin production and sporulation	Du et al. (2011)
S. *coelicolor*	*ScbA/ ScbR*	SCB1-3	Production of actinorhodin, undecylprodigiosin and coelimycin	Takano et al. (2001)
S. filipinensis	*SfbR and SfbR2*	–	Production of antifungal agent filipin and antimycin A	Barreales et al. (2020)
Streptomyces globisporus	–	Diketopiperazine MDD	Sporulation and antitumor antibiotic landomycin E production	Holden et al. (1999)

antibiotic production. Similarly, it regulates morphological development as observed in sporulation (Takano et al. 2000). The GBL-based QS signaling system and the regulated phenotypes are summarized in Table 10.1.

In *Streptomyces griseus*, the γ-butyrolactones are identified as A-factor. AfsA synthase is the enzyme involved in A-factor synthesis, and its receptor is the ArpA protein, a transcriptional factor. Evolutionary studies between AfsA and ArpA including their homologues in *Streptomyces* species showed a difference in phylogenetic tree topologies. In *Streptomyces* strain, AfsA and ArpA homologues have changed with evolution (Nishida et al. 2007). Though located adjacently on the chromosome, their evolutionary history differs in *Streptomyces* strains. Studies have shown that ArpA protein initially was a DNA-binding protein and did not act as GBL receptor. Certain plasmids with AfsA/ArpA homologues in some *Streptomyces* strains implicate their role in *afsA*/*arpA* homologue distribution (Hahn 2013). It is assumed that during evolution, on acquiring *afsA*/*arpA* homologues, the *Streptomyces* strain developed the γ-butyrolactone regulatory system for secondary metabolite production and morphogenesis (Nishida et al. 2007).

10.3.1 *Streptomyces griseus*

The intensively studied GBL is A-factor (2-isocapryloyl-3*R*-hydroxymethyl-γ-butyrolactone), a self-regulatory molecule (Chi et al. 2011). It regulates secondary metabolites such as streptomycin antibiotic production and cellular differentiation including aerial mycelium formation and sporulation in *S. griseus* as described in Fig. 10.2 (Bibb 2005; Horinouchi et al. 2010). In its absence, the receptor, a cytoplasmic protein, represses both productions of streptomycin and aerial mycelium formation (Yamazaki et al. 2004; Nishida et al. 2007; Willey and Gaskell 2011). In the A-factor-deficient mutant strain, aerial hyphae are not formed, and there is no synthesis of streptomycin. However, in the mutant, the exogenous supply of chemically synthesized A-factor (10^{-9} M) or culture filtrate of *S. griseus* wild type can restore both morphological differentiation and antibiotic production (Horinouchi 1996; Horinouchi et al. 2010). The ability of the chemical entity A-factor, to regulate physiological processes and control diverged multiple cellular functions at trace concentration in the producer strain, led to its recognition as a bacterial hormone or signaling molecule (Khokhlov et al. 1967; Hara and Beppu 1982). Further, the activity of the A-factor was specific only to *S. griseus* among *Streptomyces* species. Subsequent studies also resulted in the identification of their receptors. The GBL homologues in *Streptomyces* species are autoregulatory factors required for secondary metabolism, and all have high species specificities (Onaka et al. 1995; Kinoshita et al. 1997; Waki et al. 1997). The synthesis of the A-factor occurs in five steps with AfsA catalyzing the crucial coupling reaction between a C_3 compound and a C_{10} acyl derivative. The receptor ArpA has dual abilities. It can

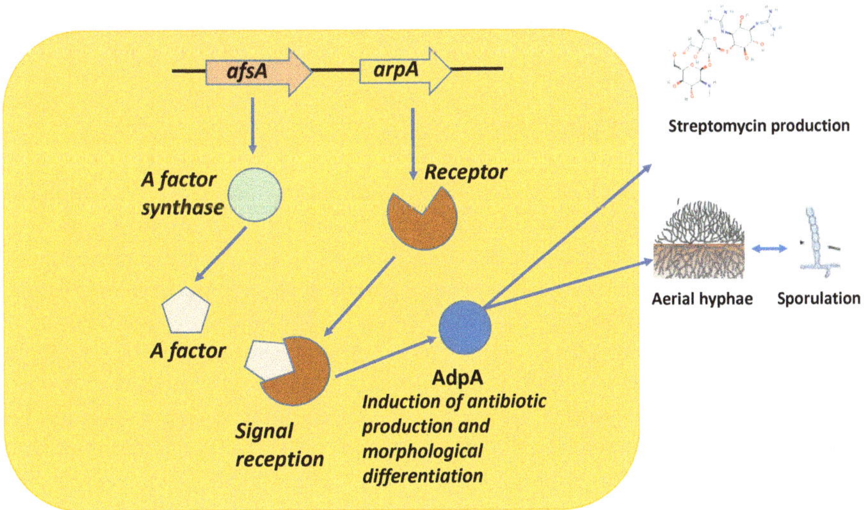

Fig. 10.2 γ-Butyrolactone signaling in *S. griseus*. A-factor autoregulator binds to ArpA repressor, releasing it from promoter of *adpA*. The derepression of *adpA* leads to morphological differentiation and secondary metabolite production including biosynthesis of streptomycin and other polyketides

bind to A-factor as a specific ligand. It can also bind to the promoter of *adpA* where it represses transcription. After binding to the signal molecule, the ArpA disassociates from the *adpA* promoter resulting in transcription and translation (Takano et al. 2001; Kitani et al. 2001). Further, genes in AdpA regulon are activated by AdpA by binding the consensus sequences located upstream of their promoters. This results in the synthesis of secondary metabolites and morphological differentiation (Horinouchi and Beppu 2007). In *S. griseus*, during the exponential growth phase, A-factor accumulates and the concentration could be around 25 to 30 ng/mL (100 nM) (Ando et al. 1997; Onaka and Horinouchi 1997). Similarly, transcription of *afsA* is constant during early growth and increases in the stationary phase (Lezhava et al. 1997).

The receptor for A-factor was identified by applying chemically synthesizing A-factor labeled with tritium (Miyake et al. 1989; Ohnishi et al. 2005). The protein was identified as ArpA and found to bind to A-factor at a 1:1 molar ratio (Beppu 1992). In mutant strains of *S. griseus* 2247, which could not produce an A-factor receptor, streptomycin production and sporulation were higher than the parental wild-type strain (Beppu 1992; Onaka and Horinouchi 1997). This study confirmed that the receptor negatively regulates antibiotic production and sporulation in *S. griseus* (Miyake et al. 1990). The A-factor is synthesized by AfsA and it diffuses through the membrane to induce signal transduction. In the early growth stage, A-factor is found at low concentration. Therefore, ArpA is bound to the promoter of its target *adpA* gene and thus represses transcription. Once A-factor concentration increases and reaches a threshold of 32 nM or higher, it binds to the receptor ArpA. The ArpA dissociates from the promoter of *adpA* resulting in the expression of genes required for secondary metabolite production used in cellular differentiation (Onaka and Horinouchi 1997; Tomono et al. 2005b; Kitani et al. 2008).

10.3.2 *Streptomyces virginiae*

The virginiae butanolide (VB) type contains a 6-α-hydroxy group and is exemplified by VB-A of *S. virginiae* (Yamada et al. 1987; Kondo et al. 1989; Kawachi et al. 2000). The structures of virginiae butanolides A, B, C, D, and E are illustrated in Fig. 10.3. VB biosynthesis can induce the production of streptogramin antibiotics, virginiamycin M and virginiamycin S in wild-type *S. virginiae* as shown in Fig. 10.4 (Takano 2006). The genes involved in the biosynthesis of VBs are *barS2* and *barX* of afsA family (Lee et al. 2008, 2010). BarX is a pleiotropic regulatory protein controlling biosynthesis of VB and virginiamycin, and in developing resistance to virginiamycin M1 (Kawachi et al. 2000). Similarly, varR gene is the transcriptional regulator for resistance to virginiamycin S (Namwat et al. 2001). The receptors for the VBs are the transcriptional regulator protein BarA and BarB (Kinoshita et al. 1997). For the biosynthesis of VB, BarX catalyzes the coupling of two malonyl-CoAs, an isovaleryl-CoA, and a dihydroxyacetone-type C3 unit to form a β-keto ester. Further, BarS2 and BarS1 are involved in conversion of β-keto ester into

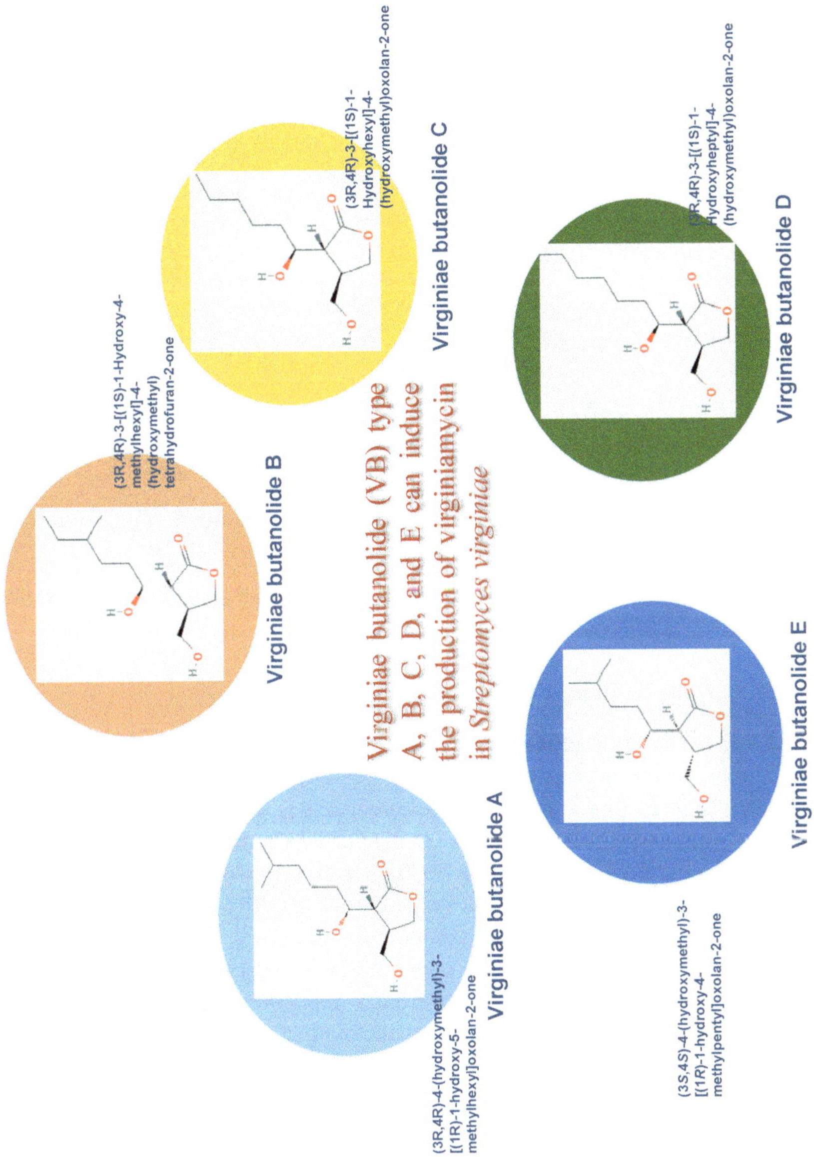

Fig. 10.3 Chemical structures of virginiae butenolide molecules A–E in *Streptomyces virginiae*

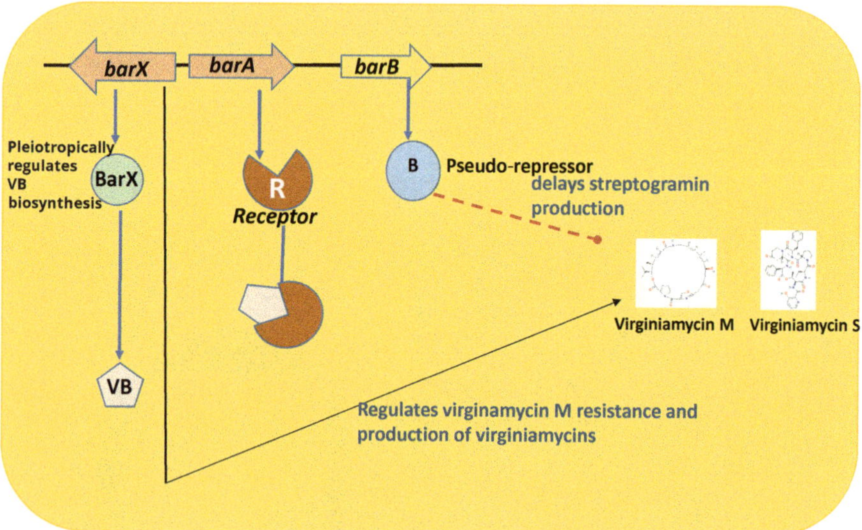

Fig. 10.4 In *S. virginiae*, *barX* pleiotropically regulates VB signal synthesis, production of virginiamycins, and resistance to virginiamycin M. Receptor protein BarA affects the transcription of *barX*. Pseudo-repressor BarB delays antibiotic synthesis until resistance to the antibiotic has been acquired

butenolide by aldol condensation, dehydration, and NAD-dependent reduction (Niu et al. 2016).

10.3.3 *Streptomyces coelicolor*

S. coelicolor produces four antibiotics at different stages of its growth. Two of the antibiotics are pigmented, namely, the polyketide actinorhodin (Act) and tri-pyrrole undecylprodigiosin (Red) (Takano et al. 2000). Antibiotic synthesis in *S. coelicolor* is regulated by the GBL SCB signaling system (Sanchez 2016).

In *S. coelicolor*, six γ-butyrolactone compounds were partly characterized but the biological activity of these compounds was not elucidated (Anisova et al. 1984).) Later, another compound IM-2-type SCB1 from *S. coelicolor* was characterized and shown to stimulate antibiotic production (Takano et al. 2000). Similar GBLs, SCB2 and SCB3, with antibiotic-inducing activity were later isolated from *S. coelicolor* (Hsiao et al. 2009).

Though the structures of γ-butyrolactones of *S. coelicolor* were studied in the 1980s, only recently, the structure of the remaining three GBLs produced by the strain is elucidated (Anisova et al. 1984; Takano 2006). Among GBLs produced in *S. coelicolor*, the butanolide 1 (SCB1) is produced in high concentration and induces Act and Red antibiotic production. The SCBs become active on attaining the threshold concentration of 0.25–0.5 mM. However, their expression and synthesis are strictly controlled. They are stabler than A-factor as they are hydroxylated at C6. Unlike the A-factor synthesis which occurs in the early exponential phase, these compounds are synthesized in later stages of the transition phase and are found even after 12 h into the stationary phase. This could be probably due to the expression of scbA at the late transition phase. The mutant deficient in scbA does not produce the SCBs (Takano et al. 2001). ScbA is not only involved in the production of SCBs but is also necessary for the expression of ScbR (Takano 2006). The receptor ScbR binds to its promoter and the adjacent scbA, and thus regulates SCB synthesis. The ScbR protein has four target repressor sites and binds to these as a dimer. One target site is a repressor site next to its own promoter (scbRp). SCB on binding to its cognate receptor ScbR relieves the repressor effect (Takano et al. 2001).

Other than ScbR, *S. coelicolor* has three receptors, namely, ScbR2, CprA, and CprB. Mutant studies have shown that mutant deficient in *cprA* produces less antibiotic and sporulates late. However, the *cprB* mutant can produce excess antibiotics, and sporulation is also observed to be early. Thus, in *S. coelicolor*, antibiotic production and sporulation are activated by CprA and repressed by CprB (Onaka et al. 1998). The crystal structure of CprB has been determined and resembles the TetR. Their dimers existing in two forms in nature have been obtained. The GBL binding region is a large cavity and found to be the regulatory domain having conserved amino acid residues. On binding to GBL, the conformation of the receptor protein changes resulting in the relocation of DNA-binding domain and dissociation of CprB (Natsume et al. 2004).

Apart from GBLs, *S. coelicolor* also produces various types of signal molecules. One of them is known as methylenomycin furans (MMFs). This signaling molecule regulates the production of methylenomycin antibiotic (Corre et al. 2008). The other is the lantipeptide antimicrobial peptide SapB produced by *S. coelicolor* which has no signaling function unlike in *Lactococcus lactis* and *Bacillus subtilis*. However, these are biosurfactants in nature and aid the aerial mycelium formation in the organism (Kodani et al. 2004; Willey et al. 2006; de Jong et al. 2012). In the lantipeptide synthesis, *ramS* codes for the SapB pre-peptide which is modified by the RamC enzyme to yield the mature peptide (Kodani et al. 2004).

10.3.4 *Streptomyces avermitilis*

The QS signal produced by *S. avermitilis* is a 4-monosubstituted butenolide known as avenolide (Kitani et al. 2011). This signaling molecule induces the production of an antibiotic avermectin known for its antihelminthic property. Avenolide is synthesized by the enzyme acyl-CoA oxidase which is coded by the *aco* gene. The gene coding for the signal receptor, AvaR1, lies between *aco* gene and the *cyp17* gene which codes for the cytochrome P450 hydroxylase (Kitani et al. 2011).

The avenolide signal has been characterized only in *S. avermitilis*. However, the genes coding for the signal have also been observed in *S. ghanaensis*, *S. griseoauranticus*, and *S. fradiae* (Kitani et al. 2011; Mutenko et al. 2014). The avenolide-like butenolides isolated from other marine *Streptomyces* species have no role in signaling but possess antibiotic property and regulate secondary metabolite production (Braun et al. 1995; Mukku et al. 2000; Cho et al. 2001; Viegelmann et al. 2014).

10.3.5 *Streptomyces rochei*

Streptomyces rochei produces two signaling molecules, SRB1 and SRB2, using an AfsA homologue SrrX. The signaling molecules induce the synthesis of antibiotics, lankacidin and lankamycin (Arakawa et al. 2012). The receptor for the signaling molecules is the protein SrrA. The SrrX/SrrA system has a complex regulatory cascade and involves genes coding for the receptor, signal synthase, regulatory proteins homologous for GBL receptors, *Streptomyces* antibiotic regulatory proteins (SARPs), and transcriptional activators organized together. This QS system regulates mainly the antibiotic production in the organism (Arakawa et al. 2007; Yamamoto et al. 2008; Suzuki et al. 2010).

10.3.6 *Streptomyces natalensis*

Streptomyces natalensis synthesizes a water-soluble signaling molecule known as PI-factor. The enzymes involved in its biosynthesis have not been elucidated. However, its biosynthesis is associated with cell growth as observed in both defined and complex media. The signaling molecule regulates the pimaricin antibiotic production. It can also induce the antibiotic pimaricin production in a non-producing strain. In a mutant non-primaricin-producing strain *S. natalensis* npi287, supplementation with A-factor or PI factor restored pimaricin biosynthesis (Recio et al. 2004).

The PI factor was structurally elucidated after obtaining its purified form from *S. natalensis* ATCC27448. This strain is known to produce the antibiotic primaricin. Chemically, PI is a 2,3-diamino-2,3-bis (hydroxymethyl)-1,4-butanediol. The PI factor at a concentration of 50 nM can initiate antibiotic production and is active at concentrations of 50–350 nM (Recio et al. 2004).

10.3.7 *Streptomyces globisporus*

Streptomyces globisporus synthesizes the signaling molecule diketopiperazine MDD ((L)-Nmethylphenylalanyl-dehydrobutyrine diketopiperazine). The signal regulates sporulation and antitumor antibiotic landomycin E production. In mutant *S. griseus* strains, the signal molecule MDD can induce the production of antibiotic streptomycin and A-factor. MDD also induces sporulation in these mutant strains. The signaling molecule can also bind to ArpA in *S. griseus* and thus mimics A-factor function and can also cause biological effect in such strains (Holden et al. 1999; Tommonaro et al. 2012; Matselyukh et al. 2015).

10.3.8 *Streptomyces albidoflavus*

In *Streptomyces albidoflavus*, the signaling molecule is a secreted protein of 324 amino acids and is known as factor C. The signal molecule induces the formation of aerial mycelium, sporulation, and production of A-factor in mutant *S. griseus* (Birkó et al. 2007; Kiss et al. 2008).

10.3.9 *Streptomyces lavendulae*

S. lavendulae produces a single signaling GBL molecule known as IM-2, which regulates the production of two types of nucleoside antibiotics, showdomycin and minimycin (Gottelt et al. 2012). IM-2 [(2R,3R,1′R)-2-1′-hydroxybutyl-3-hydroxymethyl-gamma-butanolide] prevents the production of D-cycloserine but induces production of both the nucleoside antibiotics and a blue pigment (Fig. 10.5). Here, FarA the receptor for IM-2 has a key regulatory role in the biosynthesis of IM-2 and other metabolites. Apart from farA, two more genes *farR1* and *farR2* are present in the IM-2/FarA system, whose transcription is growth-dependent. In the presence of IM-2, *farR1* and *farR2* are induced, whereas, in the absence of IM-2, FarA represses their transcription by inhibiting the RNA polymerase access (Kitani et al. 2008). *farX* of the afsA-family genes encode enzymes for GBL synthesis. FarX inhibition leads to excess production of D-

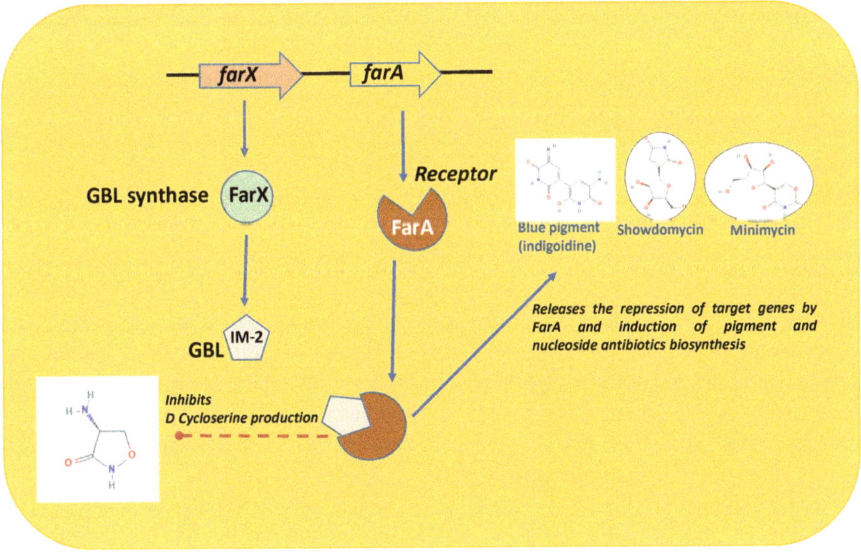

Fig. 10.5 In *S. lavendulae*, the receptor gene *farA* on binding to the signal IM-2 releases the repression of target genes by FarA, resulting in the induction of biosynthesis of pigment and nucleoside antibiotics. However, the receptor signal complex inhibits the production of the secondary metabolite D cycloserine

cycloserine, and no production of both IM-2, nucleoside antibiotics, and the blue pigment. However, exogenous supply of IM-2 restores antibiotics and pigment production implying the role of FarX in IM-2 production and farX transcription control by IM-2 (Kitani et al. 2010).

10.3.10 *Streptomyces filipinensis*

In *S. filipinensis*, GBL is involved in the biosynthesis of antifungal agent filipin. Deletion of *sfbR*, a GBL receptor, decreases filipin production, whereas deletion of pseudo-receptor, *sfbR2*, increases its production. This is due to their effect on the transcription of filipin biosynthetic genes. However, the mutations had no impact on growth and morphological development. Autoregulatory element-like sequences found in five genes of the GBL cluster show the regulatory role of SfbR and SfbR2 and its indirect control in filipin production. SfbR and SfbR2 are autoregulated and repress the transcription of each other and certain target genes. SfbR also activates its transcription and its synthesis unlike other GBL receptors, while SfbR2 has autorepression role. The pseudo-receptor binds to the antifungal antimycin A to modulate its DNA-binding activity. Both SfbR and SfbR2 inhibit the transcription of each other for filipin production. Antimycin A production is regulated by such indirect modulation of transcription of filipin biosynthetic genes. Moreover, both regulators bind to the same sites and repress the transcription of genes in the GBL cluster, including the GBL synthase gene. The role of SfbR2 as a receptor of antimycin A has also been confirmed implying the role of these receptors in antibiotic production (Barreales et al. 2020).

10.3.11 *Streptomyces chattanoogensis*

In *S. chattanoogensis* L10, GBLs regulate the production of antifungal agent natamycin. In this strain, *scgA* and *scgX* are involved in GBL production and *scgR* encodes the GBL receptor. The deletion mutants of *S. chattanoogensis* for *scgA* and *scgX* had impaired growth, morphological differentiation, and secondary metabolites. The receptor ScgR binds to the promoter region of *scgA* and represses its transcription. *scgA* is regulated by a GBL-mediated negative-feedback system. In *S. chattanoogensis*, the GBL biosynthesis is strongly regulated as GBL system is involved in the expression of more than 50 proteins and enzymes that are involved in nutrient uptake and stress response. Such tight regulation is required for the transition from primary to secondary metabolism (Du et al. 2011).

10.3.12 *Streptomyces tsukubaensis*

In *S. tsukubaensis*, the genes for synthesis and control of GBL are located in a region known as bul. The bul cluster has ten genes including two each coding for GBL receptor homologues, GBL synthetase, SARP regulatory genes, and six ARE regulatory sequences. In this organism, the receptor is BulR1 and the pseudo-receptor is BulR2. The protected region in all cases covered a 28-nt sequence with a palindromic structure. GBL regulatory system is involved in the production of the immunosuppressant tacrolimus. Deletion of the bulR1 decreased tacrolimus production, whereas bulR2 mutation did not affect its biosynthesis (Salehi-Najafabadi et al. 2014).

10.3.13 Antibiotics as Signaling Molecules

Certain antibiotics seem to have a dual role; they serve as antimicrobials at inhibitory concentrations, while they function as signals at subinhibitory concentrations. In some *Streptomyces* species, these antibiotics act as signals for utilization of nutrients, while they are also involved in intraspecific communication (Jauri et al. 2013).

Another interesting feature of some homologues of GBL receptors, known as pseudo-γ-butyrolactone receptors, is that these do not bind to GBLs but to antibiotics (Xu et al. 2010a). ScbR2 and JadR2 from *S. coelicolor* and *S. venezuelae*, respectively, are two such pseudo-GBL receptors which bind to antibiotics. ScbR2 binds to actinorhodin and undecylprodigiosin and to the promoter region of the *red* gene which encodes Red biosynthetic gene cluster activators and *adpA* (Xu et al. 2010a; Wang et al. 2014b). Similarly, JadR2 binds to chloramphenicol and jadomycin (Xu et al. 2010b). ScbR2 also binds to jadomycin, indicating interspecies communication between *S. coelicolor* and *S. venezuelae* (Wang et al. 2014b).

10.4 QS in Non-*Streptomyces* Species

The autoregulator GBLs and their cognate receptors have been intensively studied in *Streptomyces* species. But there are few studies on GBL-based QS in non-*Streptomyces* actinomycetes. The GBL regulators and their receptor proteins have been demonstrated in certain non-*Streptomyces* species including *Kitasatospora setae*, *Actinoplanes teichomyceticus*, *Amycolatopsis mediterranei*, and *Micromonospora echinospora* (Choi et al. 2003, 2004).

In the non-*Streptomyces* actinomycetes, one each among the five strains produced VB-type or IM-2-type signals. The ethyl acetate extracts of culture filtrate of non-streptomycetes were bioassayed for these signals using wild type of *S. virginiae*, *S. lavendulae* FRI-5, and *S. griseus* HH1 strains. Actinomycetes *A. teichomyceticus*, *A. mediterranei*, and *M. echinospora* had receptors for VB/SCB1, IM-2, and SCB1, respectively. *A. teichomyceticus* could bind both VB and IM-2. Its recognition of VB-type autoregulator and synthesis of VB confirms the presence of the signal and its cognate receptor in the strain. Similarly, it can be confirmed that *A. mediterranei* has an IM-2 signal/receptor pair. However, *M. echinospora* binds to SCB1 but does not produce VB- and IM-2-type signals (Choi et al. 2003).

Genome analysis has shown that *Frankia* and *Nocardia* carry homologues of AfsA and ArpA, respectively (Wuster and Madan Babu 2007; Polkade et al. 2016). The whole-genome sequencing of *N. brasiliensis* HUJEG-1 revealed the presence of a LuxR system (Vera-Cabrera et al. 2013). Other actinomycetes genera also possess LuxR-type regulators (Takano 2006; Santos et al. 2012).

B-Factor, 3′-(L-butylphosphoryl) adenosine, was initially discovered and purified from yeast extract and later identified in *Nocardia* species. B-factor synthesis is catalyzed by ribonucleases. It can induce rifamycin B production in non-producing *Nocardia* mutant, whereas curing in rifamycin producer *Nocardia* sp. resulted in a mutant unable to produce the antibiotic. Further, the effect was restored by adding yeast extract containing the B-factor. At a concentration of as low as 10 ng/mL, it could recover rifamycin B synthesis in mutant and induce its synthesis in the parental strain. The rifamycin-negative mutant of *Nocardia* showed antibiotic production at a concentration of above 2 ng/mL of B-factor and up to 30 ng/mL but not beyond that. In the presence of an adequate amount of B-factor, the rifamycin-negative mutant recovered its synthesis by 70%, whereas in the wild type its addition increased rifamycin production by 200%. In the mutant strain, the B-factor at 30 ng/mL induced rifamycin production by 1500-fold amplification. Thus, the B-factor is not only an intermediate of the antibiotic biosynthesis pathway but also is an essential stimulating factor. Another interesting observation was that, in the mutant there was no accumulation of a precursor compound ansamycin. But supplementation with the B-factor induces biosynthesis of ansamycin too. This implies the importance of B-factor in rifamycin synthesis. The *Nocardia* mutant requiring B-factor for rifamycin biosynthesis was obtained by the treatment known to cause plasmid curing. Its role is similar to that of the A-factor in *S. griseus* and *S. bikiniensis,* wherein it is involved in streptomycin biosynthesis. It is also assumed that the biosynthetic gene for B-factor is present on a plasmid or an unstable genetic determinant as is the gene for A-factor biosynthesis in *S. griseus* (Kawaguchi et al. 1984).

Structurally and functionally, B-factor resembles 3′,5′-cyclic AMP and adenosine oligophosphates which are known to regulate many cellular functions. B-factor is also implicated in regulating cellular functioning and secondary metabolism in *Nocardia* (Azuma et al. 1990).

10.5 Detection of QS Signals and Receptors in Actinomycetes

After elucidation of A-factor from *S. griseus*, 17 similar molecules were identified in other *Streptomyces* using bioassay, GC-MS, and NMR techniques (Yang et al. 2005).

Initially, GBL signaling molecules were detected by using bioassays for antibiotic synthesis. In *S. coelicolor*, the presence of SCB1 was confirmed by observing for production of the pigmented antibiotics, i.e., actinorhodin and undecylprodigiosin (Takano et al. 2000). But there were certain limitations with the bioassays, as GBLs are produced in low concentration and antibiotic production was affected by growth conditions. Hence, simple and sensitive detection methods were devised to identify new GBLs and study GBL receptor specificity (Hsiao et al. 2009). In the presence of GBL, the receptor binds to signal molecules and relieves DNA binding allowing for gene transcription (Takano et al. 2005; Pawlik et al. 2007; Hsiao et al. 2009). A reporter system of *S. coelicolor* was constructed based on the properties of ScbR. The vector used here is pIJ8 which contains *scbR* with its promoter, the *cpkO* promoter with ScbR binding site fused to promoterless kanamycin resistance gene. The plasmid pTE134 is introduced into *S. coelicolor* LW16. The constructed recombinant reporter LW16 is a double deletion mutant (*scbA* and *scbR*). Therefore, it is unable to synthesize GBL and stimulate antibiotic activity (Hsiao et al. 2009). However, it is sensitive to the antibiotic kanamycin. In this reporter system, exogenous supply of GBLs results in ScbR derepression and resistance to kanamycin. Thus, the bioassay is used for detection of GBLs having specificity to ScbR (Nodwell 2014). This bioassay has been used in characterizing two additional GBLs, i.e., SCB2 and SCB3. The detection method shows that ScbR has a preference to GBLs with 7–10 carbon side chain, a C-1′-beta-hydroxyl group, and a C-6′-methyl branch as observed in SCB3. The bioassay has been used to identify GBLs from commercial antibiotic-producing *Streptomyces* (Hsiao et al. 2009). However, again the problem associated with the bioassay is the trace concentration of GBLs synthesized by *Streptomyces* cultures resulting in the detection of few structural determinations.

Electrospray tandem mass spectrometry has been used for detecting GBLs. In *S. coelicolor*, SCB1 has been identified using His-tagged ScbR that was overexpressed in *Escherichia coli* and extracted with Ni^{2+} metal affinity beads. This technique could be used to isolate SCB1 from a small volume of 100 mL of *S. coelicolor* broth. SCB1 was identified from *S. coelicolor* using this method but not the other 2 GBLs that bind to ScbR. GBLs from non-*Streptomyces* antibiotic-producing actinomycetes using recombinant receptors have been identified. Rifamycin-producing *Amycolatopsis mediterranei* and teicoplanin producer *Actinoplanes teichomyceticus* were detected by this technique for GBL production (Yang et al. 2009).

New detection methods have to be devised to identify novel QS molecules and their cognate receptor for better understanding of their regulatory activity at the molecular level.

10.6 Interspecies Signaling in Actinomycetes

The GBLs identified in *Streptomyces* species are also involved in interspecies signaling (Nodwell 2014). The *S. griseus* signal molecule A-factor induces pimaracin antibiotic production in certain mutants of *S. natalensis*. As A-factor is not produced by *S. natalensis*, its response to exogenously supplied A-factor shows that it has a receptor to this signaling molecule (Recio et al. 2004). In *S. venezuelae*, SVB1 produced binds JadR3 protein and induces biosynthesis of jadomycin. The SVB1 found in this species is similar to SCB3 of *S. coelicolor*, an unrelated streptomycete. This suggests the probable signaling between the two streptomycetes for regulating secondary metabolites production (Zou et al. 2014). Another example involves the presence of "pseudo"-gamma-butyrolactone (GBL) receptors. These are similar to GBL receptors; however, they don't bind to GBLs but antibiotics (Wang et al. 2009). In *S. coelicolor*, the pseudo-GBL receptor ScbR are located in coelimycin gene cluster and binds to antibiotics Act and Red. At sub-MIC, the angucycline antibiotic jadomycin B from *S. venezuelae* induces sporulation and early Red antibiotic synthesis in *S. coelicolor*. JdB binds to ScbR2, relieving repression of differentiation activators and Red production (Wang et al. 2009, 2014c). These studies confirm the presence of cross talk in actinomycetes involving GBL and effector molecules for regulation of antibiotic biosynthesis and morphological differentiation.

The A-factor from *S. griseus* can restore the biosynthesis and pimaricin production in a mutant of *S. natalensis* npi287. This observation implies that apart from the PI factor, *S. natalensis* can also utilize A-factor as a signal molecule. Thus, *S. natalensis* conducts cross talk by assimilating quorum signals from other actinomycetes (Recio et al. 2004).

10.7 QS-Regulated Phenotypes in Actinomycetes

In *S. griseus*, the molecular mechanism involved in the A-factor-mediated biosynthesis of streptomycin is well understood. The gene expression for the morphological differentiation and secondary metabolism is induced by AdpA (Ohnishi and Horinouchi 2004; Bibb 2005). The AdpA transfers A-factor to the transcriptional activator genes present in the secondary metabolite biosynthesis gene cluster (Tomono et al. 2005b). In some *Streptomyces* species, the genes for the GBL synthase and receptor coding AfsA/ArpA-like protein are present in the secondary metabolite gene clusters. Many pairs of afsA-arpA are also present in certain strains (Horinouchi 2002; Hahn 2013).

The 27 genes including the positive regulatory gene *strR* for streptomycin biosynthesis are present as a big cluster on the chromosome of *S. griseus* (Distler et al. 1992; Tomono et al. 2005a). The streptomycin biosynthesis begins with the transfer of the A-factor from ArpA to AdpA and from AdpA to StrR (Horinouchi et al. 2010). StrR binds to the promoters of the nine transcriptional units resulting in their transcriptional activation (Horinouchi and Beppu 2007). During transcriptional activation, the AdpA dimer occupies the two AdpA-binding sites resulting in the complex formation and a DNA loop (Tomono et al. 2005a). Further, StrR activates all nine polycistronic transcriptional units of the biosynthesis genes, resulting in the synthesis of streptomycin from glucose (Tomono et al. 2005a). Along with *strR*, the streptomycin resistance determinant, *aphD*, which codes for streptomycin-6-phosphotransferase is also transcribed (Vujaklija et al. 1991). Thus, the *strR* and *aphD* co-transcription causes streptomycin resistance by A-factor before initiation of streptomycin synthesis (Horinouchi 2002; Horinouchi and Beppu 2007). Transcription analysis studies have shown that AdpA, the major transcriptional activator in the antibiotic biosynthesis pathway, binds the upstream activation sites at −270 and −50 nucleotide positions from *strR* (Tomono et al. 2005a).

In *S. virginiae*, virginiamycin M1 and S synthesis is activated by nanomole concentrations of VB (Yamada et al. 1987). Virginiamycin biosynthesis signal starts with VB recognition and binding to BarA (Kinoshita et al. 1997; Nakano et al. 1998). Here, a regulator-island including barA gene and other regulatory genes control both virginiamycin synthesis and resistance (Namwat et al. 2002).

In *S. natalensis*, the PI factor induces the production of pimaricin antibiotics. The antibiotic production can be stimulated by exogenously supplying the organism with up to 300 nM of purified PI factor. Thus, endogenous synthesis of PI factor is limited for pimaricin biosynthesis (Recio et al. 2004). It is also implied that the PI factor is assimilated by *S. natalensis* and initiates signal cascade at the membrane level for excess synthesis of pimaricin. The PI factor is synthesized during the growth phase of *S. natalensis* and reaches a peak at the end of the rapid growth phase. Further, the PI factor is formed prior to pimaricin biosynthesis. Thus, pimaricin production doesn't match with a high concentration of PI factor implying that it's not the PI factor level but certain biosynthetic steps that are limiting for the pimaricin production (Recio et al. 2004).

In *S. coelicolor*, various signal molecules regulate the production of many types of antibiotics. The GBLs and MMFs directly regulate coelimycin and methylenomycin production, respectively. However, the GBLs indirectly regulate the actinorhodin and undecylprodigiosin antibiotic production as biosynthetic gene clusters expression required for their production is not directly regulated by ScbR (Takano et al. 2005).

ScbR controls CPK gene cluster expression which is required for antibiotic coelimycin and pigment coelimycin P1 production (Gomez-Escribano et al. 2012). In *S. coelicolor* genome, *scbA* lies next to *scbR*. ScbR is an autorepressor and plays a crucial role in the complexly regulated SCB system (Takano et al. 2001). SCBs bind to ScbR and change its conformation, thereby preventing the binding of ScbR to the DNA. This releases repression and allows transcription of the *scbR*. In the initial stages of its growth, ScbR represses *scbA* transcription. Later in the transition phase, ScbR present at basal level binds to DNA and blocks its own transcription, while it induces its expression (Takano et al. 2001). ScbA-ScbR interaction could possibly activate *scbA* transcription (Takano et al. 2001). Induction of *scbA* expression occurs during transition from primary to secondary metabolism (Nieselt et al. 2010).

Regulation of the SCB system is a complex growth phase-dependent mechanism. In the exponential phase of growth, basal levels of ScbR exist which are sufficient to bind to the DNA in the R site, blocking its own transcription. Expression of *scbA* is activated in transition phase of growth, and γ-butyrolactones are synthesized. These signaling molecules bind to their receptor protein ScbR which provokes conformational changes on this protein that prevent its DNA-binding ability and results in transcription of *scbR*. In the same way, ScbR blocks transcription of *cpkO* in the exponential phase of growth in the absence of γ-butyrolactones. In the transition phase, expression of *cpkO* is allowed due to the presence of γ-butyrolactones, which leads to the expression of the CPK cluster (Takano et al. 2001).

10.8 Quorum Sensing Inhibition in Actinomycetes

Some actinomycetes produce AHL-degrading enzymes and certain species are capable of producing small molecules which can inhibit AHL signal synthesis (Polkade et al. 2016). The AHL-degrading or quorum quenching (QQ) enzymes and quorum sensing inhibitory molecules derived from actinobacteria have been summarized in Tables 10.2 and 10.3, respectively.

10.8.1 Quorum Quenching Enzymes in Actinomycetes

10.8.1.1 Quorum Quenching Enzymes in *Streptomyces* sp.

The first quorum quenching enzyme to be characterized from *Streptomyces* genera is *N*-acyl homoserine lactone acylase (AhlM) isolated from *Streptomyces* sp. strain M664. The acylase degrades AHL by hydrolysis of the amide bond (Park et al. 2005). AhlM deacylation activity against long acyl chain AHLs and degradation of cyclic lipopeptides has also been observed. The AhlM (20 μg/mL) has significantly reduced virulence factor production in *P. aeruginosa* PAO1 (Park et al. 2005). The gene coding for the QQ enzyme, *ahlM*, was cloned and expressed heterologously in *S. lividans*. The enzyme AhlM was a heterodimeric protein. It has 35% identity with *Ralstonia* strain XJ12B AiiD and 32% identity with *Pseudomonas aeruginosa* PAO1 PvdQ. It showed, however, 93% identity to the cyclic lipopeptide acylase from *Streptomyces* sp. strain FERM BP-5809. AhlM hydrolyzes amide bond in AHL and frees the homoserine lactone. It has high deacylation activity toward long acyl chain AHLs. However, it has broad substrate specificity and can even degrade penicillin G by deacylation. AhlM showed QQ activity in *P. aeruginosa* and reduced AHL concentration and virulence factors elastase, total protease, and LasA production. Thus, the acylase, AhlM, can be used as a QQ agent to reduce QS-mediated pathogenicity (Park et al. 2005).

Around 127 rhizospheric and 68 endophytic actinomycetes were reported for quorum quenching activity (Chankhamhaengdecha et al. 2013). An endophytic isolate *Streptomyces* LPC029 showed HSL-degrading activity. On characterization, the quorum quenching enzyme was identified as HSL-acylase which cleaves an amide bond of acyl-side chain in HSL substrate. The HSL-acylase had broad substrate specificity (C_6- to C_{12}-HSL) and best quorum quenching activity with C_{10}HSL. The HSL-acylase suppressed QS-mediated soft rot in potato by *Pectobacterium carotovorum* ssp. *carotovorum* (Chankhamhaengdecha et al. 2013).

From a library of 63 *Streptomyces* soil isolates screened for quorum quenching activity using *C. violaceum* CV026 as a biosensor, 8 of the isolates degraded synthetic hexanoyl homoserine lactone (HHL). Further, their crude extracts contained quorum quenching activity which degraded AHL signals in *P. aeruginosa* clinical isolates. An isolate *Streptomyces minutiscleroticus* St62 showed the highest activity and degraded 2 μM of HHLs in 4 h. The crude QQ enzyme had high thermal stability as it could retain 83% of its activity even at 80 °C for 60 min. It retained >90% of QQ activity at pH 8, pH 9, and pH 10 and >80% at

Table 10.2 Quorum quenching enzymes characterized from actinomycetes

Actinomycetes strain	Quorum quenching enzyme	Substrate	Target pathogen and QS-regulated phenotype inhibited	References
Streptomyces sp. strain M664	*N*-acyl homoserine lactone acylase (AhlM)	Long acyl chain AHLs and cyclic lipopeptides	Reduced virulence factor production in *P. aeruginosa* PAO1	Park et al. (2005)
Rhodococcus strains LS31 and PI33	AHL lactonase	*N*-3-oxo-hexanoyl-L-homoserine lactone	Inhibited pectate lyase activity in *Erwinia carotovora*	Park et al. (2006)
Nocardioides kongjuensis sp. nov	AHL lactonases	*N*-hexanoyl-L-homoserine lactone	–	Yoon et al. (2006)
Rhodococcus erythropolis W2	Oxidoreductase and AHL acylase activity	C6-HSL and 3-oxo derivative of acyl homoserine lactones such as 3-oxo-C6 HSL	Inhibition of violacein production in *C. violaceum*, disruption of pathogenicity in *Agrobacterium tumefaciens* and *Pectobacterium carotovorum* subsp. *carotovorum*	Leadbetter and Peter Greenberg (2000); Uroz et al. (2003); Uroz et al. (2005)
R. erythropolis	AHL lactonase QsdA	AHLs with 6–14 carbon in acyl side chains	Disrupt QS in *Pectobacterium* and control soft rot infection	Uroz et al. (2008)
Streptomyces LPC029 endophytic isolate	HSL-acylase	C_6- to C_{12}-HSL	Inhibited QS mediated soft rot in potato by *Pectobacterium carotovorum* ssp. *carotovorum*	Chankhamhaengdecha et al. (2013)
Streptomyces minutiscleroticus St62	HSL-acylase	AHLs with long acyl chains	Degraded AHL signals in *P. aeruginosa* clinical isolates and synthetic hexanoyl homoserine lactone in *C. violaceum* CV026	Sakr et al. (2015)

Table 10.3 Secondary metabolites with quorum sensing inhibitory activity derived from actinomycetes

Actinomycetes strain	QS inhibitory compound	Target pathogen and QS-regulated phenotype inhibited	References
Streptomyces sp. strain FA-70 isolated from soil	FA-70C1 (Phenylalanyl-ureido-citrullinyl-valinyl-cycloarginal)	Inhibition of Arg-gingipain (Rgp) in *Porphyromonas gingivalis*	Kadowaki et al. (2003)
Streptomyces Strain K01-0509 isolated from soil	Guadinomines A and Guadinomines B	Inhibition of type III secretion system in enteropathogenic *E. coli*	Iwatsuki et al. (2008)
Streptomyces sp. ANK313 isolated from soil	Khatmiamycin and Aloesaponarin II	Inhibition of motility in zoospores of *Plasmopara viticola*	Abdalla et al. (2011)
Streptomyces sp. BFI 250 isolated from soil	Protease	Dispersion of preformed biofilms and inhibition of biofilms formation in *Staphylococcus aureus*	Park et al. (2012)
Streptomyces sp. NIO 10068 isolated from marine invertebrate	Cinnamic acid and Linear dipeptide (Pro-Gly and *N*-amido-a-proline)	Reduction of swimming and twitching motility, inhibition of biofilm, pyocyanin, rhamnolipid, LasA protease production in *Pseudomonas aeruginosa* ATCC 27853	Naik et al. (2013)
Streptomyces sp. TOHO-Y209 and TOHO-O348 isolated from soil	Piericidin A1, 30-rhamnopiericidin A1 and Piericidin E	Inhibition of violacein production in *Chromobacterium violaceum* CV026	Ooka et al. (2013)
Nocardiopsis dassonvillei subsp. *dassonvillei* XG-8-1 from marine sediments	Nocapyrone H, Nocapyrone I and Nocapyrone M	Inhibition of QS-regulated gene expression in *P. aeruginosa* QSIS-lasI25 and *C. violaceum* CV026	Fu et al. (2013)
Streptomyces sp. CNQ343 from Seafloor sediment sample	Bahamaolide A	Inhibition of isocitrate lyase (ICL) in glyoxylate cycle in *C. albicans* ATCC 10231	Lee et al. (2014)
Streptomyces albus/ pAlnuori1aln6, a recombinant strain	Alnumycin D	Inhibition of biofilm formation in *Staphylococcus aureus* ATCC 25923	Oja et al. (2015)
Streptomyces violaceoruber Tu22 from microbial culture collection	Granaticin B	Inhibition of biofilm formation in *S. aureus* ATCC 25923	Oja et al. (2015)

(continued)

Table 10.3 (continued)

Actinomycetes strain	QS inhibitory compound	Target pathogen and QS-regulated phenotype inhibited	References
Streptomyces sp. strain MC11024 from library of culture extracts of actinomycetes	Streptorubin B	Inhibition of biofilm formation in MRSA N315	Suzuki et al. (2015)
Actinomycete strain DSW812 from marine sediment	WS9326A and WS9326B	Suppression of expression of pfoA (perfringolysin O) in *Clostridium perfringens* Inhibition of the production of hemolysin in *S. aureus* 8325–4 (type-I AIP), K12 (type-II AIP), and K9 (type-IV AIP)	Desouky et al. (2015)
Streptomyces parvulus isolated from *Codonopsis lanceolata*	Actinomycin D	Inhibition of biofilm formation, hemolysis and EPS production, and hydrophobicity in MSSA ATCC 25923, MSSA ATCC 6538, MRSA ATCC 33591	Lee et al. (2016)
Streptomyces coelicoflavus S17 isolated from soil	1H-pyrrole-2-carboxylic acid and docosanoic acid	Reduction in the production of elastase, protease, and pyocyanin Elimination of expression of las genes and rhl/pqs cascade in *P. aeruginosa* PAO1	Hassan et al. (2016)
Streptomyces xanthocidicus KPP01532 isolated from natural product library	Piericidin A and glucopiericidin A	Reduction of soft rot disease symptoms in potato caused by *Erwinia carotovora* subsp. *atroseptica* Inhibition of violacein production in *C. violaceum* CV026	Kang et al. (2016)
Streptomyces TOHO-M025 isolated from soil	Maniwamycins	Inhibition of violacein production in *C. violaceum* CV026	Fukumoto et al. (2016)
Streptomyces parvulus HY026 isolated from sea water	Actinomycin D	Inhibition of biofilm in *P. aeruginosa* PAO1 and *S. aureus*	Miao et al. (2017)

(continued)

Table 10.3 (continued)

Actinomycetes strain	QS inhibitory compound	Target pathogen and QS-regulated phenotype inhibited	References
Streptomyces sp. CCB-PSK207 isolated from marine sediment	Fatty acid methyl esters	Induction of host immunity in *C. elegans* infected with *P. aeruginosa* PA14	Fatin et al. (2017)
Streptomyces sp. AT37 isolated from desert soil	5-[(5E,7E,11E)-2,10-dihydroxy-9,11-dimethyl-5,7,11-tridecatrien-1-yl]-2-hydroxy-2-(1-hydroxyethyl)-4-methyl-3(2H)-furanone	Inhibition of biofilm formation in MRSA	Driche et al. (2017)
Streptomyces sp. MC025 isolated from a red alga	Collismycin C	Inhibition of biofilm in MRSA	Lee et al. (2017)
Streptomyces sp. OUCMDZ-3436, endobiont of *Enteromorpha prolifera*	4-Hydroxy-3-methyl-6-propylpyridin-2 (1H)-one, 3-Ethyl-4-hydroxy-6-isopropylpyridin-2 (1H)-one, 4-Hydroxy-6-isobutyl-3-methylpyridin-2 (1H)-one, and (S)-6-(sec-Butyl)-4-hydroxy-3-methylpyridin-2(1H)-one	Inhibition of QS-regulated phenotypes in *P. aeruginosa* QSIS-lasI and *P. aeruginosa* ATCC10145	Du et al. (2018)
Streptomyces sp. SCSGAA 0027 isolated from *Subergorgia suberosa*	Hygrocin C	Inhibition of biofilm formation, adhesion, EPS production, cell motility, and surface hydrophobicity in *Bacillus amyloliquefaciens* SCSGAB0082	Wang et al. (2018)
Streptomyces sp. OUCMDZ-3436 isolated from *E. prolifera*	4-hydroxy-3-methyl-6-propylpyridin-2 (1H)-one, 3-ethyl-4-hydroxy-6-isopropylpyridin-2 (1H)-one, 4-hydroxy-6-isobutyl-3-meth-ylpyridin-2 (1H)-one and (S)-6-(sec-butyl)-4-	Inhibition of QS controlled gene expression in *P. aeruginosa* QSIS-lasI biosensors	Du et al. (2018)

(continued)

Table 10.3 (continued)

Actinomycetes strain	QS inhibitory compound	Target pathogen and QS-regulated phenotype inhibited	References
	hydroxy-3-methylpyridin-2(1H)-one		
Streptomyces sp. ZL-24 isolated from wet soil	Melanin	Inhibition of biofilm formation in *P. aeruginosa* ATCC 9027 and *S. aureus* ATCC 6538	Wang et al. (2011)
Streptomyces griseoincarnatus HK12 isolated from *Callyspongia* sp.	9Z-Octadecenal, arachidic acid, erucic acid, 13Z-octadecenal, and tetracosanoic acid	Inhibition of biofilm formation in *P. aeruginosa* and *S. aureus*	Kamarudheen et al. (2019)

pH 6. On characterization, the enzyme was found to be an acylase with activity equal to 5.2 U/mg total protein with maximum activity at pH 8 and between 20–50 °C. It preferred hydrolyzing AHLs with long acyl chains than short acyl chains. The study shows that new species of *Streptomyces* need to be screened for identifying potential quorum quenching enzymes (Sakr et al. 2015).

10.8.1.2 Quorum Quenching Enzymes in *Rhodococcus* sp.

Rhodococcus sp. has three different quorum quenching enzymes including AHL lactonase, oxidoreductase, and amidase to degrade signal molecules (Park et al. 2006). *Rhodococcus* strains LS31 and PI33 were able to utilize AHLs by producing lactonases. *Rhodococcus* sp. LS31 was capable of degrading *N*-3-oxo-hexanoyl-L-homoserine lactone. It inhibited OHHL and pectate lyase activity in *Erwinia carotovora* (Park et al. 2006). AHL lactonase has been also detected in *Rhodococcus* sp. LS31 and PI33 strains (Park et al. 2006).

R. erythropolis W2 degrades C6-HSL and 3-oxo derivative of acyl-homoserine lactones such as 3-oxo-C6 HSL by oxidoreductase and AHL acylase activity (Leadbetter and Peter Greenberg 2000; Uroz et al. 2003, 2005). *Rhodococcus* sp. strain LS31 is capable of degrading AHLs of varying lengths and acyl side chains (Uroz et al. 2003, 2005). Among all the *Rhodococcus* sp., quorum quenching activity in the strains *R. erythropolis* PR4 and *Rhodococcus* RHA1 is well studied (Latour et al. 2013).

Tobacco plant rhizospheric bacteria were screened for their ability to degrade AHLs. Twenty-five isolates belonging to genera *Pseudomonas*, *Comamonas*, *Variovorax*, and *Rhodococcus* degraded AHLs. Among these isolates, *R. erythropolis* strain W2 showed in vitro quorum quenching activity as it interrupted with violacein production in *C. violaceum* and disrupted pathogenicity gene transfer in *Agrobacterium tumefaciens*. Further, in vivo studies showed that

R. erythropolis W2 disrupted QS-mediated pathogenicity of *Pectobacterium carotovorum* subsp. *carotovorum* in infected potato tubers. Thus, the rhizosphere could be a rich and diverse source of QQ enzyme-producing bacteria, which could be further developed to attenuate virulence in plant pathogens (Uroz et al. 2003). *R. erythropolis* strain W2 apart from degrading *N*-hexanoyl-L-homoserine lactone produced by other bacteria can also use AHLs as the sole carbon and energy source. *R. erythropolis* can utilize 3-oxo substituent AHLs but not unsubstituted AHLs. In the AHL-degrading pathway, AHLs with 3-oxo substituent and acyl side chains of C_8 to C_{14} are converted to 3-hydroxy derivatives. The oxidoreductase activity was also observed in *N*-(3-oxo-6-phenylhexanoyl) homoserine lactone, 3-oxododecanamide, and D- and L-isomers of *n*-(3-oxododecanoyl)-L-homoserine lactone. *R. erythropolis* cell extracts also degraded *N*-(3-oxodecanoyl)-L-homoserine lactone (3O, C10-HSL) in a temperature- and pH-dependent manner by amidolytic activity. *R. erythropolis* degrades AHLs by oxidoreductase and amidolytic reaction (Uroz et al. 2005). Screening of *R. erythropolis* strain W2A genome led to the identification of genes involved in AHL degradation. The gene *qsdA* (for *q*uorum *s*ensing *s*ignal *d*egradation) encodes AHL lactonase which is not related to either lactonase AiiA or amidohydrolase AiiD. But the lactonase is related to phosphotriesterases and could be considered to belong to a new class of lactonases. Its substrates are AHLs with an acyl chain ranging from C_6 to C_{14}, with or without substitution at carbon 3. *Rhodococcus* strains have *qsdA* gene modulate QS-regulated mechanism which could find potential application in developing quorum quenching agents (Uroz et al. 2008).

A novel class of AHL lactonase QsdA (quorum sensing signal degradation) has been reported in *R. erythropolis* strain W2 to degrade AHLs with 6–14 carbon in acyl side chains (Uroz et al. 2008). The qsdA operon has been used to assimilate lactones and also to disrupt QS signals (Latour et al. 2013).

The gamma-caprolactone (GCL) could be used as biostimulants to promote the growth of bacteria that could antagonize soft rot pathogens. For example, *R. erythropolis* utilizes various GBLs including branched-aliphatic chain GCL. The lactonase QsdA is responsible for ring-opening and *N*-acyl homoserine lactone catabolism. QsdA could also play a role in the intermediate degradation of cyclic recalcitrant molecules and synthesis of various lactones (Barbey et al. 2012). The γ-lactone catabolic pathway of *R. erythropolis* has been used to disrupt QS in *Pectobacterium* and thus control soft rot disease. The γ-lactone catabolism is induced due to QS signaling in pathogens and is seen to confer protection to plants (Barbey et al. 2013). In γ-lactone catabolic pathway, the lactone bonds of compounds with γ-butyrolactone ring coupled to an alkyl or acyl chain are hydrolyzed, and further resulting aliphatic products are activated and enter fatty acid metabolism. The pathway is induced by the presence of the γ-lactone sensed by TetR-like transcriptional regulator. Thus, biocontrol activity of *R. erythropolis* can be stimulated with γ-lactones as signal-degrading pathway gets induced (Latour et al. 2013). HiSeq transcriptomic study has revealed that no genes *R. erythropolis* were differentially expressed in the presence or absence of the *expI* mutant *P. atrosepticum.* The *expI* mutant is avirulent and unable to synthesize QS signals. However, 50 genes were differentially expressed when *R. erythropolis* was cultivated in the presence or absence of the AHL-producing virulent *P. atrosepticum*. Further, in *R. erythropolis* the expression of alkyl-sulfatase genes decreased in the presence of a virulent *P. atrosepticum*, whereas in *P. atrosepticum*, there was a lack of organic sulfur such as methionine, a key precursor for AHL synthesis (Kwasiborski et al. 2015).

10.8.1.3 Quorum Quenching Enzymes in *Nocardioides* sp.

AHL lactonases from the soil isolate *Nocardioides kongjuensis* sp. nov effectively degraded *N*-hexanoyl-L-homoserine lactone (HHL). The isolate utilized HHL as the sole carbon source. In a minimal medium supplemented with 1 mM HHL, the concentration of signaling molecules decreased with an increase in bacterial number. After 10 h of incubation, the HHL in the growth media was completely utilized, and thereafter no growth was seen (Yoon et al. 2006).

10.8.2 Quorum Sensing Inhibitory Compounds in Actinomycetes

10.8.2.1 Anti-virulence Compounds Against Oral Pathogens

Arg-gingipain (Rgp), a cysteine proteinase, is a virulence factor produced by the periodontal pathogen *Porphyromonas gingivalis*. Inhibitors of Rgp could be used as anti-virulent agents to control oral diseases. *Streptomyces* FA-70 strain was observed to produce a novel Rgp inhibitory compound, phenylalanyl-ureido-citrullinyl-L-valinyl-cycloarginal designated as FA-70C1. The compound is an antipain analogue and can also inhibit cathepsins B, L, and H. However, it had no inhibitory activity on another cysteine proteinase Lys-gingipain. FA-70C1 prevented Rgp-induced protein degradation. It also suppressed the effect of *P. gingivalis* virulence factors on polymorphonuclear leukocytes, human fibroblasts, umbilical vein endothelial cells, and vascular permeability under in vivo conditions. Thus, FA-70C1 is a potential anti-virulent agent that could be used to control *P. gingivalis* infection (Kadowaki et al. 2003).

10.8.2.2 QS and Biofilm Inhibitory Compounds Against *Vibrio* sp.

A collection of 88 actinomycetes isolated from the marine sources were screened for antibiofilm activity against *Vibrio* sp. The crude extracts of around 35 isolates at a concentration of 2.5% (v/v) exhibited biofilm inhibition in *Vibrio harveyi*, *V. vulnificus*, and *V. anguillarum*. Similarly, extracts from 33 isolates disrupted preformed mature biofilms and 6 of them were able to interfere with AHL-mediated QS in *V. harveyi*. Extracts from one of the isolates identified as *Streptomyces albus* A66 inhibited both QS signaling and biofilm formation in *Vibrio* sp. Metabolites from marine actinomycetes with potential antibiofilm activity could find application in aquaculture (You et al. 2007).

Marine actinomycetes from the Arctic were screened for novel biofilm inhibitors against *Vibrio cholerae*. Isolates belonging to *Streptomyces* sp. and *Nocardiopsis* sp. significantly inhibited biofilms of *V. cholerae*. It was observed that even the culture supernatant could inhibit biofilm formation by 80%. Solvent extract of diethyl ether fraction at 200 μg mL^{-1} inhibited biofilm formation by 60%. The study shows that marine actinomycetes are a rich source of bioactive compounds with antibiofilm activity against pathogens such as *V. cholera* (Augustine et al. 2012).

10.8.2.3 Anti-infective Compounds Against ESKAPE Pathogens

Streptomyces sp. K01-0509 produces metabolites known as guadinomines which inhibit type III secretion system (TTSS) induced hemolysis in enteropathogenic *E. coli* (EPEC). Of the six metabolites produced by the organism, guadinomines A, B, and D showed significant inhibition at IC_{50} values of 0.02, 0.007, and 8.5 μg/mL, respectively. Further, guadinomines from actinomycetes could be developed as anti-virulent agents (Iwatsuki et al. 2008).

Streptomyces sp. BFI 230 and *Kribbella* sp. BFI 1562 spent medium (1%, *v/v*) inhibited *P. aeruginosa* biofilm formation by 90%. It was observed that the spent medium components interfered with the sequestration of iron in *P. aeruginosa*. The QS inhibitors were identified to be probably extracellular peptides or proteins which could significantly reduce biofilm formation in *P. aeruginosa* (Kim et al. 2012).

Seventy-two actinomycetes isolated from marine sponge were screened for QS inhibitory activity. Methanolic extracts of 12 isolates inhibited violacein production in *C. violaceum* CV12472 and inhibited virulence factor production including swarming, biofilm formation, pyocyanin, rhamnolipid, and LasA in *P. aeruginosa* ATCC 27853. An isolate NIO 10068 exhibited the most potent anti-QS activity, and its methanol extract contained cinnamic acid and linear dipeptides proline-glycine and *N*-amido-α-proline. Among the identified metabolites, cinnamic acid could be attributed to have QS inhibitory activity (Naik et al. 2013).

A collection of 458 isolates of actinomycetes strains were screened for novel biofilm inhibitors of *S. aureus*. Of these, culture supernatants from ten isolates at 1% v/v inhibited biofilm formation in *S. aureus* by more than 80%. The biofilm inhibitor was identified as a protease which inhibited biofilm formation and detached preformed biofilms of *S. aureus*. Protease from actinomycetes could be used to inhibit and disrupt *S. aureus* biofilms (Park et al. 2012).

Three new α-pyrones, nocapyrones, H–N isolated from marine actinomycete *Nocardiopsis dassonvillei* subsp. *dassonvillei* XG-8-1 inhibited QS-regulated gene expression in *C. violaceum* CV026 and *P. aeruginosa* QSIS-*lasI* biosensors. The compounds had inhibitory activity at concentrations as low as 100 μg /mL (Fu et al. 2013). Compounds piericidin A1, 3′-rhamnopiericidin A1, and piericidin E isolated from *Streptomyces* inhibited QS mechanism in *C. violaceum* CV026 (Ooka et al. 2013).

Culture extracts of actinomycetes, when screened for activity against the agr/fsr system, revealed the presence of three cyclodepsipeptides—WS9326A, WS9326B, and cochinmicin II/III. These could inhibit the virulence gene expression of the agr system of *Staphylococcus aureus* and the fsr system of *Enterococcus faecalis*. WS9326B attenuated *S. aureus* toxicity in corneal epithelial cells. Both the compounds WS9326A and WS9326B are receptor antagonists. These cyclodepsipeptides could be used to disrupt QS in Gram-positive pathogens (Desouky et al. 2015).

Alnumycins and pyranonaphthoquinone (PNQ) polyketides from *Streptomyces* were studied for biofilm inhibitory activity in *Staphylococcus aureus*. A novel antistaphylococcal polyketide, alnumycin D, was identified and found to be highly active against both planktons and biofilms. The ribose moiety in pyranose form in alnumycins was implied for high inhibitory activity against preformed biofilms. Among the PNQ polyketides, granaticin B, kalafungin, and medermycin were tested, and granaticin B was the most potent in disrupting preformed biofilms. Thus, metabolites involved in antibiotic biosynthetic pathways and pathway intermediates could also be potential biofilm inhibitors (Oja et al. 2015). *Streptomyces* sp. MC11024 produced a compound streptorubin B which inhibited biofilms of *S. aureus*. The compound at 1 μg/mL inhibited biofilm formation in methicillin-resistant *S. aureus* (MRSA) N315 by 30% without affecting the cell growth. Streptorubin B could be developed as an anti-infective agent to control *S. aureus* biofilm-mediated infection (Suzuki et al. 2015).

Actinomycin D derived from *Streptomyces parvulus* at 0.5 μg/mL inhibited biofilm formation in three *S. aureus* strains. The antibiofilm property of the antibiotic was attributed to its ability to inhibit slime production and cause cells to be less hydrophobic. Coating actinomycin D on glass surfaces inhibited biofilm formation by *S. aureus*. The compound also reduces hemolysin activity in *S. aureus*. Thus, actinomycin D could be used to develop anti-virulence agents to treat *S. aureus* infections (Lee et al. 2016).

Streptomyces sp. TOHO-M025 produced novel compounds with an azoxy moiety, maniwamycins C–F. Maniwamycins have shown QS inhibitory activity at very low concentrations in *C. violaceum* CV026 and reduced violacein production (Fukumoto et al. 2016).

Metabolites synthesized from *Streptomyces coelicoflavus* S17, i.e., 1H-pyrrole-2-carboxylic acid and docosanoic acid, have reduced virulence factor production in *P. aeruginosa* PAO1 and thus exhibited QS inhibitory activity (Hassan et al. 2016). *Streptomyces* species S6, S12, and S17 isolated from soil inhibited QS in *C. violaceum*. The QS inhibitory compounds produced by *S. coelicoflavus* S17 was not inhibited by proteinase K. This confirmed that the compound was not a quorum quenching enzyme. The compounds were identified as behenic acid (docosanoic acid), borrelidin, and 1*H*-pyrrole-2-carboxylic acid. The compound 1*H*-pyrrole-2-carboxylic acid inhibited QS-regulated phenotypes at subinhibitory concentrations in *P. aeruginosa* PAO1 by downregulating the QS-regulated virulence gene expression. The QS inhibitory potential of *S. coelicoflavus* S17 was improved by supplementing the ISP2 medium with glucose 0.4% w/v and maintaining pH 7 (Hassan et al. 2016).

Streptomyces fradiae PE7, an estuarine sediment isolate, produces several antifouling compounds. One of the antifouling compounds PE7-C, identified as quercetin, inhibited biofilm formation between 1.6 and 25 μg/mL in about 18 biofouling bacteria. It produces more antifouling compounds in surface fermentation in comparison to submerged fermentation process (Gopikrishnan et al. 2016).

On screening 79 Micronesian marine actinomycetes, a crude extract of *Streptomyces* sp. MC025 inhibited biofilm formation in *S. aureus*. Bioactives isolated from the extract were series of 2,2′-bipyridines, i.e., collismycin B, collismycin C, pyrisulfoxin A, and pyrisulfoxin B. Collismycin C at 50 μg/mL significantly inhibited biofilm formation in MRSA by more than 90% at a concentration. Iron acquisition and the hydroxyl group in bipyridines could have a role in inhibiting biofilms (Lee et al. 2017).

A novel *Streptomyces* strain, AT37, isolated from Saharan soil and closely associated with *S. novaecaesareae* NBRC had strong inhibitory activity in MRSA ATCC 43300, clinical isolates of MRSA, and VRSA S1. The strain produced antimicrobials during the mid-stationary growth phase. Dichloromethane solvent extract contained an active compound AT37-1 identified as a derivative of furanone. In multidrug-resistant *S. aureus*, the compound had a MIC and $MBIC_{50}$ of 15–30 and 10–15 μg/mL, respectively (Driche et al. 2017).

Actinomycetes from marine sediment of Songsong Island were tested for bactericidal activities. However, methanolic extract from one of these isolates reduced infectivity of *P. aeruginosa* strain PA14 in the *Caenorhabditis elegans* animal model. Similarly, hexane extract from *Streptomyces* sp. CCB-PSK207 increased PA14-infected worm survival by 60%. This partition did not impair the feeding behavior of *C. elegans* worms. As the extract had no effect on QS-regulated activities and PA14 growth, its mechanism was studied in *lys-7*::GFP worms. The extract induced lysozyme 7 expression, an innate immunity defense which was repressed due to PA14 infection. On GC-MS analysis, the hexane extract contained methyl esters of branched saturated fatty acids. Thus, the bioactive from the marine actinomycete could rescue *C. elegans* from PA14 proving its anti-infective potential (Fatin et al. 2017).

Marine actinomycetes that are endosymbionts of sponges were capable of producing metabolites that could inhibit biofilm formation in *S. epidermidis*, *S. aureus*, and *Pseudomonas aeruginosa.* Solvent extracts from *Streptomyces* sp. SBT343 at subinhibitory concentrations prevented biofilm formation on different substrates including polystyrene, glass, and contact lens. The active component was heat stable and non-proteinaceous. The extract has the potential to be developed as an antibiofilm agent as it doesn't exert any toxicity to mammalian fibroblast, macrophage, and corneal epithelial cell lines. Such compounds could be used to prevent staphylococcal biofilm formation-mediated ocular infections (Balasubramanian et al. 2017).

Ansamycins have exhibited both inhibitions of biofilm formation and disruption of preformed biofilms. One such ansamycin is hygrocin C, extracted from *Streptomyces* sp. SCSGAA 0027, a marine isolate. Hygrocin C inhibited biofilm formation in *S. aureus* and *Bacillus amyloliquefaciens.* The compound was capable of disrupting developed and mature biofilms by its action on biofilm architecture, reducing EPS production and inhibiting cell motility and surface hydrophobicity. Transcriptomic studies showed that hygrocin C had a modulatory effect on genes associated with biofilm formation such as chemotaxis, flagellar motility, two-component QS system, and arginine and histidine synthesis. Thus, hygrocin C has the potential to be developed as a biofilm inhibitor against *S. aureus* and *B. amyloliquefaciens* (Wang et al. 2018).

Streptomyces sp. OUCMDZ-3436, an endophyte of the marine green algae *Enteromorpha prolifera*, produced novel secondary metabolites including new α-pyrones and known analogues. Based on these compounds and using diversity-oriented approach, new α-pyridones were synthesized. Some of these compounds were able to inhibit QS-regulated gene expression in *P. aeruginosa* QSIS-lasI (Du et al. 2018).

On screening marine actinomycetes from sea sediments of Andaman Sea and the Gulf of Thailand, culture supernatants from 13 strains reduced biofilm formation by 60% in the pathogens *E. coli* and *S. aureus*. The secretion of biofilm inhibitors was accompanied by protease activity. As these antibiofilm agents were found to be nontoxic, they could find application as lead molecules in developing therapeutics against *E. coli* and *S. aureus* biofilm infection (Leetanasaksakul and Thamchaipenet 2018).

A library of 185 actinomycetes was screened for antibiofilm activity against clinical strain *Staphylococcus epidermidis* 5-121-2. Culture extracts of actinomycetes species TRM 46200, TRM 41337, and TRM 46814 significantly inhibited biofilm formation in *S. epidermidis*. Hydrophilic extracellular peptides or proteins present in the extract were implied for the antibiofilm activity. The crude proteins degraded both exopolysaccharides and extracellular DNA in the biofilms and made the cells less hydrophobic. These actinomycete extracts could be developed as antibiofilm agents to control *S. epidermidis* infection (Xie et al. 2019).

Streptomyces griseoincarnatus is well known for producing alkaline protease inhibitors and antitumor agents. Crude extracts of *S. griseoincarnatus* HK 12 at 100 μg/mL inhibited biofilm activity in nosocomial pathogens *P. aeruginosa* and *S. aureus* by 82 and 78%, respectively. GC-MS analysis of the extract helped in identifying the compounds such as fatty acyls 13Z-octadecenal, 9Z-octadecenal, arachidic acid, tetracosanoic acid, and erucic acid. Docking studies revealed that 13Z-octadecenal bound to the conserved sites of substrate LasI with a binding energy of −1.90 kcal/mol confirming the QS inhibitory activity of the compound. Toxicity studies in A549 lung cancer lines showed that the compound had low toxicity of 11.5% even at 500 μg/mL. Thus, fatty acyl metabolites of *S. griseoincarnatus* HK12 could be used to develop safe biofilm inhibitors (Kamarudheen et al. 2019).

Marine *Streptomyces* isolates from Madeira Archipelago produced hybrid isoprenoids. Metabolomic profiles and molecular networking revealed a novel class of metabolites and new derivatives of napyradiomycin and marinone classes. These metabolites especially napyradiomycin SF2415B3 inhibited biofilm formation in *S. aureus* and *Marinobacter* (Bauermeister et al. 2019).

A partially purified compound 1, 4-diaza-2, 5-dioxo-3-isobutyl bicyclo[4.3.0] nonane was obtained from the marine endophytic actinomycete, *Nocardiopsis* sp. strain DMS 2 (MH900226). Both the crude extract and the compound showed the highest biofilm inhibitory activity in *Klebsiella pneumoniae*. At a minimum concentration of 300 μg/mL, the compound reduced exopolysaccharide production and modified biofilm architecture as observed by SEM and CLSM studies (Rajivgandhi et al. 2020).

Streptomyces sp. SBRK2, an endosymbiont of a marine sponge isolated from the Gulf of Mannar, produced an angucycline antibiotic, 8-*O*-metyltetrangomycin. The isolate is closely related to another actinomycete *S. longispororuber* NBRC 13488^{T}. The actinomycetes apart from antimicrobial activity were also capable of inhibiting biofilm formation in methicillin-resistant *Staphylococcus aureus* (MRSA). MIC against MRSA was observed at 2 μg/mL, while at sub-MIC, the compound reduced biofilm formation and increased the cell surface hydrophobicity index. SEM studies showed that the compound at sub-MIC disrupted cell wall structure and membrane surface resulting in cellular damage. Toxicity studies in the zebrafish embryo model showed that at up to 100 μg/mL, the compound was not lethal. Thus, angucycline compound 8-*O*-metyltetrangomycin has the potential to be developed as antibiofilm agents against drug-resistant pathogens (Mary et al. 2021).

10.8.2.4 Quorum Sensing Inhibitory Compounds Against Other Bacterial Pathogens

Actinomycetes isolated from the mucus of *Acropora digitifera* corals were screened for antibiofilm activity against biofilm-forming M serotype of *Streptococcus pyogenes*. Several isolates significantly inhibited biofilm formation by 60–80%. Of all the isolates, *S. akiyoshinensis* inhibited biofilm formation in all M serotypes. The extracts reduced cell surface hydrophobicity which is essential for biofilm development and thus inhibits biofilm formation (Nithyanand et al. 2010).

A compound WS9326A isolated from actinomycetes extract inhibited the transcription of pfoA regulated by the VirSR two-component system in *Clostridium* (Desouky et al. 2015).

Halophilic marine *Streptomyces* have been shown to produce QS and biofilm inhibitors against *Proteus mirabilis*, a pathogen known to cause biofilm-associated urinary catheters infection. Subinhibitory concentrations of the crude extract from the isolate inhibited biofilm formation by the clinical isolate *P. mirabilis* UCB4 on surfaces including urinary catheters and cover glass. The extract also inhibited the expression of QS-regulated motility, hemolysin, and urease activity (Younis et al. 2016).

Around 56 morphological different strains of marine actinomycetes were studied for quorum quenching activities. Among the actinomycetes collection, five isolates inhibited violacein production in *C. violaceum*. One of the isolates HY026 with high bioactivity was identified as *Streptomyces parvulus*. Even the spent culture medium of the isolate inhibited biofilm formation in four bacteria. The active compound was identified as actinomycin D and it constituted 26.3% of crude extracts. At a subinhibitory concentration of 12.5 μg/mL, it inhibited violacein production by 64.9%. Similarly, in *Serratia proteamaculans* 657, it inhibited prodigiosin production at 25 μg/disk. Thus, actinomycin D obtained from *S. parvulus* HY026 is a potent quorum sensing inhibitor with antibiofilm activity (Miao et al. 2017).

10.8.2.5 Quorum Sensing Inhibitory Compounds Against Plant Pathogens

Streptomyces xanthocidicus KPP01532 extracts were found to inhibit QS in *C. violaceum* CV026 strain. The quorum sensing inhibitors were identified as piericidin A and glucopiericidin A. These two compounds downregulated the expression of QS-regulated virulence genes (*pelC*, *pehA*, *celV*, and *nip*) in *Erwinia carotovora* subsp. *atroseptica*, a plant pathogen. The compounds showed similar efficacy as that of the known QS inhibitor, furanone C-30. Piericidin A or glucopiericidin A reduced soft rot diseases on potato stems and tubers caused by *E. carotovora*. Therefore, both the compounds based on their anti-virulence potential could be used as control agents for soft rot disease in potato tubers (Kang et al. 2016).

Endophytic actinomycetes from *Zingiber officinale* rhizome were screened for biofilm inhibition in *Pythium myriotylum*. An isolate, *Nocardiopsis* sp. ZoA1, showed antibiofilm activity against *Pythium* and other phytopathogens. In vivo studies showed that *Nocardiopsis* sp. extract prevented *P. myriotylum* colonization on the ginger rhizome. GC-MS analysis of the extracts showed compounds including phenol, 2,4-bis (1,1-dimethylethyl), and *trans*-cinnamic acid. Thus, bioactive metabolites from *Nocardiopsis* sp. could be applied as biocontrol agents to protect plants (Sabu et al. 2017).

10.8.2.6 Quorum Sensing Inhibitory Compounds Against Fungal Pathogens

A novel compound khatmiamycin was isolated from *Streptomyces* sp. ANK313. The compound inhibited motility in zoospores of the grapevine downy mildew pathogen *Plasmopara viticola* at 10 μg mL^{-1} by 100%. The compound also exhibited antibacterial activity against *Staphylococcus aureus* at 40 μg/disk (Abdalla et al. 2011).

Bahamaolide A is a new macrocyclic lactone isolated from the marine *Streptomyces* sp. CNQ343. It had inhibitory activity toward isocitrate lyase in *Candida albicans* with an IC_{50} of 11.82 μM. Bahamaolide A as a potent ICL inhibitor could be used for developing therapeutics against *C. albicans* infections (Lee et al. 2014).

Streptomyces isolates from soil samples were screened for biofilm inhibition in the pathogen *Candida albicans*. The most active isolate was identified as *S. toxytricini* Fz94 as it inhibited biofilm formation by 92% after 120 min at 5 g/L. Its biofilm activity was on par with the activity of Ketoconazole® which showed 90% biofilm inhibition at 2 g/L. On preformed biofilms, the crude extract at 7 g/L disrupted biofilm structure by 82% after 120 min, while Ketoconazole had no antibiofilm activity on developed biofilms. The crude extract was not toxic up to 10 g/L and thus could find potential application as anti-candidal biofilm agents (Sheir and Hafez 2017).

10.9 Conclusion and Future Prospectives

Actinomycetes are one of the most significant and natural sources of antibiotics and drugs. Hence, the emphasis for the discovery of novel bioactive compounds from actinomycetes continues. The GBL QS systems of actinomycetes have a key regulatory role in secondary metabolite production and cellular differentiation. They also have highly remarkable potential in terms of regulating biosynthetic gene clusters for the production of novel antibiotics. Thus, understanding the regulation of GBL systems and identifying cryptic BGCs that are regulated by GBLs could be useful in engineering pathways for the synthesis of commercially important antibiotics. Similarly, the quorum quenching enzymes and QS inhibitors from actinomycetes could be developed as potential anti-infectives for controlling infectious pathogens.

References

Abdalla MA, Win HY, Islam MT, Von Tiedemann A, Schüffler A, Laatsch H (2011) Khatmiamycin, a motility inhibitor and zoosporicide against the grapevine downy mildew pathogen *Plasmopara viticola* from *Streptomyces* sp. ANK313. J Antibiot 64(10):655–659

Albuquerque P, Casadevall A (2012) Quorum sensing in fungi—a review. Med Mycol 50(4):337–345

Ando N, Ueda K, Horinouchi S (1997) A *Streptomyces griseus* gene (sgaA) suppresses the growth disturbance caused by high osmolality and a high concentration of A-factor during early growth. Microbiology 143(8):2715–2723

Anisova LN, Blinova IN, Efremenkova OV, Smirnova GM, Khokhlov AS (1984) Regulators of differentiation in *Streptomyces cyaneofuscatus*. Mikrobiologiia 53(6):890–895

Arakawa K, Mochizuki S, Yamada K, Noma T, Kinashi H (2007) γ-Butyrolactone autoregulator-receptor system involved in lankacidin and lankamycin production and morphological differentiation in Streptomyces rochei. Microbiology 153(6):1817–1827

Arakawa K, Tsuda N, Taniguchi A, Kinashi H (2012) The butenolide signaling molecules SRB1 and SRB2 induce lankacidin and lankamycin production in *Streptomyces rochei*. ChemBioChem 13(10):1447–1457

Augustine N, Kerkar S, Thomas S (2012) Arctic actinomycetes as potential inhibitors of Vibrio cholerae biofilm. Curr Microbiol 64(4):338–342

Azuma M, Nishi K, Horinouchi S, Beppu T (1990) Ribonucleases catalyze the synthesis of B-factor (3′-butylphosphoryl AMP), an inducer of rifamycin production in a *Nocardia* sp. J Antibiot 43(3):321–323

Balasubramanian S, Othman EM, Kampik D, Stopper H, Hentschel U, Ziebuhr W, Oelschlaeger TA, Abdelmohsen UR (2017) Marine sponge-derived *Streptomyces* sp. SBT343 extract inhibits staphylococcal biofilm formation. Front Microbiol 8:236

Barbey C, Crépin A, Cirou A, Budin-Verneuil A, Orange N, Feuilloley M, Faure D, Dessaux Y, Burini J-F, Latour X (2012) Catabolic pathway of gamma-caprolactone in the biocontrol agent *Rhodococcus erythropolis*. J Proteome Res 11(1):206–216

Barbey C, Crépin A, Bergeau D, Ouchiha A, Mijouin L, Taupin L, Orange N et al (2013) In planta biocontrol of *Pectobacterium atrosepticum* by *Rhodococcus erythropolis* involves silencing of pathogen communication by the rhodococcal gamma-lactone catabolic pathway. PLoS One 8(6):e66642

Barreales EG, Payero TD, Jambrina E, Aparicio JF (2020) The *Streptomyces filipinensis* gamma-butyrolactone system reveals novel clues for understanding the control of secondary metabolism. Appl Environ Microbiol 86(18)

Bauermeister A, Pereira F, Grilo IR, Godinho CC, Paulino M, Almeida V, Gobbo-Neto L et al (2019) Intra-clade metabolomic profiling of MAR4 *Streptomyces* from the Macaronesia Atlantic region reveals a source of anti-biofilm metabolites. Environ Microbiol 21(3):1099–1112

Beppu T (1992) Secondary metabolites as chemical signals for cellular differentiation. Gene 115(1–2):159–165

Bhukya H, Bhujbalrao R, Bitra A, Anand R (2014) Structural and functional basis of transcriptional regulation by TetR family protein CprB from *S. coelicolo*r A3 (2). Nucleic Acids Res 42(15):10122–10133

Biarnes-Carrera M, Breitling R, Takano E (2015) Butyrolactone signalling circuits for synthetic biology. Curr Opin Chem Biol 28:91–98

Bibb MJ (2005) Regulation of secondary metabolism in streptomycetes. Curr Opin Microbiol 8(2):208–215

Bignell DRD, Bate N, Cundliffe E (2007) Regulation of tylosin production: role of a TylP-interactive ligand. Mol Microbiol 63(3):838–847

Birkó Z, Bialek S, Buzás K, Szájli E, Medzihradszky KF, Rigali S, Vijgenboom E et al (2007) The secreted signaling protein factor C triggers the A-factor response regulon in *Streptomyces griseus*: overlapping signaling routes. Mol Cell Proteomics 6(7):1248–1256

Braun D, Pauli N, Sequin U, Zähner H (1995) New butenolides from the photoconductivity screening of *Streptomyces antibioticus* (Waksman and Woodruff) Waksman and Henrici 1948. FEMS Microbiol Lett 126(1):37–42

Chankhamhaengdecha S, Hongvijit S, Srichaisupakit A, Charnchai P, Panbangred W (2013) Endophytic actinomycetes: a novel source of potential acyl homoserine lactone degrading enzymes. Biomed Res Int 2013

Chi W-J, Jin X-M, Jung S-C, Oh EA, Hong S-K (2011) Characterization of Sgr3394 produced only by the A-factor-producing *Streptomyces griseus* IFO 13350, not by the A-factor deficient mutant. J Microbiol 49(1):155–160

Cho KW, Lee H-S, Rho J-R, Kim TS, Mo SJ, Shin J (2001) New lactone-containing metabolites from a marine-derived bacterium of the genus Streptomyces. J Nat Prod 64(5):664–667

Choi S-U, Lee C-K, Hwang Y-I, Kinosita H, Nihira T (2003) γ-Butyrolactone autoregulators and receptor proteins in non-Streptomyces actinomycetes producing commercially important secondary metabolites. Arch Microbiol 180(4):303–307

Choi S-U, Lee C-K, Hwang Y-I, Kinoshita H, Nihira T (2004) Cloning and functional analysis by gene disruption of a gene encoding a γ-butyrolactone autoregulator receptor from *Kitasatospora setae*. J Bacteriol 186(11):3423–3430

Choudhary S, Schmidt-Dannert C (2010) Applications of quorum sensing in biotechnology. Appl Microbiol Biotechnol 86(5):1267–1279

Corre C, Song L, O'Rourke S, Chater KF, Challis GL (2008) 2-Alkyl-4-hydroxymethylfuran-3-carboxylic acids, antibiotic production inducers discovered by *Streptomyces coelicolor* genome mining. Proc Natl Acad Sci 105(45):17510–17515

de Jong W, Vijgenboom E, Dijkhuizen L, Wösten HAB, Claessen D (2012) SapB and the rodlins are required for development of *Streptomyces coelicolor* in high osmolarity media. FEMS Microbiol Lett 329(2):154–159

De Kievit TR, Iglewski BH (2000) Bacterial quorum sensing in pathogenic relationships. Infect Immun 68(9):4839–4849

Desouky SE, Shojima A, Singh RP, Matsufuji T, Igarashi Y, Suzuki T, Yamagaki T et al (2015) Cyclodepsipeptides produced by actinomycetes inhibit cyclic-peptide-mediated quorum sensing in Gram-positive bacteria. FEMS Microbiol Lett 362(14):fnv109

Distler J, Mansouri K, Mayer G, Stockmann M, Piepersberg W (1992) Streptomycin biosynthesis and its regulation in Streptomycetes. Gene 115(1–2):105–111

Driche EH, Sabaou N, Bijani C, Zitouni A, Pont F, Mathieu F, Badji B (2017) Streptomyces sp. AT37 isolated from a Saharan soil produces a furanone derivative active against multidrug-resistant Staphylococcus aureus. World J Microbiol Biotechnol 33(6):105

Du Y-L, Shen X-L, Yu P, Bai L-Q, Li Y-Q (2011) Gamma-butyrolactone regulatory system of *Streptomyces chattanoogensis* links nutrient utilization, metabolism, and development. Appl Environ Microbiol 77(23):8415–8426

Du Y, Sun J, Gong Q, Wang Y, Fu P, Zhu W (2018) New α-pyridones with quorum-sensing inhibitory activity from diversity-enhanced extracts of a *Streptomyces* sp. derived from marine algae. J Agric Food Chem 66(8):1807–1812

Fatin SN, Boon-Khai T, Shu-Chien AC, Khairuddean M, Abdullah AA-A (2017) A marine actinomycete rescues *Caenorhabditis elegans* from *Pseudomonas aeruginosa* infection through restitution of Lysozyme 7. Front Microbiol 8:2267

Fu P, Liu P, Gong Q, Wang Y, Wang P, Zhu W (2013) α-Pyrones from the marine-derived actinomycete *Nocardiopsis dassonvillei* subsp. *dassonvillei* XG-8-1. RSC Adv 3(43):20726–20731

Fukumoto A, Murakami C, Anzai Y, Kato F (2016) Maniwamycins: new quorum-sensing inhibitors against *Chromobacterium violaceum* CV026 were isolated from Streptomyces sp. TOHO-M025. J Antibiot 69(5):395–399

Gomez-Escribano JP, Song L, Fox DJ, Yeo V, Bibb MJ, Challis GL (2012) Structure and biosynthesis of the unusual polyketide alkaloid coelimycin P1, a metabolic product of the cpk gene cluster of *Streptomyces coelicolor* M145. Chem Sci 3(9):2716–2720

Gopikrishnan V, Radhakrishnan M, Shanmugasundaram T, Pazhanimurugan R, Balagurunathan R (2016) Antibiofouling potential of quercetin compound from marine-derived actinobacterium, *Streptomyces fradiae* PE7 and its characterization. Environ Sci Pollut Res 23(14):13832–13842

Gottelt M, Hesketh A, Bunet R, Puri P, Takano E (2012) Characterisation of a natural variant of the γ-butyrolactone signalling receptor. BMC Res Notes 5(1):1–17

Grandclément C, Tannières M, Moréra S, Dessaux Y, Faure D (2016) Quorum quenching: role in nature and applied developments. FEMS Microbiol Rev 40(1):86–116

Hahn D (2013) Induction of cryptic natural product fungicides from actinomycetes. Pest Manag Nat Prod 217–236

Hara O, Beppu T (1982) Mutants blocked in streptomycin production in *Streptomyces griseus*-the role of A-factor. J Antibiot 35(3):349–358

Hassan R, Shaaban MI, Abdel Bar FM, El-Mahdy AM, Shokralla S (2016) Quorum sensing inhibiting activity of *Streptomyces coelicoflavus* isolated from soil. Front Microbiol 7:659

Healy FG, Eaton Kevin P, Prajit L, Aldrich Joel F, Plowman Alaina K, King RR (2009) Characterization of gamma-butyrolactone autoregulatory signaling gene homologs in the angucyclinone polyketide WS5995B producer Streptomyces acidiscabies. J Bacteriol 191(15):4786–4797

Holden MTG, Chhabra SR, De Nys R, Stead P, Bainton NJ, Hill PJ, Manefield M et al (1999) Quorum-sensing cross talk: isolation and chemical characterization of cyclic dipeptides from *Pseudomonas aeruginosa* and other gram-negative bacteria. Molecular Microbiology 33(6):1254–1266

Horinouchi S (1996) *Streptomyces* genes involved in aerial mycelium formation. FEMS Microbiol Lett 141(1):1–9

Horinouchi S (2002) A microbial hormone, A-factor, as a master switch for morphological differentiation and secondary metabolism in *Streptomyces griseus*. Front Biosci 7:2045–2057

Horinouchi S, Beppu T (1992) Autoregulatory factors and communication in actinomycetes. Annu Rev Microbiol 46(1):377–392

Horinouchi S, Beppu T (2007) Hormonal control by A-factor of morphological development and secondary metabolism in *Streptomyces*. Proc Jpn Acad Ser B 83(9+10):277–295

Horinouchi S, Ueda K, Nakayama J, Ikeda T (2010) Cell-to-cell communications among microorganisms. In: Comprehensive natural products II: chemistry and biology. Elsevier Ltd, pp 283–337

Hsiao N-H, Nakayama S, Merlo ME, de Vries M, Bunet R, Kitani S, Nihira T, Takano E (2009) Analysis of two additional signaling molecules in *Streptomyces coelicolor* and the development of a butyrolactone-specific reporter system. Chem Biol 16(9):951–960

Iwatsuki M, Uchida R, Yoshijima H, Ui H, Shiomi K, Matsumoto A, Takahashi Y, Abe A, Tomoda H, Ōmura S (2008) Guadinomines, type III secretion system inhibitors, produced by *Streptomyces* sp. K01-0509. J Antibiot 61(4):222–229

Jauri PV, Bakker MG, Salomon CE, Kinkel LL (2013) Subinhibitory antibiotic concentrations mediate nutrient use and competition among soil Streptomyces. PLoS One 8(12):e81064

Kadowaki T, Kitano S, Baba A, Takii R, Hashimoto M, Katunuma N, Yamamoto K (2003) Isolation and characterization of a novel and potent inhibitor of Arg-gingipain from *Streptomyces* sp. strain FA-70. Biol Chem 384:911–920

Kamarudheen N, Bhaskara KV, Rao. (2019) Fatty acyl compounds from marine *Streptomyces griseoincarnatus* strain HK12 against two major bio-film forming nosocomial pathogens; an in vitro and in silico approach. Microb Pathog 127:121–130

Kang JE, Han JW, Jeon BJ, Kim BS (2016) Efficacies of quorum sensing inhibitors, piericidin A and glucopiericidin A, produced by *Streptomyces xanthocidicus* KPP01532 for the control of potato soft rot caused by Erwinia carotovora subsp. atroseptica. Microbiol Res 184:32–41

Kawachi R, Akashi T, Kamitani Y, Sy A, Wangchaisoonthorn U, Nihira T, Yamada Y (2000) Identification of an AfsA homologue (BarX) from *Streptomyces virginiae* as a pleiotropic regulator controlling autoregulator biosynthesis, virginiamycin biosynthesis and virginiamycin M1 resistance. Mol Microbiol 36(2):302–313

Kawaguchi T, Asahi T, Satoh T, Uozumi T, Beppu T (1984) B-factor, an essential regulatory substance inducing the production of rifamycin in a *Nocardia* sp. J Antibiot 37(12):1587–1595

Khokhlov AS, Tovarova II, Borisova LN, Pliner SA, Shevchenko LN, Ela K, Ivkina NS, Rapoport IA (1967) The A-factor, responsible for streptomycin biosynthesis by mutant strains of *Actinomyces streptomycini*. Dokl Akad Nauk SSSR 177(1):232–235

Kim, Yong-Guy, Jin-Hyung Lee, Chang-Jin Kim, Jae-Chan Lee, Yoon Jung Ju, Moo Hwan Cho, and Jintae Lee. "Antibiofilm activity of *Streptomyces* sp. BFI 230 and Kribbella sp. BFI 1562 against Pseudomonas aeruginosa." Appl Microbiol Biotechnol 96, no. 6 (2012): 1607–1617

Kinoshita H, Ipposhi H, Okamoto S, Nakano H, Nihira T, Yamada Y (1997) Butyrolactone autoregulator receptor protein (BarA) as a transcriptional regulator in *Streptomyces virginiae*. J Bacteriol 179(22):6986–6993

Kiss Z, Ward AC, Birko Z, Chater KF, Biro S (2008) Streptomyces *griseus* 45H, a producer of the extracellular autoregulator protein factor C, is a member of the species *Streptomyces albidoflavus*. Int J Syst Evol Microbiol 58(4):1029–1031

Kitani S, Kinoshita H, Nihira T, Yamada Y (1999) In vitro analysis of the butyrolactone autoregulator receptor protein (FarA) of *Streptomyces lavendulae* FRI-5 reveals that FarA acts as a DNA-binding transcriptional regulator that controls its own synthesis. J Bacteriol 181(16):5081–5084

Kitani S, Yamada Y, Nihira T (2001) Gene replacement analysis of the butyrolactone autoregulator receptor (FarA) reveals that FarA acts as a novel regulator in secondary metabolism of *Streptomyces lavendulae* FRI-5. J Bacteriol 183(14):4357–4363

Kitani S, Iida A, Izumi T-a, Maeda A, Yamada Y, Nihira T (2008) Identification of genes involved in the butyrolactone autoregulator cascade that modulates secondary metabolism in Streptomyces lavendulae FRI-5. Gene 425(1–2):9–16

Kitani S, Doi M, Shimizu T, Maeda A, Nihira T (2010) Control of secondary metabolism by farX, which is involved in the γ-butyrolactone biosynthesis of *Streptomyces lavendulae* FRI-5. Arch Microbiol 192(3):211–220

Kitani S, Miyamoto KT, Takamatsu S, Herawati E, Iguchi H, Nishitomi K, Uchida M et al (2011) Avenolide, a Streptomyces hormone controlling antibiotic production in *Streptomyces avermitilis*. Proc Natl Acad Sci 108(39):16410–16415

Kodani S, Hudson ME, Durrant MC, Buttner MJ, Nodwell JR, Willey JM (2004) The SapB morphogen is a lantibiotic-like peptide derived from the product of the developmental gene ramS in *Streptomyces coelicolor*. Proc Natl Acad Sci 101(31):11448–11453

Kondo K, Higuchi Y, Sakuda S, Nihira T, Yamada Y (1989) New virginiae butanolides from *Streptomyces virginiae*. J Antibiot 42(12):1873–1876

Kurniawan YN, Kitani S, Maeda A, Nihira T (2014) Differential contributions of two SARP family regulatory genes to indigoidine biosynthesis in *Streptomyces lavendulae* FRI-5. Appl Microbiol Biotechnol 98(23):9713–9721

Kwasiborski A, Mondy S, Chong TM, Barbey C, Chan KG, Beury-Cirou A, Latour X, Faure D (2015) Transcriptome of the quorum-sensing signal-degrading *Rhodococcus erythropolis* responds differentially to virulent and avirulent *Pectobacterium atrosepticum*. Heredity 114(5):476–484

Latour X, Barbey C, Chane A, Groboillot A, Burini J-F (2013) Rhodococcus erythropolis and its γ-lactone catabolic pathway: an unusual biocontrol system that disrupts pathogen quorum sensing communication. Agronomy 3(4):816–838

Leadbetter JR, Peter Greenberg E (2000) Metabolism of acyl-homoserine lactone quorum-sensing signals by *Variovorax paradoxus*. J Bacteriol 182(24):6921–6926

Lee YJ, Kitani S, Kinoshita H, Nihira T (2008) Identification by gene deletion analysis of barS2, a gene involved in the biosynthesis of γ-butyrolactone autoregulator in *Streptomyces virginiae*. Arch Microbiol 189(4):367–374

Lee YJ, Kitani S, Nihira T (2010) Null mutation analysis of an afsA-family gene, barX, that is involved in biosynthesis of the γ-butyrolactone autoregulator in *Streptomyces virginiae*. Microbiology 156(1):206–210

Lee S-H, Moon K, Kim H, Shin J, Oh D-C, Oh K-B (2014) Bahamaolide A from the marine-derived *Streptomyces* sp. CNQ343 inhibits isocitrate lyase in Candida albicans. Bioorg Med Chem Lett 24(17):4291–4293

Lee J-H, Kim Y-G, Lee K, Kim C-J, Park D-J, Ju Y, Lee J-C, Wood TK, Lee J (2016) Streptomyces-derived actinomycin D inhibits biofilm formation by *Staphylococcus aureus* and its hemolytic activity. Biofouling 32(1):45–56

Lee J-H, Kim E, Choi H, Lee J (2017) Collismycin C from the Micronesian marine bacterium *Streptomyces* sp. MC025 inhibits *Staphylococcus aureus* biofilm formation. Mar Drugs 15(12):387

Leetanasaksakul K, Thamchaipenet A (2018) Potential anti-biofilm producing marine actinomycetes isolated from sea sediments in Thailand. Agric Nat Resour 52(3):228–233

Lezhava A, Kameoka D, Sugino H, Goshi K, Shinkawa H, Nimi O, Horinouchi S, Beppu T, Kinashi H (1997) Chromosomal deletions in *Streptomyces griseus* that remove the afsA locus. Mol Gen Genet MGG 253(4):478–483

Martin JF, Liras P (2012) Cascades and networks of regulatory genes that control antibiotic biosynthesis. In: Reprogramming microbial metabolic pathways. pp 115–138

Mary TR, Jabila RR, Kannan AMI, Ramachandran D, Vincent SGP (2021) Cell wall distraction and biofilm inhibition of marine *Streptomyces* derived angucycline in methicillin resistant *Staphylococcus aureus*. Microb Pathog 150:104712

Mast Y, Wohlleben W (2014) Streptogramins–two are better than one! Int J Med Microbiol 304(1):44–50

Mast Y, Guezguez J, Handel F, Schinko E (2015) A complex signaling cascade governs pristinamycin biosynthesis in *Streptomyces pristinaespiralis*. Appl Environ Microbiol 81(19):6621–6636

Matselyukh B, Mohammadipanah F, Laatsch H, Rohr J, Efremenkova O, Khilya V (2015) N-methylphenylalanyl-dehydrobutyrine diketopiperazine, an A-factor mimic that restores antibiotic biosynthesis and morphogenesis in *Streptomyces globisporus* 1912-B2 and *Streptomyces griseus* 1439. J Antibiot 68(1):9–14

Matsuno K, Yamada Y, Lee C-K, Nihira T (2004) Identification by gene deletion analysis of barB as a negative regulator controlling an early process of virginiamycin biosynthesis in *Streptomyces virginiae*. Arch Microbiol 181(1):52–59

Miao L, Xu J, Yao Z, Jiang Y, Zhou H, Jiang W, Dong K (2017) The anti-quorum sensing activity and bioactive substance of a marine derived *Streptomyces*. Biotechnol Biotechnol Equipm 31(5):1007–1015

Miller MB, Bassler BL (2001) Quorum sensing in bacteria. Annu Rev Microbiol 55(1):165–199

Mingyar E, Lubomira F, Renata N, Carmen B, Jan K (2015) A γ-butyrolactone autoregulator-receptor system involved in the regulation of auricin production in Streptomyces aureofaciens CCM 3239. Appl Microbiol Biotechnol 99(1):309–325

Miyake K, Horinouchi S, Yoshida M, Chiba N, Mori K, Nogawa N, Morikawa N, Beppu T (1989) Detection and properties of A-factor-binding protein from *Streptomyces griseus*. J Bacteriol 171(8):4298–4302

Miyake K, Tomohisa K, Sueharu H, Beppu T (1990) The A-factor-binding protein of *Streptomyces griseus* negatively controls streptomycin production and sporulation. J Bacteriol 172(6):3003–3008

Mukku VJRV, Speitling M, Laatsch H, Helmke E (2000) New butenolides from two marine streptomycetes. J Nat Prod 63(11):1570–1572

Mutenko H, Makitrinskyy R, Tsypik O, Walker S, Ostash B, Fedorenko V (2014) Genes for biosynthesis of butenolide-like signalling molecules in *Streptomyces ghanaensis*, their role in moenomycin production. Russ J Genet 50(6):563–568

Naik DN, Wahidullah S, Meena RM (2013) Attenuation of *Pseudomonas aeruginosa* virulence by marine invertebrate–derived *Streptomyces* sp. Lett Appl Microbiol 56(3):197–207

Nakano H, Takehara E, Nihira T, Yamada Y (1998) Gene replacement analysis of the *Streptomyces virginiae* barA gene encoding the butyrolactone autoregulator receptor reveals that BarA acts as a repressor in virginiamycin biosynthesis. J Bacteriol 180(13):3317–3322

Namwat W, Lee C-K, Kinoshita H, Yamada Y, Nihira T (2001) Identification of the varr gene as a transcriptional regulator of virginiamycin s resistance in *Streptomyces virginiae*. J Bacteriol 183(6):2025–2031

Namwat W, Kamioka Y, Kinoshita H, Yamada Y, Nihira T (2002) Characterization of virginiamycin S biosynthetic genes from *Streptomyces virginiae*. Gene 286(2):283–290

Natsume R, Ohnishi Y, Senda T, Horinouchi S (2004) Crystal structure of a γ-butyrolactone autoregulator receptor protein in *Streptomyces coelicolor* A3 (2). J Mol Biol 336(2):409–419

Nealson KH (1977) Autoinduction of bacterial luciferase. Arch Microbiol 112(1):73–79

Nieselt K, Battke F, Herbig A, Bruheim P, Wentzel A, Jakobsen ØM, Sletta H et al (2010) The dynamic architecture of the metabolic switch in *Streptomyces coelicolor*. BMC Genomics 11(1):1–9

Nishida H, Ohnishi Y, Beppu T, Horinouchi S (2007) Evolution of γ-butyrolactone synthases and receptors in *Streptomyces*. Environ Microbiol 9(8):1986–1994

Nithyanand P, Thenmozhi R, Rathna J, Pandian SK (2010) Inhibition of *Streptococcus pyogenes* biofilm formation by coral-associated actinomycetes. Curr Microbiol 60(6):454–460

Niu G, Chater KF, Tian Y, Zhang J, Tan H (2016) Specialised metabolites regulating antibiotic biosynthesis in *Streptomyces* spp. FEMS Microbiol Rev 40(4):554–573

Nodwell JR (2014) Are you talking to me? A possible role for γ-butyrolactones in interspecies signalling. Mol Microbiol 94(3):483–485

Ohnishi Y, Horinouchi S (2004) The A-factor regulatory cascade that leads to morphological development and secondary metabolism in *Streptomyces*. Biofilms 1(4):319

Ohnishi Y, Yamazaki H, Kato J-Y, Tomono A, Horinouchi S (2005) AdpA, a central transcriptional regulator in the A-factor regulatory cascade that leads to morphological development and secondary metabolism in *Streptomyces griseus*. Biosci Biotechnol Biochem 69(3):431–439

Oja T, Galindo PSM, Taguchi T, Manner S, Vuorela PM, Ichinose K, Metsä-Ketelä M, Fallarero A (2015) Effective antibiofilm polyketides against *Staphylococcus aureus* from the pyranonaphthoquinone biosynthetic pathways of *Streptomyces* species. Antimicrob Agents Chemother 59(10):6046–6052

Onaka H, Horinouchi S (1997) DNA-binding activity of the A-factor receptor protein and its recognition DNA sequences. Mol Microbiol 24(5):991–1000

Onaka H, Ando N, Nihira T, Yamada Y, Beppu T, Horinouchi S (1995) Cloning and characterization of the A-factor receptor gene from *Streptomyces griseus*. J Bacteriol 177(21):6083–6092

Onaka H, Nakagawa T, Horinouchi S (1998) Involvement of two A-factor receptor homologues in *Streptomyces coelicolor* A3 (2) in the regulation of secondary metabolism and morphogenesis. Mol Microbiol 28(4):743–753

Ooka K, Fukumoto A, Yamanaka T, Shimada K, Ishihara R, Anzai Y, Kato F (2013) Piericidins, novel quorum-sensing inhibitors against *Chromobacterium violaceum* CV026, from *Streptomyces* sp. TOHO-Y209 and TOHO-O348. Open J Med Chem 2013

Pang X, Bertrand A, Jean-Michel G, Sophie M, Jean-Luc P, Bernard D, Pierre L (2004) Functional angucycline-like antibiotic gene cluster in the terminal inverted repeats of the Streptomyces ambofaciens linear chromosome. Antimicrob Agents Chemother 48(2):575–588

Papenfort K, Bassler BL (2016) Quorum sensing signal–response systems in Gram-negative bacteria. Nat Rev Microbiol 14(9):576

Park S-Y, Kang H-O, Jang H-S, Lee J-K, Koo B-T, Yum D-Y (2005) Identification of extracellular N-acylhomoserine lactone acylase from a *Streptomyces* sp. and its application to quorum quenching. Appl Environ Microbiol 71(5):2632–2641

Park S-Y, Hwang B-J, Shin M-H, Kim J-A, Kim H-K, Lee J-K (2006) N-acylhomoserine lactonase producing *Rhodococcus* spp. with different AHL-degrading activities. FEMS Microbiol Lett 261(1):102–108

Park J-H, Lee J-H, Kim C-J, Lee J-C, Cho MH, Lee J (2012) Extracellular protease in Actinomycetes culture supernatants inhibits and detaches *Staphylococcus aureus* biofilm formation. Biotechnol Lett 34(4):655–661

Pawlik K, Kotowska M, Chater KF, Kuczek K, Takano E (2007) A cryptic type I polyketide synthase (cpk) gene cluster in *Streptomyces coelicolor* A3 (2). Arch Microbiol 187(2):87–99

Pereira CS, Thompson JA, Xavier KB (2013) AI-2-mediated signalling in bacteria. FEMS Microbiol Rev 37(2):156–181

Polkade AV, Mantri SS, Patwekar UJ, Jangid K (2016) Quorum sensing: an under-explored phenomenon in the phylum Actinobacteria. Front Microbiol 7:131

Rajivgandhi GN, Ramachandran G, Kanisha CC, Li J-L, Yin L, Manoharan N, Alharbi NS, Kadaikunnan S, Khaled JM, Li W-J (2020) Anti-biofilm compound of 1, 4-diaza-2, 5-dioxo-3-isobutyl bicyclo [4.3. 0] nonane from marine *Nocardiopsis* sp. DMS 2 (MH900226) against biofilm forming *K. pneumoniae*. J King Saud Univ Sci 32(8):3495–3502

Recio E, Colinas Á, Rumbero Á, Aparicio JF, Martín JF (2004) PI factor, a novel type quorum-sensing inducer elicits pimaricin production in *Streptomyces natalensis*. J Biol Chem 279(40):41586–41593

Rutherford ST, Bassler BL (2012) Bacterial quorum sensing: its role in virulence and possibilities for its control. Cold Spring Harb Perspect Med 2(11):a012427

Sabu R, Soumya KR, Radhakrishnan EK (2017) Endophytic Nocardiopsis sp. from Zingiber officinale with both antiphytopathogenic mechanisms and antibiofilm activity against clinical isolates. 3 Biotech 7(2):1–13

Sakr MM, Aboshanab KMA, Aboulwafa MM, Hassouna NA-H (2015) Characterization of the quorum quenching activity of *Streptomyces minutiscleroticus*: a new approach for infection control. Afr J Microbiol Res 9(8):492–502

Salehi-Najafabadi Z, Barreiro C, Rodríguez-García A, Cruz A, López GE, Martín JF (2014) The gamma-butyrolactone receptors BulR1 and BulR2 of *Streptomyces tsukubaensis*: tacrolimus (FK506) and butyrolactone synthetases production control. Appl Microbiol Biotechnol 98(11):4919–4936

Sanchez LM (2016) Quorum sensing in Streptomyces coelicolor: regulation of the SCB signalling system that controls the synthesis of antibiotics, Thesis. University of Groningen, Groningen, 231 p

Santos CL, Correia-Neves M, Moradas-Ferreira P, Mendes MV (2012) A walk into the LuxR regulators of Actinobacteria: phylogenomic distribution and functional diversity. PLoS One 7(10):e46758

Sheir DH, Hafez MA (2017) Antibiofilm activity of *Streptomyces toxytricini* Fz94 against *Candida albicans* ATCC 10231. Microb Biosyst 2(1):26–39

Sturme MHJ, Kleerebezem M, Nakayama J, Akkermans ADL, Vaughan EE, De Vos WM (2002) Cell to cell communication by autoinducing peptides in gram-positive bacteria. Antonie Van Leeuwenhoek 81(1):233–243

Suzuki T, Mochizuki S, Yamamoto S, Arakawa K, Kinashi H (2010) Regulation of lankamycin biosynthesis in *Streptomyces rochei* by two SARP genes, srrY and srrZ. Biosci Biotechnol Biochem 74(4):819–827

Suzuki N, Ohtaguro N, Yoshida Y, Hirai M, Matsuo H, Yamada Y, Imamura N, Tsuchiya T (2015) A compound inhibits biofilm formation of *Staphylococcus aureus* from *Streptomyces*. Biol Pharm Bull 38(6):889–892

Takano E (2006) γ-Butyrolactones: *Streptomyces* signalling molecules regulating antibiotic production and differentiation. Curr Opin Microbiol 9(3):287–294

Takano E, Nihira T, Hara Y, Jones JJ, Gershater CJL, Yamada Y, Bibb M (2000) Purification and structural determination of SCB1, a γ-butyrolactone that elicits antibiotic production in *Streptomyces coelicolor* A3 (2). J Biol Chem 275(15):11010–11016

Takano E, Chakraburtty R, Nihira T, Yamada Y, Bibb MJ (2001) A complex role for the γ-butyrolactone SCB1 in regulating antibiotic production in *Streptomyces coelicolor* A3 (2). Mol Microbiol 41(5):1015–1028

Takano E, Kinoshita H, Mersinias V, Bucca G, Hotchkiss G, Nihira T, Smith CP, Bibb M, Wohlleben W, Chater K (2005) A bacterial hormone (the SCB1) directly controls the expression of a pathway-specific regulatory gene in the cryptic type I polyketide biosynthetic gene cluster of *Streptomyces coelicolor*. Mol Microbiol 56(2):465–479

Tommonaro G, Abbamondi GR, Iodice C, Tait K, De Rosa S (2012) Diketopiperazines produced by the halophilic archaeon, Haloterrigena hispanica, activate AHL bioreporters. Microb Ecol 63(3):490–495

Tomono A, Tsai Y, Yamazaki H, Ohnishi Y, Horinouchi S (2005a) Transcriptional control by A-factor of strR, the pathway-specific transcriptional activator for streptomycin biosynthesis in *Streptomyces griseus*. J Bacteriol 187(16):5595–5604

Tomono A, Tsai Y, Ohnishi Y, Horinouchi S (2005b) Three chymotrypsin genes are members of the AdpA regulon in the A-factor regulatory cascade in *Streptomyces griseus*. J Bacteriol 187(18):6341–6353

Uroz S, D'Angelo-Picard C, Carlier A, Elasri M, Sicot C, Petit A, Oger P, Faure D, Dessaux Y (2003) Novel bacteria degrading N-acylhomoserine lactones and their use as quenchers of quorum-sensing-regulated functions of plant-pathogenic bacteria. Microbiology 149(8):1981–1989

Uroz S, Chhabra SR, Camara M, Williams P, Oger P, Dessaux Y (2005) N-Acylhomoserine lactone quorum-sensing molecules are modified and degraded by *Rhodococcus erythropolis* W2 by both amidolytic and novel oxidoreductase activities. Microbiology 151(10):3313–3322

Uroz S, Oger PM, Chapelle E, Adeline M-T, Faure D, Dessaux Y (2008) A Rhodococcus qsdA-encoded enzyme defines a novel class of large-spectrum quorum-quenching lactonases. Appl Environ Microbiol 74(5):1357–1366

Vera-Cabrera L, Ortiz-Lopez R, Elizondo-Gonzalez R, Ocampo-Candiani J (2013) Complete genome sequence analysis of *Nocardia brasiliensis* HUJEG-1 reveals a saprobic lifestyle and the genes needed for human pathogenesis. PLoS One 8(6):e65425

Viegelmann C, Margassery LM, Kennedy J, Zhang T, O'Brien C, O'Gara F, Morrissey JP, Dobson ADW, Edrada-Ebel RA (2014) Metabolomic profiling and genomic study of a marine sponge-associated *Streptomyces* sp. Mar Drugs 12(6):3323–3351

Vujaklija D, Ueda K, Hong S-K, Beppu T, Horinouchi S (1991) Identification of an A-factor-dependent promoter in the streptomycin biosynthetic gene cluster of *Streptomyces griseus*. Mol Gen Genet MGG 229(1):119–128

Waki M, Nihira T, Yamada Y (1997) Cloning and characterization of the gene (farA) encoding the receptor for an extracellular regulatory factor (IM-2) from *Streptomyces* sp. strain FRI-5. J Bacteriol 179(16):5131–5137

Wang L, Tian X, Wang J, Yang H, Fan K, Xu G, Yang K, Tan H (2009) Autoregulation of antibiotic biosynthesis by binding of the end product to an atypical response regulator. Proc Natl Acad Sci 106(21):8617–8622

Wang J, Wang W, Wang L, Zhang G, Fan K, Tan H, Yang K (2011) A novel role of 'pseudo' γ-butyrolactone receptors in controlling γ-butyrolactone biosynthesis in *Streptomyces*. Mol Microbiol 82(1):236–250

Wang J-B, Zhang F, Pu J-Y, Zhao J, Zhao Q-F, Tang G-L (2014a) Characterization of AvaR1, an autoregulator receptor that negatively controls avermectins production in a high avermectin-producing strain. Biotechnol Lett 36(4):813–819

Wang T, Jiang Y, Ma K-X, Li Y-Q, Huang R, Xie X-S, Wu S-H (2014b) Two new butenolides produced by an actinomycete *Streptomyces* sp. Chem Biodivers 11(6):929–933

Wang W, Ji J, Li X, Wang J, Li S, Pan G, Fan K, Yang K (2014c) Angucyclines as signals modulate the behaviors of *Streptomyces coelicolor*. Proc Natl Acad Sci 111(15):5688–5693

Wang J, Nong X-H, Amin M, Qi S-H (2018) Hygrocin C from marine-derived *Streptomyces* sp. SCSGAA 0027 inhibits biofilm formation in *Bacillus amyloliquefaciens* SCSGAB0082 isolated from South China Sea gorgonian. Appl Microbiol Biotechnol 102(3):1417–1427

Weber T, Welzel K, Pelzer S, Vente A, Wohlleben W (2003) Exploiting the genetic potential of polyketide producing streptomycetes. J Biotechnol 106(2–3):221–232

Willey JM, Gaskell AA (2011) Morphogenetic signaling molecules of the streptomycetes. Chem Rev 111(1):174–187

Willey JM, Willems A, Kodani S, Nodwell JR (2006) Morphogenetic surfactants and their role in the formation of aerial hyphae in *Streptomyces coelicolor*. Mol Microbiol 59(3):731–742

Williams P (2007) Quorum sensing, communication and cross-kingdom signalling in the bacterial world. Microbiology 153(12):3923–3938

Wuster A, Madan Babu M (2007) Chemical molecules that regulate transcription and facilitate cell-to-cell communication. In: Wiley encyclopedia of chemical biology. pp 1–11

Xie TT, Zeng H, Ren XP, Wang N, Chen ZJ, Zhang Y, Chen W (2019) Antibiofilm activity of three Actinomycete strains against Staphylococcus epidermidis. Lett Appl Microbiol 68(1):73–80

Xu G, Wang J, Wang L, Tian X, Yang H, Fan K, Yang K, Tan H (2010a) "Pseudo" γ-butyrolactone receptors respond to antibiotic signals to coordinate antibiotic biosynthesis. J Biol Chem 285(35):27440–27448

Xu D, Seghezzi N, Esnault C, Virolle M-J (2010b) Repression of antibiotic production and sporulation in *Streptomyces coelicolor* by overexpression of a TetR family transcriptional regulator. Appl Environ Microbiol 76(23):7741–7753

Yamada Y (1999) Autoregulatory factors and regulation of antibiotic production in *Streptomyces*. Microb Signal Commun 57:177

Yamada Y, Sugamura K, Kondo K, Yanagimoto M, Okada H (1987) The structure of inducing factors for virginiamycin production in *Streptomyces virginias*. J Antibiot 40(4):496–504

Yamamoto S, He Y, Arakawa K, Kinashi H (2008) γ-butyrolactone-dependent expression of the *Streptomyces* antibiotic regulatory protein gene srrY plays a central role in the regulatory cascade leading to lankacidin and lankamycin production in *Streptomyces rochei*. J Bacteriol 190(4):1308–1316

Yamazaki H, Tomono A, Ohnishi Y, Horinouchi S (2004) DNA-binding specificity of AdpA, a transcriptional activator in the A-factor regulatory cascade in *Streptomyces griseus*. Mol Microbiol 53(2):555–572

Yang Y-H, Joo H-S, Lee K, Liou K-K, Lee H-C, Sohng J-K, Kim B-G (2005) Novel method for detection of butanolides in *Streptomyces coelicolor* culture broth, using a His-tagged receptor (ScbR) and mass spectrometry. Appl Environ Microbiol 71(9):5050–5055

Yang Y-H, Kim T-W, Park S-H, Lee K, Park H-Y, Song E, Joo H-S, Kim Y-G, Hahn J-S, Kim B-G (2009) Cell-free *Escherichia coli*-based system to screen for quorum-sensing molecules interacting with quorum receptor proteins of *Streptomyces coelicolor*. Appl Environ Microbiol 75(19):6367–6372

Yang R, Liu X, Wen Y, Song Y, Chen Z, Li J (2015) The PhoP transcription factor negatively regulates avermectin biosynthesis in *Streptomyces avermitilis*. Appl Microbiol Biotechnol 99(24):10547–10557

Yoon J-H, Lee JK, Jung SY, Kim JA, Kim HK, Oh TK (2006) *Nocardioides kongjuensis* sp. nov., an N-acylhomoserine lactone-degrading bacterium. Int J Syst Evol Microbiol 56(8):1783–1787

You JL, Xue XL, Cao LX, Lu X, Wang J, Zhang LX, Zhou SN (2007) Inhibition of *Vibrio* biofilm formation by a marine actinomycete strain A66. Appl Microbiol Biotechnol 76(5):1137–1144

Younis KM, Usup G, Ahmad A (2016) Secondary metabolites produced by marine *Streptomyces* as antibiofilm and quorum-sensing inhibitor of uropathogen *Proteus mirabilis*. Environ Sci Pollut Res 23(5):4756–4767

Zou Z, Du D, Zhang Y, Zhang J, Niu G, Tan H (2014) A γ-butyrolactone-sensing activator/repressor, JadR3, controls a regulatory mini-network for jadomycin biosynthesis. Mol Microbiol 94(3):490–505

Metabolic Engineering of Actinomycetes for Natural Product Discovery

11

Erin E. Drufva, Tien T. Sword, and Constance B. Bailey

Abstract

Actinomycetes are an incredibly prolific source of therapeutic secondary metabolites. Indeed, over 50% of currently used commercial antibiotics are of actinomycetal origin. However, discovery has slowed significantly since the so-called golden age of antibiotics in the 1940s–1960s. The expansion of genome sequencing has revealed many more undiscovered natural products which are often cryptic, or not produced under standard laboratory conditions due to their finely tuned ecological roles in their natural contexts. Advances in synthetic biology approaches have revealed new strategies to manipulate actinomycetes and perform screens to uncover some of these cryptic natural products. Additionally, such strategies provide the tools to generate analogs of natural products through manipulation of precursor flux and biosynthetic enzymes.

Keywords

Synthetic biology · Cryptic natural products · Heterologous expression · *Streptomyces*

11.1 Overview

Actinobacteria are gram-positive bacteria with high GC content known for their complex secondary metabolisms. Actinobacteria, particularly the *Streptomyces* genus, have been recognized as one of the predominant sources for microbial bioactive natural products which have applications as antibiotics,

E. E. Drufva · T. T. Sword · C. B. Bailey (✉)
Department of Chemistry, University of Tennessee-Knoxville, Knoxville, TN, USA
e-mail: edrufva@utk.edu; ttran26@vols.utk.edu; cbaile53@utk.edu

R. V. Rai, J. A. Bai (eds.), *Natural Products from Actinomycetes*,
https://doi.org/10.1007/978-981-16-6132-7_11

chemotherapeutics, immunosuppressants, and anti-parasitics for pharmaceuticals, veterinary agents, and pesticides (Li and Vederas 2011). Illustrating their prominent value as a sources for therapeutics, over 50% of clinical antibiotics are natural products produced by actinomycetes (Bérdy 2005). Most pharmaceutical natural products in use today—particularly antibiotic scaffolds—were discovered between 1940 and the early 1960s, a period which is now referred to as "the golden age of antibiotics discovery." This productive era was followed by a ~40-year innovation gap during which no new antibiotic scaffolds were introduced to the clinic—a lag in discovery of novel antibiotics that has been particularly dire given the growing emergence of antibiotic-resistant pathogens. Traditional methods of bioactivity-guided fraction and screening, now often referred to as "grind and find," facilitated the discovery of structurally diverse, potent medicinally active scaffolds for decades (Miller and Clardy 2009). However, these methods now often lead to labor-intensive rediscovery of known secondary metabolites. Due to diminishing returns, many pharmaceutical companies largely abandoned their natural products discovery pipelines. However, advances in genome sequencing revealed that many actinomycetes remain a largely untapped resource for natural product discovery (Bachmann et al. 2014). For example, while older approaches have shown *Streptomyces* as a particularly promising genus among actinomycetes for natural product discovery, the post-genomic era has revealed that we are far from finished with mining the potential of streptomycetes for secondary metabolism. A single *Streptomyces* genome usually harbors around 30 natural product biosynthetic gene clusters (BGCs), which are approximately an order of magnitude higher than the number of BGCs revealed by bioactivity screening (Li et al. 2019a). As such, the post-genomic era has uncovered biosynthetic gene clusters (BGCs) for many compounds that are not expressed under typical laboratory culture conditions, often termed "cryptic" or "silent" (Okada and Seyedsayamdost 2017). In addition to cryptic natural products, many are produced in low quantities in their physiological context, which can complicate identification and isolation when a complex mixture of secondary metabolites is produced. Thus, to fully harness the findings of this post-genomic era of natural products discovery, synthetic biology tools have been paramount for the activation and elucidation of these silent or weakly expressed BGCs. While *Streptomyces* remains one of the most prolific genera, other genera of *Actinobacteria* are also promising sources of natural products including *Saccharopolyspora*, *Saccharomonospora*, *Mycobacteria*, *Corynebacterium*, *Arthrobacteria*, *Rhodococcus*, *Salinospora*, *Amycolatopsis*, and others. This chapter primarily focuses on tools developed for *Streptomyces*, as this genus has been the target of considerable synthetic biology efforts due to its giftedness for secondary metabolism. Certainly, some of these tools can be applied to other actinomycete genera.

As natural product discovery requires genetic manipulation, unique challenges exist with actinomycetes compared to other bacteria with regard to morphology and growth properties, making them less than ideal for manipulation in the laboratory (Katz et al. 2018). While Hopwood and coworkers first devised a strategy to genetically manipulate *Streptomyces* in 1978 (Bibb et al. 1978), the development of efficient tools has lagged significantly relative to tool development for other more tractable model organisms. For many actinomycetes, *Streptomyces* in particular, the most reliable form of genetic manipulation is intergenic conjugation, often with systems largely unchanged after their development in the 1980s–1990s (Mazodier et al. 1989). The high GC content of most actinomycetes, typically exceeding 70%, can also present some challenges with DNA capture and genetic manipulation, as BGCs tend to be hundreds or thousands of kilobase pairs. For such manipulations, amplification by PCR is nontrivial, often requiring additives, specialized polymerases, and tailored thermocycling protocols. Additionally, secondary metabolite production is under the control of complex regulatory pathways managing transcription, translation, and metabolic flux (Komatsu et al. 2010; Palazzotto et al. 2019). Devising productive ways of perturbing this genetic interplay is often nontrivial. Despite all these challenges, in recent years, advancements have been made toward the application of synthetic biology tools to expand the discovery of new actinomycete natural products.

Fundamentally, there are two synthetic biology approaches to discover new natural products, most of which are cryptic. The first relies upon the discovery of novel pathways through genome mining and heterologous expression. The second relies upon manipulating the native host to reveal cryptic metabolites, or metabolites produced at such low levels that they are virtually undetectable. Finally, either heterologous or native hosts can be manipulated to create analogs of natural products via direct manipulation of the biosynthetic pathway. Taken together, the considerable advances in synthetic biology approaches to manipulate, grow, and elicit metabolites from metabolically gifted actinomycetes have facilitated the discovery and production of many bioactive compounds (Fig. 11.1).

11.2 Identification and Capture of Biosynthetic Gene Clusters (BGCs)

11.2.1 Identification

Because of the large unexplored genomic space, genome mining, which utilizes sequenced genomes and bioinformatic tools, is often the first step to identifying clusters harbored within newly isolated bacteria. The ability to analyze genomes has been greatly expedited by pipelines that allow for the identification of BGCs, typically through signature-based tools that use hidden Markov models (HMMS) or Basic Local Alignment Search Tool (BLAST). Most notably, antiSMASH (antibiotics and Secondary Metabolite Analysis Shell) (Blin et al. 2020) is one of the most extensive programs for BGC discovery. For compound identification, the Natural Products Atlas (van Santen et al. 2019) is a repository that compiles known structural information about natural products including spectra information, isolations, total synthesis, and reassignment. Both antiSMASH and NP Atlas are seamlessly integrated into the Minimum Information about a Biosynthetic Gene Cluster (MiBIG) repository which compiles antiSMASH cluster data, NP atlas structural data, and the known literature describing BGCs as a powerful means of

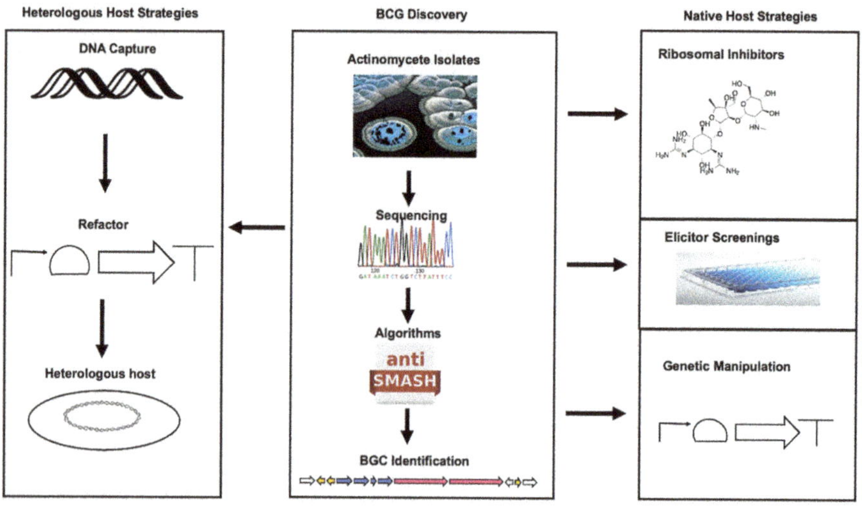

Fig. 11.1 Synthetic biology strategies for natural product discovery. Isolates are sequenced and genome mined with algorithms such as antiSMASH and clusters are identified. Then they are either captured and expressed in heterologous hosts (often with refactoring) or elicited from the native hosts either via genetic manipulation of regulatory elements, small molecules uncovered in elicitor screens, or ribosomal mutations

streamlining known information (Medema et al. 2015). Still, others further organize BGCs, such as the BiG-FAM program which organizes them into gene cluster families (Kautsar et al. 2021). Although used less extensively, other genome mining tools, for example, PRISM (Prediction Informatics for Secondary Metabolomes) (Skinnider et al. 2015), also exist. Some tools include structure prediction, including the generalized retro-biosynthesis assembly prediction engine (GRAPE) for polyketides and non-ribosomal peptides, as well as the global alignment for natural products chemoinformatics (GARLIC), developed by McGarvey and coworkers (Dejong et al. 2016). Other databases exist specific-to-specific classes of natural product biosynthetic genes (Table 11.1). The integration of these bioinformatic tools to analyze sequencing and metabolic data has enabled a multi-omics approach to natural products discovery (Figure 11.2).

11.2.2 Capture

Initial capture strategies have relied upon the development of cosmid, fosmid, and bacterial artificial chromosome (BAC) libraries. Their construction has involved genomic DNA isolation followed by partial digestion or DNA fragmentation, ligation, and transformation, typically capturing fragments of ~40 kb flanked by natural restriction sites within the genome. In theory, BACs can capture fragments up to ~100 kb. Such strategies have been employed to clone validamycin (Yu et al. 2005), spinosad (Waldron et al. 2001), borrelidin (Olano et al. 2004), and many other BGCs first elucidated from actinomycetes. Other strategies were similar in

Table 11.1 Bioinformatic tools for genome mining and multi-omics

Database	Description	Pathway type	Ref.
antiSMASH	Genome mining identification of all classes of BGCs for all classes of natural products; domain prediction; regulatory element prediction; bacterial, fungal, and plant versions	Various	Blin et al. (2020)
MiBiG	Repository compiling metadata on known BGCs including gene sequence, antiSMASH output, NP atlas link, NCBI accession numbers, and literature references	Various	Medema et al. (2015)
NP Atlas	24,594 compound repository database with an interactive Web portal for structure, structural characterization data, compound names, isolation references, total syntheses, instances of structural re-assignment, and literature references	Various	van Santen et al. (2019)
PRISM	Cluster prediction, domain prediction, metabolite structure prediction	Various	Skinnider et al. (2015)
ClustScan	Cluster prediction, domain prediction of modular biosynthetic proteins	PKS, NRPS, and PKS-NRPS hybrids	Starcevic et al. (2008)
GRAPE	Retro-biosynthetic assembly	PKS and NRPS	Dejong et al. (2016)
GARLIC	Natural products chemoinformatics	PKS and NRPS	Dejong et al. (2016)
NP.searcher	Structure prediction for modular biosynthetic pathways	PKS and NRPS	Li et al. (2009b)
NaPDoS	Phylogeny-based prediction of modular biosynthetic gene clusters	PKS and NRPS	Ziemert et al. (2012)
BiG-FAM	Gene cluster family grouping	Various	Kautsar et al. (2021)
TransATor	Structure and domain prediction	Trans-AT PKS	Helfrich et al. (2019)
RiPPMiner	Structure predictor	RiPPS	Agrawal et al. (2017)
BAGEL	Structure predictor, cluster predictor	RiPPS, especially bacteriocins	van Heel et al. (2018)
RODEO	Cluster predictor, structure predictor	RiPPS	Tietz et al. (2017)

(continued)

Table 11.1 (continued)

Database	Description	Pathway type	Ref.
RiPPMiner	Cluster predictor, structure predictor	RiPPS	Agrawal et al. (2017)
PKMiner	Cluster predictor, structure predictor	Type II PKS	Kim and Yi (2012)

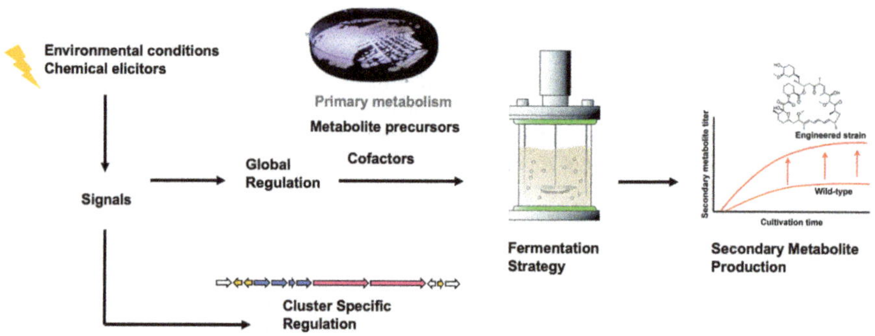

Fig. 11.2 Overview of the application of a multi-omics workflow to natural product discovery adapted from Reference (Kim et al. 2016)

approach but relied upon phage attachment sites, such as the λ EMBL3 system used to clone the erythromycin BGC (Haydock et al. 1991).

Capture strategies that are not solely reliant on restriction digestion and library formation have traditionally harnessed homologous recombination to isolate gene clusters. These strategies have typically utilized *E. coli* or *Saccharomyces cerevisiae* as "capture hosts." The "capture host" was co-transformed with linearized vector and donor DNA to capture the BGC. For *E. coli*, two parallel systems have been developed from phage recombination. One system termed "lambda red" was derived from a λ-Red prophage composed of Redα, Redβ, and Redγ. Additionally, "ET recombination" is an analogous system from the Rac prophage that involved the proteins RecE and RecT (Muyrers et al. 1999). RecE and Redα are 5′ to 3′ exonucleases, whereas RecT and Redβ are single-strand DNA-binding proteins which can facilitate complementary strand annealing. Recγ has no counterpart in the RecET system but plays a role in inhibiting the activities of RecBCD, which function as double-strand DNA and single-strand DNA exonucleases and helicases. Both the ET recombination system and the λ-Red recombination system have been used to directly capture and express DNA in various heterologous hosts. Examples of pathways captured by the λ-Red system include streptomycin (*Streptomyces griseus*), erythromycin A (*Saccharopolyspora erythraea*), cephamycin C (*Streptomyces clavuligerus*), holomycin (*Streptomyces clavuligerus*), clavulanic acid (*Streptomyces clavuligerus*), rebeccamycin (*Lechevalieria aerocolonigenes*), novobiocin (*Streptomyces analatus*), and chloramphenicol (*Streptomyces venezuelae*) (Komatsu et al. 2013). Pathways can either be captured from a large stretch of BAC, cosmid, or fosmid genomic DNA through libraries described above or directly from genomic DNA in one step.

Yeast-based transformation-associated recombination (TAR) cloning has also been applied for direct capture. However, this technique has limited utility in actinomycetes due to the frequency of mis-annealing from repetitive sequences and the high GC content of actinomycetes compared to yeast (Abbasi et al. 2020). Additionally, as the fragments are generated via PCR through TAR cloning, point mutations can be introduced which are not always easily identified due to the large size of the BGC. Notably, TAR cloning has been implemented by Brady and coworkers for the successful capture of BGCs from environmental DNA. Through TAR cloning in yeast, libraries were generated for transformation into *Streptomyces albus*. Subsequent transformants were screened for pigment formation, and through this workflow, the type II polyketide pigments, fluostatins, were uncovered (Feng et al. 2010). A final homology-based approach utilizes Gibson assembly (Gibson et al. 2009), which successfully captured the pristinamycin BGC (Li et al. 2015a). Like TAR cloning, Gibson cloning can suffer due to mismatching linker pairs and low efficiency from the high GC content of actinomycetes.

Very recently, the Clustered Regulatory Interspaced Short Palindromic Repeats/CRISPR-associated protein (CRISPR/Cas) system has been adapted for the direct capture of BGCs. Despite the great success of recombineering, the efficient cloning of large BGCs (e.g., >80 kb) remains challenging, especially considering the high GC content of actinomycetes and the NGG protospacer adjacent motif (PAM) of the commonly used *Streptococcus pyogenes* Cas9 endonuclease. To address this problem, Zhang and coworkers developed the CRISPR/Cas12a system, termed CAT-FISHING (CRISPR/Cas12a-mediated fast direct BGC cloning), which facilitated the capture of large DNA fragments from bacterial artificial chromosomes or high-GC genomic *Streptomyces* DNA. CAT-FISHING utilizes the Cas12a exonuclease rather than the more common Cas9 exonuclease as it has a T-rich protospacer adjacent motif (PAM) rather than a G-rich PAM and generates dsDNA breaks with staggered ends rather than blunt ends. The T rich PAM allows for less nonspecific cleavage in an extremely GC-rich genome, whereas the common PAM in Cas9 (NGG) can promote nonspecific cleavage. Through this effort, they were able to capture an 87-kb gene cluster encoding the surugamide and candicidin pathways (both 75% GC) and express these clusters in a heterologous host (Table 11.2) (Liang et al. 2020).

Table 11.2 BGC capture strategies

Approach	Brief description	ULCC	Limitations	Cloned BGC	Ref.
Library construction					
Cosmid	1. Genomic DNA isolation 2. Partial digestion or DNA fragmentation 3. Ligation and transformation 4. Target single clone screening	~40 kb	Time consuming; labor intensive	Validamycin, spinosad, borrelidin, etc.	Olano et al. (2004); Tan et al. (2017); Yu et al. (2005)
Fosmid		~40 kb			
BAC		~90 kb			
Yeast recombination					
TAR	1. Genomic DNA isolation 2. Transformation and recombination in yeast 3. Plasmid isolation and *E. coli* transformation	67 kb	Mis-priming or mis-annealing for high GC content or repeated sequences (e.g., polyketide synthases)	Taromycin	Yamanaka et al. (2014)
RecET/λ Red					
RecET/l λ Red	1. Genomic DNA isolation 2. DNA digestion by specific restriction enzymes 3. Transformation and application of REC	106 kb	Restriction enzyme limitation; mis-priming or mis-annealing for high GC content or repeated sequence	Salinomycin, spinosad, etc.	Song et al. (2019); Wang et al. (2016a)
Gibson assembly					
GA	1. DNA fragment preparations by PCR 2. Gibson assembly 3. Transformation	72 kb	Mutation caused by PCR; mismatched linker pairings for high GC content	Pristinamycin	Li et al. (2018a)

(continued)

Table 11.2 (continued)

Approach	Brief description	ULCC	Limitations	Cloned BGC	Ref.
CRISPR/CAS					
Cas9, λ packaging	1. Genomic DNA isolation 2. DNA digestion by CRISPR/Cas9 3. In vitro λ packaging and ligation 4. Transformation	~40 kb	Difficult for large gene cluster cloning, e.g., >50 kb	Sisomicin	Tao et al. (2019)
CATCH	1. Genomic DNA isolation 2. DNA digestion by CRISPR/Cas9 3. Gibson assembly and transformation	~100 kb	Mismatched linker pairs for high GC content sequences	N/A	Jiang and Zhu (2016); Jiang et al. (2015)
CAT-FISHING	1. Genomic DNA isolation 2. DNA digestion by CRISPR/Cas-2a	87 kb	Potential for off target cleavage	Surugamides, candicidin	Liang et al. (2020)

BAC bacterial artificial chromosome, *ULCC* upper limit of cloning capacity, *CATCH* Cas9-associated targeting of chromosome segments, *TAR* transformation-associated recombination, *GA* Gibson assembly. Adapted from Liang et al. (2020)

11.3 Manipulation and Heterologous Expression

11.3.1 Genetic Manipulation

Transformation of DNA into actinomycetes has been challenging due to their thick cell walls, high GC content, and restriction modification systems. Compared to the straightforward and reliable protocols for gram-negative bacteria like *E. coli*, transformation strategies for actinomycetes are more complex and inefficient. The first method developed in 1966 was protoplast transformation in *S. kanamyceticus* (Okanishi et al. 1966). This strategy was later modified to increase transformation efficiency via the incorporation of polyethylene glycol (Bibb et al. 1978). Although effective and widely applied to a variety of *Streptomyces* species including *S. coelicolor*, *S. griseus*, *S. venezuelae*, *S. fradiae*, *S. ambofaciens*, *S. violaceusniger*, and *S. acidomyceticus*, this process is labor-intensive and time-consuming due to the formation and regeneration of protoplasts (Bibb et al. 1978; Matsushima and Baltz 1985; Okanishi et al. 1968, 1974). Over two decades later, an electroporation protocol was developed to transform plasmid DNA (pDNA) directly into mycelial fragments, eliminating the need for protoplasts. This method was effective for certain *Streptomyces* spp. including *S. rimosus*, *S. lividans*, *S. coelicolor*, and *S. venezuelae* (Pigac and Schrempf 1995). A transformation method introduced around the same time as electroporation involved intergenic conjugation of *Streptomyces* with *E. coli.* This is the most popular transformation method today due to its applicability to many streptomycetes and high transformation efficiency compared to the protoplast method. Conjugation involves the transfer of pDNA containing *cis*-acting origin of transfer, *oriT*, from a broad-host IncP-group plasmid RK2. This plasmid is transported in trans from an *E. coli* donor strain to the *Streptomyces* recipient strain (Bierman et al. 1992; Mazodier et al. 1989). This method is straightforward compared to protoplast transformation since the only necessary steps involve growing and washing the *E. coli* donor strain, growing the *Streptomyces* spores, heat shock of the spores, and briefly mixing the two together. It is very important to use a methylation-deficient strain of *E. coli* as the donor, since many streptomycetes have restriction-modification systems that can greatly hinder transformation efficiency (Hopwood et al. 2000).

11.3.2 Vectors

Various antibiotic resistance markers have been used in the development of *Streptomyces* transformation vectors. Some of the most common resistance markers include the thiostrepton resistance gene from *S. azureus*, *tsr*; the apramycin resistance gene from *Klebsiella pneumoniae*, *apr*; the kanamycin/neomycin/gentamycin resistance gene from *E. coli*, *neo*; and the *hyg* gene from *S. hygroscopicus*, which encodes hygromycin phosphotransferase. Early studies used *tsr*, but thiostrepton unfortunately hinders morphological differentiation and interferes with antibiotic production. Alternatively, *neo*, *apr*, and *hyg* all present the advantage of selection in both *E. coli* and *Streptomyces*. However, *kan* is used to select for the popular conjugation donor strain, *E. coli* ET12567/pUZ8002, which rules out *neo* for selection of *Streptomyces* conjugants in this scenario (Hopwood et al. 2000).

Several vectors have been created for *Streptomyces* genetic engineering including replicating vectors, integrating vectors, and suicidal non-replicating vectors (Hopwood et al. 2000). Early examples of replicating vectors were constructed from native *Streptomyces* plasmids such as pIJ101 from *S. lividans* ISP5434. This plasmid is advantageous for its high copy number (about 300 copies per cell) (Kieser et al. 1982). It was used as a template for other cloning vectors, including the widely used pIJ486 and pIJ702. However, due to the absence of an *E. coli* replicating origin, these can only be used for protoplast transformation or electroporation, which were both found to be less reliable than intergenic conjugation (Hopwood et al. 2000). A different high copy number plasmid, pJV1 from *S. phaeochromogenes* (about 150 copies per cell), was modified to create several additional replicating vectors, including the stable and adaptable *Streptomyces-E. coli* shuttle cosmid vector, pFD666 (Bailey et al. 1986; Denis and Brzezinski 1992). Additionally, the low-copy vector, SCP2 from *S. coelicolor*, was used as a template to develop many robust low-copy vectors, including the *Streptomyces-E. coli* shuttle vectors pIJ698, pKC1218, and pOJ446 (Bibb et al. 1977; Bierman et al. 1992; Hopwood et al. 2000). These replicating vectors have been used to express a variety of industrially relevant proteins and enzymes including lipases, oxidoreductases, proteases, phosphatases, pharmaceutically active proteins, and others. Using these replicating vectors, most successes have occurred with different strains of *S. lividans*, although other species such as *S. coelicolor* have also been used (Berini et al. 2020).

Although replicating vectors have proved successful in some cases, they present a few disadvantages. In most cases, they are exclusively functional in native *Streptomyces* strains and a few evolutionarily related relatives. Additionally, antibiotic selection is required to maintain production of the correct plasmid. These issues can be addressed by using either integrating vectors or suicidal non-replicating vectors to express a gene of interest. Both are quite stable, especially for incorporating large genes or gene clusters up to 100 kb in size (Chen et al. 2014; Miao et al. 2005). Integrating vectors contain a DNA fragment that encodes attachment functions for site-specific integration into the *Streptomyces* genome (Smokvina et al. 1990). They can be used to heterologously produce secondary metabolites or increase the yield of a native natural product by inserting a strong promoter (Kormanec et al. 2019). On the other hand, suicidal vectors integrate into any neutral region of the *Streptomyces* genome via homologous recombination. These can be used for the stable deletion of genes or gene clusters (Knirschova et al. 2015). For instance, they can be used for the removal of negative regulatory elements in order to awaken silent genes for the generation of new natural products (*vide infra*) (Novakova et al. 2011). Furthermore, integrating and suicidal vectors can be used to decipher large sets of metagenome data related to gene clusters for novel secondary metabolites found via genome mining (Bachmann et al. 2014).

11.3.3 Recombinases

The Cre-lox and Flp classes of recombinases are tyrosine recombinases that are widely used in a range of bacteria. The Cre and Flp proteins are bidirectional recombinases that catalyze site-specific recombination at a 34-bp recognition site. For the most part, they have been utilized for simple marker removal from a chromosome. However, these systems can also be used for site-specific recombination. In a study from Hermann and coworkers, the Cre and related Dre recombinases were used at various lox sites to create multiple site mutations simultaneously. Thus, the Cre recombinase was used in the deletion of large genomic regions, including phenalinolactone, monensin, and lipomycin of *Streptomyces* sp. Tü6071, *Streptomyces cinnamonensis* A519, and *Streptomyces aureofaciens* Tü117, respectively (Herrmann et al. 2012). The Cre lox system has also been used in the actinomycete

Corynebacterium glutamicum (Suzuki et al. 2005). A separate strategy developed for recombineering in actinomycete species is the I-SceI meganuclease from *Saccharomyces cerevisiae*. This strategy employed a codon optimized variant of the yeast I-SceI endonuclease. It was used for the successful deletion of the red pigment, undecylprodigine, in *Streptomyces coelicolor* M1141 (Fernández-Martínez and Bibb 2014).

11.3.4 Phage Integration

At this time, at least eight site-specific integration systems have been developed and used to create integrating vectors for *Streptomyces*. Most are based on actinomycete bacteriophages, or actinophages, specifically from *Streptomyces* spp. (Kormanec et al. 2019) Examples of actinophages used in integrating vectors include ΦC31 (Lomovskaya et al. 1972), ΦBT1 (Gregory et al. 2003), VWB (Van Mellaert et al. 1998), TGI (Morita et al. 2009), SVI (Fayed et al. 2015), R4 (Chater and Carter 1979), ΦJoe (Fogg et al. 2017), µ1/6 (Farkašovská and Godány 2012), and pSAM2 (Boccard et al. 1989). Integrating vectors using bacteriophage functional elements, including the *attP* site, are incorporated into the host chromosome at the *attB* site via single crossover recombination. The most widely used actinophage for chromosomal integration in streptomycetes is ΦC31 derived from *S. coelicolor*. The *attP* and *int* genes encoding serine family recombinases are necessary for unidirectional integration using this system (Lomovskaya et al. 1980). Some of the earliest examples of Att/Int integrating vectors include the conjugative integrating vector pSET152, cosmid vector pOJ436, and cosmid vector pOJ444 (Bierman et al. 1992). These vectors and their future iterations were used to engineer a variety of novel *Streptomyces* strains for the successful expression of many secondary metabolite gene clusters including chlorobiocin, caprazamycin, daptomycin, moenomycin A, nikkomycin, and others. These vectors all possess cloning capacities of <50 kb (Myronovskyi and Luzhetskyy 2013). However, many secondary metabolite gene clusters are quite large, sometimes >100 kb. To address this size limitation issue, two new ΦC31-based Att/Int vectors were constructed, pStreptoBAC and pESAC13 (Miao et al. 2005; Tocchetti et al. 2018). The former is derived from the BAC, while the latter is derived from the P1 artificial chromosome, PAC. Both vectors have cloning capacities of 200 kb and can be used for *Streptomyces* conjugation with *E. coli*. These vectors, along with future derivatives, have been used for the heterologous expression of the daptomycin, *iso*-migrastatin, roseoflavin, FK506, salinomycin, anthracimycin, chaxamycin, and many other BGCs (Tocchetti et al. 2018).

Other commonly used Att/Int vectors are the ones derived from the ΦBT1 actinophage of *S. coelicolor* which is related to ΦC31. Like ΦC31, ΦBT1 integrase is a serine recombinase; however, ΦBT1 integrates into a different *attB* attachment site. This *attB* site is located within *SCO4848*, a gene encoding an integral membrane protein (Zhang et al. 2008). Integration vectors pRT801, pRT802, and pMS82, with apramycin, kanamycin, and hygromycin antibiotic resistance markers, respectively, were all constructed using this system. Since all three vectors contain *oriT*, they are suitable for *E. coli* conjugal integration into a variety of *Streptomyces* hosts (Gregory et al. 2003). Additionally, the ΦBT1 Att/Int system was used to create pSBAC, a BAC vector suitable for conjugation of large gene clusters (Liu et al. 2009). This vector was used for the efficient heterologous expression of the meridiamycin, tautomycetin, and pikromycin gene clusters in *S. lividans*, *Streptomyces* sp. CK4412, and *S. venezuelae*, respectively (Liu et al. 2009; Tocchetti et al. 2018). For the most part, vectors using the ΦBT1 Att/Int system are quite stable and versatile (Baltz 2012). However, unlike ΦC31, ΦBT1 integration is not completely unidirectional. And in some cases, ΦBT1 integrase can partially act as an excisionase, causing a low-frequency excision between *attL* and *attR* (Zhang et al. 2008).

Another example of an Att/Int integration actinophage is VWB from *S. venezuelae*. Unlike ΦC31 and ΦBT1, VWB integrase is a tyrosine recombinase, and the *attB* site is found in the putative $tRNA^{Arg}$ (AGG) gene from streptomycetes. The integration vector pKT02 was constructed using the VWB Att/Int system. Since pKT02 did not contain *oriT*, it was transformed into *S. lividans* and *S. venezuelae* via the protoplast method with a lower transformation efficiency than ΦC31 integration vectors (Van Mellaert et al. 1998). To modify this vector for *E. coli* conjugation, the derivative pSOK804 harboring *oriT* was developed. Upon integration into *S. noursei*, the transformation efficiency was twofold higher than pSET152 (Sekurova et al. 2004). Following this, the conjugative TAR vectors, pCLY10 and pCLY11, were created for the integration of large secondary metabolite gene clusters. These were constructed based on the p15a and F-factor replicons, respectively, and both were utilized to heterologously express the grecocycline BGC in *S. albus* J1074 (Bilyk et al. 2016). Additionally, pCLY10 was used for the heterologous expression of the antitumor polyketide mithramycin A in *S. lividans* with an impressive yield of 3 g/L, which is sixfold greater than the mithramycin A overproducer, *S. argillaceus* (Novakova et al. 2018) (Table 11.3).

11.3.5 Promoters

Constitutive promoters are ubiquitous in heterologous gene and gene cluster expression studies. Additionally, constitutive promoters often allow for stronger expression of natural product genes in actinomycetes than their native promoters, which frequently remain under tight regulatory control for specific environmental stimuli (Palazzotto et al. 2019). However, their efficiency can vary greatly depending on host choice, media choice, or growth stages. Prevalent examples of constitutive promoters include those based on the erythromycin resistance gene (*ermE*), *kasOP* and various derivatives, regions of the S12 ribosomal protein, and *actII orf4* and various derivatives (Myronovskyi and Luzhetskyy 2016).

In 1985, the erythromycin resistance gene (*ermE*) from *S. erythraea* was transformed into *S. lividans*. The gene was analyzed via DNA sequencing, high-resolution S1 and exonuclease VII mapping, in vitro transcription, and in vivo

Table 11.3 Phage integration strategies for actinomycete engineering

Att/Int system	Integrase	Original phage/ plasmid host	Directionality	Target gene	Vector examples	Ref.
*Φ*C31	Serine recombinase	*S. coelicolor*	Unidirectional	*SCO3798*	pSET152, pOJ444, pStreptoBAC V	Bierman et al. (1992); Combes et al. (2002); Tocchetti et al. (2018)
*Φ*BT1	Serine recombinase	*S. coelicolor*	Bidirectional	*SCO3798*	pRT801, pSBAC, pNG1	Gonzalez-Quiñonez et al. (2016); Gregory et al. (2003); Liu et al. (2009); Zhang et al. (2008)
VWB	Tyrosine recombinase	*S. venezuelae*	Unidirectional	$tRNA^{Arg}$ (AGG)	pKT02, pSOK804, pCLY10	Bilyk et al. (2016); Sekurova et al. (2004); Van Mellaert et al. (1998)
TGI	Serine recombinase	*S. cattleya*	Unidirectional	*SCO3658*	pKU462, pRF10, pFF10	Fayed et al. (2015); Morita et al. (2009)
R4	Serine recombinase	*S. albus*	Unidirectional	*SCO6196*	pLR4	Li et al. (2019b); Miura et al. (2011)
SVI	Serine recombinase	*S. venezuelae*	Unidirectional	*SCO4383*	pFB3, pLSV1	Fayed et al. (2014, 2015); Li et al. (2019b)
*Φ*Joe	Serine recombinase	*S. coelicolor*	Bidirectional	*SCO2606*	pCMF92	Fogg et al. (2017)
µ1/6	Tyrosine recombinase	*S. aureofaciens*	Unidirectional	$tRNA^{Thr}$ (ACA)	pCTint	Farkašovská and Godány (2012)
pSAM2[a]	Tyrosine recombinase	*S. ambofaciens*	Unidirectional	$tRNA^{Pro}$ (CCG)	pKC824, pKC767, pPM927	Kuhstoss et al. (1991); Raynal et al. (1998); Smokvina et al. (1990)

[a]pSAM2 is based on an integration plasmid rather than a bacteriophage

promoter probing. Through this analysis, the complicated structure of the *ermE* promoter region was elucidated. Promoters *ermEp1* and *ermEp2* were identified. *ErmEp2* initiates transcription 72 nt prior to the predicted GTG start codon, while *ermEp1* initiates at the base (G) immediately preceding the same start codon. In vivo promoter probing in *S. lividans* confirmed promoter activity. Erythromycin resistance was observed at up to 20 μg/mL (Bibb et al. 1985). The wild-type promoter containing both *ermEp1* and *ermEp2* is referred to as *ermEp*, while different mutants of the promoter have also been utilized. Trinucleotide TGG deletion within the −35 region of *ermEp1* resulted in the strong variants *ermEp1** and *ermEp**, which contains both *ermEp1** and *ermEp2*.

In *S. coelicolor*, coelimycin production is controlled by the type I PKS gene cluster, CPK (Gomez-Escribano et al. 2012). The *kasO* gene encodes the SARP transcriptional activator of this gene cluster, and its promoter, *kasOP*, is regulated by the pseudo γ-butyrolactone receptor, ScbR2 (*vide supra*) (Takano et al. 2005). *KasOP* is similar in sequence to promoters recognized by HrdB, and *S. coelicolor* was found to express HrdB during growth (Strohl 1992). One strategy to increase the robustness of *kasOP* involved removing ScbR- and ScbR2-binding sites. Upon removal of the OB binding site of ScbR2 from the 5′-end of the promoter, the transcriptional efficiency of the mutant promoter, $kasOP_3$, increased by almost 40-fold compared to the wild type. Removal of the ScbR-binding site, OA, involved sequence randomization between the −10 and −35 regions of the $kasOP_3$ promoter. This strategy generated 4 mutant promoters, $kasOP_{314}$, $kasOP_{361}$, $kasOP_{382}$, and $kasOP_{3154}$, which all displayed significantly higher transcriptional activities than the wild-type *kasOP*. $KasOP_{361}$, which was renamed *kasOP**, showed the greatest transcriptional strength and was thus subjected to further analysis. Not only was *kasOP** more effective than the original *kasOP*, but it was also more effective than both *ermE** and *SF14P* in biological and real-time qPCR assays.

Inducible promoters have also been identified for *Streptomyces* spp. They are particularly valuable for developing metabolically engineered bacterial strains or for the heterologous expression of genes that produce toxic compounds such as antibiotics. Some examples of inducible promoters for actinomycete gene expression include *PtipA*, *PnitA-NitR*, Potr, and *tcp830*, as well as promoters induced by glycerol, resorcinol, and cumate (Myronovskyi and Luzhetskyy 2016). The thiostrepton-inducible promoter, *PtipA* from *S. lividans*, has been extensively used for gene expression in streptomycetes (Takano et al. 1995). Natively, *PtipA* initiates the transcription of TipAL and TipAS, two transcriptional regulator proteins structurally related to thiopeptides that bind covalently to thiostrepton (Chiu et al. 1999). TipAL and TipAS have been shown to react with the dehydrogenase tail of thiopeptide antibiotics. Therefore, both are essential for complex antibiotic formation (Chiu et al. 1996). Protein overexpression has been achieved using *PtipA* in several heterologous hosts. For example, the glycoprotein Rv1860 of *Mycobacterium tuberculosis* was expressed in *S. lividans* under the control of *PtipA*, and SDS-PAGE analysis indicated that large quantities of Rv1860 were secreted into the culture medium at about 25% of total secreted protein (Lara et al. 2004). Although *PtipA* has been used to efficiently express a variety of genes, the use of this promoter presents a few drawbacks. First, it relies on the presence of the TipAL protein as an activator and has considerable leaky expression. Second, the presence of the resistance gene, *tsr*, is often necessary when using *PtipA* because of the high activity of thiostrepton (Murakami et al. 1989).

PnitA is another commonly used inducible promoter for heterologous gene expression in actinomycetes. *PnitA* originates from *Rhodococcus rhodochrous* J1 and is responsible for the synthesis of nitrilase. It can be strongly induced by either isovaleronitrile or the nontoxic compound, ε-caprolactam (Komeda et al. 1996). *PnitA* caught the attention of researchers because nitrilase makes up 35% of all soluble proteins in *R. rhodochrous* J1, indicating the presence of a strong promoter. Sequencing and functional analysis of the nitrilase gene, *nitA*, revealed that the transcriptional activity of *PnitA* depends on the downstream gene, *nitR*, which encodes a positive transcriptional regulator. Gene expression level using *PnitA*-NitR was revealed to be dependent on the inducer dose (ε-caprolactam), and due to its high transcriptional strength, it has facilitated many successful protein expression studies (Matsumoto et al. 2016). For example, the effectiveness of *PnitA* was tested in *S. lividans*, *S. coelicolor*, *S. avermitilis*, and *S. griseus* using the reporter genes *nitA*, *inhA*, and *xylE*. Both *inhA* and *xylE* originate from *Pseudomonas* spp. and encode isonitrile hydratase and catechol-2,3-dioxygenase, respectively. In *S. lividans*, the expression system worked very well for *nitA*, with nitrilase making up approximately 20% of the total soluble protein in cell lysate. Additionally, the maximum protein expression level was approximately 40% of total soluble protein (~400 mg/L). However, isonitrile was not expressed in *S. coelicolor* or *S. avermitilis*, and nitrilase was not expressed in *S. griseus*. Even still, *PnitA* is quite versatile and suitable for tightly controlled protein expression (Table 11.4).

11.3.6 Reporter Genes

In addition to creating a new set of stable conjugative integrating vectors, Phelan et al. tested a variety of reporter proteins for signal intensities in *S. venezuelae* including the monomeric Tag Blue Fluorescent Protein (mTagBFP), mCerulean, mTFP (teal), sfGFP (superfolder green FP), mCherry (red), mKate (red), and mCardinal (red), all of which were codon optimized for *Streptomyces*. These were expressed using the Att/Int VWB integrating system under the control of *Pgapdh (EL)*. Results indicated that mCherry was the superior FP with approximately 400-fold greater signal than the negative control (Phelan et al. 2017). An additional silent reporter gene, *bpsA*, which encodes the blue pigment producing enzyme, indigoidine synthetase, was identified in *S. albus* J1047. The *bpsA* gene was discovered in *S. albus* by inserting *ermEp** upstream, and the resulting strain produced the characteristic blue pigment (Olano et al. 2014). Similar NRPS genes were found in *S. lavendulae*, *S. aureofaciens*, *S. chromofuscus*, and *Erwinia chrysanthemi*, all of which produced the same blue pigment. *BpsA* is an attractive reporter gene for actinomycetes, since it is relatively small and naturally produced by *Streptomyces* spp. Few reporter proteins consistently produce a strong signal in

Table 11.4 Examples of promoters for actinomycete engineering. Synthetic promoters possess elements found in different hosts, and these are referred to as the "native host(s)"

Name	Gene	Engineered variants	Native host(s)	Inducer molecule (s)	Expressed metabolites	Heterologous hosts	Ref.
ermEp	*ermE*	*ermEp**, *ermEp1**, *D4-21*, *tcp830*[a]	*Saccharopolyspora erythraea*	N/A, constitutive[a]	Erythromycin, jadomycin, novobiocin, 6-*epi*-alteramides A and B, indigoidine	*S. lividans*, *S. albus*, *Salinispora tropica*, *S. venezuelae*, *S. coelicolor*	Bibb et al. (1985)); Schmitt-John and Engels (1992); Siegl et al. (2013)
SF14P	Unknown	N/A	*S. ghanaensis* phage I19	N/A, constitutive	Actinorhodin	*S. lividans*, *S. venezuelae*, *S. avermitilis*	Labes et al. (1997); Wang et al. (2013)
kasOP	*kasO*	$kasOP_{314}$, $kasOP_{382}$, $kasOP_{3154}$, *kasOP**, *SP1-44*	*S. coelicolor*	N/A, constitutive	Coelimycin, actinorhodin, spinosad, indigoidine, erythromycin	*S. lividans*, *S. venezuelae*, *S. avermitilis*, *S. albus*	Ji et al. (2019); Labes et al. (1997); Liu et al. (2019); Tan et al. (2017); Wang et al. (2013)
Pgapdh	*gapdh*	N/A	*S. griseus*, *Eggerthella lenta*	N/A, constitutive	Spectinabilin, violapyrones, methylated violapyrones, C6 and C7 ethyl ketones	*S. lividans*, *S. youssoufiensis*	Hou et al. (2018); Shao et al. (2013)
PrpsL	*rpsL*	N/A	*S. griseus*, *Cellulomonas flavigena*, *Xylanimonas cellulosilytica*	N/A, constitutive	Spectinabilin, spinosad	*S. lividans*, *S. albus*	Shao et al. (2013); Tan et al. (2017)

(continued)

Table 11.4 (continued)

Name	Gene	Engineered variants	Native host(s)	Inducer molecule (s)	Expressed metabolites	Heterologous hosts	Ref.
PtipA	*tipA*	N/A	*S. lividans*	Thiostrepton	Tendamistat, actinorhodin	*S. lividans*, *S. coelicolor*, *S. cinnamonensis*	Chiu et al. (1999); Holmes et al. (1993); Schmitt-John and Engels (1992)
PnitA	*nitA*	N/A	*Rhodococ4 us rhodochrous* J1	Isovaleronitrile, ε-caprolactam	Spectinabilin	*S. lividans*, *S. coelicolor*, *S. avermitilis*, *S. griseus*	Herai et al. (2004); Matsumoto et al. (2016); Shao et al. (2013)
tcp830	*ermE*	N/A	*Saccharopolyspora erythraea*, *E. coli*	Tetracycline, anhydrotetracycline	Novobiocin	*S. coelicolor*, *S. ambofaciens*, *S. lividans*, *S. roseosporus*, *S. griseus*, *S. venezuelae*	Dangel et al. (2010); Rodríguez-García et al. (2005)
Potr	*otrR*	*Potr**	*S. rimosus*	Oxytetracycline	Jadomycin	*S. venezuelae*, *S. coelicolor*, *S. albus*	Wang et al. (2016b)
P21-cmt	*cmR*	N/A	*Pseudomonas putida*	Cumate	Jadomycin, actinorhodin	*S. albus*, *S. coelicolor*	Horbal et al. (2014); Li et al. (2018b)

[a]*tcp830* was created from elements of *ermEp*, Tn*10*, *tetO1*, and *tetO2*

streptomycetes; therefore, indigoidine synthetase is a valuable addition to the *Streptomyces* genetic tool kit (Novakova et al. 2010).

11.3.7 CRISPR/Cas

An alternative marker-less genetic engineering strategy for microorganisms is the Clustered Regulatory Interspaced Short Palindromic Repeats/CRISPR-associated protein (CRISPR/Cas) genome editing method. Recently, this system has been revolutionary for introducing multiple simultaneous genome edits in all kingdoms of life. Essentially, CRISPR/Cas is an adaptive immune defense against external DNA or RNA. So far, three different systems have been identified: type I, type II, and type III. Type I and type III systems utilize many Cas proteins to initiate cleavage of the target DNA, while type II systems use a single Cas9 endonuclease (Selle and Barrangou 2015). Over the past 5 years, several type II systems (CRISPR/Cas9) have been developed for *Streptomyces* spp. including *S. lividans*, *S. albus*, *S. coelicolor*, *S. viridochromogenes*, *S. roseosporus*, and *S. venezuelae* (Tao et al. 2018; Tong et al. 2015). These CRISPR/Cas9 methods can perform one round of genome editing in roughly half of the time required for older methods (Huang et al. 2015). One study involved the application of CRISPR/Cas9 for targeted chromosomal deletions ranging from approximately 20 bp to 30 kb in *S. lividans*, *S. albus*, and *S. viridochromogenes*. The designed pCRISPomyces plasmids harbor a Cas9 system from *Streptococcus pyogenes* that was codon optimized for *Streptomyces*. Additional features include a *BbsI*-flanked lacZ cassette for Golden Gate assembly, an *XbaI* site for the incorporation of editing templates, machinery for *E. coli* replication and conjugation, and a temperature-sensitive pSG5 replication origin. The pCRISPomyces system resulted in the deletion of the complete undecylprodigiosin (RED) gene cluster (31.4 kb) in *S. lividans* and an uncharacterized hybrid PKS-NRPS pathway *sshg_00040/sshg_00050* (13.2 kb) in *S. albus* (Okamoto et al. 2009). Additionally, several smaller deletions were successful in all three streptomycetes, such as *phpD* in *S. viridochromogenes*, *sshg_05713* in *S. albus*, and *actVA-ORF5* in *S. lividans* (Blodgett et al. 2007; Cerdeño et al. 2001; Olano et al. 2014). Although straightforward and versatile, this system presented a couple disadvantages. In some cases, modest editing efficiency was observed (i.e., 3 positive conjugants out of 14), and in all cases, conjugation efficiency was reduced by five- to tenfold compared to traditional methods, indicating toxicity from overexpression of the Cas9 endonuclease (Cobb et al. 2015).

Another experiment involved the creation of the CRISPR/Cas9 editing plasmid pKCcas9dO for various gene deletions and point mutations in *S. coelicolor*. The plasmid pKCcas9dO contains a target-specific guide RNA (sgRNA), a *cas9* gene codon-optimized for *S. coelicolor* (*Scocas9*), two homology-directed repair templates, machinery for *E. coli* replication and conjugation, and a temperature-sensitive pSG5 replication origin. *Scocas9* expression was under the control of *PtipA* while the target-specific sgRNA was under the control of the synthetic promoter, j23119. The pKCcas9dO system resulted in the successful deletion of individual genes and gene clusters such as ACT (21.3 kb), RED (31.6 kb), and the complete Ca^{2+}-dependent antibiotic (CDA) gene cluster (82.8 kb). Additionally, a 3 bp substitution in the *rspL* gene was generated, resulting in the amino acid mutation K88E which generated streptomycin resistance in *S. coelicolor*. Like the pCRISPomyces system, editing efficiency of pKCcas9dO was modest in some cases (i.e., 10 positive conjugants out of 100). Fortunately, at least 100 conjugants grew on each of the resulting transformation plates, indicating that *S. coelicolor* growth was not hindered by this gene editing system (Huang et al. 2015).

11.3.8 Heterologous Hosts

A variety of *Streptomyces* strains have been engineered for the heterologous expression of natural product enzymes. Features relevant to heterologous expression include genome size, number of endogenous pathways that may compete for metabolic flux of precursors or other cellular resources, relevant precursor pathways, and the presence of relevant accessory genes for certain types of BGCs. Some of the most well-established species include *S. coelicolor*, *S. albus*, *S. avermitilis*, and *S. lividans*. *S. coelicolor* A3(2), the wild-type strain, is known as the model strain of actinomycete genetics (Myronovskyi and Luzhetskyy 2019). Its chromosome contains over 40 gene clusters, expressing natural products including actinorhodin, prodigiosin, and coelimycin (Bentley et al. 2002). It is one of the earliest heterologous natural product expression hosts for actinomycete BGCs (McDaniel et al. 1993). A variety of genetic tools have been developed for *S. coelicolor* including a collection of replicating and integrating vectors, conjugation and protoplast transformation protocols, promoters, terminators, and reporter genes (Li et al. 2015b). Not only has *S. coelicolor* successfully expressed heterologous proteins from other streptomycetes, but it has also expressed proteins from distantly related actinomycetes such as *Actinoallomurus fulvus*, *Saccharopolyspora erythraea*, and *Kocuria flava* (Gomez-Escribano and Bibb 2012; Hopwood et al. 2000; Inahashi et al. 2018). The first attempt to engineer *S. coelicolor* successfully deleted the actinorhodin BGC from the parent strain which lacked endogenous plasmids and housed a mutation blocking the biosynthesis of the red pigment, undecylprodigiosin. The expression of the 6-deoxyerythronolide B synthase (DEBS) gene in the resulting engineered strain yielded more than 40 mg/L of 6-deoxyerythronolide B (6-dEB) (McDaniel et al. 1993). Following this study, *S. coelicolor* strains M1152 and M1154 were engineered by deleting the ACT, RED, CDA, and CPK gene clusters. Not only were these strains shown to produce 20–40 times more chloramphenicol than the wild type, but they have also been used to heterologously express 18 different natural product gene clusters (Gomez-Escribano and Bibb 2011, 2014). These gene clusters vary greatly in structure and include aminocoumarins, RiPPS, a type I polyketide, a polyketide-terpenoid hybrid, a peptidyl nucleoside, indolocarbazoles, a nonribosomal peptide, a shikimate-derived compound, a pyrrole-amide, and a fatty-acyl nucleoside. Reported yields ranged from 0.4 to 160 mg/L (Gomez-Escribano and Bibb 2014). Strain M1152 was further engineered to heterologously express type III polyketide synthase (PKS) genes. All three endogenous type III PKS genes, *gcs*, *srsA*, and *rppA*, were deleted resulting in *S. coelicolor* M1317. *S. coelicolor* M1317 efficiently produced a variety of type III PKSs (Thanapipatsiri et al. 2015). Thus, it is evident that genetically modified strains of *S. coelicolor* are reliable hosts for the heterologous expression of a variety of classes of secondary metabolite gene clusters.

S. lividans is closely related to *S. coelicolor*; however, the genetic manipulation of *S. lividans* is easier since it accepts methylated DNA (Myronovskyi and Luzhetskyy 2019). It also has low endogenous protease activity which makes it an excellent candidate for heterologous protein expression (Rebets et al. 2017). Recently, an ordered cosmid library of *S. lividans* TK24 was constructed which allowed for the engineering of its genome. Additionally, *S. lividans* TK24 is the most frequently utilized heterologous expression host (Shima et al. 1996) and is particularly useful for the expression of NRPS and RiPPs BGCs (Myronovskyi and Luzhetskyy 2019). Many reports detail the heterologous production of several peptide products in *S. lividans* including capreomycin, daptomycin, bottromycin, and labyrinthopeptins (Ahmed et al. 2020; Felnagle et al. 2007; Penn et al. 2006). Several studies have focused on the engineering of *S. lividans* for improved heterologous protein expression. For example, three different *S. lividans* strains were engineered from TK24 by deleting various gene clusters including ACT, RED, and CPK, along with introducing additional *attB* sites into the chromosome. *S. lividans* delA9 had 9 gene clusters deleted and 1 *attB* site added, *S. lividans* delA10 had 10 gene clusters deleted and 2 *attB* sites added, and *S. lividans* delA11 had 11 gene clusters deleted and 3 *attB* sites added. These mutations resulted in the production of griseorhodin in all three engineered strains. Additionally, these strains produced deoxycinnamycin approximately threefold more efficiently than TK24 (Ahmed et al. 2020). Further engineering efforts on *S. lividans* included the stepwise integration of the global regulatory genes *nusGsc* and *afsR*; the integration of two codon-optimized multidrug efflux pump genes, *lmrA* and *mdfA*; and the deletion of the negative regulatory gene, *wblA*. This engineered strain of *S. lividans* heterologously produced piericidin, dehydrorabelomycin, and actinomycin D at significantly higher titers than the parent strain (Peng et al. 2018). Thus, *S. lividans* is a versatile, genetically tractable, and efficient chassis for heterologous protein production.

S. venezuelae has the potential to become widely used as a host for the heterologous generation of natural products. It has several advantages including a relatively high growth rate, homogenous mycelium, efficient sporulation in liquid culture, and several genetic tools including CRISPR editing (Bush et al. 2013; Pullan et al. 2011). *S. venezuelae* produces an impressive 11 PPTases (Baltz 2016). Additionally, it has a relatively small genome (8.2 Mb), six rRNA clusters, and three MbtH homologues (and accessory gene for NRPS) (Bush et al. 2013; Pullan et al. 2011). *S. venezuelae* has successfully produced many aminoglycosides, macrolactams, and flavonoids heterologously (Kim et al. 2015). For example, it was used for the heterologous expression of the phospho-glycolipid nosokomycin A2 (Lopatniuk et al. 2014). Additionally, it was genetically modified to express a plant terpene synthase, bisabolene cyclase, which was codon optimized for *Streptomyces* to make the biofuel precursor bisabolone (Phelan et al. 2015). Compared to streptomycete hosts mentioned previously, *S. venezuelae* is not as widely studied. The wild type has not been engineered to improve heterologous production yields, and as a result, titers are often very low. Furthermore, the heterologous expression of PKSs and NRPSs has not been thoroughly investigated in *S. venezuelae* despite its large number of PPTase genes (Baltz 2016).

S. albus is one of the most extensively implemented hosts with numerous genetic tools. It has one of the smallest streptomycete genomes discovered to date, with a 6.8 Mb linear chromosome (Zaburannyi et al. 2014). Many thoroughly studied promoters, terminators, and reporter genes have been developed for *S. albus* (Myronovskyi et al. 2011). Additionally, it expresses 25 natural product enzymes that synthesize many compounds including alteramides, candicidins, mansouramycins, and paulomycins (Olano et al. 2014). *S. albus* G was the first strain of interest because it naturally produces the restriction endonuclease SalI, which prevents streptomycete plaque formation (Arrand et al. 1978; Chater and Wilde 1976). The isolated mutant, *S. albus* J1074, has a mutated, inactive SalI restriction modification system and is therefore susceptible to streptomycete plaque formation and protoplast regeneration (Baltz and Matsushima 1981). Many genetic systems have been heterologously expressed in strain J1074 including the steffimycin, fredericamycin, iso-migrastatin, thiocoraline, napyradiomycin, lanomycin, griseorhodin, greocycline, moenomycin, arancimycin, and holomycin pathways (Baltz 2010, 2016). Furthermore, many secondary metabolite gene clusters that were difficult to express in other *Streptomyces* species were successfully expressed in *S. albus* J1074 (Baltz 2010). For example, *S. albus* J1074 was utilized to express the novel antibacterial compound, tetramycin A, and the novel proteome inhibitors, clarepoxins A–F and landepoxins A and B. Of the five *Streptomyces* species tested, only *S. albus* J1074 was able to produce clarepoxins and landepoxins upon chromosomal integration of cryptic BGCs from the metagenome (Owen et al. 2015).

Many other actinomycetes have been used as hosts for the heterologous production of different compounds. For example, a strain of *Salinispora tropica* was modified for heterologous protein expression by deleting the indigenous salinosporamide BGCs and adding an *attB* site for gene cluster integration (Zhang et al. 2018). Additionally, *Saccharopolyspora erythraea* has been used to produce PK-derived compounds in large quantities (Rodriguez et al. 2003). *S. roseosporus* strains with the daptomycin (Dap) biosynthetic genes removed, and *S. fradiae* strains lacking tylactone (Tyl) production, have been used for the heterologous overexpression of cyclic peptides (Baltz 2010). Furthermore, *S. cinnamonensis* C730.1 and C730.7, which produce the polyketide monensin A, have been utilized to heterologously produce the type II polyketide tetracenomycin at a yield of over 4 g/L (Li et al. 2009a). This is the highest product titer of any heterologous expression host to date, so it is likely that the *S. cinnamonensis* genome will be sequenced in the near future (Baltz 2011). For the heterologous expression of BGCs from non-streptomycete actinobacteria, it may be advantageous to select

Table 11.5 Actinomycete heterologous hosts for secondary metabolite production

Actinomycete	Genome size (Mb)	Secondary metabolite BGCs	rRNA genes	PPTase genes	Native metabolites	Heterologous metabolites	Ref.
Streptomyces coelicolor	8.7	22	3	3	Actinorhodin, kalafungin, streptorubin B, CDA	Chloramphenicol, daptomycin, germicidin, flaviolin	Choi et al. (2019); Gomez-Escribano and Bibb (2011); Gomez-Escribano et al. (2012); Thanapipatsiri et al. (2015)
Streptomyces lividans	8.3	25	18	Unknown	Actinorhodin, undecylprodigiosin, CDA	Isomigrastatin, thiocoraline, capreomycin, viomycin	Ahmed et al. (2020); Baltz (2010); Rückert et al. (2015)
Streptomyces albus	6.8	22	7	3	Fredericamycin	Steffimycin, kinamycin, isomigrastatin, thiocoraline	Baltz (2010, 2016); Rodríguez Estévez et al. (2020); Zaburannyi et al. (2014)
Streptomyces avermitilis	9.0	25	18	6	Avermictin, oligomycin, filipin, carotenoid	Isomigrastatin, streptomycin, cephamycin, pladienolide	Baltz (2010, 2016); Omura et al. (2001); Rodríguez Estévez et al. (2020)
Streptomyces ambofaciens	8.3	25	17	4	Spiramycin, netropsin	A54145	Aigle et al. (2014); Alexander et al. (2010); Bunet et al. (2014)
Streptomyces venezuelae	8.2	30	21	11	Chloramphenicol, jadomycin, pikromycin	Nosokomycin, bisabolene	He et al. (2016); Kim et al. (2020); Lee et al. (2020a); Phelan et al. (2015, 2017)
Streptomyces toyocaensis	7.3	26	3	6	A47934	Glucosyl-A47934	Kwun and Hong (2014); Solenberg et al. (1997)

(continued)

Table 11.5 (continued)

Actinomycete	Genome size (Mb)	Secondary metabolite BGCs	rRNA genes	PPTase genes	Native metabolites	Heterologous metabolites	Ref.
Streptomyces fradiae	7.7	27	18	3	Tylactone, A54145	A54145E, A54145D, CB-182,333, CB-182,350	Alexander et al. (2011); Grumaz et al. (2017)
Streptomyces filamentosus	7.8	25	24	4	Arylomycin, daptomycin,	A54145	Alexander et al. (2010); Baltz (2016); Lee et al. (2020a)
Streptomyces cinnamonensis	Unknown	Unknown	Unknown	Unknown	Monensin	Tetracenomycins A2 and C	Baltz (2016); Li et al. (2009a)
Saccharopolyspora erythraea	8.2	25	12	1	Erythromycin	Spinosad	Huang et al. (2016); Udwary et al. (2007)
Salinispora tropica	5.2	14	6	5	Salinosporamide	Thiolactomycin	Udwary et al. (2007); Zhang et al. (2018)

phylogenetically closer genera as heterologous hosts (Myronovskyi and Luzhetskyy 2019). With this in mind, other actinobacteria have recently been the subjects of genetic engineering for heterologous protein expression such as *Amycolatopsis japonicum*, which was used to generate the natural product brasilicardin (Table 11.5) (Schwarz et al. 2018).

11.4 Eliciting Production from Native Hosts

11.4.1 Coculture and Small Molecule Elicitors

Although the heterologous expression of captured BGCs can allow for the discovery of new natural products, as laid out above, the work to capture and express in a heterologous host is not always trivial. Thus, an alternative strategy has involved eliciting the production of cryptic clusters in their native hosts by optimizing culture conditions and/or culture additives. Secondary metabolites have a vast regulatory network that mediates their expression, as production of secondary metabolites is metabolically burdensome to the cell. There are a whole range of stimuli that can elicit production, many of which are typically not mimicked in culturing conditions. Some of the earliest coculture experiments provided evidence for cryptic metabolites before sequences of many *Streptomyces* spp. were elucidated. One key early example from Beppu and coworkers monitored a matrix of 76 strains of *Streptomyces* spp. in a broad binary coculture screen. Among these, 34% of streptomycetes induced antibiotic production and/or sporulation in neighboring species (Ueda et al. 2000). Additional studies with one of the interacting pairs, *Streptomyces griseus* and *Streptomyces tanashiensis*, revealed the stimulatory siderophore molecule, desferrioxamine. This suggested that secondary metabolism can be induced to mediate nutrient acquisition, such as iron. A similar phenomenon was observed when *Streptomyces lividans* was cocultured with *Myxococcus xanthus*, leading to the production of the siderophore mxyochelin, which caused iron deficiency in *Streptomyces lividans*. Iron-restrictive conditions elicited the production of the blue pigment molecule, actinorhodin (Lee et al. 2020b).

In several other examples, small molecules may be involved in signaling between metabolites to mediate competition and/or interspecies signaling. The blue pigment actinorhodin from *Streptomyces lividans* has been used extensively in coculture studies, in part due to its easily identifiable output. Studies of actinorhodin have observed activation under a broad range of environmental stressors. One of the earliest experiments involved screening supernatants of 405 species of actinomycetes. The thiazole/oxazole eliciting compound, goadsporin from *Streptomyces* TP-A054, was shown to elicit actinorhodin expression. Goadsporin was then shown to have wide-ranging elicitor properties toward secondary metabolism. Against 42 randomly selected strains, goadsporin elicited pigment production in 20 strains and sporulation in 32 strains. Strikingly, goadsporin was a relatively potent *Streptomyces*-specific antibiotic, showing no bioactivity toward *Proteobacteria*, *Firmicutes*, or fungi. Conversely, it displayed minimal inhibitory concentrations of 0.2, 3.2, and 6.4 μg/mL against *S. scabies*, *S. coelicolor*, and *S. lividans*, respectively. Thus, at subinhibitory concentrations, it served as a signaling molecule to elicit sporulation and antibiotic production, whereas at higher concentrations it served as an antibiotic, a phenomenon termed hormesis (Onaka et al. 2001).

11.4.2 Selection for Ribosomal and RNA Polymerase Mutation

One of the first attempts to elicit cryptic metabolite production was inspired by the fact that many antibiotics target the ribosome (e.g., streptomycin, kanamycin, and erythromycin). As such, ribosomal mutants are one of the major mechanisms by which resistance is conferred, and correspondingly, ribosomal mutations are selected when an organism is exposed to antibiotics that inhibit the ribosome. Compared to strain improvement through metabolic engineering, exposure to an antibiotic that affects the ribosome followed by subsequent selection is a less laborious strategy for increasing secondary metabolite production in the cell. One application of this method began with the observation that the streptomycin-resistant species, *Streptomyces lividans*, produces actinorhodin, which is cryptic in the parent strain. Thus, Ochi and coworkers pursued further investigation of ribosomal mutations and discovered that point mutations to *rpsL*, which encodes for S12 ribosomal protein, resulted in higher actinorhodin production (Shima et al. 1996). Analogous mutations in S12 were extended to additional *Streptomyces* sp. including *S. coelicolor*, *S. antibioticus*, *S. chattanoogensis*, and *S. lavendulae*, which led to ~5–6-fold overproduction of actinomycin, fredericamycin, and formycin, respectively (Hosaka et al. 2009). Ochi and coworkers expanded the selection to other antibiotics including inhibitors of RNA polymerase (RNAP). Among these experiments, exposure to rifampicin, an inhibitor of RNAP, was one of the more successful approaches. Typically, when exposed to rifampicin, mutation of the β-subunit of RNAP, *rpoB* is conferred. Rifampicin-resistant mutations in *S. lividans* resulted in an overproduction of actinorhodin of up to tenfold and undecylprodigiosin of up to fivefold. Additionally, a calcium-dependent antibiotic and three cryptic metabolites with extremely low levels of production in the wild-type strain were overexpressed (Hu et al. 2002). In the most extensive application of this approach, Ochi and coworkers screened 1068 soil actinomycete isolates. They demonstrated that 43% of antibiotic-producing streptomycetes and 6% of non-antibiotic-producing streptomycetes acquired the ability to synthesize antibacterial compounds after a selection step that generated spontaneous rifampicin or streptomycin mutations. Investigation into one of these isolates led to the identification of the new cryptic antibiotic, piperidamycin (Tanaka et al. 2017).

11.4.3 High-Throughput Elicitor Screens (HiTES)

Molecules that mediate nutrient competition (siderophores), and antibiotic resistance mechanisms that reveal cryptic production of BGCs, small molecules, and particularly antibiotics, are often major triggers to elicit secondary metabolism. Molecules like goadsporin have high levels of antibiotic activity at high concentrations but are potent elicitors at low concentrations. However, this mode of activity is clearly not generalizable, since most antibiotics tested, including streptomycin, kanamycin, thiostrepton, bacitracin, and gramicidin D, do not directly stimulate secondary metabolism of cryptic pathways. It is extremely challenging to predict which molecules may have elicitor properties and even more challenging to target a mechanism a priori that might reasonably activate cryptic BGCs. To overcome this challenge, high-throughput screening, often termed HiTES (high-throughput elicitor screens), is often employed to activate cryptic pathways.

The first of these studies was performed by Nodwell and coworkers, selecting *Streptomyces coelicolor* as an initial system. Like *Streptomyces lividans*, *Streptomyces coelicolor* synthesizes the blue pigment actinorhodin and the red prodiginine pigments, which serve as rapid readouts for secondary metabolite production, although *Streptomyces coelicolor* harbors many cryptic metabolites as well that are more challenging to monitor. Utilizing a screen of 30,569 molecules from the Canadian Compound Collection, compounds were screened to determine their ability to elicit actinorhodin or prodiginine production. From this screen, 112 compounds were identified as hits, and 4 related compounds, termed the ARC2 series, were used for further studies. While the ARC2 compounds bear little resemblance to naturally produced bacterial antibiotics, they somewhat resemble triclosan, which also induces actinorhodin biosynthesis at subinhibitory concentrations. The ARC2 series of compounds, like triclosan, appeared to target secondary metabolism as well as fatty acid metabolism likely because actinorhodin and prodiginine are both polyketide secondary metabolites that utilize methylmalonyl-CoA as precursors. Upon examining strains treated with ARC2 compounds, they elicited the production of additional secondary metabolites beyond actinorhodin. Due to the ARC2 compounds' potent elicitor properties in *Streptomyces coelicolor*, the authors treated other actinomycetes with ARC2 compounds and found them to be potent general elicitors of cryptic natural product production in *Kutzneria* sp., *S. pistinaespiralis*, and *S. peucetius*. A derivative of ARC2 was later applied to 50 additional species of *Streptomyces*, causing the induction of at least 1 compound in each strain for a total of 216 cryptic metabolites (Craney et al. 2012).

11.4.4 Refactoring

Refactoring is an effective method for the activation of silent genes or gene clusters to produce valuable natural products. For example, both the spectinabilin and lazarimide gene clusters were activated in *S. lividans* and *S. albus*, respectively (Montiel et al. 2015; Shao et al. 2013). Spectinabilin is a nitroaryl-substituted polyketide natively produced by *S. spectabilis* and *S. orinoci* with antimalarial and antiviral properties (Isaka et al. 2002). Heterologous expression studies confirmed that the gene clusters, SPN and NOR from *S. spectabilis* and *S. orinoci*, respectively, are regulated differently (Choi et al. 2010). Real-time PCR analysis revealed that the NOR cluster is expressed extremely inefficiently in a non-native environment. Nine strong constitutive promoters were used to initiate transcription of NOR while excluding *norD*, which encodes the transcriptional repressor, and *norG*, which encodes the first enzyme in the pathway. *NorG* was controlled by a separate inducible promoter, *PnitA*. The refactored biosynthetic pathway was expressed in *S. lividans* with a spectinabilin yield of 105 μg/L (Shao et al. 2013). In another experiment, the lazarimide gene cluster, LZR, was found via PCR screening of environmental cosmid libraries (Ryan and Drennan 2009). Novel tailoring enzymes expressed by genes within the LZR cluster were revealed, which implied that one or more novel indolotryptoline compounds could be encoded by LZR*lzr*. Thus, LZR was activated using synthetic promoters recognized by HrdB, which were previously developed for the homologous recombination-mediated activation of silent BGCs (Montiel et al. 2015; Seghezzi et al. 2011). Three bidirectional promoters, *P1*, *P2*, and *P3*, along with a unidirectional promoter, *P4*, were predicted to control the expression of LZR. Through a series of single insertions, *P1*, *P2*, and *P3* were replaced with the designed synthetic promoters and heterologously expressed in *S. albus*. Two of the constructs produced chromopyrrolic and indolocarbazole intermediates, while another construct produced an indolocarbazole intermediate, rather than the expected indolotryptoline intermediate. Further probing of the gene responsible for the missing step, *lzrX1*, revealed that it was nonfunctional due to one deleted base. Correcting this issue led to the production of new compounds based on lazarimides B and C (Montiel et al. 2015). The fully engineered pathway, with all four promoters replaced by the synthetic cassettes, produced one additional product, lazarimide A. Therefore, the lazarimide biosynthetic pathway was activated in *S. albus* to produce novel compounds including lazarimide A and derivatives of lazarimides B and C (Montiel et al. 2015; Ryan 2011).

11.4.5 Overexpression of Regulatory Elements

The manipulation of transcriptional regulators has also been effective for eliciting secondary metabolite production. Transcriptional regulators in the *Streptomyces* antibiotic regulatory protein (SARP) family possess a characteristic N-terminal OmpR-type winged helix-turn-helix domain. They are widely responsible for activating the transcription of genes associated with secondary metabolism (Bush et al. 2013). Additionally, TetR-like repressors that regulate secondary metabolite production have been observed, such as AveT in *Streptomyces avermitilis*. Upon overexpression of AveT, the AveR activator was turned on, which resulted in upregulation of the avermectin BGC and thus, increased titers of avermectin (Liu et al. 2015). The overexpression of activators and regulators has been shown to mediate the expression of cryptic natural products. For example, overexpression of the *papR2* SARP-type regulator from *Streptomyces pristinaespiralis* in *Streptomyces lividans* activated the silent RED gene cluster. Additionally, *papR2* activated the amicetin/plicaetin gene cluster in an environmental isolate from Indonesia, *Streptomyces* sp. SHP22-7 (Krause et al. 2020).

Finally, A-factors are hormone-like signaling molecules which function similarly to the N-acyl homoserine lactones (AHLs) in gram-negative bacteria. In streptomycetes, A-factors are typically γ-butyrolactones, although other small molecules exist that stimulate a similar function. A-factors activate downstream receptors which undergo cascades to induce the production of secondary metabolites. The first discovered signaling molecule in this category induced streptomycin production and morphological differentiation through a regulatory cascade involving the receptor ArpA, the pleiotropic regulator AdpA, and the CSR activator in *S. griseus* (Horinouchi and Beppu 2007). Thus, manipulating these small molecule-mediated bacterial hormone-like signaling pathways is of significant interest for the discovery of new cryptic natural products, especially given the success of elicitor screens through HiTES (Kong et al. 2019).

Additionally, several of the regulatory elements discussed above (such as A-factors and SARPS) have been suggested as potential elements for use as biosensors for natural product discovery (Li et al. 2019a).Transcription factor-based biosensors have been developed in some limited capacity to mine for natural products including the TetR-like repressor system (Kasey et al. 2018). Systems for lactam discovery using the NitR transcription factor (Zhang et al. 2017) as well as macrolide discovery using the MphR transcription factor (Kasey et al. 2018) have been developed for such applications.

11.5 Natural Product Analogs

The above strategies all seek to discover new natural products encoded in actinomycetes. However, because some inherent promiscuity exists in many BGCs, an alternative approach has involved the manipulation of precursor availability. This method was inspired by nature, in which a series of natural products have been isolated from the incorporation of different building blocks to varying degrees due to inherent promiscuity. For example, the avermectin PKS incorporates both isobutyryl-CoA and 2-methylbutyryl-CoA to generate avermectin B_{1b} and B_{1a}, respectively. Alternately, in the salinosporamide PKS, the major product is made from the incorporation of chloroethylmalonyl-CoA, resulting in salinosporamide A. However the minor products, salinosporamides B and C, are also produced from ethylmalonyl-CoA and methylmalonyl-CoA, respectively (Eustáquio et al. 2009). For engineering purposes, two approaches have been used. The first is mutasynthesis, a portmanteau of "mutational biosynthesis," wherein a precursor pathway was knocked out and a new precursor was supplemented, either exogenously by addition of a chemical to the media or endogenously via addition of a new precursor pathway. The second is "precursor-directed biosynthesis," which was similar in concept, but did not involve generating knockouts (Kennedy 2008). One of the earliest examples of mutasynthesis involved the perturbation of the avermectin pathway. Dutton and coworkers were able to generate several avermectin analogs with a wide range of C25 substitutions. To accomplish this, a mutant strain deficient in the BKD pathway which generates the precursors 2-methylbutyryl-CoA and isobutyryl-CoA from isoleucine and valine, respectively, were fed exogenous small carboxylic acids that could instead serve as PKS starter units. Over 36 analogs of avermectin were produced using this approach, notably including doramectin, a cyclohexyl derivative of avermectin approved for veterinary use (Dutton et al. 1991). Later, a shikimate-derived cyclohexanoyl-CoA pathway was integrated into *Streptomyces avermitilis* with the BKD mutation, resulting in the formation of doramectin without cyclohexanoic acid supplementation (Cropp et al. 2000). Similar approaches have been used to create analogs of rapamycin "rapalogs" by knocking out starter unit biosynthesis (Kendrew et al. 2013). In addition to starter units, extender unit biosynthesis has been altered. For example, a fluoro-derivative of salinosporamide A was generated via expressing genes from *Streptomyces cattleya*, a unique bacterium containing the fluoroacetate biosynthetic pathway. By replacing a chlorinase gene with the fluorinase gene, an analog of salinosporamide A with fluorine replacement was generated (Eustáquio et al. 2010).

11.6 Conclusions and Future Perspectives

While actinomycetes have historically been vital and prolific sources of bioactive natural products, we have already found the low hanging fruit from the traditional "grind and find" approaches of bioactivity-based screens and fractionation. However, the post-genomic era has revealed that actinomycetes, especially *Streptomyces* spp., remain an untapped resource, since many natural products are not produced under standard culture conditions. While not trivial to engineer, a steady development of genetic tools for these bacteria has led to new developments in the capture and heterologous expression of natural products. The discovery of new promoters, reporters, and regulatory elements has expanded the viability of expressing these cryptic clusters in both heterologous and native hosts. Furthermore, elicitor screens and ribosomal mutagenesis have facilitated cryptic metabolite formation through small molecules without genetic manipulations. This has been accomplished by hijacking signaling pathways or inducing mutations that alter secondary metabolite production. Finally, with the tools developed, known biosynthetic pathways have been manipulated through the alteration of their biosynthesis or precursor pool, in both native and heterologous hosts. With this explosion of synthetic biology strategies, we indeed have an extensive toolkit to continue to mine the metabolic giftedness of actinomycetes for the discovery and development of new drug candidates.

References

Abbasi MN, Fu J, Bian X, Wang H, Zhang Y, Li A (2020) Recombineering for genetic engineering of natural product biosynthetic pathways. Trends Biotechnol 38:715–728

Agrawal P, Khater S, Gupta M, Sain N, Mohanty D (2017) RiPPMiner: a bioinformatics resource for deciphering chemical structures of RiPPs based on prediction of cleavage and cross-links. Nucleic Acids Res 45:W80–W88

Ahmed Y, Rebets Y, Estévez MR, Zapp J, Myronovskyi M, Luzhetskyy A (2020) Engineering of Streptomyces lividans for heterologous expression of secondary metabolite gene clusters. Microb Cell Fact 19:5

Aigle B, Lautru S, Spiteller D, Dickschat JS, Challis GL, Leblond P, Pernodet J-L (2014) Genome mining of Streptomyces ambofaciens. J Ind Microbiol Biotechnol 41:251–263

Alexander DC, Rock J, He X, Brian P, Miao V, Baltz RH (2010) Development of a genetic system for combinatorial biosynthesis of lipopeptides in Streptomyces fradiae and heterologous expression of the A54145 biosynthesis gene cluster. Appl Environ Microbiol 76:6877–6887

Alexander DC, Rock J, Gu J-Q, Mascio C, Chu M, Brian P, Baltz RH (2011) Production of novel lipopeptide antibiotics related to A54145 by Streptomyces fradiae mutants blocked in biosynthesis of modified amino acids and assignment of lptJ, lptK and lptL gene functions. J Antibiot 64:79–87

Arrand JR, Myers PA, Roberts RJ (1978) A new restriction endonuclease from Streptomyces albus G. J Mol Biol 118:127–135

Bachmann BO, Van Lanen SG, Baltz RH (2014) Microbial genome mining for accelerated natural products discovery: is a renaissance in the making? J Ind Microbiol Biotechnol 41:175–184

Bailey CR, Bruton CJ, Butler MJ, Chater KF, Harris JE, Hopwood DA (1986) Properties of in vitro recombinant derivatives of pJV1, a multi-copy plasmid from Streptomyces phaeochromogenes. J Gen Microbiol 132:2071–2078

Baltz RH (2010) Streptomyces and Saccharopolyspora hosts for heterologous expression of secondary metabolite gene clusters. J Ind Microbiol Biotechnol 37:759–772

Baltz RH (2011) Strain improvement in actinomycetes in the postgenomic era. J Ind Microbiol Biotechnol 38:657–666

Baltz RH (2012) Streptomyces temperate bacteriophage integration systems for stable genetic engineering of actinomycetes (and other organisms). J Ind Microbiol Biotechnol 39:661–672

Baltz RH (2016) Genetic manipulation of secondary metabolite biosynthesis for improved production in Streptomyces and other actinomycetes. J Ind Microbiol Biotechnol 43:343–370

Baltz RH, Matsushima P (1981) Protoplast fusion in Streptomyces: conditions for efficient genetic recombination and cell regeneration. J Gen Microbiol 127:137–146

Bentley SD, Chater KF, Cerdeno-Tárraga AM, Challis GL, Thomson NR, James KD, Harris DE, Quail MA, Kieser H, Harper D et al (2002) Complete genome sequence of the model actinomycete Streptomyces coelicolor A3(2). Nature 417:141–147

Bérdy J (2005) Bioactive microbial metabolites. J Antibiot 58:1–26

Berini F, Marinelli F, Binda E (2020) Streptomycetes: attractive hosts for recombinant protein production. Front Microbiol 11:1958

Bibb MJ, Freeman RF, Hopwood DA (1977) Physical and genetical characterisation of a second sex factor, SCP2, for Streptomyces coelicolor A3(2). Mol Gen Genet 154:155–166

Bibb MJ, Ward JM, Hopwood DA (1978) Transformation of plasmid DNA into Streptomyces at high frequency. Nature 274:398–400

Bibb MJ, Janssen GR, Ward JM (1985) Cloning and analysis of the promoter region of the erythromycin resistance gene (ermE) of Streptomyces erythraeus. Gene 38:215–226

Bierman M, Logan R, O'Brien K, Seno ET, Rao RN, Schoner BE (1992) Plasmid cloning vectors for the conjugal transfer of DNA from Escherichia coli to Streptomyces spp. Gene 116:43–49

Bilyk O, Sekurova ON, Zotchev SB, Luzhetskyy A (2016) Cloning and heterologous expression of the grecocycline biosynthetic gene cluster. PLoS One 11:e0158682

Blin K, Shaw S, Kautsar SA, Medema MH, Weber T (2020) The antiSMASH database version 3: increased taxonomic coverage and new query features for modular enzymes. Nucleic Acids Res 49(D1):D639–D643

Blodgett JAV, Thomas PM, Li G, Velasquez JE, van der Donk WA, Kelleher NL, Metcalf WW (2007) Unusual transformations in the biosynthesis of the antibiotic phosphinothricin tripeptide. Nat Chem Biol 3:480–485

Boccard F, Smokvina T, Pernodet JL, Friedmann A, Guérineau M (1989) The integrated conjugative plasmid pSAM2 of Streptomyces ambofaciens is related to temperate bacteriophages. EMBO J 8:973–980

Bunet R, Riclea R, Laureti L, Hôtel L, Paris C, Girardet J-M, Spiteller D, Dickschat JS, Leblond P, Aigle B (2014) A single Sfp-type phosphopantetheinyl transferase plays a major role in the biosynthesis of PKS and NRPS derived metabolites in Streptomyces ambofaciens ATCC23877. PLoS One 9:e87607

Bush MJ, Bibb MJ, Chandra G, Findlay KC, Buttner MJ (2013) Genes required for aerial growth, cell division, and chromosome segregation are targets of WhiA before sporulation in Streptomyces venezuelae. MBio 4:e00684–e00613

Cerdeño AM, Bibb MJ, Challis GL (2001) Analysis of the prodiginine biosynthesis gene cluster of Streptomyces coelicolor A3(2): new mechanisms for chain initiation and termination in modular multienzymes. Chem Biol 8:817–829

Chater KF, Carter AT (1979) A new, wide host-range, temperate bacteriophage (R4) of Streptomyces and its interaction with some restriction-modification systems. J Gen Microbiol 115:431–442

Chater KF, Wilde LC (1976) Restriction of a bacteriophage of Streptomyces albus G involving endonuclease SalI. J Bacteriol 128:644–650

Chen C, Zhao X, Jin Y, Zhao ZK, Suh J-W (2014) Rapid construction of a Bacterial Artificial Chromosomal (BAC) expression vector using designer DNA fragments. Plasmid 76:79–86

Chiu ML, Folcher M, Griffin P, Holt T, Klatt T, Thompson CJ (1996) Characterization of the covalent binding of thiostrepton to a thiostrepton-induced protein from Streptomyces lividans. Biochemistry 35:2332–2341

Chiu ML, Folcher M, Katoh T, Puglia AM, Vohradsky J, Yun BS, Seto H, Thompson CJ (1999) Broad spectrum thiopeptide recognition specificity of the Streptomyces lividans TipAL protein and its role in regulating gene expression. J Biol Chem 274:20578–20586

Choi YS, Johannes TW, Simurdiak M, Shao Z, Lu H, Zhao H (2010) Cloning and heterologous expression of the spectinabilin biosynthetic gene cluster from Streptomyces spectabilis. Mol BioSyst 6:336–338

Choi S, Nah H-J, Choi S, Kim E-S (2019) Heterologous expression of daptomycin biosynthetic gene cluster via Streptomyces artificial chromosome vector system. J Microbiol Biotechnol 29: 1931–1937

Cobb RE, Wang Y, Zhao H (2015) High-efficiency multiplex genome editing of Streptomyces species using an engineered CRISPR/Cas system. ACS Synth Biol 4:723–728

Combes P, Till R, Bee S, Smith MCM (2002) The Streptomyces genome contains multiple pseudo-attB sites for the (phi)C31-encoded site-specific recombination system. J Bacteriol 184:5746–5752

Craney A, Ozimok C, Pimentel-Elardo SM, Capretta A, Nodwell JR (2012) Chemical perturbation of secondary metabolism demonstrates important links to primary metabolism. Chem Biol 19: 1020–1027

Cropp TA, Wilson DJ, Reynolds KA (2000) Identification of a cyclohexylcarbonyl CoA biosynthetic gene cluster and application in the production of doramectin. Nat Biotechnol 18:980–983

Dangel V, Westrich L, Smith MCM, Heide L, Gust B (2010) Use of an inducible promoter for antibiotic production in a heterologous host. Appl Microbiol Biotechnol 87:261–269

Dejong CA, Chen GM, Li H, Johnston CW, Edwards MR, Rees PN, Skinnider MA, Webster ALH, Magarvey NA (2016) Polyketide and nonribosomal peptide retro-biosynthesis and global gene cluster matching. Nat Chem Biol 12:1007–1014

Denis F, Brzezinski R (1992) A versatile shuttle cosmid vector for use in Escherichia coli and actinomycetes. Gene 111:115–118

Dutton CJ, Gibson SP, Goudie AC, Holdom KS, Pacey MS, Ruddock JC, Bu'Lock JD, Richards MK (1991) Novel avermectins produced by mutational biosynthesis. J Antibiot 44:357–365

Eustáquio AS, McGlinchey RP, Liu Y, Hazzard C, Beer LL, Florova G, Alhamadsheh MM, Lechner A, Kale AJ, Kobayashi Y et al (2009) Biosynthesis of the salinosporamide A polyketide synthase substrate chloroethylmalonyl-coenzyme A from S-adenosyl-L-methionine. Proc Natl Acad Sci U S A 106:12295–12300

Eustáquio AS, O'Hagan D, Moore BS (2010) Engineering fluorometabolite production: fluorinase expression in Salinispora tropica Yields Fluorosalinosporamide. J Nat Prod 73:378–382

Farkašovská J, Godány A (2012) Analysis of the site-specific integration system of the Streptomyces aureofaciens phage μ1/6. Curr Microbiol 64:226–233

Fayed B, Younger E, Taylor G, Smith MCM (2014) A novel Streptomyces spp. integration vector derived from the S. venezuelae phage, SV1. BMC Biotechnol 14:51

Fayed B, Ashford DA, Hashem AM, Amin MA, El Gazayerly ON, Gregory MA, Smith MCM (2015) Multiplexed integrating plasmids for engineering of the erythromycin gene cluster for expression in Streptomyces spp. and combinatorial biosynthesis. Appl Environ Microbiol 81: 8402–8413

Felnagle EA, Rondon MR, Berti AD, Crosby HA, Thomas MG (2007) Identification of the biosynthetic gene cluster and an additional gene for resistance to the antituberculosis drug capreomycin. Appl Environ Microbiol 73:4162–4170

Feng Z, Kim JH, Brady SF (2010) Fluostatins produced by the heterologous expression of a TAR reassembled environmental DNA derived type II PKS gene cluster. J Am Chem Soc 132: 11902–11903

Fernández-Martínez LT, Bibb MJ (2014) Use of the meganuclease I-SceI of Saccharomyces cerevisiae to select for gene deletions in actinomycetes. Sci Rep 4:7100

Fogg PCM, Haley JA, Stark WM, Smith MCM (2017) Genome integration and excision by a new Streptomyces bacteriophage, ϕJoe. Appl Environ Microbiol 83

Gibson DG, Young L, Chuang R-Y, Venter JC, Hutchison CA, Smith HO (2009) Enzymatic assembly of DNA molecules up to several hundred kilobases. Nat Methods 6:343–345

Gomez-Escribano JP, Bibb MJ (2011) Engineering Streptomyces coelicolor for heterologous expression of secondary metabolite gene clusters. Microb Biotechnol 4:207–215

Gomez-Escribano JP, Bibb MJ (2012) Streptomyces coelicolor as an expression host for heterologous gene clusters. Methods Enzymol 517:279–300

Gomez-Escribano JP, Bibb MJ (2014) Heterologous expression of natural product biosynthetic gene clusters in Streptomyces coelicolor: from genome mining to manipulation of biosynthetic pathways. J Ind Microbiol Biotechnol 41:425–431

Gomez-Escribano JP, Song L, Fox DJ, Yeo V, Bibb MJ, Challis GL (2012) Structure and biosynthesis of the unusual polyketide alkaloid coelimycin P1, a metabolic product of the cpk gene cluster of Streptomyces coelicolor M145. Chem Sci 3:2716

Gonzalez-Quiñonez N, López-García MT, Yagüe P, Rioseras B, Pisciotta A, Alduina R, Manteca Á (2016) New ΦBT1 site-specific integrative vectors with neutral phenotype in Streptomyces. Appl Microbiol Biotechnol 100:2797–2808

Gregory MA, Till R, Smith MCM (2003) Integration site for Streptomyces phage phiBT1 and development of site-specific integrating vectors. J Bacteriol 185:5320–5323

Grumaz C, Vainshtein Y, Kirstahler P, Luetz S, Kittelmann M, Schroer K, Eggimann FK, Czaja R, Vogel A, Hilberath T et al (2017) Draft genome sequences of three actinobacteria strains presenting new candidate organisms with high potentials for specific P450 cytochromes. Genome Announc 5

Haydock SF, Dowson JA, Dhillon N, Roberts GA, Cortes J, Leadlay PF (1991) Cloning and sequence analysis of genes involved in erythromycin biosynthesis in Saccharopolyspora erythraea: sequence similarities between EryG and a family of S-adenosylmethionine-dependent methyltransferases. Mol Gen Genet 230:120–128

He J, Sundararajan A, Devitt NP, Schilkey FD, Ramaraj T, Melançon CE (2016) Complete genome sequence of Streptomyces venezuelae ATCC 15439, producer of the methymycin/pikromycin family of Macrolide antibiotics, using PacBio technology. Genome Announc 4

Helfrich EJN, Ueoka R, Dolev A, Rust M, Meoded RA, Bhushan A, Califano G, Costa R, Gugger M, Steinbeck C et al (2019) Automated structure prediction of trans-acyltransferase polyketide synthase products. Nat Chem Biol 15:813–821

Herai S, Hashimoto Y, Higashibata H, Maseda H, Ikeda H, Omura S, Kobayashi M (2004) Hyper-inducible expression system for streptomycetes. Proc Natl Acad Sci U S A 101:14031–14035

Herrmann S, Siegl T, Luzhetska M, Petzke L, Jilg C, Welle E, Erb A, Leadlay PF, Bechthold A, Luzhetskyy A (2012) Site-specific recombination strategies for engineering actinomycete genomes. Appl Environ Microbiol 78:1804–1812

Holmes DJ, Caso JL, Thompson CJ (1993) Autogenous transcriptional activation of a thiostrepton-induced gene in Streptomyces lividans. EMBO J 12:3183–3191

Hopwood DA, Kieser T, Bibb M, Chater K (2000) Practical Streptomyces genetics | NHBS Academic & Professional Books

Horbal L, Fedorenko V, Luzhetskyy A (2014) Novel and tightly regulated resorcinol and cumate-inducible expression systems for Streptomyces and other actinobacteria. Appl Microbiol Biotechnol 98:8641–8655

Horinouchi S, Beppu T (2007) Hormonal control by A-factor of morphological development and secondary metabolism in Streptomyces. Proc Jpn Acad Ser B Phys Biol Sci 83:277–295

Hosaka T, Ohnishi-Kameyama M, Muramatsu H, Murakami K, Tsurumi Y, Kodani S, Yoshida M, Fujie A, Ochi K (2009) Antibacterial discovery in actinomycetes strains with mutations in RNA polymerase or ribosomal protein S12. Nat Biotechnol 27:462–464

Hou L, Wang S, Huang H, Li H, Wang W, Li W (2018) Generation of methylated violapyrones with improved anti-influenza A virus activity by heterologous expression of a type III PKS gene in a marine Streptomyces strain. Bioorg Med Chem Lett 28:2865–2868

Hu H, Zhang Q, Ochi K (2002) Activation of antibiotic biosynthesis by specified mutations in the rpoB gene (encoding the RNA polymerase beta subunit) of Streptomyces lividans. J Bacteriol 184:3984–3991

Huang H, Zheng G, Jiang W, Hu H, Lu Y (2015) One-step high-efficiency CRISPR/Cas9-mediated genome editing in Streptomyces. Acta Biochim Biophys Sin Shanghai 47:231–243

Huang J, Yu Z, Li M-H, Wang J-D, Bai H, Zhou J, Zheng Y-G (2016) High level of Spinosad production in the heterologous host Saccharopolyspora erythraea. Appl Environ Microbiol 82: 5603–5611

Inahashi Y, Shiraishi T, Také A, Matsumoto A, Takahashi Y, Ōmura S, Kuzuyama T, Nakashima T (2018) Identification and heterologous expression of the actinoallolide biosynthetic gene cluster. J Antibiot 71:749–752

Isaka M, Jaturapat A, Kramyu J, Tanticharoen M, Thebtaranonth Y (2002) Potent in vitro antimalarial activity of metacycloprodigiosin isolated from Streptomyces spectabilis BCC 4785. Antimicrob Agents Chemother 46:1112–1113

Ji C-H, Kim H, Kang H-S (2019) Synthetic inducible regulatory systems optimized for the modulation of secondary metabolite production in Streptomyces. ACS Synth Biol 8:577–586

Jiang W, Zhu TF (2016) Targeted isolation and cloning of 100-kb microbial genomic sequences by Cas9-assisted targeting of chromosome segments. Nat Protoc 11:960–975

Jiang W, Zhao X, Gabrieli T, Lou C, Ebenstein Y, Zhu TF (2015) Cas9-assisted targeting of CHromosome segments CATCH enables one-step targeted cloning of large gene clusters. Nat Commun 6:8101

Kasey CM, Zerrad M, Li Y, Cropp TA, Williams GJ (2018) Development of transcription factor-based designer macrolide biosensors for metabolic engineering and synthetic biology. ACS Synth Biol 7:227–239

Katz L, Chen YY, Gonzalez R, Peterson TC, Zhao H, Baltz RH (2018) Synthetic biology advances and applications in the biotechnology industry: a perspective. J Ind Microbiol Biotechnol 45: 449–461

Kautsar SA, Blin K, Shaw S, Weber T, Medema MH (2021) BiG-FAM: the biosynthetic gene cluster families database. Nucleic Acids Res 49:D490–D497

Kendrew SG, Petkovic H, Gaisser S, Ready SJ, Gregory MA, Coates NJ, Nur-E-Alam M, Warneck T, Suthar D, Foster TA et al (2013) Recombinant strains for the enhanced production of bioengineered rapalogs. Metab Eng 15:167–173

Kennedy J (2008) Mutasynthesis, chemobiosynthesis, and back to semi-synthesis: combining synthetic chemistry and biosynthetic engineering for diversifying natural products. Nat Prod Rep 25:25–34

Kieser T, Hopwood DA, Wright HM, Thompson CJ (1982) pIJ101, a multi-copy broad host-range Streptomyces plasmid: functional analysis and development of DNA cloning vectors. Mol Gen Genet 185:223–228

Kim J, Yi G-S (2012) PKMiner: a database for exploring type II polyketide synthases. BMC Microbiol 12:169

Kim EJ, Yang I, Yoon YJ (2015) Developing Streptomyces venezuelae as a cell factory for the production of small molecules used in drug discovery. Arch Pharm Res 38:1606–1616

Kim HU, Charusanti P, Lee SY, Weber T (2016) Metabolic engineering with systems biology tools to optimize production of prokaryotic secondary metabolites. Nat Prod Rep 33:933–941

Kim W, Lee N, Hwang S, Lee Y, Kim J, Cho S, Palsson B, Cho B-K (2020) comparative genomics determines strain-dependent secondary metabolite production in Streptomyces venezuelae strains. Biomolecules 10

Knirschova R, Novakova R, Mingyar E, Bekeova C, Homerova D, Kormanec J (2015) Utilization of a reporter system based on the blue pigment indigoidine biosynthetic gene bpsA for detection of promoter activity and deletion of genes in Streptomyces. J Microbiol Methods 113:1–3

Komatsu M, Uchiyama T, Omura S, Cane DE, Ikeda H (2010) Genome-minimized Streptomyces host for the heterologous expression of secondary metabolism. Proc Natl Acad Sci U S A 107: 2646–2651

Komatsu M, Komatsu K, Koiwai H, Yamada Y, Kozone I, Izumikawa M, Hashimoto J, Takagi M, Omura S, Shin-ya K et al (2013) Engineered Streptomyces avermitilis host for heterologous expression of biosynthetic gene cluster for secondary metabolites. ACS Synth Biol 2:384–396

Komeda H, Hori Y, Kobayashi M, Shimizu S (1996) Transcriptional regulation of the Rhodococcus rhodochrous J1 nitA gene encoding a nitrilase. Proc Natl Acad Sci U S A 93:10572–10577

Kong D, Wang X, Nie J, Niu G (2019) Regulation of antibiotic production by signaling molecules in Streptomyces. Front Microbiol 10:2927

Kormanec J, Rezuchova B, Homerova D, Csolleiova D, Sevcikova B, Novakova R, Feckova L (2019) Recent achievements in the generation of stable genome alterations/mutations in species of the genus Streptomyces. Appl Microbiol Biotechnol 103:5463–5482

Krause J, Handayani I, Blin K, Kulik A, Mast Y (2020) Disclosing the potential of the SARP-type regulator PapR2 for the activation of antibiotic gene clusters in Streptomycetes. Front Microbiol 11:225

Kuhstoss S, Richardson MA, Rao RN (1991) Plasmid cloning vectors that integrate site-specifically in Streptomyces spp. Gene 97:143–146

Kwun MJ, Hong H-J (2014) Genome sequence of Streptomyces toyocaensis NRRL 15009, producer of the glycopeptide antibiotic A47934. Genome Announc 2

Labes G, Bibb M, Wohlleben W (1997) Isolation and characterization of a strong promoter element from the Streptomyces ghanaensis phage I19 using the gentamicin resistance gene (aacC1) of Tn 1696 as reporter. Microbiology (Reading, Engl) 143(Pt 5):1503–1512

Lara M, Servín-González L, Singh M, Moreno C, Cohen I, Nimtz M, Espitia C (2004) Expression, secretion, and glycosylation of the 45- and 47-kDa glycoprotein of Mycobacterium tuberculosis in Streptomyces lividans. Appl Environ Microbiol 70:679–685

Lee N, Kim W, Hwang S, Lee Y, Cho S, Palsson B, Cho B-K (2020a) Thirty complete Streptomyces genome sequences for mining novel secondary metabolite biosynthetic gene clusters. Sci Data 7:55

Lee N, Kim W, Chung J, Lee Y, Cho S, Jang K-S, Kim SC, Palsson B, Cho B-K (2020b) Iron competition triggers antibiotic biosynthesis in Streptomyces coelicolor during coculture with Myxococcus xanthus. ISME J 14:1111–1124

Li JW-H, Vederas JC (2011) [Drug discovery and natural products: end of era or an endless frontier?]. Biomed Khim 57:148–160

Li C, Hazzard C, Florova G, Reynolds KA (2009a) High titer production of tetracenomycins by heterologous expression of the pathway in a Streptomyces cinnamonensis industrial monensin producer strain. Metab Eng 11:319–327

Li MHT, Ung PMU, Zajkowski J, Garneau-Tsodikova S, Sherman DH (2009b) Automated genome mining for natural products. BMC Bioinformatics 10:185

Li L, Zhao Y, Ruan L, Yang S, Ge M, Jiang W, Lu Y (2015a) A stepwise increase in pristinamycin II biosynthesis by Streptomyces pristinaespiralis through combinatorial metabolic engineering. Metab Eng 29:12–25

Li S, Wang J, Li X, Yin S, Wang W, Yang K (2015b) Genome-wide identification and evaluation of constitutive promoters in streptomycetes. Microb Cell Fact 14:172

Li L, Jiang W, Lu Y (2018a) A modified Gibson assembly method for cloning large DNA fragments with high GC contents. Methods Mol Biol 1671:203–209

Li S, Wang J, Xiang W, Yang K, Li Z, Wang W (2018b) An autoregulated fine-tuning strategy for titer improvement of secondary metabolites using native promoters in Streptomyces. ACS Synth Biol 7:522–530

Li L, Liu X, Jiang W, Lu Y (2019a) Recent advances in synthetic biology approaches to optimize production of bioactive natural products in actinobacteria. Front Microbiol 10:2467

Li L, Wei K, Liu X, Wu Y, Zheng G, Chen S, Jiang W, Lu Y (2019b) aMSGE: advanced multiplex site-specific genome engineering with orthogonal modular recombinases in actinomycetes. Metab Eng 52:153–167

Liang M, Liu L, Wang W, Zeng X, Liu J, Loganathan K, Zhu G, Bai L, Hou C, Chen X et al (2020) Simple cloning of large natural product biosynthetic gene cluster by CRISPR/Cas12a-mediated fast direct capturing strategy. BioRxiv

Liu H, Jiang H, Haltli B, Kulowski K, Muszynska E, Feng X, Summers M, Young M, Graziani E, Koehn F et al (2009) Rapid cloning and heterologous expression of the meridamycin biosynthetic gene cluster using a versatile Escherichia coli-Streptomyces artificial chromosome vector, pSBAC. J Nat Prod 72:389–395

Liu W, Zhang Q, Guo J, Chen Z, Li J, Wen Y (2015) Increasing avermectin production in Streptomyces avermitilis by manipulating the expression of a novel TetR-family regulator and its target gene product. Appl Environ Microbiol 81:5157–5173

Liu Y, Ren C-Y, Wei W-P, You D, Yin B-C, Ye B-C (2019) A CRISPR-Cas9 strategy for activating the Saccharopolyspora erythraea erythromycin biosynthetic gene cluster with knock-in bidirectional promoters. ACS Synth Biol 8:1134–1143

Lomovskaya ND, Mkrtumian NM, Gostimskaya NL, Danilenko VN (1972) Characterization of temperate actinophage phi C31 isolated from Streptomyces coelicolor A3(2). J Virol 9:258–262

Lomovskaya ND, Chater KF, Mkrtumian NM (1980) Genetics and molecular biology of Streptomyces bacteriophages. Microbiol Rev 44:206–229

Lopatniuk M, Ostash B, Luzhetskyy A, Walker S, Fedorenko V (2014) Generation and study of the strains of streptomycetes—heterologous hosts for production of moenomycin. Russ J Genet 50: 360–365

Matsumoto M, Hashimoto Y, Saitoh Y, Kumano T, Kobayashi M (2016) Development of nitrilase promoter-derived inducible vectors for Streptomyces. Biosci Biotechnol Biochem 80:1230–1237

Matsushima P, Baltz RH (1985) Efficient plasmid transformation of Streptomyces ambofaciens and Streptomyces fradiae protoplasts. J Bacteriol 163:180–185

Mazodier P, Petter R, Thompson C (1989) Intergeneric conjugation between Escherichia coli and Streptomyces species. J Bacteriol 171:3583–3585

McDaniel R, Ebert-Khosla S, Hopwood DA, Khosla C (1993) Engineered biosynthesis of novel polyketides. Science 262:1546–1550

Medema MH, Kottmann R, Yilmaz P, Cummings M, Biggins JB, Blin K, de Bruijn I, Chooi YH, Claesen J, Coates RC et al (2015) Minimum Information about a Biosynthetic Gene cluster. Nat Chem Biol 11:625–631

Miao V, Coëffet-LeGal M-F, Brian P, Brost R, Penn J, Whiting A, Martin S, Ford R, Parr I, Bouchard M et al (2005) Daptomycin biosynthesis in Streptomyces roseosporus: cloning and analysis of the gene cluster and revision of peptide stereochemistry. Microbiology (Reading, Engl) 151:1507–1523

Miller SJ, Clardy J (2009) Beyond grind and find. Nat Chem 1:261–263

Miura T, Hosaka Y, Yan-Zhuo Y, Nishizawa T, Asayama M, Takahashi H, Shirai M (2011) In vivo and in vitro characterization of site-specific recombination of actinophage R4 integrase. J Gen Appl Microbiol 57:45–57

Montiel D, Kang H-S, Chang F-Y, Charlop-Powers Z, Brady SF (2015) Yeast homologous recombination-based promoter engineering for the activation of silent natural product biosynthetic gene clusters. Proc Natl Acad Sci U S A 112:8953–8958

Morita K, Yamamoto T, Fusada N, Komatsu M, Ikeda H, Hirano N, Takahashi H (2009) The site-specific recombination system of actinophage TG1. FEMS Microbiol Lett 297:234–240

Murakami T, Holt TG, Thompson CJ (1989) Thiostrepton-induced gene expression in Streptomyces lividans. J Bacteriol 171:1459–1466

Muyrers JP, Zhang Y, Testa G, Stewart AF (1999) Rapid modification of bacterial artificial chromosomes by ET-recombination. Nucleic Acids Res 27:1555–1557

Myronovskyi M, Luzhetskyy A (2013) Genome engineering in actinomycetes using site-specific recombinases. Appl Microbiol Biotechnol 97:4701–4712

Myronovskyi M, Luzhetskyy A (2016) Native and engineered promoters in natural product discovery. Nat Prod Rep 33:1006–1019

Myronovskyi M, Luzhetskyy A (2019) Heterologous production of small molecules in the optimized Streptomyces hosts. Nat Prod Rep 36:1281–1294

Myronovskyi M, Welle E, Fedorenko V, Luzhetskyy A (2011) Beta-glucuronidase as a sensitive and versatile reporter in actinomycetes. Appl Environ Microbiol 77:5370–5383

Novakova R, Odnogova Z, Kutas P, Feckova L, Kormanec J (2010) Identification and characterization of an indigoidine-like gene for a blue pigment biosynthesis in Streptomyces aureofaciens CCM 3239. Folia Microbiol (Praha) 55:119–125

Novakova R, Rehakova A, Feckova L, Kutas P, Knischova R, Kormanec J (2011) Genetic manipulation of pathway regulation for overproduction of angucycline-like antibiotic auricin in Streptomyces aureofaciens CCM 3239. Folia Microbiol (Praha) 56:276–282

Novakova R, Núñez LE, Homerova D, Knirschova R, Feckova L, Rezuchova B, Sevcikova B, Menéndez N, Morís F, Cortés J et al (2018) Increased heterologous production of the antitumoral polyketide mithramycin A by engineered Streptomyces lividans TK24 strains. Appl Microbiol Biotechnol 102:857–869

Okada BK, Seyedsayamdost MR (2017) Antibiotic dialogues: induction of silent biosynthetic gene clusters by exogenous small molecules. FEMS Microbiol Rev 41:19–33

Okamoto S, Taguchi T, Ochi K, Ichinose K (2009) Biosynthesis of actinorhodin and related antibiotics: discovery of alternative routes for quinone formation encoded in the act gene cluster. Chem Biol 16:226–236

Okanishi M, Utahara R, Okami Y (1966) Infection of the protoplasts of Streptomyces kanamyceticus with deoxyribonucleic acid preparation from actinophage PK-66. J Bacteriol 92:1850–1852

Okanishi M, Hamana K, Umezawa H (1968) Factors affecting infection of protoplasts with deoxyribonucleic acid of actinophage PK-66. J Virol 2:686–691

Okanishi M, Suzuki K, Umezawa H (1974) Formation and reversion of Streptomycete protoplasts: cultural condition and morphological study. J Gen Microbiol 80:389–400

Olano C, Wilkinson B, Sánchez C, Moss SJ, Sheridan R, Math V, Weston AJ, Braña AF, Martin CJ, Oliynyk M et al (2004) Biosynthesis of the angiogenesis inhibitor borrelidin by Streptomyces parvulus Tü4055: cluster analysis and assignment of functions. Chem Biol 11:87–97

Olano C, García I, González A, Rodriguez M, Rozas D, Rubio J, Sánchez-Hidalgo M, Braña AF, Méndez C, Salas JA (2014) Activation and identification of five clusters for secondary metabolites in Streptomyces albus J1074. Microb Biotechnol 7:242–256

Omura S, Ikeda H, Ishikawa J, Hanamoto A, Takahashi C, Shinose M, Takahashi Y, Horikawa H, Nakazawa H, Osonoe T et al (2001) Genome sequence of an industrial microorganism Streptomyces avermitilis: deducing the ability of producing secondary metabolites. Proc Natl Acad Sci U S A 98:12215–12220

Onaka H, Tabata H, Igarashi Y, Sato Y, Furumai T (2001) Goadsporin, a chemical substance which promotes secondary metabolism and morphogenesis in streptomycetes. I. Purification and characterization. J Antibiot 54:1036–1044

Owen JG, Charlop-Powers Z, Smith AG, Ternei MA, Calle PY, Reddy BVB, Montiel D, Brady SF (2015) Multiplexed metagenome mining using short DNA sequence tags facilitates targeted discovery of epoxyketone proteasome inhibitors. Proc Natl Acad Sci U S A 112:4221–4226

Palazzotto E, Tong Y, Lee SY, Weber T (2019) Synthetic biology and metabolic engineering of actinomycetes for natural product discovery. Biotechnol Adv 37:107366

Peng Q, Gao G, Lü J, Long Q, Chen X, Zhang F, Xu M, Liu K, Wang Y, Deng Z et al (2018) Engineered Streptomyces lividans strains for optimal identification and expression of cryptic biosynthetic gene clusters. Front Microbiol 9:3042

Penn J, Li X, Whiting A, Latif M, Gibson T, Silva CJ, Brian P, Davies J, Miao V, Wrigley SK et al (2006) Heterologous production of daptomycin in Streptomyces lividans. J Ind Microbiol Biotechnol 33:121–128

Phelan RM, Sekurova ON, Keasling JD, Zotchev SB (2015) Engineering terpene biosynthesis in Streptomyces for production of the advanced biofuel precursor bisabolene. ACS Synth Biol 4: 393–399

Phelan RM, Sachs D, Petkiewicz SJ, Barajas JF, Blake-Hedges JM, Thompson MG, Reider Apel A, Rasor BJ, Katz L, Keasling JD (2017) Development of next generation synthetic biology tools for use in Streptomyces venezuelae. ACS Synth Biol 6:159–166

Pigac J, Schrempf H (1995) A simple and rapid method of transformation of Streptomyces rimosus R6 and other Streptomycetes by electroporation. Appl Environ Microbiol 61:352–356

Pullan ST, Chandra G, Bibb MJ, Merrick M (2011) Genome-wide analysis of the role of GlnR in Streptomyces venezuelae provides new insights into global nitrogen regulation in actinomycetes. BMC Genomics 12:175

Raynal A, Tuphile K, Gerbaud C, Luther T, Guérineau M, Pernodet JL (1998) Structure of the chromosomal insertion site for pSAM2: functional analysis in Escherichia coli. Mol Microbiol 28:333–342

Rebets Y, Kormanec J, Luzhetskyy A, Bernaerts K, Anné J (2017) Cloning and expression of metagenomic DNA in Streptomyces lividans and subsequent fermentation for optimized production. Methods Mol Biol 1539:99–144

Rodríguez Estévez M, Myronovskyi M, Rosenkränzer B, Paululat T, Petzke L, Ristau J, Luzhetskyy A (2020) Novel fredericamycin variant overproduced by a streptomycin-resistant Streptomyces albus subsp. chlorinus strain. Mar Drugs 18

Rodriguez E, Hu Z, Ou S, Volchegursky Y, Hutchinson CR, McDaniel R (2003) Rapid engineering of polyketide overproduction by gene transfer to industrially optimized strains. J Ind Microbiol Biotechnol 30:480–488

Rodríguez-García A, Combes P, Pérez-Redondo R, Smith MCA, Smith MCM (2005) Natural and synthetic tetracycline-inducible promoters for use in the antibiotic-producing bacteria Streptomyces. Nucleic Acids Res 33:e87

Rückert C, Albersmeier A, Busche T, Jaenicke S, Winkler A, Friðjónsson ÓH, Hreggviðsson GÓ, Lambert C, Badcock D, Bernaerts K et al (2015) Complete genome sequence of Streptomyces lividans TK24. J Biotechnol 199:21–22

Ryan KS (2011) Biosynthetic gene cluster for the cladoniamides, bis-indoles with a rearranged scaffold. PLoS One 6:e23694

Ryan KS, Drennan CL (2009) Divergent pathways in the biosynthesis of bisindole natural products. Chem Biol 16:351–364

Schmitt-John T, Engels JW (1992) Promoter constructions for efficient secretion expression in Streptomyces lividans. Appl Microbiol Biotechnol 36:493–498

Schwarz PN, Roller L, Kulik A, Wohlleben W, Stegmann E (2018) Engineering metabolic pathways in Amycolatopsis japonicum for the optimization of the precursor supply for heterologous brasilicardin congeners production. Synth Syst Biotechnol 3:56–63

Seghezzi N, Amar P, Koebmann B, Jensen PR, Virolle M-J (2011) The construction of a library of synthetic promoters revealed some specific features of strong Streptomyces promoters. Appl Microbiol Biotechnol 90:615–623

Sekurova ON, Brautaset T, Sletta H, Borgos SEF, Jakobsen M ØM, Ellingsen TE, Strøm AR, Valla S, Zotchev SB (2004) In vivo analysis of the regulatory genes in the nystatin biosynthetic gene cluster of Streptomyces noursei ATCC 11455 reveals their differential control over antibiotic biosynthesis. J Bacteriol 186:1345–1354

Selle K, Barrangou R (2015) Harnessing CRISPR-Cas systems for bacterial genome editing. Trends Microbiol 23:225–232

Shao Z, Rao G, Li C, Abil Z, Luo Y, Zhao H (2013) Refactoring the silent spectinabilin gene cluster using a plug-and-play scaffold. ACS Synth Biol 2:662–669

Shima J, Hesketh A, Okamoto S, Kawamoto S, Ochi K (1996) Induction of actinorhodin production by rpsL (encoding ribosomal protein S12) mutations that confer streptomycin resistance in Streptomyces lividans and Streptomyces coelicolor A3(2). J Bacteriol 178:7276–7284

Siegl T, Tokovenko B, Myronovskyi M, Luzhetskyy A (2013) Design, construction and characterisation of a synthetic promoter library for fine-tuned gene expression in actinomycetes. Metab Eng 19:98–106

Skinnider MA, Dejong CA, Rees PN, Johnston CW, Li H, Webster ALH, Wyatt MA, Magarvey NA (2015) Genomes to natural products PRediction Informatics for Secondary Metabolomes (PRISM). Nucleic Acids Res 43:9645–9662

Smokvina T, Mazodier P, Boccard F, Thompson CJ, Guérineau M (1990) Construction of a series of pSAM2-based integrative vectors for use in actinomycetes. Gene 94:53–59

Solenberg PJ, Matsushima P, Stack DR, Wilkie SC, Thompson RC, Baltz RH (1997) Production of hybrid glycopeptide antibiotics in vitro and in Streptomyces toyocaensis. Chem Biol 4:195–202

Song C, Luan J, Cui Q, Duan Q, Li Z, Gao Y, Li R, Li A, Shen Y, Li Y et al (2019) Enhanced heterologous Spinosad production from a 79-kb synthetic multioperon assembly. ACS Synth Biol 8:137–147

Starcevic A, Zucko J, Simunkovic J, Long PF, Cullum J, Hranueli D (2008) ClustScan: an integrated program package for the semi-automatic annotation of modular biosynthetic gene clusters and in silico prediction of novel chemical structures. Nucleic Acids Res 36:6882–6892

Strohl WR (1992) Compilation and analysis of DNA sequences associated with apparent streptomycete promoters. Nucleic Acids Res 20:961–974

Suzuki N, Nonaka H, Tsuge Y, Inui M, Yukawa H (2005) New multiple-deletion method for the Corynebacterium glutamicum genome, using a mutant lox sequence. Appl Environ Microbiol 71:8472–8480

Takano E, White J, Thompson CJ, Bibb MJ (1995) Construction of thiostrepton-inducible, high-copy-number expression vectors for use in Streptomyces spp. Gene 166:133–137

Takano E, Kinoshita H, Mersinias V, Bucca G, Hotchkiss G, Nihira T, Smith CP, Bibb M, Wohlleben W, Chater K (2005) A bacterial hormone (the SCB1) directly controls the expression of a pathway-specific regulatory gene in the cryptic type I polyketide biosynthetic gene cluster of Streptomyces coelicolor. Mol Microbiol 56:465–479

Tan G-Y, Deng K, Liu X, Tao H, Chang Y, Chen J, Chen K, Sheng Z, Deng Z, Liu T (2017) Heterologous biosynthesis of Spinosad: an omics-guided large polyketide synthase gene cluster reconstitution in Streptomyces. ACS Synth Biol 6:995–1005

Tanaka Y, Izawa M, Hiraga Y, Misaki Y, Watanabe T, Ochi K (2017) Metabolic perturbation to enhance polyketide and nonribosomal peptide antibiotic production using triclosan and ribosome-targeting drugs. Appl Microbiol Biotechnol 101:4417–4431

Tao W, Yang A, Deng Z, Sun Y (2018) CRISPR/Cas9-Based Editing of Streptomyces for Discovery, Characterization, and Production of Natural Products. Front Microbiol 9:1660

Tao W, Chen L, Zhao C, Wu J, Yan D, Deng Z, Sun Y (2019) In vitro packaging mediated one-step targeted cloning of natural product pathway. ACS Synth Biol 8:1991–1997

Thanapipatsiri A, Claesen J, Gomez-Escribano J-P, Bibb M, Thamchaipenet A (2015) A Streptomyces coelicolor host for the heterologous expression of Type III polyketide synthase genes. Microb Cell Factories 14:145

Tietz JI, Schwalen CJ, Patel PS, Maxson T, Blair PM, Tai H-C, Zakai UI, Mitchell DA (2017) A new genome-mining tool redefines the lasso peptide biosynthetic landscape. Nat Chem Biol 13: 470–478

Tocchetti A, Donadio S, Sosio M (2018) Large inserts for big data: artificial chromosomes in the genomic era. FEMS Microbiol Lett 365

Tong Y, Charusanti P, Zhang L, Weber T, Lee SY (2015) CRISPR-Cas9 based engineering of actinomycetal genomes. ACS Synth Biol 4:1020–1029

Udwary DW, Zeigler L, Asolkar RN, Singan V, Lapidus A, Fenical W, Jensen PR, Moore BS (2007) Genome sequencing reveals complex secondary metabolome in the marine actinomycete Salinispora tropica. Proc Natl Acad Sci U S A 104:10376–10381

Ueda K, Kawai S, Ogawa H-O, Kiyama A, Kubota T, Kawanobe H, Beppu T (2000) Wide distribution of interspecific stimulatory events on antibiotic production and sporulation among Streptomyces species. J Antibiot 53:979–982

van Heel AJ, de Jong A, Song C, Viel JH, Kok J, Kuipers OP (2018) BAGEL4: a user-friendly web server to thoroughly mine RiPPs and bacteriocins. Nucleic Acids Res 46:W278–W281

Van Mellaert L, Mei L, Lammertyn E, Schacht S, Anné J (1998) Site-specific integration of bacteriophage VWB genome into Streptomyces venezuelae and construction of a VWB-based integrative vector. Microbiology (Reading, Engl) 144(Pt 12):3351–3358

van Santen JA, Jacob G, Singh AL, Aniebok V, Balunas MJ, Bunsko D, Neto FC, Castaño-Espriu L, Chang C, Clark TN et al (2019) The natural products atlas: an open access knowledge base for microbial natural products discovery. ACS Cent Sci 5:1824–1833

Waldron C, Matsushima P, Rosteck PR, Broughton MC, Turner J, Madduri K, Crawford KP, Merlo DJ, Baltz RH (2001) Cloning and analysis of the spinosad biosynthetic gene cluster of Saccharopolyspora spinosa. Chem Biol 8:487–499

Wang W, Li X, Wang J, Xiang S, Feng X, Yang K (2013) An engineered strong promoter for streptomycetes. Appl Environ Microbiol 79:4484–4492

Wang H, Li Z, Jia R, Hou Y, Yin J, Bian X, Li A, Müller R, Stewart AF, Fu J et al (2016a) RecET direct cloning and Redαβ recombineering of biosynthetic gene clusters, large operons or single genes for heterologous expression. Nat Protoc 11:1175–1190

Wang W, Yang T, Li Y, Li S, Yin S, Styles K, Corre C, Yang K (2016b) Development of a synthetic oxytetracycline-inducible expression system for Streptomycetes using de novo characterized genetic parts. ACS Synth Biol 5:765–773

Yamanaka K, Reynolds KA, Kersten RD, Ryan KS, Gonzalez DJ, Nizet V, Dorrestein PC, Moore BS (2014) Direct cloning and refactoring of a silent lipopeptide biosynthetic gene cluster yields the antibiotic taromycin A. Proc Natl Acad Sci U S A 111:1957–1962

Yu Y, Bai L, Minagawa K, Jian X, Li L, Li J, Chen S, Cao E, Mahmud T, Floss HG et al (2005) Gene cluster responsible for validamycin biosynthesis in Streptomyces hygroscopicus subsp. jinggangensis 5008. Appl Environ Microbiol 71:5066–5076

Zaburannyi N, Rabyk M, Ostash B, Fedorenko V, Luzhetskyy A (2014) Insights into naturally minimised Streptomyces albus J1074 genome. BMC Genomics 15:97

Zhang L, Ou X, Zhao G, Ding X (2008) Highly efficient in vitro site-specific recombination system based on Streptomyces phage phiBT1 integrase. J Bacteriol 190:6392–6397

Zhang J, Barajas JF, Burdu M, Ruegg TL, Dias B, Keasling JD (2017) Development of a transcription factor-based lactam biosensor. ACS Synth Biol 6:439–445

Zhang JJ, Moore BS, Tang X (2018) Engineering Salinispora tropica for heterologous expression of natural product biosynthetic gene clusters. Appl Microbiol Biotechnol 102:8437–8446

Ziemert N, Podell S, Penn K, Badger JH, Allen E, Jensen PR (2012) The natural product domain seeker NaPDoS: a phylogeny based bioinformatic tool to classify secondary metabolite gene diversity. PLoS One 7:e34064

Application of CRISPR/Cas9 Editing for Production of Secondary Metabolites in Actinomycetes

12

Jamuna A. Bai and Ravishankar V. Rai

Abstract

The CRISPR (clustered regularly interspaced short palindromic repeats)/Cas9-based RNA-guided DNA editing technique is used to make double-stranded breaks in genomes. Further, it can be directed to introduce either site-specific insertions or deletions and to replace genes in bacteria. It has been successfully used as a genome editing tool in *Escherichia coli*, *Streptococcus pneumonia,* and *Lactobacillus reuteri*. The application of CRISPR/Cas9 for industrially relevant strains especially the actinomycetes, which are not amenable to conventional editing, is of paramount importance. The advantage of using CRISPR/Cas9 for gene editing is that it is easy, less laborious, and has rapid application. It can be used in genetically difficult to access microbes especially those with the potential to produce diverse secondary metabolites. In this chapter, the focus is on understanding the mechanism of CRISPR/Cas9 editing in bacteria, and the design and construction of CRISPR/Cas9-based editing platforms in actinomycetes. We have also discussed the advantages and limitations of the four CRISPR/Cas9-based toolkits in *Streptomyces*. The potential application of CRISPR/Cas9 genome editing in actinomycetes for the detection of cryptic BGCs and genes encoding novel secondary metabolites to increase the production of secondary metabolites has also been explored. Finally, the advances made using the variants in CRISPR/Cas9-based genome editing techniques have been described.

J. A. Bai
Department of Microbiology, School of Life Sciences, JSS AHER, Mysore, Karnataka, India

R. V. Rai (✉)
Department of Studies in Microbiology, University of Mysore, Mysore, Karnataka, India

R. V. Rai, J. A. Bai (eds.), *Natural Products from Actinomycetes*,
https://doi.org/10.1007/978-981-16-6132-7_12

Keywords

CRISPR/Cas9 · Actinomycetes · *Streptomyces* · Genome editing · Biosynthetic gene cluster · Secondary metabolites

12.1 Introduction to CRISPR/Cas9 Editing in Bacteria

The CRISPR/Cas9 genome editing process involves cleaving a target double-stranded DNA (dsDNA) by endonuclease enzyme Cas9 (Tong et al. 2015). Firstly, the enzyme Cas9 interacts with a dual-RNA structure. This dual-RNA structure is made of two molecules, one is the CRISPR-RNA (crRNA) and the other is a transactivating crRNA (tracrRNA) (Alberti and Corre 2019). Secondly, a synthetic guide RNA (sgRNA) is generated by combining crRNA and tracrRNA. Further, a protospacer sequence (20 bp) in the crRNA and protospacer adjacent motif (PAM) in the target DNA are used to recognize specific target sites for double-stranded breaks (DSB) (Fig. 12.1). The DSB is located three bp upstream to PAM sequence in target dsDNA. It is repaired either through the nonhomologous end-joining (NHEJ) pathway or homology-directed repair (HDR). NHEJ is error-prone, whereas HDR is considered to be highly efficient and accurate and used for precise genome editing. HDR involves two mechanisms. One is the addition of a custom-designed editing template. The second involves the use of homologous recombination arms (Alberti and Corre 2019).

The use of CRISPR/Cas9 for the first time in introducing DSBs in the genome of *Streptococcus pyogenes,* paved way for its application as a genome editing tool in other microbes including *Methanosarcina acetivorans, Escherichia coli,* and *Saccharomyces cerevisiae* (Tong et al. 2015). CRISPR/Cas9 gene editing technology has also been successfully used for genome engineering in higher organisms

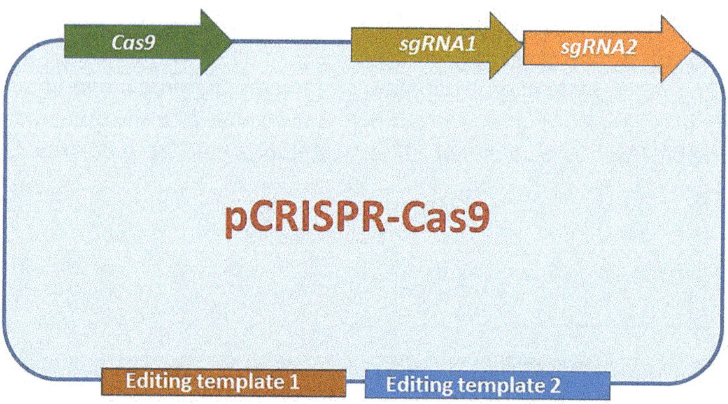

Fig. 12.1 Components of a simple pCRISPR/Cas9 genome editing tool include Cas9, shortguide RNA clusters, and two editing templates

including the plants *Arabidopsis thaliana* and *Nicotiana benthamiana* as well as in animal models of zebrafish and mouse and in human cell lines. Such widespread application is possible because CRISPR/Cas-based editing tool has no requisite for preintegration of a unique enzyme recognition sequence into the genome, as it uses a transcribed synthetic guide RNA or only crRNA to direct Cas proteins to the target sites (Knott and Doudna 2018).

In bacteria, conventional genome editing techniques were multistep and laborious. One such technique, involves the use of suicide plasmids. However, the discovery of CRISPR-Cas phenomenon in bacteria and its easy and programmable applications have transformed genome editing technique. In the initial days, due to certain shortcomings, it was not of much use in bacterial genome editing. Nonetheless, research on overcoming these problems by using alternatives such as CRISPR nickases, Cas12a, base editors, CRISPR-associated transposases, prime-editing, endogenous CRISPR systems, and the use of premade ribonucleoprotein complexes of Cas proteins and guide RNAs has made CRISPR-Cas a potential genome editing tool in bacteria (Alberti and Corre 2019; Arroyo-Olarte et al. 2021). In this chapter, the mechanism of CRISPR/Cas9 and its associated advanced tools, the advantages and limitations for their application in genome editing in actinomycetes, with special reference to streptomycetes have been discussed. The four toolkits based on CRISPR/Cas9 developed for genome editing in *Streptomyces* sp. have been described. The chapter focuses on the application of CRISPR/Cas9-based tools for genome editing in actinomycetes to discover hidden biosynthetic gene clusters (BGCs) for identifying new bioactive secondary metabolites, and in in vivo engineering of industrially important strains to increase the production of certain secondary metabolites having therapeutic value.

12.2 Conventional Genome Editing in Actinomycetes

The conventional technique used for genome editing in actinomycetes is by using homologous recombination, involving single or double crossover (Table 12.1). Gene disruption is carried out by inserting a selectable marker using a single crossover. But when a whole plasmid is inserted, it will disrupt the expression of genes located downstream. Similarly, if suitable selective markers are not used, there could be a reversion of the genotype to its earlier form. Streptomycetes have few antibiotic markers which include aminoglycosides such as kanamycin, apramycin, and hygromycin. Gene editing exploiting double crossovers is a multistep tedious process, but it results in the generation of stable mutations. In the first step, a single crossover event ensures integration of the DNA construct into the target locus, located between the two homologous recombination arms. Double crossover occurs in the next integration step in the vector. Further, the simple replica plating technique helps in identifying preferred genotype and distinguishing mutants with from the wild-type reverted strains (Alberti and Corre 2019).

The Cre-loxP recombination system has also been used for gene editing in actinomycetes. It involves introducing two loxP sites in the identical direction to

Table 12.1 Genome editing techniques applied in actinomycetes for secondary metabolite production and strain improvement

Techniques	Construction and features	Application	Reference
PCR-targeting system	Uses high-efficiency λ Red recombination system	For gene knockout in *Escherichia coli* and nonpolar and in-frame deletion of genes or gene clusters in *Streptomyces coelicolor*	Gust et al. (2003)
Cre-loxP recombination system	Involves introduction of two loxP sites in the genome and the expression of the Cre recombinase	To knockout large fragments of DNA in in the industrial strain, *Streptomyces avermitilis* and other *Streptomyces* sp.	Komatsu et al. (2010)
I-SceI Meganuclease-promoted recombination system	Use of I-SceI meganuclease to recognize unique sequence and cause DNA double-strand breaks	Deletion of actinorhodin and undecylprodigiosin BGC in *S. coelicolor*	Lu et al. (2010)
CRISPR/Cas-based genome editing	Employs a transcribed sgRNA, crRNA, and tracrRNA, or only crRNA, to direct Cas proteins to any site on the genome	To generate CRISPR/Cas-based genome editing toolbox in *Streptomyces*	Li et al. (2018)
Cas9-based genome editing	pCRISPomyces-1 and pCRISPomyces-2	Precise deletion of various DNA sizes including individual genes, double genes simultaneously, and single antibiotic BGCs with efficiency of 21–100% in three different *Streptomyces* species	Cobb et al. (2015)
	pKCcas9dO in *S. coelicolor*	Deletion of individual genes (*actII-orf4*, *redD*, and *glnR*) and single antibiotic BGCs with 60–100% efficiency; simultaneous deletion of two genes (*actII-orf4* and *redD*) and two BGCs; to introduce point mutations into the *rpsL* gene	Huang et al. (2015)
	pCRISPR-Cas9-ScaligD (contains the ligD gene from *Streptomyces carneus*)	Precise deletion of individual and multiple genes by HDR in *S. coelicolor*	Tong et al. (2015)
	pWHU2653 based on a plasmid with pIJ101 and counter-selection marker CodA(sm)	Improved plasmid curing by 95 %	Zeng et al. (2015)

(continued)

Table 12.1 (continued)

Techniques	Construction and features	Application	Reference
	pMWCas9 with thiostrepton-inducible promoter tipAp replacing the ermEp* promoter of Cas9	Used to delete highly repetitive DNA sequences, such as the *eryA*III gene from erythromycin PKS in *Saccharopolyspora erythraea*	Mo et al. (2019)
	pWHU2653 (pWHU2653-TRMA) by modulating Cas9 activity at multiple levels including use of inducible promoter tipAp and introducing the theophylline-inducible riboswitch and Mag-based blue light-inducible system pKC1139-TRMA based on pKC1139	Reduction in toxicity and increased DNA transformation efficiency by 250-fold in *S. coelicolor* Individual deletion of *actII-orf4* and *redD* in *S. coelicolor* with 35–80 % efficiencies ranging from 35 to 80 %	Wang et al. (2019)
	dual-functional chromogenic-screening CRISPR/Cas9 tool (pQS-gusA and pQS-idgS) based on pWHU2653 by replacing CodA(sm) with GusA and IdgS reporters	Reporter systems with improved genome editing and plasmid curing efficiency by 100 % Successfully applied in genetically recalcitrant and slow-growing rare strain *Verrucosispora* sp. MS100137; Used for the deletion of carotenoid and the abyssomicin with 100 % efficiency	Wang et al. (2020)
Cpf1-assisted genome editing	Cpf1 from *Francisella novicida* (FnCpf1) that recognizes T-rich PAM sequences (5′-TTV-3)	For precise deletion of single and double genes simultaneously with 75–95 % efficiency in *S. coelicolor* and for multiplex genome editing	Li et al. (2018)
Alternative CRISPR/Cas-based	Cas9 from *Streptococcus thermophilus* CRISPR1 (Sth1Cas9, PAM: NNAGAA and NNGGAA), Cas9 from *Staphylococcus aureus* (SaCas9, PAM: NNGRRT), and Cpf1 from *F. novicida* (FnCpf1)	To edit strains that cannot be edited by Cas9 from *S. pyogenes*, such as *Streptomyces* sp. NRRL S-244	Yeo et al. (2019)
dCas-based transcriptional repression (CRISPRi)	Inducible CRISPRi based on pCRISPR-Cas9 by replacing Cas9 with dCas9, the replicative plasmid pGM1190 with pSG5 replicon, is used to	For efficient repression of single genes upon induction	Tong et al. (2015)

(continued)

Table 12.1 (continued)

Techniques	Construction and features	Application	Reference
	express the dCas9/sgRNA complex and dCas9 gene is regulated by tipAp		
	pSET152, an integrative plasmid for dCas9/ sgRNAs or ddCpf1/ crRNAs complex expression	For high efficiency multiplexed gene repression	Zhao et al. (2018); Li et al. (2018)
Base editors based on the Cas9 variants (dCas9 or Cas9n)	CRISPR-cBEST (belonging to CBE) and CRISPR-aBEST (belonging to ABE), by fusing rat APOBEC1 (rAPOBEC1) cytidine deaminase and the adenosine deaminase ecTadA to the N-terminus of the codon-optimized Cas9n	CRISPR-cBEST is used for point mutation (C to T) with 100 % frequencies in *S. coelicolor*; used for introduction of STOP codons in *S. griseofuscus* at 60–100 % frequency and for simultaneous targeted mutagenesis of kirN in *S. collinus* Tu365. The Csy4-based RNA processing system can be used for simultaneous editing of three different sites at 100 % frequencies	Tong et al. (2019)
	CBE system dCas9-CDA-ULstr comprises dCas9, PmCDA1 [an activation-induced cytidine deaminase (AID), UGI, and the degradation tag (LVA)]	It is used for single-, double- and triple-point mutations in *S. coelicolor* and for highly efficient base editing in the industrial strain *S. rapamycinicus*	Zhao et al. (2019)

flank the DNA fragment that needs to be deleted by two single-crossover events. It is followed by induction of Cre recombinase expression for deletion of genes between the loxP sites. This system can be used for deleting large DNA fragments as observed with the successful deletion of DNA fragment of 1.4 Mb in *Streptomyces avermitilis* (Komatsu et al. 2010).

The application of Red/ET-based recombineering technique for gene editing in streptomycetes has been highly popular. In this rapid method of gene editing, cosmid clones are initially modified by λ-Red-mediated homologous recombination in the host *E. coli* and later transferred to streptomycetes. The limitation of this method is the lengthy process of generating and screening double cross-over events in streptomycete genome (Gust et al. 2003). Double crossovers can be selected by inserting unique recognition sites of 18 bp for the enzyme meganuclease I-SceI in the DNA construct (Lu et al. 2010). When I-SceI is expressed, integrated DNA becomes susceptible to DSBs and DNA break repair occurs due to homologous recombination and leads to the selection of double crossover mutants. Similarly, *gusA* blue-white screening can also be used for the single and double crossover selection processes. Here, expression of the *gusA* gene that encodes for β-glucuronidase in a single crossover configuration and in presence of X-Gluc results in the formation of a blue product. In a double crossover event, the plasmid backbone is disrupted and desired mutant colonies appear white. Though this method uses visual screening, but it requires exconjugants to be passaged 5–10 times in broth for the enrichment of cells having double crossovers (Alberti and Corre 2019).

12.3 CRISPR/Cas9 Editing in Actinomycetes

Actinomycetes are routinely screened for secondary metabolites and bioactive compounds for pharmacological and industrial application. However, they are not amenable to gene manipulation, and consequently to metabolic engineering. The use of CRISPR/Cas9 geome editing tool overcomes the dearth of rapid and effective genome engineering techniques, and expedites natural product discovery in actinomycetes especially streptomycetes. In the recent years, various CRISPR/Cas9-based protocols and toolkits have been established for actinomycetes (Table 12.1). These in vivo editing tools are now finding significant application for studying biosynthesis and regulation of natural products (Culp et al. 2019).

A CRISPR-Cas9 system of high efficiency has been developed to remove genes, replace genes, and control gene expression in *Streptomyces coelicolor* A3. Using this tool, two genes actIORF1 (SCO5087) and actVB (SCO5092) in the actinorhodin BGC were targeted in the organism. The CRISPR-Cas9 system disrupted the target genes and the DSBs introduced were repaired through the NHEJ pathway. As it is an error-prone repair system, a deletion library of variable sizes was obtained. However, in presence of templates for HDR, the removal of the target gene was precise and 100 % frequency was obtained. Similarly, CRISPRi was developed using variant of Cas9, dCas9 which is dead or inactive in catalytic functions and this system could efficiently and reversibly control target gene expression (Tong et al. 2015).

In *S. coelicolor* M145 strain, a genome editing plasmid pKCCas9dO was introduced into the organism through one-step intergeneric transfer. This CRISPR/Cas9 system consisted of three components, the guide RNA with Cas9 along with the two HDR templates. Up to 60–100 % gene editing could be obtained (Zhang et al. 2020). Single gene deletion could be carried out for actII-orf4 encoded actinorhodin, redD encoded undecylprodigiosin, and glnR genes. Only deletion of the large size antibiotic BGCs of actinorhodin, undecylprodigiosin, and Ca(2+)-dependent antibiotic was possible. The actII-orf4 of actinorhodin and red of undecylprodigiosin BGCs deletion were also obtained at the same time. The CRISPR/Cas9 system could perform point mutations in the gene rpsL, resulting in streptomycin-resistant mutants. This established CRISPR/Cas9 genome editing platform is extremely rapid and extremely efficient than the conventional methods for genome editing in *Streptomyces* (Huang et al. 2015).

Actinoplanes sp. SE50/110 is used for the production of acarbose, a drug used for diabetes type II treatment. CRISPR/Cas9 technology has been used in this strain to improve its functionality. The genome editing tool was used to delete MelC gene encoding tyrosinase. The enzyme tyrosinase is involved in the synthesis of eumelanin. The resulting *Actinoplanes* sp. SE50/110 ΔmelC2 mutant was deficient in pigment production and consequently the supernatant did not darken. The plasmid coding for Cas9 protein has temperature-sensitive replication. Thus, increasing temperature can denature and remove the plasmid (Wolf et al. 2016).

One of the limitations of using CRISPR-Cas9-assisted genome editing tool in the industrially relevant *Streptomyces* strains is the toxicity due to the expression of *Sp*Cas9. The other problem is the prerequisite of intricate expression constructs that can target numerous genomic loci. These limitations were overcome by developing a highly efficient CRISPR-Cpf1 platform derived from another bacteria *Francisella novicida*. CRISPR-Cpf1 could be used for both engineering complex genome and repressing transcription in *Streptomyces*. It utilizes one editing plasmid with HDR and with high precision deletes single/double genes by 75–95 % in *S. coelicolor*. In absence of HDR templates, *Fn*Cpf1-induced DSB repairs using a reconstructed NHEJ, generates random-sized deletions. For repression of multiple genes, a platform consisting of an integrated customized crRNA array and Dnase-deactivated Cpf1 (ddCpf1) CRISPRi can be used. The strains in which *Sp*Cas9 cannot be used, such as in *S. hygroscopicus* SIPI-KF known for 5-oxomilbemycin production, the *Fn*Cpf1 can be used for HDR-mediated gene deletion. Thus, *Fn*CRISPR-Cpf1 can be considered to be a potential substitute for the conventional CRISPR-Cas9 from *S. pyogenes* (Li et al. 2018).

During the screening for novel antimicrobials, 50% of actinomycete strains are rejected due to the detection of previously identified or known antibiotics. In such cases, CRISPR-Cas9 can be used to knockout genes encoding known antibiotics. This approach has been used in 11 actinomycete strains to delete genes for biosynthesis of well-known antibiotic streptothricin or streptomycin. The application of this simple technique has resulted in the discovery of rare and new variants of thiolactomycin, amicetin, phenanthroviridin, and 5-chloro-3-formylindole antibiotics (Culp et al. 2019).

An all-inclusive toolkit containing variants of CRISPR-Cas9 systems, CRISPRi, and CRISPR-base editing systems (CRISPR-BEST) has been developed for genome editing in *Streptomyces* sp. The toolkit can be used for engineering CRISPR plasmids and introducing them in important strains. The other application of this toolkit in streptomycetes is to synthesize a library consisting of arbitrary sized deletions, delete target genes, regulate gene transcription, and replace base pairs in genome. The toolkit also includes a Csy4-based multiplexing which can be used for numerous editing in one attempt. In comparison to conventional methods, this toolkit can be used to edit and delete a gene in less than 10 days, and hence is considered to be faster (Tong et al. 2018, 2019).

CRISPR/Cas9-based editing tool pMWCas9 has been developed by modifying pWHU2653. The *ermE*p* promoter of Cas9 is replaced with the thiostrepton-inducible promoter *tipA*p to enhance transformation efficiency. This tool successfully deleted highly repetitive *eryAIII* gene DNA sequence from erythromycin PKS in *Saccharopolyspora erythraea* (Mo et al. 2019). Another pWHU2653-modified tool functioning as a dual-functional chromogenic-screen is the pQS-*gusA* and pQS-*idgS*. It has two reporters GusA and IdgS for chromogenic screening, and its inclusion increased genome editing and plasmid curing efficiency by 100 %. It has been successful used in editing genetically recalcitrant strain *Verrucosispora* sp. MS100137. It showed 100 % efficiency in deleting carotenoid and the abyssomicin BGC in these rare and slow-growing actinomycetes (Wang et al. 2020). pWHU2653-TRMA is also constructed by modifying WHU2653. It uses inducible promoter *tipA*p to regulate the Cas9 expression and contains theophylline-inducible riboswitch and mag-based blue light-inducible reconstitution system to regulate Cas9 activity (Wang et al. 2019). Similalry, pKC1139-TRMA based on pKC1139 has been used for deletion of *actII-orf4* and *redD* in *S. coelicolor*. These tools WHU2653-TRMA and pKC1139-TRMA can be used for editing *Streptomyces* species with low transformation efficiency (Wang et al. 2019). And for those *Streptomyces* strains in which *S. pyogenes* Cas9 cannot be applied, pCRISPomyces-2 system can be used. It is constructed using Sth1Cas9 from *Streptococcus thermophilus* CRISPR1, SaCas9 from *Staphylococcus aureus,* and Cpf1 from *F. novicida* (FnCpf1). This tool efficiently carries out HDR-mediated DNA knock-in and gene deletion in several streptomycetes. It was successfully used in *Streptomyces* sp. NRRL S-244 strain (Yeo et al. 2019). These diverse CRISPR/Cas tools have potential for discovering novel natural products and to overproduce desired secondary metabolites in actinomycetes.

12.4 Design and Construction of CRISPR/Cas9 Toolkits in *Streptomyces*

The recently developed and commonly used CRISPR-Cas9 toolkits in *Streptomyces* for in vivo genome editing are pCRISPomyces-2, pKCcas9dO, pCRISPR-Cas9, and pWHU2653 (Alberti and Corre 2019; Zhao et al. 2020). The CRISPR/Cas-based technologies for genome editing in *Streptomyces* are presented in Table 12.2 and Fig. 12.2.

The earliest developed CRISPR/Cas9-based toolkit for streptomycetes was composed of two plasmids, i.e., pCRISPomyces-1 and pCRISPomyces-2. The plasmid pCRISPomyces-1 contains crRNA and tracrRNA cassettes as two entities and pCRISPomyces-2, sgRNA consists of tracrRNA and the crRNA fused together. The temperature-sensitive pSG5 replication region helps in the maintenance and selection of plasmids in cells and their clearance after genome editing. Similarly, streptomycetes codon-optimized *cas9* gene and sgRNA cassette are placed under strong promoters *rpsLp* (CF, from *Cellulomonas flavigena*) and *gapdhp* (EL, from *Eggerthella lenta*), respectively. These promoters are responsible for the efficient expression in *S. lividans*, and tenfold higher activity than the routinely used promoter *ermE*p*. These promoters have shown high efficiency in other streptomycetes such as *S. albus* J1074, *S. venezuelae* ISP5230, *S. coelicolor* M1146, and *S. avermitilis* SUKA16. pCRISPomyces-1 generates short deletions only in *S. lividans* and has lower efficiency, while pCRISPomyces-2 generates both short and large deletions in

Table 12.2 CRSIPR/Cas9-based plasmid toolkits for genome editing in *Streptomyces* species

CRSIPR/Cas9-based plasmid toolkits	*cas9* expression promoter	Guide RNA expression promoter	Replicons	Species tested	References
pCRISPomyces-1	*rpsLp*	*gapdhp* for crRNA and *rpsLp* for tracrRNA	pSG5	*S. lividans*	Cobb et al. (2015)
pCRISPomyces-2	*rpsLp*	*gapdhp* for sgRNA	pSG5	*S. lividans*, *S. viridochromogenes*, *S. albus*, *S. formicae*, *S. rimosus*, *S. showdoensis*, *S. roseosporus*, *S. venezuelae*, and *Actinoplanes* sp. SE50/110	Cobb et al. (2015)
pKCcas9dO	*tipAp*	*j23199p* for sgRNA	pSG5	*S. coelicolor* M145, *S. pristinaespiralis*, and *S. cinnamonensis*	Huang et al. (2015)
pCRISPR-Cas9	*tipAp*	ermE*p for sgRNA	pSG5	*S. coelicolor* A3(2), *Streptomyces* SD-85, and *Micromonospora chersina*	Tong et al. (2015)
pCRISPR-dCas9	*tipAp*	ermE*p for sgRNA	pSG5	*S. coelicolor* A3(2)	Tong et al. (2015)
pWHU2653	*aac(3)IVp*	ermE*p for sgRNA	*pIJ101*	*S. coelicolor* M145	Zeng et al. (2015)

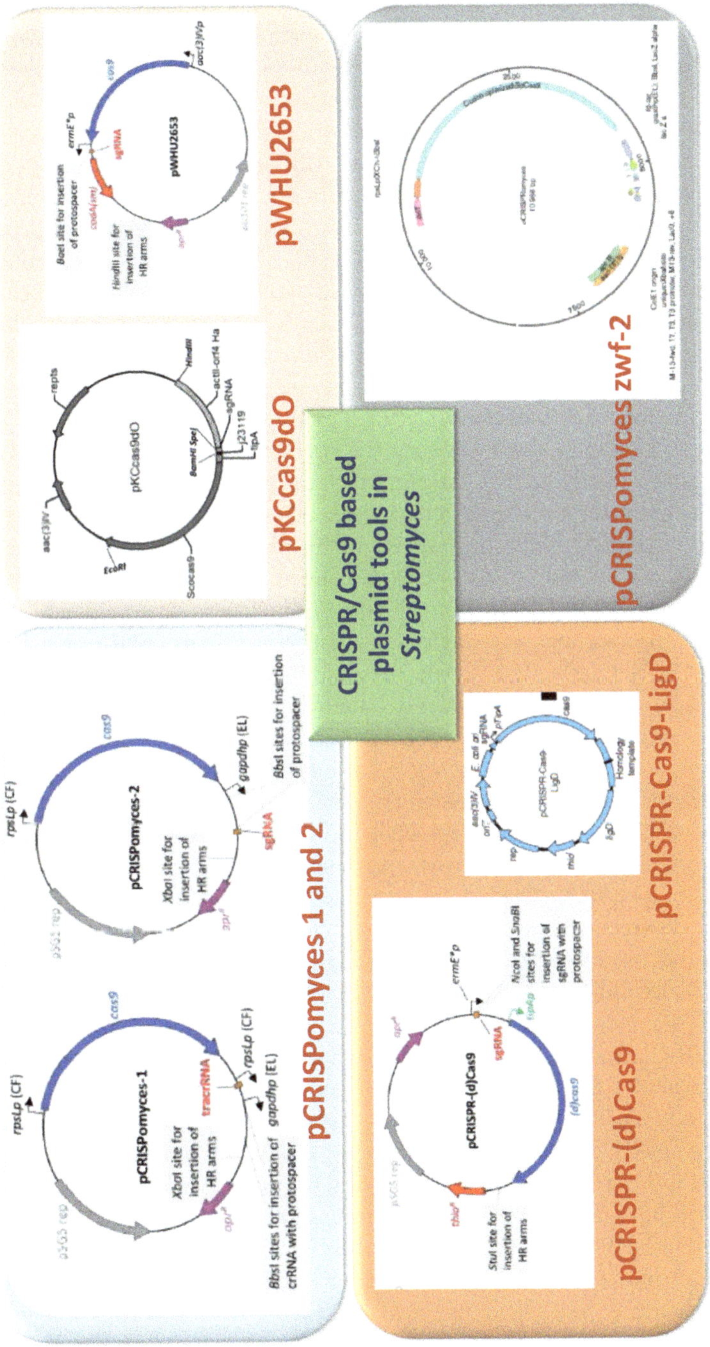

Fig. 12.2 CRISPR/Cas-based toolkits and technologies for genome editing in *Streptomyces* species

S. lividans, *S. albus,* and *S. viridochromogenes* with an efficiency of 67–100 %. Further, pCRISPomyces-2 an cause simultaneous short deletions in two different loci and thus, it is useful for multiplex genome editing. The major drawback in CRISPR/Cas9 editing is the off-target effects of the Cas9 activity. However, such effects can be minimized by using protospacer saving unique last 12 bp plus PAM. pCRISPomyces-2 has been successfully used for gene editing in both established hosts and in new strains. In *S. formicae*, pCRISPomyces-2 was used for deleting both one gene and whole clusters which led to the identification of formicamycin BGC and halogenase ForV. In *S. rimosus,* it was used to cause point mutations and deletions for oxytetracycline overproduction. In *S. showdoensis*, it has been used to understand showdomycin antibiotic glycosylation. pCRISPomyces-2 has been used for activating silent BGCs in *S. albus*, *S. lividans*, *S. roseosporus*, *S. venezuelae,* and *S. viridochromogenes* by inserting single/bidirectional heterologous promoters (Cobb et al. 2015). However, the *cas9* gene and sgRNA promoters need to be optimized in other streptomycetes for wider applications.

Another CRISPR/Cas9 toolkit used for streptomycete genome editing uses a plasmid with a chimeric sgRNA cassette which has a synthetic constitutive promoter j23119. The promoter has been evaluated in *E. coli* (Huang et al. 2015). *Streptomyces coelicolor* codon-optimized *cas9* gene in this plasmid is controlled by *tipAp*, an inducible promoter. Here, pKCcas9dO was initially tested for editing genome of *S. coelicolor* M145 and showed a 29–100 % efficiency for single and multiple deletions, while deletions for single and multiple antibiotic BGC were between 38 and 100 %. Similarly, 100 % deletion was observed in actinorhodin-containing BGC but at low efficiency for undecylprodigiosin BGC (29–71 %) (Li et al. 2018).

sgRNA was used for concurrent double deletion by causing cleavage in BGCs. In *S. coelicolor*, it was also used to generate deletions and point mutations in *rpsL* gene which codes for resistance to streptomycin. This gene editing was useful in studying the role of AfsQ1/Q2 and GlnR regulation in coelimycin P2 and antibiotic production, respectively. The plasmid when used in *S. pristinaespiralis* showed 94 % efficacy in a 25 kb BGC deletion and generated point mutations, which was useful in studying the regulation of pristinamycin I biosynthesis. Further, antibiotic production could be increased by removing a set of transcriptional repressors. In *S. cinnamonensis,* it was used for deletion of gene involved in regulation of DasR in the antibiotic monensin synthesis (Huang et al. 2015).

pCRISPR-Cas9 is a toolkit used in streptomycetes to induce *cas9* gene expression with the promoter thiostrepton responsive activator *tipAp*. But it works only in streptomycetes such as *S. coelicolor,* as its genome has a copy of *TipA* promoter required for sgRNA expression (Tong et al. 2015). In *S. coelicolor* A3(2), the pCRISPR-Cas9 was tested for targeting six positions in *actIORF1* and *actVB*. These genes have a role in actinorhodin biosynthesis. DSBs generated can be repaired using NHEJ pathway or HDR. NHEJ repairs DSB resulting in insertions and deletions in the desired site. Its efficiency could be increased by introducing *LigD* gene which encodes heterologous DNA Ligase D as it increased deletion by 37–77 %. pCRISPR-Cas9 plasmids containing 2.2 kb homologous recombination arms targeted the same two genes of the actinorhodin pathway, causing 100 % deletion and no off-target effects. It was also used to generate genomic deletions by CRISPR interference (CRISPRi). It involved the inactivation of the RuvC1 and HNH nuclease domains in *cas9* gene. The targeting of single amino acid resulted in generation of a Cas9 variant lacking endonuclease function. Hence, this catalytically inactive or dead enzyme is known as dCas9. The enzyme disrupts gene expression by steric hindrance resulting in loss of actinorhodin production in the presence of sgRNA. This pCRISPR-Cas9 toolkit has been used to delete and down-regulate genes in *Corynebacterium glutamicum* for targeting γ-aminobutyric acid biosynthetic genes. Similarly, it has been used in a mangrove streptomycete strain for identifying BGC coding for the polyene macrolactam, sceliphrolactam (Alberti and Corre 2019). In nonstreptomycete species, it was used to remove cytochrome P450 genes and certain PKS genes in *Micromonospora chersina* genome to identify PKS for anthraquinone portion of the dynemicin A antitumor antibiotic.

pWHU2650, CRISPR-Cas9 toolkit uses the constitutive promoter *ermE*p* for controlling streptomycete codon-optimized *cas9* gene expression, while sgRNA has the *aac(3)IVp* promoter control. This promoter is from the gene coding for an aminoglycoside 3-*N*-acetyltransferase type IV, a commonly used marker in the shuttle vectors *E. coli* and *Streptomyces*. It confers resistance to the antibiotic apramycin and hence is used as a marker for all the CRISPR/Cas9-based toolkits in streptomycetes (Zeng et al. 2015). In *S. coelicolor* M145, pWHU2650 was used for targeting actinorhodin biosynthesis genes. An initial study with the use of plasmid-containing sgRNA and without template for HDR, generated seven exconjugants of *S. coelicolor* lacking any mutations. However, the use of HR arms (2 kb) increased *actI-ORF2* gene deletion by 93–99 %. Whereas, on using plasmid without sgRNA but containing both *cas9* gene and the homologous arms had an editing efficiency of only 4 %. Thus, Cas9-generated DSB for homology-directed repair has a crucial role in the process. Introduction of *codA(sm)* gene encoding cytosine deaminase generated both plasmid pWHU2653 and pWHU2659 with sgRNA and HR arms which could act on *actI-ORF2* gene. Inclusion of *codA(sm)* in the plasmid improvised screening and increased genome editing rounds. The exconjugants containing CodA could catalyze the conversion of 5-fluorocytosine to 5-fluorouracil. Therefore, it can also be used as a counter marker. Whereas, the progeny which is marker-free could undergo another round of gene editing. The use of a counter-selectable marker is useful in plasmid clearance and in in vivo engineering which exploits temperature-sensitive replication region in pSG5 gene. Through this technique, 94 % of exconjugants showed preferred loss of editing plasmid and no off-target Cas9 activity (Zeng et al. 2015).

12.5 Application of CRISPR/Cas9 in Actinomycetes

The CRISPR/Cas9 editing tools can find potential application in actinomycetes for biosynthetic pathway reconstruction and metabolic engineering to detect new bioactive compounds and increase the production of certain secondary metabolites (Fig. 12.3).

12.5.1 Detection of Cryptic Genes Encoding Novel Secondary Metabolites

Most of the antibiotics discovered so far and in use are from actinomycetes. Their genome contains numerous biosynthetic gene clusters (BGCs) involved in the synthesis of secondary metabolites. However, only a few BGCs are characterized. The setback for the discovery of novel antibiotics from actinomycetes is due to the lack of use of dereplication tools which leads to the rediscovery of known compounds during screening processes.

The antimicrobial screening process is abandoned if the actinomycete strain produces previously identified antibiotics and it is not explored further for novel compounds. Thus, the rediscovery of a known antibiotic hampers the screening for new compounds. In such scenarios, with the advent of CRISPR-Cas9 editing tool, genome engineering can be performed to knockout genes involved in the synthesis of known antibiotics. This technique has been successfully applied to knockout streptothricin and streptomycin antibiotic-coding genes in an actinomycete screening program. As a result, previously masked rare antibiotics were discovered and new variants of thiolactomycin, amicetin, phenanthroviridin, and 5-chloro-3-formylindole were identified (Culp et al. 2019). The experiment involved developing

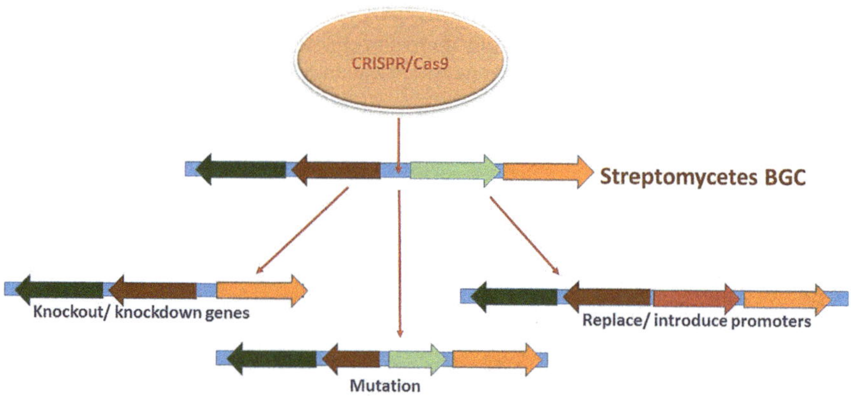

Fig. 12.3 Application of CRISPR/Cas9 editing tools in actinomycetes BGCs for biosynthetic pathway reconstruction. Gene knockout techniques, mutagenesis, and by replacing/introducing of promoters in actinomycetes novel metabolites can be identified and the secondary metabolites production increased.

a CRISPR-Cas9 tool for inactivating antibiotic production by targeting significant genes involved in their biosynthesis. Initially, highly conserved single guide (sg) RNA target sequences in a given BGC were identified. This helped in the preparation of CRISPR-based targeting constructs which could be used in various strains even when BGC sequences were not characterized. Conserved target sites were identified in 20 BGCs of streptothricin with a Python script. Further, sgRNA sites were refined resulting in the identification of *orf15* and *orf17* targets. The use of the editing tool resulted in the identification of amicetin and thiolactomycin producers. Thus, CRISPR/Cas9-targeted inactivation of known BGCs could aid in large-scale genome mining of available strain collections for discovering novel antibiotics (Culp et al. 2019).

12.5.2 Increased Production of Secondary Metabolites

CRISPR/Cas tools can be applied to increase the production of certain bioactive compounds. This tool has been applied in *Streptomyces roseosporus* for increased production of the cyclic lipopeptide antibiotic daptomycin. Three nonribosomal peptide synthetase (NRPS) genes, *dptA*, *dptBC,* and *dptD* are involved in daptomycin biosynthesis, wherein *dptA* gene encodes the first subunit of the NRPS (Miao et al. 2005). In *S. roseosporus*, an increase in the production of daptomycin can be achieved by introducing *kasO*p* in front of the *dptA* gene. However, to introduce a strong promoter in NRPS gene using *Fn*Cas12a WT, TTN PAM sequences suitable for *Fn*Cas12a recognition were not available. Therefore, a CRISPR-Cas12a3 system was developed with the *Fn*Cas12a mutant EP16 (Wang et al. 2019). The *Fn*Cas12a mutant EP16 can recognize YN, TAC, and CAA PAMs. *Fn*Cas12a variant EP16 unlike the WT *Fn*Cas12a worked with the three spacers adjacent to CCG, CCA, and ATC PAMs with good *kasO*p* promoter insertion efficiencies. Thus, CRISPR-*Fn*Cas12a3 system can be used to select suitable PAMs for precise editing in GC-rich genome. Introduction of *kasO*p* followed by plasmid clearance of *S. roseosporus*/PkasA resulted in increased transcription levels of the *dptA*, *dptBC,* and *dptD* genes and consequently increased the production of daptomycin. Though, the editing efficiency of *Fn*Cas12a3 system is low, yet such a Cas12a variant could be used to overcome the requirement of TTN PAM for editing in *Streptomyces* and in applying for a broad range of PAMs (Zhang et al. 2020).

FnCRISPR-Cpf1 system, an alternative to *S. pyogenes* CRISPR-Cas9 system is highly efficient and promising as it in combination with HDR or NHEJ, enables gene (s) deletion with high efficiency. The FnCpf1-based editing tool has been used in genetic modification of an industrially important strain *Streptomyces hygroscopicus* SIPI-KF mainly used for 5-oxomilbemycin production. Using *Fn* CRISPR-Cpf1 system, HDR-mediated precise genome editing was performed in *S. hygroscopicus* SIPI-KF. A TetR-family regulatory gene, *SBI00792*, adjacent to the 5-oxomilbemycin BGC was knocked out. Homologous arms (HAs) of 2.5 kb were used to delete *SBI00792*. Further, the introduction of the plasmid pKCCpf1-*SBI00792*-HA2.5 with two 2.5-kb HAs generated more than 20 exconjugants. Thus, FnCRISPR-Cpf1 genome editing can be used for HDR-mediated gene deletion in specific *Streptomyces* species which cannot utilize the SpCRISPR-Cas9 system. Further, this tool was developed into a simple ddCpf1-CRISPRi platform which can be used for the multiplex transcriptional repression in *Streptomyces* species (Li et al. 2018).

12.5.3 Genome Editing for Elucidation of Biosynthetic Pathways

An endophytic novel species *Streptomyces formicae* isolated from *Tetraponerapenzigi* produces novel pentacyclic polyketides with antimicrobial activity against MRSA and VRE, and also 16 other novel bioactive molecules. Using CRISPR/Cas9 genome editing, it was identified that these compounds were encoded by a single type 2 PKS BGC in *S. formicae*. Further, reinsertion of this BGC led to restoration of formicamycin biosynthesis. Deletion of a gene, *forV* that codes for flavin-dependent halogenase resulted in the loss of halogenated compounds and disrupted the biosynthetic pathway at the fasamycin congener stage. This revealed that halogenation is crucial for post-PKS modification for the formicamycin scaffold. Thus, using the genome editing tool, formicamycin BGC and its biosynthetic pathway were identified (Qin et al. 2017).

The type II CRISPR/Cas9 system has been used for targeted gene editing in the oxytetracycline antibiotic producer *Streptomyces rimosus*. The genome of *S. rimosus* varies from other *Streptomyces* in the terminal and core regions. The genes in this region are involved in the biosynthesis of important metabolites. In *S. rimosus*, *zwf2* and *devB* target genes were edited separately by three different techniques. Using sgRNA-1 or sgRNA-2, single-site mutation was carried out for changing GG to CA, GC to AT, and GG to CC. Similarly, sgRNA-1 and sgRNA-2 were used to make double-site mutations involving deletions and/or point mutations. Gene fragment disruptions were also carried out to edit the genome. It was observed that mutations occurred in the gRNA sequence regions and deletion led to a loss of eight bases, of which three were located in the PAM sequence, its upstream and two bases downstream. This genome editing led to the development of an oxytetracycline high yielding mutant strain (zwf2-devB-). In comparison to the wild-type *S. rimosus*, it was observed that the mutant produced increased antibiotic, that is by 36.8 %. Thus, CRISPR/Cas9 can be used for targeted genome editing and increased production of desired secondary metabolites such as antibiotics (Jia et al. 2017).

WGS of *Streptomyces* sp. SD85, a mangrove sediment isolate revealed that 21.2 % of its 8.6-Mb genome comprises of 52 BGCs. The isolate produced sceliphrolactam, a 26-membered polyene macrolactam. However, its biosynthetic pathway is not known. Genome mining studies showed that BGC of sceliphrolactam (sce) codes for the type I PKS, enzymes for biosynthesis of β-amino acid starter, transporters, and transcriptional regulators. Further, gene knockout with CRISPR/Cas9 tool revealed the importance of sce BGC in sceliphrolactam biosynthesis and

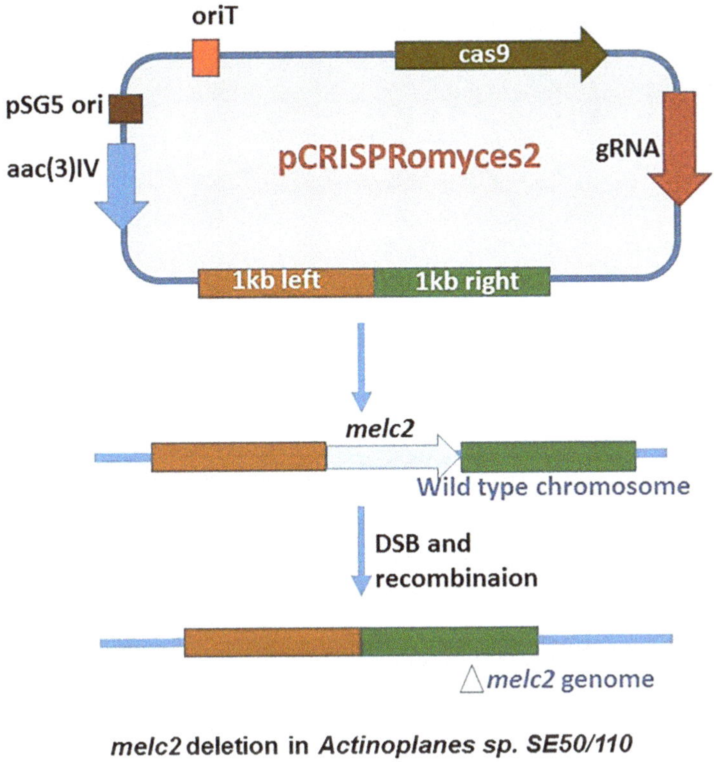

Fig. 12.4 CRISPR/Cas9-based pCRISPomyces system used in the industrial strain *Actinoplanes* sp. SE50/110 for *melC2* deletion to disrupt eumelanin pigment production

the role of sce-encoded PKS in the assembly of the macrolactam skeleton (Low et al. 2018).

CRISPR/Cas9-based pCRISPomyces tool was used for genome editing of industrially relevant *Actinoplanes* sp. SE50/110. This strain is used for the production of therapeutic molecule acarbose which used for the treatment of type II diabetes. Here, genome editing tool was used to stop the production of a dark pigment eumelanin in the strain. CRISPR/Cas9 genome editing was used to delete the gene encoding the tyrosinase MelC, which is involved in the biosynthesis of eumelanin. The resulting ΔmelC2 mutant did not produce the pigment and thus, the culture supernatant was not dark (Fig. 12.4). Further, the Cas9 protein-coding plasmid had temperature-sensitive replication and could be easily removed after editing by increasing the growth temperature. Thus, CRISPR/Cas9-based genome editing was used to create mutation, delete undesirable biosynthesis of metabolites and improve the features of industrially important microorganisms (Wolf et al. 2016). The pCRISPomyces expression system has also been used in successful deletion of redN, from the undecylprodigiosin gene cluster in *S. lividans* genome (Fig 12.5) (Cobb et al.

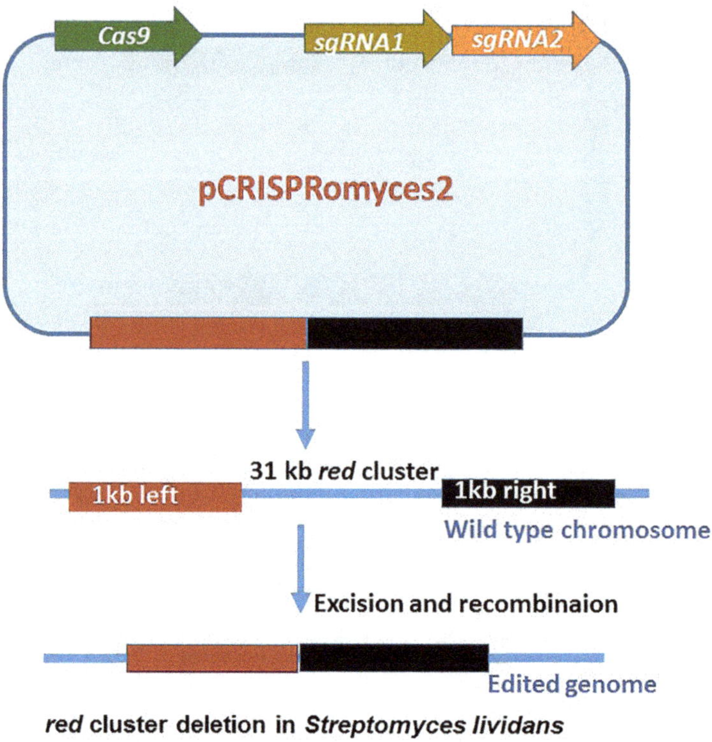

Fig. 12.5 pCRISPomyces plasmids for targeted editing of *red* BGC in *Streptomyces lividans*

2015). Thus, pCRISPomyces system can be used in various *Streptomyces* sp. for targeted editing natural product gene clusters in the wild types and to identify diverse secondary metabolites.

Similarly, CRISPR/Cas9 editing tool has been used to identify the BGCs in the actinobacteria *Micromonospora chersina*. The organism *M. chersina* is known to produce an antitumor antibiotic dynemicin A. This antibiotic belongs to enediynes antibiotics and is known to contain two compounds of polyketide origin fused to each other, i.e., a ten-membered carbocycle and an anthraquinone. However, sequencing the dynemicin BGC revealed only enediyne PKS, whereas anthraquinone PKS was not found. It was concluded that anthraquinone biosynthesis was not part of dynemicin cluster. Therefore, the CRISPR-Cas9 editing tool was used to create mutants and identify the genes coding for anthraquinone. Initially, CRISPR-Cas9 mutants retained dynemicin production. Further, deletion of two cytochromes P450 in the dynemicin BGC led to the identification of DynE8 in dynemicin enediyne PKS, as the one involved in anthraquinone biosynthesis. DynE8, of the enediyne PKS, produces the enediyne and anthraquinone core scaffolds which is required for the synthesis of dynemicin A octaketide (Cohen and Townsend 2018).

12.6 Conclusion and Future Perspectives

The CRISPR/Cas9 application has transformed genome engineering in streptomycetes. The in vivo CRISPR/Cas9-based toolkits have been used for deletion of the single gene as well as BGCs, for the introduction of point mutations and genes. Especially, pCRISPomyces-2 has been used extensively in streptomycetes and non-streptomycetes with the exception of *S. coelicolor*. On using the CRISPR/Cas9-based toolkit and with the inducible expression of *cas9*, the chances of getting exconjugants before Cas9 inactivation are high. The reversible gene knockdowns engineering through CRISPRi, helps in elucidating BGCs and biosynthetic pathways for obtaining the natural products with wide potential of applicability. CRISPR/Cas system derived from *Prevotella* and *Francisella* 1 is known as called Cpf1. It can complement CRISPR/Cas9-based plasmids with T-rich PAM optionality and prerequisite of only a crRNA. The Cpf1 toolkit can be used for CRISPRi-based gene repression. This toolkit works in streptomycete strains such as *S. hygroscopicus* SIPI-KF, where Cas9 is nonfunctional (Alberti and Corre 2019).

CRISPR/Cas9-based tools reduce the time required for genome engineering in comparison to the conventional homologous recombination methods. It can increase editing rounds especially with pWHU2653 plasmid as it has markers for counter selection for loss of plasmid after editing. Similarly, plasmid clearance in other CRISPR/Cas9-based toolkits is done by subculturing repeatedly at 37–39 °C, but it increases the time required for gene editing. The main advantage of CRISPR over Red/ET-based engineering is the avoidance of preparing a cosmid library as the strain of interest can be directly used. The technique is also more efficient than double crossover using genome editing. However, it requires the use of cosmid clones or BGC-captured plasmids especially with CRISPR/Cas9-based ICE in vitro toolkit and with genetically intractable strains. The other limitations with CRISPR/Cas9-based toolkits could be the production of mixed phenotypes strains and Cas9 off-target results as it is tolerant to protospacer sequence mismatches. However, it has been proposed that use of CRISPy-web tools can help in the selection of the best protospacer and the use of multiple protospacer for targets can reduce undesirable events. Further, the application of WGS for mutants can avoid off-target editing in CRISPR/Cas9-based and Red/ET-based genome editing (Tong et al. 2015; Alberti and Corre 2019).

References

Alberti F, Corre C (2019) Editing streptomycete genomes in the CRISPR/Cas9 age. Nat Prod Rep 36(9):1237–1248

Arroyo-Olarte RD, Bravo Rodríguez R, Morales-Ríos E (2021) Genome editing in bacteria: CRISPR-Cas and beyond. Microorganisms 9(4):844. https://doi.org/10.3390/microorganisms9040844. PMID: 33920749; PMCID: PMC8071187

Cobb RE, Wang Y, Zhao H (2015) High-efficiency multiplex genome editing of *Streptomyces* species using an engineered CRISPR/Cas system. ACS Synth Biol 4:723–728. https://doi.org/10.1021/sb500351f

Cohen DR, Townsend CA (2018) A dual role for a polyketide synthase in dynemicin enediyne and anthraquinone biosynthesis. Nat Chem 10(2):231–236. https://doi.org/10.1038/nchem.2876

Culp EJ, Yim G, Waglechner N et al (2019) Hidden antibiotics in actinomycetes can be identified by inactivation of gene clusters for common antibiotics. Nat Biotechnol 37:1149–1154. https://doi.org/10.1038/s41587-019-0241-9

Gust B, Challis GL, Fowler K, Kieser T, Chater KF (2003) PCR-targeted Streptomyces gene replacement identifies a protein domain needed for biosynthesis of the sesquiterpene soil odor geosmin. Proc Natl Acad Sci U S A 100(4):1541–1546. https://doi.org/10.1073/pnas.0337542100. Epub 2003 Jan 31. PMID: 12563033; PMCID: PMC149868

Huang H, Zheng G, Jiang W, Hu H, Lu Y (2015) One-step high-efficiency CRISPR/Cas9-mediated genome editing in *Streptomyces*. Acta Biochim Biophys Sin (Shanghai) 47(4):231–243. https://doi.org/10.1093/abbs/gmv007. Epub 2015 Mar 3. PMID: 25739462

Jia H, Zhang L, Wang T, Han J, Tang H, Zhang L (2017) Development of a CRISPR/Cas9-mediated gene editing tool in *Streptomycesrimosus*. Microbiology (Reading) 163(8):1148–1155. https://doi.org/10.1099/mic.0.000501. Epub 2017 Jul 18. PMID: 28742008

Knott GJ, Doudna JA (2018) CRISPR-Cas guides the future of genetic engineering. Science (New York, N.Y.) 361(6405):866–869. https://doi.org/10.1126/science.aat5011

Komatsu M, Uchiyama T, Ōmura S, Cane DE, Ikeda H (2010) Genome-minimized *Streptomyces* host for the heterologous expression of secondary metabolism. Proc Natl Acad Sci U S A 107: 2646–2651. https://doi.org/10.1073/pnas.0914833107

Li L, Wei K, Zheng G, Liu X, Chen S, Jiang W, Lu Y (2018) CRISPR-Cpf1-assisted multiplex genome editing and transcriptional repression in *Streptomyces*. Appl Environ Microbiol 84(18): e00827–e00818. https://doi.org/10.1128/AEM.00827-18. PMID: 29980561; PMCID: PMC6121969

Low ZJ, Pang LM, Ding Y, Cheang QW, Le Mai Hoang K, Thi Tran H, Li J, Liu XW, Kanagasundaram Y, Yang L, Liang ZX (2018) Identification of a biosynthetic gene cluster for the polyene macrolactam sceliphrolactam in a *Streptomyces* strain isolated from mangrove sediment. Sci Rep 8(1):1594. https://doi.org/10.1038/s41598-018-20018-8. PMID: 29371699; PMCID: PMC5785472

Lu Z, Xie P, Qin Z (2010) Promotion of markerless deletion of the actinorhodin biosynthetic gene cluster in Streptomyces coelicolor. Acta Biochim Biophys Sin (Shanghai) 42(10):717–721. https://doi.org/10.1093/abbs/gmq080. Epub 2010 Sep 1 PMID: 20810535

Miao V, Coeffet-LeGal M-F, Brian P, Brost R, Penn J, Whiting A, Martin S, Ford R, Parr I, Bouchard M, Silva CJ, Wrigley SK, Baltz RH (2005) Daptomycin biosynthesis in Streptomyces roseosporus: cloning and analysis of the gene cluster and revision of peptide stereochemistry. Microbiology 151(Pt 5):1507–1523

Mo J, Wang S, Zhang W, Li C, Deng Z, Zhang L, Qu X (2019) Efficient editing DNA regions with high sequence identity in actinomycetal genomes by a CRISPR-Cas9 system. Synth Syst Biotechnol 4:86–91. https://doi.org/10.1016/j.synbio.2019.02.004

Qin Z, Munnoch JT, Devine R, Holmes NA, Seipke RF, Wilkinson KA, Wilkinson B, Hutchings MI (2017) Formicamycins, antibacterial polyketides produced by *Streptomyces* formicae isolated from African Tetraponera plant-ants. Chem Sci 8(4):3218–3227. https://doi.org/10.1039/c6sc04265a

Tong Y, Charusanti P, Zhang L, Weber T, Lee SY (2015) CRISPR-Cas9-based engineering of actinomycetal genomes. ACS Synth Biol 4:1020–1029. https://doi.org/10.1021/acssynbio.5b00038

Tong Y, Robertsen HL, Blin K, Weber T, Lee SY (2018) CRISPR-Cas9 toolkit for actinomycete genome editing. In: Jensen MK, Keasling JD (eds) Synthetic metabolic pathways. Methods in molecular biology, vol 1671. Humana Press, New York. https://doi.org/10.1007/978-1-4939-7295-1_11

Tong Y, Whitford CM, Robertsen HL, Blin K, Jørgensen TS, Klitgaard AK, Gren T, Jiang X, Weber T, Lee SY (2019) Highly efficient DSB-free base editing for streptomycetes with

CRISPR-BEST. Proc Natl Acad Sci U S A 116:20366–20375. https://doi.org/10.1073/pnas.1913493116

Wang K, Zhao Q-W, Liu Y-F, Sun C-F, Chen X, Burchmore R, Burgess K, Li Y-Q, Mao X-M (2019) Multi-layer controls of Cas9 activity coupled with ATP synthase over-expression for efficient genome editing in Streptomyces. Front Bioeng Biotechnol 7:304. https://doi.org/10.3389/fbioe.2019.00304

Wang Q, Xie F, Tong Y, Habisch R, Yang B, Zhang L, Müller R, Fu C (2020) Dual-function chromogenic screening-based CRISPR/Cas9 genome editing system for actinomycetes. Appl Microbiol Biotechnol 104:225–239. https://doi.org/10.1007/s00253-019-10223-4

Wolf T, Gren T, Thieme E, Wibberg D, Zemke T, Pühler A, Kalinowski J (2016) Targeted genome editing in the rare actinomycete Actinoplanes sp. SE50/110 by using the CRISPR/Cas9 System. J Biotechnol 231:122–128. https://doi.org/10.1016/j.jbiotec.2016.05.039. Epub 2016 Jun 1. PMID: 27262504

Yeo WL, Heng E, Tan LL, Lim YW, Lim YH, Hoon S, Zhao H, Zhang MM, Wong FT (2019) Characterization of Cas proteins for CRISPR-Cas editing in streptomycetes. Biotechnol Bioeng 116:2330–2338. https://doi.org/10.1002/bit.27021

Zeng H, Wen S, Xu W, He Z, Zhai G, Liu Y, Deng Z, Sun Y (2015) Highly efficient editing of the actinorhodin polyketide chain length factor gene in *Streptomyces* coelicolor M145 using CRISPR/Cas9-CodA (sm) combined system. Appl Microbiol Biotechnol 99:10575–10585. https://doi.org/10.1007/s00253-015-6931-4

Zhang J, Zhang D, Zhu J, Liu H, Liang S, Luo Y (2020) Efficient multiplex genome editing in *Streptomyces* via engineered CRISPR-Cas12a systems. Front Bioeng Biotechnol 8:726. https://doi.org/10.3389/fbioe.2020.00726

Zhao Y, Li L, Zheng G, Jiang W, Deng Z, Wang Z, Lu Y (2018) CRISPR/dCas9-mediated multiplex gene repression in *Streptomyces*. Biotechnol J 13:1800121. https://doi.org/10.1002/biot.201800121

Zhao Y, Tian J, Zheng G, Chen J, Sun C, Yang Z, Zimin AA, Jiang W, Deng Z, Wang Z (2019) Multiplex genome editing using a dCas9-cytidine deaminase fusion in *Streptomyces*. Sci China Life Sci 63(7):1053–1062. https://doi.org/10.1007/s11427-019-1559-y

Zhao Y, Li G, Chen Y, Lu Y (2020) Challenges and advances in genome editing technologies in *Streptomyces*. Biomol Ther 10(5):734. https://doi.org/10.3390/biom10050734

Synthetic Biology in Actinomycetes for Natural Product Discovery

13

Shuqing Ning, Tingting Wu, Yushuang Ren, and Yunzi Luo

Abstract

Actinomycetes are outstanding natural products (NPs) producers. As the development of Next Generation Sequencing (NGS) technologies, genomic analysis reveals that a large number of gene clusters in actinomycetes encode rich bioactive substances to expand the natural products library. However, most biosynthetic gene clusters (BGCs) are silent or poorly expressed in native hosts under common laboratory conditions. What is more, traditional fermentation methods always repeatedly identified known product, which is one of the main bottlenecks of NPs discovery. Nowadays, synthetic biology strategies driven by genomics are

Shuqing Ning and TingtingWu contributed equally to this work.

S. Ning
Frontier Science Center for Synthetic Biology and Key Laboratory of Systems Bioengineering (Ministry of Education), School of Chemical Engineering and Technology, Tianjin University, Tianjin, China
e-mail: 2019207579@tju.edu.cn

T. Wu · Y. Ren
Department of Gastroenterology, State Key Laboratory of Biotherapy, West China Hospital, Sichuan University and Collaborative Innovation Center of Biotherapy, Chengdu, China

Y. Luo (✉)
Frontier Science Center for Synthetic Biology and Key Laboratory of Systems Bioengineering (Ministry of Education), School of Chemical Engineering and Technology, Tianjin University, Tianjin, China

Department of Gastroenterology, State Key Laboratory of Biotherapy, West China Hospital, Sichuan University and Collaborative Innovation Center of Biotherapy, Chengdu, China

Collaborative Innovation Center of Chemical Science and Engineering (Tianjin), Tianjin University, Tianjin, China
e-mail: yunzi.luo@tju.edu.cn

R. V. Rai, J. A. Bai (eds.), *Natural Products from Actinomycetes*,
https://doi.org/10.1007/978-981-16-6132-7_13

widely applied in silent BGCs characterization and expression, providing a theoretical and practical technique for the discovery of novel NPs. In this chapter, we summarized the effective synthetic biology strategies for silent gene clusters activation in actinomycetes. We look forward to the NPs "explosion" stage in the near future, ushering by the effective and broad-spectrum synthetic biology platform.

Keywords

Actinomycetes · Synthetic biology · Natural products · Silent BGCs · Pathway engineering · In situ activation · Heterologous expression · CRISPR/Cas9

13.1 Introduction

Natural products (NPs) are valuable sources of inspiration for drug design and development (Davison and Brimble 2019). Actually, NPs are widely applied in the medical and agricultural fields given that they possess plentiful structural diversity and extensive bioactivity. Actinomycetes have been recognized as crucial producers of NPs in the golden age of antibiotics discovery in 1940s to 1960s (Hwang et al. 2014; Robertsen and Musiol-Kroll 2019). In this stage, a number of drugs were isolated from actinomycetes, such as erythromycin A, tylosin, rifamycin, and tetracyclines (Robertsen and Musiol-Kroll 2019). However, antimicrobial resistance (AMR) caused by antibiotics abuse becomes a public health problem seriously threatening human health (https://www.cdc.gov/drugresistance/intl-activities/amr-challenge.html). Thus, discovering bioactive NPs with new skeleton is urgently needed to overcome existing clinical drug resistance (Rutledge and Challis 2015).

Actinomycetes were able to produce a variety of secondary metabolites, including polyketides, lactams, nonribosomal peptides, and terpenes (Ghai et al. 2012; Hwang et al. 2014; Medema and Fischbach 2015). Initially, precious NPs can be easily obtained by simple fermentation, separation, and purification. However, as early as 20–30 years ago, traditional strategies started to receive fewer and fewer returns (Foulston 2019). Recently, powerful bioinformatics tools based on Next Generation Sequencing Technologies (NGST) and genome mining have advanced accurate and effective analysis of actinomycetes genomes (Robertsen and Musiol-Kroll 2019). For instance, antiSMASH can be used to predict various types of BGCs such as polyketide synthase (PKS), nonribosomal peptide synthetase (NRPS), and ribosomally synthesized and post-translationally modified peptide (RiPP). RxnFinder is a search engine using artificial intelligence to collect biochemical pathway information from literature database (Hu et al. 2011). Undoubtedly, the bioinformatics tools promote the discovery and identification of NPs, and discover that about 90% of mysterious NPs are encoded by silent BGCs (Walsh and Fischbach 2010). It is worth noting that, the number of silent BGCs greatly exceed that of the active BGCs (Mao et al. 2018), and few of them can be activated in nonselective fermentation broths (Zhu et al. 2014). To address this issue, efficient strategies to awake the expression of silent BGCs are urgently needed to obtain more bioactive compounds (Baltz 2016; Mao et al. 2018). Meanwhile, it is indispensable to improve high-efficient strains for the reserves of therapeutic molecules (Baltz 2016; Mao et al. 2018).

Generally, synthetic biology whose characteristics are modularization and standardization can be defined as the science of using biotechnology to design a new system or to improve an existing system (Medema and Zhao 2016; Wang et al. 2019; Zhao and Medema 2016; Palazzotto et al. 2019). At present, synthetic biology has been widely used in medical, chemical, food, and agricultural industries. Besides, it is also revolutionizing our understanding of fundamental life sciences. In response to the large number of unexplored resources in actinomycetes genome, synthetic biology techniques are utilized to activate silent BGCs for NPs discovery (Begani et al. 2018). Commonly, the main steps in NPs discovery are as follows: firstly, bioinformatics tools are used to identify the potential BGCs in the genome and infer the potential product molecules. Secondly, DNA assembly technologies are used to clone the related genes and essential regulatory elements to reconstruct the silent BGCs. Thirdly, the biosynthetic pathway is transferred and integrated into the heterologous hosts (Cobb et al. 2012). Finally, analysis tools such as LC-MS and NMR are used to the identification and characterization of NPs (Fig. 13.1).

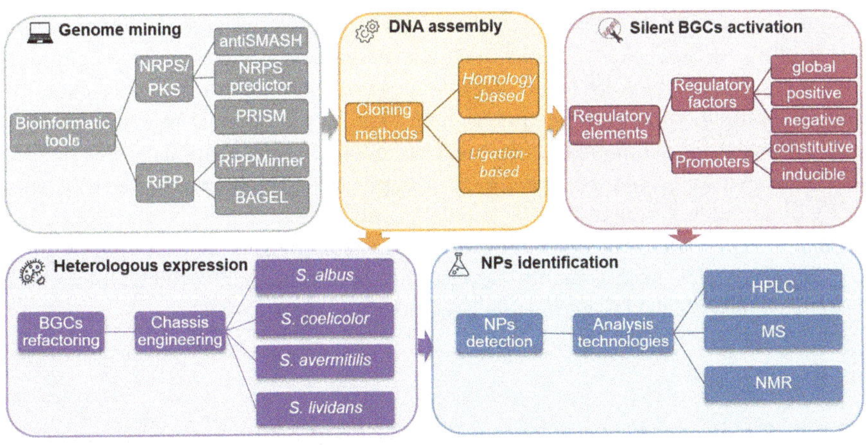

Fig. 13.1 Workflow of natural product discovery in actinomycetes. Firstly, the cryptic BGCs are identified by bioinformatics tools like antiSMASH, NRPSpredictor, and BAGEL after extracting and sequencing the genome of the microorganism of interest; Secondly, the cryptic BGCs are cloned by DNA assembly technologies; Thirdly, regulatory elements are inserted into the original pathway to reconstruct a biosynthetic pathway. Fourthly, the modified pathway is transferred into the heterologous hosts. Finally, the products are purified and identified by HPLC, LC-MS, and NMR

This chapter summarized and discussed two aspects involved in NPs' study via synthetic biology. Firstly, we summarized the current synthetic biology methods including in vivo genome editing tools and in vitro DNA assembly methods. Then we concluded the popular synthetic biology strategies in new NPs discovery, including in situ activation strategies and heterologous expression strategies. As far as we are concerned, synthetic biology technology could provide a new direction for the synthesis of NPs and lay the basis for further expansion of natural product libraries. We are looking forward that novel NPs could overcome the multidrug resistance of microorganisms and provide more options for the treatment of infectious diseases.

13.2 Synthetic Biology Strategies

13.2.1 Definition of Synthetic Biology

For more than half a century, the ability to reasonably design microorganisms has been a long-term goal for biological experts (Cameron et al. 2014). For this purpose, Szybalski explicitly stated a vision for synthetic biology for the first time in 1974 (Zhao 2013). Later, synthetic biology was formally defined in 1990s with the genomics revolution and the rise of systems biology. It is a multidisciplinary subject, combining methods and characteristics of metabolic engineering, bioinformatics, protein engineering, functional genomics, chemical biology, systems biology, and many other research areas. Along with the rapid development in biotechnology industry, synthetic biology has undergone considerable growth in scope, expectation, and output, which has become a main branch of biological research (Alberts 2011). Nowadays, synthetic biology has been applied in many applications, including cost-effective DNA synthesis, protein engineering, pathway engineering, and many other fields (Liang et al. 2011). To avoid confusion, this chapter would define synthetic biology as "An integration of biology and engineering that applies engineering principles to reasonably design and build complex artificial biological systems with predictable and tunable behaviors".

13.2.2 Synthetic Biology Technologies Applied in Actinomycetes

13.2.2.1 Pathway Engineering

Actinomycetes are prokaryotic organisms with complex regulatory cascades and networks (Yin et al. 2019), making them adapt to changeable environments and produce characteristic secondary metabolites, especially potential antibiotics (Barka et al. 2016). It is crucial for discovering new antibiotics given that the synthetic biology strategy led by pathway engineering is conducive to elucidate the regulation mechanism of the biosynthetic pathway (Zhu et al. 2014; Yin et al. 2019).

13.2.2.1.1 Biosynthetic Pathway Regulation via Synthetic Biology Tools

In actinomycetes, the biosynthesis of NPs is highly regulated as gene clusters often remain silent until suitable conditions are met. Regulators usually fall into three groups (Yin et al. 2019). Firstly, cluster-situated regulators (CSRs) located in the biosynthetic gene clusters mainly regulate the production of antibiotics. Secondly, pleiotropic regulators situated outside of BGCs control the production of various antibiotics or the development of bacterial morphology. Thirdly, global regulators which are scattered throughout the chromosomes regulate central metabolic genes and pleiotropic regulatory genes or CSRs (Wei et al. 2018). Expressing the pathway-specific regulator under a well-characterized promoter is an effective approach to trigger the expression of silent BGCs. Researchers created many efficient tools to regulate the expression of crucial genes or regulators in targeted BGCs. For example, Shao et al. developed a simple synthetic biology approach that decouples pathway expression from complex native regulations. In this strategy, strong promoters are confirmed in the target expression host under the selected culture condition to ensure the transcription of downstream genes (Shao et al. 2013). Similarly, Zhang et al. extended the CRISPR/Cas9 technology to perform strategic promoter knock-in for the activation of silent BGCs in native *Streptomyces* hosts (Zhang et al. 2017). In those *Streptomyces* strains that CRISPR/Cas9 does not work, CRISPR/FnCas12a system can be applied to introduce promoters in front of the regulators or BGCs (Zhang et al. 2020). In other conditions, regulators inhibit the expression of some biosynthetic genes and CRISPR/dCas9-based interference system (CRISPRi) was designed to repress their functions (Zhao et al. 2018).

13.2.2.1.2 Biosynthetic Pathway Construction via Synthetic Biology Tools

For many pathways which are difficult to manipulate in situ, reconstruction and modification of corresponding BGCs with DNA assembly techniques would be an efficient activation method. The different composition of assembly line contributes to the diversity of polyketides, therefore modification of individual or combined domains, replacement of modules sourced from other PKS systems, and generation of novel combinations of subunits comprising single or multiple modules are common routines to reprogram PKS clusters (Weissman 2016; Kudo et al. 2020; Staunton 1998).

Decades of research and efforts, synthetic biology has led to the development of plentiful novel DNA assembly strategies. Modern DNA assembly techniques can roughly be classified into two large groups, those based on homology recombination and those based on digestion-ligation (Table 13.1 and Fig. 13.2).

Homology-Based DNA Assembly Methods

In homology recombination-based methods, splicing can be performed by annealing with extending homologous ends in vitro or by homologous recombination in vivo. One of the most well-known in vitro homology-based methods is "Gibson assembly" (Gibson et al. 2009), which is an isothermal and single-reacted method for assembling multiple overlapping DNA molecules by the concerted action of a DNA polymerase, a 5′ exonuclease, and a DNA ligase. This method is broadly used to

Table 13.1 DNA assembly techniques

Method	Description	References
Homology-based *in vitro* assembly methods		
Gibson assembly	An isothermal, single-reaction method for assembling multiple overlapping DNA molecules	Gibson et al. (2009)
CATCH	A method combined CRISPR/Cas9 with Gibson assembly	Jiang et al. (2015)
PIPE	Polymerase incomplete primer extension method	Klock et al. (2008)
SLIC	A method for Sequence- and Ligation-Independent Cloning	Li and Elledge (2012)
USER	An improved method based on ligasefree UDG-mediated cloning	Bitinaite et al. (2007)
CPEC	A method based on polymerase overlap extension	Quan and Tian (2011)
CAT-FISHING	A method combined CRISPR/Cas12a with Gibson assembly	Liang et al. (2020)
Homology-based *in vivo* assembly methods		
DNA assembler	An *in vivo* genetic method for rapid construction of biochemical pathways	Shao et al. (2009)
TAR	A method for constructing new plasmids that relies on interchanging parts of plasmids by homologous recombination	Ma et al. (1987)
mCRISTAR	Multiplexed-CRISPR-TAR	Kang et al. (2016)
miCASTAR	Multiplex in vitro Cas9-TAR	Kim et al. (2019)
MAGIC	A method relies on bacterial mating, *in vivo* site-specific endonuclease cleavage and homologous recombination	Li and Elledge (2005)
SLiCE	A method utilizes bacterial cell extracts to assemble multiple DNA fragments into recombinant DNA molecules	Zhang et al. (2012)
SSRTA	A tandem assembly method based on *Streptomyces* phage wBT1 integrase	Zhang et al. (2011)
Red/ET	A method combined RecET direct cloning with Redαβ recombineering	Wang et al. (2016a)
Ligation-based assembly methods		
Biobrick™	A sequential assembly method based on iterative cycles of restriction digestion and ligation reactions	Shetty et al. (2008)
BglBrick	A method Modified from Biobrick™ standard	Anderson et al. (2010)
golden gate	A one-step, one-pot assembly method relies on type IIs restriction enzymes	Engler et al. (2008)
Start-Stop assembly	A particularly efficient, unbiased, multi-part, hierarchical Golden Gate type assembly method	Taylor et al. (2019)
MASTER	A ligation method for seamless DNA assembly relies on type II restriction enzyme MspJI	Chen et al. (2013)

(continued)

Table 13.1 (continued)

Method	Description	References
LCR	A one-step, scarless DNA assembly method based on ligase cycling reaction	de Kok et al. (2014)
DATEL	A Scarless and Sequence-Independent DNA Assembly Method	Jin et al. (2016)

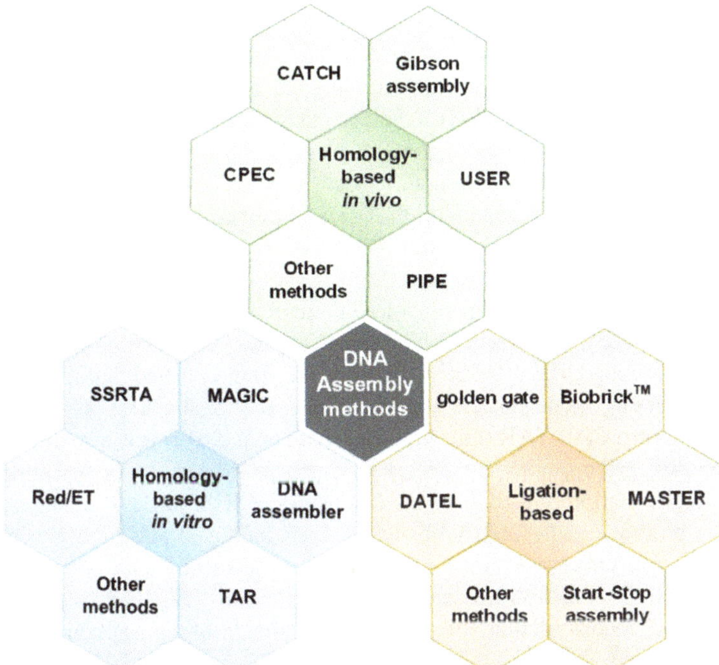

Fig. 13.2 An overview of DNA assembly methods. CATCH Cas9-assisted targeting of chromosome segments, CPEC circular polymerase extension cloning, PIPE the polymerase incomplete primer extension, USER uracil-specific excision reagent friendly DNA engineering, MAGIC mating-assisted genetically integrated cloning, SSRTA the site-specific recombination-based tandem assembly method, Red/ET RecET direct cloning and Redαβ recombineering, TAR transformation-associated recombination, DATEL DNA assembly with thermostable exonuclease and ligase, MASTER methylation-assisted tailorable ends rational ligation

seamlessly assemble natural or synthetic genes, genetic pathways, and entire genomes. Apart from Gibson assembly, there are also many previously developed in vitro assembly techniques. Uracil-Specific Excision Reagent (USER) friendly DNA Engineering (Bitinaite et al. 2007) utilizes uracil-containing primers along with a uracil-specific glycosylase and endonuclease to generate defined single-stranded 3′ ends. The Polymerase Incomplete Primer Extension (PIPE) method (Klock et al. 2008) employs normal PCR reactions to generate a population of partially single-stranded DNA fragments resulting from incomplete primer extension. Circular polymerase extension cloning (CPEC) (Quan and Tian 2011) is based on polymerase overlap extension and is therefore free of restriction digestion, ligation or single-stranded homologous recombination. Sequence- and Ligation-Independent Cloning (SLIC) (Li and Elledge 2012) uses exonuclease and T4 DNA polymerase to generate single-stranded DNA overhangs in insert and vector sequences.

As for in vivo recombination, one prominent representative is the DNA Assembler (Shao et al. 2009) method relying on the homologous recombination mechanism in *S. cerevisiae*. Sharing the similar mechanism, another study constructed a complete synthetic *Mycoplasma genitalium* using 25 overlapping DNA fragments in *S. cerevisiae* (Gibson et al. 2008). Another classical method for constructing new plasmids relying on interchanging parts of plasmids by homologous recombination in *S. cerevisiae* is called transformation-associated recombination (TAR). By this method, large portions of genomic DNA can be cloned via simultaneous transformation of *S. cerevisiae* with genomic DNA containing the target sequence and a receiver vector (Ma et al. 1987).

In order to improve the efficiency of gene manipulation, several DNA assembly methods are usually combined. For instance, via the construction of mCRISTAR platform combining CRISPR/Cas9 with TAR, the single-labeled multiple promoters engineering of large gene clusters was realized in yeast (Kang et al. 2016). Recently, multiplex in vitro Cas9-TAR (miCASTAR) (Kim et al. 2019) was also developed to enable multiplexed promoter engineering with single-marker. Those two approaches involve fragmentation of a BGC of interest using a CRISPR/Cas9 system targeting the native promoter sequences, followed by TAR mediated reassembly to incorporate synthetic promoters. Moreover, the site-specific recombination-based tandem assembly (SSRTA) method (Zhang et al. 2011) utilizes *Streptomyces* phage φBT1 integrase for reconstruction of biological parts. Mating-assisted genetically integrated cloning (MAGIC) (Li and Elledge 2005) uses bacterial mating, in vivo site-specific endonuclease cleavage, and homologous recombination to catalyze the transfer of a DNA fragment between a donor vector and a recipient plasmid. Seamless Ligation Cloning Extract (SLiCE) (Zhang et al. 2012) utilizes bacterial cell extracts to assemble multiple DNA fragments into recombinant DNA molecules in a single in vitro recombination reaction. Red/ET is a useful strategy combining RecE/T direct cloning with Redαβ recombineering (Wang et al. 2016a; Zhang et al. 2003), which can be applied in directly cloning large DNA regions from genomes via *Escherichia coli*. Most recently, CRISPR/Cas12a-mediated fast direct biosynthetic gene cluster cloning (CAT-FISHING) is reported to successfully capture a 87 kb gene cluster with a relatively high efficiency (Liang et al. 2020).

Ligation-Based DNA Assembly Methods

Although numerous powerful homology recombination-based assembly methods are developed, assembly techniques that do not rely on homologous recombination are still needed. Among these methods, BioBrick assembly (Shetty et al. 2008) and its upgraded version BglBrick assembly (Anderson et al. 2010) were created to generate standard biological parts and automate DNA assembly process. Simultaneously, golden gate assembly (Engler et al. 2008) simplifies simultaneous assemble multiple fragments in the desired configuration. Recent years, modular DNA assembly methods based on golden gate assembly were established successively, such as Modular Idempotent DNA Assembly System (MIDAS) (van Dolleweerd et al. 2018) and Start-Stop Assembly (Taylor et al. 2019). According to the methods mentioned above, all DNA recognition sites of the selected type IIS restriction endonuclease within the fragments will be assembled. Different from ligation-based DNA assembly methods mentioned above, methylation-assisted tailorable ends rational (MASTER) ligation (Chen et al. 2013) applies restriction endonuclease MspJI which shares both type IIM and type IIS properties. It only recognizes the methylation-specific 4 bp sites and cuts DNA outside of the recognition sequences.

Apart from sequence homology-based and restriction endonuclease-based methods, DNA assembly method relying on ligase cycling reaction (LCR) was developed (de Kok et al. 2014). In this method, single-stranded bridging oligos complementary to the ends of DNA parts are used. Inspired by the concept of the scarless LCR, DNA assembly with thermostable exonuclease and ligase (DATEL) was created. Besides, the nicking-ligation activity of DNA ligase was designed and developed (Jin et al. 2016).

13.2.2.2 Genome Editing

Decades ago, it was difficult to imagine that people could site-specifically edit the genomes of animals, plants, and microorganisms. With the emerging and developing of genome editing tools, targeted genome editing was successfully implemented, including zinc-finger nucleases (ZFNs), transcription activator-like effector nucleases (TALENs), and clustered regularly interspaced short palindromic repeats-CRISPR-Associated 9 (CRISPR/Cas9). Sharing similar mechanisms, these methods enable a broad range of genetic modifications by introducing DNA double-strand breaks (DSBs) that stimulate error-prone nonhomologous end joining (NHEJ) or homology-directed repair (HDR) at specific genomic locations (Gaj et al. 2013).

CRISPR/Cas system functions as adaptive immune system in numerous bacteria and archaea, of which RNAs harboring "spacer" sequence from previously exposed bacteriophages help Cas proteins recognize and cleave the specific exogenous DNA (Tao et al. 2018). The best studied example is the CRISPR/Cas9 system of *Streptococcus pyogenes*, which was initially used as a genome editing tool for editing bacterial genomes and even human cells (Tong et al. 2019a). The general CRISPR/Cas9-based genome editing process is often initiated by the Cas9 protein combining with sgRNA to form a Cas9-sgRNA complex. Then, the complex searches the 5′-NGG-3′ PAM in the whole genome by random collisions. When a PAM is found, the complex starts to bind to this PAM and interrogates the flanking DNA for spacer complementarity. After that, Cas9 undergoes a series of conformational changes that trigger the nuclease activity, inducing DNA DSB. Finally, the DSB is repaired via the NHEJ repair pathway or through the more efficient and accurate HDR (Fig. 13.3).

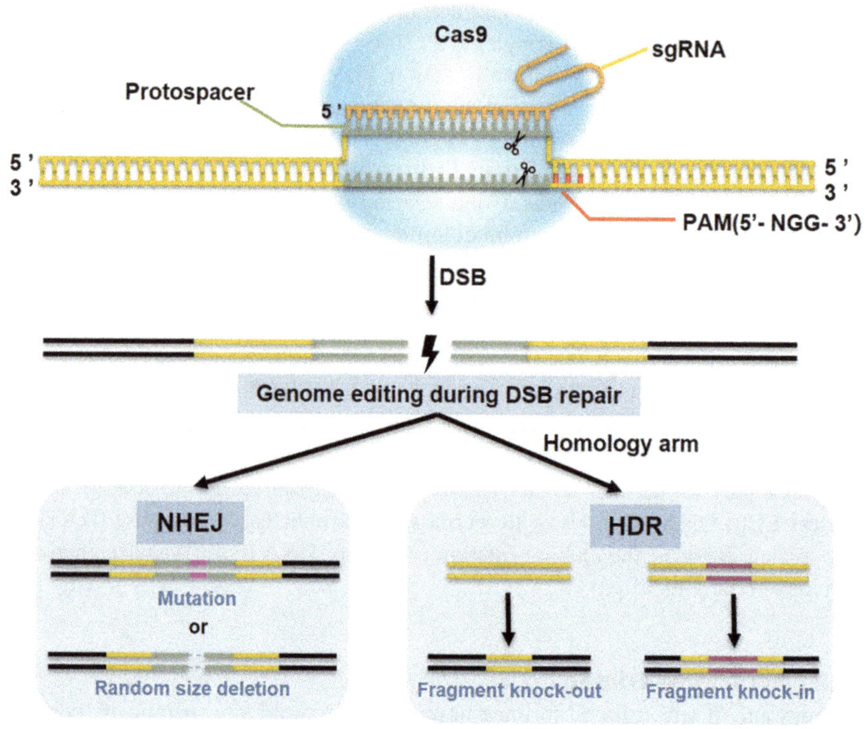

Fig. 13.3 The general CRISPR/Cas9-based genome editing process: (1) the conformation of Cas9-sgRNA complex, (2) the search for 5′-NGG-3′ PAMs in the whole genome under the guidance of Cas9-sgRNA complex, (3) DNA double-strand breaks caused by Cas9 nuclease, (4) the stimulation of nonhomologous end joining or homology-directed repair

CRISPR/Cas9 has become a widely used synthetic biology tool for microbial biosynthetic gene clusters engineering (Li et al. 2017a). Recent years, the emergence of engineered Cas9 homologs and other CRISPR/Cas RNA-guided enzymes, underscores the utility of CRISPR/Cas technology for genome editing and other applications including gene regulation and base editing, driving genome-oriented NPs mining of actinomycetes (Tong et al. 2019a; Doudna 2020). When it comes to actinomycetes genome engineering, compared to single-crossover insertion and other strategies, CRISPR/Cas9 system-based genome editing is more flexible and capable for producing seamless and markerless mutations in a reduced timeframe (Alberti and Corre 2019).

13.2.2.2.1 CRISPR/Cas9

The clustered regularly interspaced short palindromic repeats (CRISPR)/CRISPR-associated (Cas) protein 9 system is a tool for DNA editing at specific sites (Zhang et al. 2014). Nowadays, it has been successfully applied to bacterial, fungi, plants, mammalian, and human cells.

The CRISPR/Cas9 technology originates from type II CRISPR/Cas systems which provide bacteria with adaptive immunity to viruses and plasmids. There are two components of the system: a protein with DNA-cutting activity called Cas9 and a RNA molecule known as the guide RNA. These two components bound together to form a complex that can identify and cut specific sections of target DNA.

The simplicity of CRISPR/Cas9 programming along with a unique DNA double-strain break and repair mechanism makes it a great tool for genetic editing. Researchers obtained a dCas9 mutant without cutting ability by introducing the D10A and H840A mutations. Then, by fusing dCas9 with the reporter gene and regulatory gene, they successfully realized CRISPRi and CRISPRa (Bikard et al. 2013). Moreover, dCas9 fused with pyrimidine deaminase forms "base editors", which can convert the base pairs of C:G to T:A and A:T to G:C (Tong et al. 2019b; Li et al. 2020).

In a word, the simplicity of CRISPR/Cas9 programming, the capacity for multiplexed target recognition, and the existence of many natural or engineered CRISPR/Cas9 system, have enabled remarkable developments for precise and efficient targeting, editing, modification, regulation, and marking genomic loci of a wide array of cells and organisms (Doudna and Charpentier 2014; Doudna 2020).

13.3 Synthetic Biology for Silent Gene Clusters Activation in Actinomycetes

13.3.1 In Situ Activation of BGCs

Signaling factors, pathway-specific regulatory factors, global regulatory factors, and antibiotic feedback regulation compose the complex and elaborate biosynthetic pathways' networks (Xia et al. 2020), which help to balance the dynamic production of metabolites in actinomycetes. Whereas, traditional methods of NPs discovery (like nondirected cultivation and screening optimization) are hard to balance the metabolic network, making it time-consuming and labor-intensive to activate certain BGCs (Xia et al. 2020; Baltz 2016). Consequently, it is necessary to reasonably design or modify natural regulatory elements, such as regulatory factors for coordinating the regulatory network and optimizing the biological metabolic pathways (Hwang et al. 2014; Zhu et al. 2014; Liang et al. 2011). Therefore, this section mainly focuses on using synthetic biology to manipulate key regulatory elements in the native hosts for activating silent BGCs.

13.3.1.1 Promoter Engineering

Promoters occupy an indisputably important position because they are responsible for efficiently starting the transcription process, the first stage of gene expression. Based on the principle of "plug and play", strong promoters are integrated into target BGCs to perform activation functions (Fig. 13.4).

One of the most popular promoter engineering strategies is to directly substitute native promoters with well-characterized strong promoters. Olano et al. inserted the

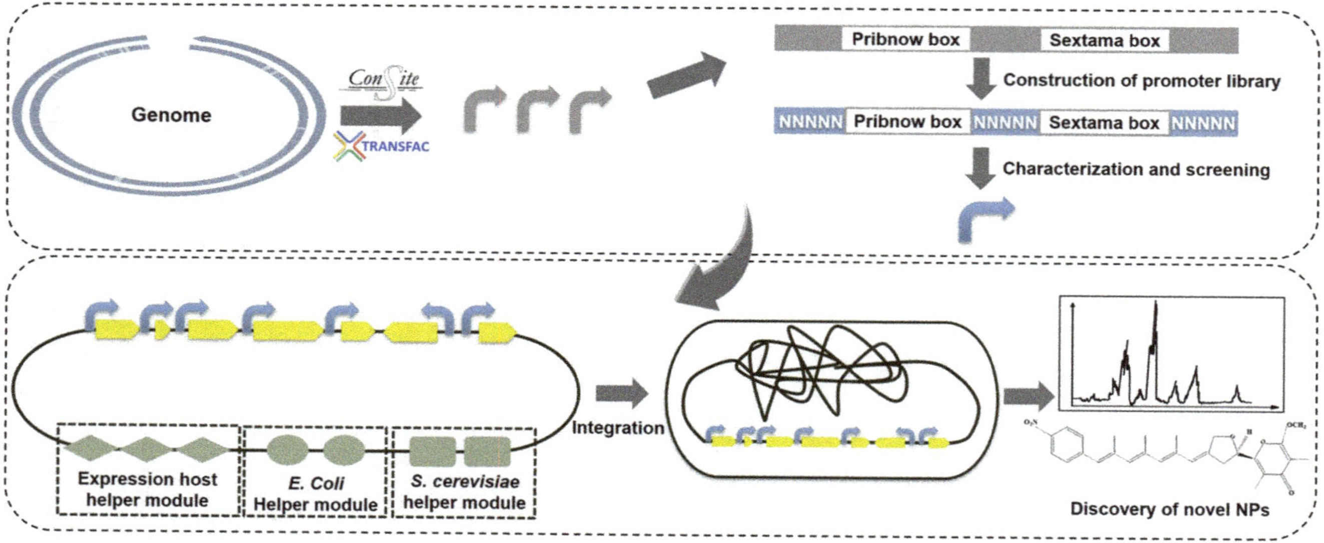

Fig. 13.4 Using promoter engineering to reconstruct and activate BGCs. Firstly, bioinformatics tools are utilized to analyze and predict the promoters of the genome. Then, through random mutation and mutation library screening, the natural promoters are transformed to obtain strong synthetic promoters. Finally, based on plug and play, strong promoters are used for biological pathway reconstruction and heterologous expression to activate BGCs and obtain new NPs

constitutive promoter *ermEp** upstream of the *sshg_00313* gene of a NRPS gene cluster in *S. albus* J1074, successfully activating the biosynthesis of blue pigment indigoidine (Olano et al. 2014). Similarly, inserting *ermEp** in front of the *sshg_05712* and *sshg_05713* genes in the PKS-NRPS hybrid gene cluster successfully activated the biosynthesis of 6-epi-alteramide A and B (Olano et al. 2014). Shao et al. used the "plug and play" scaffolds to reconstruct the entire silent biosynthetic pathway, awakening the spectinabilin pathway of *S. orinoci* (Shao et al. 2013). Luo et al. reconstructed the polycyclic tetramate macrolactams (PTM) biosynthetic genes SGR810-815 from *S. griseus* by inserting six constitutive promoters upstream of the *SGR* genes. Through integration of all modules into *S. lividans*, they found two major products and three minor products in the metabolites (Luo et al. 2013). More commonly, CRISPR/Cas9 system was utilized to knock-in the composite promoter *Pj23119-PkasO* into the *Saccharopolyspora erythraea* genome, activating the expression of the erythromycin BGCs (Liu et al. 2019c). Kim et al. adopted the multiplasmid method, in which each plasmid carries one or two unique gRNAs and a unique auxotrophic marker to construct multiple plasmids-based CRISPR/Cas9 and TAR (mpCRISTAR). The method can simultaneously replace four promoters of the actinomycin BGC with an efficiency close to 100% (Kim et al. 2020a).

To meet the urgent needs of strong promoters in BGCs activation and overproduction, Siegl et al. reasonably modified the consensus sequences in −10 (Pribnow box) and −35 (Sextama box) of promoter *ermEp1* and generated a synthetic promoter library. The sequences located upstream, midstream, and downstream of the consensus sequences were randomized and cloned into the integrated plasmid upstream of the *gusA* reporter gene. As a result, a series of promoters with an intensity between 2 and 319% compared to *ermEp1* were created (Siegl et al. 2013). Similarly, Wang et al. engineered the *kasOp*$_3$ promoter, which is strictly controlled by two regulators, ScbR and ScbR2, based on random mutations and screening mutant libraries. Through eliminating ScbR and ScbR2 binding sites in the core region, they obtained the strong promoter *kasOp** (Wang et al. 2013). In another case, Luo et al. applied RNAseq analysis to characterize 32 candidate promoters identified from *S. albus* J1074. The strength of the identified strong promoters varies from 200 to 1300% of the well-known *ermEp** (Luo et al. 2015).

Fusion of the promoter and the reporting system can be used to screen the activation of silent BGCs. For example, reporter-guided mutant selection (RGMS) was established for targeted activation of silent BGCs, which requires the promoter cloned in front of the *xylE-neo* box to generate the reporter plasmid. Neo reporter molecules are mutagenized by ultraviolet light to produce qualified RGMS, which relied on *xylE* reporter molecular to achieve targeted screening. Using this method, the silent *pga* gene cluster in *Streptomyces* was successfully activated, revealing two new anthraquinone aminoglycosides, gaudimycin D and E (Guo et al. 2015). Similarly, a high-throughput screening ("HiTES") method was invented using a large library of NPs for discovering small molecule initiators of silent BGCs. Two different *S. albus* reporter strains were created using eGFP reporter to activate the *sur* (surugamide) cluster. Ultimately, they identified small molecule products of the *sur* cluster induced by etoposide and ivermectin, such as surugamides G-J, acyl-surugamides A-E, and albucyclones A-F (Okada and Seyedsayamdost 2017; Xu et al. 2017).

13.3.1.2 Regulatory Factors

Regulatory factors are proteins that can directly or indirectly bind cis-element, participating in the regulation of transcription activity. Actinomycetes microorganisms have abundant and diverse regulatory factors (Table 13.2) (Flinspach et al. 2014; Yang et al. 2015b; Jiang et al. 2017; Mao et al. 2017; Som et al. 2017; Barreales et al. 2018; Lu et al. 2018; Wu et al. 2018; Guo et al. 2018; Hou et al. 2018; Wang et al. 2018; Liu et al. 2019a; Martin et al. 2019; Cui et al. 2016). Among them, the presence of some regulatory factors can promote the expression of BGCs, known as positive regulators. Commonly and importantly, up-regulating regulators have been wildly applied to enhance the transcription of certain silent genes, resulting in yielding of novel products (Zhou et al. 2020).

LuxR family proteins are frequently detected in PKS gene clusters (Wilson et al. 2001; Kuscer et al. 2007). LuxR proteins can recognize and combine N-acyl homoserine lactone (AHL) which is a quorum sensing (QS) inducer, to form an active complex. The complex is able to enhance the strength of the promoter *pLux*, and then up-regulate the expression of downstream genes (Subramoni and Venturi 2009). In a previous study, a large ATP-binding LuxR (LAL)-encoding gene *samR0484* in *S. ambofaciens* ATCC 23877 was placed downstream of the constitutive promoter *ermEp** for overexpression. Subsequently, four 51-membered glycosylated macrolides were discovered, named stambomycins A–D (Laureti et al. 2011). Besides, the overexpression of PimM (LuxR family), which comes from the pimaricin cluster in *S. natalensis*, activated a type I PKS biosynthetic gene cluster, contributing to the production of candicidins and antimycins, simultaneously (Olano et al. 2014). Apart from the LuxR family, *Streptomyces* antibiotic regulatory protein (SARP)-type regulators have been revealed to activate transcription by binding to a set of heptameric direct repeats around the −35 region of their cognate promoters, which are important pathway-specific activators of pyramidal cascades of regulation in actinomycetes (Garg and Parry 2010). For example, KyaL is a SARP type regulatory factor. The constitutive promoter drives the transcription of the genes *kyaR1* (a gene encoding KyaL) and *kyaL* (a gene encoding a methylase), bypassing the natural regulatory mechanism of *Saccharopolyspora* and triggering the appearance of kyamicin (Vikeli et al. 2020). Jin et al. characterized seven SARP-family regulators, SARP1-7 in *Streptomyces sp*. Through overexpression of SARP-7, the silent BGCs of this strain were successfully activated, and a new amide-containing polyene was obtained (Du et al. 2016).

In the process of regulating biosynthetic pathways, the regulatory factors that play a repressed role are called negative regulatory factors. Inhibiting or deleting negative regulatory factors can reduce their influence on the expression of BGCs, even obtaining novel NPs. Gottelt et al. identified and deleted the negative regulatory factor coding gene *scbR2* in the type I PKS gene cluster in *S. coelicolor* M145, obtaining a novel yellow compound named yCPK (yellow coelicolor polyketide)

Table 13.2 Regulatory factors in *Actinomycetes*

Year	Hosts	Regulatory factors	Types	Targeted BGCs	References
2014	*Planobispora rosea ATCC 53733*	PbtR	TetR family-like	GE2270(*pbt*) BGCs	Flinspach et al. (2014)
2015	*Streptomyces bottropensis*	Txn9 Txn10	SARP family Two-component response regulator	Trioxacarcin BGCs	Yang et al. (2015b)
2016	*Streptomyces ahygroscopicus*	TtmR I-III	LAL family	Tetramycin BGCs	Cui et al. (2016)
2017	*Streptomyces autolyticus CGMCC0516*	GdmR III	TetR family	Geldanamycin BGCs	Jiang et al. (2017)
2017	*Streptomyces roseosporus*	DepR2	ArsR family	Daptomycin BGCs	Mao et al. (2017)
2017	*Streptomyces colicolor A3(2)*	MtrAB	Two-component response regulator	Actinorbodin and undecylprodigiosyn BGCs	Som et al. (2017)
2018	*Streptomyces natalensis*	PimR	SARP-LAL family	Pimaricin BGCs	Barreales et al. (2018)
2018	*Streptomyces pactum*	PtmE, PtmF PhoR-PhoP	Two-component response regulator	Pactamycin BGCs	Lu et al. (2018)
2018	*Streptomyces actuosus*	NosP	SARP family	Nosiheptide BGCs	Wu et al. (2018)
2018	*Streptomyces avermitilis*	SAV4189	MarR family	Avermectin BGCs	Guo et al. (2018)
2018	*Streptomyces lincolnensis*	LmbU	–	Lincomycin BGCs	Hou et al. (2018)
2018	*Streptomyces ansochromogenes*	SabR1, SabR2	–	Nikkomycin BGCs	Wang et al. (2018)
2019	*Streptomyces avermitilis*	AveI	TetR family	Avermectin and oligomycin BGCs	Liu et al. (2019b)
2019	*Streptomyces hygroscopicus*	GdmR I-III	TetR family	Geldanamycin BGCs	Martin et al. (2019)

(Gottelt et al. 2010). The pathway-specific negative regulatory factor AlpW is a member of the TetR regulatory factor family, which has DNA-binding activity for specific motifs of promoters in BGCs. Mutating gene *alpW* to yield inactivated protein AlpW stimulated the potential of *S. ambofaciens* to produce kinamycins

(Bunet et al. 2011). The *arp* (argimycins P) cluster-coded streptazone E and argimycins P (PVII to PXII) bear a single piperidine ring which was characterized in *S. argillaceus*. Ye et al. believed that inhibiting the negative regulator ArpRII (TetR-like regulatory factor) while overexpressing the positive regulatory factor ArpRI (SARP-like regulatory factor) can significantly increase the expression of *arp* cluster (Ye et al. 2018). Silent chromomycin biosynthetic gene cluster was identified in *S. reseiscleroticus*. Overexpression of the SARP-type activator SrcmRI as well as synergistic inhibition of the PadR-like repressor SrcmRII effectively turned on the biosynthetic pathway of chromomycin, obtaining the titers of chromomycins A3 and A2 to 145.1 ± 15.3 and 158.3 ± 15.4 mg/L in liquid fermentation, respectively (Sun et al. 2018).

Based on existing genetic elements, cleverly designed transcriptional regulation tools can effectively activate silent BGCs. Wang et al. selected and recharacterized the oxytetracycline (OTC)-inducible regulator OtrR, operon *otrO,* and promoter *otrBp* in *S. rimosus*. Using three genetic parts, they developed a tightly induced expression system Potr*, which was induced by OTC under certain concentration range. Potr* was used to activate silent jadomycin BGC and optimize the production of this antibiotic in *S. venezuelae* ISP5230 which demonstrated the usefulness of the system in triggering the discovery of secretive NPs in *Streptomyces* (Wang et al. 2016b). Transcription factor decoy (TFD), a DNA molecule, is similar to the regulatory DNA that binds to the regulator to interfere with gene transcription, thus preventing the regulator from binding to its homologous DNA target. In this study, TFD was designed to bind to negative regulatory factors in silent gene clusters, thus bypassing natural regulatory mechanisms and ensuring smooth biosynthetic pathways. The authors successfully induced and activated eight large silent PKS and NRPS gene clusters (50–134 kb), characterizing a new oxazole family compound produced by a 98 kb BGC (Wang et al. 2019). As a gene editing tool, the CRISPR/Cas system can be used for gene expression regulation. The esterase encoded by the *SACE_1765* inactivates erythromycin by hydrolyzing the macrolactone ring. Through reasonable design, CRISPR/Cas9 was used to knockout the *SACE_1765* gene in *Saccharopolyspora erythraea*, activating a small amount of erythromycin production which has not been characterized before (Liu et al. 2018) (Fig. 13.5).

13.3.2 Heterologous Expression Using *Streptomyces* as Hosts

As mentioned above, optimizing the metabolic regulation in situ is an effective option to obtain new NPs. Usually, the native strains are relative complexity of metabolites' networks, and may be difficult to cultivate. Considering the factors involved in NPs biosynthesis, such as precursors and genetic manipulation, it would be better to activate silent gene clusters by heterologous expression strategy in optimal chassis cells.

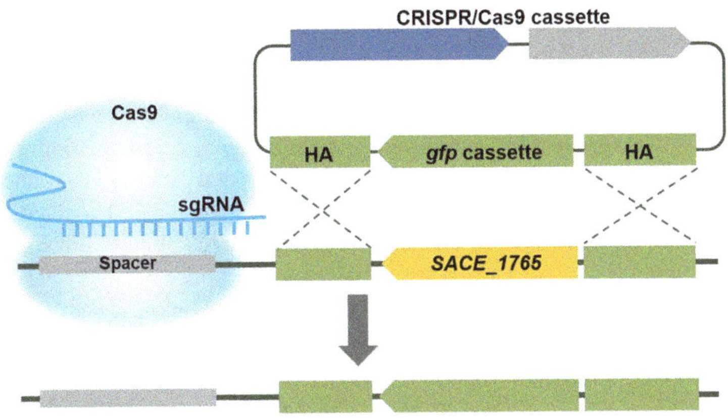

Fig. 13.5 Under the mediation of the CRISPR/Cas9 system, the SACE_1765 gene is knocked out. The gfp gene is inserted as a selection marker. HA homology arm

13.3.2.1 BGCs Construction

The premise of achieving heterologous production is to accurately clone the corresponding BGCs from the original hosts. However, the size of BGCs in actinomycetes ranges from dozens kilobase to hundreds kilobase, which is quite a challenge for traditionally PCR-based cloning, especially with high GC content. Contemporary cloning techniques have been introduced in the second part, such as TAR (Lee et al. 2015) and DNA assembler (Shao et al. 2009) based on homologous recombination of *Saccharomyces cerevisiae*, Red/ET based on homologous recombination of *E. coli* (Wang and Zhang 2005; Fu et al. 2012), and site-specific recombinase (SSR) (Li et al. 2019). These techniques are often applied to effectively clone large BGCs of actinomycetes.

TAR can directly capture large genomic loci (up to 250 kb) with an efficiency of 0.5–2%. Yamanaka et al. used TAR to clone a silent NRPSBGC (67 kb) from a marine actinomycete *Saccharomonospora sp*. CNQ-490. Taking *S. coelicolor* as the host, they obtained taromycinA, a new fatty peptide antibiotic similar to the clinical antibiotic daptomycin, with the yield of 1 mg/L (Yamanaka et al. 2014). Shao et al. reported the DNA assembler method to achieve one-step BGCs cloning of NPs. They constructed the antibiotic aureothin biosynthetic pathway (29.1 kb) and performed its heterologous expression in *S. lividans*, with a yield of 6.1 + 0.2 mg/L (Shao et al. 2011). Heterologous expression can not only activate silent BGCs, but also characterize the function of the genes related to the biosynthetic pathways. They constructed three mutant strains to elucidate that *spnK*, *spnL*, and *spnM* were not required for the spectinabilin biosynthesis, which was consistent with the bioinformatics analysis results (Choi et al. 2010).

Currently, Cas9 has also been utilized in the cloning of large DNA fragments. The combination of Cas9-mediated DNA capture technology and Gibson assembly improves the accuracy of DNA fragment capture. Jiang et al. used Cas9-assisted targeting of chromosome segments (CATCH) to complete one-step cloning of the jadomycin BGC (36 kb) from *S. venezuelan* and chlortetracycline (*ctc*) BGC (32 kb) from *Staphylococcus aureus* (Jiang et al. 2015).

In addition to homologous recombination, SSR system is also an effective tool for cloning large DNA fragments. Almost all site-specific integrases can be divided into two groups: tyrosine recombination enzymes and serine recombination enzymes. Cre, Dre, and Flp are tyrosine recombination enzymes that catalyze DNA interaction and recombination at *loxP*, *rox,* and *FRT* sites, respectively. φC31 and φBT1 from *Streptomyces* gentler phages 1 belong to the serine recombinant enzyme family which catalyzes the site-specific recombination at the phage attachment site (*attP*) and the bacterial attachment site (*attB*) to form two hybridization sites (*attL* and *attR*) (Gregory et al. 2003). Du et al. developed a new strategy of site-specific recombination mediated by phage φBT1 integrase and successfully cloned actinorhodin gene cluster (25 kb) from *S. coelicolor* M145, napsamycin gene cluster (45 kb), and daptomycin gene cluster (157 kb) of *S. roseosporus* NRRL 15998 with positive-clone frequencies above 80% (Du et al. 2015). What is more, an important application of SSR is to delete DNA fragments from the genome, even to construct and modify chassis cells.

13.3.2.2 Refactoring of BGCs

In the process of heterologous activation, the complex regulatory network remains an insurmountable barrier (Kang et al. 2016). Therefore, except for optimizing the cultivation of pleiotropic conditions, collaborative regulatory network is essential for the maintenance of balanced metabolism (Myronovskyi et al. 2016). In transcriptomics, large amounts of mRNA data which are sequenced at transcriptional and translational levels by RNA-seq are elucidated to reflect metabolic profile (Kim et al. 2020b; Bauer et al. 2017). Nowadays, several libraries of regulatory elements have been established, including various constitutive and inducible promoters, ribosome switches, transcriptional terminators, and ribosomal binding sites. Different regulatory elements will lead to unequal regulatory levels. It is significant to select the most suitable element based on quantitative data from transcriptional analysis (Kim et al. 2020b, Bauer et al. 2017). In order to control the yield of novobiocin, an aminocoumarin antibiotic produced by *Streptomyces spheroids*, a synthetic inducible promoter *tcp830* was inserted into the upstream of the BGC (23.4 Kb) and achieved its heterogeneous expression in *S. coelicolor*. In the later research, the *tcp830* promoter was also used for the production of bioactive compounds through heterologous expression (Dangel et al. 2010).

13.3.2.3 Engineering Chassis Cells

In addition to the aspects mentioned above, accurate selection and optimization of chassis cells are the basis for achieving heterologous activation of silent BGCs. It is a widely held view that, except *E. coli* and *S.cerevisiae*, *Streptomyces* is also an outstanding chassis for heterologous expression. According to the statistics, about 30% BGCs of native *Streptomyces* and 10% BGCs of non-*Streptomyces* actinomycetes could be expressed in heterologous *Streptomyces* hosts. The most common chassis cells used for heterologous expression include *S. albus*, *S. coelicolor*, *S. avermitilis*, and *S. lividans* (Table 13.3) (Luzhetskyy et al. 2008;

Table 13.3 Engineered chassis cells for NPs discovery

Name	Engineering strategies	Improvements	Genome size	References
S. albus J1074	Yielded by deleting an active SalI restriction modification system of *S. albus* G genome	Used for cloning and expression of *Streptomyces* spp. genes; Available genetic toolkit; Simple protocols for conjugation	6.79 Mb	Luzhetskyy et al. (2008)
S. albus Del14	Yielded by deleting 15 endogenous BGCs from *S. albus* J1074	Easier to detect and purify the heterologous NPs by simplifying the metabolite profile	6.29 Mb	Myronovskyi et al. (2018)
S. albus B2P1	Yielded by adding additional phage φ31 *attB* sites	Used to improve the titres of heterologous compounds by inserting additional BGC copies	6.79 Mb	Myronovskyi et al. (2018)
*S. coelicolor*A3(2)	Wild type strain	Prominent at heterologous expression of BGCs related antibacterial	8.67 Mb	Yang et al. (2015a)
*S. coelicolor*M1152	Yielded by deleting the Act, Red, CDA, and Cpk biosynthetic gene clusters and mutating *rpoB* (S433L)	High yields than the wild type strain	8.40 Mb	Gomez-Escribano and Bibb (2012)
*S. coelicolor*ZM12	Yielded by deleting 10 PKS and NRPS gene clusters from the genome of *S. coelicolor*M145 and a telomeric region around of 900 kb	Producing more actinorhodin than the wild-type strain	7.53 Mb	Liu et al. (2015)
*S. lividans*TK24	Yielded by mutating *rpsL* [K88E] to enhances the level of NPs production	The most frequently used host of all *S. lividans* for	8.34 Mb	Novakova et al. (2018)

(continued)

Table 13.3 (continued)

Name	Engineering strategies	Improvements	Genome size	References
		heterologous expression		
*S. lividans*ΔYA9	Yielded by deleting 9 gene clusters (178.5 kb)	Producing7 potential NPs with a clean metabolic background	8.11 Mb	Ahmed et al. (2020)
*S. avermitilis*SUKA17	Yielded by deleting 1.4 Mb sub-telomeric region genes	Used to offer a heterologous expression platform of small molecules	7.35 Mb	(Komatsu et al. (2010)

Ahmed et al. 2020; Komatsu et al. 2010; Yang et al. 2015a; Liu et al. 2015; Novakova et al. 2018). Chromosome of *S. albus* J1074 was shortened to 6.8 Mb, so it grows very fast with life cycle of only 4 days (Zaburannyi et al. 2014; Chater and Wilde 1976). As such, *S. albus* J1074 has an outstanding ability to synthesize novel products that are unequalled by other hosts (Winn et al. 2018; Jiang et al. 2018). For example, *S.albus* J1074 is rich in the precursor substances needed by the synthesis of polyketides, thus it is more suitable for the synthesis of PKSs than NRPSs and RiPPs (Myronovskyi et al. 2014; Olano et al. 2014). In order to increase the biosynthetic capacity of NRPSs and RiPPs, 15 endogenous BGCs of *S. albus* J1074 were deleted and generated a new strain *S. albus* Del14. Studies have shown that polyketides (such as tunicamycin, didesmethylmensacarcin, pyridinopyrone) and RiPPs (such as cinnamycin) yielded in *S. albus* Del14 are higher than those in *S. albus* J1074 (Rodriguez Estevez et al. 2019). In addition, the mutant *S. albus* B2P1 was obtained by adding three integrase loci to the genome, which was conducive to increase the copy number of introduced BGCs and thus achieving the goal of increasing product yield (Myronovskyi et al. 2018). Synthetic biology tools simplified the engineering of actinomycetes chassis cells. *S. coelicolor*, with a clear genetic background and simplicity of genetic modification, has been used as a model strain for the study of heterologous expression of BGCs from non-*Streptomyces actinomycetes*, such as *Nocardiopsis* and *Micromonospora* (Gomez-Escribano and Bibb 2012; Widdick et al. 2018). Based on the model strain *S. coelicolor* M154, *S. coelicolor* M1146 is constructed by deleting the *act, red, cpk, and cda* gene clusters. A recent research found two novel compounds in *S. coelicolor* M1146 and *S. lividans* SBT18 by performing the heterologous expression of *asm* gene cluster from marine *S. seoulensis* A01 in the two hosts (Liu, Wang et al. 2019). In addition, on the basis of *S. coelicolor* M1146, *S. coelicolor* M1152 was obtained by mutating *rpoB* (S433L), while *S. coelicolor* M1154 was yielded by mutating *rpoB* (S433L) and *rpsL* (K88E). *RpsL* is known to increase the production of NPs because of inducting protein production in the stationary growth phase, while *rpoB* increased infinity of RNA polymerase. As it was reported, *S. coelicolor* M1152 and

S. coelicolor M1154 could produce higher yields of chloramphenicol and congocidine than the parent strain (Gomez-Escribano and Bibb 2011). Similarly, *S. coelicolor* M1317 was obtained by removing all three endogenous of type III PKS genes (*gcs*, *srsA*, *rppA*) to express exogenous type III PKS genes (Thanapipatsiri et al. 2015). The deletion of ten PKS and NRPS gene clusters on *S. coelicolor* M154 yielded *S. coelicolor* ZM12, which produced more actinomycetin than the wild type strain (Zhou et al. 2012).

S. lividans is also one of the commonly used hosts, which is characterized by low protease activity (Busche et al. 2018). It is the reason why *S. lividans* is suitable for synthesis of peptides, such as capreomycin, daptomycin, viomycin, and labyrinthopeptins (Felnagle et al. 2007; Krawczyk et al. 2013; Barkei et al. 2009; Penn et al. 2006). *S. lividans* TK24 was obtained by mutating the *rpsL* (K88E) gene, and *S. lividans* K4-114 was obtained by deleting the *act* gene. In recent studies, the host capacity was optimized by progressively integrating the global regulatory genes (*nusGsc* and *afsR*), eliminating the negative regulatory genes *wblA* and multidrug efflux pump genes (*lmrA* and *mdfA*) (Rebets et al. 2018).

S. avermitilis has attracted considerable attention for industrial production of avermectin, but its ability to synthesize other NPs has not been explored. One of the outstanding derived strain is *S. avermitilis* SUKA17 yielded by minimizing the chromosomes to construct an expression platform for small molecules (Komatsu et al. 2013).

In addition to the above-mentioned strains, *S. venezuelae*, *S. roseosporus,* and *S. toyocaensis* have also been developed as high-output and low-background chassis cells. *Streptomyces chattanoogensis* L321 was yielded by deleting 0.7 Mb nonessential genomic regions from the genome of *Streptomyces chattanoogensis* L10, which could serve as an efficient chassis for the production of polyketides (Bu et al. 2019). In summary, the engineered chassis cells could follow these principles: (1) Simplify the metabolite profile. Importantly, minimizing the genome size to generate clean metabolic backgrounds was applied in many chassis, such as *S. albus* Del14 and *S. coelicolor* ZM12. By knocking out the endogenous BGCs of *Streptomyces,* the competitive pathways are weakened and the difficulty of product separation and purification in the subsequent process is also simplified. (2) Enrich the precursors. Heterologous hosts prefer to produce certain compounds relying on different precursors. For example, acetyl-CoA is abundant in *S. albus*, so it is a preferred host for polyketides biosynthesis. (3) Coordinate the regulatory system. If a strong promoter is inserted to overexpress a single-positive regulator, an imbalance may be caused, leading to the production of a large number of useless byproducts. Therefore, it is particularly important to design the regulatory system in combination with transcriptome analysis. (4) Be well tolerant to antibiotic. Compared with *S. cerevisiae* and *E. coli*, *Streptomyces* has stronger antibiotic tolerance, thus more suitable in antibiotic biosynthesis. (5) Possess high conjugal transfer efficiency. Due to the different morphology and structure of actinomycetes, the efficiency of conjugal transfer is inconsistent. Thus, it is necessary to explore the optimal conjugation condition in practice. (6) Possess mature gene editing tools. The genome of actinomycetes is rich in GC bases, increasing the difficulty of gene editing.

Therefore, the development of gene editing tools for actinomycetes is also a very popular research direction at present.

13.4 Conclusion and Perspectives

Since the 1990s, the utilization of synthetic biology strategies in the activation of silent BGCs has led to the discovery of plentiful valuable bioactive compounds such as gaudimycin and jadomycin (Myronovskyi and Luzhetskyy 2016; Tao et al. 2018; Zhang et al. 2020; Li et al. 2017b; Zhao et al. 2019; Guo et al. 2015; Wang et al. 2016b). However, discovery of new NPs drugs from talented actinomycetes is still on a declining period. There are several factors that should be taken into consideration. Firstly, the current bioinformatics database lacks integrity and comprehensiveness, due to the limited sequenced genome and well-characterized BGCs. Thus, some BGCs cannot be predicted and annotated with the present bioinformatics database. On the other hand, it is comparatively difficult for researchers to study rare BGCs or the BGCs corresponding to produce completely novel compounds. Recently, the popular used website antiSMASH combined several Comprehensive Data System which provides a significant help in genome mining for potential bioactive components (Blin et al. 2013, 2017a, b, 2019; Weber et al. 2015; Medema et al. 2011). However, the latest antiSMASH 5.0 is deficient in accurate gene functional annotation and detailed structure prediction (Blin et al. 2019). Secondly, structural details of truly novel compounds cannot be predicted accurately by existing bioinformatics tools and the function of many genes in known BGCs is unknown. Thirdly, the available gene manipulation tools cannot meet all cases in actinomycetes for low transformation efficiency and the toxicity of effector proteins. For example, no study on efficiently gene editing tools in rare actinomycetes *gerenzanensis* ATCC 39727 has been reported (Yushchuk et al. 2020). Hopefully, new synthetic gene editing strategies are under studying. Lately, CRISPR-Base Editing System (CRISPR-BEST), an efficient system combining CRISPR/Cas9-associated base editing and site-specific recombination was successfully used to inactivate genes of hygromycin B, resulting further proposed biosynthetic routes for hygromycin B. Besides, the well-characterized CRISPR systems stimulate us to explore new CRISPR platforms or investigate native CRISPR systems to accommodate uneditable actinomycetes (Tong et al. 2019a; Li et al. 2020).

In all, it is of great importance to develop efficient pathway refactoring and synthetic biology tools to breed robust strains of actinomycetes for new NPs drug exploitation. On the other hand, further study on biosynthesis mechanism and machine learning would greatly promote the prediction of uncovered BGCs and corresponding structures as well as annotation of key metabolic pathways. More importantly, combined with accurate bioinformation and reasonable synthetic biology tools may direct us to a second "Golden Age for Antibiotics".

Acknowledgments This work was supported by the National Key R&D Program of China (2018YFA0903300), the Natural Science Foundation of Tianjin Province (19JCYBJC24200), and the National Natural Science Foundation of China (32071426).

References

Ahmed Y, Rebets Y, Estevez MR, Zapp J, Myronovskyi M, Luzhetskyy A (2020) Engineering of Streptomyces lividans for heterologous expression of secondary metabolite gene clusters. Microb Cell Fact 19(1):5. https://doi.org/10.1186/s12934-020-1277-8

Alberti F, Corre C (2019) Editing streptomycete genomes in the CRISPR/Cas9 age. Nat Prod Rep 36(9):1237–1248. https://doi.org/10.1039/c8np00081f

Alberts B (2011) A grand challenge in biology. Science 333(6047):1200. https://doi.org/10.1126/science.1213238

Anderson JC, Dueber JE, Leguia M, Wu GC, Goler JA, Arkin AP, Keasling JD (2010) BglBricks: a flexible standard for biological part assembly. J Biol Eng 4(1):1. https://doi.org/10.1186/1754-1611-4-1

Baltz RH (2016) Genetic manipulation of secondary metabolite biosynthesis for improved production in Streptomyces and other actinomycetes. J Ind Microbiol Biotechnol 43(2–3):343–370. https://doi.org/10.1007/s10295-015-1682-x

Barka EA, Vatsa P, Sanchez L, Gaveau-Vaillant N, Jacquard C, Meier-Kolthoff JP, Klenk HP, Clement C, Ouhdouch Y, van Wezel GP (2016) Taxonomy, physiology, and natural products of Actinobacteria. Microbiol Mol Biol Rev 80(1):1–43. https://doi.org/10.1128/MMBR.00019-15

Barkei JJ, Kevany BM, Felnagle EA, Thomas MG (2009) Investigations into viomycin biosynthesis by using heterologous production in Streptomyces lividans. Chembiochem 10(2):366–376. https://doi.org/10.1002/cbic.200800646

Barreales EG, Vicente CM, de Pedro A, Santos-Aberturas J, Aparicio JF (2018) Promoter engineering reveals the importance of heptameric direct repeats for DNA binding by Streptomyces antibiotic regulatory protein-large ATP-binding regulator of the LuxR family (SARP-LAL) regulators in Streptomyces natalensis. Appl Environ Microbiol 84(10). https://doi.org/10.1128/AEM.00246-18

Bauer JS, Fillinger S, Forstner K, Herbig A, Jones AC, Flinspach K, Sharma C, Gross H, Nieselt K, Apel AK (2017) dRNA-seq transcriptional profiling of the FK506 biosynthetic gene cluster in Streptomyces tsukubaensis NRRL18488 and general analysis of the transcriptome. RNA Biol 14(11):1617–1626. https://doi.org/10.1080/15476286.2017.1341020

Begani J, Lakhani J, Harwani D (2018) Current strategies to induce secondary metabolites from microbial biosynthetic cryptic gene clusters. Ann Microbiol 68(7):419–432. https://doi.org/10.1007/s13213-018-1351-1

Bikard D, Jiang W, Samai P, Hochschild A, Zhang F, Marraffini LA (2013) Programmable repression and activation of bacterial gene expression using an engineered CRISPR-Cas system. Nucleic Acids Res 41(15):7429–7437. https://doi.org/10.1093/nar/gkt520

Bitinaite J, Rubino M, Varma KH, Schildkraut I, Vaisvila R, Vaiskunaite R (2007) USERTM friendly DNA engineering and cloning method by uracil excision. Nucleic Acids Res 35(6):1992–2002. https://doi.org/10.1093/nar/gkm041

Blin K, Medema MH, Kazempour D, Fischbach MA, Breitling R, Takano E, Weber T (2013) antiSMASH 2.0—a versatile platform for genome mining of secondary metabolite producers. Nucleic Acids Res 41(Web Server issue):W204–W212. https://doi.org/10.1093/nar/gkt449

Blin K, Medema MH, Kottmann R, Lee SY, Weber T (2017a) The antiSMASH database, a comprehensive database of microbial secondary metabolite biosynthetic gene clusters. Nucleic Acids Res 45(D1):D555–D559. https://doi.org/10.1093/nar/gkw960

Blin K, Wolf T, Chevrette MG, Lu X, Schwalen CJ, Kautsar SA, Suarez Duran HG, de Los Santos ELC, Kim HU, Nave M, Dickschat JS, Mitchell DA, Shelest E, Breitling R, Takano E, Lee SY, Weber T, Medema MH (2017b) antiSMASH 4.0-improvements in chemistry prediction and gene cluster boundary identification. Nucleic Acids Res 45(W1):W36–W41. https://doi.org/10.1093/nar/gkx319

Blin K, Shaw S, Steinke K, Villebro R, Ziemert N, Lee SY, Medema MH, Weber T (2019) antiSMASH 5.0: updates to the secondary metabolite genome mining pipeline. Nucleic Acids Res 47(W1):W81–W87. https://doi.org/10.1093/nar/gkz310

Bu QT, Yu P, Wang J, Li ZY, Chen XA, Mao XM, Li YQ (2019) Rational construction of genome-reduced and high-efficient industrial Streptomyces chassis based on multiple comparative genomic approaches. Microb Cell Fact 18(1):16. https://doi.org/10.1186/s12934-019-1055-7

Bunet R, Song L, Mendes MV, Corre C, Hotel L, Rouhier N, Framboisier X, Leblond P, Challis GL, Aigle B (2011) Characterization and manipulation of the pathway-specific late regulator AlpW reveals Streptomyces ambofaciens as a new producer of Kinamycins. J Bacteriol 193(5): 1142–1153. https://doi.org/10.1128/JB.01269-10

Busche T, Tsolis KC, Koepff J, Rebets Y, Ruckert C, Hamed MB, Bleidt A, Wiechert W, Lopatniuk M, Yousra A, Anne J, Karamanou S, Oldiges M, Kalinowski J, Luzhetskyy A, Economou A (2018) Multi-omics and targeted approaches to determine the role of cellular proteases in Streptomyces protein secretion. Front Microbiol 9:1174. https://doi.org/10.3389/fmicb.2018.01174

Cameron DE, Bashor CJ, Collins JJ (2014) A brief history of synthetic biology. Nat Rev Microbiol 12(5):381–390. https://doi.org/10.1038/nrmicro3239

Chater KF, Wilde LC (1976) Restriction of a bacteriophage of Streptomyces albus G involving endonuclease SalI. J Bacteriol 128(2):644–650

Chen WH, Qin ZJ, Wang J, Zhao GP (2013) The MASTER (methylation-assisted tailorable ends rational) ligation method for seamless DNA assembly. Nucleic Acids Res 41(8):e93. https://doi.org/10.1093/nar/gkt122

Choi YS, Johannes TW, Simurdiak M, Shao Z, Lu H, Zhao H (2010) Cloning and heterologous expression of the spectinabilin biosynthetic gene cluster from Streptomyces spectabilis. Mol Biosyst 6(2):336–338. https://doi.org/10.1039/b923177c

Cobb RE, Si T, Zhao HM (2012) Directed evolution: an evolving and enabling synthetic biology tool. Curr Opin Chem Biol 16(3–4):285–291

Cui H, Ni X, Liu S, Wang J, Sun Z, Ren J, Su J, Chen G, Xia H (2016) Characterization of three positive regulators for tetramycin biosynthesis in Streptomyces ahygroscopicus. FEMS Microbiol Lett 363(12). https://doi.org/10.1093/femsle/fnw109

Dangel V, Westrich L, Smith MCM, Heide L, Gust B (2010) Use of an inducible promoter for antibiotic production in a heterologous host. Appl Microbiol Biotechnol 87(1):261–269. https://doi.org/10.1007/s00253-009-2435-4

Davison EK, Brimble MA (2019) Natural product derived privileged scaffolds in drug discovery. Curr Opin Chem Biol 52:1–8. https://doi.org/10.1016/j.cbpa.2018.12.007

de Kok S, Stanton LH, Slaby T, Durot M, Holmes VF, Patel KG, Platt D, Shapland EB, Serber Z, Dean J, Newman JD, Chandran SS (2014) Rapid and reliable DNA assembly via ligase cycling reaction. ACS Synth Biol 3(2):97–106. https://doi.org/10.1021/sb4001992

Doudna JA (2020) The promise and challenge of therapeutic genome editing. Nature 578(7794): 229–236. https://doi.org/10.1038/s41586-020-1978-5

Doudna JA, Charpentier E (2014) Genome editing. The new frontier of genome engineering with CRISPR-Cas9. Science 346(6213):1258096. https://doi.org/10.1126/science.1258096

Du DY, Wang L, Tian YQ, Liu H, Tan HR, Niu GQ (2015) Genome engineering and direct cloning of antibiotic gene clusters via phage phi BT1 integrase-mediated site-specific recombination in Streptomyces. Sci Rep 5:8740. https://doi.org/10.1038/srep08740

Du D, Katsuyama Y, Onaka H, Fujie M, Satoh N, Shin-Ya K, Ohnishi Y (2016) Production of a novel amide-containing polyene by activating a cryptic biosynthetic gene cluster in Streptomyces sp. MSC090213JE08. Chembiochem 17(15):1464–1471. https://doi.org/10.1002/cbic.201600167

Engler C, Kandzia R, Marillonnet S (2008) A one pot, one step, precision cloning method with high-throughput capability. PLoS One 3(11):e3647. https://doi.org/10.1371/journal.pone.0003647

Felnagle EA, Rondon MR, Berti AD, Crosby HA, Thomas MG (2007) Identification of the biosynthetic gene cluster and an additional gene for resistance to the antituberculosis drug capreomycin. Appl Environ Microbiol 73(13):4162–4170. https://doi.org/10.1128/AEM.00485-07

Flinspach K, Kapitzke C, Tocchetti A, Sosio M, Apel AK (2014) Heterologous expression of the thiopeptide antibiotic GE2270 from Planobispora rosea ATCC 53733 in Streptomyces coelicolor requires deletion of ribosomal genes from the expression construct. PLoS One 9(3): e90499. https://doi.org/10.1371/journal.pone.0090499

Foulston L (2019) Genome mining and prospects for antibiotic discovery. Curr Opin Microbiol 51: 1–8. https://doi.org/10.1016/j.mib.2019.01.001

Fu J, Bian X, Hu S, Wang H, Huang F, Seibert PM, Plaza A, Xia L, Muller R, Stewart AF, Zhang Y (2012) Full-length RecE enhances linear-linear homologous recombination and facilitates direct cloning for bioprospecting. Nat Biotechnol 30(5):440–446. https://doi.org/10.1038/nbt.2183

Gaj T, Gersbach CA, Barbas CF 3rd. (2013) ZFN, TALEN, and CRISPR/Cas-based methods for genome engineering. Trends Biotechnol 31(7):397–405. https://doi.org/10.1016/j.tibtech.2013.04.004

Garg RP, Parry RJ (2010) Regulation of valanimycin biosynthesis in Streptomyces viridifaciens: characterization of VlmI as a Streptomyces antibiotic regulatory protein (SARP). Microbiology 156(Pt 2):472–483. https://doi.org/10.1099/mic.0.033167-0

Ghai R, McMahon KD, Rodriguez-Valera F (2012) Breaking a paradigm: cosmopolitan and abundant freshwater actinobacteria are low GC. Environ Microbiol Rep 4(1):29–35. https://doi.org/10.1111/j.1758-2229.2011.00274.x

Gibson DG, Benders GA, Axelrod KC, Zaveri J, Algire MA, Moodie M, Montague MG, Craig Venter J, Smith HO, Hutchison CA (2008) One-step assembly in yeast of 25 overlapping DNA fragments to form a complete synthetic Mycoplasma genitalium genome. PNAS 105:20404–20409

Gibson DG, Young L, Chuang R-Y, Craig Venter J, Hutchison CA, Smith HO (2009) Enzymatic assembly of DNA molecules up to several hundred kilobases. Nat Methods 6(5):343–345. https://doi.org/10.1038/nmeth.1318

Gomez-Escribano JP, Bibb MJ (2011) Engineering Streptomyces coelicolor for heterologous expression of secondary metabolite gene clusters. Microb Biotechnol 4(2):207–215. https://doi.org/10.1111/j.1751-7915.2010.00219.x

Gomez-Escribano JP, Bibb MJ (2012) Streptomyces coelicolor as an expression host for heterologous gene clusters. Nat Prod Biosynth Microorgan Plants Pt C 517:279–300. https://doi.org/10.1016/B978-0-12-404634-4.00014-0

Gottelt M, Kol S, Gomez-Escribano JP, Bibb M, Takano E (2010) Deletion of a regulatory gene within the cpk gene cluster reveals novel antibacterial activity in Streptomyces coelicolor A3(2). Microbiology 156(Pt 8):2343–2353. https://doi.org/10.1099/mic.0.038281-0

Gregory MA, Till R, Smith MC (2003) Integration site for Streptomyces phage phiBT1 and development of site-specific integrating vectors. J Bacteriol 185(17):5320–5323. https://doi.org/10.1128/jb.185.17.5320-5323.2003

Guo F, Xiang S, Li L, Wang B, Rajasarkka J, Grondahl-Yli-Hannuksela K, Ai G, Metsa-Ketela M, Yang K (2015) Targeted activation of silent natural product biosynthesis pathways by reporter-guided mutant selection. Metab Eng 28:134–142. https://doi.org/10.1016/j.ymben.2014.12.006

Guo J, Zhang X, Lu X, Liu W, Chen Z, Li J, Deng L, Wen Y (2018) SAV4189, a MarR-family regulator in Streptomyces avermitilis, activates Avermectin biosynthesis. Front Microbiol 9: 1358. https://doi.org/10.3389/fmicb.2018.01358

Hou B, Lin Y, Wu H, Guo M, Petkovic H, Tao L, Zhu X, Ye J, Zhang H (2018) The novel transcriptional regulator LmbU promotes lincomycin biosynthesis through regulating expression of its target genes in Streptomyces lincolnensis. J Bacteriol 200(2):e00447–e00417. https://doi.org/10.1128/JB.00447-17

Hu QN, Deng Z, Hu H, Cao DS, Liang YZ (2011) RxnFinder: biochemical reaction search engines using molecular structures, molecular fragments and reaction similarity. Bioinformatics 27(17): 2465–2467. https://doi.org/10.1093/bioinformatics/btr413

Hwang KS, Kim HU, Charusanti P, Palsson BO, Lee SY (2014) Systems biology and biotechnology of Streptomyces species for the production of secondary metabolites. Biotechnol Adv 32(2):255–268. https://doi.org/10.1016/j.biotechadv.2013.10.008

Jiang W, Zhao X, Gabrieli T, Lou C, Ebenstein Y, Zhu TF (2015) Cas9-Assisted Targeting of CHromosome segments CATCH enables one-step targeted cloning of large gene clusters. Nat Commun 6:8101. https://doi.org/10.1038/ncomms9101

Jiang M, Yin M, Wu S, Han X, Ji K, Wen M, Lu T (2017) GdmRIII, a TetR family transcriptional regulator, controls geldanamycin and elaiophylin biosynthesis in Streptomyces autolyticus CGMCC0516. Sci Rep 7(1):4803. https://doi.org/10.1038/s41598-017-05073-x

Jiang G, Zhang Y, Powell MM, Zhang P, Zuo R, Zhang Y, Kallifidas D, Tieu AM, Luesch H, Loria R, Ding Y (2018) High-yield production of herbicidal thaxtomins and thaxtomin analogs in a nonpathogenic Streptomyces strain. Appl Environ Microbiol 84(11). https://doi.org/10.1128/AEM.00164-18

Jin P, Ding W, Du G, Chen J, Kang Z (2016) DATEL: a Scarless and sequence-independent DNA assembly method using thermostable exonucleases and ligase. ACS Synth Biol 5(9): 1028–1032. https://doi.org/10.1021/acssynbio.6b00078

Kang HS, Charlop-Powers Z, Brady SF (2016) Multiplexed CRISPR/Cas9- and TAR-mediated promoter engineering of natural product biosynthetic gene clusters in yeast. ACS Synth Biol 5(9):1002–1010. https://doi.org/10.1021/acssynbio.6b00080

Kim SH, Lu W, Ahmadi MK, Montiel D, Ternei MA, Brady SF (2019) Atolypenes, tricyclic bacterial sesterterpenes discovered using a multiplexed in vitro Cas9-TAR gene cluster refactoring approach. ACS Synth Biol 8(1):109–118. https://doi.org/10.1021/acssynbio.8b00361

Kim H, Ji CH, Je HW, Kim JP, Kang HS (2020a) mpCRISTAR: multiple plasmid approach for CRISPR/Cas9 and TAR-mediated multiplexed refactoring of natural product biosynthetic gene clusters. ACS Synth Biol 9(1):175–180. https://doi.org/10.1021/acssynbio.9b00382

Kim W, Hwang S, Lee N, Lee Y, Cho S, Palsson B, Cho BK (2020b) Transcriptome and translatome profiles of Streptomyces species in different growth phases. Sci Data 7(1):138. https://doi.org/10.1038/s41597-020-0476-9

Klock HE, Koesema EJ, Knuth MW, Lesley SA (2008) Combining the polymerase incomplete primer extension method for cloning and mutagenesis with microscreening to accelerate structural genomics efforts. Proteins 71(2):982–994. https://doi.org/10.1002/prot.21786

Komatsu M, Uchiyama T, Omura S, Cane DE, Ikeda H (2010) Genome-minimized Streptomyces host for the heterologous expression of secondary metabolism. Proc Natl Acad Sci U S A 107(6):2646–2651. https://doi.org/10.1073/pnas.0914833107

Komatsu M, Komatsu K, Koiwai H, Yamada Y, Kozone I, Izumikawa M, Hashimoto J, Takagi M, Omura S, Shin-ya K, Cane DE, Ikeda H (2013) Engineered Streptomyces avermitilis host for heterologous expression of biosynthetic gene cluster for secondary metabolites. ACS Synth Biol 2(7):384–396. https://doi.org/10.1021/sb3001003

Krawczyk JM, Voller GH, Krawczyk B, Kretz J, Bronstrup M, Sussmuth RD (2013) Heterologous expression and engineering studies of labyrinthopeptins, class III lantibiotics from Actinomadura namibiensis. Chem Biol 20(1):111–122. https://doi.org/10.1016/j.chembiol.2012.10.023

Kudo K, Hashimoto T, Hashimoto J, Kozone I, Kagaya N, Ueoka R, Nishimura T, Komatsu M, Suenaga H, Ikeda H, Shin-Ya K (2020) In vitro Cas9-assisted editing of modular polyketide synthase genes to produce desired natural product derivatives. Nat Commun 11(1):4022. https://doi.org/10.1038/s41467-020-17769-2

Kuscer E, Coates N, Challis I, Gregory M, Wilkinson B, Sheridan R, Petkovic H (2007) Roles of rapH and rapG in positive regulation of rapamycin biosynthesis in Streptomyces hygroscopicus. J Bacteriol 189(13):4756–4763. https://doi.org/10.1128/JB.00129-07

Laureti L, Song L, Huang S, Corre C, Leblond P, Challis GL, Aigle B (2011) Identification of a bioactive 51-membered macrolide complex by activation of a silent polyketide synthase in Streptomyces ambofaciens. Proc Natl Acad Sci U S A 108(15):6258–6263. https://doi.org/10.1073/pnas.1019077108

Lee NC, Larionov V, Kouprina N (2015) Highly efficient CRISPR/Cas9-mediated TAR cloning of genes and chromosomal loci from complex genomes in yeast. Nucleic Acids Res 43(8):e55. https://doi.org/10.1093/nar/gkv112. %/ Published by Oxford University Press on behalf of Nucleic Acids Research 2015. This work is written by (a) US Government employee(s) and is in the public domain in the US

Li MZ, Elledge SJ (2005) MAGIC, an in vivo genetic method for the rapid construction of recombinant DNA molecules. Nat Genet 37(3):311–319. https://doi.org/10.1038/ng1505

Li MZ, Elledge SJ (2012) SLIC: a method for sequence- and ligation-independent cloning. Methods Mol Biol 852:51–59. https://doi.org/10.1007/978-1-61779-564-0_5

Li L, Jiang W, Lu Y (2017a) New strategies and approaches for engineering biosynthetic gene clusters of microbial natural products. Biotechnol Adv 35(8):936–949. https://doi.org/10.1016/j.biotechadv.2017.03.007

Li L, Zheng G, Chen J, Ge M, Jiang W, Lu Y (2017b) Multiplexed site-specific genome engineering for overproducing bioactive secondary metabolites in actinomycetes. Metab Eng 40:80–92. https://doi.org/10.1016/j.ymben.2017.01.004

Li L, Liu X, Wei K, Lu Y, Jiang W (2019) Synthetic biology approaches for chromosomal integration of genes and pathways in industrial microbial systems. Biotechnol Adv 37(5): 730–745. https://doi.org/10.1016/j.biotechadv.2019.04.002

Li S, Liu Q, Zhong Z, Deng Z, Sun Y (2020) Exploration of hygromycin B biosynthesis utilizing CRISPR-Cas9-associated base editing. ACS Chem Biol 15(6):1417–1423. https://doi.org/10.1021/acschembio.0c00071

Liang J, Luo Y, Zhao H (2011) Synthetic biology: putting synthesis into biology. Wiley Interdiscip Rev Syst Biol Med 3(1):7–20. https://doi.org/10.1002/wsbm.104

Liang M, Liu L, Wang W, Zeng X, Liu J, Karthik L, Zhu G, Bai L, Hou C, Chen X, Ouyang L, Liu X, Hu B, Xia X, Tong Y, Lou C, Alterovitz G, Tan GY, Zhang LX (2020) Simple cloning of large natural product biosynthetic gene cluster by CRISPR/Cas12a-mediated fast direct capturing strategy. https://doi.org/10.1101/2020.06.25.170191

Liu C, Zhang J, Lu C, Shen Y (2015) Heterologous expression of galbonolide biosynthetic genes in Streptomyces coelicolor. Antonie Van Leeuwenhoek 107(5):1359–1366. https://doi.org/10.1007/s10482-015-0415-5

Liu Y, Wei WP, Ye BC (2018) High GC content Cas9-mediated genome-editing and biosynthetic gene cluster activation in Saccharopolyspora erythraea. ACS Synth Biol 7(5):1338–1348. https://doi.org/10.1021/acssynbio.7b00448

Liu L, Cheng Y, Lyu M, Zhao X, Wen Y, Li J, Chen Z (2019a) AveI, an AtrA homolog of Streptomyces avermitilis, controls avermectin and oligomycin production, melanogenesis, and morphological differentiation. Appl Microbiol Biotechnol 103(20):8459–8472. https://doi.org/10.1007/s00253-019-10062-3

Liu SH, Wang W, Wang KB, Zhang B, Li W, Shi J, Jiao RH, Tan RX, Ge HM (2019b) Heterologous expression of a cryptic giant type I PKS gene cluster leads to the production of ansaseomycin. Org Lett 21(10):3785–3788. https://doi.org/10.1021/acs.orglett.9b01237

Liu Y, Ren CY, Wei WP, You D, Yin BC, Ye BC (2019c) A CRISPR-Cas9 strategy for activating the Saccharopolyspora erythraea erythromycin biosynthetic gene cluster with Knock-in bidirectional promoters. ACS Synth Biol 8(5):1134–1143. https://doi.org/10.1021/acssynbio.9b00024

Lu W, Alanzi AR, Abugrain ME, Ito T, Mahmud T (2018) Global and pathway-specific transcriptional regulations of pactamycin biosynthesis in Streptomyces pactum. Appl Microbiol Biotechnol 102(24):10589–10601. https://doi.org/10.1007/s00253-018-9375-9

Luo Y, Huang H, Liang J, Wang M, Lu L, Shao Z, Cobb RE, Zhao H (2013) Activation and characterization of a cryptic polycyclic tetramate macrolactam biosynthetic gene cluster. Nat Commun 4:2894. https://doi.org/10.1038/ncomms3894

Luo Y, Zhang L, Barton KW, Zhao H (2015) Systematic identification of a panel of strong constitutive promoters from Streptomyces albus. ACS Synth Biol 4(9):1001–1010. https://doi.org/10.1021/acssynbio.5b00016

Luzhetskyy A, Hoffmann J, Pelzer S, Wohlert S-E, Vente A, Bechthold A (2008) Aranciamycin analogs generated by combinatorial biosynthesis show improved antitumor activity. Appl Microbiol Biotechnol 80(1):15–19. https://doi.org/10.1007/s00253-008-1515-1

Ma H, Kunes S, Schatz PJ, Botstein D (1987) Plasmid construction by homologous recombination in yeast. Gene 58:201–216

Mao XM, Luo S, Li YQ (2017) Negative regulation of daptomycin production by DepR2, an ArsR-family transcriptional factor. J Ind Microbiol Biotechnol 44(12):1653–1658. https://doi.org/10.1007/s10295-017-1983-3

Mao D, Okada BK, Wu Y, Xu F, Seyedsayamdost MR (2018) Recent advances in activating silent biosynthetic gene clusters in bacteria. Curr Opin Microbiol 45:156–163. https://doi.org/10.1016/j.mib.2018.05.001

Martin JF, Ramos A, Liras P (2019) Regulation of geldanamycin biosynthesis by cluster-situated transcription factors and the master regulator PhoP. Antibiotics (Basel) 8(3). https://doi.org/10.3390/antibiotics8030087

Medema MH, Fischbach MA (2015) Computational approaches to natural product discovery. Nat Chem Biol 11(9):639–648. https://doi.org/10.1038/nchembio.1884

Medema MH, Zhao HM (2016) Editorial: synthetic biology and bioinformatics. Nat Prod Rep 33(8):913–914

Medema MH, Blin K, Cimermancic P, de Jager V, Zakrzewski P, Fischbach MA, Weber T, Takano E, Breitling R (2011) antiSMASH: rapid identification, annotation and analysis of secondary metabolite biosynthesis gene clusters in bacterial and fungal genome sequences. Nucleic Acids Res 39(Web Server issue):W339–W346. https://doi.org/10.1093/nar/gkr466

Myronovskyi M, Luzhetskyy A (2016) Native and engineered promoters in natural product discovery. Nat Prod Rep 33(8):1006–1019. https://doi.org/10.1039/c6np00002a

Myronovskyi M, Tokovenko B, Brotz E, Ruckert C, Kalinowski J, Luzhetskyy A (2014) Genome rearrangements of Streptomyces albus J1074 lead to the carotenoid gene cluster activation. Appl Microbiol Biotechnol 98(2):795–806. https://doi.org/10.1007/s00253-013-5440-6

Myronovskyi M, Brotz E, Rosenkranzer B, Manderscheid N, Tokovenko B, Rebets Y, Luzhetskyy A (2016) Generation of new compounds through unbalanced transcription of landomycin A cluster. Appl Microbiol Biotechnol 100(21):9175–9186. https://doi.org/10.1007/s00253-016-7721-3

Myronovskyi M, Rosenkranzer B, Nadmid S, Pujic P, Normand P, Luzhetskyy A (2018) Generation of a cluster-free Streptomyces albus chassis strains for improved heterologous expression of secondary metabolite clusters. Metab Eng 49:316–324. https://doi.org/10.1016/j.ymben.2018.09.004

Novakova R, Nunez LE, Homerova D, Knirschova R, Feckova L, Rezuchova B, Sevcikova B, Menendez N, Moris F, Cortes J, Kormanec J (2018) Increased heterologous production of the antitumoral polyketide mithramycin A by engineered Streptomyces lividans TK24 strains. Appl Microbiol Biotechnol 102(2):857–869. https://doi.org/10.1007/s00253-017-8642-5

Okada BK, Seyedsayamdost MR (2017) Antibiotic dialogues: induction of silent biosynthetic gene clusters by exogenous small molecules. FEMS Microbiol Rev 41(1):19–33. https://doi.org/10.1093/femsre/fuw035

Olano C, Garcia I, Gonzalez A, Rodriguez M, Rozas D, Rubio J, Sanchez-Hidalgo M, Brana AF, Mendez C, Salas JA (2014) Activation and identification of five clusters for secondary metabolites in Streptomyces albus J1074. Microb Biotechnol 7(3):242–256. https://doi.org/10.1111/1751-7915.12116

Palazzotto E, Tong YJ, Lee SY, Weber T (2019) Synthetic biology and metabolic engineering of actinomycetes for natural product discovery. Biotechnol Adv 37(6):107366. https://doi.org/10.1016/j.biotechadv.2019.03.005

Penn J, Li X, Whiting A, Latif M, Gibson T, Silva CJ, Brian P, Davies J, Miao V, Wrigley SK, Baltz RH (2006) Heterologous production of daptomycin in Streptomyces lividans. J Ind Microbiol Biotechnol 33(2):121–128. https://doi.org/10.1007/s10295-005-0033-8

Quan J, Tian J (2011) Circular polymerase extension cloning for high-throughput cloning of complex and combinatorial DNA libraries. Nat Protoc 6(2):242–251. https://doi.org/10.1038/nprot.2010.181

Rebets Y, Tsolis KC, Guethmundsdottir EE, Koepff J, Wawiernia B, Busche T, Bleidt A, Horbal L, Myronovskyi M, Ahmed Y, Wiechert W, Ruckert C, Hamed MB, Bilyk B, Anne J, Friethjonsson O, Kalinowski J, Oldiges M, Economou A, Luzhetskyy A (2018) Characterization of sigma factor genes in Streptomyces lividans TK24 using a genomic library-based approach for multiple gene deletions. Front Microbiol 9:3033. https://doi.org/10.3389/fmicb.2018.03033

Robertsen HL, Musiol-Kroll EM (2019) Actinomycete-derived polyketides as a source of antibiotics and lead structures for the development of new antimicrobial drugs. Antibiotics (Basel) 8(4). https://doi.org/10.3390/antibiotics8040157

Rodriguez Estevez M, Gummerlich N, Myronovskyi M, Zapp J, Luzhetskyy A (2019) Benzanthric acid, a novel metabolite from Streptomyces albus Del14 expressing the nybomycin gene cluster. Front Chem 7:896. https://doi.org/10.3389/fchem.2019.00896

Rutledge PJ, Challis GL (2015) Discovery of microbial natural products by activation of silent biosynthetic gene clusters. Nat Rev Microbiol 13(8):509–523. https://doi.org/10.1038/nrmicro3496

Shao Z, Zhao H, Zhao H (2009) DNA assembler, an in vivo genetic method for rapid construction of biochemical pathways. Nucleic Acids Res 37(2):e16. https://doi.org/10.1093/nar/gkn991

Shao Z, Luo Y, Zhao H (2011) Rapid characterization and engineering of natural product biosynthetic pathways via DNA assembler. Mol Biosyst 7(4):1056–1059. https://doi.org/10.1039/c0mb00338g

Shao Z, Rao G, Li C, Abil Z, Luo Y, Zhao H (2013) Refactoring the silent spectinabilin gene cluster using a plug-and-play scaffold. ACS Synth Biol 2(11):662–669. https://doi.org/10.1021/sb400058n

Shetty RP, Endy D, Knight TF Jr (2008) Engineering BioBrick vectors from BioBrick parts. J Biol Eng 2:5. https://doi.org/10.1186/1754-1611-2-5

Siegl T, Tokovenko B, Myronovskyi M, Luzhetskyy A (2013) Design, construction and characterisation of a synthetic promoter library for fine-tuned gene expression in actinomycetes. Metab Eng 19:98–106. https://doi.org/10.1016/j.ymben.2013.07.006

Som NF, Heine D, Holmes N, Knowles F, Chandra G, Seipke RF, Hoskisson PA, Wilkinson B, Hutchings MI (2017) The MtrAB two-component system controls antibiotic production in Streptomyces coelicolor A3(2). Microbiology 163(10):1415–1419. https://doi.org/10.1099/mic.0.000524

Staunton J (1998) Combinatorial biosynthesis of erythromycin and complex polyketides. Curr Opin Chem Biol 2(3):339–345. https://doi.org/10.1016/S1367-5931(98)80007-0

Subramoni S, Venturi V (2009) LuxR-family 'solos': bachelor sensors/regulators of signalling molecules. Microbiology 155(Pt 5):1377–1385. https://doi.org/10.1099/mic.0.026849-0

Sun L, Zeng J, Cui P, Wang W, Yu D, Zhan J (2018) Manipulation of two regulatory genes for efficient production of chromomycins in Streptomyces reseiscleroticus. J Biol Eng 12:9. https://doi.org/10.1186/s13036-018-0103-x

Tao W, Yang A, Deng Z, Sun Y (2018) CRISPR/Cas9-based editing of Streptomyces for discovery, characterization, and production of natural products. Front Microbiol 9:1660. https://doi.org/10.3389/fmicb.2018.01660

Taylor GM, Mordaka PM, Heap JT (2019) Start-Stop Assembly: a functionally scarless DNA assembly system optimized for metabolic engineering. Nucleic Acids Res 47(3):e17. https://doi.org/10.1093/nar/gky1182

Thanapipatsiri A, Claesen J, Gomez-Escribano JP, Bibb M, Thamchaipenet A (2015) A Streptomyces coelicolor host for the heterologous expression of Type III polyketide synthase genes. Microb Cell Fact 14:145. https://doi.org/10.1186/s12934-015-0335-0

Tong Y, Weber T, Lee SY (2019a) CRISPR/Cas-based genome engineering in natural product discovery. Nat Prod Rep 36(9):1262–1280. https://doi.org/10.1039/c8np00089a

Tong Y, Whitford CM, Robertsen HL, Blin K, Jorgensen TS, Klitgaard AK, Gren T, Jiang X, Weber T, Lee SY (2019b) Highly efficient DSB-free base editing for streptomycetes with CRISPR-BEST. Proc Natl Acad Sci U S A 116(41):20366–20375. https://doi.org/10.1073/pnas.1913493116

van Dolleweerd CJ, Kessans SA, Van de Bittner KC, Bustamante LY, Bundela R, Scott B, Nicholson MJ, Parker EJ (2018) MIDAS: a modular DNA assembly system for synthetic biology. ACS Synth Biol 7(4):1018–1029. https://doi.org/10.1021/acssynbio.7b00363

Vikeli E, Widdick DA, Batey SFD, Heine D, Holmes NA, Bibb MJ, Martins DJ, Pierce NE, Hutchings MI, Wilkinson B (2020) In situ activation and heterologous production of a cryptic lantibiotic from an African plant ant-derived Saccharopolyspora species. Appl Environ Microbiol 86(3). https://doi.org/10.1128/AEM.01876-19

Walsh CT, Fischbach MA (2010) Natural products version 2.0: connecting genes to molecules. J Am Chem Soc 132(8):2469–2493. https://doi.org/10.1021/ja909118a

Wang JP, Zhang YM (2005) [The application of Red/ET recombination to high efficient gene-targeting vector construction]. Yi Chuan 27(6):953–958

Wang W, Li X, Wang J, Xiang S, Feng X, Yang K (2013) An engineered strong promoter for streptomycetes. Appl Environ Microbiol 79(14):4484–4492. https://doi.org/10.1128/AEM.00985-13

Wang H, Li Z, Jia R, Hou Y, Yin J, Bian X, Li A, Muller R, Stewart AF, Fu J, Zhang Y (2016a) RecET direct cloning and Redalphabeta recombineering of biosynthetic gene clusters, large operons or single genes for heterologous expression. Nat Protoc 11(7):1175–1190. https://doi.org/10.1038/nprot.2016.054

Wang W, Yang T, Li Y, Li S, Yin S, Styles K, Corre C, Yang K (2016b) Development of a synthetic oxytetracycline-inducible expression system for Streptomycetes using de novo characterized genetic parts. ACS Synth Biol 5(7):765–773. https://doi.org/10.1021/acssynbio.6b00087

Wang W, Zhang J, Liu X, Li D, Li Y, Tian Y, Tan H (2018) Identification of a butenolide signaling system that regulates nikkomycin biosynthesis in Streptomyces. J Biol Chem 293(52): 20029–20040. https://doi.org/10.1074/jbc.RA118.005667

Wang B, Guo F, Dong SH, Zhao H (2019) Activation of silent biosynthetic gene clusters using transcription factor decoys. Nat Chem Biol 15(2):111–114. https://doi.org/10.1038/s41589-018-0187-0

Weber T, Blin K, Duddela S, Krug D, Kim HU, Bruccoleri R, Lee SY, Fischbach MA, Muller R, Wohlleben W, Breitling R, Takano E, Medema MH (2015) antiSMASH 3.0-a comprehensive resource for the genome mining of biosynthetic gene clusters. Nucleic Acids Res 43(W1): W237–W243. https://doi.org/10.1093/nar/gkv437

Wei J, He L, Niu G (2018) Regulation of antibiotic biosynthesis in actinomycetes: perspectives and challenges. Synth Syst Biotechnol 3(4):229–235. https://doi.org/10.1016/j.synbio.2018.10.005

Weissman KJ (2016) Genetic engineering of modular PKSs: from combinatorial biosynthesis to synthetic biology. Nat Prod Rep 33(2):203–230. https://doi.org/10.1039/c5np00109a

Widdick D, Royer SF, Wang H, Vior NM, Gomez-Escribano JP, Davis BG, Bibb MJ (2018) Analysis of the tunicamycin biosynthetic gene cluster of Streptomyces chartreusis reveals new insights into tunicamycin production and immunity. Antimicrob Agents Chemother 62(8). https://doi.org/10.1128/AAC.00130-18;e00130-18

Wilson DJ, Xue Y, Reynolds KA, Sherman DH (2001) Characterization and analysis of the PikD regulatory factor in the pikromycin biosynthetic pathway of Streptomyces venezuelae. J Bacteriol 183(11):3468–3475. https://doi.org/10.1128/JB.183.11.3468-3475.2001

Winn M, Francis D, Micklefield J (2018) De novo biosynthesis of "non-natural" thaxtomin phytotoxins. Angew Chem Int Ed Engl 57(23):6830–6833. https://doi.org/10.1002/anie.201801525

Wu X, Jin L, Zhang H, Tong R, Ma M, Chen Y (2018) Identification of truncated form of NosP as a transcription factor to regulate the biosynthesis of nosiheptide. FASEB J 32(1):453–465. https://doi.org/10.1096/fj.201700556R

Xia H, Li X, Li Z, Zhan X, Mao X, Li Y (2020) The application of regulatory cascades in Streptomyces: yield enhancement and metabolite mining. Front Microbiol 11:406. https://doi.org/10.3389/fmicb.2020.00406

Xu F, Nazari B, Moon K, Bushin LB, Seyedsayamdost MR (2017) Discovery of a cryptic antifungal compound from Streptomyces albus J1074 using high-throughput elicitor screens. J Am Chem Soc 139(27):9203–9212. https://doi.org/10.1021/jacs.7b02716

Yamanaka K, Reynolds KA, Kersten RD, Ryan KS, Gonzalez DJ, Nizet V, Dorrestein PC, Moore BS (2014) Direct cloning and refactoring of a silent lipopeptide biosynthetic gene cluster yields the antibiotic taromycin A. Proc Natl Acad Sci U S A 111(5):1957–1962. https://doi.org/10.1073/pnas.1319584111

Yang C, Huang C, Zhang W, Zhu Y, Zhang C (2015a) Heterologous expression of fluostatin gene cluster leads to a bioactive heterodimer. Org Lett 17(21):5324–5327. https://doi.org/10.1021/acs.orglett.5b02683

Yang K, Qi LH, Zhang M, Hou XF, Pan HX, Tang GL, Wang W, Yuan H (2015b) The SARP family regulator Txn9 and two-component response regulator Txn11 are key activators for trioxacarcin biosynthesis in Streptomyces bottropensis. Curr Microbiol 71(4):458–464. https://doi.org/10.1007/s00284-015-0868-9

Ye S, Brana AF, Gonzalez-Sabin J, Moris F, Olano C, Salas JA, Mendez C (2018) New insights into the biosynthesis pathway of polyketide alkaloid Argimycins P in Streptomyces argillaceus. Front Microbiol 9:252. https://doi.org/10.3389/fmicb.2018.00252

Yin H, Wang W, Fan K, Li Z (2019) Regulatory perspective of antibiotic biosynthesis in Streptomyces. Sci China Life Sci 62(5):698–700. https://doi.org/10.1007/s11427-019-9497-5

Yushchuk O, Andreo-Vidal A, Marcone GL, Bibb M, Marinelli F, Binda E (2020) New molecular tools for regulation and improvement of A40926 glycopeptide antibiotic production in Nonomuraea gerenzanensis ATCC 39727. Front Microbiol 11:8. https://doi.org/10.3389/fmicb.2020.00008

Zaburannyi N, Rabyk M, Ostash B, Fedorenko V, Luzhetskyy A (2014) Insights into naturally minimised Streptomyces albus J1074 genome. BMC Genomics 15:97. https://doi.org/10.1186/1471-2164-15-97

Zhang Y, Muyrers JP, Rientjes J, Stewart AF (2003) Red/ET is a useful stragtey combined RecE/T direct cloning with Redαβ recombineering. BMC Mol Biol 4(1):1

Zhang L, Zhao G, Ding X (2011) Tandem assembly of the epothilone biosynthetic gene cluster by in vitro site-specific recombination. Sci Rep 1:141. https://doi.org/10.1038/srep00141

Zhang Y, Werling U, Edelmann W (2012) SLiCE: a novel bacterial cell extract-based DNA cloning method. Nucleic Acids Res 40(8):e55. https://doi.org/10.1093/nar/gkr1288

Zhang F, Wen Y, Guo X (2014) CRISPR/Cas9 for genome editing: progress, implications and challenges. Hum Mol Genet 23(R1):R40–R46. https://doi.org/10.1093/hmg/ddu125

Zhang MM, Wong FT, Wang Y, Luo S, Lim YH, Heng E, Yeo WL, Cobb RE, Enghiad B, Ang EL, Zhao H (2017) CRISPR–Cas9 strategy for activation of silent Streptomyces biosynthetic gene clusters. Nat Chem Biol 13(6):607–609. https://doi.org/10.1038/nchembio.2341

Zhang J, Zhang D, Zhu J, Liu H, Liang S, Luo Y (2020) Efficient multiplex genome editing in Streptomyces via engineered CRISPR-Cas12a systems. Front Bioeng Biotechnol 8. https://doi.org/10.3389/fbioe.2020.00726

Zhao H (2013) Synthetic biology: tools and applications. Elsevier, London

Zhao HM, Medema MH (2016) Standardization for natural product synthetic biology. Nat Prod Rep 33(8):920–924

Zhao Y, Li L, Zheng G, Jiang W, Deng Z, Wang Z, Lu Y (2018) CRISPR/dCas9-mediated multiplex gene repression in Streptomyces. Biotechnol J 13(9):e1800121. https://doi.org/10.1002/biot.201800121

Zhao Q, Wang L, Luo Y (2019) Recent advances in natural products exploitation in Streptomyces via synthetic biology. Eng Life Sci 19(6):452–462. https://doi.org/10.1002/elsc.201800137

Zhou M, Jing X, Xie P, Chen W, Wang T, Xia H, Qin Z (2012) Sequential deletion of all the polyketide synthase and nonribosomal peptide synthetase biosynthetic gene clusters and a 900-kb subtelomeric sequence of the linear chromosome of Streptomyces coelicolor. FEMS Microbiol Lett 333(2):169–179. https://doi.org/10.1111/j.1574-6968.2012.02609.x

Zhou Q, Ning S, Luo Y (2020) Coordinated regulation for nature products discovery and overproduction in Streptomyces. Synth Syst Biotechnol 5(2):49–58. https://doi.org/10.1016/j.synbio.2020.04.002

Zhu H, Sandiford SK, van Wezel GP (2014) Triggers and cues that activate antibiotic production by actinomycetes. J Ind Microbiol Biotechnol 41(2):371–386. https://doi.org/10.1007/s10295-013-1309-z

Endophytic Actinomycetes: Secondary Metabolites and Genomic Approaches

14

Nattakorn Kuncharoen and Somboon Tanasupawat

Abstract

Endophytic actinomycetes, Gram-positive filamentous bacteria, live inside the plant tissues of different organs including roots, stems, leaves, flowers, fruits, and seeds. They not only harm the living plants, but also have proper effects on their host plants such as promoting plant growth and defending phytopathogens. Numerous classes of bioactive natural compounds, polyketides, macrolides, alkaloids, peptides, and terpenes, with various bioactivities: antibacterial, anti-fungal, anti-phytopathogens, immunosuppressant, anticancer, antioxidant, and anti-inflammatory, were produced by *Streptomyces kebangsaanensis*, *Streptomyces albidoflavus, Actinoallomurus fulvus, Micromonospora lupini*, *M. endophytica*, *Polymorphospora rubra*, and *Streptosporangium oxazolinicum* strains. Moreover, their novel compounds are also continued to discover and develop for preclinical stages. Thus, endophytic actinomycetes are the most key sources of potential bioactive metabolites, as well as a reservoir to mine for novel chemical structures. Moreover, with the progress in high-throughput genome sequencing techniques, many genome mining methods has been developed and helped to link the discovered compounds to their biosynthetic gene clusters (BGCs). This chapter highlights bioactive compounds derived from endophytic actinomycetes on their biological activity and biotechnological potential together

N. Kuncharoen
Department of Plant Pathology, Faculty of Agriculture, Kasetsart University, Bangkok, Thailand
e-mail: fagrnoku@ku.ac.th

S. Tanasupawat (✉)
Department of Biochemistry and Microbiology, Faculty of Pharmaceutical Sciences, Chulalongkorn University, Bangkok, Thailand
e-mail: somboon.T@chula.ac.th

R. V. Rai, J. A. Bai (eds.), *Natural Products from Actinomycetes*, https://doi.org/10.1007/978-981-16-6132-7_14

with the modern disciplines, and genome mining tools for discovery of the novel compounds.

Keywords

Bioactive metabolites · Biological control · Biological potential · Biosynthetic gene clusters (BGCs) · Endophytic actinomycetes · Genome mining · In silico · Drug discovery · Genome sequences · Secondary metabolite biosynthetic gene clusters (smBGCs)

14.1 Introduction

During the past few decades, numerous natural products have continuously played a highly important role in drug discovery and developmental process (Newman and Cragg 2007). Over 22,000 bioactive compounds have been obtained from microorganisms. Among them, 45% were produced by actinomycetes, particularly in members of the genus *Streptomyces* (Bérdy 2005). Actinomycetes have made an extraordinary contribution to the health and well-being of people around the world (Qin et al. 2011). The dawning of antibiotics resistance has been developed in pathogenic bacteria and the recent rising in the number of new diseases and pathogens, acquired immunodeficiency syndrome (AIDS), severe acute respiratory syndrome (SARS), and flu from influenza virus, has caused a revival of interest in finding new bioactive metabolites for the drug discovery. Nevertheless, for many years of intensive screening of soil microorganisms, the frequency of discovering novel compounds is seemingly shrunk (Bérdy 2005; Bérdy 2012). Consequently, the undiscovered and novel microbial habitats need to be explored for microbial resources, which served as a reservoir of useful biological active compounds.

Actinomycetes are Gram-positive, filamentous bacteria that formed true branched mycelia and contained genomic guanine (G) plus cytosine (C) content over 55 mol% belonged to the phylum *Actinobacteria*. They are saprophytic organisms, which played an important role in decomposing complex organic matters such as dead plants, animals, algae, and fungi, contributing to the formation of humus (Lechevalier and Lechevalier 1967). Actinomycetes are a large group of microorganisms, which are well known as a key producer of bioactive compounds, especially antimicrobial drugs: avermectin, chloramphenicol, gentamicin, erythromycin, kanamycin, rifampicin tetracycline, and vancomycin (Bérdy 2005; Bérdy 2012).

The actinomycetes are generally inhabited in both terrestrial and aquatic ecosystems. One virtually overlooked and promising niche is plant endosphere (El-Shatoury et al. 2013). Previous reports illustrated that some actinomycetes could form intimate related to plants and colonize inside their tissues. The genus *Frankia* and some species of the genus *Streptomyces*, *S. scabies*, can enter their host and establish either endophytic or pathogenic relationships (Benson and Silvester 1993; Doumbou et al. 1998). The actinomycetes, which inhabit inside the plant endosphere, do not apparently ruin the living plants, and have appropriate effects on the host plants, are known as endophytic actinomycetes (Quadt-Hallmann et al.

1997; Hasegawa et al. 2006; Rosenblueth and Martínez-Romero 2006). These organisms exist in different host plant's organs including roots, stems, leaves, flowers, fruits, and seeds, primarily in inter- or intracellular spaces. It is particularly noteworthy that, of approximately 300,000 different plant species on the earth, each species is considered to host one or more kind of endophytes (Strobel and Daisy 2003; Qin et al. 2011), initiating a gigantic biodiversity. Nonetheless, plant-associated actinomycetes have been studied, and they were found to fall within narrow genera; for example, *Streptomyces* is a major genus, and *Micromonospora, Microbispora, Nocardia, Nocardioides, Pseudonocardia,* and *Streptosporangium* were also recognized as common genera (Qin et al. 2009). According to the discovery in the narrow genera, it is pointing out the favorable chance to find many interesting species and their numerous natural products among myriads of plants in different niches and ecosystems is great. Several current reports revealed an enormous abundance of endophytic actinomycete species and diverse structures of metabolites with different biological functions (Coombs and Franco 2003; Ryan et al. 2008; Inahashi et al. 2015). Additionally, they act as biological control agents (Cao et al. 2005), enhance plant growth, and promote plant formation under unfriendly environment (Igarashi et al. 2002; Rungin et al. 2012; Hasegawa et al. 2006). These actinomycetes are relatively unexplored and potential reservoirs of new natural products for taking advantages in pharmaceutics, agriculture and industry (Strobel et al. 2004).

Since the whole-genome sequence of *S. coelicolor* A3 had been proposed (Bentley et al. 2002), more than 100 genomes of actinomycetes were sequenced and annotated. These data significantly showed that the genomes encode many biosynthetic gene clusters (BGCs) for diverse secondary metabolite production. Nevertheless, only a few bioclusters are revealed at standard conditions; hence, the potent bioclusters of numerous actinomycetes remain unexplicit (Mitousis et al. 2020). Therefore, this chapter mainly focused on bioactive compounds and biological potential as well as the methods for discovery of novel bioactive natural compounds.

14.2 Bioactive Metabolites, Biological Activity, and Biotechnological Potential

In view of the drug discovery, new actinomycetes are very interesting because they are prone to have novel genes in theory and held promising for novel metabolites; therefore, the prospect of finding new bioactive compounds from endophytic actinomycetes is extraordinarily considerable. Endophytic actinomycetes derived from medicinal plants in ethnobotanical uses, especially in the tropical zone, could be a rich belt of various functional metabolites and continued as the attention for many researchers around the world (Strobel et al. 2004; Qin et al. 2011). Compounds obtained from them have been validated to be advantage in the area of medicine, industry, and agriculture. Most of the metabolites were classified into various classes: polyketides, terpenes, macrolides, aminoglycosides and peptides (Genilloud 2017). These bioactive compounds, biological activity, and biotechnological potential (Table 14.1) are described below.

Table 14.1 Bioactive compounds from endophytic actinomycetes, their biological activity and biotechnological potential

Compounds	Chemical class	Species	Activity	References
Actinoallolides A-E	Macrolide	*Actinoallomurus flavus*	Anti-trypanosomal	Inahashi et al. (2015)
Alnumycin	Naphthoquinone	*Streptomyces* sp. DSM 11575	Antibacterial	Bieber et al. (1998)
Antimycin A_{18}	Macrolide	*S. albidoflavus*	Antifungal	Yan et al. (2010)
Bacaryolanes A-C	Sesquiterpene	*Streptomyces* sp.	Antibacterial	Ding et al. (2015)
Bafilomycins	Macrolide	*Streptomyces* sp.	Cytotoxic	Yu et al. (2011)
Cedarmycins A-B	Butyrolactone	*Streptomyces* TP-A0456	Antifungal	Sasaki et al. (2001)
Celastramycins A-B	Polyketide	*S. sampsonii* MaB-QuH-8	Antibacterial	Pullen et al. (2002)
Coronamycin	Peptide	*Streptomyces* sp.	Antifungal	Ezra et al. (2004)
24-Demethylbafilomycin C1	Macrolide	*Streptomyces* sp. CS	Antitumor	Lu and Shen (2003, 2004)
5, 7-Dimethoxy-4-phenylcoumarin 5, 7-Dimethoxy-4-*p*-methoxylphenylcoumarin	Flavonoid	*S. aureofaciens* CMUAc130	Antitumor Anti-inflammatory Antifungal	Taechowisan et al. (2005, 2007)
Dimeric dinactin	Macrolide	*Streptomyces* sp. Is9131	Antibacterial Anticancer	Zhao et al. (2005)
1-Hydroxy-β-carboline	Alkaloid	*Micromonospora endophytica* 161,111	Anti-influenza (H1N1)	Wang et al. (2014)
Kakadumycin A	Peptide	*Streptomyces* sp.	Antibacterial	Castillo et al. (2003)
Kandenols A-E	Sesquiterpene	*Streptomyces* sp.	Antibacterial	Ding et al. (2012)
Lumichrome	Flavonoid	*S. pseudovenezuelae* SKH1-2	Anticancer	Kuncharoen et al. (2019), Chantarawong et al. (2019)
Lupinacidins A-C	Anthraquinone	*M. lupini* Lupac 08	Antitumor	Igarashi et al. (2007), Igarashi et al. (2011)

Munumbicins A-D	Polyketide (actinomycin family)	*Streptomyces* sp. NRRL 30562	Antibacterial Antifungal Antimalarial	Castillo et al. (2002)
Munumbicins E-4, E-5	Polyketide (actinomycin family)	*Streptomyces* sp. NRRL 30562	Antibacterial Antifungal	Castillo et al. (2006)
Naphthomycin K	Naphthoquinone	*Streptomyces* sp. CS	Anticancer	Lu and Shen (2007)
Pteridic acids A-B	Spirocyclic octaketide	*S. hygroscopicus* TP-A0451	Plant growth-promoting (auxin-like hormone)	Igarashi et al. (2002)
Pterocidin	Polyketide	*S. hygroscopicus* TP-A0451	Anticancer	Igarashi et al. (2006)
Saadamycin	Polyketide	*Streptomyces* sp. Hedaya48	Antifungal	El-Tarabily et al. (2010)
Salaceyins A-B	Polyketide	*S. laceyi* MS53	Anticancer Antifungal	Kim et al. (2006)
Trehangelins A-C	Polyketide	*Polymorphospora rubra*	Photo-oxidative hemolysis inhibitor	Nakashima et al. (2013)

14.2.1 Antibiotics from Endophytic Actinomycetes

Numerous plant-derived actinomycetes, especially those from medicinal and indigenous plants, possess the ability to suppress or kill a wide range of pathogenic bacteria, fungi, and viruses. Consequently, there is tremendous value to discover and develop antimicrobial drugs from endophytic actinomycetes. Currently, many novel antibiotics and antimicrobial agents have been isolated, for example, alnumycin (Bieber et al. 1998), celastramycins A-B (Pullen et al. 2002), munumbicins A-D (Castillo et al. 2002), kakadumycins (Castillo et al. 2003), and demethylnovobiocins (Igarashi 2004).

From a cultured broth of *Streptomyces* strain TP-A0456, two new antibiotics, cedarmycins A and B, were isolated and purified. The structure of cedarmycins was an α, β-unsaturated butyrolactone with a fatty acid side chain. These two compounds exhibited potential to inhibit the growth of *Candida glabrata* with the MIC value of 0.4 μg/mL (Sasaki et al. 2001). In 2002, four novel wide-spectrum antibiotics, munumbicins A, B, C and D, were isolated from *Streptomyces* strain NRRL 30562, which was derived from the medicinal plant snake vine (*Kennedia nigriscans*), native to the Northern Territory of Australia. The compounds exhibited broad-spectrum activity against numerous human and plant pathogenic fungi and bacteria as well as a *Plasmodium* sp. For example, munumbicin B had ability to inhibit a methicillin-resistant strain of *Staphylococcus aureus* (MRSA) with the MIC of 2.5 μg/mL, while munumbicin A was not active. Although the munumbicins generally showed activity against Gram-stain-positive bacteria: *Bacillus anthracis* and multidrug-resistant *Mycobacterium tuberculosis*, the most impressive biological activity of any of the munumbicins was that of munumbicin D against the malarial parasite, *P. falciparum*, with the IC_{50} of 4.5 ± 0.07 ng/mL (Castillo et al. 2002).

In 2003, a complex of new peptide antibiotics, coronamycin, was obtained from a verticillate *Streptomyces* strain MSU-2110 isolated from *Monstera* sp. This compound illustrated activity against the pythiaceous fungi, *Pythium ultimum* and *Phytophthora cinnamomi*, with MIC values ranged from 2 to 16 μg/mL and the human fungal pathogen *Cryptococcus neoformans* with MIC of 4 μg/mL. The compound was also active against *P. falciparum*, with an IC_{50} of 9.0 ng/mL (Ezra et al. 2004).

After the discovery of munumbicins, Castillo et al. (2006) still studied the isolation and purification of secondary metabolites from the endophytic *Streptomyces* NRRL 30562 and found two novel peptide antibiotics, munumbicins E-4 and E-5. Both compounds exhibited a wide-spectrum activity against Gram-stain-positive and Gram-stain-negative bacteria. Additionally, *Pythium ultimum*, a plant pathogenic fungus, was sensitive to both munumbicins at the concentration of 5.0 μg/mL. The munumbicins E-4 and E-5 also had capability to inhibit *P. falciparum* with the IC_{50} values of 0.5 ± 0.08 and 0.87 ± 0.026 μg/mL, respectively. It appears that other bioactive compounds, related to E-4 and E-5, are also produced making it the most biologically active endophytic *Streptomyces* strains.

Extraction of the culture medium of endophytic *Streptomyces* strain Hedaya48 furnished a new antimycotic compound, saadamycin (Fig. 14.1), which showed

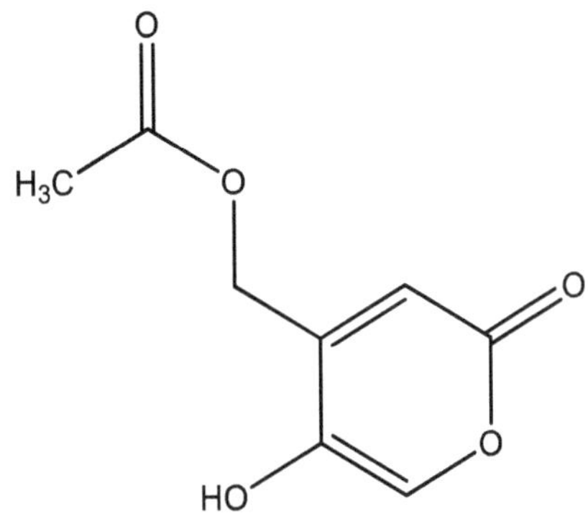

Fig. 14.1 Chemical structure of saadamycin

antifungal activity against dermatophytes and other clinical fungi such as *Aspergillus niger*, *A. fumigatus*, *Fusarium oxysporum*, *Candida albicans*, and *Cryptococcus humicolus* with lower MIC and minimum fungicidal concentration (MFC) values of 1–5.16 μg/mL and 1.25–10 μg/mL, respectively (El-Gendy and El-Bondkly 2010).

Streptomyces kebangsaanensis SUK12^{T} isolated from a Malaysian ethnomedicinal plant, *Portulaca oleracea* L. (local name, Gelang pasir), was proposed as a novel species in 2013. This strain produced tubermycin B (phenazine-1-carboxylic acid), which was active against Gram-positive bacteria and some filamentous fungi, for instance, *B. subtilis*, *A. fumigatus*, *Geotrichum candidum*, and *Rhizoctonia solani* (Tan et al. 2013). In addition, *S. kebangsaanensis* SUK12^{T} contained the biosynthetic gene clusters encoding for a novel bioactive phenazine compound, 6-((2-hydroxy-4-methoxyphenoxy) carbonyl) phenazine-1-carboxylic acid (HCPCA), which exhibited a wide-spectrum antimicrobial activity (Remali et al. 2017). After the discovery of a wide-spectrum antimicrobial activity of *S. kebangsaanensis* SUK12^{T}, the Universiti Kebangsaan Malaysia continuously investigated the endophytic *Streptomyces* strain SUK25 isolated from the root of *Zingiber spectabile,* which produced four active diketopiperazines and an acetamide derivative including *cyclo*-(L-Val-L-Pro), *cyclo*-(L-Leu-L-Pro), *cyclo*-(L-Phe-L-Pro), *cyclo*-(L-Val-L-Phe), and *N*-(7-hydroxy-6-methyl-octyl)-acetamide. All of the compounds were active against methicillin-resistant *S. aureus* ATCC 43300 and *Enterococcus raffinosus*, with low cytotoxicity against human hepatoma HepaRG cells (Alshaibani et al. 2017).

Recently, the overuse of antibiotics has been highly contributed to the increase in antibiotic resistance in microbes; hence, the finding and development of new and effective natural antibiotics are still important. Endophytic actinomycetes are still interesting as a rich reservoir of novel and known antimicrobial drugs. Endophytic *Streptomyces parvulus* Av-R5 derived from root of *Aloe vera* exhibited high potential of antimicrobial activity. This strain produced actinomycin D and actinomycin $X_{0\beta}$ as major compounds, which inhibited several antibiotic-resistant pathogenic bacteria and fungi consisting of multidrug-resistant *S. aureus* JNMC-3, *S. epidermidis* JNMC-4, *Klebsiella pneumoniae* MTCC-3384, *K. pneumoniae*

JNMC-6, *Pseudomonas aeruginosa* MTCC-741, *Proteus vulgaris* JNMC-7, *C. albicans* MTCC-183, and *A. niger* MTCC-872 (Chandrakar and Gupta 2019).

At this point, the endophytic actinomycetes are prone to be an important source of antibiotics. Furthermore, the whole-genome sequence analysis is effortlessly carried out and lots of biosynthetic gene clusters of functional molecules have been identified. Consequently, there is merit to cultivation of microbes to help wholly discover more high potential bioactive metabolites that they hold.

14.2.2 Anticancers, Anti-inflammatory and Other Pharmaceutical Agents from Endophytic Actinomycetes

Currently, there are significantly increasing of anticancer drugs from endophytic actinomycetes. They can produce the compounds as same as host plants that they lived. So, it could be possibly assumed that genes involved in the biosynthesis of natural products could be exchanged through horizontal gene transfer (HGT) between plants and microorganisms, contributing to plant-derived compounds production: *Kitasatospora* sp. isolated from *Taxus baccata* in Italy produced paclitaxel, which was the first report of the taxol production from the endophytic actinomycetes (Janso and Carter 2010; Qin et al. 2011).

Maytansinoids (19-membered macrocyclic lactams related to ansamycin antibiotics) are extremely potent antitumor drugs, which were originally produced by members of the higher plant families: Celastraceae, Euphorbiaceae, and Rhamnaceae (Kupchan et al. 1972; Powel et al. 1982) as well as some mosses (Suwanborirux 1990). The compounds were also isolated from plant-associated actinomycete *Actinosynnema pretiosum* (Higashide et al. 1977). Notably, naphthomycin K (Fig. 14.2), a novel chlorine-containing ansamycin, was isolated from an endophytic *Streptomyces* strain CS derived from *Maytenus hookeri*. This compound evidently exhibited cytotoxicity against P388 (leukemia) and A-549 (a carcinomic human alveolar basal epithelial cells) cell lines at IC_{50} values of 0.07 and 3.17 μM, respectively (Lu and Shen 2007).

Interestingly, the endophytic *Streptomyces* strain CS was an extraordinary bioactive metabolite producer. The strain produced a new member of the bafilomycin subfamily, 24-demethylbafilomycin C1 (Fig. 14.3a), and two more new bafilomycin derivatives, which showed a strong potent antitumor activity (Lu and Shen 2003, 2004). As a high active potent of the bafilomycin family, Li et al. (2010b) had continued to isolate the metabolites from *Streptomyces* sp. CS and discovered five novel 16-membered macrolides, belonging to the family of bafilomycin. All of the new compounds demonstrated cytotoxic activity against MDA-MB-435, a melanoma cell line, in vitro. From the extraction and purification of the fermentation broth of *Streptomyces* strain ls9131 associated with *M. hookeri*, two new macrolides -dimeric dinactin and dimeric nonactin, were discovered. Biological activity assay showed that dimeric dinactin (Fig. 14.3b) had ability to inhibit Gram-positive bacteria, *S. aureus* and *M. tuberculosis*, as well as the growth of cancer cell lines including HL60 (leukemia), A549 (lung cancer), SGC7901

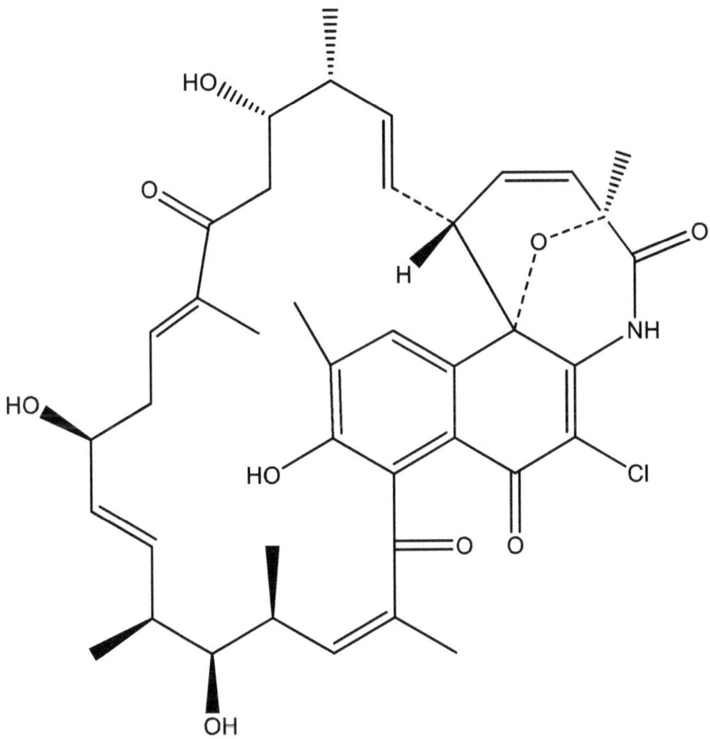

Fig. 14.2 Chemical structure of naphthomycin K

(gastric cancer), and BEL7401 (liver cancer) cell lines with IC_{50} at 0.26, 1.28, 1.80, and 2.16 μM, respectively (Zhao et al. 2005).

One of the endophytic actinomycete genera that produced many bioactive compounds is the genus *Micromonospora*. *M. lupini* Lupac 08 was a novel actinomycete associated with root nodules of *Lupinus angustifolius*, which produced three novel anthraquinones, lupinacidins A, B, and C. These compounds had capability to inhibit the invasion of murine colon (26-L5) carcinoma cells without inhibiting cell growth (Igarashi et al. 2007, 2011).

Members in the genus *Streptomyces* were still a potent source of anticancer agents. Salaceyins A and B, two novel 6-alkylsalicylic acids, isolated from the endophytic *Streptomyces laceyi* MS53, were active against a human breast cancer cell line (SKBR3) with IC_{50} values of 3.0 and 5.5 μg/mL, respectively (Kim et al. 2006). Pterocidin, a novel cytotoxic compound, obtained from *Streptomyces hygroscopicus* TP-A0451 illustrated cytotoxicity against many human cancer cell lines with IC_{50} ranged from 2.9 to 7.1 μM (Igarashi et al. 2006). *Streptomyces aureofaciens* CMUAc130 isolated from roots of *Zingiber officinale* Rosc. produced two metabolites: 5, 7-dimethoxy-4-phenylcoumarin and 5, 7-dimethoxy-4-*p*-methoxylphenylcoumarin, which affected not only the generation of nitric oxide

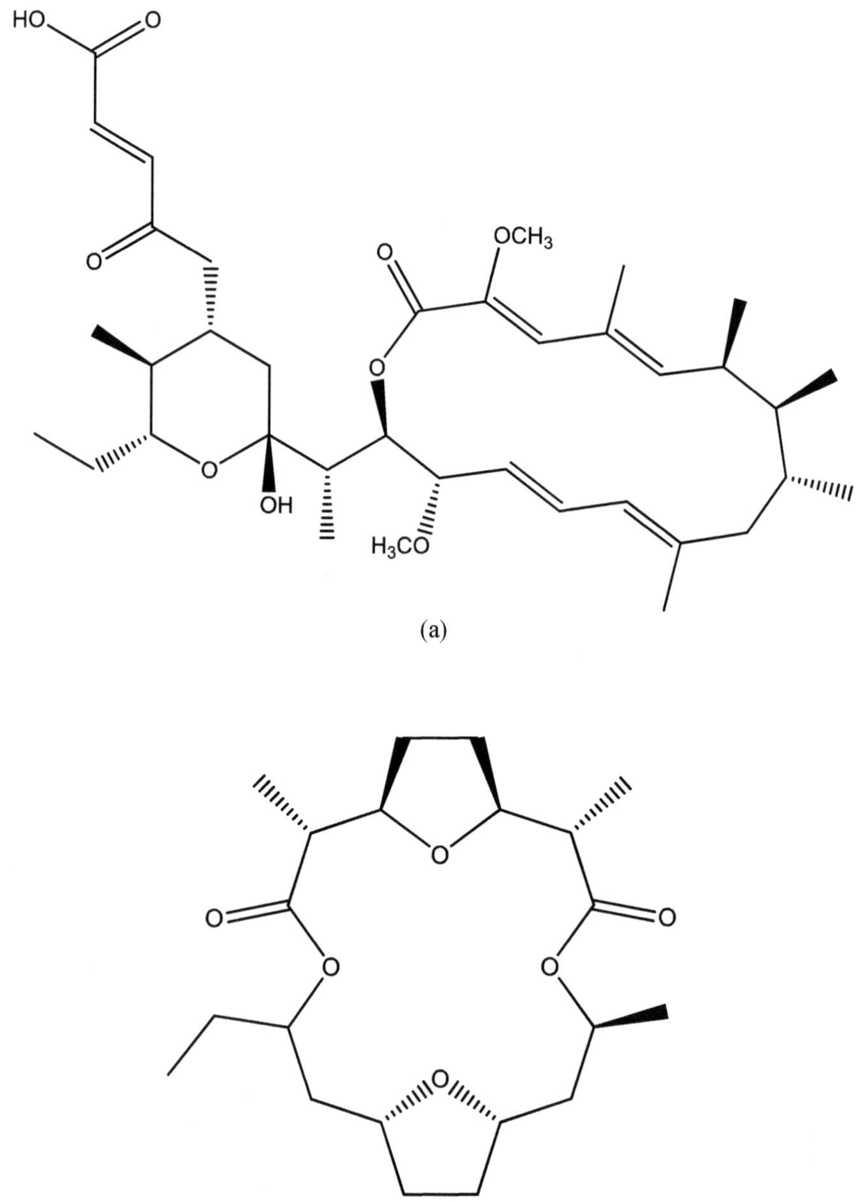

Fig. 14.3 Chemical structure of 24-demethylbafilomycin C1 (**a**) and dimeric dinactin (**b**)

Fig. 14.4 Chemical structure of lumichrome

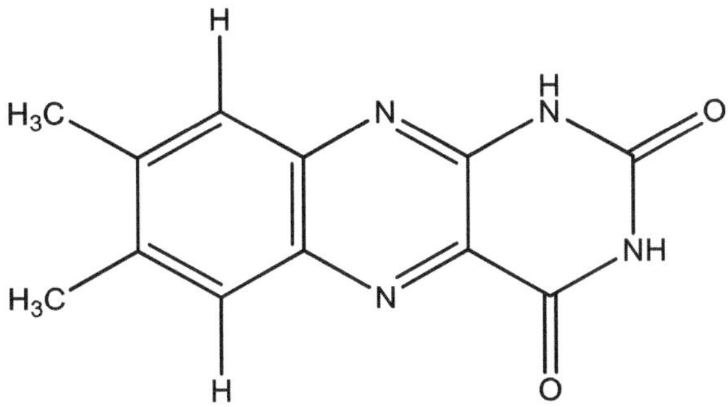

(NO), prostaglandin E2 (PGE2), and tumor necrosis factor (TNF-α) but also affected the inductive of nitric oxide synthase and cyclooxygenase-2 in lipopolysaccharide (LPS)-induced murine macrophage RAW 264.7 cells. Moreover, the compounds significantly lowered the formation of TNF-α. These results indicated the application of the compounds as anti-inflammatory agent (Taechowisan et al. 2007). *Streptomyces* strain neau-D50A produced a novel prenylated indole derivative 3-acetonylidene-7-prenylindolin-2-one (1) together with four known compounds of hybrid isoprenoids: 7-isoprenylindole-3-carboxylic acid (2), 3-cyanomethyl-6-prenylindole (3), 6-isoprenylindole-3-carboxylic acid (4), and 7,4′-dihydroxy-5-methoxy-8-(γ,γ-dimethylallyl)-flavanone (5). Compounds 1 and 2 were active against human lung adenocarcinoma cell line A549 with IC_{50} values of 3.3 and 5.1 μg/mL, respectively (Zhang et al. 2014).

Recently, two biphenyls, 3′-hydroxy-5-methoxy-3,4-methylenedioxybiphenyl (1) and 3′-hydroxy-5,5′-dimethoxy-3,4-methylenedioxybiphenyl (2), were produced by the endophytic *Streptomyces* strain BO-07, which was endosymbiotic with the root of *Boesenbergia rotunda* (L.) Mansf A. Compounds 1 and 2 not only exhibited the highest DPPH antioxidant activity with a scavenging concentration (SC_{50}) value of 85.8 and 88.3 μg/mL, respectively, but also strongly active against all three cancer cell lines: HeLa, HepG2, and Huh7, at the IC_{50} values of 3.04–20.30 μg/mL. Furthermore, the compounds 1 and 2 were less cytotoxicity on normal cells (L929) than on the tested cancer cell lines (Taechowisan et al. 2017). These findings supported the application of the compounds as antioxidant and anticancer drugs.

Lumichrome (Fig. 14.4), a major derivative of riboflavin, isolated from the endophytic *Streptomyces pseudovenezuelae* SKH1-2 (Kuncharoen et al. 2019), could inhibit lung cancer cell growth and reduce survival in both normal and anchorage-independent conditions. Moreover, this compound also induced apoptosis in lung cancer cells via a p53-dependent mitochondrial mechanism with substantial selectivity, shown by its lower cytotoxicity to the normal primary dermal papilla cells (Chantarawong et al. 2019).

Many endophytic actinomycetes exhibited anti-trypanosomal activity. Three novel alkaloid antibiotics, spoxazomicins A, B, and C, have been isolated from an endophytic *Streptosporangium oxazolinicum* K07-0460 (Inahashi et al. 2011). Spoxazomicin A was highly active against *T. b. brucei* strain GUT at 3.1 (causative agent of Nagana disease in animals) with IC_{50} at 0.11 μg/mL, which was more potent than the clinical drugs from 14- to 21-fold. Spoxazomicin B was also active against *T. b. brucei* strain GUT at 3.1 with IC_{50} at 0.55 μg/mL, whereas spoxazomicin C exhibited weakly anti-trypanosomal activity against *T. b. brucei* strain GUT at 3.1 with IC_{50} value at 3 μg/mL.

An endophytic *Actinoallomurus fulvus* MK10-036 produced five new anti-trypanosomal macrolide compounds, actinoallolides A, B, C, D and E. Actinoallolide A showed strong potent anti-trypanosomal activity against *T. b. brucei* strain GUT at 3.1 and *T. cruzi* Tulahuen C4C8 strain with the IC_{50} values of 0.0049 and 0.226 μg/mL, respectively, with no toxicity against MRC-5 cell. This suggested that actinoallolide A could be further developed as medicine for sleeping sickness and Chagas disease (Inahashi et al. 2015).

Four metabolites including lumichrome (1), perlolyrine (2), 1-hydroxy-β-carboline (3), and 1*H*-indole-3-carboxaldehyde (4), which were extracted and isolated from the fermentation broth of the endophytic *Micromonospora endophytica* 161111, were active against the influenza A virus subtype H1N1 at IC_{50} values of 39.7, 38.3, 25.0, and 45.9 μg/mL, respectively (Wang et al. 2014). Compound 3 was potentially active against H1N1 and thus further developed as the anti-influenza viral drug.

Briefly, endophytic actinomycetes still remain a relatively untapped source of novel bioactive natural products, presumed to drive forward the borderline of drug discovery and benefit the development of the compound for management of the diseases.

14.2.3 Biological Control and Plant Growth-Promoting Agents

Endophytic actinomycetes were also used to promote growth of plants by implementation with a wide range of mechanisms of actions. The possible mechanisms of this activity included (1) the ability to produce plant hormone; (2) nitrogen fixation; (3) siderophore production; (4) solubilization of phosphates and other minerals; and (5) production and represent for biocontrol agents (Cattelan et al. 1999).

In terms of plant growth regulator, Igarashi et al. (2002) had extracted and purified two auxin-like compounds, pteridic acids A (Fig. 14.5a) and B (Fig. 14.5b), from the endophytic *Streptomyces hygroscopicus* TP-A0451 (derived from a stem of *Pteridium aquilinum*). These two compounds were structurally related to azalomycin B and exhibited as a natural plant growth hormone by inducing the formation of adventitious roots in hypocotyl of kidney beans with a concentration of 1 nM.

Shutsrirung et al. (2013) discovered that the endophytic actinomycete-associated *Citrus reticulata* L. can build up shoot height, fresh shoot weight, and fresh root weight of the seedlings with the values ranged 20.2–49.1, 14.9–53.6, and 1.6–102%, respectively. In addition, Dochhil et al. (2013) also found that two strains of endophytic *Streptomyces* CA10 and CA26, derived from *Centella asiatica,* can

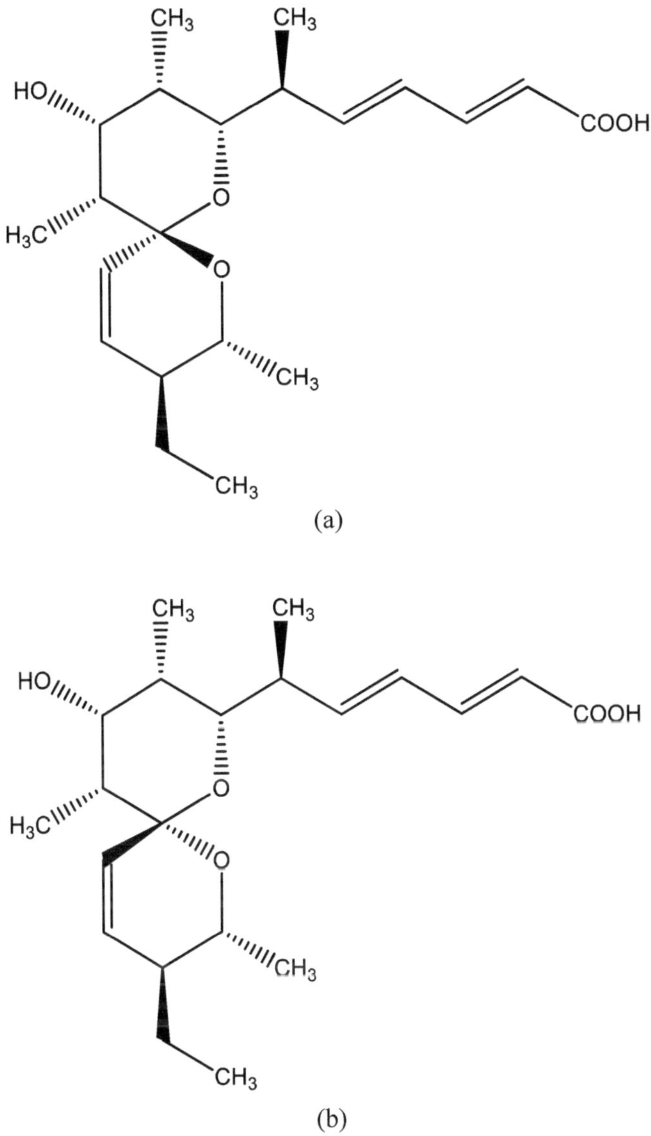

Fig. 14.5 Chemical structure of pteridic acid A (**a**) and pteridic acid B (**b**)

promote the germination of seed and seedling growth of *Phaseolus vulgaris* by producing a high concentration of indole-3-acetic acid (IAA) from 71 to 197 μg/mL.

Plant diseases anthracnose of banana and wilt of wheat were commonly caused by *C. musae* and *F. oxysporum*, respectively. The endophytic *S. aureofaciens* CMUAc130 isolated from plant roots in the Zingiberaceae family produced two compounds, 5, 7-dimethoxy-4-phenylcoumarin and 5, 7-dimethoxy-4-*p*-

methoxylphenylcoumarin, which affected the tested fungi: *C. musae* and *F. oxysporum* with the MICs of 120 and 150 μg/mL. Growth of *C. musae* mycelia was inhibited completely by 150–180 μg/mL of the compounds (Taechowisan et al. 2005). *Pythium aphanidermatum* was a filamentous fungus, which usually caused the dumping-off disease of the seedlings and roots as well as the crown rots of cucumber (*Cucumis sativus*) and affected the commercial production of cucumber in the United Arab Emirates (UAE). Three endophytic actinomycetes including *S. spiralis*, *M. chalcea*, and *Actinoplanes campanulatus* were active to inhibit that pathogenic fungus; thus, these endophytes might be further developed and used as biocontrol agents instead of metalaxyl, a fungicide widely used for the control of Pythium disease in UAE (El-Tarabily et al. 2010).

Presently, the novel prenylated indole derivative 3-acetonylidene-7-prenylindolin-2-one and three known metabolites of hybrid isoprenoids, 7-isoprenylindole-3-carboxylic acid, 3-cyanomethyl-6-prenylindole, and 6-isoprenylindole-3-carboxylic acid, were extracted and isolated from the cultured broth of the endophytic *Streptomyces* sp. neau-D50A. These compounds showed potentially antifungal activity against many phytopathogenic fungi: *Colletotrichum orbiculare*, *Phytophthora capsici*, *Corynespora cassiicola* and *F. oxysporum*, with the IC_{50} values ranging from 36.92 to 90.61 μg/mL (Zhang et al. 2014).

The impact relationship between the endophytic actinomycetes and plants and bioactive natural metabolite production provided a unique opportunity to discover potent compounds in plant protection and biological control. Moreover, abilities of plant growth-promoting endophytic actinomycetes and valuable understanding of several mechanisms offered further studies for potential application in agriculture.

14.3 Genome Mining for Search and Discovery of Bioactive Compounds

A classical approach for identifying the biosynthetic gene clusters (BGCs) in the endophytic actinomycetes related to the identification of secondary metabolites generally used physicochemical-based methods. The compounds were isolated from the cultured broth of the actinobacterial endophytes, elucidated the chemical structures using mass spectrometry, UV-visible spectroscopy and nuclear magnetic resonance (NMR), then characterized the corresponding BGCs by gene deletion or mutagenesis, and finally screened the nonproducing clones (Ikeda et al. 1987). However, this method was very tedious and high labor-intensive. Recently, the genome technologies and bioinformatic tools were in high progress. Therefore, the new method called "genome mining" was developed.

Genome mining is the in silico methods which is not only used to identify the biosynthetic pathway and linked to bioactive compounds, but also possibly predicted functional and chemical interactions (Ziemert et al. 2016). Additionally, the genome mining included the identification of formerly uncharacterized BCGs within the bacterial whole genomes of sequences, sequence analysis of the coding enzymes by the gene clusters and the product identification of the gene clusters (Trivella and de Felicio 2018).

Genome mining solely relied on computation technology and bioinformatic platforms. At this point, a large amount of data including genome sequences and their annotations are now publicly available in accessible databases: NCBI, DDBJ, and EMBL. So, when all genes within a new genome were identified, they could be compared with those of known functions in those public databases (Albarano et al. 2020). This approach brought the feasibility to compare the target gene clusters to known gene clusters, which are convenient for the prediction of their function and structure using different associated web databases (Boddy 2014). Previously, the genome mining was granted to find and identify the gene clusters accountable for the production of natural products; nonetheless, web-based tools and databases have been combined to improve the performance of the approach in the last decade (Ziemert et al. 2016).

A scientific progress has allowed the development of important web-based tools including (1) antibiotics and Secondary Metabolite Analysis Shell (antiSMASH), a web server for the identification of gene clusters with a series of specific algorithms for metabolite analysis (Blin et al. 2013). Consequently, this tool implemented the sequence prediction and furnished a more detailed analysis of identified gene clusters involving the stereochemistry structure of amino acid (Boddy 2014); (2) PRISM, PRediction Informatics for Secondary Metabolomes, was an open web-based tool, which contained a genomic prediction of secondary metabolomes. This platform gave the prospect of obtaining a correspondence between known natural compounds and probable novel ones (Skinnider et al. 2015); and (3) Integrated Microbial Genomes Atlas of Biosynthetic gene Clusters (IMG/ABC) tool is a huge open database website of known predicted microbial BGCs that able to analyze both BGCs and secondary metabolites. In this database, it provided the capability of finding similar function between BGCs in database and BGCs to be identified (Hadjithomas et al. 2017).

In this section, we focused on the significance of genome mining for identification of novel natural compounds, the advantages, and disadvantages of this method, in silico tools for genome mining, characterization of secondary metabolite biosynthetic gene clusters (smBGC) and use of genome mining in endophytic actinomycetes.

14.3.1 Importance of Genome Mining in Drug Discovery

Half of the clinical approved medicines including antibiotics were directly extracted and purified from whole cell and fermentation broth of microorganisms. Currently, the development of new bioinformatics and genetic analytic tools gave many new strategies for the discovery of natural bioactive compounds known as "combinatorial biosynthesis approaches" (Albarano et al. 2020). These methods and bioinformatic tools have shown the ability of microbes to produce bioactive metabolites, which have been underestimated (Nett et al. 2009). These microbes have been widely searched by the sequencing of their genome and the application of genome mining (Olano et al. 2014). Genomic analysis has released the presence of numerous BCGs, which could be covered in the biosynthesis of other secondary metabolites designed cryptic or orphan for their unknown origin (Nett et al. 2009).

The genome mining aimed to predict the gene clusters encoding for novel compounds of biotechnological interests using several bioinformatic platforms (Albarano et al. 2020). The significance of genome mining depends on the demanding need to explore new drugs owing to the high incidence of severe diseases and the reduced efficiency of existing medicines (Mody and Haldar 2015). Moreover, the BCGs comprised the elements, which might be used to increase the production of both natural and engineered products by promoting cost reduction and their commercial use (Olano et al. 2014).

14.3.2 Advantages and Disadvantages of Genome Mining

One of the strengths of using genome mining is to advance the detection of a huge amount of biological active natural metabolites (Wohlleben et al. 2016). Furthermore, the genome mining technique is quite cheap and simple to perform in laboratory, and it required no special skills or experience of the researchers (Scheffler et al. 2013). Integration of the genome mining with genetic engineering could make it possible to accomplish maximum diversity of metabolites (Zerikly and Challis 2009). This bioinformatic tool was granted to predict the chemical structure of bioactive compounds; however, the forecasts are regularly difficult to draw up (Zerikly and Challis 2009; Scheffler et al. 2013).

Although the genome mining approach provided many advantages, a tremendous disadvantage of the genome mining is only known biosynthetic gene clusters that can be predicted (Zerikly and Challis 2009; Wohlleben et al. 2016). In addition, with this platform, it could not predict the biological activities of the identified metabolites (Olano et al. 2014). Nevertheless, genome mining is still performed as an expanded tool (Albarano et al. 2020). Recently, many scientists are trying to upgrade the bioinformatic platforms in order to reduce the limitation. The advantages and disadvantages of the genome mining are concluded in Table 14.2.

Table 14.2 Advantages and disadvantages of genome mining

Advantages	Disadvantages
1. Cheap and easy to perform the experiments in laboratory	1. Not to identify the biotechnological function of the natural metabolites
2. Well support the prediction of chemical structures of bioactive metabolites	2. Difficult to generate the chemical structures and only applicable for known biosynthetic gene clusters
3. Specific skills and/or experiences are not required for the users	3. Too new approach that required to be deepened

14.3.3 In Silico Tools for Mining smBGCs

Improvement of nucleotide or protein sequence alignment platforms including BLAST, Diamond and HMMer, as well as the high-throughput genome sequencing allowed researchers to mine for new smBGCs in databases without time-consuming and labor-intensive of operating a Southern blot. DECIPHER, constructed by Ecopia Biosciences Inc., was the first proprietary database of microbial natural product biosynthetic loci for the in silico genome mining of smBGCs (Zazopoulos et al. 2003). Although various free databases and tools for smBGC identification: BAGEL3 (van Heel et al. 2013), ClustScan (Starcevic et al. 2008), CLUSEAN (Weber et al. 2009) and NP.searcher (Li et al. 2009) have been developed and launched, most of them are limited to the discovery of specific classes of secondary metabolites, PKS and NRPS.

PRISM 3 (Skinnider et al. 2017) and antiSMASH 5.0 (Blin et al. 2019) are currently typical in silico tools for forecasting various types of smBGCs. These equipments predicted types of smBGC by using a sequence alignment-based data in a hidden Markov model (HMM) of genes, which are definite for certain smBGC types. For instance, antiSMASH identified smBGCs relied on the highly conserved core biosynthetic enzymes and determined the results using a set of manually curated BGC rules, subsequently eliminating false positives using negative models, and fatty acid synthases are homologous to PKSs. The antiSMASH also consisted of a "Cluster Blast" and a "SubCluster Blast" that allowed the conserved operons detection within a gene cluster, which is responsible for the synthesis of specific building blocks of the natural compounds (Fig. 14.6). Moreover, the graphical output and the annotation could be downloaded with different file formats (Weber 2014).

PRISM 3 can predict 22 different smBGC types, while antiSMASH version 5 can predict up to 52 different smBGC types. Both tools are user-friendly web servers, which gave fast gene annotation submission of the FASTA format of bacterial genome sequences, hence making them as well-known tools in current mining investigations. Presently, smBGC mining approaches have utilized machine learning strategies, ClusterFinder and DeepBGC, for allowing the prediction of unknown smBGCs (Hannigan et al. 2019; Cimermancic et al. 2014). Nonetheless, machine learning-based tools have a higher rate of false-positive than the rule-based tools. Moreover, the combination of a set of known clusters (MIBiG database) and a set of clusters predicted by one of the rule-based tools (antiSMASH) has already been implemented in smBGCs mining; therefore, it is still challenging to determine completely novel smBGCs.

14.3.4 Characterization of the Identified smBGCs Based on Genome Mining

Although genome mining approaches highlighted the completed biosynthetic potential of some actinomycetes (i.e., *Streptomyces* sp. and *Micromonospora* sp.), it had

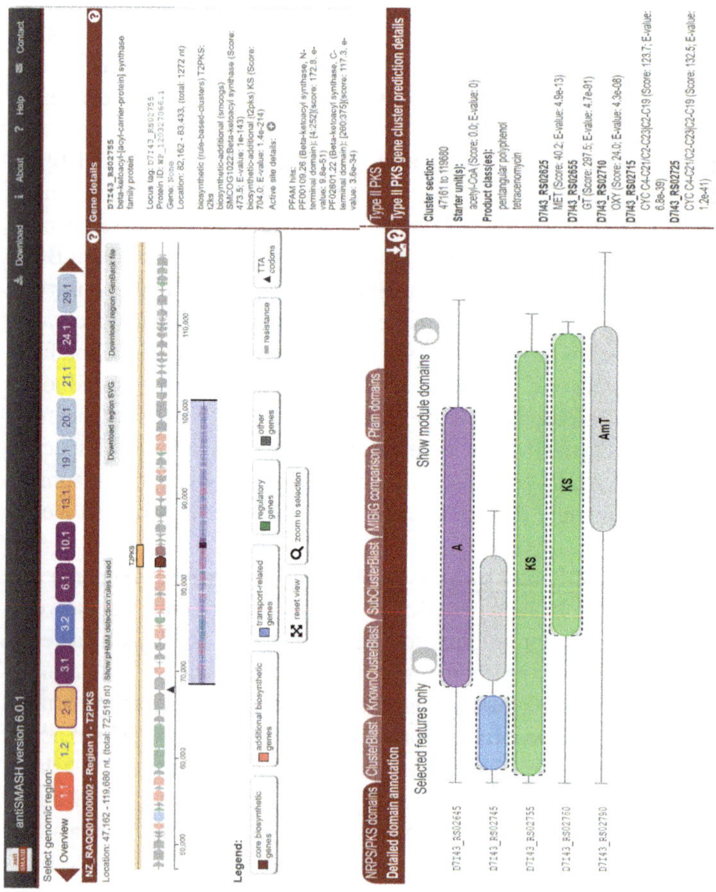

Fig. 14.6 A screenshot of the antiSMASH version 5.0/6.0 result page. The numbered circles represent the identified gene clusters in the genome

not linked the predicted smBGCs to their product. In this part, we explained some samples of the genome mining tools that connected to various metabolites with their corresponding bioclusters using reverse (metabolites to genes) or forward (genes to metabolites) approaches.

The reverse approach was the method that linked the secondary metabolites to their biosynthetic gene clusters. This method was started after knowing the compounds and then identifying the smBGCs, which were associated with their biosynthetic pathway. The forward approach was the procedure that connects the BGCs to their biosynthetic compounds. This approach was beginning with the genome sequence, analyzing genes of gene clusters, and then predicting the biosynthetic metabolites (Cacho et al. 2014; Khater et al. 2016). The two approaches for deciphering the relationship between secondary metabolites and their gene clusters are shown in Fig. 14.7.

In current years, advances in genome mining tools allowed users to apply a reverse approach for detecting the BGCs of known secondary metabolites produced by many genera of actinomycetes, especially the genus *Streptomyces*. A reverse approach allowed the isolated bioactive natural products to know the BCGs (Fig. 14.7). These efforts also facilitated us to predict and elucidate the biosynthetic pathways of various important secondary metabolites much faster and more effective than conventional randomized mutagenesis-based methods, which sometimes had no stability in the production processes. The smBGC data obtained from the reverse paths have enhanced databases of smBGC and increased the precision of the genome mining tools together with the number of predictable types of smBGC (Lee et al. 2020).

Advance in the genome mining tools also gave opportunities for the forward approach to the identification of smBGCs, which permitted the researchers to identify the new smBGCs from the genome, and then predict the products of these smBGCs (Fig. 14.7). There are two things that are significantly required for successful forward approaches in the smBGC genome mining comprising a predictable draft structure of the final product and a high expression level of smBGCs, which are enough to produce detectable secondary metabolites (Lee et al. 2020).

There are many computational platforms and databases (Table 14.3) to predict the smBGC putative products, particularly for the classes of PKS and NRPS. These methods employed the basic rules of structure prediction, which is regarded to the substrate specificity of the catalytic domains of PKS and the modules of NRPs to generate the backbone structure of the products, subsequently by the identification of conforming domains to evaluate further cyclization or modifications of the metabolites. These results are drawn back to the database providing the idea of the secondary metabolite produced by their unknown smBGC to researchers. Global genome mining approaches, PRISM and antiSMASH, also furnished the prediction of chemical structure of putative metabolites from unknown smBGCs (Khater et al. 2016). The precision of chemical structure prediction depends on the methods and the databases that are used to identify the enzyme catalytic domains and the substrate specificity. The tools, PRISM and AntiSMASH, contained a function of chemistry prediction, which can create a wide range of combinatorial libraries of predicted

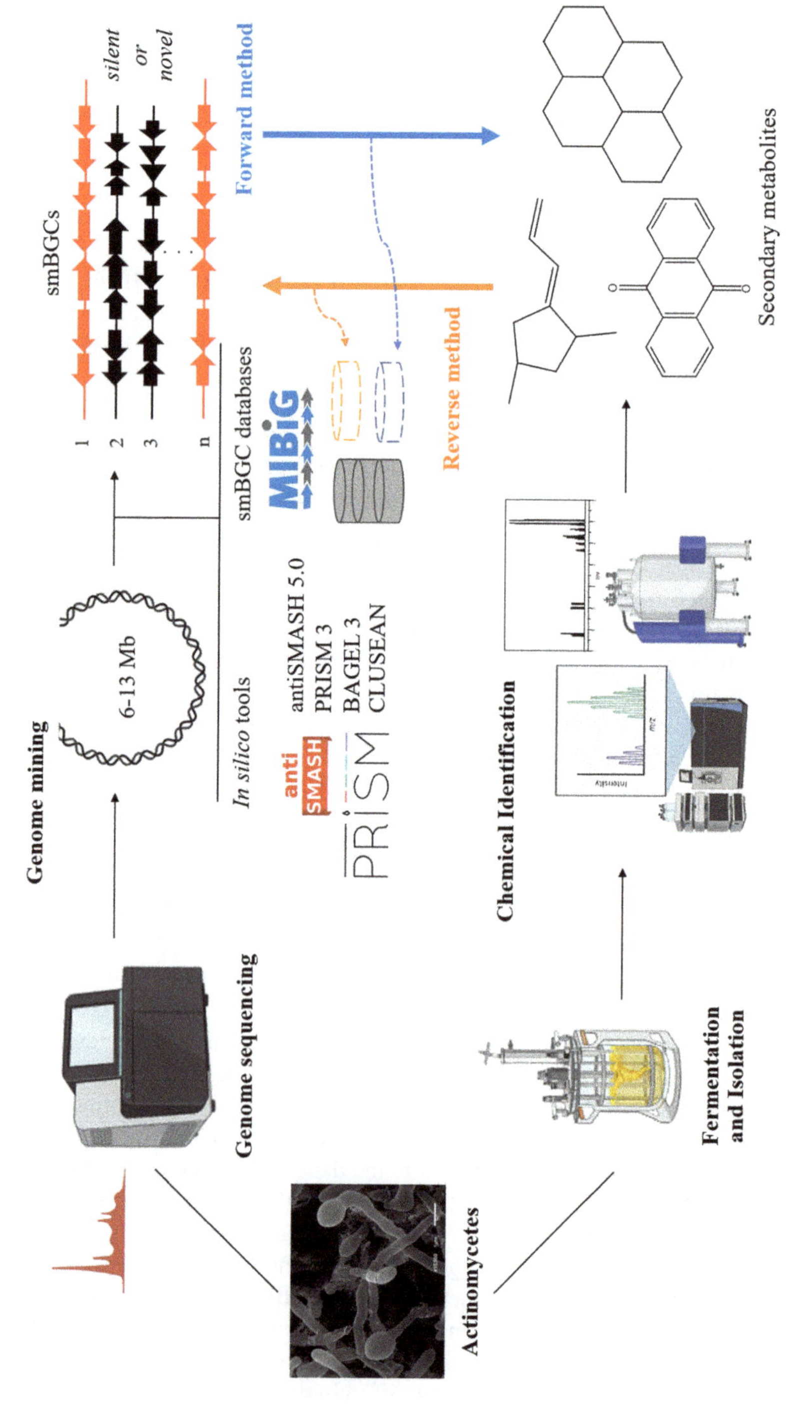

Fig. 14.7 Overview of genome mining method to identify smBGCs in actinomycetes (modified from Lee et al. 2020)

Table 14.3 Computational tools and databases associated with the genome mining

Software/database	Website	Reference
Databases for gene clusters		
ClustScan Database	http://csdb.bioserv.pbf.hr/csdb/ClustScanWeb.html	Starcevic et al. (2008), Diminic et al. (2013)
ClusterMine360	http://www.clustermine360.ca/	Conway and Boddy (2013)
Databases for bioactive natural metabolites		
PubChem	http://pubchem.ncbi.nlm.nih.gov/	Li et al. (2010a)
KNApSAcK database	http://kanaya.aist-nara.ac.jp/KNApSAcK/	Nakamura et al. (2014)
StreptomeDB	https://www.pharmaceutical-bioinformatics.de/streptomedb/	Lucas et al. (2013)
Platforms for the prediction of substrate specificities		
NRPSpredictor2	http://nrps.informatik.uni-tuebingen.de	Rottig et al. (2011)
PKS/NRPS Web Server/ Predictive Blast Server	http://nrps.igs.umaryland.edu/nrps/	Bachmann and Ravel (2009)
PKSIIIexplorer	http://type3pks.in/tsvm/pks3/	Vijayan et al. (2011)
NRPSSP	http://www.nrpssp.com/	Prieto et al. (2012)
Tools for the genome mining of smBGCs		
antiSMASH version 5.0	http://antismash.secondarymetabolites.org	Blin et al. (2019)
BAGEL3	http://bagel2.molgenrug.nl/	van Heel et al. (2013)
CLUSEAN	https://bitbucket.org/antismash/clusean	Weber et al. (2009)
PRISM 3	http://magarveylab.ca/prism/	Skinnider et al. (2017)
PKMiner	http://pks.kaist.ac.kr/pkminer/	Kim and Yi (2012)

chemical structures (Blin et al. 2019). However, the accuracy of chemical structure prediction of the recent versions of antiSMASH version 5.0 and PRISM version 3 has never been compared, hence, it is proper to employ both of them in accordance with the purposes of the research.

To accomplish the second requirement of forward approaches, numerous other technologies were combined into the genome mining to raise the production of secondary metabolites or activate the silent smBGCs. Many smBGCs of actinomycetes are muted under the condition of fermentation, so modifying the level of smBGC expression to produce enough quantities of metabolites must come before connecting the secondary metabolites to the corresponding bioclusters (Lee et al. 2020). This protocol is based on the treatment of fermentation culture with elicitors to high the expression of smBGCs as seen in the discovery of curacozole from *Streptomyces curacoi* (Kaweewan et al. 2019). In addition, although genome engineering is a suitable procedure to induce the silent smBGCs, it is not always useful because actinomycetes had slow growth rates and high G + C content in the genomes (Lee et al. 2020). It could be highlighted that the forward method significantly seems to be more challenging than the reverse approach when it came to chemical characterization of secondary metabolites. Furthermore, the presence of a large number of completed unknown genes encoding catalytic enzymes for the products strongly supported the accuracy of chemical structure predictions for the forward approach (Lee et al. 2020). Thus, the forward method has the most potent for the isolation and elucidation of new bioactive metabolites from actinomycetes.

14.3.5 Application of Genome Mining Approach in Endophytic Actinomycetes

Endophytic actinomycetes are widespread and symbiosis with plant physiological benefits; however, their genomes and secondary metabolites remain broadly unidentified. Ceapă et al. (2018) have isolated *S. scabrisporus* NF3 from the *Amphipterygium adstringens*, an endemic plant of Mexico, and studied the potent bioactive molecules together with genes and metabolites comprised in host interactions using genome mining tools: antiSMASH, PRISM, BAGEL3, and InterProScan 5. As a result, the endophytic *S. scabrisporus* NF3 exhibited outstanding biotechnological capabilities, with functional attributes in the interaction between microbe and plant. The genome mining illustrated an advancement in metabolic pathways associated with amino acid and protein synthesis and carbohydrate degradation together with the existence of over 50 biosynthetic gene clusters encoding potentially active molecules against plant pathogenic microorganisms such as griseobactin, abyssomicin, streptomycin, kanamycin, sanglifehrin, and polyoxypeptin (Ceapă et al. 2018). According to this study, the genome mining approaches not only confirmed the production of potential secondary metabolites and plant growth-promoting agents, but also provided invaluable ways to improve agricultural practices and reduced the use and persistence of chemicals currently used for plant growth.

Streptomyces strain YIM 130001 isolated from the lichen showed 100% similarity of the 16S rRNA gene sequence to the endophytic *Streptomyces* strain KLBMP1330 isolated from *Dendranthema indicum* (Xing et al. 2014). The strain was analyzed for searching and isolation of novel thiopeptide antibiotics based on the genome mining and metabolomic methods. The data obtained from antiSMASH version 4.0 demonstrated that the draft genome of *Streptomyces* strain YIM 130001 contained 18 putative BGCs encoding for the class of thiopeptide secondary metabolites. An active compound was isolated and purified by implementation with the high-performance liquid chromatography (HPLC) chromatogram, and then elucidated a structure using nuclear magnetic resonance (NMR). The compound was identified as geninthiocin B, which exhibited high similarity to several 35-membered macrocyclic thiopeptides geninthiocin, Val-geninthiocin and berninamycin A. The integration of spectroscopic data and bioinformatic analysis of the biocluster of geninthiocin B exhibited that it was closely related to berninamycins (Schneider et al. 2018). As reported in this study, it could be affirmed that the genome mining is a powerful tool for the discovery of bioactive natural products.

14.4 Conclusion and Future Prospectives

Over the past decades, it has been adequately increased in information for bioactive metabolites derived from known and novel species of the endophytic actinomycetes, especially in the genus *Streptomyces*. Many actinomycete metabolites consisting of antibacterial, antifungal, anticancer, anti-inflammatory and antioxidant drugs have been isolated and developed for using in the clinical field and the agricultural field in order to promote plant growth and defend host plants from phytopathogens. However, the study on bioactive compounds produced by the plant-derived actinomycetes is still in the early stage. Further success is not only continued to isolate and evaluate the biological activities, but modern strategies are also applied.

Currently, there are great advances in molecular and computational biology in addition to many actinomycete genomes have been continuously sequenced and publicly available in the databases. Genome mining represents a good method to powerfully force the identification of biosynthetic gene clusters, which encoded for enzymes involving in the natural product biosynthesis based on the reverse and forward algorithms. These highly civilized technologies contribute to the development of new generation for the studies in the discovery of bioactive natural products and diversity of endophytic actinomycetes. Furthermore, the developed methods also provide a more understanding of host–microbe interactions such as plant growth-promoting and plant defense mechanisms.

In the near future, machine learning and synthetic biology will be integrated to the genome mining for powerful analysis of microbiome datasets, especially in plants, animals and humans, which were still unexplored in genomic BGC resource. The datasets consisting of associations with disease phenotypes and characterization of novel chemical compounds including antibiotics, anticancer and immunomodulatory drugs could have important clinical effects for translating microbiome information to therapeutic interventions. We consequently suggest that the genome mining approach together with all of the techniques in molecular and computational biology could be represented as novel challenging platforms in the drug discovery and in the finding of the relationship among disease symptoms, pathogens, and therapeutic methods.

References

Albarano L, Esposito R, Ruocco N, Costantini M (2020) Genome mining as new challenge in natural products discovery. Mar Drugs 18(4):199

Alshaibani MM, MohamadZin N, Jalil J, Sidik NM, Ahmad SJ, Kamal N, Edrada-Ebel R (2017) Isolation, purification, and characterization of five active diketopiperazine derivatives from endophytic *Streptomyces* SUK 25 with antimicrobial and cytotoxic activities. J Microbiol Biotechnol 27(7):1249–1256

Bachmann BO, Ravel J (2009) Chapter 8. Methods for *in silico* prediction of microbial polyketide and nonribosomal peptide biosynthetic pathways from DNA sequence data. Methods Enzymol 458:181–217

Benson DR, Silvester WB (1993) Biology of *Frankia* strains, actinomycete symbionts of actinorhizal plants. Microbiol Mol Biol Rev 57(2):293–319

Bentley SD, Chater KF, Cerdeño-Tárraga AM, Challis GL, Thomson NR, James KD, Harris DE, Quail MA, Kieser H, Harper D (2002) Complete genome sequence of the model actinomycete *Streptomyces coelicolor* A3 (2). Nature 417(6885):141–147

Bérdy J (2005) Bioactive microbial metabolites. J Antibiot 58(1):1–26

Bérdy J (2012) Thoughts and facts about antibiotics: where we are now and where we are heading. J Antibiot 65(8):385–395

Bieber B, Nüske J, Ritzau M, Gräfe U (1998) Alnumycin a new naphthoquinone antibiotic produced by an endophytic *Streptomyces* sp. J Antibiot 51(3):381–382

Blin K, Medema MH, Kazempour D, Fischbach MA, Breitling R, Takano E, Weber T (2013) antiSMASH 2.0—a versatile platform for genome mining of secondary metabolite producers. Nucleic Acids Res 41(W1):W204–W212

Blin K, Shaw S, Steinke K, Villebro R, Ziemert N, Lee SY, Medema MH, Weber T (2019) antiSMASH 5.0: updates to the secondary metabolite genome mining pipeline. Nucleic Acids Res 47(W1):W81–W87

Boddy CN (2014) Bioinformatics tools for genome mining of polyketide and non-ribosomal peptides. J Ind Microbiol Biotechnol 41(2):443–450

Cacho RA, Tang Y, Chooi YH (2014) Next-generation sequencing approach for connecting secondary metabolites to biosynthetic gene clusters in fungi. Front Microbiol 5:774

Cao L, Qiu Z, You J, Tan H, Zhou S (2005) Isolation and characterization of endophytic streptomycete antagonists of *Fusarium* wilt pathogen from surface-sterilized banana roots. FEMS Microbiol Lett 247(2):147–152

Castillo UF, Strobel GA, Ford EJ, Hess WM, Porter H, Jensen JB, Albert H, Robison R, Condron MMA, Teplow DB (2002) Munumbicins, wide-spectrum antibiotics produced by *Streptomyces* NRRL 30562, endophytic on *Kennedia nigriscans*. Microbiology 148(9):2675–2685

Castillo U, Harper JK, Strobel GA, Sears J, Alesi K, Ford EJ, Lin J, Hunter M, Maranta M, Ge H (2003) Kakadumycins, novel antibiotics from *Streptomyces* sp. NRRL 30566, an endophyte of *Grevillea pteridifolia*. FEMS Microbiol Lett 224(2):183–190

Castillo UF, Strobel GA, Mullenberg K, Condron MM, Teplow DB, Folgiano V, Gallo M, Ferracane R, Mannina L, Viel S (2006) Munumbicins E-4 and E-5: novel broad-spectrum antibiotics from *Streptomyces* NRRL 3052. FEMS Microbiol Lett 255(2):296–300

Cattelan AJ, Hartel PG, Fuhrmann JJ (1999) Screening for plant growth–promoting rhizobacteria to promote early soybean growth. Soil Sci Soc Am J 63(6):1670–1680

Ceapă CD, Vázquez-Hernández M, Rodríguez-Luna SD, Cruz Vázquez AP, Jiménez Suárez V, Rodríguez-Sanoja R, Alvarez-Buylla ER, Sánchez S (2018) Genome mining of *Streptomyces scabrisporus* NF3 reveals symbiotic features including genes related to plant interactions. PLoS One 13(2):e0192618

Chandrakar S, Gupta AK (2019) Actinomycin-producing endophytic *Streptomyces parvulus* associated with root of aloe vera and optimization of conditions for antibiotic production. Probiot Antimicrob Proteins 11(3):1055–1069

Chantarawong W, Kuncharoen N, Tanasupawat S, Chanvorachote P (2019) Lumichrome inhibits human lung cancer cell growth and induces apoptosis via a p53-dependent mechanism. Nutr Cancer 71(8):1390–1402

Cimermancic P, Medema MH, Claesen J, Kurita K, Brown LCW, Mavrommatis K, Pati A, Godfrey PA, Koehrsen M, Clardy J (2014) Insights into secondary metabolism from a global analysis of prokaryotic biosynthetic gene clusters. Cell 158(2):412–421

Conway KR, Boddy CN (2013) ClusterMine360: a database of microbial PKS/NRPS biosynthesis. Nucleic Acids Res 41(Database issue):D402–D407

Coombs JT, Franco CMM (2003) Visualization of an endophytic *Streptomyces* species in wheat seed. Appl Environ Microbiol 69(7):4260–4262

Diminic J, Zucko J, Ruzic IT, Gacesa R, Hranueli D, Long PF, Cullum J, Starcevic A (2013) Databases of the thiotemplate modular systems (CSDB) and their in silico recombinants (r-CSDB). J Ind Microbiol Biotechnol 40(6):653–659

Ding L, Maier A, Fiebig HH, Lin WH, Peschel G, Hertweck C (2012) Kandenols A-E, Eudesmenes from an endophytic *Streptomyces* sp. of the mangrove tree *Kandelia candel*. J Nat Prod 75: 2223–2227

Ding L, Goerls H, Dornblut K, Lin W, Maier A, Fiebig H-H, Hertweck C (2015) Bacaryolanes A-C, rare bacterial caryolanes from a mangrove endophyte. J Nat Prod 78:2663–2967

Dochhil H, Dkhar MS, Barman D (2013) Seed germination enhancing activity of endophytic *Streptomyces* isolated from indigenous ethno-medicinal plant *Centella asiatica*. Int J Pharm Biol Sci 4(1):256–262

Doumbou CY, Akimov V, Beaulieu C (1998) Selection and characterization of microorganisms utilizing Thaxtomin A, a Phytotoxin produced by *Streptomyces scabies*. Appl Environ Microbiol 64(11):4313–4316

El-Gendy MMA, El-Bondkly AMA (2010) Production and genetic improvement of a novel antimycotic agent, saadamycin, against dermatophytes and other clinical fungi from endophytic *Streptomyces* sp. Hedaya48. J Ind Microbiol Biotechnol 37(8):831–841

El-Shatoury SA, El-Kraly OA, Trujillo ME, El-Kazzaz WM, El-Din ESG, Dewedar A (2013) Generic and functional diversity in endophytic actinomycetes from wild Compositae plant species at South Sinai—Egypt. Res Microbiol 164(7):761–769

El-Tarabily KA, Hardy GE, Sivasithamparam K (2010) Performance of three endophytic actinomycetes in relation to plant growth promotion and biological control of *Pythium aphanidermatum*, a pathogen of cucumber under commercial field production conditions in the United Arab Emirates. Eur J Plant Pathol 128(4):527–539

Ezra D, Castillo UF, Strobel GA, Hess WM, Porter H, Jensen JB, Condron MAM, Teplow DB, Sears J, Maranta M (2004) Coronamycins, peptide antibiotics produced by a verticillate *Streptomyces* sp. (MSU-2110) endophytic on *Monstera* sp. Microbiology 150(4):785–793

Genilloud O (2017) Actinomycetes: still a source of novel antibiotics. Nat Prod Rep 34(10):1203–1232

Hadjithomas M, Chen I-MA, Chu K, Huang J, Ratner A, Palaniappan K, Andersen E, Markowitz V, Kyrpides NC, Ivanova NN (2017) IMG-ABC: new features for bacterial secondary metabolism analysis and targeted biosynthetic gene cluster discovery in thousands of microbial genomes. Nucleic Acids Res 45(D1):D560–D565

Hannigan GD, Prihoda D, Palicka A, Soukup J, Klempir O, Rampula L, Durcak J, Wurst M, Kotowski J, Chang D (2019) A deep learning genome-mining strategy for biosynthetic gene cluster prediction. Nucleic Acids Res 47(18):e110–e110

Hasegawa S, Meguro A, Shimizu M, Nishimura T, Kunoh H (2006) Endophytic actinomycetes and their interactions with host plants. Actinomycetologica 20(2):72–81

Higashide E, Asai M, Ootsu K, Tanida S, Kozai Y, Hasegawa T, Kishi T, Sugino Y, Yoneda M (1977) Ansamitocin, a group of novel maytansinoid antibiotics with antitumour properties from Nocardia. Nature 270(5639):721–722

Igarashi Y (2004) Screening of novel bioactive compounds from plant-associated actinomycetes. Actinomycetologica 18(2):63–66

Igarashi Y, Iida T, Yoshida R, Furumai T (2002) Pteridic acids A and B, novel plant growth promoters with auxin-like activity from *Streptomyces hygroscopicus* TP-A0451. J Antibiot 55(8):764–767

Igarashi Y, Miura SS, Fujita T, Furumai T (2006) Pterocidin, a cytotoxic compound from endophytic *Streptomyces hygroscopicus*. J Antibiot 59:193–195

Igarashi Y, Trujillo ME, Martínez-Molina E, Yanase S, Miyanaga S, Obata T, Sakurai H, Saiki I, Fujita T, Furumai T (2007) Antitumor anthraquinones from an endophytic actinomycete *Micromonospora lupini* sp. nov. Bioorg Med Chem Lett 17(13):3702–3705

Igarashi Y, Yanase S, Sugimoto K, Enomoto M, Miyanaga S, Trujillo ME, Saiki I, Kuwahara S (2011) Lupinacidin C, an inhibitor of tumor cell invasion from *Micromonospora lupini*. J Nat Prod 74(4):862–865

Ikeda H, Kotaki H, Omura S (1987) Genetic studies of avermectin biosynthesis in *Streptomyces avermitilis*. J Bacteriol 169:5615–5621

Inahashi Y, Iwatsuki M, Ishiyama A, Namatame M, Nishihara-Tsukashima A, Matsumoto A, Hirose T, Sunazuka T, Yamada H, Otoguro K (2011) Spoxazomicins A–C, novel antitrypanosomal alkaloids produced by an endophytic actinomycete, *Streptosporangium oxazolinicum* K07-0460 T. J Antibiot 64(4):303–307

Inahashi Y, Iwatsuki M, Ishiyama A, Matsumoto A, Hirose T, Oshita J, Sunazuka T, Panbangred W, Takahashi Y, Kaiser M (2015) Actinoallolides A–E, new anti-trypanosomal macrolides, produced by an endophytic actinomycete, *Actinoallomurus fulvus* MK10-036. Org Lett 17(4):864–867

Janso JE, Carter GT (2010) Biosynthetic potential of phylogenetically unique endophytic actinomycetes from tropical plants. Appl Environ Microbiol 76(13):4377–4386

Kaweewan I, Komaki H, Hemmi H, Hoshino K, Hosaka T, Isokawa G, Oyoshi T, Kodani S (2019) Isolation and structure determination of a new cytotoxic peptide, curacozole, from *Streptomyces curacoi* based on genome mining. J Antibiot 72(1):1–7

Khater S, Anand S, Mohanty D (2016) *In silico* methods for linking genes and secondary metabolites: the way forward. Synth Syst Biotechnol 1(2):80–88

Kim J, Yi GS (2012) PKMiner: a database for exploring type II polyketide synthases. BMC Microbiol 12:169

Kim N, Shin JC, Kim W, Hwang BY, Kim BS, Hong YS, Lee D (2006) Cytotoxic 6-alkylsalicylic acids from the endophytic *Streptomyces laceyi*. J Antibiot 59(12):797–800

Kuncharoen N, Fukasawa W, Iwatsuki M, Mori M, Shiomi K, Tanasupawat S (2019) Characterisation of two polyketides from *Streptomyces* sp. SKH1-2 isolated from roots of *Musa* (ABB) cv. 'Kluai Sao Kratuep Ho'. Int Microbiol 22(4):451–459

Kupchan SM, Komoda Y, Thomas GJ, Smith RM, Karim A, Gilmore CJ, Haltiwanger RC, Bryan RF (1972) Maytansine, a novel antileukemic ansa macrolide from *Maytenus ovatus*. J Am Chem Soc 94(4):1354–1356

Lechevalier HA, Lechevalier MP (1967) Biology of actinomycetes. Annu Rev Microbiol 21:71–100

Lee N, Hwang S, Kim J, Cho SH, Palsson B, Cho BK (2020) Mini review: genome mining approaches for the identification of secondary metabolite biosynthetic gene clusters in *Streptomyces*. Comput Struct Biotechnol J 18:1548–1556

Li MHT, Ung PMU, Zajkowski J, Garneau-Tsodikova S, Sherman DH (2009) Automated genome mining for natural products. BMC Bioinform 10(1):1–10

Li Q, Cheng T, Wang Y, Bryant SH (2010a) PubChem as a public resource for drug discovery. Drug Discov Today 15(23–24):1052–1057

Li J, Lu C, Shen Y (2010b) Macrolides of the bafilomycin family produced by *Streptomyces* sp. CS. J Antibiot 63(10):595–599

Lu C, Shen Y (2003) A new macrolide antibiotic with antitumor activity produced by *Streptomyces* sp. CS, a commensal microbe of *Maytenus hookeri*. J Antibiot 56(4):415–418

Lu C, Shen Y (2004) Two new macrolides produced by *Streptomyces* sp. CS. J Antibiot 57(9):597–600

Lu C, Shen Y (2007) A novel ansamycin, naphthomycin K from *Streptomyces* sp. J Antibiot 60(10):649–653

Lucas X, Senger C, Erxleben A, Gruning BA, Doring K, Mosch J, Flemming S, Gunther S (2013) StreptomeDB: a resource for natural compounds isolated from *Streptomyces* species. Nucleic Acids Res 41(Database issue):D1130–D1136

Mitousis L, Thoma Y, Musiol-Kroll EM (2020) An update on molecular tools for genetic engineering of Actinomycetes—the source of important antibiotics and other valuable compounds. Antibiotics 9(8):494

Mody KH, Haldar S (2015) Genome mining for bioactive compounds. In: Springer handbook of marine biotechnology. Springer, pp 531–539

Nakamura Y, Afendi FM, Parvin AK, Ono N, Tanaka K, Hirai Morita A, Sato T, Sugiura T, Altaf-Ul-Amin M, Kanaya S (2014) KNApSAcK Metabolite Activity Database for retrieving the relationships between metabolites and biological activities. Plant Cell Physiol 55(1):e7

Nakashima T, Okuyama R, Kamiya Y, Matsumoto A, Iwatsuki M, Inahashi Y, Yamaji K, Takahashi Y, Omura S (2013) Trehangelins A, B and C, novel photo-oxidative hemolysis inhibitors produced by an endophytic actinomycete, *Polymorphospora rubra* K07-0510. J Antibiot 66:311–317

Nett M, Ikeda H, Moore BS (2009) Genomic basis for natural product biosynthetic diversity in the actinomycetes. Nat Prod Rep 26(11):1362–1384

Newman DJ, Cragg GM (2007) Natural products as sources of new drugs over the last 25 years. J Nat Prod 70(3):461–477

Olano C, García I, González A, Rodriguez M, Rozas D, Rubio J, Sánchez-Hidalgo M, Braña AF, Méndez C, Salas JA (2014) Activation and identification of five clusters for secondary metabolites in *Streptomyces albus* J 1074. Microb Biotechnol 7(3):242–256

Powel RG, Weisleder D, Smith CR, Kozlowski J, Rohwedder WK (1982) Treflorine, trenudine, and Nmethyltrenudone: novel maytansinoids tumour inhibitors containing two fused macrocyclic rings. J Am Chem Soc 104:4929–4934

Prieto C, Garcia-Estrada C, Lorenzana D, Martin JF (2012) NRPSsp: non-ribosomal peptide synthase substrate predictor. Bioinform 28(3):426–427

Pullen C, Schmitz P, Meurer K, Bamberg DD, Lohmann S, Franca S, Groth I, Schlegel B, Möllmann U, Gollmick F (2002) New and bioactive compounds from *Streptomyces* strains residing in the wood of Celastraceae. Planta 216(1):162–167

Qin S, Li J, Chen HH, Zhao GZ, Zhu WY, Jiang CL, Xu LH, Li WJ (2009) Isolation, diversity, and antimicrobial activity of rare actinobacteria from medicinal plants of tropical rain forests in Xishuangbanna, China. Appl Environ Microbiol 75(19):6176–6186

Qin S, Xing K, Jiang JH, Xu LH, Li WJ (2011) Biodiversity, bioactive natural products and biotechnological potential of plant-associated endophytic actinobacteria. Appl Microbiol Biotechnol 89(3):457–473

Quadt-Hallmann A, Kloepper JW, Benhamou N (1997) Bacterial endophytes in cotton: mechanisms of entering the plant. Can J Microbiol 43(6):577–582

Remali J, Sarmin NM, Ng CL, Tiong JJL, Aizat WM, Keong LK, Zin NM (2017) Genomic characterization of a new endophytic *Streptomyces kebangsaanensis* identifies biosynthetic pathway gene clusters for novel phenazine antibiotic production. PeerJ 5:e3738

Rosenblueth M, Martínez-Romero E (2006) Bacterial endophytes and their interactions with hosts. Mol Plant Microbe Interact 19(8):827–837

Rottig M, Medema MH, Blin K, Weber T, Rausch C, Kohlbacher O (2011) NRPSpredictor2—a web server for predicting NRPS adenylation domain specificity. Nucleic Acids Res 39(Web Server issue):W362–W367

Rungin S, Indananda C, Suttiviriya P, Kruasuwan W, Jaemsaeng R, Thamchaipenet A (2012) Plant growth enhancing effects by a siderophore-producing endophytic streptomycete isolated from a Thai jasmine rice plant (*Oryza sativa* L. cv. KDML105). Antonie Van Leeuwenhoek 102(3):463–472

Ryan RP, Germaine K, Franks A, Ryan DJ, Dowling DN (2008) Bacterial endophytes: recent developments and applications. FEMS Microbiol Lett 278(1):1–9

Sasaki T, Igarashi Y, Saito N, Furumai T (2001) Cedarmycins A and B, new antimicrobial antibiotics from *Streptomyces* sp. TP-A0456. J Antibiot 54(7):567–572

Scheffler RJ, Colmer S, Tynan H, Demain AL, Gullo VP (2013) Antimicrobials, drug discovery, and genome mining. Appl Microbiol Biotechnol 97(3):969–978

Schneider O, Simic N, Aachmann FL, Rückert C, Kristiansen KA, Kalinowski J, Jiang Y, Wang L, Jiang CL, Lale R (2018) Genome mining of *Streptomyces* sp. YIM 130001 isolated from lichen affords new thiopeptide antibiotic. Front Microbiol 9:3139

Shutsirung A, Chromkaew Y, Pathom-Aree W, Choonluchanon S, Boonkerd N (2013) Diversity of endophytic actinomycetes in mandarin grown in northern Thailand, their phytohormone production potential and plant growth promoting activity. Soil Sci Plant Nutr 59(3):322–330

Skinnider MA, Dejong CA, Rees PN, Johnston CW, Li H, Webster ALH, Wyatt MA, Magarvey NA (2015) Genomes to natural products prediction informatics for secondary metabolomes (PRISM). Nucleic Acids Res 43(20):9645–9662

Skinnider MA, Merwin NJ, Johnston CW, Magarvey NA (2017) PRISM 3: expanded prediction of natural product chemical structures from microbial genomes. Nucleic Acids Res 45(W1):W49–W54

Starcevic A, Zucko J, Simunkovic J, Long PF, Cullum J, Hranueli D (2008) ClustScan: an integrated program package for the semi-automatic annotation of modular biosynthetic gene clusters and in silico prediction of novel chemical structures. Nucleic Acids Res 36(21):6882–6892

Strobel GA, Daisy B (2003) Bioprospecting for microbial endophytes and their natural products. Microbiol Mol Biol Rev 67(4):491–502

Strobel GA, Daisy B, Castillo U, Harper J (2004) Natural products from endophytic microorganisms. J Nat Prod 67(2):257–268

Suwanborirux K (1990) Cytotoxic constituents from the higher plants *Spathelia sorbifolia, Psorospermum febrifugum* and the mosses *Claopodium crispifolium, Anomodon attenuatus.* Purdue University

Taechowisan T, Lu C, Shen Y, Lumyong S (2005) Secondary metabolites from endophytic *Streptomyces aureofaciens* CMUAc130 and their antifungal activity. Microbiology 151(5):1691–1695

Taechowisan T, Lu C, Shen Y, Lumyong S (2007) Antitumor activity of 4-arylcoumarins from endophytic *Streptomyces aureofaciens* CMUAc130. J Cancer Res Ther 3(2):86

Taechowisan T, Chaisaeng S, Phutdhawong WS (2017) Antibacterial, antioxidant and anticancer activities of biphenyls from *Streptomyces* sp. BO-07: an endophyte in *Boesenbergia rotunda* (L.) Mansf A. Food Agric Immunol 28(6):1330–1346

Tan GYA, Franco CMM, Edrada-Ebel R, Latip J, Zin NM (2013) *Streptomyces kebangsaanensis* sp. nov., an endophytic actinomycete isolated from an ethnomedicinal plant, which produces phenazine-1-carboxylic acid. Int J Syst Evol Microbiol 63(10):3733–3738

Trivella DBB, de Felicio R (2018) The tripod for bacterial natural product discovery: genome mining, silent pathway induction, and mass spectrometry-based molecular networking. mSystems 3(2):e00160–e00117

van Heel AJ, de Jong A, Montalban-Lopez M, Kok J, Kuipers OP (2013) BAGEL3: automated identification of genes encoding bacteriocins and (non-) bactericidal posttranslationally modified peptides. Nucleic Acids Res 41(W1):W448–W453

Vijayan M, Chandrika SK, Vasudevan SE (2011) PKSIIIexplorer: TSVM approach for predicting Type III polyketide synthase proteins. Bioinformation 6(3):125–127

Wang P, Kong F, Wei J, Wang Y, Wang W, Hong K, Zhu W (2014) Alkaloids from the mangrove-derived actinomycete *Jishengella endophytica* 161111. Mar Drugs 12(1):477–490

Weber T (2014) *In silico* tools for the analysis of antibiotic biosynthetic pathways. Int J Med Microbiol 304(3–4):230–235

Weber T, Rausch C, Lopez P, Hoof I, Gaykova V, Huson DH, Wohlleben W (2009) CLUSEAN: a computer-based framework for the automated analysis of bacterial secondary metabolite biosynthetic gene clusters. J Biotechnol 140(1–2):13–17

Wohlleben W, Mast Y, Stegmann E, Ziemert N (2016) Antibiotic drug discovery. Microb Biotechnol 9(5):541–548

Xing K, Qin S, Zhang WD, Cao CL, Ruan JS, Huang Y, Jiang JH (2014) *Glycomyces phytohabitans* sp. nov., a novel endophytic actinomycete isolated from the coastal halophyte in Jiangsu, East China. J Antibiot 67(8):559–563

Yan LL, Han NN, Zhang YQ, Yu LY, Chen J, Wei YZ, Li QP, Tao L, Zheng GH, Yang SE (2010) Antimycin A 18 produced by an endophytic *Streptomyces albidoflavus* isolated from a mangrove plant. J Antibiot 63(5):259–261

Yu Z, Zhao LX, Jiang CL, Duan Y, Wong L, Carver KC, Schuler LA, Shen B (2011) Bafilomycins, produced by an endophytic actinomycete *Streptomyces* sp. YIM 56209. J Antibiot 64:159–162

Zazopoulos E, Huang K, Staffa A, Liu W, Bachmann BO, Nonaka K, Ahlert J, Thorson JS, Shen B, Farnet CM (2003) A genomics-guided approach for discovering and expressing cryptic metabolic pathways. Nat Biotechnol 21(2):187–190

Zerikly M, Challis GL (2009) Strategies for the discovery of new natural products by genome mining. ChemBioChem 10(4):625–633

Zhang J, Wang JD, Liu CX, Yuan JH, Wang XJ, Xiang WS (2014) A new prenylated indole derivative from endophytic actinobacteria *Streptomyces* sp. neau-D50. Nat Prod Res 28(7):431–437

Zhao PJ, Fan LM, Li GH, Zhu N, Shen YM (2005) Antibacterial and antitumor macrolides from *Streptomyces* sp. Is9131. Arch Pharm Res 28(11):1228–1232

Ziemert N, Alanjary M, Weber T (2016) The evolution of genome mining in microbes—a review. Nat Prod Rep 33(8):988–1005

Mining for NRPS and PKS Genes in *Actinobacteria* Using Whole-Genome Sequencing and Bioinformatic Tools

15

Heidi El-Gawahergy, Dina H. Amin, and Alaa F. Elsayed

Abstract

Antimicrobial resistance of pathogens against known antibiotics is a growing concern. It is a call of emergency to find an effective cure for pathogens. *Actinobacteria* are great producers of antibiotic, antifungal, and anticancer agents. Non-ribosomal peptide synthetase and polyketide synthase are multi-enzymes responsible to produce complex polyketides and peptides of significant pharmaceutical importance. Many NPPS and PKS gene clusters in *Actinobacteria* are not normally expressed under laboratory conditions, but they can be only revealed through the next-generation sequencing technologies supported by bioinformatic analysis. In this chapter, we highlight the role of genome assembly and annotation processes in determining the structural and functional part of secondary metabolites genes. We also shed the light on some bioinformatic software tools for targeting NRPS and PKS gene clusters. Additionally, we referred to the developments in genomics and synthetic biology, which can strongly trigger the discovery of new antibiotics from *Actinobacteria*.

Keywords

Actinobacteria · Next-generation sequencing · PKS and NRPS genes · Bioinformatic tools

H. El-Gawahergy
Department of Microbiology, Egyptian Drug Authority (EDA), Cairo, Egypt

D. H. Amin (✉) · A. F. Elsayed
Department of Microbiology, Faculty of Science, Ain Shams University, Cairo, Egypt
e-mail: dina.hatem@sci.asu.edu.eg; alaafayez@sci.asu.edu.eg

R. V. Rai, J. A. Bai (eds.), *Natural Products from Actinomycetes*,
https://doi.org/10.1007/978-981-16-6132-7_15

15.1 Introduction

The phylum *Actinobacteria* is one of the largest taxonomic units recognized within the b*acteria* domain (Ludwig et al. 2012). The majority of the *Actinobacteria* are free-living saprophytic organisms that are widely distributed in broad range of ecological environments: They are present in both terrestrial and aquatic (including marine) ecosystems. They can be heterotrophic or chemoautotrophic, but most are chemoheterotrophic and capable of consuming a wide range of nutrients, including complex polysaccharides (Macagnan et al. 2006). *Actinobacteria* produce a wide range of secondary metabolites with significant medicinal and commercial value. *Actinobacteria,* particularly the genus *Streptomyces,* have produced a number of antibiotics since the discovery of actinomycin. They are also responsible for the decomposition of humus (a soil-resistant substance) and the earthy smell of freshly ploughed soils, as well as the production of nitrogen and a number of important antibiotics like streptomycin, terramycin, and aureomycin.

Actinobacteria are Gram-positive filamentous bacteria with a high guanine and cytosine (G_C) content in their genomes. Tip extension and hyphae branching are used to grow them. This is the origin of their name, which comes from the Greek word "ray" (aktis or aktin) (Lechevalier and Lechevalier 1965; Zimmerman 1980). Therefore, *Actinomycetes* were considered transitional forms between fungi and bacteria. Mycelium is produced by *Actinobacteria*, just like filamentous fungi. Actinomycetes, like all bacteria, have a thin unicellular chromosome structured in a prokaryotic nucleoid and a peptidoglycan cell wall; moreover, the cells are vulnerable to antibacterial treatments (Chaudhary et al. 2013; Aftab et al. 2015; Weber et al. 2015; Barka et al. 2016; Behie et al. 2017; Ser et al. 2017).

15.2 Classification of *Actinomycetes*

The phylum *Actinobacteria* is considered as one of the largest phyla among the 30 major phyla known within the domain bacteria until 1983, *Actinobacteria* phylum. It was divided into five subclasses, six orders, and 14 suborders, making it one of the largest taxonomic units in the bacteria domain (Goodfellow and Williams 1983). The sequencing and amplification of the 16S rRNA gene from microbiotic communities, on the other hand, have become a standard for comparing communities. The 16S ribosomal DNA sequences provided actinomycetologists with a phylogenetic tree that allowed them to investigate Actinomycete evolution and identify them (Burke and Darling 2016), and this leads to six classes (*Actinobacteria*, *Acidimicrobiia*, *Coriobacteriia*, *Nitriliruptoria*, *Rubrobacteria*, and *Thermoleophilia*), 16 orders (*Actinomycetales*, *Actinopolysporales*, *Bifidobacteriales*, *Catenulisporales*, *Corynebacteriales*, *Frankiales*, *Glycomycetales*, *Jiangellales*, *Kineosporiales*, *Micrococcales*, *Micromonosporales*, *Propionibacteriales*, *Pseudonocardiales*, *Streptomycetales*, *Streptosporangiales*, and *Incertaesedis*), 43 families, and 130 genera of *Actinobacteria* (Zhi et al. 2009; Ranjani et al. 2016).

15.3 General Characteristics of *Actinobacteria*:-

It is mainly composed of aerial mycelium that is usually thicker than the substrate mycelium and exhibits enough differentiation that an assortment of isolates may be divided into several groups with similar morphological traits under fixed conditions. This includes structure (cottony, velvety, or powdery), creation of rings or concentric zones, and coloring, and is one of the most essential criteria for the classification of *Actinobacteria*. *Actinobacteria* substrate mycelium varies in size, shape, and thickness. It might be white or nearly colorless, or it can be yellow, brown, red, pink, orange, green, or black (Holt et al. 1994).

15.4 Habitat of *Actinobacteria*

Actinobacteria are a widespread group found in natural ecosystems, which implies they can be found in a variety of settings. They are the most common type of microorganism found in soil, but they can also be found in aquatic habitats including rivers, oceans, and lakes. They are also found in the gastrointestinal tracts of some mammals and are part of symbiotic relationships with plants and animals (Kumar et al. 2014; Rana and Salam 2014; Barka et al. 2016).

Desert soil, marine sponge, and radon-containing thermal fonts were among the unusual habitats where new species were discovered. *Actinomycetes* capacity to live in a wide range of environments is due to their ability to create extracellular hydrolytic enzymes, and they are responsible for the degradation of organic matter, making them crucially important organisms in the carbon cycle. Some species can break complex and recalcitrant compounds, for example, *Rhodococcus* species, which can degrade nitrophenol, dinitrophenol, pyridine, and nitroaromatic compounds (Ul-Hassan and Wellington 2009; Kumar et al. 2014).

The most important habitat for *Actinobacteria* is soil, with *Streptomycetes* making up a large part of the population. They are usually found in concentrations of 106–109 cells per gram of soil (Goodfellow and Williams 1983). Temperature, pH, and soil moisture are all elements that influence *Actinobacteria* growth. They are generally mesophilic, growing best at temperatures between 25 and 30 °C. Thermophilic *Actinobacteria,* on the other hand, can grow at temperatures as high as 60 °C (Edwards 1993). Low humidity promotes *Actinobacteria* vegetative growth, especially when the spores are submerged in water. In dry soils, on the other hand, where the moisture tension is higher, growth is restricted. The majority of *Actinobacteria* thrive in soils with a pH of 7. They thrive in a pH range of 6–9, with maximal growth occurring near neutrality. A few *Streptomyces* strains have been isolated from acidic soils, though (pH 3.5) (Kim et al. 2003).

Actinobacteria are widely distributed in aquatic habitats. Although it has been long believed that *Actinobacteria* isolated from water are of soil origin, they do not develop in the aquatic environment, having inactive state in the form of spores (Goodfellow and Williams 1983). The molecular methods independent of cultivation, mainly such as fluorescent cell labeling and PCR analysis of 16S rRNA gene sequences, revealed that uncultivated members of *Actinobacteria* are numerous and cosmopolitan inhabitants of freshwater ecosystems, constituting the dominant fraction of heterotrophic bacterioplankton (Zwart et al. 2002; Warnecke et al. 2004; Allgaier and Grossart 2006; Newton et al. 2011).

15.5 Next-Generation Sequencing and *Actinobacteria*

Nucleic acid sequencing is a technique used for determining the nucleotide sequence of a DNA or RNA molecule. As the capacity to sequence has become more available to academic and clinical laboratories over the last century, the usage of nucleic acid sequencing has blasted. The Human Genome Project, a $3 billion, 13-year initiative finished in 2003, was the first major project for DNA sequencing. First-generation sequencing, also known as Sanger sequencing, was used to complete the Human Genome Project. Sanger sequencing is a method of determining the order of the chain termination method, and Edward Sanger's invention in 1975 was the gold standard for nucleic acid sequencing for the next two and a half decades (Sanger et al. 1977).

Sanger sequencing (first-generation sequencing) has become widely employed in molecular biology research, leading to the development of various PCR-based approaches for studying microbial communities. To investigate uncultured microbial communities in various contexts, denaturing gradient gel electrophoresis (DGGE) or terminal restriction fragment length polymorphism (T-RFLP) has been created (Muyzer et al. 1993; Cancilla et al. 1992). Following the completion of the first human genome sequence, there has been a significant surge in demand for cheaper and quicker sequencing methods. Second-generation sequencing (SGS) technologies, often known as next-generation sequencing, were developed in response to this need (NGS). NGS platforms do massive parallel sequencing, which involves sequencing millions of DNA fragments from a single sample. High-throughput sequencing is made possible by massive parallel sequencing technology, which allows a complete genome to be sequenced in less than a day (Kircher and Kelso 2010; Rastogi and Sani 2011).

SGS (second-generation sequencing) was one of the earliest NGS methods to arrive, and it relied on cycles of terminating DNA polymerization and recording the included nucleotides in each cycle. In 2005, 454 pyrosequencing was the first SGC technology to be commercialized by 454 Life sciences company, which is now a subsidary of Roche Diagnostics (Margulies et al. 2005). In 2006, the first sequencer was launched by Solexa/Illumina company. It was based on reversible terminator chemistry sequencing approach and it was widely used by scientists because of it's low cost, high-throughput level, and high accuracy (Loman et al. 2012). Several disciplines of research and medical diagnostics have become the first choice sequencing technology. Other SGS technologies have been commercialized, but none have achieved the same level of success as these two. For example, SOLiD (issued in 2006 by Applied Biosystems Inc., now Thermo Fisher Scientific/Life

Technologies) and Ion Torrent (launched in 2006 by Thermo Fisher Scientific/Life Technologies) have been utilized for Actinobacteria genome sequencing.

Actinobacteria becoming more important in various disciplines coupled with advances in molecular biology, particularly in this post-genomic era, can help us gain a better knowledge of these organisms by analyzing their genome. The availability of NGS technologies and -omics methodologies (metagenomics, metaproteomic) has tremendously aided *Actinobacteria* biosynthesis research (Qin et al. 2016). The availability of huge collections of *Actinobacterial* genome sequences is expected to help us better comprehend *Actinobacteria* phylogeny and aid in the discovery of medically helpful novel natural compounds (Ventura et al. 2007).

Actinobacteria have broadly diverse genomic sizes, ranging from 1 to 12 Mb (Větrovský and Baldrian 2013), where bioactive compounds are genetically coded by biosynthetic gene clusters (BGCs). The advancement of NGS technologies boosted the understanding of secondary metabolite biosynthesis possibilities of *Actinobacteria* (Nouioui et al. 2018). Far of these BGCs are quiet or minimally expressed under conventional laboratory circumstances, according to the genome research, which revealed *Actinobacteria* capable of generating many more chemicals than were detected in in vitro culture.

The large G + C content of *Actinomycete* genomes creates challenges not only for sequencing technology but also for the computing algorithms employed in genome assembly (Nakamura et al. 2011). Improvements in library preparation have reduced many of the mistakes and sequencing biases caused by mol percent G + C bias (Kozarewa et al. 2009; Quail et al. 2012). The linear chromosome and plasmids of many significant *Actinobacteria*, such as the *Streptomycetes*, provide a more specific difficulty, with long terminal inverted repeats that can reach over one megabase, with today's sequencing methods, and this is impossible to resolve. Furthermore, extracting high molecular weight DNA of the high-quality necessary for NGS library creation is not simple and is currently impossible in many circumstances, particularly from Actinobacteria that are difficult to culture and resistant to cell wall breakdown (Weaver et al. 2004).

15.6 Genome Assembly

Genome assembly is the process of piecing together a huge number of small DNA sequences to construct a representation of the chromosomes from whence the DNA came. Millions of tiny DNA fragments are "read" by automated sequencing equipment, which can read up to 1000 nucleotides. A genome assembly method works by matching all of the parts together and recognizing overlapping sections of the reads. The operation can be continued when these overlapping reads are combined (Paszkiewicz and Studholme 2010).

Because many genomes contain huge numbers of identical sequences, known as repeats, genome assembly is a difficult computing challenge. These repeats can be thousands of nucleotides long, and some can be found in many locations in plants' and animals' genomes. The resultant genome is called a draft genome sequence; it is produced by merging the information sequenced together in the correct orientation and order, and then connecting them to generate scaffolds (Paszkiewicz and Studholme 2010). Scaffolds are bigger DNA pieces that are positioned along the chromosome's physical map. Contigs collected from sequencing genomes can also be built using the database's most similar reference genome to fill in the gaps (Darling et al. 2011).

15.7 Genome Annotation

The process of identifying the elements of the genome and attaching biological information to these elements is known as genome annotation (Stein 2001). Gene annotation is required after genome assembly to determine the structural and functional identity of those genes (Kisand and Lettieri 2013). Instead of manual annotation, which requires human skill, automatic annotation technologies attempt to accomplish all of this through computer analysis. Rapid annotation using subsystem technology (RAST), a popular online automated annotation system, is an automated service for annotating whole or almost complete bacterial genomes with high-quality genome annotations for these genomes in phylogenetic trees (Aziz et al. 2008). The integrated microbial genomes based on BLAST p (IMG) system is a resource for comparative analysis of publicly available genomes in a broad integrated context for both draft and complete microbial genomes (Markowitz et al. 2009) and Prokaryotic Genomes Automatic Annotation Pipeline (PGAAP) established at the National Center for Biotechnology Information (NCBI) that is based on homology-based approaches and gene prediction algorithms. PGAAP annotates both complete genomes and draft genomes covering multiple contigs (Tatusova et al. 2016).

15.8 Locating Secondary Metabolite Gene Clusters Using Bioinformatic Tools

Thousands of compounds have been isolated from *Actinobacteria* to date, with a wide range of structure and bioactivities; yet, with the advent of microbial resistance to commonly used antibiotics, the need for novel bioactive chemicals remains critical (Arnison et al. 2013; WHO 2017). Traditional bioactivity screening approaches, unfortunately, result in the rediscovery of previously identified compounds. The availability of bacterial genome sequences, as well as ever-improving algorithms for computer prediction of bacterial secondary metabolites, pave the way for genome mining (Medema and Fischbach 2015), which leads to the identification of secondary metabolite gene clusters (SMGCs) within genomic data. Understanding the composition and regulation of SMGCs can help scientists design experiments that isolate molecules more precisely, saving time and money in the process of discovering new chemicals. Furthermore, genetic data can aid in the activation of "silent" gene clusters that are not expressed under typical laboratory growth conditions. Furthermore, understanding regulatory systems can aid in the optimization of heterologous gene expression laboratory settings. However, identifying SMGCs and predicting their products are a difficult task. It is crucial to understand various SMGC types, typical genes, functional domains, and assembly methods in order to comprehend the basic mechanisms of secondary metabolite biosynthesis (Spellberg et al. 2008; Weissman 2015).

Complex regulatory networks mediated by biotic and abiotic stimuli present in the bacteria's natural habitat influence the expression of these clusters (Craney et al. 2013). As a result, only a small percentage of secondary metabolites can be synthesized in laboratory cultures, especially when we do not know the exact environmental cues that would trigger their synthesis. As a result, developing a set of bioinformatics tools to annotate and mine these genomes has become increasingly important.

15.9 NRPS and PKS Gene Clusters in *Actinobacteria*

Both NRPS and PKS are considered to be main classes of secondary metabolites that possess different chemical structures and a precious source of essential pharmaceutical molecules. They are a collection of enzymes, coded by genes in charge of production of key antibiotic groups. The existence of NRPS and PKS genes in an *Actinobacterial* strain is strongly linked to their biosynthetic ability (Marahiel et al. 1997; Amin et al. 2020).

Both PKSs and NRPs are biosynthesized through the thiotemplate mechanism, where the PK/NRP chain is collected on enzyme templates and the biosynthetic intermediates are covalently linked to the templates as thioesters (Hopwood 1997; Marahiel et al. 1997). PKs are built from small building blocks like acetate and other short carboxylic acids through sequential decarboxylative condensations similar to fatty acid biosynthesis, and this process is catalyzed by PKSs (Hopwood 1997; Staunton and Weissman 2001). NRPs are a family of structurally complex natural products that are majorly present in microorganisms. They are 2–48 amino acid residues long, usually with very different structures. Over 300 amino acids, including D-configured, N-methylated, and other unusual non-proteinogenic residues, are present in the products. The NRPSs are organized in a modular structure, where each module is a relatively separate functional block that fulfills a cycle of peptide elongation (Cane et al. 1998).

During the past decade, modular PKSs and NRPSs have been a source of attraction for a large number of bioinformatic studies. These multi-enzymes are molecular assembly lines for the production of strongly complex polyketides and peptides, respectively. Both groups of secondary metabolites are of remarkable pharmaceutical importance. PKS/NRPS megasynthases act sequentially, as each module is responsible for one round of chain elongation (Staunton and Weissman 2001; Finking and Marahiel 2004).

15.10 PKS and NRPS Databases

With the increasing number of both characterized and uncharacterized modular NRPS and PKS sequences in databases, it is not only essential to develop knowledge-based in silico approaches for correlating sequence and domain organization, but also to compile the existing data in searchable databases. Yadav et al. (2003a) were the first to follow this idea and have organized a searchable database for modular PKS (Yadav et al. 2003b). Subsequently, the group extended that database and added NRPS, iterative PKS, and chalcone synthases. The entire platform is designated NRPS–PKS (http://www.nii.res.in/nrps-pks.html). Overall, the group has summarized knowledge about 20 PKS, 17 NRPS, and 5 NRPS/PKS hybrid gene clusters. The database provides links to chemical structures, to references describing experimental characterization of the clusters, and shows the organization of each cluster into modules, as well as the respective domain organization and their functionality. The database is being updated continuously and currently includes data of 41 characterized PKS pathways. Several bioinformatic tools have been developed, including BAGEL (De Jong et al. 2006), ClustScan (Starcevic et al. 2008), CLUSEAN (Weber et al. 2009), NP.searcher (Li et al. 2009), PRISM (Skinnider et al. 2017), and antiSMASH (Blin et al. 2019), to identify SMBGCs within the genome, with most of these technologies relying on the highly conserved sequences within the SMBGCs to map their location. Below, we review several open-source and commercial software tools that are used for identifying metabolic gene clusters.

15.10.1 NP.searcher

Natural Product searcher (NP.searcher) (Li et al. 2009) is an open-source software program intended to scan microbial genomes for the identification of NRPS and PKS or hybrid NRPS-PKS gene clusters and the produced putative chemical structures of candidate NRP and PK specialized metabolites. Initially, the query sequences are aligned against NRPS/PKS sequences through the algorithm by using BLAST (http://www.ncbi.nlm.nih.gov/blast/) (Altschul et al. 1990) as shown in Fig. 15.1. To identify gene clusters, key catalytic domains of these signature enzymes are located, namely adenylation ("A") domain for NRPS and acyltransferase ("AT") domain for PKS. After that, sequence motifs are identified using BLAST again within the "A" and "AT" domains to predict substrate specificity by querying against an internal database of "signature" sequences derived from published literature. BLAST is used twice more NP.searcher to identify auxiliary domains that perform functions such as epimerization, reduction, and methylation during collection of the core product, and identify additional enzymes within the cluster, which further tailor the product through catalyzing reactions like glycosylation, heterocyclization, hydroxylation, and halogenation. After substrate specificity is predicted, structures of the produced NRP or PK molecules are predicted.

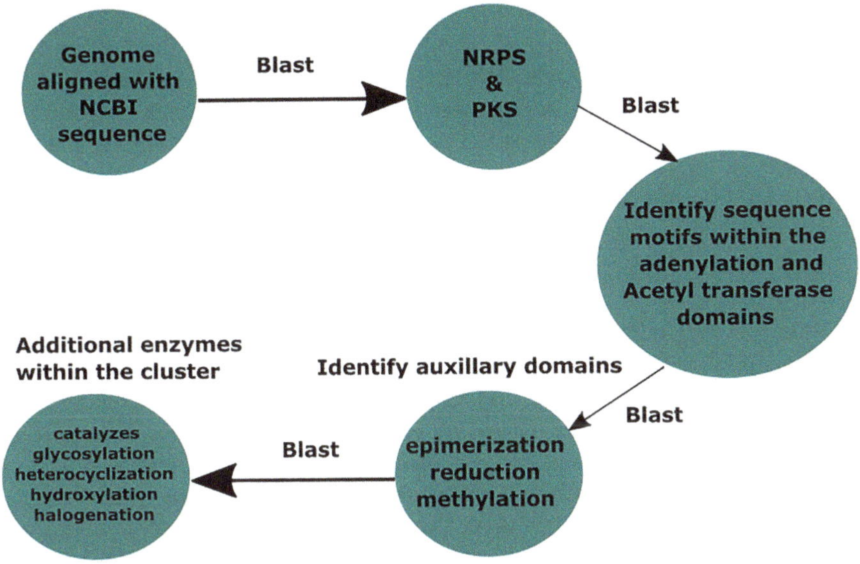

Fig. 15.1 NP searcher uses BLAST in genome mining for natural products

The NP.searcher internal database of 187 NRPS and 18 PKS sequences is used to locate substrate specificity of catalytic domains from query sequences. Unknown signature enzyme sequences that do not match previously characterized sequences and corresponding substrate specificities contained in the internal database can result in false predictions. Furthermore, if the catalytic domains of signature enzymes are separated by >15 kbp, the algorithm distinguishes them as two clusters. This factor is mostly required to be modified for specific genomes where metabolic clusters may span larger physical distances (Li et al. 2009).

15.10.2 antiSMASH

In 2011, a Web server for genomic determination and analysis of BGCs of any type called antiSMASH (antibiotics and Secondary Metabolite Analysis Shell) was introduced; thus, it simplifies rapid genome annotation of a wide range of bacterial and fungal strains (Blin et al. 2013). antiSMASH is able to annotate extensive chemical structures of secondary metabolites. However, it is still restricted to annotate peptides and polyketides coded by modular assembly lines only. Annotation of chemical compounds coded by cyclization and tailoring reactions is still limited. To overcome this limitation, multiple possible end product compound strategy can be applied. This is essential to avoid the replication of existing compounds for efficient drug discovery and comparative analysis of unknown and known gene clusters (Weber et al. 2015).

A distinct advantage of antiSMASH over NP.searcher is wide coverage of specialized metabolite compound classes. While comparing the performance between the two tools, 47 NRPS/PKS gene clusters were detected by antiSMASH as opposed to 31 detected by NP.searcher, and all predicted clusters by NP.searcher were also recovered by antiSMASH (Medema et al. 2011). antiSMASH and NP. searcher are like in defining a fixed cluster boundary. While NP.searcher differentiates two clusters if the catalytic domains of their signature enzymes are separated by >15 kbp, antiSMASH identifies signature gene within a 10 kbp distance to belong to the same cluster (Medema et al. 2011). However, antiSMASH also allows merging of predicted clusters; tailoring genes are incorporated by the algorithm through implementing a greedy approach and extending a cluster on either side by a distance of 5, 10, or 20 kbp (Medema et al. 2011). This permits overlaps among clusters. In this case, the algorithm assumes the creation of a supercluster. Thus, superclusters could be formed by either two separate gene clusters being closely spaced or by the construction of a hybrid cluster, which produces a compound derived from two or more chemical scaffolds (Medema et al. 2011). As with NP.searcher, antiSMASH implementation on more complex genomes requires alteration of parameters for spacing of clusters.

Secondary metabolite gene clusters in rare *Actinobacteria* are predicted by bioinformatic analysis using antiSMASH 3.0. Amin et al.'s research group used antiSMASH server to mine the whole genomic sequence of *Micromonospora* sp. Rc5 isolated from the Egyptian desert. Reads of 33 potential secondary metabolite gene clusters including PKS, NRPS, hybrid polyketide synthases, terpenes, lantipeptides, saccharides, siderophore, bacteriocin, aryl polyene, and unidentified clusters were demonstrated in this study (Amin et al. 2019). The annotation of the draft genome sequence of *Micromonospora* sp. DSW705 was reported by another study using antiSMASH analysis that predicts 3 PKS gene clusters, 1 NRPS gene cluster, and 3 hybrid PKS/NRPS gene clusters responsible for antitumor rakicidin synthesis (Komaki et al. 2016).

15.10.3 ClustScan

Cluster Scanner (ClustScan) is a cluster identification approach developed for annotation of NRPS, PKS, and hybrid NRPS-PKS enzymes and prediction of chemical structures of generated products (Starcevic et al. 2008). It is a commercial software package convenient for bacterial DNA sequences. It employs a top-down approach based on profile HMMs to annotate BGCs encoding modular biosynthetic enzymes, allowing the semi-automatic structural prediction of the products. The predicted chemical structures of products can be exported in a SMILES/SMARTS format for further analysis by standard chemistry programs.

15.10.4 CLUSEAN (Cluster Sequence Analyzer)

CLUSEAN is a BioPerl-based toolkit developed for genome annotation for identification of BGC region of secondary metabolite synthesis in bacteria (Weber et al. 2009). Analysis section for automated homology search, enzyme classification, and specificity predictions of conserved protein domains in NRPS, PKS, and NRPS A domains is found in CLUSEAN. The annotation results of CLUSEAN are exported as EMBL or MS Excel files.

15.10.5 PRISM

Prediction Informatics for Secondary Metabolomes (PRISM) is a Java application and Web server for chemical structure prediction of genetically encoded non-ribosomal peptides (NRPs) and type I and type II polyketides (PKs) (Skinnider et al. 2015). A microbial nucleotide sequence in FASTA or GenBank format is taken by PRISM as input. PRISM then searches the sequence with a library of HMMs associated with secondary metabolism, clusters the identified biosynthetic genes, and shows identified biosynthetic data for structure prediction. Because the exact site of tailoring process is not always predictable, combinatorial libraries of predicted structures are generated PRISM to account for variability in the action of tailoring enzymes or in the permutation of monomers that constitute the natural product backbone (Skinnider et al. 2016).

15.11 Manipulation of NRPS and PKS Genes

Since the end of the twentieth century, isolation of the natural product has remarkably decreased with the rise of combinatorial synthesis and high-throughput screening (Shen 2015). Additionally, low quantities, slow or poor growth of the producers, or the inability to cultivate the desired bacterial strains within laboratory conditions often interrupt the detection of natural products (Wenzel and Müller 2005; Zhang et al. 2008). Two complementary approaches have been mostly considered.

15.12 Activating the Gene Cluster in the Native Producer

A huge number of actinobacterial genomes have recently been sequenced, revealing the presence of multiple gene clusters that can control the manufacture of new specialized metabolites. However, the metabolic products of such cryptic gene clusters are rarely identified, because they are not produced in laboratory circumstances. An example of activation of gene clusters is *Streptomyces venezuelae* ATCC 1071; despite its reputation as a chloramphenicol producer, it was unable to manufacture this antibiotic in laboratory under a variety of growth conditions. However, deletion of *bldM*, which encodes an atypical response regulator needed for morphological development, triggered transcription of the chloramphenicol biosynthesis gene (Challis 2008; Al-Bassam et al. 2014; Fernández-Martínez et al. 2014).

Another example indicates that genetic manipulation of the regulatory genes activated a 92-kb silent hybrid polyketide and non-ribosomal peptide gene cluster in marine-derived *Streptomyces pactum* SCSIO 02999, by knocking out two negative regulators (*totR5* and *totR3*) and overexpressing a positive regulator *totR1*, to direct the production of known totopotensamides (A (1) and B (3)) and a novel sulfonate-containing analog (C (2)) (Chen et al. 2017).

15.13 Expressing the Gene Cluster in a Heterologous Model Host

The rapidly expanding genome databases have revealed a slew of previously unknown secondary metabolite biosynthesis gene clusters. The heterologous production of these gene clusters in an engineered host strain is a significant strategy for studying them. Actinomycetes gene clusters are frequently expressed in a *Streptomyces* host strain to identify and research the chemicals they produce. However, heterologous expression is frequently accompanied by obstacles that impact secondary metabolite synthesis rates. The choice of an appropriate expression vector, host strain, is thus the initial step, suitability of pathway manipulation, methods for stable maintenance, and optimization. Once production has been established, there are a number of ways to boost compound yields, including media screening, overexpression of regulatory or transport genes, or by introduction of constitutive or inducible promoters (Bekiesch et al. 2016; Teijaro et al. 2019). However, this does not also guarantee the functional expression of the complex heterologous enzyme(s) or successful production of the product of the pathway. As a result, it is hard to predict the success degree for heterologous expression (Fu et al. 2008; Ongley et al. 2013).

Amin et al. (2017) clarified the capture of NRPS genes in fosmid *E. coli* clones for further manipulation. However, it did not show any expression in *E. coli* clones. The chosen host strain can often limit the efficient expression of entire biosynthetic gene clusters. The use of conventional microorganisms like *E. coli* or *Saccharomyces cerevisiae* as hosts for secondary metabolism gene expression is limited due to the high GC content of actinomycetes genes and differences in regulatory components of gene expression (Rebets et al. 2014). The heterologous synthesis of PKS/NRPS hybrid yersiniabactin by *Yersinia pestis* in a genetically modified *E. coli* host was described by Pfeifer et al. (2003). The ability to modify biosynthetic gene clusters in a different host was clarified in this study. After that, the changed biosynthetic gene clusters can be heterologously expressed in a suitable host organism to generate new natural compounds. Another example is the manipulation of the oxytetracycline (OTC) gene cluster in *Streptomyces venezuelae* WVR2006, a heterologous host was reported by modulating the expression of two cluster-situated regulators (CSR) OtcR and OtrR, as well as the precursor supply, the OTC production level in this heterologous host was considerably boosted in 48 h, from 75 to 431 mg/L, a level comparable to the native producer *Streptomyces rimosus* M4018 in 8 days (Yin et al. 2016).

15.14 Conclusion and Future Prospective

Actinobacteria are proving to have a great potential benefits for humans as sources of antibiotic, antifungal, anticancer agents, and other secondary metabolites that are used in medicine. However, there still remains area for more study to guarantee *Actinobacteria* as producers of new bioactive compounds for the benefit of human well-being. The biology of the *Actinobacteria* is a strongly recommended research field with the rapid developments in the fields of genomics, synthetic biology, and the urgent requirement for new antimicrobial compounds to combat antimicrobial resistance. It is expected to see many new advances in this field in the upcoming years. It is a call of emergency to discover an effective cure for pathogens (Ziemert et al. 2016; Blin et al. 2019).

Recently, a powerful tool and useful information are provided to support the search of novel bioactive metabolites for drug development with the advancement of next-generation sequencing and accumulation of high-quality whole-genome data. Currently, these genome data of *Actinobacteria* reveal the existence of many biosynthetic gene clusters of secondary metabolites and reconfirm that *Actinobacteria* are prolific producers of bioactive compounds. However, these gene clusters are not normally expressed under laboratory conditions. Many secondary metabolites encoded by these gene clusters are still unidentified in fermentation broth (Scherlach and Hertweck 2009; Ren et al. 2017). Therefore, the challenge lies in our ability to activate these silent gene clusters. Recently, certain biological and chemical stimuli namely exposure to antibiotics, metals, and mixed microbial culture were successfully employed to activate secondary metabolite production in *Actinobacteria* (Covington et al. 2018). This would allow the emergence of a set of new guiding rules for engineered production of natural products and high producing chassis hosts.

Declarations **Competing Interests** The authors declare that they have no competing interests.

Ethical Consideration Not available.

Author Contributions HE wrote the first draft of manuscript, DHA revised and managed the work, and HE and DHA agreed with the manuscript's results and conclusions, and confirmed that they have contacted. AFE wrote the final manuscript.

References

Aftab U, Zechel DL, Sajid I (2015) Antitumor compounds from *Streptomyces* sp. KML-2, isolated from Khewra salt mines, Pakistan. Biol Res 48(1):1–10

Al-Bassam MM, Bibb MJ, Bush MJ, Chandra G, Buttner MJ (2014) Response regulator heterodimer formation controls a key stage in Streptomyces development. PLoS Genet 10(8): e1004554

Allgaier M, Grossart HP (2006) Diversity and seasonal dynamics of Actinobacteria populations in four lakes in northeastern Germany. Appl Environ Microbiol 72(5):3489–3497

Altschul SF, Gish W, Miller W, Myers EW, Lipman DJ (1990) Basic local alignment search tool. J Mol Biol 215(3):403–410

Amin DH, Borsetto C, Tolba S, Abolmaaty A, Abdallah NA, Wellington EM (2017) Phylogenic analysis of NRPS and PKS genes associated with antagonistic Micromonospora Rc5 and Streptomyces Ru87 isolates. J Adv Biol Biotechnol 1–22

Amin DH, Abolmaaty A, Borsetto C, Tolba S, Abdallah NA, Wellington EM (2019) In silico genomic mining reveals unexplored bioactive potential of rare *Actinobacteria* isolated from Egyptian soil. Bull Natl Res Cent 43(1):1–9

Amin DH, Abdallah NA, Abolmaaty A, Tolba S, Wellington EM (2020) Microbiological and molecular insights on rare *Actinobacteria* harboring bioactive prospective. Bull Natl Res Cent 44(1):1–12

Arnison PG, Bibb MJ, Bierbaum G et al (2013) Ribosomally synthesized and post-translationally modified peptide natural products: overview and recommendations for a universal nomenclature. Nat Prod Rep 30(1):108–160

Aziz RK, Bartels D, Best AA, DeJongh M, Disz T, Edwards RA et al (2008) The RAST Server: rapid annotations using subsystems technology. BMC Genomics 9(1):1–15

Barka EA, Vatsa P, Sanchez L, Gaveau-Vaillant N, Jacquard C, Klenk HP et al (2016) Taxonomy, physiology, and natural products of *Actinobacteria*. Microbiol Mol Biol Rev 80(1):1–43

Behie SW, Bonet B, Zacharia VM, McClung DJ, Traxler MF (2017) Molecules to ecosystems: *Actinomycete* natural products in situ. Front Microbiol 7:2149

Bekiesch P, Basitta P, Apel AK (2016) Challenges in the heterologous production of antibiotics in Streptomyces. Arch Pharm Chem Life Sci 349:594–601. https://doi.org/10.1002/ardp.201600058

Bentley DR, Balasubramanian S, Swerdlow HP, Smith GP, Milton J, Brown CG et al (2008) Accurate whole human genome sequencing using reversible terminator chemistry. Nature 456(7218):53–59

Blin K, Medema MH, Kazempour D, Fischbach MA, Breitling R, Takano E, Weber T (2013) antiSMASH 2.0—a versatile platform for genome mining of secondary metabolite producers. Nucleic Acids Res 41(1):204–212

Blin K, Shaw S, Steinke K, Villebro R, Ziemert N, Lee SY et al (2019) antiSMASH 5.0: updates to the secondary metabolite genome mining pipeline. Nucleic Acids Res 47(W1):W81–W87

Burke CM, Darling AE (2016) A method for high precision sequencing of near full-length 16S rRNA genes on an Illumina MiSeq. PeerJ 4:2492

Cancilla MR, Powell IB, Hillier AJ, Davidson BE (1992) Rapid genomic fingerprinting of *Lactococcus lactis* strains by arbitrarily primed polymerase chain reaction with 32P and fluorescent labels. Appl Environ Microbiol 58(5):1772–1775

Cane DE, Walsh CT, Khosla C (1998) Harnessing the biosynthetic code: combinations, permutations, and mutations. Science 282(5386):63–68

Challis GL (2008) Mining microbial genomes for new natural products and biosynthetic pathways. Microbiology 154(6):1555–1569

Chaudhary HS, Soni B, Shrivastava AR, Shrivastava S (2013) Diversity and versatility of *Actinomycetes* and its role in antibiotic production. J Appl Pharm Sci 3(8):S83–S94

Chen R, Zhang Q, Tan B, Zheng L, Li H, Zhu Y, Zhang C (2017) Genome mining and activation of a silent PKS/NRPS gene cluster direct the production of totopotensamides. Org Lett 19(20):5697–5700

Covington BC, Spraggins JM, Ynigez-Gutierrez AE, Hylton ZB, Bachmann BO (2018) Response of secondary metabolism of hypogean *Actinobacterial* genera to chemical and biological stimuli. Appl Environ Microbiol 84(19):e01125–e01118

Craney A, Ahmed S, Nodwell J (2013) Towards a new science of secondary metabolism. J Antibiot 66(7):387–400

Darling AE, Tritt A, Eisen JA, Facciotti MT (2011) Mauve assembly metrics. Bioinformatics 27(19):2756–2757

De Jong A, van Hijum SA, Bijlsma JJ, Kok J, Kuipers OP (2006) BAGEL: a web-based *bacteriocin* genome mining tool. Nucleic Acids Res 34(Suppl 2):W273–W279

Edwards C (1993) Isolation properties and potential applications of thermophilic *Actinomycetes*. Appl Biochem Biotechnol 42(2):161–179

Fernández-Martínez LT, Borsetto C, Gomez-Escribano JP, Bibb MJ, Al-Bassam MM, Chandra G, Bibb MJ (2014) New insights into chloramphenicol biosynthesis in Streptomyces venezuelae ATCC 10712. Antimicrob Agents Chemother 58(12):7441

Finking R, Marahiel MA (2004) Biosynthesis of nonribosomal peptides. Annu Rev Microbiol 58: 453–488

Fu J, Wenzel SC, Perlova O, Wang J, Gross F, Tang Z et al (2008) Efficient transfer of two large secondary metabolite pathway gene clusters into heterologous hosts by transposition. Nucleic Acids Res 36(17):e113–e113

Goodfellow M, Williams ST (1983) Ecology of *Actinomycetes*. Annu Rev Microbiol 37(1):189–216

Holt JG, Krieg NR, Sneath PH, Staley JT, Williams ST (1994) Bergey's manual of determinative bacteriology, 9th edn. Williams & Wilkins, Baltimore, MA, USA

Hopwood DA (1997) Genetic contributions to understanding polyketide synthases. Chem Rev 97(7):2465–2498

Kim SB, Lonsdale J, Seong CN, Goodfellow M (2003) *Streptacidiphilus gen.* nov., acidophilic *Actinomycetes* with wall chemotype I and emendation of the family *Streptomycetaceae* (Waksman and Henrici (1943) AL) emend. Rainey et al. 1997. Antonie Van Leeuwenhoek 83(2):107–116

Kircher M, Kelso J (2010) High-throughput DNA sequencing–concepts and limitations. Bioessays 32(6):524–536

Kisand V, Lettieri T (2013) Genome sequencing of bacteria: sequencing, de novo assembly and rapid analysis using open source tools. BMC Genomics 14(1):1–11

Komaki H, Ichikawa N, Hosoyama A, Hamada M, Harunari E, Ishikawa A, Igarashi Y (2016) Draft genome sequence of *Micromonospora* sp. DSW705 and distribution of biosynthetic gene clusters for depsipeptides bearing 4-amino-2, 4-pentadienoate in *Actinomycetes*. Stand Genomic Sci 11(1):1–10

Kozarewa I, Ning Z, Quail MA, Sanders MJ, Berriman M, Turner DJ (2009) Amplification-free Illumina sequencing-library preparation facilitates improved mapping and assembly of (G+ C)-biased genomes. Nat Methods 6(4):291–295

Kumar V, Naik B, Gusain O, Bisht GS (2014) An *Actinomycete* isolate from solitary wasp mud nest having strong antibacterial activity and kills the *Candida* cells due to the shrinkage and the cytosolic loss. Front Microbiol 5:446

Lechevalier H, Lechevalier MP (1965) Classification des *Actinomycètesaérobies*basée sur leurmorphologie et leur composition chimique. In: ANNALES DE L INSTITUT PASTEUR (Vol. 108, No. 5, pp. 662-+). 21 STREET CAMILLE DESMOULINS, ISSY, 92789 MOULINEAUX CEDEX 9, FRANCE: MASSON EDITEUR

Li MH, Ung PM, Zajkowski J, Garneau-Tsodikova S, Sherman DH (2009) Automated genome mining for natural products. BMC Bioinformatics 10(1):1–10

Loman NJ, Misra RV, Dallman TJ, Constantinidou C, Gharbia SE, Wain J, Pallen MJ (2012) Performance comparison of benchtop high-throughput sequencing platforms. Nat Biotechnol 30(5):434–439

Ludwig W, Euzéby J, Schumann P, Busse HJ, Trujillo ME, Kämpfer P, Whitman WB (2012) Road map of the phylum *Actinobacteria*. In: Bergey's manual of systematic bacteriology. Springer, New York, pp 1–28

Macagnan D, Romeiro RDS, de Souza JT, Pomella AW (2006) Isolation of Actinomycetes and endospore-forming bacteria from the cacao pod surface and their antagonistic activity against the witches' broom and black pod pathogens. Phytoparasitica 34(2):122–132

Marahiel MA, Stachelhaus T, Mootz HD (1997) Modular peptide synthetases involved in nonribosomal peptide synthesis. Chem Rev 97(7):2651–2674

Margulies M, Egholm M, Altman WE, Attiya S, Bader JS, Bemben LA et al (2005) Genome sequencing in microfabricated high-density picolitre reactors. Nature 437(7057):376–380

Markowitz VM, Mavromatis K, Ivanova NN, Chen IMA, Chu K, Kyrpides NC (2009) IMG ER: a system for microbial genome annotation expert review and curation. Bioinformatics 25(17):2271–2278

Medema MH, Fischbach MA (2015) Computational approaches to natural product discovery. Nat Chem Biol 11(9):639

Medema MH, Blin K, Cimermancic P, de Jager V, Zakrzewski P, Fischbach MA et al (2011) antiSMASH: rapid identification, annotation and analysis of secondary metabolite biosynthesis gene clusters in bacterial and fungal genome sequences. Nucleic Acids Res 39(2):339–346

Muyzer G, De Waal EC, Uitterlinden AG (1993) Profiling of complex microbial populations by denaturing gradient gel electrophoresis analysis of polymerase chain reaction-amplified genes coding for 16S rRNA. Appl Environ Microbiol 59(3):695–700

Nakamura K, Oshima T, Morimoto T, Ikeda S, Yoshikawa H, Shiwa Y et al (2011) Sequence-specific error profile of Illumina sequencers. Nucleic Acids Res 39(13):e90–e90

Newton RJ, Jones SE, Eiler A, McMahon KD, Bertilsson S (2011) A guide to the natural history of freshwater lake bacteria. Microbiol Mol Biol Rev 75:14–49

Nouioui I, Carro L, García-López M, Meier-Kolthoff JP, Woyke T, Kyrpides NC et al (2018) Genome-based taxonomic classification of the phylum *Actinobacteria*. Front Microbiol 9:2007

Ongley SE, Bian X, Zhang Y, Chau R, Gerwick WH, Müller R, Neilan BA (2013) High-titer heterologous production in *E. coli* of *lyngbyatoxin*, a protein kinase C activator from an uncultured marine *cyanobacterium*. ACS Chem Biol 8(9):1888–1893

Paszkiewicz K, Studholme DJ (2010) De novo assembly of short sequence reads. Brief Bioinform 11(5):457–472

Pfeifer BA, Wang CC, Walsh CT, Khosla C (2003) Biosynthesis of yersiniabactin, a complex polyketide-nonribosomal peptide, using Escherichia coli as a heterologous host. Appl Environ Microbiol 69(11):6698

Qin S, Li WJ, Dastager SG, Hozzein WN (2016) *Actinobacteria* in special and extreme habitats: diversity, function roles, and environmental adaptations. Front Microbiol 7:1415

Quail MA, Smith M, Coupland P, Otto TD, Harris SR, Connor TR et al (2012) A tale of three next generation sequencing platforms: comparison of Ion Torrent, Pacific Biosciences and Illumina MiSeq sequencers. BMC Genomics 13(1):1–13

Rana S, Salam MD (2014) Antimicrobial potential of *Actinomycetes* isolated from soil samples of Punjab, India. J Microbiol Exp 1(2):00010

Ranjani A, Gopinath PM, Rajesh K, Dhanasekaran D, Priyadharsini P (2016) Diversity of silver nanoparticle synthesizing *Actinobacteria* isolated from marine soil, Tamil Nadu, India. Arab J Sci Eng 41(1):25–32

Rastogi G, Sani RK (2011) Molecular techniques to assess microbial community structure, function, and dynamics in the environment. In: Microbes and microbial technology. Springer, New York, pp 29–57

Rebets Y, Brötz E, Tokovenko B, Luzhetskyy A (2014) Actinomycetes biosynthetic potential: how to bridge in silico and in vivo? J Ind Microbiol Biotechnol 41(2):387–402

Ren H, Wang B, Zhao H (2017) Breaking the silence: new strategies for discovering novel natural products. Curr Opin Biotechnol 48:21–27

Sanger F, Nicklen S, Coulson AR (1977) DNA sequencing with chain-terminating inhibitors. Proc Natl Acad Sci 74(12):5463–5467

Scherlach K, Hertweck C (2009) Triggering cryptic natural product biosynthesis in microorganisms. Org Biomol Chem 7(9):1753–1760

Ser HL, Tan LTH, Law JWF, Chan KG, Duangjai A, Saokaew S et al (2017) Focused review: cytotoxic and antioxidant potentials of mangrove-derived *Streptomyces*. Front Microbiol 8:2065

Shen B (2015) A new golden age of natural products drug discovery. Cell 163(6):1297–1300

Skinnider MA, Dejong CA, Rees PN, Johnston CW, Li H, Webster AL et al (2015) Genomes to natural products prediction informatics for secondary metabolomes (PRISM). Nucleic Acids Res 43(20):9645–9662

Skinnider MA, Johnston CW, Edgar RE, Dejong CA, Merwin NJ, Rees PN, Magarvey NA (2016) Genomic charting of ribosomally synthesized natural product chemical space facilitates targeted mining. Proc Natl Acad Sci 113(42):E6343–E6351

Skinnider MA, Merwin NJ, Johnston CW, Magarvey NA (2017) PRISM 3: expanded prediction of natural product chemical structures from microbial genomes. Nucleic Acids Res 45(W1):W49–W54

Spellberg B, Guidos R, Gilbert D, Bradley J, Boucher HW, Scheld WM, Bartlett JG, Edwards J Jr, Infectious Diseases Society of America (2008) The epidemic of antibiotic-resistant infections: a call to action for the medical community from the Infectious Diseases Society of America. Clin Infect Dis 46(2):155–164

Starcevic A, Zucko J, Simunkovic J, Long PF, Cullum J, Hranueli D (2008) ClustScan: an integrated program package for the semi-automatic annotation of modular biosynthetic gene clusters and in silico prediction of novel chemical structures. Nucleic Acids Res 36(21):6882–6892

Staunton J, Weissman KJ (2001) Polyketide biosynthesis: a millennium review. Nat Prod Rep 18(4):380–416

Stein L (2001) Genome annotation: from sequence to biology. Nat Rev Genet 2(7):493–503

Tatusova T, DiCuccio M, Badretdin A, Chetvernin V, Nawrocki EP, Zaslavsky L et al (2016) NCBI prokaryotic genome annotation pipeline. Nucleic Acids Res 44(14):6614–6624

Teijaro CN, Adhikari A, Shen B (2019) Challenges and opportunities for natural product discovery, production, and engineering in native producers versus heterologous hosts. J Ind Microbiol Biotechnol 46(3–4):433–444

Ul-Hassan A, Wellington EMH (2009) Actinobacteria. Encycl Microbiol 3:25–44

Ventura M, Canchaya C, Tauch A, Chandra G, Fitzgerald GF, Chater KF, van Sinderen D (2007) Genomics of *Actinobacteria*: tracing the evolutionary history of an ancient phylum. Microbiol Mol Biol Rev 71(3):495–548

Větrovský T, Baldrian P (2013) The variability of the 16S rRNA gene in bacterial genomes and its consequences for bacterial community analyses. PLoS One 8(2):e57923

Warnecke F, Amann R, Pernthaler J (2004) *Actinobacterial* 16S rRNA genes from freshwater habitats cluster in four distinct lineages. Environ Microbiol 6(3):242–253

Weaver D, Karoonuthaisiri N, Tsai HH, Huang CH, Ho ML, Gai S et al (2004) Genome plasticity in Streptomyces: identification of 1 Mb TIRs in the S. coelicolor A3 (2) chromosome. Mol Microbiol 51(6):1535–1550

Weber T, Rausch C, Lopez P, Hoof I, Gaykova V, Huson DH, Wohlleben W (2009) CLUSEAN: a computer-based framework for the automated analysis of bacterial secondary metabolite biosynthetic gene clusters. J Biotechnol 140(1–2):13–17

Weber T, Blin K, Duddela S, Krug D, Kim HU, Bruccoleri R et al (2015) antiSMASH 3.0—a comprehensive resource for the genome mining of biosynthetic gene clusters. Nucleic Acids Res 43(W1):W237–W243

Weissman KJ (2015) The structural biology of biosynthetic megaenzymes. Nat Chem Biol 11(9):660

Wenzel SC, Müller R (2005) Recent developments towards the heterologous expression of complex bacterial natural product biosynthetic pathways. Curr Opin Biotechnol 16(6):594–606

World Health Organization (2017) Antibacterial agents in clinical development: an analysis of the antibacterial clinical development pipeline, including tuberculosis (No. WHO/EMP/IAU/2017.11). World Health Organization

Yadav G, Gokhale RS, Mohanty D (2003a) Computational approach for prediction of domain organization and substrate specificity of modular polyketide synthases. J Mol Biol 328(2):335–363

Yadav G, Gokhale RS, Mohanty D (2003b) SEARCHPKS: a program for detection and analysis of polyketide synthase domains. Nucleic Acids Res 31(13):3654–3658

Yin S, Li Z, Wang X et al (2016) Heterologous expression of oxytetracycline biosynthetic gene cluster in Streptomyces venezuelae WVR2006 to improve production level and to alter fermentation process. Appl Microbiol Biotechnol 100:10563–10572. https://doi.org/10.1007/s00253-016-7873-1

Zhang H, Wang Y, Pfeifer BA (2008) Erratum: bacterial hosts for natural product production. Mol Pharm 5(2):212–225

Zhi XY, Li WJ, Stackebrandt E (2009) An update of the structure and 16S rRNA gene sequence-based definition of higher ranks of the class *Actinobacteria*, with the proposal of two new suborders and four new families and emended descriptions of the existing higher taxa. Int J Syst Evol Microbiol 59(3):589–608

Ziemert N, Alanjary M, Weber T (2016) The evolution of genome mining in microbes—a review. Nat Prod Rep 33(8):988–1005

Zimmerman W (1980) Degradation of lignin by bacteria. J Biotechnol 13:199–130

Zwart G, Crump BC, Kamst-van Agterveld MP, Hagen F, Han SK (2002) Typical freshwater bacteria: an analysis of available 16S rRNA gene sequences from plankton of lakes and rivers. Aquat Microb Ecol 28(2):141–155

Glycopeptide Antibiotics: Genetics, Chemistry, and New Screening Approaches 16

Oleksandr Yushchuk and Bohdan Ostash

Abstract

Glycopeptide antibiotics (GPAs) that target bacterial cell wall remain one of the most successful families of antibiotics. GPAs are produced by Actinobacteria, a class of Gram-positive microorganisms with complex life cycle and sizable genomes. The elucidation of biosynthesis of GPAs is important for several reasons. First, GPA producers are likely a primary source of respective antimicrobial resistance (AMR) genes; their study will inform us of possible ways for AMR evolution among pathogens. Second, the biosynthetic studies yield information about regulatory mechanisms governing the production of GPAs, which can be used in biotechnology. Third, manipulation of GPA biosynthetic genes is a viable approach to generate novel GPAs of practical value. Lastly, the information on GPA genes can be used to find new compounds through computational analyses of publicly available genomic information. In this chapter, we outline a comprehensive picture of GPAs, focusing on genetics and chemistry behind their biosynthetic pathways. Examples of the use of the gained knowledge for strain improvement and discovery of novel compounds are given. A concise outlook of bioactivities exhibited by GPAs will conclude the chapter.

Keywords

Glycopeptide antibiotics · Lipid II binders · Actinobacteria · Biosynthetic gene clusters · Non-ribosomal peptide synthetase · Transcriptional regulation · Resistance to glycopeptide antibiotics

O. Yushchuk · B. Ostash (✉)
Department of Genetics and Biotechnology, Ivan Franko National University of Lviv, Lviv, Ukraine
e-mail: oleksandr.yushchuk@lnu.edu.ua; bohdan.ostash@lnu.edu.ua

R. V. Rai, J. A. Bai (eds.), *Natural Products from Actinomycetes*,
https://doi.org/10.1007/978-981-16-6132-7_16

16.1 Introduction

16.1.1 Bacterial Cell Wall: Organization, Biosynthesis, and Inhibitors

The cell membrane of the Gram-positive bacteria is covered with a thick (10–40 nm) layer of peptidoglycan (murein, PG), which is decorated with lipoteichoic acids (anchored in the cell membrane, LTA) and wall-teichoic acids (WTAs, submerged into PG). Such an organization of the cell wall is characteristic for many pathogens, such as staphylococci. The cell envelope of Gram-negative bacteria contains a thinner (3–6 nm) layer of PG, located between two membranes—the inner cytoplasmic membrane and the outer membrane. Unlike a symmetric phospholipid bilayer of the cytoplasmic membrane, the outer membrane is asymmetrically composed with PG-adjacent phospholipid layer and the external lipopolysaccharide (LPS) layer. Corynebacteria and mycobacteria possess much more complicated cell envelope, although belonging to a typically Gram-positive phylum Actinobacteria (Silhavy et al. 2010).

Chains of PG consist of repeating building blocks—*N*-acetylglucosamine (Glc*N*Ac) linked to *N*-acetylmuramic acid (Mur*N*Ac) via β 1 → 4 glycosidic bonds; pentapeptide side chains are attached to Mur*N*Ac. Most commonly, these side chains are formed with L-Ala–D-Gln–L-Lys–D-Ala–D-Ala in Gram-positive bacteria. PG biosynthesis is similar across bacteria; this topic is extensively reviewed (Barreteau et al. 2008; Vollmer et al. 2008; Silhavy et al. 2010; Liu and Breukink 2016) and summarized in Fig. 16.1.

PG biosynthesis is vital for bacteria and remains the most sought-after target for antibiotics (Sarkar et al. 2017). All inhibitors of PG biosynthesis fall into one of the two main groups: direct inhibitors of biosynthetic enzymes and binders of PG intermediates, such as lipid I and lipid II, and pentapeptide stems of PG monomers (see Fig. 16.1). All compounds that belong to the second group act on cell surface, and thus do not require mechanism(s) for cell entry.

16.1.2 Why Are GPAs So Important?

Among all PG-targeting antibiotics, glycopeptides (GPAs, *sensu* related to vancomycin) are the most thoroughly studied and clinically successful compounds (rivaled probably only by β-lactams). Resistance to GPAs among pathogens emerges very slowly (Zeng et al. 2016), perhaps due to the peculiar nature of their principal target, lipid II. The latter consists of Glc*N*Ac-Mur*N*Ac-pentapeptide attached to undecaprenyl lipid chain and acts as a molecular handle for PG-building units. GPAs bind D-Ala-D-Ala terminus of pentapeptide stem, forming five hydrogen bonds (Barna and Williams 1984). This interferes with the upstream transpeptidase and transglycosylase reactions, preventing the formation of cross-linked PG. The fraction of lipid II is extremely small among all cell membrane phospholipids (less than 1%, de Kruijff et al. 2008). At the same time, lipid II is heavily used; export of more than 5000 molecules of lipid II to the outer side of cell membrane is necessary

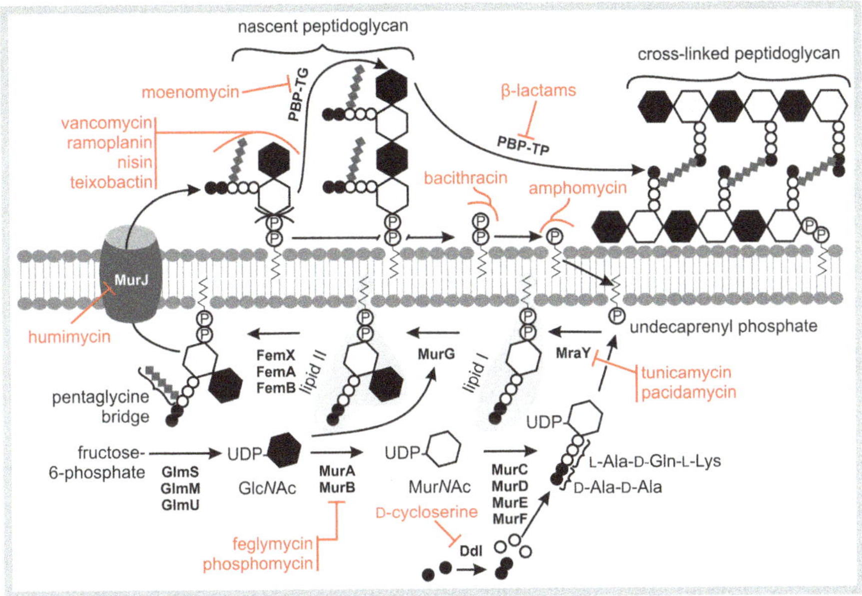

Fig. 16.1 Basic scheme of PG biosynthesis indicating most common PG biosynthesis inhibitors (typed in red), acting on different stages of the biosynthesis. PBP-TG and PBP-TP stand for penicillin-binding protein transglycosylase and transpeptidase domains, respectively

to maintain physiological rates of PG synthesis in exponential phase of bacterial growth (Manat et al. 2014). The importance of lipid II is reflected also in conservation of its structure among different bacteria; enzymes for lipid II biosynthesis are also very conserved. No other membrane phospholipids can replace the loss of lipid II. Mutations in genes for lipid II production would impair cell vitality and thus are eliminated by natural selection. This leaves a very limited space for the variability of lipid II (in terms of quantity and structure) and, consequently, for the emergence of GPA resistance mutations.

GPA resistance in pathogens does not appear de novo, but is mediated by horizontal gene transfer of GPA resistance genes—so-called *van* genes—from GPA producers and other environmental bacteria (Binda et al. 2014; Yushchuk et al. 2020b). Molecular and genetic background of GPA resistance was reviewed exhaustively in recent years (Binda et al. 2014; Zeng et al. 2016; Marcone et al. 2018; Yushchuk et al. 2020b). In brief, pathogens and the producer Actinobacteria use two ways of cell wall remodeling to avoid GPA binding. One way involves replacement of terminal D-Ala with either D-lactate (D-Lac, Arthur et al. 1992) or D-serine (D-Ser, occurs only in pathogens, Arthur and Courvalin 1993; Marshall et al. 1997, 1998; Courvalin 2006). Installation of D-Lac is mediated by three enzymes: VanH (D-Lac dehydrogenase, supplier of the D-Lac pool), VanX (D-alanyl-D-alanine dipeptidase), and VanA (D-Ala-D-Lac ligase). Similarly, D-Ser is installed by D-Ser racemase, D-alanyl-D-alanine dipeptidase, and D-Ala-D-Ser ligase. In Actinobacteria and in pathogens, triads of *van* genes usually form one operon. GPAs exhibit decreased affinity to D-Ala-D-Lac and D-Ala-D-Ser termini, forming only four hydrogen bonds, not enough for stable interaction. Alternatively, GPA resistance is achieved with the help of VanY, a D,D-carboxypeptidase cleaving the terminal D-Ala from pentapeptide stem. Such PG can still be cross-linked, but GPAs do not form stable complexes with it anymore. VanY is rather accessory GPA resistance factor in pathogens, but it might be a single resistance determinant in GPA-producing Actinobacteria (Marcone et al. 2010). Expression of *van* genes is usually regulated with VanRS two-component regulatory circuit, which consists of a sensor histidine kinase (VanS) and response regulator (VanR). VanS, upon sensing GPA, phosphorylates VanR, which upregulates *van* operon. VanS from different species often recognizes only certain GPA.

Today, two natural GPAs (the first generation)—vancomycin and teicoplanin, as well as three semisynthetic GPAs (second generation)—telavancin, oritavancin, and dalbavancin—are in clinical practice (Zeng et al. 2016). Vancomycin, produced by different *Amycolatopsis* (*Am.*) *orientalis* strains, was the first GPA marketed in 1958. Only 30 years later, the first vancomycin-resistant clinical isolate (*Enterococcus faecium* strain) was reported. Nowadays, unfortunately, vancomycin and teicoplanin resistance is widespread among enterococci and staphylococci. The spread of GPA resistance does not render GPAs worthless. This is due to several reasons. First, nature offers a remarkable chemical variety of GPAs (Nicolaou et al. 1999), which expands ever since the discovery of vancomycin and exploded in modern "post-genomic" era. Second, biosynthesis of GPAs is modular, which allows to use enzymes from one GPA biosynthetic pathways to modify another, leading to unprecedented possibilities for combinatorial biosynthesis. Therefore, VanS sensors, acting in pathogens, might be "bluffed" with minor modifications of the existing drugs with relative ease (Kalan et al. 2013; Yim et al. 2018). Finally, deep knowledge of genetics behind the GPA production allows to develop GPA overproducers much faster than before (Horbal et al. 2014; Lo Grasso et al. 2015; Yushchuk et al. 2020a). These advances allow to counter the spread of GPA-resistant pathogens with novel GPAs. Recent years were especially fruitful for new discoveries of unusual GPAs and underlying biosynthetic routes, which merit a detailed review. In this chapter, we portray an updated picture of biosynthesis of natural GPAs, going from genes to products.

16.1.3 Definitions

Historically, to define GPAs one took into account structure, origin, and mode of action. That is, GPAs are non-ribosomally synthesized molecules whose peptide core consists of seven amino acids, often non-proteinogenic; the core can be decorated with sugar moieties, halogen atoms, sulfate groups, or acyl chains; they are produced by Actinobacteria, and they bind D-Ala-D-Ala termini of PG of Gram-positive bacteria. In this chapter, we discuss GPA as they are denoted above. However, in a broad chemical sense, term "glycopeptide" can be attributed to any glycosylated peptide molecule, such as bleomycin, that does not target PG (Hecht 2000).

All GPAs (*sensu* structurally related to vancomycin) are classified into five types according to their structural features (Nicolaou et al. 1999). Type I antibiotics have aliphatic amino acids in positions 1 and 3 of their aglyca (e.g., vancomycin); type II features amino acids with aromatic side chains at these positions, uncross-linked (e.g., keratinimicin); types III and IV have aromatic amino acids at 1 and 3 aglycone positions and they are cross-linked (e.g., ristocetin); finally, type V is different from all the rest by carrying tryptophan residue at the position 2 of aglycone (e.g., complestatin). Type I–IV GPAs bind D-Ala-D-Ala. Some GPAs, like A47934 (Boeck and Mertz 1986; Pootoolal et al. 2002), are D-Ala-D-Ala binders whose aglycones are not glycosylated. Recent discoveries of novel type V compounds (Cheng and Wuest 2020; Culp et al. 2020; Xu et al. 2020) revealed that at least some of them might have nine amino acid peptide core (or even more, Waglechner et al. 2019). Thus, historical definition of all these compounds as GPAs is quite confusing. There was another early term for types I-IV—dalbaheptides (Parenti and Cavalleri 1989; Cavalleri and Parenti 2000), derived from abbreviation of "***D-Al***a-D-Ala-***b***inding ***a***ntibiotics with ***heptapeptide*** cores." This term includes both glycosylated and nonglycosylated members of types I–IV and might be further extended to glycodalbaheptides and lipoglycodalbaheptides. However, type V compounds are definitely not "dalba-" and often not "-heptides." Further on, we will often use term dalbaheptide referring to type I–IV GPAs. It is important, in our opinion, to propose a term, which would distinguish both dalbaheptides and complestatin-related antibiotics from the other "glycopeptides," *sensu lato*.

16.2 Structural Classification of GPAs with Recent Updates

GPAs are currently represented by dozens of natural compounds (Nicolaou et al. 1999). Majority of known GPAs are composed of oxidatively cross-linked heptapeptide aglyca, which may undergo various modifications. Seven amino acids are usually denoted as *aa-1–aa-7* from N- to C-terminus of the oligopeptide (Fig. 16.2). In rare cases (discussed further), the position of *aa-1* could be actually occupied with a keto acid. Vancomycin, the archetypical GPA (produced by *Am. orientalis* strains, Fig. 16.2), harbors amino acids with aliphatic side chains in *aa-1* and *aa-3* positions; positions 2 and 4 to 7 are occupied with aromatic amino acid. These five aryl rings (denoted with letters, A to E, Fig. 16.2) are involved in the formation of macrocycle structures by oxidative cross-linking, giving one biaryl (AB) and two bisaryl ethers (C-*O*-D and D-*O*-E). Such combination of features is characteristic for the other vancomycin-like GPAs (balhimycin, chloroeremomycin) and recently discovered type I GPA pekiskomycin (Thaker et al. 2013).

Aglyca of type II GPAs are composed exclusively of aromatic amino acids, thus adding F and G aryl rings at amino acid positions 1 and 3 (Fig. 16.2). They have the same macrocycles as type I GPAs (Fig. 16.2). Among antibiotics in this rather scanty type are recently discovered keratinimicins (Fig. 16.2, Xu et al. 2019), produced by

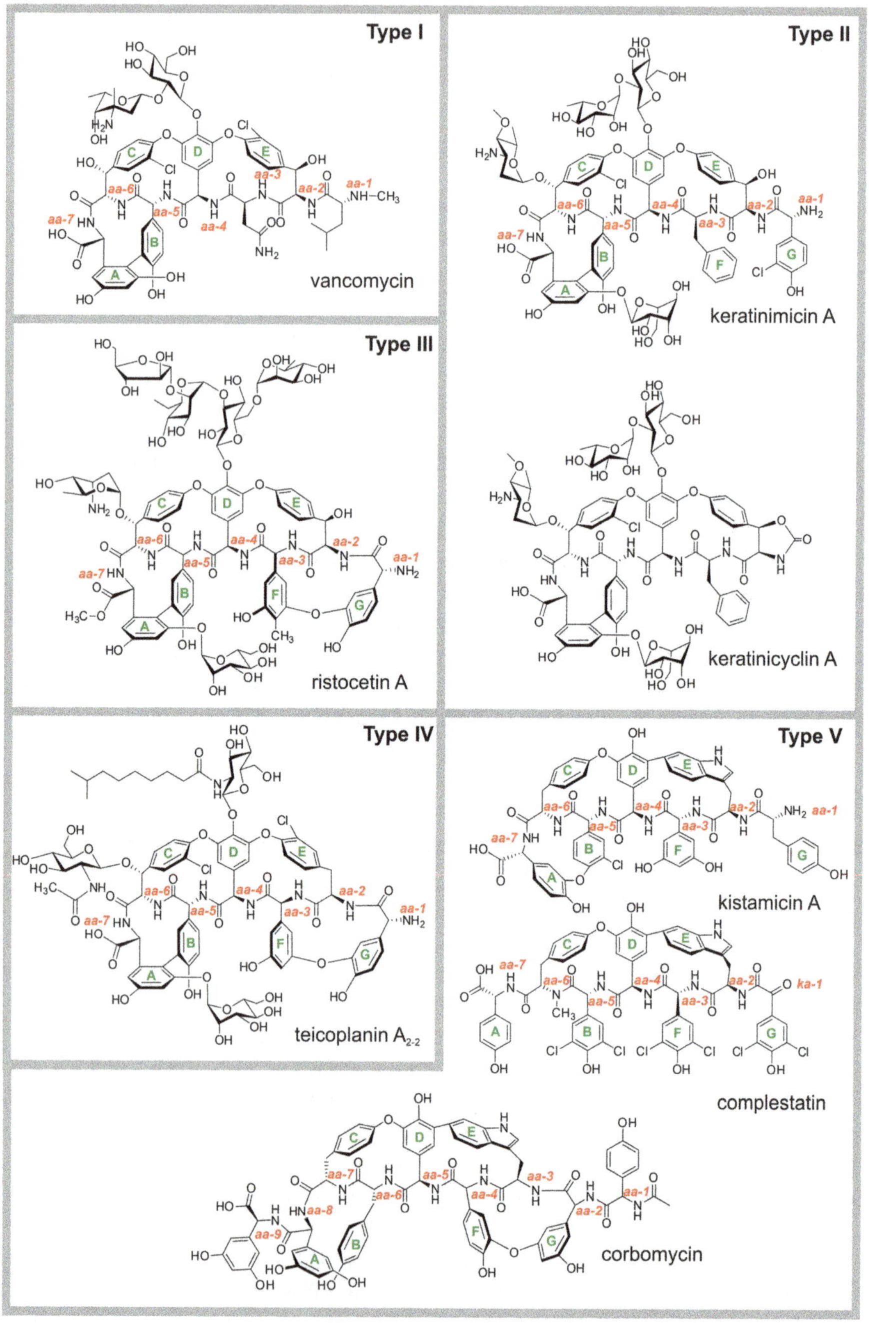

Fig. 16.2 Structural classification of GPAs into five types according to Nicolaou et al. (1999); each class shown with typical representative. Refer to the main text for more details

Am. keratiniphila NRRL B-24117. The latter also produces new natural derivatives of keratinimicins, known as keratinicyclines (minor fraction, Xu et al. 2019). In their aglycone, *aa-1* is substituted with N-terminal 2-oxazolidinone ring (Fig. 16.2). Thus, keratinicyclines combine features of dalbaheptides with the active site of oxazolidinone antibiotics that inhibit translation. However, such combination might not augment biological activity of these compounds, since targets of oxazolidinones are inside the cell, whereas dalbaheptides act on PG. Mechanisms behind the formation of keratinicyclines are not yet known. It is likely that the other type II GPAs also undergo such "-cyclinization."

Type III and IV GPAs feature an additional F-*O*-G macrocycle (Fig. 16.2). Type III GPAs are often hyperglycosylated (like ristocetin and actaplanin), some are aglyca (A47934). The presence of fatty acid side chain is a marker of type IV GPAs (such as teicoplanin and A40926, Fig. 16.2). Teicoplanin and A40926 are GPAs of clinical value (Alduina et al. 2018; Yushchuk et al. 2020c).

Type V GPAs are not limited to compounds with heptapeptide cores. For example, recently discovered GP6738 (Xu et al. 2020) and corbomycin (Culp et al. 2020) are nonapeptides. Oxidative cross-linking patterns are also variable in type V GPAs. Although D-E biaryl and C-*O*-D bisaryl ether bonds are present in all known type V GPAs (and might be the only cross-links, as in complestatin), they can also feature additional A-*O*-B (kistamicin) or A-B and F-*O*-G macrocycles (corbomycin). Type V GPAs are not glycosylated.

Members of all five types are exclusively Actinobacteria-derived antibiotics. GPAs are most abundant in "rare" actinobacterial families such as *Pseudonocardiaceae* (Adamek et al. 2018; Xu et al. 2020), *Micromonosporaceae, and Streptosporangiaceae* (Goldstein et al. 1987; Naruse et al. 1993a). Until very recently, GPAs were believed to be rather rare in genus *Streptomyces,* but now it becomes evident that streptomycetes might be abundant source of type I (Thaker et al. 2013), type III (Pootoolal et al. 2002), and especially type V GPAs (Waglechner et al. 2019; Culp et al. 2020; Xu et al. 2020).

16.3 Organization of GPA BGCs and Where (How) to Find Them

Multiple GPAs were discovered in course of microbiological screening programs, yielding remarkable chemical diversity of this class (Nicolaou et al. 1999). Genetic studies of producers of well-known GPAs yielded first GPA BGCs. These included BGCs for chloroeremomycin (*cep* from *Am. orientalis*), later shown to be in fact Kibdelosporangium aridum A82846, Waglechner et al. 2019), complestatin (*com* from *Streptomyces (S.) lavendulae* SANK 60477, Chiu et al. 2001), A47934 (*sta* from *S. toyocaensis* NRRL 15009, Pootoolal et al. 2002), teicoplanin (*tei* from *Actinoplanes (A.) teichomyceticus* NRRL B-16726, Sosio et al. 2003; Li et al. 2004), A40926 (*dbv* from *Nonomuraea (N.) gerenzanensis* ATCC 39727, Sosio et al. 2003), and balhimycin (*bal* from *Am. balhimycina* DSM 5908, Shawky et al. 2007). Later on, when genome sequencing became routine, *auk* and *kis* (kistamycin) BGCs were discovered in UK-68,597 producer *Actinoplanes* sp. ATCC 53533

(Yim et al. 2014) and *Nonomuraea* sp. ATCC 55076 (Nazari et al. 2017; Greule et al. 2019).

Later, analysis of metagenomic data showed the abundance of GPA-like BGCs. These were TEG and VEG BGCs (Banik and Brady 2008); CA37, CA878, and CA915 BGCs (Banik et al. 2010); GQ475282 (Kim et al. 2010) and esnapd15 (Owen et al. 2013). Not a single natural host of these BGCs was cultivated, and their final products are unknown (although predicted); however, these BGCs became a source of novel GPA tailoring enzymes. In the same time, accumulation of knowledge about the genetics of GPA production and resistance allowed to develop screening pipelines tailored specifically for discovering GPAs. One such pipeline yielded pekiskomycin (and *pek* BGCs coming from *Streptomyces* spp. WAC1420 and WAC4229, Thaker et al. 2013). Here, a collection of cultivable soil actinomycetes was "sifted" through a several "sieves." First, these strains were cultivated in the presence of vancomycin, and resistant colonies were selected. The vancomycin-resistant strains were assumed to be GPA producers. The presence of *van* genes in the selected resistant strains was verified via PCR. Strains that passed these two "filters" were further PCR-screened for the most conserved genes for cross-linking P450 monooxygenases involved in GPA biosynthesis. Positive strains were tested for GPA production and underwent shotgun genome sequencing to determine corresponding BGCs. Similar approach, involving GPA-sensitive microbiological assay system, was utilized to discover novel ristocetin BGC in *Amycolatopsis* sp. MJM2582 (Truman et al. 2014). Some of other GPA BGCs were silent and required activation (Spohn et al. 2014; Xu et al. 2019).

The recent dawn of post-genomic era had a significant impact on the search of novel GPA BGCs. It turned out that genomic databases harbor many GPA BGC-encoding genomes. This was revealed in a large phylogenomic attempt to reconstruct the evolution of GPA BGCs (Waglechner et al. 2019). Interestingly, one subset of GPA BGCs lacked typical GPA resistance genes, leading to BGC for a novel nonapeptide type V GPA—corbomycin (from *Streptomyces* sp. WAC01529, *crb* BGC, Culp et al. 2020). The latter has an unprecedented mode of action against Gram-positive bacteria (discussed further). A GPA BGC expression platform was then created, allowing to express in a heterologous host BGCs silent in native hosts (Xu et al. 2020).

GPA BGCs are large and have complex transcriptional ensembles. Certain common features of their organization could be deduced. In all cases, non-ribosomal peptide synthetase (NRPS) genes are oriented in one direction and immediately followed be cross-linking oxygenase genes (Fig. 16.3). A single exception is *dbv*, which is bizarrely rearranged for unknown reasons (Fig. 16.3g). Peculiarly, a recently described BGC (*noc*) for a novel A40926-related GPA from *N. coxensis* DSM 45129 is almost identical to *dbv,* but has normal arrangement of NRPS genes (Yushchuk et al. 2021). Majority of structural genes are located downstream of NRPS genes, at least in BGCs from *Actinoplanes* spp. and *Amycolatopsis* spp. BGCs from *Streptomyces* spp. (including type V GPA BGCs) tend to have structural genes both up- and downstream of NRPS genes (Fig. 16.3d, e, i). Another characteristic feature of all dalbaheptide BGCs is the presence of *van* genes (see above); their

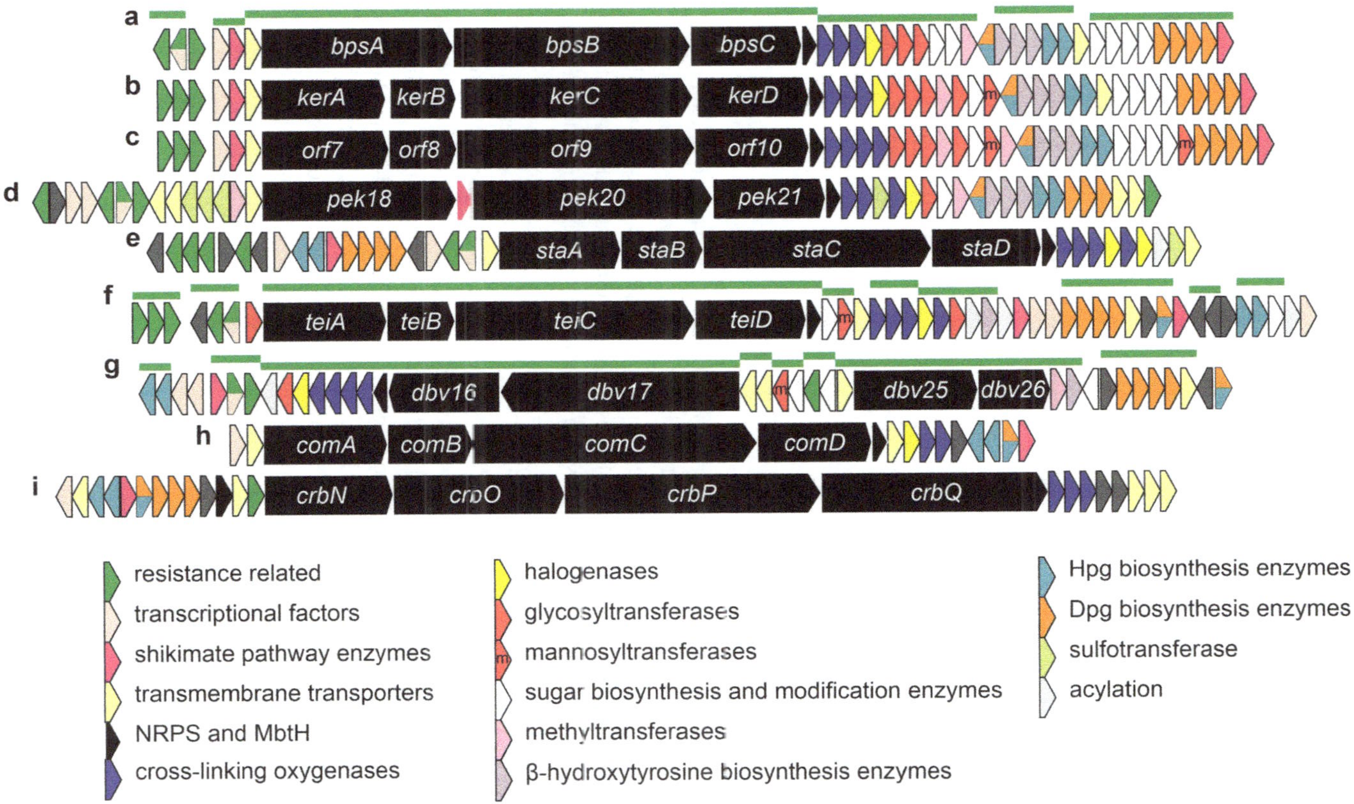

Fig. 16.3 Comparison of the genetic organization of some GPA BGCs, namely (**a**) *bal* for balhimycin; (**b**) *ker* for keratinimicin; (**c**) BGC for ristocetin; (**d**) *pek* for pekiskomycin; (**e**) *sta* for A47934; (**f**) *tei* for teicoplanin; (**g**) *dbv* for A40926; (**h**) *com* for complestatin; and (**i**) *crb* for corbomycin. For (**a**), (**f**), and (**g**), operon structure was established experimentally; operons are shown above the genes with solid green lines. Gene sizes are given not to scale

organization in different BGCs was recently discussed in detail (Yushchuk et al. 2020b).

Semi-quantitative reverse transcription PCR was used to elucidate transcriptional organization for *bal*, *tei,* and *dbv* BGCs. The first of them has rather straightforward organization with six polygenic transcriptional units (Fig. 16.3a, Shawky et al. 2007). Since other *Amycolatopsis*-derived GPA BGCs are rather syntenic with *bal*, it might be construed that they are organized similarly. In *tei* BGCs, nine polygenic transcriptional units were found (Fig. 16.3f, Yushchuk et al. 2019), 8—in *dbv* (Fig. 16.3g, Alduina et al. 2007). In all these examples, genes for Hpg and Dpg biosynthesis form operons.

16.4 Biosynthetic Machinery Behind the Production of GPAs

Biosynthesis of all GPAs consists of three stages: (I) biosynthesis of non-proteinogenic amino acids; (II) non-ribosomal synthesis of peptide core, together with the halogenation and oxidative cross-linking; and (III) further tailoring steps of the aglycone, which may include a combination of methylation, sulfation, acylation, or glycosylation reactions.

16.4.1 "Secondary Metabolism" of Tyrosine

GPA aglyca usually require a pool of certain non-proteinogenic amino acids for their biosyntheses. These are L-β-hydroxytyrosine (β-Ht), L-4-hydroxyphenylglycine (Hpg), and L-3,5-dihydroxyphenylglycine (Dpg). Respective enzymes could be found in all GPA BGCs. First two of these non-proteinogenic amino acids are directly derived from tyrosine (Tyr), which also acts as the amino donor in Hpg and Dpg biosyntheses.

A majority of characterized to date GPA BGCs contain genes for key enzymes of the shikimate pathway, leading to the tyrosine: 3-deoxy-D-arabino-heptulosonate 7-phosphate (DAHP) synthases, chorismate mutases (CMs), and the prephenate dehydrogenases (PDHs) (Thykaer et al. 2010; Waglechner et al. 2019; Yushchuk et al. 2020c). Roles of these enzymes are summarized in Fig. 16.4a. Knockout of the DAHP synthase gene from balhimycin BGC in *Am. balhimycina* DSM 5908 led to decreased antibiotic production, whereas overexpression had an opposite effect. Neither knockout nor overexpression of PDH gene influenced balhimycin production (Thykaer et al. 2010). Not a single GPA BGC-encoded CM was studied to date experimentally.

16.4.2 Formation of β-Hydroxytyrosine

All GPAs contain a β-hydroxytyrosine as a part of aglycone. In A40926 biosynthetic pathway of *N. gerenzanensis*, nonheme dioxygenase (β-hydroxylase) Dbv28 is

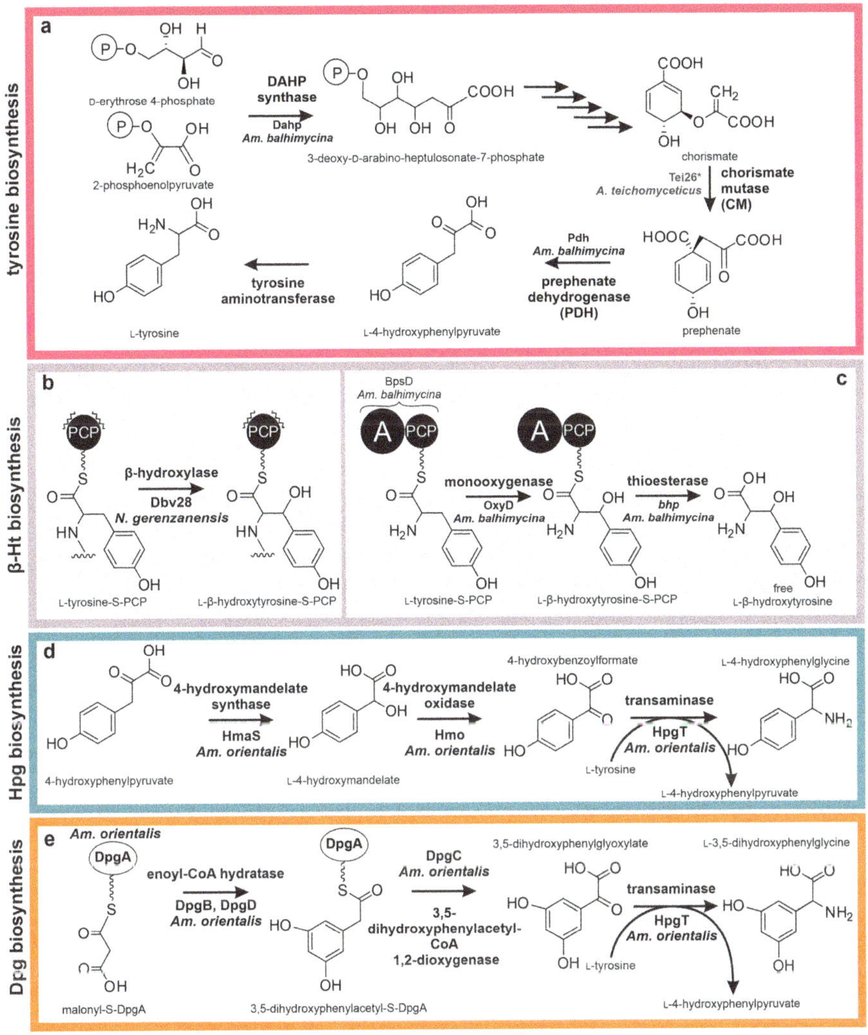

Fig. 16.4 Biosynthesis of tyrosine and non-proteinogenic amino acids, catalyzed by GPA BGC-encoded enzymes. The depicted enzymatic routes are taken from experimentally studied examples. (**a**) Shikimate pathway, at least three enzymes (DAHP synthase, PDH, and CM) are encoded within GPA BGCs. DAHP and PDH were studied experimentally in *Am. balhimycina*. CM Tei28* comes from *tei* BGC, but was not studied experimentally. There are two different pathways to L-β-hydroxytyrosine in A40926 biosynthesis (**b**) and chloroeremomycin biosynthesis (**c**). (**d**, **e**) Biosynthesis of Hpg and Dpg, respectively. Refer to the main text for more details

involved (Fig. 16.4b), acting on a tyrosine in a NRPS-bound aglycone. A different strategy is observed in balhimycin biosynthetic pathway, and most likely in the biosyntheses of the other related GPAs (Stegmann and Frasch 2010). Here, β-hydroxytyrosine is synthesized in free form in cytoplasm and only then is incorporated in nascent peptide core (Mulyani et al. 2010). Three-gene operon—*bhp, bpsD,* and *oxyD*—is involved in this process (Mulyani et al. 2010; Recktenwald et al. 2002; Puk et al. 2004). L-tyrosine is first loaded on NRPS module BpsD and then hydroxylated by monooxygenase OxyD; thioesterase Bhp releases L-β-hydroxytyrosine from BpdD PCP domain (Fig. 16.4c).

16.4.3 Biosynthesis of Hpg and Dpg

Enzymes involved in Hpg biosynthesis were studied in *Am. orientalis*, chloroeremomycin producer (Hubbard et al. 2000; Li et al. 2001). HmaS converts 4-hydroxyphenylpyruvate (likely product of cluster-encoded PDH) into L-4-hydroxymandelate. The latter is oxidized by Hmo into L-4-hydroxybenzoylformate. HpgT uses an amino group of L-tyrosine to convert L-4-hydroxybenzoylformate into L-4-hydroxyphenylglycine (with L-4-hydroxyphenylpyruvate as a byproduct of this reaction, Fig. 16.4d).

Dpg biosynthesis was studied in chloroeremomycin and balhimycin producers (Chen et al. 2001; Pfeifer et al. 2001). Here, biosynthetic enzymes include DpgA, DpgB, DpgC, DpgD, and, once again, HpgT. DpgA is a type III polyketide synthase that forms a complex with the enoyl-CoA hydratases DpgB and DpgD, making 3,5-dihydroxyphenylacetate (tethered to DpgA) out of malonyl-CoA units (Fig. 16.4e). 3,5-dihydroxyphenylacetyl-CoA 1,2-dioxygenase DpgC converts 3,5-dihydroxyphenylacetate into 3,5-dihydroxyphenylglyoxylate. The final reaction of L-3,5-dihydroxyphenylglycine biosynthesis is catalyzed by a HpgT transaminase (Fig. 16.4e, Pfeifer et al. 2001; Sandercock et al. 2001). As in Hpg biosynthesis, L-tyrosine here serves as an amino donor and L-4-hydroxyphenylpyruvate is formed as a by-product (Fig. 16.4e). Some GPA BGCs, such as pekiskomycin ones, lack genes for DpgD (Thaker et al. 2013). This enzyme is probably dispensable for Dpg biosynthesis and is accessory to DpgB.

All aforementioned biosynthetic pathways are tightly intertwined. Products of shikimate pathway, L-tyrosine and L-4-hydroxyphenylpyruvate, serve either as direct precursors of non-proteinogenic amino acids, or take part in their biosynthesis. The final step of Dpg and Hpg biosyntheses also yields L-4-hydroxyphenylpyruvate, which can enter either the biosynthesis of L-tyrosine or Hpg. Finally, L-tyrosine itself is a part of the aglyca of GPAs such as teicoplanin and A40926. Tyrosine and Hpg/Dpg biosynthesis intermediates are also found within aglyca of some GPAs: 4-hydroxyphenylpyruvate in UK-68,597 (Yim et al. 2014) or 4-hydroxybenzoylformate in keratinimicins C and D (Xu et al. 2019) and complestatin (Chiu et al. 2001).

16.4.4 Non-ribosomal Biosynthesis of GPA Cores

NRPSs involved in the biosynthesis of dalbaheptides and type V GPAs exhibit typical modular organization (Fig. 16.5). Substrate specificities of corresponding

Fig. 16.5 Patterns in the organization of NRPSs (and corresponding genes) from GPA BGCs. Domains believed to be nonfunctional are shadowed. Refer to the main text for more details

adenylation (A) domains could be confidently predicted using in silico tools (Bachmann and Ravel 2009; Röttig et al. 2011).

Heptapeptide aglyca of dalbaheptides are genetically "imprinted" within seven modules of NRPS (Fig. 16.5a–h). These are encoded with three or four genes. In first case, e.g., in type I GPA balhimycin biosynthesis (Fig. 16.5a), genes *bpsA, bpsB,* and *bpsD* code 1st–3rd, 4th–6th, and 7th modules of NRPS, respectively (Shawky et al. 2007); in second, like in type II GPA keratinimicin biosynthesis (Fig. 16.5c), genes *kerA-D* (Xu et al. 2019) encode 1st–2nd, 3rd, 4th–6th, and 7th NRPS modules, respectively. One exception is the organization of NRPS genes in metagenome-derived CA915 BGC, where first three NRPS modules are encoded within single gene each (Fig. 16.5f, Banik et al. 2010). NRPS genes are located on one strand collinearly to the order of their function within NRPS assembly line. The *dbv* gene cluster is exception here: NRPS genes are on different strands and are separated by the other biosynthetic genes (Fig. 16.5h, Sosio et al. 2003). In *pek* BGC (Thaker et al. 2013), PDH gene (*pek19*) separates first two NRPS genes (*pek18* and *pek20*, Fig. 16.5b).

The domain organization of each module in dalbaheptide biosynthesis NRPSs remains quite conserved: 1st—(A)-(peptidyl carrier protein, PCP); 2nd—(condensation domain, C)-(A)-(PCP)-(epimerization domain, E); 3rd—(C)-(A)-(PCP); 4th—(C)-(A)-(PCP)-(E); 5th—(C)-(A)-(PCP)-(E); 6th—(C)-(A)-(PCP)-(E); and 7th—(C)-(A)-(PCP)-(X)-(thioesterase domain, Te).

Such an arrangement of E-domains should yield NH_2-L-D-L-D-D-L-L-COOH stereochemical configuration of the heptapeptide. However, this is not in agreement with experimentally solved NH_2-D-D-L-D-D-L-L-COOH configuration of all known dalbaheptides (e.g., Schäfer et al. 1996, 1998; Han et al. 2014). Such epimerization may arise from some orphan E-domain or epimerase, encoded outside of BGCs. However, recent in vitro experiments have shown that module 1 C-domain is likely involved in the epimerization of *aa-1* (Kaniusaite et al. 2021). A nonfunctional E-domain might sometimes be found in NRPS third module, like in A47934 NRPS StaB (Fig. 16.5e, Pootoolal et al. 2002).

Among all dalbaheptides, A-domain specificities in modules 2 (Tyr/Bht) and 4–7 (Hpg-Hpg-Tyr/Bht-Dpg) are the most conserved. On the contrary, A-domains of modules 1 and 3 might be responsible for the introduction of Leu and Asn (vancomycin, balhimycin), Ala and Gln (pekiskomycin), Hpg and Phe (keratinimicin), and Hpg and Dpg (type III and IV GPAs). In rare cases, *aa-1* position is occupied with a keto acid, whereas an amino acid is predicted in silico as a substrate for the corresponding A-domain. It might be due to substrate promiscuity of corresponding A-domain, or, alternatively, aglycone *aa-1* might be deaminated later on, although the exact mechanism is still to be unraveled.

Modular organizations of type V GPA NRPSs are more variable. For instance, E-domains are more common, giving different stereochemical patterns of aglyca in different compounds. An unusual truncated module 1, lacking A-domain, was found in kistamicin NRPS (Fig. 16.5j, Greule et al. 2019). It is so far unknown how the loading of first amino acid is achieved in kistamicin biosynthesis. Corbomycin NRPS starts with C-domain (Culp et al. 2020). Since corbomycin carries acetyl group attached to N-terminus of its aglycone, it was proposed that as-yet-unknown polyketide synthase module initiates the biosynthesis of this antibiotic. It makes module 1 of CorA actually a hybrid PKS-NRPS, explaining the presence of C-domain. Complestatin NRPS features a unique methylation domain in module 6, responsible for the methylation of the amino group of tyrosine (Chiu et al. 2001). Finally, last modules of NRPSs of nonapeptide type V GPAs carry an additional, likely nonfunctional, A-domain (Fig. 16.5k, l).

16.4.5 Oxidative Cross-Linking and Halogenation of GPA Aglyca

Nascent GPA aglycone undergoes oxidative cross-linking while still being attached to the PCP domain of the last module of NRPS. There is invariant X-domain within all GPA NRPS (see Fig. 16.5) that plays a crucial role in this. It orchestrates the activity of P450 monooxygenases (Haslinger et al. 2015; Peschke et al. 2016b, c). For years, this domain was believed to be a nonfunctional C-domain. However, recent studies show that X-domain is required for the recruitment of cross-linking oxygenases to the non-ribosomal assembly line. The X-domain probably descended from an ancestral C-domain (Peschke et al. 2016b), which became catalytically inactive, but now serves as a docking site for the P450 monooxygenases (Haslinger et al. 2015; Peschke et al. 2016a, c). Indeed, Te domains of GPA NRPSs (Fig. 16.5) are specific toward cross-linked aglyca (Peschke et al. 2018), assisting the dissociation of fully processed aglycone from NRPS assembly line.

Type I and II GPAs feature three cross-links (Fig. 16.2). As investigated in *Am. balhimycina*, formation of these cross-links is achieved with P450 monooxygenases OxyA, OxyB, and OxyC (Fig. 16.6a, Stegmann et al. 2006). OxyB acts first, installing C-*O*-D cross-link, followed by OxyA (D-*O*-E) and OxyC (A-B). Type III and IV GPAs have fourth F-*O*-G cross-link, furnished by an additional P450 monooxygenase OxyE. Studies of oxygenases from *sta* and *tei* BGCs showed that OxyE acts after OxyB and before OxyA (Fig. 16.6b, Hadatsch et al. 2007; Cryle et al. 2011; Li et al. 2011). Thus, the number of cross-linking oxygenases encoded in dalbaheptide BGCs equals to the number of cross-links.

This, however, is different for type V GPAs: *kis* BGC, for instance, features two oxygenase genes, *kisN* and *kisO*, while the GPA has three cross-links (Nazari et al. 2017; Greule et al. 2019). KisN and KisO are related to dalbaheptide P450 oxygenases OxyA and OxyC, respectively (Fig. 16.6c). KisN installs D-E cross-link (although here it is biaryl bond, not bisaryl ether bond, Fig. 16.6c). KisO generates C-*O*-D and A-*O*-B cross-links (Fig. 16.6c). The sequence of the reactions is as follows: C-*O*-D > D-E > A-*O*-B (Greule et al. 2019).

Early works on *Am. balhimycina* (Puk et al. 2004) hinted that halogenation of GPA aglyca is coupled with non-ribosomal synthesis. This was supported with recent in vitro assays that have shown halogenases, such as Tei8* (responsible for the chlorination of *aa-2* and *aa-6*), acting only upon amino acids loaded onto the PCPs of the NRPS (Kittilä et al. 2017). This confirmed the aforementioned

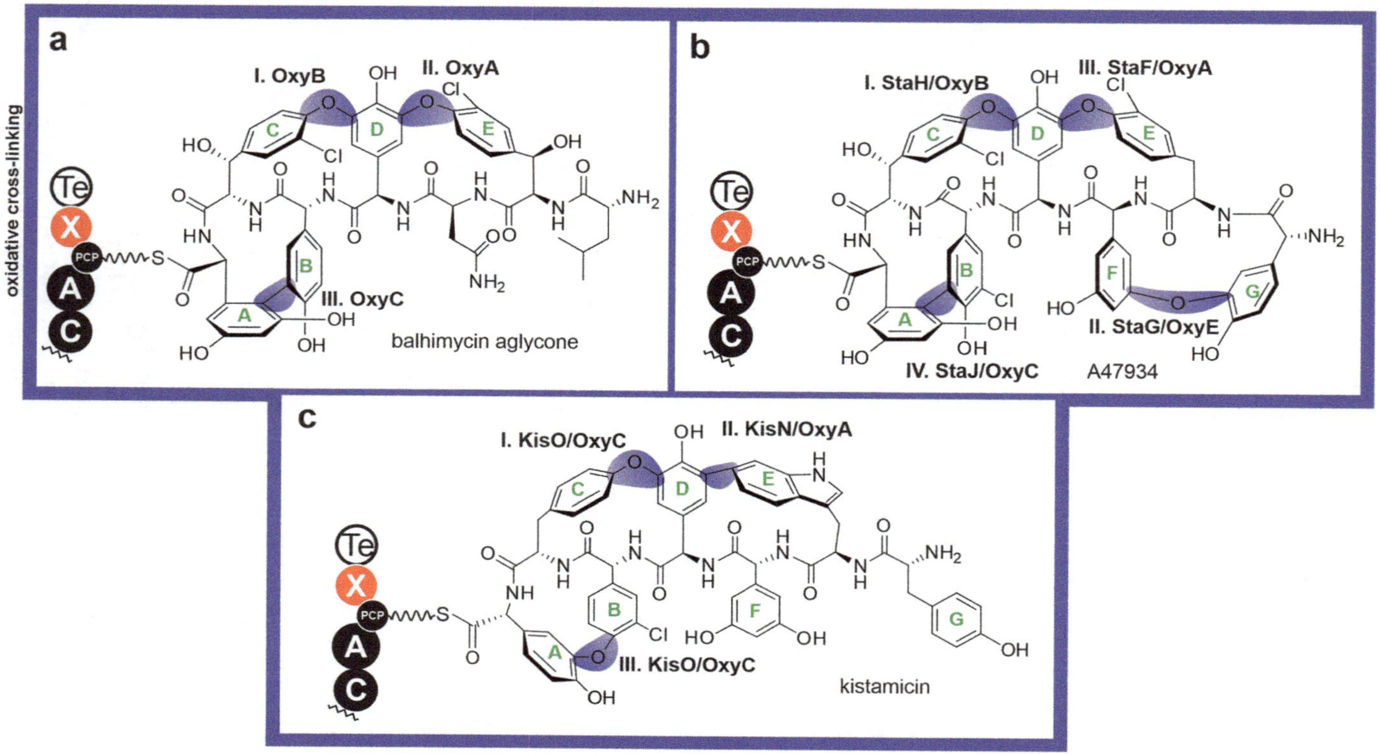

Fig. 16.6 Strategies of the oxidative cross-linking of NRPS-tethered GPA aglyca: (**a**) cross-linking of type I GPA, balhimycin; P450 monooxygenases act in following order: OxyB > OxyA > OxyC; (**b**) oxidative cross-linking of type III–IV GPAs exemplified with A47934, here one additional oxygenase acts between OxyB and OxyA; and (**c**) oxidative cross-linking of type V GPA kistamicin, featuring a unique dual-action P450 monooxygenase KisO

suggestion. Furthermore, a preference for chlorinated substrates was found for teicoplanin P450 monooxygenases (Peschke et al. 2017).

Most often among dalbaheptides, aglycones *aa-2* and *aa-6* are halogenated; corresponding BGCs carry a single halogenase gene responsible for this, like *bhaA* from *bal* (Puk et al. 2002) or *tei8** from *tei* (Li et al. 2004). The A40926 BGC carries a single *dbv10* halogenase gene sharing 81% of amino acid sequence identity to *bhaA* at protein level. Yet, A40926 carries chlorines on *aa-3* and *aa-6* (Sosio et al. 2003). A single halogenase gene within *pek* gene cluster (Thaker et al. 2013) is even more similar to BhaA (92%), but pekiskomycin is halogenated only at *aa-6*. The A47934 and *auk* BGCs carry two genes for halogenases, and corresponding antibiotics are halogenated at amino acids 2, 5, and 6 and 1, 2, 5, and 6, respectively. Recent work has shown that in A47934 biosynthetic pathway, StaK is inactive, and StaI alone carries out three chlorinations (Yim et al. 2018). Among type V GPAs, complestatin harbors six chlorines installed onto *aa-1, aa*-3, and *aa-5*. Such a "hyperchlorinated" pattern is likely achieved with a single BGC-encoded halogenase ComH. Other type V GPA, kistamicin, carries a single chlorine on *aa-5*. At the same time, *kis*-encoded halogenase—KisU—is 72% identical to ComH. Thus, specificities of GPA halogenases might be inferred from in silico analysis very cautiously.

GPA halogenases might also decorate GPA aglyca with halogens other than chlorine; feeding of *Am. balhimycina* with bromine yielded bromobalhimycin (Bister et al. 2003). Some GPAs (ristocetin, corbomycin, GP6738) lack halogen atoms. Corresponding BGCs have no halogenase genes (Spohn et al. 2014; Truman et al. 2014; Culp et al. 2020; Xu et al. 2020).

16.4.6 Methylation and Sulfation of GPA Aglyca

Aglyca of dalbaheptides are often, but not necessarily, methylated. Most commonly, N-terminal end of oligopeptide is methylated with an (*S*)-adenosyl-L-methionine (SAM)-dependent *N*-methyltransferases, initially studied on the basis of MtfA from chloroeremomycin BGC (O'Brien et al. 2000; Shi et al. 2009). MtfA has a very low affinity to a non-cross-linked heptapeptide, suggesting that *N*-methylation occurs on mature aglycone as a substrate (O'Brien et al. 2000; Brieke et al. 2016). Some *N*-methyltransferases methylate N-terminus of heptapeptide twice (like Pek28), or even three times (possibly in demethylvancomycin producer WAC4169, Thaker et al. 2013). *N*-methyltransferase genes are present in BGCs for vancomycin-like antibiotics, pekiskomycin and A40926. The *tei* and *sta* gene clusters, however, lack such genes. Yet, *N*-methyltransferases from *dbv, pek,* and *cem* BGCs are promiscuous enough to *N*-methylate A40926 and teicoplanin aglyca (Shi et al. 2009; Brieke et al. 2016). Interestingly, Pek28 was unable to methylate N-terminus of these aglyca twice, losing substrate specificity after the first round (Brieke et al. 2016). Metagenome-derived VEG BGC contains two genes for *N*-methyltransferases—*veg24* and *veg25* (Banik and Brady 2008).

Methylation of GPA aglyca is not limited to N-terminus of heptapeptide. Ristocetin aglycone features C-methylated *aa-3* (Dpg) and *O*-methylated C-terminus of the heptapeptide. Orf19 and Orf23 methyltransferases were predicted to methylate these sites (Truman et al. 2014). Function of Orf19 and Orf23 was confirmed later via heterologous expression in *S. coelicolor*, carrying *sta* gene cluster (Yim et al. 2016). Both enzymes are unique for GPAs; Orf19 has Veg18 as the only ortholog from metagenome-derived VEG BGC. Accordingly, Veg18 is able to methylate *aa-3* of A47934 in vitro (Banik and Brady 2008).

Sulfations of the GPA aglyca are much rarer than methylations. Among naturally occurring GPAs, only A47934, UK-68,897, and pekiskomycin are sulfated at *aa-1, aa-3,* and *aa-6,* respectively. Sulfotransferase genes are abundant in metagenome-derived GPA BGCs, mainly in TEG BGC (Banik and Brady 2008). Many of them were investigated in vitro and in vivo. StaL sulfates teicoplanin aglycone in vitro (Lamb et al. 2006) at the *aa-1* (as it naturally happens in A47934 biosynthesis). Auk20 sulfates A47934 at *aa-3* in vivo and vancomycin- and teicoplanin-derived compounds in vitro (Yim et al. 2014). Pek25 sulfates A47934 at *aa-6* in vivo (Yim et al. 2016). TEG-derived sulfotransferases offered more interesting case, being able to sulfate teicoplanin aglycone at *aa-3* (Teg12), *aa-4* (Teg14), and *aa-6* (Teg13). Crystal structures of StaL, Teg14, and Teg12 were published (Bick et al. 2010; Shi et al. 2012). We are not aware of type V GPAs carrying sulfate group naturally or as a result of genetic engineering.

16.4.7 Glycosylation and Acylation of Dalbaheptides

As mentioned above, eponymous tailoring steps are limited to type I-IV GPAs, and only type IV compounds carry aliphatic side chains. All dalbaheptides (with a very few exceptions, like A47934) are glycosylated. Each amino acid in aglycone could be glycosylated (except, probably, *aa-5,* Nicolaou et al. 1999). The astonishing variability of glycosylation patterns remains poorly understood at the genetic background. Nevertheless, glycosylation of some dalbaheptides was studied in great detail.

Among type I dalbaheptides, mechanisms of glycosylation are best known for chloroeremomycin, vancomycin, and balhimycin (Fig. 16.7a). All three antibiotics are glucosylated at *aa-4*. Chloroeremomycin also carries two residues of exotic amino sugar L-epivancosamine—one at *aa-6* and second attached to D-glucose; balhimycin has another amino sugar, L-4-oxovancosamine, at *aa-6*. The *aa-6* position of vancomycin is not glycosylated; L-vancosamine residue is appended to D-glucose (Fig. 16.7a).

Corresponding BGCs harbor different sets of glycosyltransferase genes: three in *cep* gene cluster—GtfA, GtfB, and GtfC; and three in *bal* gene cluster—BgtfA, BgtfB, and BgtfC. Vancomycin BGC encodes two glycosyltransferases—GtfD and GtfE. GtfB/GtfE transfers D-glucose residue onto *aa-4*, and then, GtfC/D attach L-epivancosamine/L-vancosamine to D-glucose, respectively (Fig. 16.7a, Losey et al. 2001, 2002; Mulichak et al. 2004). GtfA installs L-epivancosamine onto *aa-6*

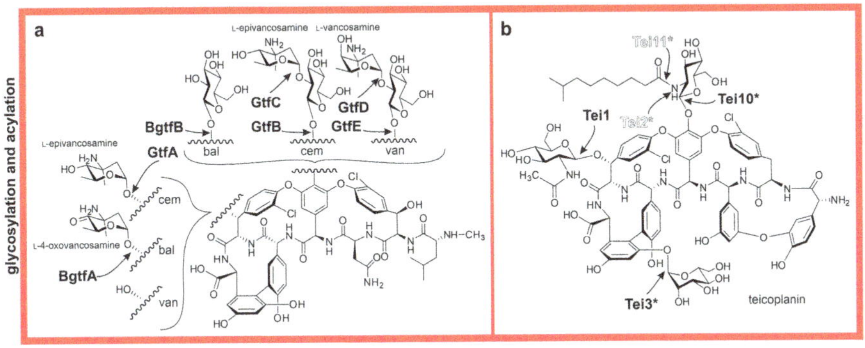

Fig. 16.7 Some of the best studied GPA glycosylation (and acylation) cascades: (**a**) glycosylation cascades for type I GPAs (balhimycin—bal, chloroeremomycin—cem, and vancomycin—van); (**b**) coupled glycosylation/acylation of type IV GPA teicoplanin. Refer to the main text for more details

(Fig. 16.7a, Mulichak et al. 2003; Lu et al. 2004). All these enzymes are quite promiscuous with regard to carbohydrate donor substrates and acceptor aglyca, albeit retaining strict regiospecificity. This allows combinatorial glycosylation of vancomycin and teicoplanin aglyca. Function of BgtfB was studied in vivo. The *bgtfB* knockout mutant fails to produce any glycosylated GPA, implying that BgtfB catalyzes initial reaction of glycosylation cascade (Fig. 16.7a, Pelzer et al. 1999). BgtfC is practically unable to work on balhimycin scaffold (Pelzer et al. 1999; Stegmann and Frasch 2010).

Experimental data on glycosyltransferases from type II dalbaheptides are absent, although these GPAs are often decorated at unusual positions, like avoparcin (glycosylated at *aa-1* and *aa-2*), or carry exotic amino sugars, such as L-acosamine in actinoidin (Nicolaou et al. 1999).

Glycosylation is ubiquitous among type III dalbaheptides. A good example is ristocetin, decorated with six sugar moieties, including common D-arabinose, D-glucose, L-rhamnose, two D-mannose residues, and exotic amino sugar L-ristosamine (Truman et al. 2014). The ristocetin BGC carries six genes for glycosyltransferases (Truman et al. 2014). Functions of some genes were investigated in a heterologous strain *S. coelicolor* producing A47934 and desulfo-A47934 (Yim et al. 2016), showing that Orf22 is responsible for D-mannosylation of *aa-7*.

Glycosylation of type IV dalbaheptides, such a teicoplanin, is coupled with acylation. The teicoplanin aglycone carries three sugar residues and one lipid chain (Fig. 16.7b); correspondingly, three glycosyltransferases (Tei1, Tei3*, and Tei10*) and one acyltransferase (Tei11*) are encoded within *tei* BGC. Glycosylation and acylation cascade in teicoplanin biosynthesis begins with Tei10* transferring Glc*N*Ac to *aa-4* (Fig. 16.7b, Li et al. 2004). Glc*N*Ac is then deacetylated by Tei2* yielding a glucosaminyl moiety (Truman et al. 2006). Tei11* then transfers an acyl chain to the free amino group of this glucosaminyl moiety (Fig. 16.7b, Li et al. 2004; Kruger et al. 2005; Lyu et al. 2014). Tei11* is quite flexible toward the donor substrates, cell membrane lipids. It leads to the production of different teicoplanin congeners, varying in length and branching of lipid chains (Lazzarini et al. 1997). Inactivation of *tei11** leads to accumulation of a teicoplanin derivative lacking *N*-acetylglucosamine on *aa-6* and D-mannose on *aa-7* (Yushchuk et al. 2016). In vitro characterization of Tei1 showed that it is responsible for transferring Glc*N*Ac moiety to *aa-6* (Fig. 16.7b, Li et al. 2004), while knockout of *tei3** showed that Tei3* attaches a D-mannose to *aa-7* (Fig. 16.7b, Yushchuk et al. 2016). Another type IV dalbaheptide, A40926, undergoes tailoring similarly to teicoplanin. Here, Dbv9 and Dbv20 are homologues of Tei10* and Tei3*, attaching Glc*N*Ac to *aa-4* and mannose to *aa-7*, respectively. Glc*N*Ac is then deacetylated by Dbv21 (Tei2* homologue) and acylated by Dbv8 (Tei11* homologue). However, two additional enzymes, absent in teicoplanin biosynthesis pathway, are involved in further modifications of A40926 sugars. These are Glc*N*Ac oxidase Dbv29 (which forms carboxylic group on Glc*N*Ac moiety), and Dbv23, which acetylates D-mannose residue.

The biosynthesis of unusual amino sugars (mentioned above) is directed by BGC-situated genes and was thoroughly elucidated on the example of L-epivancosamine production in chloroeremomycin pathway (Fig. 16.8, Chen et al. 2000). This pathway consists of five (EvaA-E) enzymes and serves as a reference to understand amino sugar biosynthesis for all GPAs (Fig. 16.8).

16.5 Transcriptional Regulation of the GPA Biosynthetic Gene Expression

All described to date BGCs for dalbaheptides and complestatin-like antibiotics possess a gene for a StrR-like transcriptional regulator (Waglechner et al. 2019; Yushchuk et al. 2019). These regulators act as a key transcriptional activators at lowest—pathway-specific—regulatory level. There is deep understanding of StrR homologues from teicoplanin, balhimycin, and A40926 BGCs (Shawky et al. 2007; Horbal et al. 2014; Lo Grasso et al. 2015). StrR family regulators are not the only transcriptional factors (TFs) encoded within GPA BGCs. The *tei* and *dbv* gene clusters carry also regulatory gene for a LuxR-like TF. The latter play central role in GPA biosynthesis. Type V GP6738 (Xu et al. 2020) BGC also contains a *lnbU*-like TF, overexpression of which increases the production of GP6738. Majority of GPA BGCs also harbor *vanR-vanS* genes involved only in regulation of self-resistance (not discussed here).

16.5.1 Pathway-Specific Regulation in Balhimycin BGC (and Related BGCs)

In *Am. balhimycina*, balhimycin BGC encodes StrR-like protein Bbr (Shawky et al. 2007). Bbr DNA-binding sites (BSs) are found within the promoter regions of *tba*, *oxyA*, *orf7*, *dvaA*, and *bbr* itself, and consensus sequence is [**GTCCAa** $(N)_{17}$**TtGGAC**] (base pairs corresponding to the inverted repeats here and further

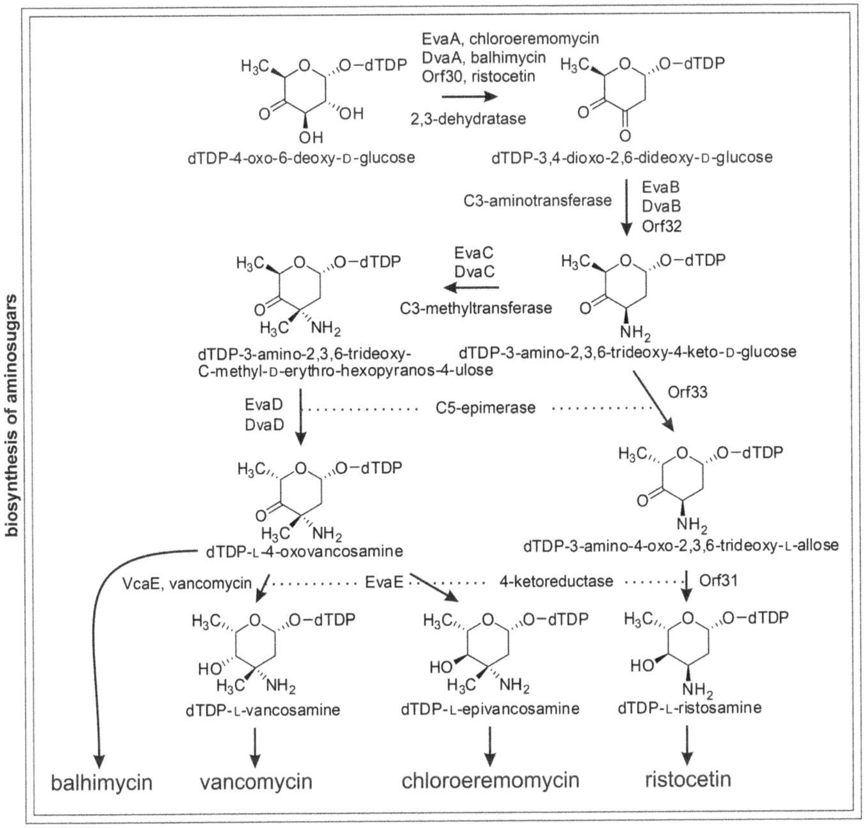

Fig. 16.8 Biosynthetic pathways that enable the production of some GPA amino sugars. The biosynthesis of L-epivancosamine was investigated experimentally in chloroeremomycin pathway; enzymes, required for the biosynthesis of L-ristosamine and others, were not studied experimentally, but their functions were predicted by analogy

are given in bold, Shawky et al. 2007). Bbr does not bind *vanY* promoter region, leaving the expression of biosynthetic and resistance genes without a link.

Balhimycin BGC is a model for studying the biosynthesis of clinically relevant vancomycin, since the producer of the latter poses formidable challenge for genetic manipulations. Recent work on the regulation of norvancomycin biosynthesis in *Am. orientalis* NCPC 2–48 (Li et al. 2021) provides collateral insight into Bbr-mediated regulation of balhimycin production. The norvancomycin BGC—*nvcm*—was described to be regulated by Bbr ortholog AoStrR1. Balhimycin BGC and *nvcm* are very similar. However, DNA-binding properties of AoStrR1 were drastically different from that of Bbr. Particularly, AoStrR1 was shown to bind at least 8 promoter regions within *nvcm*: *AoStrR1* (corresponding to *bbr* promoter region, where Bbr binds), *oxyA* (also recognized by Bbr), *oxyB* (Bbr binding to the homologous sequence was not observed), *vasA* (*dvaA* in balhimycin BGC, where Bbr binds), *vanY* (not recognized by Bbr), *vhp* (corresponds to *bhp* in balhimycin BGC; Bbr does not bind), *hmaS* (same in balhimycin BGC; Bbr does not bind), and *ald* (corresponding to *dahp* in balhimycin BGC, where no binding was observed for Bbr). Some of them contain almost perfect BSs matching the consensus described for Bbr, some have rather degenerate sites. The consensus BS of AoStrR1 was thus much more degenerate than the one for Bbr—[**(G/T)NNCNN**$(N)_{17}$**(T/C)N(G/C)(G/C)(A/G)C**] (Li et al. 2021). Given the 84% amino acid sequence identity between Bbr and AoStrR1, the promiscuity of the latter is mind boggling. Perhaps, some AoStrR1 BSs are artificial (especially the most degenerate ones) and the binding occurs only under conditions of the in vitro assay. One of such nonphysiological BSs might be in the region upstream of *oxyB*, which is most likely a part of *oxyA-B-C-vhaA* polycistronic transcription unit (by analogy to the *oxyA-B-C-bhaA* operon in balhimycin BGC).

The same work (Li et al. 2021) also described another regulator, found on the outer boundary of *nvcm* BGC to be involved in norvancomycin biosynthesis. This TF contains a LuxR-like DNA-binding domain (completely different from the ones found in *dbv/tei*-derived LuxR-like TFs). Its overexpression led to the increase in norvancomycin production.

16.5.2 Pathway-Specific Regulation of tei BGC

Transcriptional regulation of teicoplanin biosynthesis in *A. teichomyceticus* was recently discussed in Yushchuk et al. (2020c). In brief, *tei* BGC contains three genes, coding for TFs: *tei15** (StrR-like), *tei16** (LuxR-like), and *tei31** (AfsR-like) (Li et al. 2004; Horbal et al. 2014; Yushchuk et al. 2019). Tei15* and Bbr are orthologs sharing 48% of amino acid sequence identity. Knockout of *tei15** canceled teicoplanin production (Horbal et al. 2014), and overexpression of the same gene improved teicoplanin titers up to 40-fold (Horbal et al. 2013, 2014). As Bbr, DNA-binding properties of Tei15* were studied in vitro. Tei15* BSs were found upstream *tei2–3* (*vanRS* orthologs), *teiA* (first NRPS gene), *tei2** (Glc*N*Ac deacetylase), *tei5** (OxyA), *tei16** (LuxR-like TF), *tei17** (DpgA), *tei23** (HpgT), *tei26** (type II CM), *tei28** (HmaS), *tei30** (type II thioesterase), and *tei31** (AfsR-like TF). Six of these promoter regions (*teiA, tei2*, tei16*, tei17*, tei27**, and *tei31**) were bound in vitro by Tei15*; the others were not tested. Tei15* BS consensus sequence is [**(G/C)NN(C/G)NN**$(N)_{17}$**NT(G/C)G(A/G)C**] (Horbal et al. 2014), being more degenerate than the one for Bbr. Expression of *tei1, teiA-1** (one operon, Tei15* binding verified), *tei2*-3** (operon, Tei15* binding verified), *tei4*, tei5*-7** (operon, Tei15* BS predicted in silico), *tei8*-12** (operon, Tei15* BS predicted in silico), *tei13*, tei14*, tei24**, and *tei28*-30** (operon, Tei15* BS predicted in silico) is decreased in Δ*tei15** mutant. At the same time, the expression of *tei2–4, tei17*-23** operons, and *tei16** was rather intact. Corresponding promoter regions contained either predicted in silico or verified in vitro Tei15* BS. Some genes having ample Tei15* BS in their promoter regions were not expressed at all in the wild type under teicoplanin production conditions. These results make the picture of Tei15*-mediated regulation of teicoplanin biosynthesis quite puzzling, highlighting the gap in our understanding of DNA-binding properties of TFs under in vitro and in vivo conditions.

Tei16* is a second pathway-specific regulator of teicoplanin biosynthesis. The knockout of *tei16** completely abolishes the production of teicoplanin (Horbal et al. 2014); the overexpression of this gene boosts teicoplanin titers more that it was observed for *tei15** overexpression (Horbal et al. 2013, 2014). Studies of Tei16* DNA-binding properties in vitro shed no light on its targets within *tei* BGC (Horbal et al. 2014). Tei16* likely requires some partner proteins or ligands to bind DNA (like other LuxR-like TFs, de Schrijver and de Mot 1999). The transcriptional changes within *tei* gene cluster of Δ*tei16** and Δ*tei15** mutants were similar, including the complete cessation of *tei15** expression (Yushchuk et al. 2019). The *tei15** is likely to be a single target of Tei16*, thus placing the latter at the top of *tei* pathway-specific regulatory hierarchy. As Tei15* binds *tei16** promoter region in vitro, some feedback mechanism may exist, which so far remains unstudied. Knockouts of both *tei15** and *tei16** have no impact on *tei2–4* and *tei7–5* operons for teicoplanin resistance, similarly to what was reported for balhimycin BGC.

TF Tei31* appears to be dispensable for teicoplanin biosynthesis under teicoplanin production conditions (Yushchuk et al. 2019). However, Tei15* binds *tei31** promoter in vitro; so, Tei31* might have some effect on teicoplanin biosynthesis under certain conditions.

16.5.3 Pathway-Specific Regulation of dbv BGC

The A40926 (*dbv*) BGC of *N. gerenzanensis* encodes Dbv4 TF, which is Bbr/Tei15* ortholog (Sosio et al. 2003). LuxR-like Dbv3 is also found in *dbv* BGC (Sosio et al. 2003). A two-component regulatory system consisting of a sensor histidine kinase Dbv22 and response TF Dbv6 was very recently shown to be involved in A40926 production as well (Alduina et al. 2020).

Dbv4 possesses rather small regulon in *dbv* BGC as compared to Bbr and Tei15* regulons. Dbv4 directly controls the expression of *dbv30–35* operon (Dpg biosynthesis genes) and *dbv14–8* suboperon (cross-linking oxygenase and some tailoring enzyme genes). This was shown both by studying Dbv4 DNA-binding properties in vitro (Alduina et al. 2007) and by the analysis of the *dbv* gene expression in Δ*dbv4* mutant (Lo Grasso et al. 2015). DNA-binding properties of Dbv4 and Bbr are remarkably similar (Alduina et al. 2007), underscoring overall identity of these proteins. The promoter regions of *dbv30* and *dbv14* contain BSs identical to consensus BSs described for Bbr; moreover, Dbv4 was able to bind Bbr-regulated promoter regions within balhimycin BGC. Although the regulon of Dbv4 is smaller than the regulons of Bbr and Tei15*, it governs crucial A40926 biosynthetic genes, as *dbv4* knockout blocks A40926 production (Lo Grasso et al. 2015), while overexpression improves the latter (Yushchuk et al. 2020a).

Dbv3 is a second major TF in *dbv* BGC and a crucial one for A40926 biosynthesis, because *dbv3* knockout abolishes the production of this GPA (Lo Grasso et al.

2015), while overexpression improves the yield (Yushchuk et al. 2020a). Dbv3 targets within *dbv* BGC are known only through gene expression studies (Lo Grasso et al. 2015). Following *dbv* operons are not expressed in Δ*dbv3* mutant (Lo Grasso et al. 2015): *dbv1–2* (Hpg biosynthetic enzymes), *dbv15–17* (NRPS modules 4–7 and MbtH), *dbv21–21* (mannosyltransferase and mannose *O*-acetylase), and *dbv24–28* (NRPS modules 1–3, A40926 export, and some tailoring enzymes). The expression of *dbv29* (hexose oxidase), *dbv36* (type II thioesterase), and *dbv37* (HpgT) genes is absent in Δ*dbv3* mutant, as well as the expression of *dbv4*. Thus, unlike Tei16*, Dbv3 has a multiple *dbv* targets, except *dbv4*. It is not clear whether Dbv3 regulates the expression of *dbv* genes alone, or some yet-unknown intermediate player is involved.

A40926 BGC encodes a two-component regulatory system—*dbv6* and *dbv22*. Given its similarity to VanR/S regulatory circuit, Dbv6/22 was previously believed to be orthologs of the latter (Yushchuk et al. 2020b). When *dbv6* was first knocked out, no impact was observed on either A40926 production or other physiological properties of *N. gerenzanensis* (Lo Grasso et al. 2015). Dbv6 and Dbv22 were shown recently to be accessory regulators of A40926 biosynthesis (Alduina et al. 2020). Dbv6 binds promoter regions of *dbv4, dbv3,* and *dbv24–28* operons (also controlled by Dbv3), and *dbv22–23* operon (controlled by neither Dbv4 nor Dbv3). Moreover, it somehow binds *dbv7* upstream region, although the latter gene was previously shown to be part of polycistronic *dbv5–6–7* operon (Alduina et al. 2007). At least under some growth conditions (in RARE3 medium, but not in R3 medium, Lo Grasso et al. 2015; Alduina et al. 2020), Dbv6 and Dbv22 act as a A40926 biosynthesis repressors. At the same time, Δ*dbv22* mutant (but not Δ*dbv6*) was slightly more sensitive to A40926 and its precursor—*O*-acetyl-A40926. Additionally, A40926 (but not *O*-acetyl-A40926) was shown to stimulate the expression of *dbv* genes, maybe through Dbv22. In this scenario, Dbv22 acts as a sensor histidine kinase, which represses certain *dbv* genes (likely through Dbv6); Dbv22 displays greater affinity to A40926 than to *O*-acetyl-A40926 and loses its repressing function upon A40926 binding. Although solid, evidence-based scenario is still unavailable, one thing is certain—Dbv6/Dbv22 regulatory pair is involved in fine-tuning of the *dbv* gene expression. Being accessory to Dbv4 and Dbv3, Dbv6/Dbv22 likely couples antibiotic biosynthesis and self-resistance. Kistamicin BGC also carries *dbv6/dbv22* orthologs, in addition to the ones for *dbv3/dbv4* (Nazari et al. 2017).

To sum up, there currently are three "archetypal" models for pathway-specific regulation of dalbaheptide biosynthesis. The best understood model applies to *Amycolatopsis*-derived antibiotics. Here, a StrR family Bbr-like TF governs the biosynthesis of the majority of biosynthetic genes and operons. Its binding sites are firmly identified. All Bbr-like regulators are very similar and function across different biosynthetic pathways (Spohn et al. 2014). In contrast, the regulation of teicoplanin and, most likely, other *Actinoplanes*-derived GPAs (Yim et al. 2014) involves two regulators. One is StrR/Bbr type, exemplified by Tei15*, and the latter is a subject to action of LuxR-like TF Tei16*. Tei15* binds multiple promoters within *tei* BGC, recognizing, however, a more degenerate binding sequence than Bbr does. Gene *tei15** is likely a single target of Tei16*, but might upregulate the expression of *tei16** as well. Factors contributing to the function of Tei16* are unknown. In *dbv* biosynthesis regulation, a StrR and LuxR regulators Dbv4 and Dbv3 are also focal. However, here LuxR-like TF Dbv3 is not an orthologue of Tei16*. Dbv4 is closely related to Bbr, but is responsible for the activation of only two *dbv* operons.

16.6 Biological Activities of GPAs: Beyond D-Ala-D-Ala Binding

In contrast to the other GPAs, type V GPAs, such as complestatin and corbomycin, do not bind D-Ala-D-Ala termini but are still active against Gram-positive bacteria, most likely because they target autolysins (Culp et al. 2020). Autolysins are a large group of lytic enzymes, involved in cell wall remodeling necessary for growth and division (Smith and Blackman 2000). Pharmacological properties of both dalbaheptides and complestatin-like antibiotics are not limited to the antimicrobial activities, as described below.

Anticancer properties Most of them show certain cytotoxic activities against cancer cells in vitro, but clinical prospects are uncertain (e.g., Balzarini et al. 2003; Xu et al. 2019). Vancomycin was found to potentiate antitumor immune response to the radiotherapy, thus inhibiting tumor growth (Uribe-Herranz et al. 2020). This effect was achieved not by vancomycin per se, but by depletion of vancomycin-sensitive gut microbiota.

Antiviral properties Complestatin inhibits the development of HIV-1 in different models in vitro (Momota et al. 1991). Complestatin appears to prevent the adsorption of viral particles on the cell surface. Antiviral properties were also described for kistamicin (Naruse et al. 1993a, b). Other type V GPAs, chloropeptins I and II, possess activity against HIV (Tanaka et al. 1997). Among dalbaheptides, only those carrying aliphatic side chain (type IV)—teicoplanin and A40926—exhibit antiviral properties (Balzarini et al. 2003). Heavily glycosylated dalbaheptides tended to lose antiviral potency and vice versa—vancomycin aglycone has weak antiviral effect (Balzarini et al. 2003). Teicoplanin inhibits the infection of cell lines with Ebolavirus in vitro (Wang et al. 2016; Zhou et al. 2016). Teicoplanin was effective only before viral entry, hinting that the cell surface, not the Ebola virus itself, is drug target. Teicoplanin is completely inefficient against non-enveloped human respiratory syncytial virus (hSCV) (Wang et al. 2016). It is now known that teicoplanin targets cathepsin L protease, which is located in the host cell membrane. Cathepsin L is involved in the proteolysis of the envelope glycoproteins, which is necessary for the entry of Ebolavirus into the host cell (e.g., Hood et al. 2010). Activity of teicoplanin against a number of other enveloped viruses was shown. Hydrophobic groups present in teicoplanin, dalbavancin, oritavancin, and telavancin are probably necessary for the proper anchoring of the antibiotic within host cell membrane, which therefore delivers the pharmacophore (probably located in aglycone) to cathepsin L. Activity of teicoplanin and related drugs against coronaviruses like SARS-CoV-1 and MERS-CoV is noticeable in times of pandemics caused by SARS-CoV-1 and SARS-CoV-2 (Wu et al. 2020).

Activity of dalbaheptides against SARS-CoV-2 might not be only cathepsin-mediated but also ACE2-dependent (Wang et al. 2021); the latter protein is the entry receptor for the SARS-CoV-2 (Shang et al. 2020). ACE2 tropism was shown for dalbavancin and not vancomycin. The presence of the aliphatic side chain in the dalbavancin is the most salient structural difference between abovementioned GPAs.

Usage of ristocetin in von Willebrand disease diagnostics Ristocetin induces thrombophlebitis and thrombocytopenia when administered as antibacterial, which prompted its withdrawal from clinical usage. These properties, however, find their use in the diagnostics of von Willebrand disease (James and Goodeve 2011), as well as rarer Bernard–Soulier syndrome (Lanza 2006). Ristocetin causes agglutination of thrombocytes mediated by von Willebrand factor multimers. Thus, addition of the ristocetin to a blood sample, lacking this factor (or corresponding receptor on the thrombocyte surface, as it is in the case of Bernard–Soulier syndrome), will not cause agglutination of thrombocytes, which could be detected (Coller and Gralnick 1977).

Some other "exotic properties" of complestatin Complestatin is one of the most potent inhibitors of sensitized erythrocytes hemolysis, caused by a complement system. This inhibition is irreversible and, most likely, is achieved due to the inhibition of C3 convertase (Kaneko et al. 1989). This GPA also prevents neuronal excitotoxicity caused by *N*-methyl-D-aspartate (NMDA) and α-amino-3-hydroxy-5-methyl-4-isoxazolepropionic (AMPA) acid. Thus, complestatin protects cortical neurons from prolonged oxygen and glucose starvation (Seo et al. 2001). Finally, complestatin has certain antiapoptotic properties (Kim et al. 2004).

16.7 Conclusions and Future Prospects

In this chapter, we have focused on a rather "mechanistic" description of the GPA family. However, many other aspects of the biology of this peculiar compounds merit detailed discussion and more research as well. One of such aspects is relationship of GPAs with other peptide inhibitors of PG synthesis, such as feglymycin (Gonsior et al. 2015) and ramoplanin (Farver et al. 2005). Feglymycin BGC—*feg*—has some genetic elements highly related to GPA BGCs, especially *tei.* These include two regulatory genes for StrR- and LuxR-like orthologs of Tei15* and Tei16*. Indeed, the discovery of corbomycin and GP6738 makes it easier to imagine how a GPA-like BGC might have evolved into *feg*-like one through the rearrangement of NRPS genes and loss of cross-linking oxygenases. Alternatively, *feg* might represent a sister evolutionary line to GPAs, coming from common ancestral pool of BGCs carrying basic NRPS genes, Dpg and Hpg biosynthetic genes, regulatory elements, and transporters. Ramoplanin BGC, at the same time, codes for a mannosyltransferase with a striking similarity to mannosyltransferases involved in the GPA biosynthesis, as well as some GPA-like amino acid synthesis genes. Recent works on the phylogeny of GPAs (Waglechner et al. 2019) clearly highlight these links. However, a bigger picture of the evolution of GPAs together with related compounds still remains elusive. There is no doubt that further reconstruction of this "bigger picture" will yield novel classes of antibiotics related to GPAs, novel tailoring enzymes, and data on the biosynthetic regulatory mechanisms. Altogether, this will contribute to our arsenal against multidrug-resistant pathogens.

References

Adamek M, Alanjary M, Sales-Ortells H et al (2018) Comparative genomics reveals phylogenetic distribution patterns of secondary metabolites in *Amycolatopsis* species. BMC Genomics 19: 426. https://doi.org/10.1186/s12864-018-4809-4

Alduina R, Lo Piccolo L, D'Alia D et al (2007) Phosphate-controlled regulator for the biosynthesis of the dalbavancin precursor A40926. J Bacteriol 189:8120–8129. https://doi.org/10.1128/JB.01247-07

Alduina R, Sosio M, Donadio S (2018) Complex regulatory networks governing production of the glycopeptide A40926. Antibiotics (Basel) 7(2):30. https://doi.org/10.3390/antibiotics7020030

Alduina R, Tocchetti A, Costa S et al (2020) A two-component regulatory system with opposite effects on glycopeptide antibiotic biosynthesis and resistance. Sci Rep 10:6200. https://doi.org/10.1038/s41598-020-63257-4

Arthur M, Courvalin P (1993) Genetics and mechanisms of glycopeptide resistance in enterococci. Antimicrob Agents Chemother 37(8):1563–1571. https://doi.org/10.1128/aac.37.8.1563

Arthur M, Molinas C, Bugg TDH et al (1992) Evidence for *in vivo* incorporation of D-lactate into peptidoglycan precursors of vancomycin-resistant enterococci. Antimicrob Agents Chemother 36(4):867–869. https://doi.org/10.1128/aac.36.4.867

Bachmann BO, Ravel J (2009) Methods for *in silico* prediction of microbial polyketide and nonribosomal peptide biosynthetic pathways from DNA sequence data. Methods Enzymol 458:181–217. https://doi.org/10.1016/S0076-6879(09)04808-3

Balzarini J, Pannecouque C, De Clercq E et al (2003) Antiretroviral activity of semisynthetic derivatives of glycopeptide antibiotics. J Med Chem 46:2755–2764. https://doi.org/10.1021/jm0300882

Banik JJ, Brady SF (2008) Cloning and characterization of new glycopeptide gene clusters found in an environmental DNA megalibrary. Proc Natl Acad Sci U S A 105(45):17273–17277. https://doi.org/10.1073/pnas.0807564105

Banik JJ, Craig JW, Calle PY, Brady SF (2010) Tailoring enzyme-rich environmental DNA clones: a source of enzymes for generating libraries of unnatural natural products. J Am Chem Soc 132: 15661–15670. https://doi.org/10.1021/ja105825a

Barna JC, Williams DH (1984) The structure and mode of action of glycopeptide antibiotics of the vancomycin group. Annu Rev Microbiol 38:339–357. https://doi.org/10.1146/annurev.mi.38.100184.002011

Barreteau H, Kovač A, Boniface A et al (2008) Cytoplasmic steps of peptidoglycan biosynthesis. FEMS Microbiol Rev 32(2):168–207. https://doi.org/10.1111/j.1574-6976.2008.00104.x

Bick MJ, Banik JJ, Darst SA, Brady SF (2010) The 2.7 Å resolution structure of the glycopeptide sulfotransferase Teg14. Acta Crystallogr D Biol Crystallogr 66(Pt 12):1278–1286. https://doi.org/10.1107/S0907444910036681

Binda E, Marinelli F, Marcone GL (2014) Old and new glycopeptide antibiotics: action and resistance. Antibiotics (Basel) 3(4):572–594. https://doi.org/10.3390/antibiotics3040572

Bister B, Bischoff D, Nicholson GJ et al (2003) Bromobalhimycin and chlorobromobalhimycins—illuminating the potential of halogenases in glycopeptide antibiotic biosyntheses. ChemBioChem 4(7):658–662. https://doi.org/10.1002/cbic.200300619

Boeck LD, Mertz FP (1986) A47934, a novel glycopeptide-aglycone antibiotic produced by a strain of *Streptomyces toyocaensis* taxonomy and fermentation studies. J Antibiot (Tokyo) 39:1533–1540. https://doi.org/10.7164/antibiotics.39.1533

Brieke C, Yim G, Peschke M et al (2016) Catalytic promiscuity of glycopeptide: N-methyltransferases enables bio-orthogonal labelling of biosynthetic intermediates. Chem Commun (Camb) 52(94):13679–13682. https://doi.org/10.1039/c6cc06975d

Cavalleri B, Parenti F (2000) Glycopeptides (Dalbaheptides). In: Kirk-Othmer encyclopedia of chemical technology. https://doi.org/10.1002/0471238961.0712250303012201.a01

Chen H, Thomas MG, Hubbard BK et al (2000) Deoxysugars in glycopeptide antibiotics: enzymatic synthesis of TDP-L-epivancosamine in chloroeremomycin biosynthesis. Proc Natl Acad Sci U S A 97(22):11942–11947. https://doi.org/10.1073/pnas.210395097

Chen H, Tseng CC, Hubbard BK, Walsh CT (2001) Glycopeptide antibiotic biosynthesis: enzymatic assembly of the dedicated amino acid monomer (S)-3,5-dihydroxyphenylglycine. Proc Natl Acad Sci U S A 98(26):14901–14906. https://doi.org/10.1073/pnas.221582098

Cheng AV, Wuest WM (2020) Phylogeny-guided approach yields glycopeptides with unique action. Trends Pharmacol Sci 41(5):297–299. https://doi.org/10.1016/j.tips.2020.03.002

Chiu HT, Hubbard BK, Shah AN et al (2001) Molecular cloning and sequence analysis of the complestatin biosynthetic gene cluster. Proc Natl Acad Sci U S A 98(15):8548–8553. https://doi.org/10.1073/pnas.151246498

Coller BS, Gralnick HR (1977) Studies on the mechanism of ristocetin induced platelet agglutination. Effects of structural modification of ristocetin and vancomycin. J Clin Invest 60:302–312. https://doi.org/10.1172/JCI108778

Courvalin P (2006) Vancomycin resistance in gram-positive cocci. Clin Infect Dis 42(Suppl 1):S25–S34. https://doi.org/10.1086/491711

Cryle MJ, Staaden J, Schlichting I (2011) Structural characterization of CYP165D3, a cytochrome P450 involved in phenolic coupling in teicoplanin biosynthesis. Arch Biochem Biophys 507(1):163–173. https://doi.org/10.1016/j.abb.2010.10.017

Culp EJ, Waglechner N, Wang W et al (2020) Evolution-guided discovery of antibiotics that inhibit peptidoglycan remodelling. Nature 578:582–587. https://doi.org/10.1038/s41586-020-1990-9

de Kruijff B, van Dam V, Breukink E (2008) Lipid II: a central component in bacterial cell wall synthesis and a target for antibiotics. Prostaglandins Leukot Essent Fatty Acids 79(3–5):117–121. https://doi.org/10.1016/j.plefa.2008.09.020

de Schrijver A, de Mot R (1999) A subfamily of MalT-related ATP-dependent regulators in the LuxR family. Microbiology 145:1287–1288. https://doi.org/10.1099/13500872-145-6-1287

Farver DK, Hedge DD, Lee SC (2005) Ramoplanin: a lipoglycodepsipeptide antibiotic. Ann Pharmacother 39(5):863–868. https://doi.org/10.1345/aph.1E397

Goldstein BP, Selva E, Gastaldo L et al (1987) A40926, a new glycopeptide antibiotic with anti-*Neisseria* activity. Antimicrob Agents Chemother 31:1961–1966. https://doi.org/10.1128/AAC.31.12.1961

Gonsior M, Mühlenweg A, Tietzmann M et al (2015) Biosynthesis of the peptide antibiotic feglymycin by a linear nonribosomal peptide synthetase mechanism. ChemBioChem 16(18):2610–2614. https://doi.org/10.1002/cbic.201500432

Greule A, Izoré T, Iftime D et al (2019) Kistamicin biosynthesis reveals the biosynthetic requirements for production of highly crosslinked glycopeptide antibiotics. Nat Commun 10(1):2613. https://doi.org/10.1038/s41467-019-10384-w

Hadatsch B, Butz D et al (2007) The biosynthesis of teicoplanin-type glycopeptide antibiotics: assignment of P450 mono-oxygenases to side chain cyclizations of glycopeptide A47934. Chem Biol 14(9):1078–1089. https://doi.org/10.1016/j.chembiol.2007.08.014

Han S, Le BV, Hajare HS et al (2014) X-ray crystal structure of teicoplanin A_{2-2} bound to a catalytic peptide sequence via the carrier protein strategy. J Org Chem 79:8550–8556. https://doi.org/10.1021/jo501625f

Haslinger K, Peschke M, Brieke C et al (2015) X-domain of peptide synthetases recruits oxygenases crucial for glycopeptide biosynthesis. Nature 521:105–109. https://doi.org/10.1038/nature14141

Hecht SM (2000) Bleomycin: new perspectives on the mechanism of action. J Nat Prod 63 (1):158–168. https://doi.org/10.1021/np990549f

Hood CL, Abraham J, Boyington JC et al (2010) Biochemical and structural characterization of cathepsin L-processed Ebola Virus glycoprotein: implications for viral entry and immunogenicity. J Virol 84:2972–2982. https://doi.org/10.1128/jvi.02151-09

Horbal L, Kobylyanskyy A, Yushchuk O et al (2013) Evaluation of heterologous promoters for genetic analysis of *Actinoplanes teichomyceticus*—producer of teicoplanin, drug of last defense. J Biotechnol 168:367–372. https://doi.org/10.1016/j.jbiotec.2013.10.018

Horbal L, Kobylyanskyy A, Truman AW et al (2014) The pathway-specific regulatory genes, *tei15** and *tei16**, are the master switches of teicoplanin production in *Actinoplanes teichomyceticus*. Appl Microbiol Biotechnol 98:9295–9309. https://doi.org/10.1007/s00253-014-5969-z

Hubbard BK, Thomas MG, Walsh CT (2000) Biosynthesis of L-*p*-hydroxyphenylglycine, a non-proteinogenic amino acid constituent of peptide antibiotics. Chem Biol 7(12):931–942. https://doi.org/10.1016/s1074-5521(00)00043-0

James PD, Goodeve AC (2011) Von Willebrand disease. Genet Med 13(5):365–376. https://doi.org/10.1097/GIM.0b013e3182035931

Kalan L, Perry J, Koteva K et al (2013) Glycopeptide sulfation evades resistance. J Bacteriol 195 (1):167–171. https://doi.org/10.1128/JB.01617-12

Kaneko I, Kamoshida K, Takahashi S (1989) Complestain, a potent anti-complement substance produced by *Streptomyces lavendulae*. I. Fermentation, isolation and biological characterization. J Antibiot (Tokyo) 42:236–241. https://doi.org/10.7164/antibiotics.42.236

Kaniusaite M, Tailhades J, Kittilä T et al (2021) Understanding the early stages of peptide formation during the biosynthesis of teicoplanin and related glycopeptide antibiotics. FEBS J 288:507–529. https://doi.org/10.1111/febs.15350

Kim EC, Yun BS, Ryoo IJ et al (2004) Complestatin prevents apoptotic cell death: inhibition of a mitochondrial caspase pathway through AKT/PKB activation. Biochem Biophys Res Commun 313:193–204. https://doi.org/10.1016/j.bbrc.2003.11.104

Kim JH, Feng Z, Bauer JD et al (2010) Cloning large natural product gene clusters from the environment: piecing environmental DNA gene clusters back together with TAR. Biopolymers 93(9):833–844. https://doi.org/10.1002/bip.21450

Kittilä T, Kittel C, Tailhades J et al (2017) Halogenation of glycopeptide antibiotics occurs at the amino acid level during non-ribosomal peptide synthesis. Chem Sci 8(9):5992–6004. https://doi.org/10.1039/c7sc00460e

Kruger RG, Lu W, Oberthür M et al (2005) Tailoring of glycopeptide scaffolds by the acyltransferases from the teicoplanin and A-40,926 biosynthetic operons. Chem Biol 12 (1):131–140. https://doi.org/10.1016/j.chembiol.2004.12.005

Lamb SS, Patel T, Koteva KP, Wright GD (2006) Biosynthesis of sulfated glycopeptide antibiotics by using the sulfotransferase StaL. Chem Biol 13(2):171–181. https://doi.org/10.1016/j.chembiol.2005.12.003

Lanza F (2006) Bernard-Soulier syndrome (Hemorrhagiparous thrombocytic dystrophy). Orphanet J Rare Dis 1:46. https://doi.org/10.1186/1750-1172-1-46

Lazzarini A, Borghi A, Zerilli LF et al (1997) Novel teicoplanins by directed biosynthesis. J Antibiot 50(2):180–183. https://doi.org/10.7164/antibiotics.50.180

Li TL, Choroba OW, Charles EH et al (2001) Characterisation of a hydroxymandelate oxidase involved in the biosynthesis of two unusual amino acids occurring in the vancomycin group of antibiotics. Chem Commun (Camb) 18:1752–1753. https://doi.org/10.1039/b103548g

Li TL, Huang F, Haydock SF et al (2004) Biosynthetic gene cluster of the glycopeptide antibiotic teicoplanin: characterization of two glycosyltransferases and the key acyltransferase. Chem Biol 11:107–119. https://doi.org/10.1016/S1074-5521(04)00002-X

Li Z, Rupasinghe SG, Schuler MA, Nair SK (2011) Crystal structure of a phenol-coupling P450 monooxygenase involved in teicoplanin biosynthesis. Proteins 79(6):1728–1738. https://doi.org/10.1002/prot.22996

Li X, Zhang C, Zhao Y et al (2021) Comparative genomics and transcriptomics analyses provide insights into the high yield and regulatory mechanism of norvancomycin biosynthesis in *Amycolatopsis orientalis* NCPC 2-48. Microb Cell Fact 20(1):28. https://doi.org/10.1186/s12934-021-01521-6

Liu Y, Breukink E (2016) The membrane steps of bacterial cell wall synthesis as antibiotic targets. Antibiotics (Basel) 5(3):28. https://doi.org/10.3390/antibiotics5030028

Lo Grasso L, Maffioli S, Sosio M et al (2015) Two master switch regulators trigger A40926 biosynthesis in *Nonomuraea* sp. strain ATCC 39727. J Bacteriol 197:2536–2544. https://doi.org/10.1128/JB.00262-15

Losey HC, Peczuh MW, Chen Z et al (2001) Tandem action of glycosyltransferases in the maturation of vancomycin and teicoplanin aglycones: novel glycopeptides. Biochemistry 40 (15):4745–4755. https://doi.org/10.1021/bi010050w

Losey HC, Jiang J, Biggins JB et al (2002) Incorporation of glucose analogs by GtfE and GtfD from the vancomycin biosynthetic pathway to generate variant glycopeptides. Chem Biol 9 (12):1305–1314. https://doi.org/10.1016/s1074-5521(02)00270-3

Lu W, Oberthür M, Leimkuhler C et al (2004) Characterization of a regiospecific epivancosaminyl transferase GtfA and enzymatic reconstitution of the antibiotic chloroeremomycin. Proc Natl Acad Sci U S A 101(13):4390–4395. https://doi.org/10.1073/pnas.0400277101

Lyu SY, Liu YC, Chang CY et al (2014) Multiple complexes of long aliphatic N-acyltransferases lead to synthesis of 2,6-diacylated/2-acyl-substituted glycopeptide antibiotics, effectively killing vancomycin-resistant enterococcus. J Am Chem Soc 136(31):10989–10995. https://doi.org/10.1021/ja504125v

Manat G, Roure S, Auger R et al (2014) Deciphering the metabolism of undecaprenyl-phosphate: the bacterial cell-wall unit carrier at the membrane frontier. Microb Drug Resist 20(3):199–214. https://doi.org/10.1089/mdr.2014.0035

Marcone GL, Beltrametti F, Binda E et al (2010) Novel mechanism of glycopeptide resistance in the A40926 producer *Nonomuraea* sp. ATCC 39727. Antimicrob Agents Chemother 54:2465–2472. https://doi.org/10.1128/AAC.00106-10

Marcone GL, Binda E, Berini F, Marinelli F (2018) Old and new glycopeptide antibiotics: from product to gene and back in the post-genomic era. Biotechnol Adv 36(2):534–554. https://doi.org/10.1016/j.biotechadv.2018.02.009

Marshall CG, Broadhead G, Leskiw BK, Wright GD (1997) D-Ala-D-Ala ligases from glycopeptide antibiotic-producing organisms are highly homologous to the enterococcal vancomycin-resistance ligases VanA and VanB. Proc Natl Acad Sci U S A 94(12):6480–6483. https://doi.org/10.1073/pnas.94.12.6480

Marshall CG, Lessard IAD, Park IS, Wright GD (1998) Glycopeptide antibiotic resistance genes in glycopeptide-producing organisms. Antimicrob Agents Chemother 42(9):2215–2220. https://doi.org/10.1128/AAC.42.9.2215

Momota K, Kaneko I, Kimura S et al (1991) Inhibition of human immunodeficiency virus type-1-induced syncytium formation and cytopathicity by complestatin. Biochem Biophys Res Commun 179:243–250. https://doi.org/10.1016/0006-291X(91)91361-F

Mulichak AM, Losey HC, Lu W et al (2003) Structure of the TDP-epi-vancosaminyltransferase GtfA from the chloroeremomycin biosynthetic pathway. Proc Natl Acad Sci U S A 100 (16):9238–9243. https://doi.org/10.1073/pnas.1233577100

Mulichak AM, Lu W, Losey HC et al (2004) Crystal structure of vancosaminyltransferase GtfD from the vancomycin biosynthetic pathway: interactions with acceptor and nucleotide ligands. Biochemistry 43(18):5170–5180. https://doi.org/10.1021/bi036130c

Mulyani S, Egel E, Kittel C et al (2010) The thioesterase Bhp is involved in the formation of β-hydroxytyrosine during balhimycin biosynthesis in *Amycolatopsis balhimycina*. ChemBioChem 11:266–271. https://doi.org/10.1002/cbic.200900600

Naruse N, Oka M, Konishi M, Oki T (1993a) New antiviral antibiotics, kistamicins a and b II. Structure determination. J Antibiot (Tokyo) 46:1812–1818. https://doi.org/10.7164/antibiotics.46.1812

Naruse N, Tenmyo O, Kobaru S et al (1993b) New antiviral antibiotics, kistamicins a and b I. taxonomy, production, isolation, physico-chemical properties and biological activities. J Antibiot (Tokyo) 46:1804–1811. https://doi.org/10.7164/antibiotics.46.1804

Nazari B, Forneris CC, Gibson MI et al (2017) *Nonomuraea* sp. ATCC 55076 harbours the largest actinomycete chromosome to date and the kistamicin biosynthetic gene cluster. Medchemcomm 8:780–788. https://doi.org/10.1039/c6md00637j

Nicolaou KC, Boddy CNC, Bräse S, Winssinger N (1999) Chemistry, biology, and medicine of the glycopeptide antibiotics. Angew Chem Int Ed Engl 38(15):2096–2152. https://doi.org/10.1002/(sici)1521-3773(19990802)38:15<2096::aid-anie2096>3.0.co;2-f

O'Brien DP, Kirkpatrick PN, O'Brien SW et al (2000) Expression and assay of an N-methyltransferase involved in the biosynthesis of a vancomycin group antibiotic. Chem Commun 1:103–104. https://doi.org/10.1039/a907953j

Owen JG, Reddy BVB, Ternei MA et al (2013) Mapping gene clusters within arrayed metagenomic libraries to expand the structural diversity of biomedically relevant natural products. Proc Natl Acad Sci U S A 110(29):11797–11802. https://doi.org/10.1073/pnas.1222159110

Parenti F, Cavalleri B (1989) Proposal to name the vancomycin-ristocetin like glycopeptides as dalbaheptides. J Antibiot 42:1882–1883. https://doi.org/10.7164/antibiotics.42.1882

Pelzer S, Süßmuth R, Heckmann D et al (1999) Identification and analysis of the balhimycin biosynthetic gene cluster and its use for manipulating glycopeptide biosynthesis in *Amycolatopsis mediterranei* DSM5908. Antimicrob Agents Chemother 43(7):1565–1573. https://doi.org/10.1128/AAC.43.7.1565

Peschke M, Brieke C, Cryle MJ (2016a) F-*O*-G ring formation in glycopeptide antibiotic biosynthesis is catalysed by OxyE. Sci Rep 6:35584. https://doi.org/10.1038/srep35584

Peschke M, Gonsior M, Süssmuth RD, Cryle MJ (2016b) Understanding the crucial interactions between cytochrome P450s and non-ribosomal peptide synthetases during glycopeptide antibiotic biosynthesis. Curr Opin Struct Biol 41:46–53. https://doi.org/10.1016/j.sbi.2016.05.018

Peschke M, Haslinger K, Brieke C et al (2016c) Regulation of the P450 oxygenation cascade involved in glycopeptide antibiotic biosynthesis. J Am Chem Soc 138:6746–6753. https://doi.org/10.1021/jacs.6b00307

Peschke M, Brieke C, Goode RJA et al (2017) Chlorinated glycopeptide antibiotic peptide precursors improve cytochrome p450-catalyzed cyclization cascade efficiency. Biochemistry 56(9):1239–1247. https://doi.org/10.1021/acs.biochem.6b01102

Peschke M, Brieke C, Heimes M, Cryle MJ (2018) The thioesterase domain in glycopeptide antibiotic biosynthesis is selective for cross-linked aglycones. ACS Chem Biol 13:110–120. https://doi.org/10.1021/acschembio.7b00943

Pfeifer V, Nicholson GJ, Ries J et al (2001) A polyketide synthase in glycopeptide biosynthesis. The biosynthesis of the non-proteinogenic amino acid (S)-3,5-dihydroxyphenylglycine. J Biol Chem 276:38370–38377. https://doi.org/10.1074/jbc.M106580200

Pootoolal J, Thomas MG, Marshall CG et al (2002) Assembling the glycopeptide antibiotic scaffold: the biosynthesis of A47934 from *Streptomyces toyocaensis* NRRL15009. Proc Natl Acad Sci U S A 99:8962–8967. https://doi.org/10.1073/pnas.102285099

Puk O, Huber P, Bischoff D et al (2002) Glycopeptide biosynthesis in *Amycolatopsis mediterranei* DSM5908: function of a halogenase and a haloperoxidase/perhydrolase. Chem Biol 9 (2):225–235. https://doi.org/10.1016/s1074-5521(02)00101-1

Puk O, Bischoff D, Kittel C et al (2004) Biosynthesis of chloro-β-hydroxytyrosine, a nonproteinogenic amino acid of the peptidic backbone of glycopeptide antibiotics. J Bacteriol 186:6093–6100. https://doi.org/10.1128/JB.186.18.6093-6100.2004

Recktenwald J, Shawky R, Puk O et al (2002) Nonribosomal biosynthesis of vancomycin-type antibiotics: a heptapeptide backbone and eight peptide synthetase modules. Microbiology (Reading) 148(Pt 4):1105–1118. https://doi.org/10.1099/00221287-148-4-1105

Röttig M, Medema MH, Blin K et al (2011) NRPSpredictor2—a web server for predicting NRPS adenylation domain specificity. Nucleic Acids Res 39:W362–W367. https://doi.org/10.1093/nar/gkr323

Sandercock AM, Charles EH, Scaife W et al (2001) Biosynthesis of the di-meta-hydroxyphenylglycine constituent of the vancomycin-group antibiotic chloroeremomycin. Chem Biol 7(12):931–942. https://doi.org/10.1016/s1074-5521(00)00043-0

Sarkar P, Yarlagadda V, Ghosh C, Haldar J (2017) A review on cell wall synthesis inhibitors with an emphasis on glycopeptide antibiotics. Medchemcomm 8(3):516–533. https://doi.org/10.1039/c6md00585c

Schäfer M, Schneider TR, Shieldrick GM (1996) Crystal structure of vancomycin. Structure 4:1509–1515. https://doi.org/10.1016/S0969-2126(96)00156-6

Schäfer M, Sheldrick GM, Schneider TR, Vértesy L (1998) Structure of balhimycin and its complex with solvent molecules. Acta Crystallogr D Biol Crystallogr 54:175–183. https://doi.org/10.1107/S0907444997008895

Seo SY, Yun BS et al (2001) Complestatin is a noncompetitive peptide antagonist of N-methyl-D-aspartate and α-amino-3-hydroxy-5-methyl-4-isoxazolepropionic acid/kainate receptors: secure blockade of ischemic neuronal death. J Pharmacol Exp Ther 299(1):377–384

Shang J, Ye G, Shi K et al (2020) Structural basis of receptor recognition by SARS-CoV-2. Nature 581:221–224. https://doi.org/10.1038/s41586-020-2179-y

Shawky RM, Puk O, Wietzorrek A et al (2007) The border sequence of the balhimycin biosynthesis gene cluster from *Amycolatopsis balhimycina* contains *bbr*, encoding a StrR-like pathway-specific regulator. J Mol Microbiol Biotechnol 13:76–88. https://doi.org/10.1159/000103599

Shi R, Lamb SS, Zakeri B et al (2009) Structure and function of the glycopeptide N-methyltransferase MtfA, a tool for the biosynthesis of modified glycopeptide antibiotics. Chem Biol 16(4):401–410. https://doi.org/10.1016/j.chembiol.2009.02.007

Shi R, Munger C, Kalan L et al (2012) Sulfonation of glycopeptide antibiotics by sulfotransferase StaL depends on conformational flexibility of aglycone scaffold. Proc Natl Acad Sci U S A 109 (29):11824–11829. https://doi.org/10.1073/pnas.1205377109

Silhavy TJ, Kahne D, Walker S (2010) The bacterial cell envelope. Cold Spring Harb Perspect Biol 2(5):a000414. https://doi.org/10.1101/cshperspect.a000414

Smith TJ, Blackman SA (2000) Autolysins of *Bacillus subtilis*: multiple enzymes with multiple functions. Microbiology 146:249–262. https://doi.org/10.1099/00221287-146-2-249

Sosio M, Stinchi S, Beltrametti F et al (2003) The gene cluster for the biosynthesis of the glycopeptide antibiotic A40926 by *Nonomuraea* species. Chem Biol 10:541–549. https://doi.org/10.1016/S1074-5521(03)00120-0

Spohn M, Kirchner N, Kulik A et al (2014) Overproduction of ristomycin a by activation of a silent gene cluster in *Amycolatopsis japonicum* MG417-CF17. Antimicrob Agents Chemother 58:6185–6196. https://doi.org/10.1128/AAC.03512-14

Stegmann E, Frasch HJ (2010) Glycopeptide biosynthesis in the context of basic cellular functions. Curr Opin Microbiol 13:595–602. https://doi.org/10.1016/j.mib.2010.08.011

Stegmann E, Pelzer S, Bischoff D et al (2006) Genetic analysis of the balhimycin (vancomycin-type) oxygenase genes. J Biotechnol 124(4):640–653. https://doi.org/10.1016/j.jbiotec.2006.04.009

Tanaka H, Matsuzaki K, Nakashima H et al (1997) Chloropeptins, new anti-HIV antibiotics inhibiting GP120-CD4 binding from *Streptomyces* sp. I. Taxonomy, fermentation, isolation, and physico-chemical properties and biological activities. J Antibiot 50:58–65. https://doi.org/10.7164/antibiotics.50.58

Thaker MN, Wang W, Spanogiannopoulos P et al (2013) Identifying producers of antibacterial compounds by screening for antibiotic resistance. Nat Biotechnol 31:922–927. https://doi.org/10.1038/nbt.2685

Thykaer J, Nielsen J, Wohlleben W et al (2010) Increased glycopeptide production after overexpression of shikimate pathway genes being part of the balhimycin biosynthetic gene cluster. Metab Eng 12:455–461. https://doi.org/10.1016/j.ymben.2010.05.001

Truman AW, Robinson L, Spencer JB (2006) Identification of a deacetylase involved in the maturation of teicoplanin. ChemBioChem 7(11):1670–1675. https://doi.org/10.1002/cbic.200600308

Truman AW, Kwun MJ, Cheng J et al (2014) Antibiotic resistance mechanisms inform discovery: identification and characterization of a novel *Amycolatopsis* strain producing ristocetin. Antimicrob Agents Chemother 58:5687–5695. https://doi.org/10.1128/AAC.03349-14

Uribe-Herranz M, Rafail S, Beghi S et al (2020) Gut microbiota modulate dendritic cell antigen presentation and radiotherapy-induced antitumor immune response. J Clin Invest 130:466–479. https://doi.org/10.1172/JCI124332

Vollmer W, Blanot D, De Pedro MA (2008) Peptidoglycan structure and architecture. FEMS Microbiol Rev 32(2):149–167. https://doi.org/10.1111/j.1574-6976.2007.00094.x

Waglechner N, McArthur AG, Wright GD (2019) Phylogenetic reconciliation reveals the natural history of glycopeptide antibiotic biosynthesis and resistance. Nat Microbiol 4:1862–1871. https://doi.org/10.1038/s41564-019-0531-5

Wang Y, Cui R, Li G et al (2016) Teicoplanin inhibits Ebola pseudovirus infection in cell culture. Antivir Res 125:1–7. https://doi.org/10.1016/j.antiviral.2015.11.003

Wang G, Yang ML, Duan ZL et al (2021) Dalbavancin binds ACE2 to block its interaction with SARS-CoV-2 spike protein and is effective in inhibiting SARS-CoV-2 infection in animal models. Cell Res 31:17–24. https://doi.org/10.1038/s41422-020-00450-0

Wu F, Zhao S, Yu B et al (2020) A new coronavirus associated with human respiratory disease in China. Nature 579:265–269. https://doi.org/10.1038/s41586-020-2008-3

Xu F, Wu Y, Zhang C et al (2019) A genetics-free method for high-throughput discovery of cryptic microbial metabolites. Nat Chem Biol 15:161–168. https://doi.org/10.1038/s41589-018-0193-2

Xu M, Wang W, Waglechner N et al (2020) GPAHex-A synthetic biology platform for Type IV–V glycopeptide antibiotic production and discovery. Nat Commun 11(1):5232. https://doi.org/10.1038/s41467-020-19138-5

Yim G, Kalan L, Koteva K et al (2014) Harnessing the synthetic capabilities of glycopeptide antibiotic tailoring enzymes: characterization of the UK-68, 597 biosynthetic cluster. ChemBioChem 15:2613–2623. https://doi.org/10.1002/cbic.201402179

Yim G, Wang W, Thaker MN et al (2016) How to make a glycopeptide: a synthetic biology approach to expand antibiotic chemical diversity. ACS Infect Dis 2(9):642–650. https://doi.org/10.1021/acsinfecdis.6b00105

Yim G, Wang W, Pawlowski AC, Wright GD (2018) Trichlorination of a teicoplanin-type glycopeptide antibiotic by the halogenase StaI evades resistance. Antimicrob Agents Chemother 62(12):e01540–e01518. https://doi.org/10.1128/AAC.01540-18

Yushchuk O, Ostash B, Pham TH et al (2016) Characterization of the post-assembly line tailoring processes in teicoplanin biosynthesis. ACS Chem Biol 11:2254–2264. https://doi.org/10.1021/acschembio.6b00018

Yushchuk O, Horbal L, Ostash B et al (2019) Regulation of teicoplanin biosynthesis: refining the roles of *tei* cluster-situated regulatory genes. Appl Microbiol Biotechnol 103:4089–4102. https://doi.org/10.1007/s00253-019-09789-w

Yushchuk O, Andreo-Vidal A, Marcone GL et al (2020a) New molecular tools for regulation and improvement of A40926 glycopeptide antibiotic production in *Nonomuraea gerenzanensis* ATCC 39727. Front Microbiol 11:8. https://doi.org/10.3389/fmicb.2020.00008

Yushchuk O, Binda E, Marinelli F (2020b) Glycopeptide antibiotic resistance genes: distribution and function in the producer actinomycetes. Front Microbiol 11:1173. https://doi.org/10.3389/fmicb.2020.01173

Yushchuk O, Ostash B, Truman AW et al (2020c) Teicoplanin biosynthesis: unraveling the interplay of structural, regulatory, and resistance genes. Appl Microbiol Biotechnol 104: 3279–3291. https://doi.org/10.1007/s00253-020-10436-y

Yushchuk O, Vior NM, Andreo-Vidal A et al (2021) Genomic-led discovery of a novel glycopeptide antibiotic by *Nonomuraea coxensis* DSM 45129. ACS Chem Biol 16(5):915–928. https://doi.org/10.1021/acschembio.1c00170

Zeng D, Debabov D, Hartsell TL et al (2016) Approved glycopeptide antibacterial drugs: mechanism of action and resistance. Cold Spring Harb Perspect Med 6(12):a026989. https://doi.org/10.1101/cshperspect.a026989

Zhou N, Pan T, Zhang J et al (2016) Glycopeptide antibiotics potently inhibit cathepsin l in the late endosome/lysosome and block the entry of ebola virus, middle east respiratory syndrome coronavirus (MERS-CoV), and severe acute respiratory syndrome coronavirus (SARS-CoV). J Biol Chem 291:9218–9232. https://doi.org/10.1074/jbc.M116.716100

17 Biotechnological Aspects of Siderophore Biosynthesis by Actinobacteria

Artur Maier, Carolin Mügge, and Dirk Tischler

Abstract

Actinobacteria (Actinomycetes) comprise diverse and ancient phyla of the bacteria with important properties for biotechnology. Large genomes and (mega)-plasmids pose a gigantic reservoir of genetic information resulting in interesting and often powerful catabolic and anabolic pathways. Hence, many actinobacteria are able to utilize all kinds of carbon sources and have either evolved specific pathways or perform co-metabolism. In addition, their anabolic capabilities are enormous. Interestingly, natural products, some already being exploited by industries and produced in large amounts, are primary or secondary metabolites of actinobacteria. Among those natural products, the class of siderophores is gaining increasing attention in recent years. These compounds have the ability to coordinate iron or other metal and metalloid ions, earning themselves the name metallophores. Their natural role is to mobilize iron and selected other metal ions to supply the producing and secreting organism with nutrients. Thus, they allow microbes to colonize new habitats or maintain their metabolic activity even under limiting conditions. Siderophores provide access to various applications, and herein, we will highlight and discuss some prominent but also emerging examples, such as phytomining, soil remediation, and medical or imaging applications.

Keywords

Metallophore · Chelator · Natural products · Metal mobilization · Actinomycetes · Enzymology · *N*-hydroxylase · Monooxygenase · Decarboxylase

A. Maier · C. Mügge · D. Tischler (✉)
Microbial Biotechnology, Ruhr-Universität Bochum, Bochum, Germany
e-mail: artur.maier@rub.de; carolin.muegge@rub.de; dirk.tischler@rub.de

R. V. Rai, J. A. Bai (eds.), *Natural Products from Actinomycetes*,
https://doi.org/10.1007/978-981-16-6132-7_17

Abbreviations

2,3-DHBL	2,3-dihydroxybenzoate-AMP-ligase
2,3-diDHB	2,3-dihydro-2,3-dihydroxybenzoate
2,3-diHA	2,3-dihydro-3-hydroxyanthranilate
ACAD	Acyl-CoA dehydrogenase
ACS	Acyl-CoA synthase
ADIC	2-amino-2-deoxyisochorismate
ArCP	Aryl carrier protein
AS	Anthranilate synthase
DFO	Desferrioxamine
DHB	2,3-dihydroxybenzoate
FAD	Flavin adenine dinucleotide
FRET	Fluorescence resonance energy transfer
HA	3-hydroxyanthranilate
HAC	*N*-hydroxy-*N*-acetylcadaverine
hCad	*N*-hydroxycadaverine
hOrn	*N*-hydroxyornithine
HSC	*N*-hydroxy-*N*-succinylcadaverine
IC	Isochorismatase
ICS	Isochorismate synthase
IM	Inner membrane
LDC	Lysine decarboxylase
MA-DFO	*N*-methylanthranyl desferrioxamine
MIC	Minimum inhibitory concentration
MM	Minimal medium
$NAD(P)^+$	Nicotinamide adenine dinucleotide (phosphate), oxidized form
NAD(P)H	Nicotinamide adenine dinucleotide (phosphate), reduced form
NIS	NRPS-independent synthase
NMO	*N*-hydroxylating monooxygenase
NRPS domains	A: Adenylation; C: Condensation; E: Epimerization; TE: Thioesterase; PCP: Peptidyl carrier protein
NRPS	Non-ribosomal peptide synthase
OM	Outer membrane
PCR	Polymerase chain reaction
PET	Positron emission spectroscopy
PKS domains	ACP: Acyl carrier protein; AT: Acyltransferase; DH: Dehydratase; ER: Enoyl reductase; KR: β-ketoreductase; KS: Ketosynthase
PKS	Polyketide synthase
PLP	Pyridoxal phosphate
proCP	Proline carrier protein
ROS	Reactive oxygen species
SCI	Spinal cord injuries

17.1 Siderophores as Natural Products from Actinobacteria

Actinobacteria, also designated as Actinomycetes, are a diverse and ancient phylum of bacteria (Tischler et al. 2019). Often, the representatives are recognized by their beauty of colony morphology or spectacular growth properties. Due to the large diversity and long evolutionary history, it is just reasonable that actinobacteria have conquered various habitats and thus have developed manifold lifestyles and a huge arsenal of metabolic activities. Hence, we refer to them as "metabolic powerhouses." This broad spectrum of physiological properties is based on a highly complex and fluid genomic background, represented by frequently found huge genomes, megaplasmids, gene redundancy, and horizontal gene transfer (Bottacini et al. 2015; McLeod et al. 2006; Tischler et al. 2009, 2010, 2013; Roberts et al. 2011; Riebel et al. 2012; Gröning et al. 2014; Riedel et al. 2015a, b; Nguyen et al. 2017; Chen et al. 2018; Gran-Scheuch et al. 2018; Ventura et al. 2007). The mostly Gram-positive or Gram-variable microorganisms have typically a high GC content above 60% (ranges from 50 to more than 70%) (Stackebrandt and Schumann 2006; Ventura et al. 2007; Lawson 2018). This can make it hard to work directly with them on a molecular biological approach or provide issues during (meta)genome and transcriptome studies. Especially, the large linear chromosomes or megaplasmids can provide serious hurdles whenever the needed information is located at one of the termini, which are hardly available for PCR or related methods. However, especially these megaplasmids function as "storage device" and provide access to often redundant genetic information (König et al. 2004; Medema et al. 2010; Wagenknecht et al. 2010; Bottacini et al. 2015; Lewin et al. 2016). Among those, many complex-degradative or biosynthetic routes are described. Hence, the metabolic power stays in direct connection to large genomes and especially to DNA-based storage devices. This plethora of genetic diversity provides access to many biotechnologically relevant systems such as enzymes (Chen et al. 2018; Tischler et al. 2009, 2010, 2013, 2019; Riebel et al. 2012; Riedel et al. 2015a, b; Zimmerling et al. 2017; Roberts et al. 2011; Hou et al. 2012; Jin et al. 2007), amino acid production capabilities (Poetsch et al. 2011; Goldbeck et al. 2018; Pérez-García et al. 2018), various secondary metabolites and pigments (Niu et al. 2016; Rostami et al. 2016; Senges et al. 2018; Retamal-Morales et al. 2018a, b, 2021), biosurfactants (Retamal-Morales et al. 2018a; Franzetti et al. 2010; Tischler et al. 2013; Pacheco et al. 2010), and polymer and organic pollutant-degrading members (Linos et al. 1999; Heine et al. 2018; McLeod et al. 2006; Kim et al. 2018; Kumaran et al. 2020). Among those, the prediction and use of secondary metabolite-related gene clusters became more and more important or prominent in the last years. Due to increasing incidences of pathogenic microorganisms developing (multiple) resistances against various antibiotics, the demand for novel pharmaceutical bioactive compounds to treat infections is rising. Hence, this is of industrial interest and an ongoing field of research. Nevertheless, we provide here a view on a class of those secondary metabolites, especially from actinobacteria: siderophores.

17.1.1 Secondary Metabolites from Actinobacteria

Secondary metabolites are defined as low molecular weight natural products and synthesized by plants, fungi, and microorganisms. They all have in common that they are not produced all the time and are not essential for the organisms' survival. Rather, their production is initiated by certain external or internal triggers, such as limitation or stress. Hence, growth-affecting conditions (excess of carbon source, N or S limitations, metal starvation, presence of toxic compounds or elements) are employed to (over)produce these molecules directly with the wild-type strains. Often, their metabolic or ecological role is unclear and not well understood. In the last years, with the vast sequencing surveys of (micro)organisms and environmental DNA, respective gene clusters encoding for such secondary metabolite production machineries are detected or predicted. The products thereof can mostly only be proposed and were described in few cases. This presents us with a large playground to uncover novel molecules. Furthermore, many of the secondary metabolites are of industrial interest (Table 17.1). Here, especially the powerful antibiotics produced by *Streptomyces* species have to be mentioned.

Often, these molecules have multiple roles in the environmental context and thus can have different applications in industrial biotechnology as well. For example, some antibiotics produced by actinobacteria are siderophores/metallophores and alter, for example, the iron availability in order to fight against other organisms in the same habitat. Hence, they are important for iron uptake for the producers and to stress competitors at the same time. In some cases, they even have an antimicrobial activity (Ribeiro and Simões 2019). In consequence, we can classify siderophores as secondary metabolites and in some cases as bioactive compounds.

17.1.2 Siderophores

Iron and some other metals or metalloids are essential to maintain cellular functionality and are available in many habitats for organisms to utilize. However, due to environmental changes or competition between various organisms, these important elements can become a limited resource. This typically triggers a cellular response as described above for secondary metabolites in general. Organisms have evolved various strategies to acquire metals from various sources such as minerals, metallo-organic complexes, or even other organisms. A common strategy allowing to fight iron limitation is the formation of designated secondary metabolites, siderophores (Hider and Kong 2010; Hofmann et al. 2020; Kramer et al. 2020). The term is based on the Greek words "sidero" and "phore" for iron and carrying, respectively. These molecules (MW: ~500 to 1500 daltons) are typically formed in the absence of iron, or, if specificity is variable, other elements can trigger siderophore synthesis, too. They have common features such as variable backbone structures and chelating functional groups, e.g., phenolate, catecholate, hydroxamate, or carboxylate. Some siderophores carry various functional groups. The combination of backbone and functional groups defines the possible interaction

Table 17.1 Examples of secondary metabolites from actinobacteria

Molecule(s)	Example(s)	Application(s)	Reference(s)
β-Diketones	Asenjonamides Gilvocarcin	Bioactives, antimicrobial activities	Abdelkader et al. (2018); Hou et al. (2012)
Spiroketals	Calcimycin Spirofungin Rubromycins	Antimicrobial, antitumoral, and ionophoric activities	Wu et al. (2011); Chen et al. (2014); Sperry et al. (2010)
Lactones	Actinomycin	Various bioactivities	Rateb et al. (2011); Khalil et al. (2017); Rathod et al. (2018); Compton et al. (2013); Dhakal et al. (2019)
Siderophores	Desferrioxamines	Various, see below	He and Xie (2011); Hider and Kong (2010); Hofmann et al. (2021); Kramer et al. (2020)
Spirotetronates	Abyssomycins Maklamicin	Antimicrobial and antitumoral activities	Braddock and Theodorakis (2019); Lacoske and Theodorakis (2015); Igarashi et al. (2011); Niu et al. (2011)
Aminoglycosides	Streptomycin Kanamycin	Antimicrobial activities	Vakulenko and Mobashery (2003); Chang et al. (2010)
Terpenoids	Albaflavenone Isozizaene Pentalenolactone	Bioactives	Gürtler et al. (1994); Citron et al. (2012); Singh and Dubey (2020); Hussein and El-Anssary (2019)
Trehalose lipids		Biosurfactants in remediation and oil industry	Franzetti et al. (2010)

with ions and thus the chelation power. Siderophores can often bind iron and other ions. In few cases, they prefer other elements over iron; hence, these compounds are either referred to as metallophores (general term) or with more specific terms like zincophores. Siderophores are formed by all kinds of bacteria, plants, and fungi. However, especially actinobacteria comprise large genomes and often multiple sets of genes to allow the production siderophores, and thus, often a single strain produces multiple molecules acting as natural metal/metalloid chelator.

17.1.2.1 Selected Siderophores from Actinobacteria

The diversity of actinobacteria and their enormous genetic potential provide access to numerous secondary metabolites and so also toward various natural metal/metalloid chelators. A list of prominent and well-studied metallophores is given in Table 17.2, and some related structures are presented in Fig. 17.1.

The structural versatility of siderophore molecules reflects the diversity of siderophore-producing organisms. The individual enzyme machineries of the producing organisms' biosynthetic routes and also their varying substrate tolerance result in numerous structural motifs that are combined to manifold molecules. Here, the NRPS (non-ribosomal peptide synthetases), NIS (NRPS-independent

Table 17.2 Described siderophores/metallophores from actinobacteria

Siderophore	Structural characteristics	Pathway	Organisms	Source
Amychelin	Hydroxamate–phenolate	NRPS	*Amycolatopsis* sp.	Seyedsayamdost et al. (2011)
Attinimicin	Hydroxamate–phenolate	NRPS	*Pseudonocardia* spp.	Fukuda et al. (2021)
Cahuitamycin A–C	Hydroxamate–phenolate	NRPS	*Streptomyces griseus*	Park et al. (2016)
Calcimycin	Benzoxazole–pyrrole	PKS	*Streptomyces chartreusis*	Wu et al. (2011)
Coelichelin	Hydroxamate	NRPS	*Streptomyces coelicolor*	Challis and Ravel (2000)
Desferrioxamine B	Hydroxamate	NIS	*Streptomyces coelicolor, Streptomyces pilosus, Gordonia rubripertincta Nocardioides simplex*	Barona-Gómez et al. (2004); Hofmann et al. (2021); Schupp et al. (1987); Schwabe et al. (2020); Ronan et al. (2018)
Desferrioxamine G_1 and E	Hydroxamate	NIS	*Streptomyces coelicolor, Gordonia rubripertincta Nocardioides simplex*	Hofmann et al. (2021); Imbert et al. (1995); Schwabe et al. (2020)
Enterobactin	Catecholate	NRPS	*Streptomyces* sp.	Fiedler et al. (2001)
Foroxymithine	Hydroxamate	NRPS	*Streptomyces nitrosporeus*	Dolence and Miller (1991)
Griseobactin	Catecholate	NRPS	*Streptomyces griseus*	Patzer and Braun (2010)
Heterobactin A	Hydroxamate–catecholate	NRPS	*Rhodococcus erythropolis*	Bosello et al. (2013); Retamal-Morales et al. (2021)
Madurastatins	Hydroxamate	NRPS	*Actinomadura* sp.	Mazzei et al. (2012); Yan et al. (2019)
Mycobactins Exochelins	Hydroxamate–phenolate	NRPS/ PKS	*Mycobacterium tuberculosis, Mycobacterium marinum*	McMahon et al. (2012); Knobloch et al. (2020); Gobin et al. (1995); Harrison et al. (2006)
Nocardichelin A and B	Hydroxamate–phenolate	NIS	*Nocardia* spp., *Gordonia* sp.	Schneider et al. (2007); Santos et al. (2019)
Rhequichelin	Hydroxamate	NRPS	*Rhodococcus equi*	Miranda-CasoLuengo et al. (2012)
Rhodobactin	Hydroxamate–catecholate	NRPS	*Rhodococcus rhodochrous*	Dhungana et al. (2007)
Rhodochelin	Hydroxamate–catecholate	NRPS	*Rhodococcus jostii*	Bosello et al. (2011)

(continued)

Table 17.2 (continued)

Siderophore	Structural characteristics	Pathway	Organisms	Source
Salinichelins	Hydroxamate	NRPS	*Salinispora* sp.	Bruns et al. (2018)
Thermochelin	Hydroxamate	NRPS	*Thermocrispum agreste*	Heine et al. (2017)

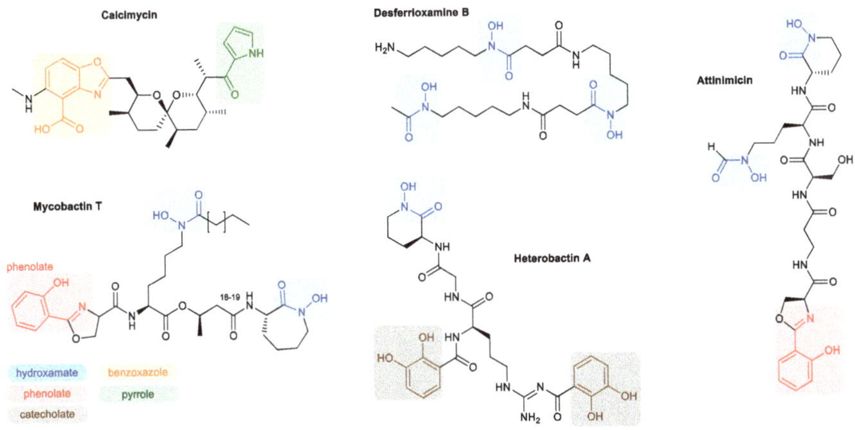

Fig. 17.1 Structural examples of actinobacterial siderophores, calcimycin (Wu et al. 2011), desferrioxamine B (Ronan et al. 2018), mycobactin T (Harrison et al. 2006), heterobactin A (Bosello et al. 2013), and attinimicin (Fukuda et al. 2021). Despite their structural versatility, they share common structural motifs, as highlighted by color code

siderophore), and PKS (polyketide synthases) routes have to be mentioned. In few cases, e.g., mycobactin and exochelin, some elements of two biosynthetic routes are combined. In the following, we exemplarily highlight the three major routes along with the most important, i.e., industrially relevant, siderophore examples.

17.1.2.2 Heterobactin Biosynthesis via Non-ribosomal Peptide Synthetases (NRPSs)

Most of the siderophore classes are synthesized by non-ribosomal peptide synthetases (NRPSs), which are essentially an assembly line of specified domains. These normally comprise an adenylation (A) domain, peptidyl carrier protein (PCP) domain, and condensation (C) domain, linking amino acids through thioester intermediates (Frueh et al. 2008). Additional domains can be present in specific modules for further modification of the peptides such as epimerization, cyclization, methylation, reduction, or oxidation (Hider and Kong 2010). Heterobactin A (and related variants), a mixed-type siderophore (catecholate–hydroxamate) produced by *Rhodococcus erythropolis* among others, is the product of an NRPS pathway involving a specified epimerization (E) domain. The NRPS for heterobactin consists of three modules, each comprising an A, PCP, and C domain. Each A domain is

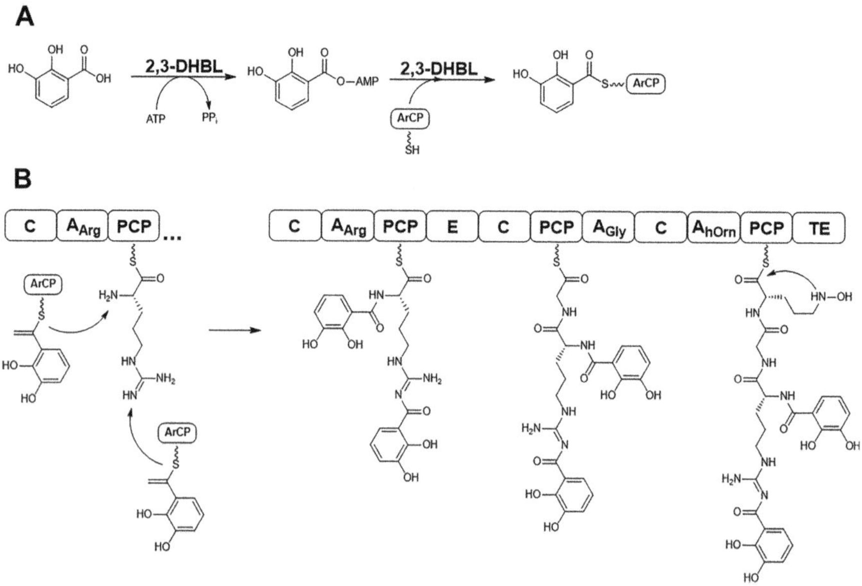

Fig. 17.2 Proposed biosynthesis pathway of heterobactin. (**a**) Activation of DHB by 2,3-dihydroxybenzoate-AMP ligase (2,3-DHBL). (**b**) NRPS-mediated assembly of the heterobactin backbone and TE domain-mediated elimination of the compound (Bosello et al. 2013)

specific for one amino acid, in this case Arg, Gly, and *N*-hydroxyornithine (hOrn), respectively. The latter is synthesized from ornithine by a flavin-dependent monooxygenase (NMO) (Mügge et al. 2020; Bufkin and Sobrado 2017). In the initial step, 2,3-dihydroxybenzoate (DHB) is activated by 2,3-dihydroxybenzoate-AMP ligase (2,3-DHBL) and transferred to the aryl carrier protein (ArCP). At the first module, Arg is bound to the PCP domain and linked through condensation with two molecules of DHB, forming amide bonds. Afterward, the Arg side chain is epimerized by the adjacent E domain. The second module elongates the tripeptide by incorporation of Gly. Additionally, it exhibits an unusual C-PCP-A arrangement instead of the more common C-A-PCP order, which is thought to orient the tripeptide and the Gly-S-PCP intermediate in closer proximity to the C domain. Eventually, the last module attaches the final hOrn building block to the tetrapeptide before the thioesterase domain (TE) cleaves the assembled product (Bosello et al. 2013). The general procedure is well known; however, there are still some open questions regarding the true nature of heterobactin A, since alternative structures were proposed (Carrano et al. 2001) and various heterobactin-like structures were reported, questioning the previous results (Retamal-Morales et al. 2021) (Fig. 17.2).

17.1.2.3 Desferrioxamine Biosynthesis via NRPS-Independent Siderophore Pathways (NISs)

In contrast to the previously mentioned NPRSs, some siderophores are the product of the NRPS-independent siderophore pathways (NISs) (Fig. 17.3). One of these siderophores is desferrioxamine (DFO), a hydroxamate type consisting of either three *N*-hydroxy-*N*-succinylcadaverine (HSC) units (DFO-G_1 and DFO-E) or two HSC and one *N*-hydroxy-*N*-acetylcadaverine (HAC) unit (DFO-B), linked by amide bonds. HSC and HAC are synthesized by an acyltransferase, which uses succinyl-CoA or acetyl-CoA, respectively, to acylate *N*-hydroxycadaverine (hCad). Subsequently, the NRPS-independent siderophore synthetase catalyzes the adenylation of the carboxylic group of HSC, followed by condensation with the amino group of another HSC. This intermediate is further adenylated and condensed with an additional HAC to give desferrioxamine B, or with an HSC to give desferrioxamine G_1. Through an additional cycle of adenylation and condensation, the latter can be converted to DFO-E (Ronan et al. 2018).

17.1.2.4 Calcimycin Biosynthesis via Polyketide Synthases (PKSs)

Calcimycin is a natural antibiotic with ionophoric properties, which is produced by *Streptomyces chartreusis* (Wu et al. 2011). In contrary to the most known siderophores, it exhibits the highest affinities toward calcium and magnesium. The compound belongs to the family of pyrrole polyethers, with a pyrrole moiety, a spiroketal ring, and a benzoxazole heterocycle. The main part of its biosynthesis is catalyzed by polyketide synthases, multimeric proteins built up from individual modules. A module contains minimum a ketosynthase (KS) domain, an acyltransferase (AT) domain, and an acyl carrier protein (ACP) domain (Khosla et al. 2014). Additionally, dehydratase (DH) domains, enoyl reductase (ER) domains, and β-ketoreductase (KR) domains can be present for further

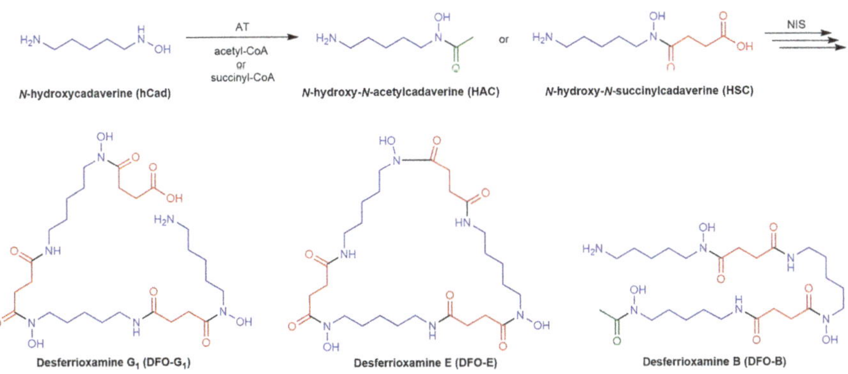

Fig. 17.3 DFO biosynthesis by NISs. *N*-hydroxycadaverine (hCad) is acylated by an acyltransferase (AT) with acetyl-CoA or succinyl-CoA to give *N*-hydroxy-*N*-acetylcadaverine (HAC) or *N*-hydroxy-*N*-succinylcadaverine (HSC), respectively. NIS-catalyzed condensations yield desferrioxamine G1 (DFO-G1), desferrioxamine E (DFO-E), or desferrioxamine B (DFO-B) (Ronan et al. 2018)

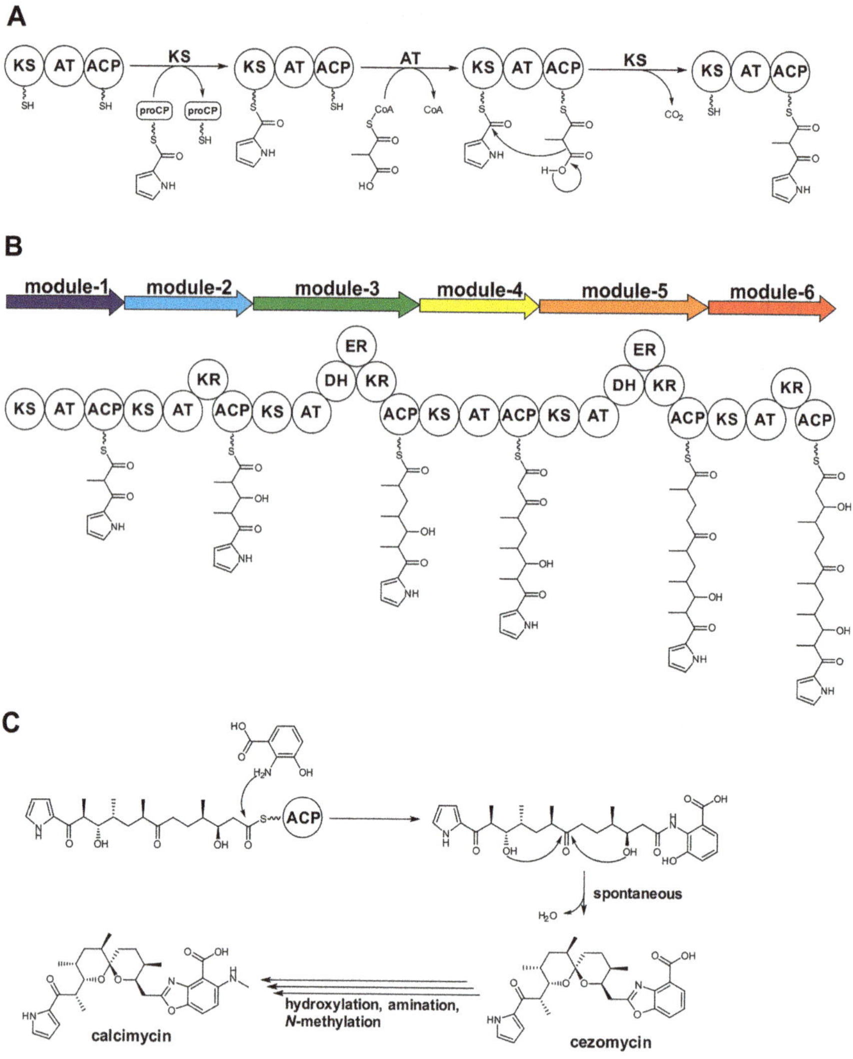

Fig. 17.4 Calcimycin biosynthesis pathway. (**a**) Initial step of the PKS-mediated calcimycin synthesis. (**b**) Module compositions and assembly process of the backbone. (**c**) Final maturation steps to yield calcimycin (Wu et al. 2011)

modification (Nivina et al. 2019). In the case of calcimycin, the responsible PKS consists of six modules of which modules 2 and 6 bear an additional KR domain and modules 3 and 5 each bear additional KR, DH, and ER domains (Fig. 17.4b). The assembly of the polyketide backbone is initiated by transfer of the pyrrole–carboxylic acid unit to the KS domain from a proline carrier protein (proCP) by transthiolation. The unit is bound to the active site cysteine residue. Next, the AT domain catalyzes the condensation reaction of methylmalonyl-CoA with the SH group of the ACP domains' pantetheinyl residue (Dunn et al. 2013). Subsequently, the KS domain initiates the decarboxylative condensation between both units to create the polyketide (Fig. 17.4a). This process is carried on by the PKS domains, which are located downstream to form the polyketide backbone (Fig. 17.4b). Depending on the presence and composition of the optional domains, the incorporated unit can be in different reduction states: The KR domain reduces the initial keto group to a hydroxy group, the DH domain creates a *trans* double bond through dehydrogenation, and the NADPH-dependent ER domain reduces the double bond to a saturated CH_2-CH_2 unit (Reid et al. 2003; Wu et al. 2011). Terminal nucleophilic attack of 3-hydroxyanthranilate on the polyketide releases the polyketide chain, and the benzoxazole moiety is generated through an acetalization reaction. Formation of the spiroketal presumably proceeds spontaneously to create the intermediate cezomycin. Further modification including hydroxylation, amination, and *N*-methylation (Wu et al. 2013, 2018) gives the final compound calcimycin (Fig. 17.4c). Overall, the structure and procedure of the PKSs resemble the previously described NRPSs (Wu et al. 2011).

17.1.2.5 Enzymes Initiating Siderophore Biosynthesis

Before the designated domains or respective enzymes can perform the siderophore biosynthesis, various enzymes have to provide the necessary precursors. These are created by conversion of common compounds such as lysine, ornithine, proline, and chorismate. One central process is the hydroxylation of amines, catalyzed by flavin-dependent *N*-hydroxylating monooxygenases (NMOs), as it occurs in the conversion of cadaverine (Cad) and ornithine (Orn) to *N*-hydroxycadaverine (hCad) and *N*-hydroxyornithine (hOrn), respectively (Fig. 17.5) (Mügge et al. 2020). These products play an important role in the NRPS and NIS pathways. Cad is produced from lysine by a pyrodoxal phosphate (PLP)-dependent lysine decarboxylase (LDC).

The compounds 2,3-dihydroxybenzoate (DHB) and 3-hydroxyanthranilate (HA), which are important building blocks for siderophores synthesized via NRPS and

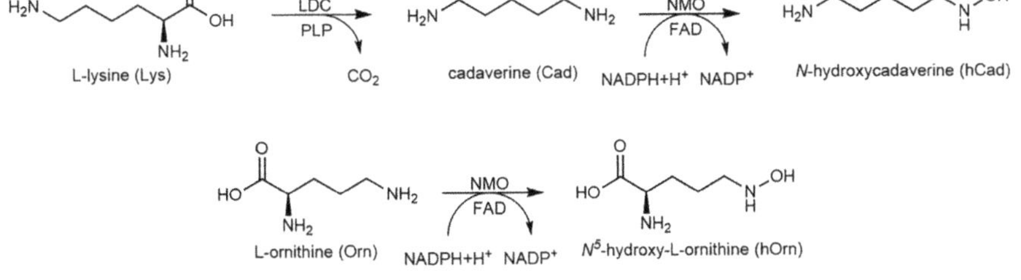

Fig. 17.5 Biosynthesis of *N*-hydroxycadaverine (hCad) and *N*-hydroxyornithine (hOrn) from L-lysine (Lys) and L-ornithine (Orn), respectively. Decarboxylation of Lys by the pyridoxal phosphate (PLP)-dependent lysine decarboxylase (LDC) gives cadaverine (Cad). *N*-hydroxylation of Cad by a FAD- and NADPH-dependent *N*-hydroxylating monooxygenase (NMO) gives hCad. Some NMOs can directly *N*-hydroxylate Orn to hOrn (Mügge et al. 2020)

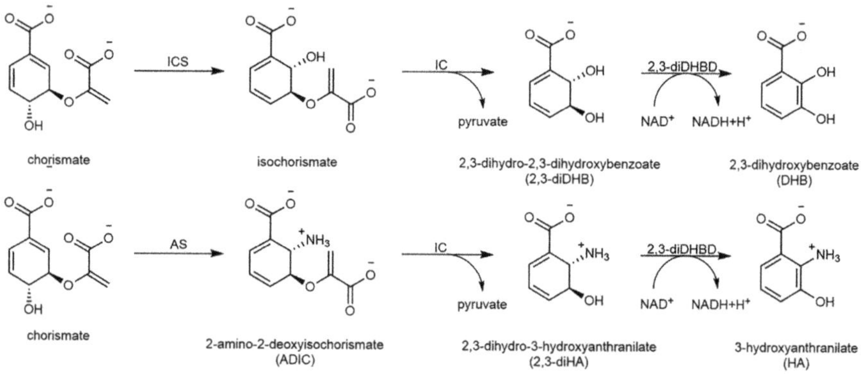

Fig. 17.6 Biosynthesis of 2,3-dihydroxybenzoate and 3-hydroxyanthranilate from chorismate. Isochorismate and 2-amino-2deoxychorismate (ADIC) are converted from chorismate by isochorismate synthase (ICS) and anthranilate synthase (AS), respectively. Elimination of pyruvate by isochorismatase (IC) and subsequent reduction by 2,3-dihydro-2,3-dihydroxybenzoate dehydrogenase (2,3-diDHBD) yields 2,3-dihydroxybenzoate (DHB) and 3-hydroxyanthranilate (HA) (Bosello et al. 2011; Wu et al. 2011; Rowland et al. 1996)

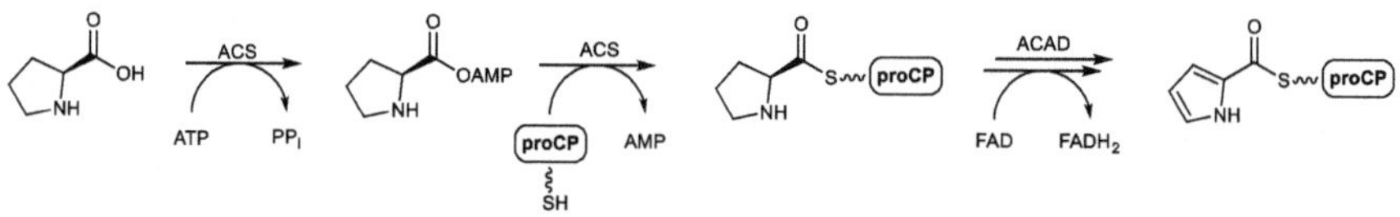

Fig. 17.7 Ketopyrrole biosynthesis from proline. Acyl-CoA synthase links the proline to the proline carrier protein (proCP) via thiolation. Oxidation by acyl-CoA dehydrogenase (ACAD) gives the ketopyrrole moiety (Wu et al. 2011)

PKS, respectively, show structural similarities and are both provided from chorismate. The pathways differ only in the initial step, which is carried out by isochorismate synthase (ICS) for DHB or by anthranilate synthase (AS) for HA. Subsequent release of pyruvate catalyzed by isochorismatase (IC) and NADH-dependent reduction in the intermediate by 2,3-dihydro-2,3-dihydroxybenzoate dehydrogenase (2,3-diDHBD) (Rowland et al. 1996) gives the final product DHB or HA (Fig. 17.6).

The ketopyrrole moiety, which is involved in PKS-mediated pathways, is synthesized from proline. Initially, acyl-CoA synthase (ACS) adenylates proline and links the intermediate to the proCP via thiolation. Final oxidation to yield the ketopyrrole moiety is catalyzed by FAD-dependent acyl-CoA dehydrogenase (Wu et al. 2011) (Fig. 17.7).

17.2 Overproduction of Siderophores

Siderophores are mostly obtained by aid of naturally producing wild-type strains. Upon iron deficiency, respective gene clusters are transcribed, leading to active biosynthetic clusters. Hence, media with low iron content or media containing chemical iron chelators are typically used to produce siderophores (Saha et al. 2016; Khan et al. 2018). There is no obvious correlation between strain, gene cluster, medium composition, cultivation conditions, and siderophore titer. Therefore, the parameters need to be studied and optimized for each strain individually. An alternative approach would be the heterologous production either in wild-type strains by artificially changed regulatory elements or by transferring gene clusters into new host strains. Only little information is available on these kinds of recombinant siderophore production routes to date. The following sections will outline the production of siderophores typically found among Actinomycetes.

Evaluation of media-dependent siderophore production in actinobacteria revealed a correlation between media richness and complexity and quantity of the produced siderophores. Production of rhodochelin and heterobactins by *Rhodococcus* spp. was induced by iron-depleted minimal medium with glucose as optional carbon source (Bosello et al. 2011, 2013). For *Streptomyces coelicolor,* various medium compositions were tested toward the production of desferrioxamines. In minimal medium, the smaller DFO-B was the most abundant, whereas in nutritional-rich medium, the more complex DFO-E had the predominant role. This was attributed to either the availability of nutrition and precursors or to an adaption to the nutritional-rich environment, in which siderophores could fulfill yet unknown functions (Senges et al. 2018). As for many siderophore-producing organisms, iron deprivation was the main factor inducing siderophore production.

A similar approach was chosen for the elucidation of siderophore production in *Gordonia rubripertincta* CWB2, which is known to produce siderophores like DFO-B among others. In complex casamino acid-based media with various supplements including $MgSO_4$, calcium, vitamin B_1, and others, siderophore concentrations of 50–85 μM were observed. The yield was further increased by the use of succinate as DFO precursor and carbon source, and avoidance of other supplements by using minimal medium. This resulted in siderophore levels of up to 400 μM. Addition of iron inhibited siderophore production (Schwabe et al. 2020). Furthermore, the impact of various metals and metalloids at a concentration of 100 μM on siderophore production was evaluated. Compared to the iron control, supplementation of Al, Co, Gd, Ge, Nd, and Zn promoted siderophore production, but growth inhibition was observed for V, Ga, and Ge. Only Nd promoted both siderophore production and growth of *Gordonia rubripertincta* (Schwabe et al. 2020).

Recombinant DFO-E and related siderophore production was successfully performed in *Escherichia coli* (*E. coli*) by introduction of the respective genes. Three genes coding for the enzymes responsible for the initial steps were taken from a marine metagenomic DNA, and the fourth gene, coding for a NIS, was from the terrestrial bacterium *Erwinia amylovora*. DFO-E titers of up to 23 mg/L were achieved with standard medium and growth conditions typically used for *E. coli* expressions (Fujita and Sakai 2013).

The impact of the pH value on siderophore production is correlating with iron solubility, since it is decreasing with rising pH due to formation of iron hydroxide on the one hand, and with optimal bacterial growth conditions and on the other. Experiments showed that marine actinobacteria *Salinispora tropica* produces more complex siderophores like DFO-B and DFO-E with increasing pH (up to pH 9.0) instead of smaller analogs, since DFO-B and DFO-E show higher binding constants to iron (Ejje et al. 2013). However, if the ideal pH for cell growth is taken into account, a neutral pH is more favorable for maximal siderophore production (Yu et al. 2017).

17.3 Application of Siderophores

With respect to siderophore application, we need to discriminate the applications in types of utilizing the pure, isolated form or those employing siderophores in a non-purified manner (Table 17.3).

In the following sections, a brief introduction is given to the most important applications of siderophores. These applications can typically be achieved with siderophores produced by various organisms. Here, we focus on those produced in actinobacteria.

17.3.1 Desferal

Targeted siderophore utilization began in the late 1960s, when desferrioxamine B, produced from *Streptomyces pilosus* among others, was approved under the name desferal in the USA for the treatment of iron and aluminum intoxication, especially

Table 17.3 Classification and examples of siderophore applications

Siderophore	Application	Mode	Reference
Calcimycin	Biochemical and pharmacological tool	Isolated	Kozian et al. (2005); Wu et al. (2011)
Desferrioxamine B	Medicine, imaging, biosensors	Isolated	Shander et al. (2009)
Desferrioxamines	Bioremediation, plant growth support, phytomining, phytoremediation	Secreted from active growing cultures	Wiche et al. (2017); Trögl et al. (2018)
Heterobactins	Bioremediation, detoxification	Secreted	Hofmann et al. (2021); Retamal-Morales et al. (2021)
Various	Metal winning	Crude extract in immobilized form	Hofmann et al. (2021)

for people suffering from thalassemia or other erythropoiesis disorders and depending on blood transfusion (Shander et al. 2009). Recently, desferal was considered as a potential treatment for spinal cord injuries (SCIs), since a newly discovered iron-dependent cell death pathway, called ferroptosis, was identified to play a major role in SCI pathologies. This is particularly interesting since there is no effective cure for SCI so far (Yao et al. 2019). Moreover, desferal is listed on the World Health Organization's list of essential medicines, underlining the importance of siderophores for the medical sector.

17.3.2 Imaging

Imaging techniques are a fundamental part of clinical diagnostics and are continuously being refined. Nevertheless, there is still potential for further improvements as the detection of bacterial infections, among others, remains a diagnostic challenge. Latest research results show that siderophores, loaded with radiolabeled metal ions, can be utilized to identify infection centers in the organism using positron emission spectroscopy (PET). It was demonstrated *in vivo* that ^{67}Ga-radiolabeled desferrioxamines were selectively taken up by *Staphylococcus aureus* in a mouse model, allowing identification and location of the infection (Ioppolo et al. 2017). Besides desferrioxamines, other siderophores were elucidated for this purpose and showed promising results (Petrik et al. 2020). For even further improvements, the conjugation of siderophores with an additional fluorescent dye was investigated to give a combination of PET and optical imaging. It was demonstrated in a mouse model that infections with *Pseudomonas aeruginosa* can be visualized by PET/CT imaging. One major drawback is the rapid elimination of labeled siderophores through the renal system, which can to some extent be overcome by structural adaption, but still limiting the diagnostics in the involved structures (Ioppolo et al. 2017; Petrik et al. 2020, 2021).

The use of siderophores for imaging is not only limited to bacterial or fungal infections. DFO radiolabeled with ^{89}Zr and conjugated to the monoclonal antibody elotuzumab gave rise to a new potential diagnostic tool for multiple myeloma. This FDA-approved antibody selectively binds to the glycoprotein CS1, which is overexpressed in multiple myeloma cells, guiding the linked and radiolabeled siderophore to the target. Affected cells can be identified and localized by PET imaging. First experiments showed the feasibility of the DFO-^{89}Zr-elotozumab as a companion diagnostic in preclinical models (Perk et al. 2010; Ghai et al. 2020).

17.3.3 Mineral Dissolution and Bioremediation

The versatile application and importance of heavy metals for industry, involving sectors like the electric vehicle market (Cusenza et al. 2019), lead to an increasing demand for metals like cobalt (Co) and manganese (Mn), which resulted in a worldwide distribution and partial contamination with rising health concerns

(He et al. 2005; Tchounwou et al. 2012). Due to their chelating abilities, which are not limited to iron, siderophores can be versatilely used for either bioremediation, metal recovery, or metal extraction by mineral dissolution (Hofmann et al. 2020).

In experiments, DFO-mediated dissolution of Co and Mn from minerals such as heterogenite, goethite, and manganite was observed (Bi et al. 2010; Parker et al. 2004). Two mechanisms, the ligand-promoted and the reductive dissolution process, are discussed. The first leads to Mn(III)/Co(III) DFO-B complexes without change in the oxidation state. The second leads to Mn(II) and Co(II) complexes through reduction in the trivalent metal centers and subsequent release of the metal ions due to the instability of the Mn(II)/Co(II) DFO-B complexes (Akafia et al. 2014).

A number of other toxic elements are being bound by siderophores and as a consequence not taken up into the bacterial cell of the respective siderophore producers (Schalk et al. 2011). Thus, they endow the organisms with a tolerance toward elements such as arsenic, lead, or mercury, as was shown for *Pseudomonas* sp., *Rhodococcus* sp., and *Streptomyces* sp., respectively (Braud et al. 2009; Dimkpa et al. 2008; Drewniak et al. 2008). However, problems arise as these mobilized toxic elements might be taken up by other bacteria, fungi, or plants, which are not producing siderophores. Hence, siderophores add to the availability of those elements, but at the same time negatively affect the environment and thus, for example, bioremediation.

Asbestos is a group of six fibrous minerals, which have been widely used in the past on industrial scale and are known to cause serious health issues as asbestosis and lung cancer upon exposure. Because of this, asbestos was banned in most developed countries. Nevertheless, many developing countries still use it on large scale. Chrysotile, belonging to the group of asbestos, is one of the commonly used and the cause of contamination of various places. The presence of iron impurities in these structures enhances the generation of reactive oxygen species (ROS) in lung tissue, potentially inducing inflammation processes. Extraction of these iron impurities by DFO siderophores significantly reduced toxicity of chrysotile (Mohanty et al. 2018).

17.3.4 Phytoremediation and Phytomining

Typically, metallophores are produced by organisms to overcome a metal limitation in their habitat or environment, and hence, they like to take or bind the metal-loaded molecules themselves. Metallophores do mobilize and chelate various metals and metalloids and therefore improve their bioavailability. It is well established that other organisms hijack these little treasures to not waste energy by generating metallophores themselves (Miethke and Marahiel 2007). This is true even over boundaries of genera and phyla or domains of life. Thus, it is not surprising that siderophores from soil bacteria are taken up by other microorganisms and by plants. Plants can sequester metals and metalloids and enrich those in their biomass through these shuttle molecules, resulting in a growth benefit (Wang et al. 1993; Palaniyandi et al. 2013). The supportive bacteria or molecules are defined as plant growth-promoting bacteria/compounds, respectively. This phenomenon can also be utilized in a more direct fashion. Certain plants have hyper-accumulating properties and bind more metal than they need for their metabolism or take up elements even if they are not of any biological need (Wiche et al. 2017). This so-called phytoextraction mediated by metallophores has two potential applications, which are currently studied by multiple laboratories: phytoremediation and phytomining (Hofmann et al. 2020).

Phytoremediation is an emerging application and makes use of the above described processes in the rhizosphere. Here, plants are seen as efficient metal traps and at the same time as carbon sink. Fixing CO_2 during photosynthesis uses sunlight in a low-energy demanding biomass production system to create energy and molecular oxygen. Thus, phytoremediation can be considered to be an environmentally friendly and cost-efficient strategy to absorb, bind, and even accumulate toxic pollutants from contaminated sites (Rajkumar et al. 2010; Prasad et al. 2010). If the biomass is collected and processed properly, it allows to clean contaminated sites from pollutants such as heavy metals and can therefore be considered as environmental detoxification strategy to improve soil quality. The plant biomass and raised microbial diversity in the treated soil lead to better growth conditions also for secondary biomass. Along with this, often metal stress is reduced and plants become more stable and are less often attacked or affected by plant pathogens.

Phytomining is in principle the same as phytoremediation from a biological perspective. However, the scope of application and also the follow-up utilization are different. Important and rare elements in soil can be accumulated by plants as described above. By selecting a suitable plant species, metal accumulation can be triggered and certain target elements, like strategic rare earth elements, are accumulated in their biomass. This can further be improved by bacteria and their secondary metabolites. As soon as the plants have accumulated a certain element amount, they can be harvested. The target metals/metalloids can then be extracted directly or from the ash of previously combusted biomass. This is an indirect mining approach toward rare elements in soil or from heaps of former mining activities (Heilmeier and Wiche 2020).

17.3.5 Biosensors

The detection of extremely small traces of bioavailable iron in aqueous systems is becoming more and more important, as iron plays a crucial role in ocean health and productivity and thus CO_2 binding (Chisholm 2000). Various chemical and analytical techniqus such as colorimetry, flame photometry, UV–Vis spectrophotometry, atomic absorption spectrometry, or electron microprobe analysis were developed in the past decades for the quantification of metal ions. However, most of these methods are expensive and time-consuming and require complex instruments (Nosrati et al. 2018). The use of siderophores as biosensors represents a promising alternative. One approach is the indicator displacement assay based on fluorescence resonance energy transfer (FRET). Here, the chemosensors' core structure consists of a fluorophore as antenna, a spacer, and a siderophore-based indicator. Upon excitation, the antenna transfers energy to the indicator, which gives a fluorescence signal as long as lanthanide metals are bound to the siderophore subunit. Displacement of the lanthanide metal by the far more stable-bound analyte, here iron, results in quenching of the fluorescence emission signal. The semisynthetic siderophore *N*-methylanthranyl-desferrioxamine (MA-DFO) loaded with a terbium ion (Tb^{3+}) allowed detection of Fe^{3+} at concentrations as low as 0.28 ng/mL (Orcutt et al. 2010).

In another approach, gold nanoparticles were coated with siderophore-inspired polymers containing a catechol terminus meant to coordinate metal ions. In a colorimetric approach, the selective detection of Fe^{3+} over Fe^{2+} and other metals was possible in the range of 8–25 μM in a simple 96-well optical readout setup (Phillips et al. 2015).

17.3.6 Bioactives (Naturally or Artificially)

Siderophores can be coupled with various bioactives such as antibiotics to overcome resistances or to reduce the minimum inhibitory concentration (MIC) (Ribeiro and Simões 2019). Microorganisms like *Streptomyces*, among others, already utilized this possibility by forming siderophore–antibiotic conjugates, so-called sideromycins. Their mechanism of action is based on the "Trojan horse" strategy. Here, the need of bacteria for Fe(III) is exploited by linking siderophores with antibacterial compounds, easing their entrance into the bacterial cell (Fig. 17.8). The first representative of this group, albomycin, was already discovered in 1947 and consists of a ferrichrome bound to an antimicrobial thionucleoside moiety (Lin et al. 2018), which was identified to be a potent seryl tRNA synthase inhibitor (Stefanska et al. 2000). In the target organism, the antibiotic is released by peptidase-mediated cleavage of the linker (Braun et al. 2009; Negash et al. 2019). Moreover, albomycin showed activity against various bacteria such as *Staphylococcus aureus* and *Streptococcus pneumoniae* (Pramanik et al. 2007). This could be especially important for the treatment of infections caused by Gram-negative bacteria such as of the genus *Acinetobacter*, *Klebsiella,* and *Pseudomonas*, as these bacteria possess an additional cell membrane, which impairs the uptake of several antibiotics.

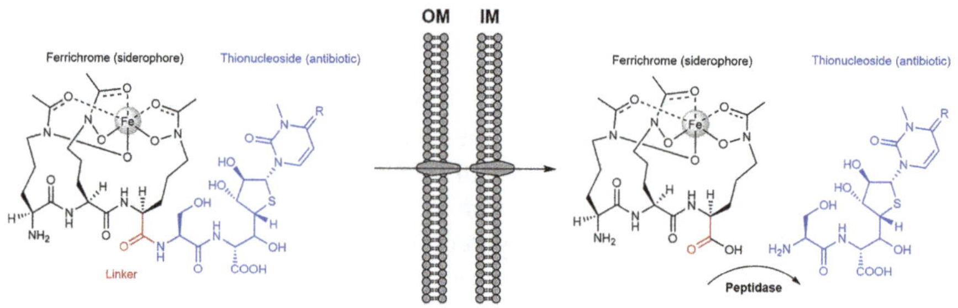

Fig. 17.8 Simplified mode of action of albomycin—"Trojan horse" (Negash et al. 2019)

Cefiderocol is the first artificial siderophore–antibiotic conjugate, which was approved by the FDA in 2019 and the European Union in 2020 for the treatment of infections caused by multidrug-resistant Gram-negative bacteria such as *Pseudomonas aeruginosa* (Parsels et al. 2021). It is a combination of a catechol-type siderophore and a cephalosporin, which belongs to the group of β-lactam antibiotics and inhibits cell wall synthesis. Furthermore, the structure is chemically enhanced by addition of oxime and dimethyl groups for β-lactamase stability and aminothiazole and pyrrolidinium to improve antibacterial activity (Sato and Yamawaki 2019).

Recently, a new siderophore, namely attinimicin, was discovered. It is produced by *Pseudonocardia* spp., which are living in symbioses with fungus-growing ant colonies in the South American rainforest. These bacteria protect the fungal food source of the ants against pathogens like *Escovopsis* spp. by secretion of various antifungal agents such as attinimicin. Attinimicin was found to have antifungal activity *in vivo* but negligible activity *in vitro*, which is attributed to its iron dependency. Surprisingly, attinimicin was only active when not bound to iron, which is uncommon among siderophores as their antimicrobial or antifungal activity is based, among others, on iron binding and iron deprivation of the target organism (Fukuda et al. 2021).

Despite its antibacterial and antifungal activity, calcimycin is primarily exploited for its ionophoric properties. Contrary to most of the other siderophores, calcimycin exhibits affinity to divalent cations like Ca^{2+} and Mg^{2+} besides Fe^{2+} and even to monovalent cations like Li^{+}, K^{+}, Na^{+}, and Rb^{+} (Pal et al. 2018). Therefore, it is widely used as a biochemical and pharmacological tool due to its activities on various cell signaling pathways (Kozian et al. 2005; Ebner et al. 2015; Torrente et al. 2014).

17.4 Conclusions and Future Perspectives

From the above outlined details, it is clear that siderophores, or better metallophores, are important secondary metabolites not only in an environmental context but also in an industrial context. Especially, the diversity of metallophores from actinobacteria leads to a plethora of applications and potentials. While already a huge diversity of these molecules is described, even more novel siderophores are continuously reported. Screening for siderophores is quite simple as it can be done on agar plates. In combination with LC-MS and database clustering, novel structural data are frequently elucidated (Hofmann et al. 2021; Mehnert et al. 2017; Alexander and Zuberer 1991; Schwyn and Neilands 1987). Often, structures and metal(loid) affinities are known, and in order to gain this information, studies in analytical scale are sufficient and do not require large quantities (Hider and Kong 2010; Hofmann et al. 2020).

The major issue with all these compounds is, however, their deliberate production, defined by often unsatisfying space time yields, which makes them unattractive or not competitive. The production of large quantities is and remains problematic and hampers or limits application or even respective studies. So far, only a few metallophores such as desferrioxamines and heterobactins can be produced at significant titers (Hofmann et al. 2021; Chiani et al. 2010) and in case of siderophores we speak of few mg per L broth. Mostly, cultivation and process parameters are optimized in order to increase these small amounts to a certain maximum. However, it is difficult to achieve higher levels. The often complex biosynthetic pathways (see Sect. 17.1.2) do not simply allow modifications of the genome (e.g., promoter regions) or recombinant cloning and (heterologous) expression. But the latest developments in genetic engineering bring forth the possibility to alter the genomic structure of actinobacteria and thus manipulate metallophore production, e.g., by CRISPR/Cas tools (Tong et al. 2018). Thus, we hope to see the targeted manipulation of good producers to efficient cellular production strains in the future. The available genetic tools can be used, for example, to provide controllable promoters, to simplify operons or combine relevant gene clusters with respect to localization and transcription control, to manipulate sensors, which (down)-regulate the level/activity, to enhance secretion while lowering uptake of metallophores, and to destroy siderophore degradation routes, among many other options.

In addition, the development of tools to modify metallophores or pathways to provide access towards new compounds, even molecule families, represents an emerging field. As it was stated above, mostly the biosynthetic routes naturally provide access to a spectrum of molecules and with suitable feeding strategies one can alter the composition to some extent. Nevertheless, there is still room for further developments. Even combination of the existing routes combined with artificial enzymes or chemical modifications can provide more versatile compounds and thus empower the use of siderophores in more fields. Here, especially, the Trojan horse strategy to create new bioactives or labels for imaging applications will become important. One probably does not need large titers for such applications, and thus, standard production followed by modification can be sufficient.

With respect to the environment, we learn more and more about the complex networks nature has evolved. With the better understanding of how secondary metabolites affect their environment, like in metal mobilization for plants by siderophores, we will learn to make use of these complex systems. This can be beneficial for agriculture in terms of increasing yields of biomass, to sustain healthy plants to fight pathogens or to colonize difficult habitats. Further, it can provide routes to reduce mobilization of toxic elements and thus to maintain healthy ecosystems. And also, phytomining is affected by these compounds so that either toxic elements or precious target elements can be collected via metallophores in higher plants for winning purposes.

With the recent progress in the siderophore research and the approval of the first siderophore conjugate for the treatment of infections, new opportunities are opened up for the pharmaceutical sector as these concepts proved to be viable at various levels. Especially, the versatility of the siderophores and their linked compounds will draw more attention and fuel further research as the potential is far from being exhausted. Hence, siderophores will become a more prominent role in pharmaceutical development.

Acknowledgements All authors were supported by the Federal Ministry for Innovation, Science and Research of North Rhine–Westphalia (PtJ-TRI/1141ng006). We thank the DECHEMA for providing a Max-Buchner Scholarship to Dirk Tischler (MBFSt 3646). Further, the research in this direction was supported by Junior Research Grant from the German Federal Ministry of Education and Research (BakSolEx 033R147).

References

Abdelkader MSA, Philippon T, Asenjo JA, Bull AT, Goodfellow M, Ebel R et al (2018) Asenjonamides A-C, antibacterial metabolites isolated from Streptomyces asenjonii strain KNN 42.f from an extreme-hyper arid Atacama Desert soil. J Antibiot 71(4):425–431. https://doi.org/10.1038/s41429-017-0012-0

Akafia MM, Harrington JM, Bargar JR, Duckworth OW (2014) Metal oxyhydroxide dissolution as promoted by structurally diverse siderophores and oxalate. Geochim Cosmochim Acta 141: 258–269. https://doi.org/10.1016/j.gca.2014.06.024

Alexander DB, Zuberer DA (1991) Use of chrome azurol S reagents to evaluate siderophore production by rhizosphere bacteria. Biol Fert Soils 12(1):39–45. https://doi.org/10.1007/BF00369386

Barona-Gómez F, Wong U, Giannakopulos AE, Derrick PJ, Challis GL (2004) Identification of a cluster of genes that directs desferrioxamine biosynthesis in Streptomyces coelicolor M145. J Am Chem Soc 126(50):16282–16283. https://doi.org/10.1021/ja045774k

Bi Y, Hesterberg DL, Duckworth OW (2010) Siderophore-promoted dissolution of cobalt from hydroxide minerals. Geochim Cosmochim Acta 74(10):2915–2925. https://doi.org/10.1016/j.gca.2010.02.028

Bosello M, Robbel L, Linne U, Xie X, Marahiel MA (2011) Biosynthesis of the siderophore rhodochelin requires the coordinated expression of three independent gene clusters in Rhodococcus jostii RHA1. J Am Chem Soc 133(12):4587–4595. https://doi.org/10.1021/ja1109453

Bosello M, Zeyadi M, Kraas FI, Linne U, Xie X, Marahiel MA (2013) Structural characterization of the heterobactin siderophores from Rhodococcus erythropolis PR4 and elucidation of their biosynthetic machinery. J Nat Prod 76(12):2282–2290. https://doi.org/10.1021/np4006579

Bottacini F, O'connell Motherway M, Casey E, McDonnell B, Mahony J, Ventura M, van Sinderen D (2015) Discovery of a conjugative megaplasmid in Bifidobacterium breve. Appl Environ Microbiol 81(1):166–176. https://doi.org/10.1128/AEM.02871-14

Braddock AA, Theodorakis EA (2019) Marine spirotetronates: biosynthetic edifices that inspire drug discovery. Mar Drugs 17(4):232. https://doi.org/10.3390/md17040232

Braud A, Hannauer M, Mislin GLA, Schalk IJ (2009) The Pseudomonas aeruginosa pyochelin-iron uptake pathway and its metal specificity. J Bacteriol 191(11):3517–3525. https://doi.org/10.1128/JB.00010-09

Braun V, Pramanik A, Gwinner T, Köberle M, Bohn E (2009) Sideromycins: tools and antibiotics. Biometals 22(1):3–13. https://doi.org/10.1007/s10534-008-9199-7

Bruns H, Crüsemann M, Letzel AC, Alanjary M, McInerney JO, Jensen PR, Schulz S, Moore BS, Ziemert N (2018) Function-related replacement of bacterial siderophore pathways. ISME J 12:320–329. https://doi.org/10.1038/ismej.2017.137

Bufkin K, Sobrado P (2017) Characterization of the ornithine hydroxylation step in albachelin biosynthesis. Molecules (Basel, Switzerland) 22(10):1652. https://doi.org/10.3390/molecules22101652

Carrano CJ, Jordan M, Drechsel H, Schmid DG, Winkelmann G (2001) Heterobactins: a new class of siderophores from Rhodococcus erythropolis IGTS8 containing both hydroxamate and catecholate donor groups. Biometals 14(2):119–125. https://doi.org/10.1023/A:1016633529461

Challis GL, Ravel J (2000) Coelichelin, a new peptide siderophore encoded by the Streptomyces coelicolor genome: structure prediction from the sequence of its non-ribosomal peptide synthetase. FEMS Microbiol Lett 187(2):111–114. https://doi.org/10.1111/j.1574-6968.2000.tb09145.x

Chang C-WT, Fosso M, Kawasaki Y, Shrestha S, Bensaci MF, Wang J et al (2010) Antibacterial to antifungal conversion of neamine aminoglycosides through alkyl modification. Strategy for reviving old drugs into agrofungicides. J Antibiot 63(11):667–672. https://doi.org/10.1038/ja.2010.110

Chen Y-P, Liu Q, Gao H, Lin H-P, Tian H-Y, Hong K et al (2014) Streptospirodienoic acids A and B, 6,6-spiroketal polyketides from Streptomyces sp. RSC Adv 4(108):63324–63327. https://doi.org/10.1039/C4RA10672E

Chen B-S, Médici R, van der Helm MP, van Zwet Y, Gjonaj L, van der Geest R et al (2018) Rhodococcus strains as source for ene-reductase activity. Appl Microbiol Biotechnol 102(13):5545–5556. https://doi.org/10.1007/s00253-018-8984-7

Chiani M, Akbarzadeh A, Farhangi A, Mazinani M, Saffari Z, Emadzadeh K, Mehrabi MR (2010) Optimization of culture medium to increase the production of desferrioxamine B (Desferal) in Streptomyces pilosus. Pak J Biol Sci 13(11):546–550. https://doi.org/10.3923/pjbs.2010.546.550

Chisholm SW (2000) Stirring times in the Southern Ocean. Nature 407(6805):685–686. https://doi.org/10.1038/35037696

Citron CA, Gleitzmann J, Laurenzano G, Pukall R, Dickschat JS (2012) Terpenoids are widespread in actinomycetes: a correlation of secondary metabolism and genome data. Chembiochem 13(2):202–214. https://doi.org/10.1002/cbic.201100641

Compton CL, Schmitz KR, Sauer RT, Sello JK (2013) Antibacterial activity of and resistance to small molecule inhibitors of the ClpP peptidase. ACS Chem Biol 8(12):2669–2677. https://doi.org/10.1021/cb400577b

Cusenza MA, Bobba S, Ardente F, Cellura M, Di Persio F (2019) Energy and environmental assessment of a traction lithium-ion battery pack for plug-in hybrid electric vehicles. J Clean Prod 215:634–649. https://doi.org/10.1016/j.jclepro.2019.01.056

Dhakal D, Sohng JK, Pandey RP (2019) Engineering actinomycetes for biosynthesis of macrolactone polyketides. Microb Cell Fact 18(1):137. https://doi.org/10.1186/s12934-019-1184-z

Dhungana S, Michalczyk R, Boukhalfa H, Lack JG, Koppisch AT, Fairlee JM et al (2007) Purification and characterization of rhodobactin: a mixed ligand siderophore from Rhodococcus rhodochrous strain OFS. Biometals 20(6):853–867. https://doi.org/10.1007/s10534-006-9079-y

Dimkpa C, Svatos A, Merten D, Büchel G, Kothe E (2008) Hydroxamate siderophores produced by Streptomyces acidiscabies E13 bind nickel and promote growth in cowpea (Vigna unguiculata L.) under nickel stress. Can J Microbiol 54(3):163–172. https://doi.org/10.1139/w07-130

Dolence EK, Miller MJ (1991) Synthesis of foroxymithine, a microbial fermentation product and angiotensin 1 converting enzyme inhibitor. J Org Chem 56(2):492–499. https://doi.org/10.1021/jo00002a005

Drewniak L, Styczek A, Majder-Lopatka M, Sklodowska A (2008) Bacteria, hypertolerant to arsenic in the rocks of an ancient gold mine, and their potential role in dissemination of arsenic pollution. Environ Pollut (Barking, Essex: 1987) 156(3):1069–1074. https://doi.org/10.1016/j.envpol.2008.04.019

Dunn BJ, Cane DE, Khosla C (2013) Mechanism and specificity of an acyltransferase domain from a modular polyketide synthase. Biochemistry 52(11):1839–1841. https://doi.org/10.1021/bi400185v

Ebner T, Oppelt P, Wöber M, Staples P, Mayer RB, Sonnleitner U et al (2015) Treatment with Ca2+ ionophore improves embryo development and outcome in cases with previous developmental problems: a prospective multicenter study. Hum Reprod (Oxford, England) 30(1):97–102. https://doi.org/10.1093/humrep/deu285

Ejje N, Soe CZ, Gu J, Codd R (2013) The variable hydroxamic acid siderophore metabolome of the marine actinomycete Salinispora tropica CNB-440. Metallomics 5(11):1519–1528. https://doi.org/10.1039/c3mt00230f

Fiedler HP, Krastel P, Müller J, Gebhardt K, Zeeck A (2001) Enterobactin: the characteristic catecholate siderophore of Enterobacteriaceae is produced by Streptomyces species.(1). FEMS Microbiol Lett 196(2):147–151. https://doi.org/10.1111/j.1574-6968.2001.tb10556.x

Franzetti A, Gandolfi I, Bestetti G, Smyth TJP, Banat IM (2010) Production and applications of trehalose lipid biosurfactants. Eur J Lipid Sci Technol 112(6):617–627. https://doi.org/10.1002/ejlt.200900162

Frueh DP, Arthanari H, Koglin A, Vosburg DA, Bennett AE, Walsh CT, Wagner G (2008) Dynamic thiolation-thioesterase structure of a non-ribosomal peptide synthetase. Nature 454(7206):903–906. https://doi.org/10.1038/nature07162

Fujita MJ, Sakai R (2013) Heterologous production of desferrioxamines with a fusion biosynthetic gene cluster. Biosci Biotechnol Biochem 77(12):2467–2472. https://doi.org/10.1271/bbb.130597

Fukuda TTH, Helfrich EJN, Mevers E, Melo WGP, van Arnam EB, Andes DR et al (2021) Specialized metabolites reveal evolutionary history and geographic dispersion of a multilateral symbiosis. ACS Cent Sci 7(2):292–299. https://doi.org/10.1021/acscentsci.0c00978

Ghai A, Zheleznyak A, Mixdorf M, O'Neal J, Ritchey J, Rettig M et al (2020) Development of 89ZrDFO-elotuzumab for immunoPET imaging of CS1 in multiple myeloma. Eur J Nucl Med Mol Imaging 48(5):1302–1311. https://doi.org/10.1007/s00259-020-05097-y

Gobin J, Moore CH, Reeve JR, Wong DK, Gibson BW, Horwitz MA (1995) Iron acquisition by Mycobacterium tuberculosis: isolation and characterization of a family of iron-binding exochelins. Proceedings of the National Academy of Sciences 92(11):5189. https://doi.org/10.1073/pnas.92.11.5189

Goldbeck O, Eck AW, Seibold GM (2018) Real time monitoring of NADPH concentrations in Corynebacterium glutamicum and Escherichia coli via the genetically encoded sensor mBFP. Front Microbiol 9:2564. https://doi.org/10.3389/fmicb.2018.02564

Gran-Scheuch A, Trajkovic M, Parra L, Fraaije MW (2018) Mining the genome of Streptomyces leeuwenhoekii: two new type I Baeyer-Villiger monooxygenases from atacama Desert. Front. Microbiol 9:1609. https://doi.org/10.3389/fmicb.2018.01609

Gröning JAD, Eulberg D, Tischler D, Kaschabek SR, Schlömann M (2014) Gene redundancy of two-component (chloro)phenol hydroxylases in Rhodococcus opacus 1CP. FEMS Microbiol Lett 361(1):68–75. https://doi.org/10.1111/1574-6968.12616

Gürtler H, Pedersen R, Anthoni U, Christophersen C, Nielsen PH, Wellington EM et al (1994) Albaflavenone, a sesquiterpene ketone with a zizaene skeleton produced by a streptomycete with a new rope morphology. J Antibiot 47(4):434–439. https://doi.org/10.7164/antibiotics.47.434

Harrison A, Yu M, Gårdenborg T, Middleditch M, Ramsay R, Baker E, Lott J (2006) The structure of MbtI from Mycobacterium tuberculosis, the first enzyme in the biosynthesis of the siderophore mycobactin, reveals it to be a salicylate synthase. J Bacteriol 188:6081–6091. https://doi.org/10.1128/JB.00338-06

He J-l, Xie J-p (2011) Advances in Mycobacterium siderophore-based drug discovery. Acta Pharm Sin B 1(1):8–13. https://doi.org/10.1016/j.apsb.2011.04.008

He ZL, Yang XE, Stoffella PJ (2005) Trace elements in agroecosystems and impacts on the environment. J Trace Elem Med Biol 19(2–3):125–140. https://doi.org/10.1016/j.jtemb.2005.02.010

Heilmeier H, Wiche O (2020) The PCA of phytomining: principles, challenges and achievements. Carpath J Earth Environ Sci 15(1):37–42. https://doi.org/10.26471/cjees/2020/015/106

Heine T, Mehnert M, Schwabe R, Tischler D (2017) Thermochelin, a hydroxamate siderophore from thermocrispum agreste DSM 44070. SSP 262:501–504. https://doi.org/10.4028/www.scientific.net/SSP.262.501

Heine T, Zimmerling J, Ballmann A, Kleeberg SB, Rückert C, Busche T et al (2018) On the enigma of glutathione-dependent styrene degradation in Gordonia rubripertincta CWB2. Appl Environ Microbiol 84(9):e00154–e00118. https://doi.org/10.1128/AEM.00154-18

Hider RC, Kong X (2010) Chemistry and biology of siderophores. Nat Prod Rep 27(5):637–657. https://doi.org/10.1039/b906679a

Hofmann M, Retamal-Morales G, Tischler D (2020) Metal binding ability of microbial natural metal chelators and potential applications. Nat Prod Rep 37(9):1262–1283. https://doi.org/10.1039/c9np00058e

Hofmann M, Heine T, Malik L, Hofmann S, Joffroy K, Senges C et al (2021) Screening for microbial metal-chelating siderophores for the removal of metal ions from solutions. Microorganisms 9:111. https://doi.org/10.3390/microorganisms9010111

Hou J, Liu P, Qu H, Fu P, Wang Y, Wang Z et al (2012) Gilvocarcin HE: a new polyketide glycoside from Streptomyces sp. J Antibiot 65(10):523–526. https://doi.org/10.1038/ja.2012.61

Hussein RA, El-Anssary AA (2019) Chapter 2: Plants secondary metabolites: the key drivers of the pharmacological actions of medicinal Plants. In: Builders PF (ed) Herbal medicine. IntechOpen, Rijeka

Igarashi Y, Ogura H, Furihata K, Oku N, Indananda C, Thamchaipenet A (2011) Maklamicin, an antibacterial polyketide from an endophytic Micromonospora sp. J Nat Prod 74(4):670–674. https://doi.org/10.1021/np100727h

Imbert M, Bchet M, Blondeau R (1995) Comparison of the main siderophores produced by some species of Streptomyces. Curr Microbiol 31(2):129–133. https://doi.org/10.1007/BF00294289

Ioppolo JA, Caldwell D, Beiraghi O, Llano L, Blacker M, Valliant JF, Berti PJ (2017) 67Ga-labeled deferoxamine derivatives for imaging bacterial infection: preparation and screening of functionalized siderophore complexes. Nucl Med Biol 52:32–41. https://doi.org/10.1016/j.nucmedbio.2017.05.010

Jin J, Mazon H, van den Heuvel RHH, Janssen DB, Fraaije MW (2007) Discovery of a eugenol oxidase from Rhodococcus sp. strain RHA1. FEBS J 274(9):2311–2321. https://doi.org/10.1111/j.1742-4658.2007.05767.x

Khalil ZG, Salim AA, Vuong D, Crombie A, Lacey E, Blumenthal A, Capon RJ (2017) Amycolatopsins A-C: antimycobacterial glycosylated polyketide macrolides from the Australian soil Amycolatopsis sp. MST-108494. J Antibiot 70(12):1097–1103. https://doi.org/10.1038/ja.2017.119

Khan A, Singh P, Srivastava A (2018) Synthesis, nature and utility of universal iron chelator—siderophore: a review. Microbiol Res 212–213:103–111. https://doi.org/10.1016/j.micres.2017.10.012

Khosla C, Herschlag D, Cane DE, Walsh CT (2014) Assembly line polyketide synthases: mechanistic insights and unsolved problems. Biochemistry 53(18):2875–2883. https://doi.org/10.1021/bi500290t

Kim D, Choi KY, Yoo M, Zylstra GJ, Kim E (2018) Biotechnological potential of Rhodococcus biodegradative pathways. J Microbiol Biotechnol 28(7):1037–1051. https://doi.org/10.4014/jmb.1712.12017

Knobloch P, Koliwer-Brandl H, Arnold FM, Hanna N, Gonda I, Adenau S et al (2020) Mycobacterium marinum produces distinct mycobactin and carboxymycobactin siderophores to promote growth in broth and phagocytes. Cell Microbiol 22(5):e13163. https://doi.org/10.1111/cmi.13163

König C, Eulberg D, Gröning J, Lakner S, Seibert V, Kaschabek SR, Schlömann M (2004) A linear megaplasmid, p1CP, carrying the genes for chlorocatechol catabolism of Rhodococcus opacus 1CP. Microbiology (Reading, England) 150(Pt 9):3075–3087. https://doi.org/10.1099/mic.0.27217-0

Kozian D, Proulle V, Nitsche A, Galitzine M, Martinez M-C, Schumann B et al (2005) Identification of genes involved in Ca2+ ionophore A23187-mediated apoptosis and demonstration of a high susceptibility for transcriptional repression of cell cycle genes in B lymphoblasts from a patient with Scott syndrome. BMC Genomics 6:146. https://doi.org/10.1186/1471-2164-6-146

Kramer J, Özkaya Ö, Kümmerli R (2020) Bacterial siderophores in community and host interactions. Nat Rev Microbiol 18(3):152–163. https://doi.org/10.1038/s41579-019-0284-4

Kumaran S, Ngo ACR, Schultes FPJ, Tischler D (2020) Draft genome sequence of Kocuria indica DP-K7, a methyl red degrading actinobacterium. 3 Biotech 10(4):175. https://doi.org/10.1007/s13205-020-2136-3

Lacoske MH, Theodorakis EA (2015) Spirotetronate polyketides as leads in drug discovery. J Nat Prod 78(3):562–575. https://doi.org/10.1021/np500757w

Lawson PA (2018) The phylum Actinobacteria. In: Mattarelli P, Biavati B, Holzapfel WH, Wood BJB (eds) The bifidobacteria and related organisms. Biology, taxonomy, applications. Elsevier/Academic, London, pp 1–8

Lewin GR, Carlos C, Chevrette MG, Horn HA, McDonald BR, Stankey RJ et al (2016) Evolution and ecology of Actinobacteria and their bioenergy applications. Annu Rev Microbiol 70(1):235–254. https://doi.org/10.1146/annurev-micro-102215-095748

Lin Z, Xu X, Zhao S, Yang X, Guo J, Zhang Q et al (2018) Total synthesis and antimicrobial evaluation of natural albomycins against clinical pathogens. Nat Commun 9(1):3445. https://doi.org/10.1038/s41467-018-05821-1

Linos A, Steinbüchel A, Spröer C, Kroppenstedt RM (1999) Gordonia polyisoprenivorans sp. nov., a rubber-degrading actinomycete isolated from an automobile tyre. Int J Syst Bacteriol 49 (Pt 4):1785–1791. https://doi.org/10.1099/00207713-49-4-1785

Mazzei E, Iorio M, Maffioli SI, Sosio M, Donadio S (2012) Characterization of madurastatin C1, a novel siderophore from Actinomadura sp. J Antibiot 65(5):267–269. https://doi.org/10.1038/ja.2012.10

McLeod MP, Warren RL, Hsiao WWL, Araki N, Myhre M, Fernandes C et al (2006) The complete genome of Rhodococcus sp. RHA1 provides insights into a catabolic powerhouse. Proc Natl Acad Sci U S A 103(42):15582–15587. https://doi.org/10.1073/pnas.0607048103

McMahon MD, Rush JS, Thomas MG (2012) Analyses of MbtB, MbtE, and MbtF suggest revisions to the mycobactin biosynthesis pathway in Mycobacterium tuberculosis. J Bacteriol 194(11):2809–2818. https://doi.org/10.1128/JB.00088-12

Medema MH, Trefzer A, Kovalchuk A, van den Berg M, Müller U, Heijne W et al (2010) The sequence of a 1.8-Mb bacterial linear plasmid reveals a rich evolutionary reservoir of secondary metabolic pathways. Genome Biol Evol 2:212–224. https://doi.org/10.1093/gbe/evq013

Mehnert M, Retamal-Morales G, Schwabe R, Vater S, Heine T, Levicán GJ et al (2017) Revisiting the chrome Azurol S assay for various metal ions. SSP 262:509–512. https://doi.org/10.4028/www.scientific.net/SSP.262.509

Miethke M, Marahiel MA (2007) Siderophore-based iron acquisition and pathogen control. Microbiol Mol Biol Rev 71(3):413–451. https://doi.org/10.1128/MMBR.00012-07

Miranda-CasoLuengo R, Coulson G, Miranda-CasoLuengo A, Vázquez-Boland J, Hondalus M, Meijer W (2012) The hydroxamate siderophore rhequichelin is required for virulence of the pathogenic actinomycete Rhodococcus equi. Infect Immun 80:4106–4114. https://doi.org/10.1128/IAI.00678-12

Mohanty SK, Gonneau C, Salamatipour A, Pietrofesa RA, Casper B, Christofidou-Solomidou M, Willenbring JK (2018) Siderophore-mediated iron removal from chrysotile: implications for asbestos toxicity reduction and bioremediation. J Hazard Mater 341:290–296. https://doi.org/10.1016/j.jhazmat.2017.07.033

Mügge C, Heine T, Baraibar AG, van Berkel WJH, Paul CE, Tischler D (2020) Flavin-dependent N-hydroxylating enzymes: distribution and application. Appl Microbiol Biotechnol 104(15):6481–6499. https://doi.org/10.1007/s00253-020-10705-w

Negash KH, Norris JKS, Hodgkinson JT (2019) Siderophore-antibiotic conjugate design: new drugs for bad bugs? Molecules 24(18):3314. https://doi.org/10.3390/molecules24183314

Nguyen Q-T, Trinco G, Binda C, Mattevi A, Fraaije MW (2017) Discovery and characterization of an F420-dependent glucose-6-phosphate dehydrogenase (Rh-FGD1) from Rhodococcus jostii RHA1. Appl Microbiol Biotechnol 101(7):2831–2842. https://doi.org/10.1007/s00253-016-8038-y

Niu S, Li S, Chen Y, Tian X, Zhang H, Zhang G et al (2011) Lobophorins E and F, new spirotetronate antibiotics from a South China Sea-derived Streptomyces sp. SCSIO 01127. J Antibiot 64(11):711–716. https://doi.org/10.1038/ja.2011.78

Niu G, Chater KF, Tian Y, Zhang J, Tan H (2016) Specialised metabolites regulating antibiotic biosynthesis in Streptomyces spp. FEMS Microbiol Rev 40(4):554–573. https://doi.org/10.1093/femsre/fuw012

Nivina A, Yuet KP, Hsu J, Khosla C (2019) Evolution and diversity of assembly-line polyketide synthases. Chem Rev 119(24):12524–12547. https://doi.org/10.1021/acs.chemrev.9b00525

Nosrati R, Dehghani S, Karimi B, Yousefi M, Taghdisi SM, Abnous K et al (2018) Siderophore-based biosensors and nanosensors; new approach on the development of diagnostic systems. Biosens Bioelectron 117:1–14. https://doi.org/10.1016/j.bios.2018.05.057

Orcutt KM, Jones WS, McDonald A, Schrock D, Wallace KJ (2010) A lanthanide-based chemosensor for bioavailable Fe3+ using a fluorescent siderophore: an assay displacement approach. Sensors (Basel, Switzerland) 10(2):1326–1337. https://doi.org/10.3390/s100201326

Pacheco GJ, Ciapina EMP, Gomes E d B, Junior NP (2010) Biosurfactant production by rhodococcus erythropolis and its application to oil removal. Braz J Microbiol 41(3):685–693. https://doi.org/10.1590/S1517-83822010000300019

Pal S, Manjunath B, Ghorai S, Sasmal S (2018) Benzoxazole alkaloids: occurrence, chemistry, and biology. Alkaloids Chem Biol 79:71–137. https://doi.org/10.1016/bs.alkal.2017.12.002

Palaniyandi SA, Yang SH, Zhang L, Suh J-W (2013) Effects of actinobacteria on plant disease suppression and growth promotion. Appl Microbiol Biotechnol 97(22):9621–9636. https://doi.org/10.1007/s00253-013-5206-1

Park SR, Tripathi A, Wu J, Schultz PJ, Yim I, McQuade TJ et al (2016) Discovery of cahuitamycins as biofilm inhibitors derived from a convergent biosynthetic pathway. Nat Commun 7:10710. https://doi.org/10.1038/ncomms10710

Parker DL, Sposito G, Tebo BM (2004) Manganese(III) binding to a pyoverdine siderophore produced by a manganese(II)-oxidizing bacterium. Geochim Cosmochim Acta 68(23):4809–4820. https://doi.org/10.1016/j.gca.2004.05.038

Parsels KA, Mastro KA, Steele JM, Thomas SJ, Kufel WD (2021) Cefiderocol: a novel siderophore cephalosporin for multidrug-resistant Gram-negative bacterial infections. J Antimicrob Chemother 76(6):1379–1391. https://doi.org/10.1093/jac/dkab015

Patzer SI, Braun V (2010) Gene cluster involved in the biosynthesis of griseobactin, a catechol-peptide siderophore of Streptomyces sp. ATCC 700974. J Bacteriol 192(2):426–435. https://doi.org/10.1128/JB.01250-09

Pérez-García F, Jorge JMP, Dreyszas A, Risse JM, Wendisch VF (2018) Efficient production of the dicarboxylic acid glutarate by Corynebacterium glutamicum via a novel synthetic pathway. Front Microbiol 9:2589. https://doi.org/10.3389/fmicb.2018.02589

Perk LR, Vosjan MJWD, Visser GWM, Budde M, Jurek P, Kiefer GE, van Dongen GAMS (2010) p-Isothiocyanatobenzyl-desferrioxamine: a new bifunctional chelate for facile radiolabeling of monoclonal antibodies with zirconium-89 for immuno-PET imaging. Eur J Nucl Med Mol Imaging 37(2):250–259. https://doi.org/10.1007/s00259-009-1263-1

Petrik M, Pfister J, Misslinger M, Decristoforo C, Haas H (2020) Siderophore-based molecular imaging of fungal and bacterial infections-current status and future perspectives. J Fungi (Basel, Switzerland) 6(2):73. https://doi.org/10.3390/jof6020073

Petrik M, Umlaufova E, Raclavsky V, Palyzova A, Havlicek V, Pfister J et al (2021) 68Ga-labelled desferrioxamine-B for bacterial infection imaging. Eur J Nucl Med Mol Imaging 48(2):372–382. https://doi.org/10.1007/s00259-020-04948-y

Phillips DJ, Davies G-L, Gibson MI (2015) Siderophore-inspired nanoparticle-based biosensor for the selective detection of Fe3. J Mater Chem 3(2):270–275. https://doi.org/10.1039/c4tb01501k

Poetsch A, Haußmann U, Burkovski A (2011) Proteomics of corynebacteria: from biotechnology workhorses to pathogens. Proteomics 11(15):3244–3255. https://doi.org/10.1002/pmic.201000786

Pramanik A, Stroeher UH, Krejci J, Standish AJ, Bohn E, Paton JC et al (2007) Albomycin is an effective antibiotic, as exemplified with Yersinia enterocolitica and Streptococcus pneumoniae. Int J Med Microbiol 297(6):459–469. https://doi.org/10.1016/j.ijmm.2007.03.002

Prasad MNV, Freitas H, Fraenzle S, Wuenschmann S, Markert B (2010) Knowledge explosion in phytotechnologies for environmental solutions. Environ Pollut 158(1):18–23. https://doi.org/10.1016/j.envpol.2009.07.038

Rajkumar M, Ae N, Prasad MNV, Freitas H (2010) Potential of siderophore-producing bacteria for improving heavy metal phytoextraction. Trends Biotechnol 28(3):142–149. https://doi.org/10.1016/j.tibtech.2009.12.002

Rateb ME, Houssen WE, Arnold M, Abdelrahman MH, Deng H, Harrison WTA et al (2011) Chaxamycins A-D, bioactive ansamycins from a hyper-arid desert Streptomyces sp. J Nat Prod 74(6):1491–1499. https://doi.org/10.1021/np200320u

Rathod BB, Korasapati R, Sripadi P, Reddy Shetty P (2018) Novel actinomycin group compound from newly isolated Streptomyces sp. RAB12: isolation, characterization, and evaluation of antimicrobial potential. Appl Microbiol Biotechnol 102(3):1241–1250. https://doi.org/10.1007/s00253-017-8696-4

Reid R, Piagentini M, Rodriguez E, Ashley G, Viswanathan N, Carney J et al (2003) A model of structure and catalysis for ketoreductase domains in modular polyketide synthases. Biochemistry 42(1):72–79. https://doi.org/10.1021/bi0268706

Retamal-Morales G, Heine T, Tischler JS, Erler B, Gröning JAD, Kaschabek SR et al (2018a) Draft genome sequence of Rhodococcus erythropolis B7g, a biosurfactant producing actinobacterium. J Biotechnol 280:38–41. https://doi.org/10.1016/j.jbiotec.2018.06.001

Retamal-Morales G, Mehnert M, Schwabe R, Tischler D, Zapata C, Chávez R et al (2018b) Detection of arsenic-binding siderophores in arsenic-tolerating Actinobacteria by a modified CAS assay. Ecotoxicol Environ Saf 157:176–181. https://doi.org/10.1016/j.ecoenv.2018.03.087

Retamal-Morales G, Senges CHR, Stapf M, Olguín A, Modak B, Bandow JE et al (2021) Isolation and characterization of arsenic-binding siderophores from Rhodococcus erythropolis S43: role of heterobactin B and other heterobactin variants. Appl Microbiol Biotechnol 105(4):1731–1744. https://doi.org/10.1007/s00253-021-11123-2

Ribeiro M, Simões M (2019) Advances in the antimicrobial and therapeutic potential of siderophores. Environ Chem Lett 17(4):1485–1494. https://doi.org/10.1007/s10311-019-00887-9

Riebel A, Dudek HM, de Gonzalo G, Stepniak P, Rychlewski L, Fraaije MW (2012) Expanding the set of rhodococcal Baeyer-Villiger monooxygenases by high-throughput cloning, expression and substrate screening. Appl Microbiol Biotechnol 95(6):1479–1489. https://doi.org/10.1007/s00253-011-3823-0

Riedel A, Heine T, Westphal AH, Conrad C, Rathsack P, van Berkel WJH, Tischler D (2015a) Catalytic and hydrodynamic properties of styrene monooxygenases from Rhodococcus opacus 1CP are modulated by cofactor binding. AMB Express 5(1):112. https://doi.org/10.1186/s13568-015-0112-9

Riedel A, Mehnert M, Paul CE, Westphal AH, van Berkel WJH, Tischler D (2015b) Functional characterization and stability improvement of a 'thermophilic-like' ene-reductase from Rhodococcus opacus 1CP. Front Microbiol 6:1073. https://doi.org/10.3389/fmicb.2015.01073

Roberts JN, Singh R, Grigg JC, Murphy MEP, Bugg TDH, Eltis LD (2011) Characterization of dye-decolorizing peroxidases from Rhodococcus jostii RHA1. Biochemistry 50(23):5108–5119. https://doi.org/10.1021/bi200427h

Ronan JL, Kadi N, McMahon SA, Naismith JH, Alkhalaf LM, Challis GL (2018) Desferrioxamine biosynthesis: diverse hydroxamate assembly by substrate-tolerant acyl transferase DesC. Philos Trans R Soc B Biol Sci 373(1748):20170068. https://doi.org/10.1098/rstb.2017.0068

Rostami H, Hamedi H, Yolmeh M (2016) Some biological activities of pigments extracted from Micrococcus roseus (PTCC 1411) and Rhodotorula glutinis (PTCC 5257). Int J Immunopathol Pharmacol 29(4):684–695. https://doi.org/10.1177/0394632016673846

Rowland BM, Grossman TH, Osburne MS, Taber HW (1996) Sequence and genetic organization of a Bacillus subtilis operon encoding 2,3-dihydroxybenzoate biosynthetic enzymes. Gene 178 (1–2):119–123. https://doi.org/10.1016/0378-1119(96)00349-6

Saha M, Sarkar S, Sarkar B, Sharma BK, Bhattacharjee S, Tribedi P (2016) Microbial siderophores and their potential applications: a review. Environ Sci Pollut Res Int 23(5):3984–3999. https://doi.org/10.1007/s11356-015-4294-0

Santos JD, Vitorino I, de La Cruz M, Díaz C, Cautain B, Annang F et al (2019) Bioactivities and extract dereplication of Actinomycetales isolated from marine sponges. Front Microbiol 10:727. https://doi.org/10.3389/fmicb.2019.00727

Sato T, Yamawaki K (2019) Cefiderocol: discovery, chemistry, and in vivo profiles of a novel siderophore cephalosporin. Clin Infect Dis 69(Suppl 7):S538–S543. https://doi.org/10.1093/cid/ciz826

Schalk IJ, Hannauer M, Braud A (2011) New roles for bacterial siderophores in metal transport and tolerance. Environ Microbiol 13(11):2844–2854. https://doi.org/10.1111/j.1462-2920.2011.02556.x

Schneider K, Rose I, Vikineswary S, Jones AL, Goodfellow M, Nicholson G et al (2007) Nocardichelins A and B, siderophores from Nocardia strain Acta 3026. J Nat Prod 70(6):932–935. https://doi.org/10.1021/np060612i

Schupp T, Waldmeier U, Divers M (1987) Biosynthesis of desferrioxamine B in Streptomyces pilosus : evidence for the involvement of lysine decarboxylase. FEMS Microbiol Lett 42 (2–3):135–139. https://doi.org/10.1111/j.1574-6968.1987.tb02060.x

Schwabe R, Senges CHR, Bandow JE, Heine T, Lehmann H, Wiche O et al (2020) Cultivation dependent formation of siderophores by Gordonia rubripertincta CWB2. Microbiol Res 238: 126481. https://doi.org/10.1016/j.micres.2020.126481

Schwyn B, Neilands JB (1987) Universal chemical assay for the detection and determination of siderophores. Anal Biochem 160(1):47–56. https://doi.org/10.1016/0003-2697(87)90612-9

Senges CHR, Al-Dilaimi A, Marchbank DH, Wibberg D, Winkler A, Haltli B et al (2018) The secreted metabolome of Streptomyces chartreusis and implications for bacterial chemistry. Proc Natl Acad Sci U S A 115(10):2490–2495. https://doi.org/10.1073/pnas.1715713115

Seyedsayamdost MR, Traxler MF, Zheng S-L, Kolter R, Clardy J (2011) Structure and biosynthesis of amychelin, an unusual mixed-ligand siderophore from amycolatopsis sp. AA4. J Am Chem Soc 133(30):11434–11437. https://doi.org/10.1021/ja203577e

Shander A, Cappellini MD, Goodnough LT (2009) Iron overload and toxicity: the hidden risk of multiple blood transfusions. Vox Sang 97(3):185–197. https://doi.org/10.1111/j.1423-0410.2009.01207.x

Singh R, Dubey AK (2020) Isolation and characterization of a new endophytic Actinobacterium Streptomyces californicus strain ADR1 as a promising source of anti-bacterial, anti-biofilm and antioxidant metabolites. Microorganisms 8(6):929. https://doi.org/10.3390/microorganisms8060929

Sperry J, Wilson ZE, Rathwell DCK, Brimble MA (2010) Isolation, biological activity and synthesis of benzannulated spiroketal natural products. Nat Prod Rep 27(8):1117–1137. https://doi.org/10.1039/b911514p

Stackebrandt E, Schumann P (2006) Introduction to the taxonomy of Actinobacteria. In: Dworkin M, Falkow S, Rosenberg E, Schleifer K-H, Stackebrandt E (eds) The prokaryotes. A handbook on the biology of bacteria. Springer, New York, pp 297–321

Stefanska AL, Fulston M, Houge-Frydrych CS, Jones JJ, Warr SR (2000) A potent seryl tRNA synthetase inhibitor SB-217452 isolated from a Streptomyces species. J Antibiot 53(12):1346–1353. https://doi.org/10.7164/antibiotics.53.1346

Tchounwou PB, Yedjou CG, Patlolla AK, Sutton DJ (2012) Heavy metal toxicity and the environment. Exp Suppl 101:133–164. https://doi.org/10.1007/978-3-7643-8340-4_6

Tischler D, Eulberg D, Lakner S, Kaschabek SR, van Berkel WJH, Schlömann M (2009) Identification of a novel self-sufficient styrene monooxygenase from Rhodococcus opacus 1CP. J Bacteriol 191(15):4996–5009. https://doi.org/10.1128/JB.00307-09

Tischler D, Kermer R, Gröning JAD, Kaschabek SR, van Berkel WJH, Schlömann M (2010) StyA1 and StyA2B from Rhodococcus opacus 1CP: a multifunctional styrene monooxygenase system. J Bacteriol 192(19):5220–5227. https://doi.org/10.1128/JB.00723-10

Tischler D, Niescher S, Kaschabek SR, Schlömann M (2013) Trehalose phosphate synthases OtsA1 and OtsA2 of Rhodococcus opacus 1CP. FEMS Microbiol Lett 342(2):113–122. https://doi.org/10.1111/1574-6968.12096

Tischler D, van Berkel WJH, Fraaije MW (2019) Editorial: Actinobacteria, a source of biocatalytic tools. Front Microbiol 10:800. https://doi.org/10.3389/fmicb.2019.00800

Tong Y, Robertsen HL, Blin K, Weber T, Lee SY (2018) CRISPR-Cas9 toolkit for actinomycete genome editing. Methods Mol Biol (Clifton, N.J.) 1671:163–184. https://doi.org/10.1007/978-1-4939-7295-1_11

Torrente D, Mendes-da-Silva RF, Lopes AAC, González J, Barreto GE, Guedes RCA (2014) Increased calcium influx triggers and accelerates cortical spreading depression in vivo in male adult rats. Neurosci Lett 558:87–90. https://doi.org/10.1016/j.neulet.2013.11.004

Trögl J, Esuola C, Kříženecká S, Kuráň P, Seidlová L, Veronesi-Dáňová P et al (2018) Biodegradation of high concentrations of aliphatic hydrocarbons in soil from a petroleum refinery: implications for applicability of new actinobacterial strains. Appl Sci 8(10):1855. https://doi.org/10.3390/app8101855

Vakulenko SB, Mobashery S (2003) Versatility of aminoglycosides and prospects for their future. Clin Microbiol Rev 16(3):430–450. https://doi.org/10.1128/cmr.16.3.430-450.2003

Ventura M, Canchaya C, Tauch A, Chandra G, Fitzgerald GF, Chater KF, van Sinderen D (2007) Genomics of Actinobacteria: tracing the evolutionary history of an ancient phylum. Microbiol Mol Biol Rev 71(3):495–548. https://doi.org/10.1128/MMBR.00005-07

Wagenknecht M, Dib JR, Thürmer A, Daniel R, Farías ME, Meinhardt F (2010) Structural peculiarities of linear megaplasmid, pLMA1, from Micrococcus luteus interfere with pyrosequencing reads assembly. Biotechnol Lett 32(12):1853–1862. https://doi.org/10.1007/s10529-010-0357-y

Wang Y, Brown HN, Crowley DE, Szaniszlo PJ (1993) Evidence for direct utilization of a siderophore, ferrioxamine B, in axenically grown cucumber. Plant Cell Environ 16(5):579–585. https://doi.org/10.1111/j.1365-3040.1993.tb00906.x

Wiche O, Tischler D, Fauser C, Lodemann J, Heilmeier H (2017) Effects of citric acid and the siderophore desferrioxamine B (DFO-B) on the mobility of germanium and rare earth elements in soil and uptake in Phalaris arundinacea. Int J Phytoremediation 19(8):746–754. https://doi.org/10.1080/15226514.2017.1284752

Wu Q, Liang J, Lin S, Zhou X, Bai L, Deng Z, Wang Z (2011) Characterization of the biosynthesis gene cluster for the pyrrole polyether antibiotic calcimycin (A23187) in Streptomyces chartreusis NRRL 3882. Antimicrob Agents Chemother 55(3):974. https://doi.org/10.1128/AAC.01130-10

Wu Q, Gou L, Lin S, Liang J, Yin J, Zhou X et al (2013) Characterization of the N-methyltransferase CalM involved in calcimycin biosynthesis by Streptomyces chartreusis NRRL 3882. Biochimie 95(7):1487–1493. https://doi.org/10.1016/j.biochi.2013.03.014

Wu H, Liang J, Wang J, Liang W-J, Gou L, Wu Q et al (2018) Cezomycin is activated by CalC to its ester form for further biosynthesis steps in the production of calcimycin in Streptomyces chartreusis NRRL 3882. Appl Environ Microbiol 84(12). https://doi.org/10.1128/AEM.00586-18

Yan J-X, Chevrette MG, Braun DR, Harper MK, Currie CR, Bugni TS (2019) Madurastatin D1 and D2, oxazoline containing siderophores isolated from an Actinomadura sp. Org Lett 21(16):6275–6279. https://doi.org/10.1021/acs.orglett.9b02159

Yao X, Zhang Y, Hao J, Duan H-Q, Zhao C-X, Sun C et al (2019) Deferoxamine promotes recovery of traumatic spinal cord injury by inhibiting ferroptosis. Neural Regen Res 14(3):532–541. https://doi.org/10.4103/1673-5374.245480

Yu S, Teng C, Bai X, Liang J, Song T, Dong L et al (2017) Optimization of siderophore production by bacillus sp. PZ-1 and its potential enhancement of phytoextraction of Pb from soil. J Microbiol Biotechnol 27(8):1500–1512. https://doi.org/10.4014/jmb.1705.05021

Zimmerling J, Tischler D, Großmann C, Schlömann M, Oelschlägel M (2017) Characterization of aldehyde dehydrogenases applying an enzyme assay with in situ formation of phenylacetaldehydes. Appl Biochem Biotechnol 182(3):1095–1107. https://doi.org/10.1007/s12010-016-2384-1

An Overview of Biomedical, Biotechnological, and Industrial Applications of *Actinomycetes*

18

H. A. D. Ruwandeepika, G. C. P. Fernando, and T. S. P. Jayaweera

Abstract

Actinobacteria are the largest taxonomic group within the domain bacteria and one of the largest phyla. They are cosmopolitan organisms with the capability of living in an array of environmental and ecological conditions ranging from tranquil forest soils to ecological extremities of volcanoes and marine hydrothermal vents. With this nature of being omnipresent, they play several vital roles in the ecosystem such as recycling of substances, synthesis of bioactive molecules, and degradation of complex polymers. They are considered to be the treasure house of secondary metabolites as they are the main source of preponderant amount of naturally derived, modern-day antibiotics, and many other antifungal, antiparasitic, anticancer, antitumor, and immunomodulator compounds. Furthermore, they are being espoused in different biotechnologies to produce numerous industrially vital enzymes, organic acids, amino acids, pigments, vitamins, and toxins. Consequently, attributed to the massive diversity and the survival ability under harsh exogenous insults, Actinobacteria have also being adopted in ecosystem reclamation, biotransformation, and bioremediation schemes. This chapter is written with the core objective of discussing the involvements of diverse actinomycetes species in the fields of biotechnology, biomedicine, agriculture, environment, and other industries as underpinned by the researches of the international pioneers of the respective fields.

H. A. D. Ruwandeepika (✉) · G. C. P. Fernando · T. S. P. Jayaweera
Department of Livestock Production, Faculty of Agricultural Sciences, Sabaragamuwa University of Sri Lanka, Belihuloya, Sri Lanka
e-mail: sanjeewapj@agri.sab.ac.lk

R. V. Rai, J. A. Bai (eds.), *Natural Products from Actinomycetes*,
https://doi.org/10.1007/978-981-16-6132-7_18

Keywords

Actinobacteria · Antibiotics · Bioactive molecules · Biomedical · Biotechnical · Industrial

18.1 Introduction

Actinobacteria are morphologically and physiologically diverse bacterial group representing one of the largest bacterial groups of the bacterial domain having large number of genera. They are ubiquitous in nature and spread in wide ranges of niches including soil, rhizosphere, and aquatic (marine and freshwater) environments (Lawson 2018). They are capable of tolerating extreme environment, and they live in more extreme locations, such as deep-sea sediments and hyperarid desert soils. Some of the members of this group play immensely important role by producing several biologicals, which are important in biomedical, industrial, biotechnical fields, and some of these compounds are antibiotics, enzymes, signaling molecules, immunomodulators, enzyme inhibitors, etc. It has shown a high potential for the production of secondary metabolites (15–25 secondary metabolites) by single strain, and it has been proven that 45% of all bioactive microbial metabolites discovered are from Actinobacteria counting about 10,000 antibiotics (Santos et al. 2020). Actinobacteria represent the normal microbial population in human body (from skin to mucosal surfaces, gastrointestinal tract, etc.) maintaining a balanced micro-ecosystem in the human body. Members of the genera Corynebacterium, Propionibacterium, Rothia, Actinomyces, and Bifidobacterium are the most important groups found in healthy individuals. Although the majority of Actinobacteria are saprophytes having ability of decomposition of plant and animal debris, there are some members of this group that is pathogenic to humans and animals (ul Hassan and Shaikh 2017).

They have the characteristic of high guanine and cytosine contents in DNA, and also characteristic of filamentous morphology. This phylum consists of bacteria with different morphologies such as coccoid, rod–coccoid, hyphal, and branched forms. Additionally, they are with variety of physiological and metabolic capabilities viz. spore or nonspore-forming, production of extracellular enzymes, metabolic products, and antibiotics (Stackebrandt and Schumann 2006).

The phylum Actinobacteria is divided into six classes, namely Actinobacteria, Acidimicrobiia, Coriobacteriia, Nitriliruptoria, Rubrobacteria, and Thermoleophilia. The class Actinobacteria is further divided into 16 orders that are Actinopolysporales, Actinomycetales, Bifidobacteriales, Catenulisporales, Corynebacteriales, Frankiales, Glycomycetales, Jiangellales, Kineosporiales, Micrococcales, Micromonosporales, Propionibacteriales, Pseudonocardiales, Streptomycetales, Streptosporangiales, and Incertae sedis (Puttaswamygowda et al. 2019). Order Actinomycetales contains the family Actinomycetaceae, and the family consists of genera Actinomyces, Actinobaculum, Arcanobacterium, Mobiluncus, Trueperella, and Varibaculum. Members of this family are Gram-positive, non-acid-fast, non-spore-forming, nonmotile, facultatively anaerobic, predominantly diphtheroid cells that tend to form branched filaments. The actinomycetes are extremely diverse, with at least 350 genera reported to date (Yassin 2014).

Actinomyces species are frequently found as commensal members of the normal microflora of human mucous membranes, and also, they do occur on the buccal and nasopharyngeal mucous membranes of several animal species such as cattle, sheep, pig, dogs, and horses. Additionally, they are reported in the nature as well. Members of the genus Arcanobacterium are known as obligate human pathogen, but they are reported to be present in animals as well. Trueperella species occur as commensals of the mucous membranes of many domestic animals such as dairy and beef cattle, sheep, swine, and goats. Actinobaculum is a well-recognized commensal flora of the mucosal surfaces and skin of animals and humans, whereas the genus Varibaculum is a commensal in human skin. Though the distribution of genus Mobiluncus is doubtful, they have been isolated from humans. Members of the family Actinomycetaceae are generally considered to be of relatively low virulence.

There is enormous use of Actinobacteria identified up to date in several fields including biomedical, biotechnological, and industrial. They are widely known for their vital role in decomposing organic debris ranging from chitin to cellulose and in humus formation. They support in keeping the wheels of the material cycle of almost all the ecosystems in motion by accelerating the organic matter turnover rate and carbon cycle, recharging the soil nutrient levels. Actinobacteria were recently put under the spotlight by their ability to produce a range of secondary metabolites with high pharmacological and industrial value. Recent discoveries have unveiled that the activity of Actinobacteria in soils, where they produce a sundry of antibiotics compounds such as aureomycin, terramycin, and streptomycin, is responsible for the earthy smell elicited when surface soil is being ploughed. When focusing on biotic resources on earth, most organisms produce both primary and secondary metabolites. Of these bioactive compounds, secondary metabolites often have diverse and unusual chemical structures with comparatively lower molecular weight, and unlike primary metabolites, they have no direct influence on the lifecycle of the organisms. Molecular biological tools and high-throughput screening techniques have unveiled over a million of natural compounds originating in different ecosystems. While the majority of these compounds are of plant origin, around 500,000 compounds have been recognized to have a microbial origin. Out of these, ~250,000 have been proven to have bioactive characteristics and approximately 22,500 compounds are directly synthesized by microbes. In this array of microbial metabolites, around 45% (~10,100) are produced by Actinobacteria, while ~17% and ~38% are coming from unicellular bacteria and fungi, respectively (Takahashi and Nakashima 2018). These Actinobacterial bioactive metabolites have been proven as potential sources of many novel compounds with multifarious industrial importance ranging from pigments, enzymes, biosurfactants, probiotic compounds and vitamins, antibiotics, antifungal agents, herbicides, insecticides, plant-protective compounds, nematicides and larvicides, anthelmintic and antiparasitic compounds, antitumor and anticancer compounds, immunomodulator compounds, nanoparticle compounds, and organoleptic modifiers. Furthermore, about 75% of bioactive metabolites discovered to date come from the genus *Streptomyces*. Actinobacteria have also been proven highly productive in bioremediation/organic waste biodegradation schemes where their decomposing and material cycling capacity are being exploited and Actinobacteria, their cellular machinery, and genetic components are currently being used for developing molecular tools that aid for the progress of modified enzyme research, researches on isolation of bioactive compounds from biotic sources, molecular tools and protocol development practices, novel and rare antimicrobial compound researches, bioreactor and fermentation technology development, etc. (Grasso et al. 2016). This chapter focuses on biomedical, technological, and industrial applications of *Actinomycetes*.

18.2 Antimicrobials of Actinomycetes

Vast majority (nearly two thirds) of all known clinically used antibiotics are produced by Actinobacteria. Of the 34,000 microbial bioactive compounds that have been reported until 2010, 40% are produced by Actinobacteria. About 80% of the antibiotics produced by actinomycetes are made by members of the genus Streptomyces. Antibiotics are the substances that selectively inhibit the bacterial growth or kill bacteria.

From an ecosystem's perspective, most of these antibiotic synthesizers are soil-dwelling and Actinobacterial antibiotics include different peptides, beta-lactams (cephalosporins) and lactams (cephamycins), anthracyclines (daunorubicin), lactamase inhibitor clavulanic acid, ansamycins (rifamycin, geldanamycin), chloramphenicol, tetracyclines, oxazolidinones, streptogramins (streptogramin), glutarimides (cycloheximide), nucleosides, polyenes, polyketides/macrolides (clarithromycin, erythromycin, tylosin, clarithromycin), glycopeptides (vancomycin, teicoplanin), angucyclines (landomycin and moromycin), lipopeptides (daptomycin), and polyesters and aminoglycosides (neomycin, kanamycin, streptomycin) (Gohain et al. 2020). *Streptomyces* species hold the throne as the predominant genus of antibiotic production, synthesizing around 7600 bioactive secondary metabolite compounds where majority of them function as potent antibiotics (Table 18.1).

Until about 1980, exploitation of bioactive compounds was strictly limited to *Nocardia* and *Streptomyces* species. Later, only genera like *Ampullariella, Actinoplanes, Actinomadura, Actinosynnema,* and *Dactylosporangium* were researched in order to discover antibiotics. Actinobacteria produce bacteriocins, which are ribosomally synthesized antimicrobial peptides. They produce several types of bacteriosins with different capabilities such as antibacterial, antiallergic, antitumor, and antinociceptive activities. Actinomycetes research has made recent breakthroughs in the class of lantibiotics as well (Raja and Prabakarana 2011; Wink et al. 2017).

Table 18.1 Common antimicrobial agents produced by *Actinobacteria*

Compound	Microorganism
Actinomycins, 2-allyloxyphenol, amphomycin, aspartocins, avermectin, fluorometabolites, bisanthraquinone, bisoxazolomycin, bonactin, carbomycin, chloramphenicol, chlorodihydroquinones, cycloheximide, cycloserine, daptomycin, essramycin, frigocyclinone, glaciapyrroles, gutingimycin, himalomycins, hygromycin, iminimycins A and B, gentamicin, kanamycin, lajollamycin, leucomycin, lincomycin, marinopyrroles, neomycins, novobiocin, oleandomycin, oxytetracycline, pacificanones A and B, pristinamycin, retamycin, resistoflavin, ansamycins (rifapentine, rifampicin, rifabutin), spiramycin, staphylomycin, stendomycin, streptolydigin, streptomycin, tetracyclines (chlortetracycline and oxytetracycline), tirandamycin, valinomycin, vancomycin, penicillin N (penicillin group), cephalosporin C (cephalosporin group), thienamycin (cephalosporin type β-lactam), daptomycin	*Streptomyces* spp.
Norresistomycin	*Schisandra chinensis*
Abyssomicin, proximicins	*Verrucosispora* spp.
Anthracycline, clostomicins, diazepinomicin	*Micromonospora* spp.
Arenimycin	*Salinispora arenicola*
Dipyrimicins A and B, ristomycins A and B	*Amycolatopsis* spp.
Erythromycin (macrolides), telithromycin	*Saccharopolyspora* spp.
Marinomycin	*Marinispora* spp.
Mumiamicin	*Mumia* spp.
Natamicin	*Micromonospora* spp.
Pyrizomicins A and B	*Lechevalieria aerocolonigenes*
Ristocetin, nocardicin (monocyclic β-lactam),	*Nocardia* spp.
Fidaxomicin, teicoplanin (Tcp)	*Actinoplanes* spp.

18.3 Antifungals of Actinomycetes

The majority of antifungal compounds produced by *Actinomycetes* species come under polyene macrolactones group. It is an extensive family of natural bioactive agents produced mostly by soil-borne Actinobacterial species. These compounds are generally known for their potent antifungal nature, but some compounds have also demonstrated antitumor, antiparasitic, antiviral, and anti-prion properties. Despite their effectiveness, polyene macrolactones are generally highly toxic to mammalian cells, and therefore, continuous researches are being carried out for identification and structural modification of polyene macrolides to expand their general uses. Recent researches on biosynthetic pathways of polyene macrolides and genetic alteration of those pathways have opened new doors for creating new analogous compounds. Recently, a disaccharide-containing NPP (nystatin-like pseudonocardia polyene) was identified indicating the possibility of a novel class of polyene antifungals. In general, polyene macrolides are composed of polyhydroxylated macrocyclic lactones. The common means of antifungal activity of these compounds are exerted by interfering with cell wall integrity. In antifungal therapy, only amphotericin B is considered as systemically applicable antifungal compound. This is extracted from the fermentation of *Streptomyces nodosus* and usually considered as a frontline drug for treating *Histoplasma, Cryptococcus*, and zygomycetes species-borne infections and blastomycosis (mostly skin infections). Nystatin, produced by soil-borne *Streptomyces noursei,* is another antifungal agent used in treating immune-compromised cancer patients receiving chemotherapy. It has been proven effective in controlling fungal infections in AIDS patients and is one of the most prominent antifungal agents used in laboratories. Compared to other antifungal therapeutics, the resistance development against polyene macrolides is extremely rare, but the usage of these drugs is often limited because of the extreme toxicity (Lee et al. 2012).

Sceliphrolactam is a novel antifungal compound isolated from *Streptomyces* sp. In 2011, it consists of polyunsaturated and polyoxygenated 26-membered macrocyclic lactam. Sceliphrolactam is a potent antifungal activity against amphotericin B-resistant *C. albicans* (Jakubiec-Krzesniak et al. 2018). 15-glycidylfilipin III; 16α, 17α-epoxyfilipin V; and 16β, 17β-epoxyfilipin V are other three novel antifungals of the polyene macrolide class, and they were isolated from the cultures of a soil actinomycete, *S. lavenduligriseus,* in 2016 (Yang et al. 2016). Six more new antifungal compounds (cyclic octadepsipeptides, enduspeptides A-F) were identified from *Streptomyces* sp. strain isolated from a soil sample collected from a coal mine of China in 2017 (Chen et al. 2017). Cet1p RNA 5′-triphosphatase inhibitors, designated as kribellosides A-D, were isolated from *Kribella* MI481-42F6 in soil sample (in Japan) having antifungal effect. There is a report on purifying another antifungal compounds, mohangamides A and B from marine *Streptomyces* sp. SNM55 (in Korea), whereas neomaclafungin A was isolated from *Actinoalloteichus* sp. NPS702 (from marine sediment in Japan). There are many other compounds of Actinobacteria with antifungal effect is reported, such as actinomycins, amphotericin B, bonactin, blasticidin, candicidin, carbomycin, chandrananimycin, fungichromin, galbonolides, guanidylfungin, jadomycin, kasugamycin, kitamycin, natamycin, oligomycin, phenylacetate, polyoxin B, resistomycin, and anthraquinones synthesized by *Streptomyces anulatus, Streptomyces nodosus, Streptomyces* spp., *Streptomyces griseochromogenes, Streptomyces griseus, Streptomyces* spp., *Actinomadura* sp., *Streptomyces padanus, Streptomyces galbus, Streptomyces violaceusniger, Streptomyces venezuelae, Streptomyces kasugaensis, Streptomyces* spp., *Streptomyces natalensis, Streptomyces diastatochromogenes, Streptomyces humidus, Streptomyces cacaoi, Streptomyces canus, and Micromonospora* spp., respectively. Asterobactin, borrelidin, diazepinomicin, lomaiviticins, lupinacidins, and nikkomycin are also antifungal compounds derived from several Actinobacteria spp. such as *Nocardia asteroids, Streptomyces* spp., *Micromonospora* spp., *Micromonospora* spp., *Micromonospora* spp., and *Streptomyces tendae,* respectively (Igarashi et al. 2012; Sato et al. 2012; Bae et al. 2015).

18.4 Antivirals of Actinobacteria

A sundry of antibiotic compounds produced by Actinobacteria demonstrates antiviral properties. For instance, amphotericin, doxorubicin, and pepstatin have been identified to exert activity against HIV like retroviral infections and both hygromycin and kanamycin have antiviral activity on influenza viruses (family Orthomyxoviridae). Furthermore, novobiocin and daunomycin are known to exhibit antiviral activity against herpes virus. However, neither of these compounds are not potent enough to be developed as commercial antiviral drugs and compared to antiviral agents like acyclovir, zanamivir, peramivir, and baloxavir marboxil, and therefore, the attention given remains at a much lower ebb. Recently, a novel class of lantibiotics produced by an *Actinomadura* species came under the spotlight due to their antiviral activity, and namely, these compounds are known as "Labyrinthopeptins." Although the history of these compounds dates back to late 1980s, it was only given attention after about 20 years of its discovery (Wink et al. 2017).

Salmon pink actinobacterium, *Actinomadura namibiensis,* found in harsh environmental niche of Namib Desert, Namibia, reported to produce a novel peptide that exhibited weak antiviral activity against influenza, herpes viruses, and adenoviruses. However, the exploitation process came into an abrupt stop because of the difficulties of solubility and structure determination. But back in 2008, a research group at Sanofi-Aventis Pharmaceuticals restarted the project and discovered 3 structural variants of labyrinthopeptins (A1-A3). Labyrinthopeptin A1 was the most powerful out of the three, and the compound itself and chemically modified variants of it are now being tested for its effectiveness for controlling HIV (human immunodeficiency virus) and HSV (herpes simplex viruses). The research for antiviral agents has become an area of utmost importance because most of the antivirals available to date are failing at their game. Recent surveys have unveiled that most antiviral therapeutics are not able to control progressed infections and sometimes infections at the early stages (Wink et al. 2003, 2017).

In 2003, a promising producer of novel bioactive compounds belonging to the genera *Streptomyces* was discovered from a Brazilian tropical forest soil. This has proven to inhibit the propagation of an acyclovir-resistant herpes simplex virus type 1 strain on HEp-2 cells at non-cytotoxic concentration. Marine Actinobacteria are also found to be great sources of novel broad-spectrum antiviral agents. A marine actinobacterium *Streptomyces kaviengensis,* a novel actinomycetes isolated from marine sediments, was discovered to produce another potent antiviral agent, antimycin A. This compound has proven effective against western equine encephalitis virus in cultured cells. Moreover, antimycin A has proven effective against RNA viruses of Togaviridae, Flaviviridae, Bunyaviridae, Picornaviridae, and Paramyxoviridae families. A study done in Brazil, in 2015, has also unveiled extracts from two termite-associated strains of *Streptomyces chartreusis* with in vitro antiviral activity against bovine viral diarrhea virus (BVDV). A Korean research has also unveiled several other promising antiviral agents (xiamycins C-E) from fermentation of *Streptomyces* sp. HK18 from the topsoil of a Korean solar saltern. Xiamycin D, of these compounds, has indicated the most predominant activity against porcine epidemic diarrhea virus (PEDV). Liu and his colleagues discovered a novel inhibitor of HIV-1 protease, 4862F derived from the fermentation of *Streptomyces albosporus* I03A-04862. This has been proved as a novel therapeutic tool against HIV infection. Furthermore, in 2018 a linear peptide and a novel pyrrolidine derivative Ahmpatinin iBu was identified from the fermentation of *Streptomyces* sp. CPCC 202950, which is another novel HIV-1 protease inhibitor (Sacramento et al. 2004; Liu et al. 2013; Kim et al. 2015; Padilla et al. 2015; Chen et al. 2018).

The quartromycins produced by a strain of Amycolatopsis orientalis is reported to be effective against herpes simplex virus type 1 and influenza virus A. A study done with five strains of extremophilic actinomycetes isolated from the unique ecosystems of Kazakhstan has shown the extracts of the Actinomycetes with antiviral properties against the influenza viruses, Sendai virus, and Newcastle disease virus. Astonishingly, these activities were comparable to those shown by Rimantadine and Tamiflu drugs, and "Virospan" and "Flavovir" plant preparations and structural characterization of exact active components are currently being done in order to assess the feasibility as commercial antivirals. Extract of marine Actinobacteria, *Nocardia alba,* has also shown antiviral effect against Newcastle virus and infectious bursal disease virus of poultry. Benzastatin C, a 3-chloro-tetrahydroquinolone alkaloid isolated from Streptomyces nitrosporeus, has also shown an antiviral potential against herpes simplex virus type 1 (HSV-1) and herpes simplex virus type 2 (HSV2). Additionally, anti-adenoviral property of butenolides 1a, 1b, 2, 3, and 4 was obtained from crude extracts of Streptomyces sp. A novel compound abyssomicin 2 isolated from marine action bacteria Streptomyces sp. was examined in in vitro potential as viral re-activator (Sanglier et al. 1993; Lee et al. 2007; Janardhan et al. 2018; Berezin et al. 2019).

18.5 Antiparasitics of Actinobacteria

Actinobacteria are also known to produce several potent antiparasitic compounds as well. They have shown substantial effect against parasitic diseases caused, especially by Plasmodium sp., Leishmania sp., Trypanosoma brucei, and helminthes. The avermectins and milbemycins, which are currently used as anthelminthics/insecticides in veterinary practice, both have actinobacterial origins and belong to the 16-ring macrolactone chemical group. This compound group was first detected as a fermentation by-product of *Streptomyces avermectinius.* Milbemycins are also 16-ring macrolides and fermentation products of a *Streptomyces* strain. To date, approximately 13 different natural milbemycins have been discovered. That is, *Streptomyces hygroscopicus* subsp. *Aureolacrimosus* is known to produce milbemycin oxime. In the recent years, two other novel antiparasitic compounds were detected in *S. nanchangensis,* which are nanchangmycin, which was proven effective against coccidial parasites and meilingmycin, which has a chemical profile similar to avermectins. *S. axinella* PO1001 were also detected to produce antagonists inhibiting *Trypanosoma brucei,* and antibiotics such as valinomycin (isolated from *S. flavissimus, S. roseochromogenes, S. griseus* var. *flexipartum*), staurosporine (isolated from *Streptomyces* sp. strain 11), and butanolides have been detected to exhibit antitrypanasomal and anti-leishmaniasis activity. Recently, another two antibiotics, sinefungin VA and dehydrosinefungin V, have been isolated from *Streptomyces* sp. K05-0178 with proven activity against *Trypanosoma brucei* (Solecka et al. 2012).

As an antimalarial agent, cyclic heptapeptide compound cyclomarin A has been isolated from Streptomyces sp. Activity of this compound against Plasmodium falciparum is reported to be due to the inhibition of the plasmodial enzyme diadenosine triphosphate hydrolase (PfAp3Aase). Also, the trioxacarcins A, B, and C were obtained from Streptomyces ochraceus and *Streptomyces bottropensis* as antimalarial agents and also it exhibits antitumor activity potent antibacterial activity against *Staphylococcus aureus*. Cyclic depsipeptide valinomycin and indolocarbazole alkaloid staurosporine were obtained from Streptomyces sp. isolated from Mediterranean sponge reported to exhibit antiparasitic activities specifically against Leishmania major against Trypanosoma brucei for staurosporine. In another study, 90 actinobacterial strains isolated from marine sponges collected from offshore Egypt also exhibited the antifungal, antibacterial, and antiparasitic activity including against *Leishmania major* and *Trypanosoma brucei*. Another antiparasitic drug against *Trypanosoma brucei*, i.e., diazepinomicin, is isolated from marine Actinobacteria (Maskey et al. 2004; Cheng et al. 2015; Bürstner et al. 2015; ul Hassan and Shaikh 2017).

Several anti helminthic compounds also have been identified from Actinobacteria. Larvicidal and acaricidal compound ((2S,5R,6R)-2-hydroxy-3,5,6-trimethyloctan-4-one) has been isolated from Streptomyces with marine origin. The compound is shown to have larvicidal efficacy against larvae of *Rhipicephalus microplus*, *Anopheles subpictus*, and *Culex quinquefasciatus*. Polyketide biomolecule isolated from Streptomyces sp. of marine origin has shown a remarkable larvicidal, antifeedant, and growth-arresting activity against polyphagous pests, antifeedant activity of against *Spodoptera litura* and *Helicoverpa armigera,* and larvicidal activity against *Spodoptera litura* and *Helicoverpa armigera* (Deepika and Kannabiran 2010; Arasu et al. 2013).

18.6 Antitumor/Anticancer Compounds of Actinobacteria

Many of the antitumor drugs reported are from marine Actinobacteria, and the first of this type of drugs identified in Actinobacteria is Actinomycin produced by *Streptomyces antibioticus* subsp. *Antibioticus*. Actinomycin D, a transcription inhibitor (originated from *Streptomyces parvulus*) of this class, demonstrates excellent anticancer activity and is still in use of antitumor. Daunomycin, Daunorubicin (*Streptomyces peucetius*), and Adriamycin belonging to the class of anthracyclines are some of the other common antitumor drugs originating from Actinobacteria.

Diazepinomicin is a farnesylated dibenzodiazepinone alkaloid produced by a Micromonospora strain with strong antitumor, antimicrobial, and anti-inflammatory therapeutic potential. Streptochlorin isolated from marine Streptomyces roseolilacinus proved itself as antiangiogenic, antitumor, anti-inflammatory, and antibiotic. Salinosporamide A is an unusual bicyclic beta-lactone gamma-lactam produced by Salinispora tropica with anticancer and antimalarial effect. A novel 16-membered macrolide Chalcomycin, Manumycins, Luminacin, Strepsesquitriol, Thiocoraline, and Carcinoids obtained from Streptomyces sp. also have shown the antitumor effect. Compounds such as 1,8-dihydroxy-2-ethyl-3-methyl anthraquinone, 1-hydroxy-1-norresistomycin, anthracyclines, arenimycin, butenolides, carbomycin, chinikomycins, daryamides, marinomycins A–D, mechercharmycins, mitomycin C, piericidins, proximicins, saliniketal, salinosporamide A, streptokordin, elaiomycins B and C, *N*-[2-hydroxyphenyl)-2-phenazinamine (NHP), chromomycins B, A2, and A3, anthraquinones, asterobactin, borrelidin, diazepinomicin, lomaiviticins, and lupinacidins derived from different actinobacterial spp. are known to exhibit antiviral properties. Epirubicin is another anthracycline group compound that has a better therapeutic profile with much less toxication reports than doxorubicin. It was approved for clinical usage in 1999 by the FDA and is used in the treatment of breast cancer, ovarian cancer, lung cancer, and leukemia (Charan et al. 2004; Prudhomme et al. 2008; Lee et al. 2013).

Novel anthracyclines like tetracenoquinocin, 5-iminoaranciamycin, cytotoxic polyketide, and mayamycin produced by *Streptomyces* sp. exert anticancer effects on human cancer cell lines (Solecka et al. 2012). Apart from anthracyclines, bleomycin and mitomycin C coming under the glycopeptide group of antibiotics are two other anticancer drug isolated from *Streptomyces verticillus* and *Streptomyces caespitosus*, respectively. Streptozotocin, which is a fermentation by-product of *Streptomyces achromogenes*, has been also identified to demonstrate selective toxicity on pancreatic β-cells and therefore adopted for treating pancreatic cancers (Solecka et al. 2012). Several other actinobacterial anticancer compounds were also discovered during the past decade, i.e., 11-methoxy-17-formyl-17-demethoxy-18-*O*-21-*O*-dihydrogeldanamycin originated from *S. hygroscopicus* A070101, which is an analog of geldanamycin (proved effective in treating breast cancer MCF-7, lung carcinoma COR-L23, skin melanoma SK-MEL-2), resistoflavin, a quinone-related antibiotic biosynthesized by *S. chibaensis* $AUBN_1/7$ (effective in treating HMO2 (gastric adenocarcinoma) and HePG2 (hepatic carcinoma)), marinomycin A isolated from marine Actinobacteria, daryamide C, lucentamycins (A-D) isolated from *Nocardiopsis lucentensis* (proven effective in destroying HCT-116 cells of human colon carcinoma), mansouramycins A-D isolated from *Streptomyces* sp. Mei37 (antitumor activity against lung cancer, breast cancer, melanoma, and prostate cancer cells), carbomycin isolated from *Streptomyces* sp. NTK 937 (effective in treating gastric adenocarcinoma cell lines (AGS), hepatocellular carcinoma (HepG2), and breast carcinoma (MCF7)), phenazine-type secondary metabolites like dermacozines extracted from the fermentation of *Dermacoccus abyssi* (effective in treating leukemia), teleocidin analog, and JBIR-31 produced by *Streptomyces* sp. NBRC 105896 that shows weak cytotoxicity in human cervical carcinoma

Table 18.2 Common antitumor/anticancer compounds of actinobacterial origin

Compound	Microorganism
1,8-Dihydroxy-2-ethyl-3 m. anthraquinone, anthracyclines, carbomycin, chinikomycins, streptokordin, daryamides, mitomycin C, piericidins, elaiomycins B and C, chromomycins B, A2, and A3, borrelidin	*Streptomyces* spp.
1-Hydroxy-1-norresistomycin	*Schisandra chinensis*
Arenimycin	*Salinispora arenicola*
Butenolides	*Streptoverticillium luteoverticillatum*
Marinomycins A–D	*Marinispora*
Mechercharmycins	*Thermoactinomyces* spp.
Proximicins	*Verrucosispora* spp.
Saliniketal, salinosporamide A	*Salinispora* spp.
ZHD-0501, IB-00208	*Actinomadura* spp.
N-[2-hydroxyphenyl)-2-phenazinamine (NHP), asterobactin	*Nocardia* spp.
Anthraquinones, diazepinomicin	*Micromonospora* spp.
LL-E33288 complex	Micromonospora spp.
Lomaiviticins	
Lupinacidins	

HeLa cells and human malignant pleural mesothelioma (MPM) ACC-MESO-1 cells, JBIR-69 (a novel compounds of soil Actinobacteria *Streptomyces* sp. OG05 having anticancer properties against human acute myelogenous leukemia HL-60 cells), angucycline representative, JBIR-88, and a butenolide, JBIR-89, are biosynthesized, respectively, by *Streptomyces* sp. RI104-LiC106 and RI104-LiB101 (exhibit cytotoxicity against HeLa and ACC-MESO-1 cells) and tetralones like tetralone D (antitumor agents). Cancer drug formulation is usually done by combining toxins to carrier antibodies, and when to antibody–antigen reaction occurs at the tumor/cancer cells, the immunotoxin enters the cell, and the active compound (toxin) kills the tumor. The enediyne class of cytostatics is one of the prominent drugs in this group. Esperamicins (originated from *Actinomadura verrucosospora*) and calicheamicins (originated from *Micromonospora echinospora*) are two of the enediyne class drugs, which are highly effective antitumor agents. Apart from aforementioned, leptomycin (isolated from *Streptomyces canosus*) is another anticancer immunoconjugate, which is known to inhibit the chromosomal region maintenance (CRMI) of cells (Kudo et al. 1998; Galm et al. 2005; Kane and Hanes 2017) (Table 18.2).

18.7 Immunomodulator Compounds of Actinobacteria

Actinobacteria are known to produce a range of immunomodifiers that could alter the immune response in both humans and animals. These compounds behave either as therapeutic immunosuppressants or as immunoconjugates in drug delivering.

Amastatin, phenicine, and bestatin originated from *Streptomyces* spp. have been identified to enhance the immune response of the mouse models, and compounds like rapamycin, a product of *Streptomyces rapamycinicus*, sirolimus (originated from *S. hygroscopicus*), tacrolimus (FK-506), and FR-900506 produced by *S. tsukubaensis* are well-recognized immunosuppressants originating from Actinobacteria. Usually, these immunosuppressants are used in conjunction with ciclosporin and corticosteroids after transplantations to minimize the risk of organ/tissue rejection. Several other variants of rapamycin were recently identified in actinoplanes, and FK 506 on the other hand is an inhibitor of interleukin/IL-2, c-myc, IL-3, TNF-α, IFN-ϒ production, mixed lymphocyte reaction, and other interferons. It blocks the cytotoxic T cells by binding to a cytosolic receptor, an immunophilin (FKBP12) within the target cell, thereby immobilizing the serine–threonine phosphatase calcineurin resulting in minimization of the activity of the immune system on the transplanted organs. FK 506 could also activate factor C induction cascade of the cells. Cyclic peptides and thalassospiramides A and B were isolated from a new member of the marine α-proteobacterium Thalassospira with immunosuppressant activities (Kino et al. 1987; Huang et al. 1995; Kumar and Goodfellow 2008; Oh et al. 2008; Wink et al. 2017).

There are several anti-inflammatory compounds that have been identified from Actinobacteria. Salinamides A-F are bicyclic depsipeptides produced by marine-derived Streptomyces sp. isolated from the jelly fish Cassiopeia xamachana. Pyrazine-type bioactive molecule griseusrazin A produced from marine Streptomyces griseus inhibits the inflammatory mediator's production. Quinoline alkaloids actinoquinolines A and B, produced by marine-derived Streptomyces sp., were isolated from California sediment sample. Axinelline A derived from Streptomyces axinellae exhibited a potent anti-inflammatory activity. Marine Streptomyces derivatives of phenazines 1 and 2 also have anti-inflammatory properties in addition to the anticancerous properties (Kondratyuk et al. 2012; Hassan et al. 2015, 2016; Lee et al. 2016).

18.8 Role of Actinobacteria in Industrial Enzyme Production

Apart from the production of bioactive agents such as antibiotics, antivirals, and anticancer compounds, they are also the main driving force behind the modern-day white biotechnology. An array of both terrestrial and marine Actinobacteria is currently being used for the production of different enzymes, organic acids, and recently for the production of biofuel. Attributed to their involvement of actinomycetes in these fields, the load of fossil fuels being inputted to the processes has exponentially reduced, and despite the ecological benefits, their involvement has demonstrated elevated efficiency and purity of most of the end products in contrast to chemical industries of the early age.

Enzymes are considered as one of the most important catalysts of biological origin that are key components in the research disciplines of microbiology, biochemistry, molecular genetics, and vital biological machines for most of the linked industries (i.e., food industry, fermentation, and textile to paper industries). Actinobacteria are known to synthesize a variety of bio-catalytic enzymes, and in statistical terms, they account for more than 50% of the industrial enzymes. The reasons behind this are quite simple. Most of the actinobacterial species are the end organisms of the material cycle with a natural ability to degrade and biotransform organic compounds and bio-convert both urban and agricultural organic wastes into novel products. Due to their diversity, they are a repertoire of enzymes involved in decomposing and bio-converting a range of substrates such as lignocelluloses, lignin, cellulose, and plant residues. As per the BRENDA database, around 66,900 enzymes of industrial importance have been utilized from various genera of Actinobacteria including cellulases, proteases, lipases, xylanases, pectinases, amylases, phytases, keratinases, glucose oxidases, and chitinases (Shivlata and Satyanarayana 2017).

18.9 Amylases

Amylases have the ability to hydrolyze starch into dextrins and much smaller polymeric glucose units, and majority of amylases are being utilized in bakery, alcohol brewing, juice, textile, laundry detergent production, pulp and paper production, fructose syrup production industries, and production of cyclodextrins for the pharmaceutical industry. They are grouped into endo- and exo-types, and they account for over 30% of the global enzyme production with most of the industrially important amylases extracted from the Actinobacteria genus. *Streptomyces* and *Nocardia* are commonly considered as a potential source of amylolytic enzymes. Currently, research is being carried out in order to exploit for thermostable amylolytic enzymes and Syed et al. (2009) reported optimized extracellular α-amylase production from alkali-thermo-tolerant strain *Streptomyces gulbargensis* DAS. Recently, discovered amylases of *Streptomyces hygroscopicus* and *Streptomyces praecox* could be effectively used for high maltose sugar syrup production and amylases of *Nocardiopsis* spp. and *Streptomyces erumpens* high-temperature industrial processes (>70 °C). Haloalkaliphilic *Saccharopolyspora* spp. strain A9 producing surfactants and oxidant stable, calcium ion-independent and detergent stable 471 amylase has been reported (Chakraborty et al. 2011). In a study done in Indonesia, the researchers were able to isolate amylolytic Actinobacteria from the rumen fluid of ruminants. Several other Actinobacterial spp. such as *Streptomyces hygroscopicus, Streptomyces limosus, Streptomyces praecox, Streptomyces erumpens, Nocardiopsis* spp., *Thermomonospora curvata, and Thermobifida fusca* are known to produce amylases (Syed et al. 2009; Wink et al. 2017; Salwan and Sharma 2018; Ratnakomala and Perwitasari 2020).

18.10 Lignocellulolytic Enzymes (Cellulases and Laccases)

Lignocelluloses are the main components of plant biomass, and these include polymeric compounds such as lignin, cellulose, and hemicellulose. These complex polysaccharides are hydrolyzed by the activity of cellulases, hemicellulases, and ligninolytic enzymes of microorganisms, and one of the most industrially important lignocellulolytic enzymes that Actinobacteria produce is cellulases. Genomic studies have unveiled an array of putative cellulose-degrading enzymes belonging to families GH5, GH6, GH8, GH9, GH12, GH48, and GH51 in the genomes of the Actinobacteria. These gene expression products/enzymes are exogenously secreted by actinobacterial species that grow on cellulosic substrates, and three main groups of cellulases have been identified, which are endo-(1, 4)-β-D-glucanase (digests cellulose to produce glucan chains through random catalysis of glycosidic bonds), exo-(1, 4)-β-D-glucanase (cellulose is digested to produce β-cellobiose), and β-glucosidases (digests β-cellobiose disaccharides to produce glucose). These different variants of cellulases have varying catalytic mechanisms following different reaction kinetics resulting in different end products. Cellulases are generally adopted as bio-detergents controlling color brightening and softening, fabric softeners/textile finishing compounds, especially in denim fabrics/biostoning of jeans, bioethanol/biofuel production, color extractions of juices, paper and pulp deinking/fiber modifier compounds/pretreatment of biomass that contains cellulose to improve nutritional quality of forage, and pretreatment of industrial wastes; bio-converters/laboratory applications; etc. Majority of cellulase-producing Actinobacteria belongs to mesophilic category (*Streptomyces antibioticus, Streptomyces flavogriseus, and Acidothermus cellulolyticus*), and some belongs to thermophilic group (*Streptomyces transformant* T3-1, *Thermomonospora curvata*). Furthermore, several researches have unveiled alkali-tolerant cellulase variants in *Streptomyces* and *Thermoactinomyces* spp. and species with the capability of producing cellulase from agro-industrial residues (*Streptomyces noboritoensis, Streptomyces viridobrunneus*). Additionally, *Thermomonospora* spp., *Streptomyces antibioticus, Streptomyces flavogriseus, Acidothermus cellulolyticus, Streptomyces transformant, Thermomonospora curvata, Thermobifida fusca, Thermobifida halotolerans, Streptomyces thermodiastaticus,* and *Streptomyces ruber*are reported to produce cellulases (Techapun et al. 2003; Karmakar and Ray 2011; Kuhad et al. 2011; Palaniyandi et al. 2014; Wink et al. 2017).

Hydrolysis of lignin is sequentially catalyzed by laccases or laccase-like multicopper oxidases, manganese peroxidases, and lignin peroxidases. Laccases producing Actinobacteria include several species of *Streptomyces: Streptomyces cyaneus, S. lavendulae, S. griseus, S. coelicolor, S. ipomoea, S. psammoticus,* and other *Streptomyces* spp. To date, there are more than ten laccase enzymes with industrial significance have been recognized in *Streptomyces* spp. and major fraction of these enzymes are being adopted in stain removal from fabrics, removal of unwanted phenolic compounds from beer, wine, juices, and bakery products, deoxygenating certain packed food items, pulp/paper bleaching, and for bioremediation purposes (Fernandes et al. 2014; Salwan and Sharma 2018).

18.11 Xylanases

Another commercially important type of hydrolytic enzymes from Actinobacteria is xylanase (composed of D-xylosyl residues and arabinosyl, acetyl, and glucuronosyl units) and is the second most abundant type of polysaccharide. Xylanase production from various species of actinomycetes has been reported such as production of β-1, 4-endoxylanases, xylosidases, β-1,3-glucanases, xyloglucanases, α-*N*-arabinofuranosidases, and α-L-arabinofuranosidases (by *T. bifida*), endo- and exo-xylanases, β-mannanase, mannosidase, and xel74 (by Cellulomonas fimi), xylanases (by *Cellulomonas flavigena*), β-1,4-glucan glucanohydrolase (by *Streptomyces flavogriseus*), and β-mannosidase, β-xylanase, β-xylanase, and acetylxylan esterase (by *Streptomyces* sp. SirexAA-E). Most of the xylanases are extracted from *Streptomyces* species, and almost all of them behave as endo-types (Morosoli et al. 1986; Salwan and Sharma 2018).

Xylan is hydrolyzed into xylose by the activity of xylanases, and the complete breakdown of xylan is mediated by a combined action of α-L-arabinofuranosidases, endo-β-1, 4-xylanases, α-glucuronidases, β-D-xylosidases, ferulic acid esterases, and acetyl xylan esterases. Xylanses are widely used as food modifier in confectionaries' sector. Furthermore, it is also used in biofuel production from organic debris, animal feed modification, paper and pulp milling industries, and as a bioleaching agent. Similar to cellulases, xylanases are also found in the form of enzyme complexes, which exert a substrate-specific function on different polymeric forms of xylan. These are usually mesophilic enzymes and have an optimal working temperature ranging from 40 to 60 °C. Xylanases extracted from *Actinomadura* spp., *Thermomonospora* spp., and *Nonomuraeaflexuosa* are on the other hand highly heat-stable and have an optimal function at temperatures ranging from 60 to 70 °C. *Thermomonospora fusca* is another actinobacterium that can produce β-1,4-endoxylanases, xylosidases, β-1,3-glucanases, xyloglucanases, α-*N*-arabinofuranosidases, and α-L-arabinofuranosidases, and these enzymes could withstand both high temperatures and pH levels from 5 to 8. Similar studies done using *Cellulomonas fimi* have reported endo- and exo-xylanases, β-mannanase, mannosidase, and xel74. Furthermore, *Cellulomonas flavigena* has been reported for the production of xylanases belonging to family GH10, GH11, GH13, GH16, GH26, GH30, GH43, GH51, and GH81, while parallel studies have characterized SirexAA-E for β-mannosidase, β-xylanase, β-xylanase, and acetylxylan esterase in *Streptomyces* spp. Recently, trials done using xylanases of *Streptomyces* spp. and *Kocuria* spp. proved that they could be effectively used as animal feed modifiers (McCarthy et al. 1985; Christopherson et al. 2013; Takasuka et al. 2013; Gomez del Pulgar et al. 2014; Wink et al. 2017).

18.12 Proteases

Proteases account for the largest group of commercially utilized enzymes and are used mainly in the silk, leather dehiding, processing and tanning, food and beverage, detergent, laboratory reagent production, cheese, and dairy and alcohol brewing industries. Maong, the bacterial proteases, and actinobacterial proteases (from members of Actinomycetes such as Streptomyces, Nocardia, and Nocardiopsis) have been discovered from many ecosystems. Thirty culturable Actinobacteria (belonged to eight genus in the phylum Actinobacteria: Arthrobacter, Brevibacterium, Curtobacterium, Janibacter, Knoellia, Rhodococcus, Streptomyces, and Thermoleophilum) were reported with the ability of producing extracellular enzymes (protease, gelatinase, and cellulase enzymes) with their biotechnological potential. Leather industries use proteases of *Nocardiopsis prasina* and *Saccharomonospora viridis*, and *Streptomyces* spp. for degradation of animal skins for further processing and as a detergent additive. *Streptomyces* proteases isolated from *S. albus*, *Streptomyces* strain BA7 and *S. pactum*, *S. fradiae*, *S. thermoviolaceus*, and *S. sclerotialus* have keratinolytic activity, and therefore, they are used in agricultural industries like feather meal production and as bioremediation components in textile, medicine, cosmetic, leather, feed, and poultry processing industries. Production of alkaline proteases from different Actinobacteria (*Nesterenkonia* sp., *Nocardiopsis* spp., *Actinopolyspora* spp.) is reported. Also, it is reported to produce alkaline serine proteases by alkalophilic actinomycete, *Nocardiopsis dassonvillei* (Moreira et al. 2001; Korkmaz et al. 2003; De Azeredo et al. 2004; Mitra and Chakrabartty 2005; Esawy 2007; Brandelli 2008; Shata and Farid 2012; Suthindhiran et al. 2014; Lamilla et al. 2017).

18.13 Pectinases

Pectinases are widely being used in food industries, and these enzymes are oriented toward modification of fruit-based products, where general applications are clarification of fruit juices, wines, and texture modification of fruit pulps, jams, cordials, and chutneys. The pectinases of *Streptomyces* species hold a predominant role in these industries. The endophytic actinobacterium isolated from tomato plant roots (*Lycopersicon esculentum*) have shown the ability of producing pectinases in addition to other enzymes such as amylase, cellulose, lipase, esterase, caseinase, gelatinase, and catalase. Hydrolytic depolymerases, which is a kind of pectinases, have been isolated, purified, and characterized from *Streptomyces lydicus* (Niladevi and Prema 2008; Minotto et al. 2014).

18.14 Chitinases

Chitin is the second most common biopolymer, which is an essential component of both fungi (cell wall component), and crustaceans and insects (exoskeleton). Chitinases are the enzymes that hydrolyze the chitin, and chitin is available in three polymeric forms. Chitinases play an important role in recycling the wastes generated by the seafood industries (i.e., shrimp processing) and attributed to the degradation capability of fungal cell walls, and this enzyme is also used in production of antifungal formulation. Chitinases are produced by many bacteria, plants, and animals among the bacteria. Actinobacteria play a major role in chitin breakdown. Actinobacteria like *Nocardiopsis prasina, Streptomyces* spp., *S. coelicolor*, and *S. aureofaciens* are well-known chitinase producers, and *S. griseoloalbus*, *S. clauifer*, *S. anulatus,* and *S. griseus* spp. produced chitinase. Terrestrial Actinobacteria are the best adapted prokaryotes to use chitin, and many Streptomyces (*S. Lividans*, *S. Olivaceoviridis*, *S. coelicolor,* etc.) are among them. In a study aiming to explore the microbial potential of a semi-arid sandy soil from south-central Algeria in order to isolate new chitinolytic Actinobacteria, they found that isolates were belonging to Streptomyces or Micromonospora genera with seven possibly new bacterial species. A study done by Kawase and his group investigated Actinobacteria (from the Japan Collection of Microorganisms) for the production of chitinase (Kawase et al. 2004; Lacombe-Harvey et al. 2018; Gasmi et al. 2019).

18.15 Lipolytic Enzymes, Lipases, and Phospholipases

Lipases enzymes have the capability to hydrolyze ester bonds in fats/triglycerides to form diglycerides, monoglycerides, fatty acids, and glycerol. Phospholipases are capable of selective cleavage of ester bonds in glycerophosphatides. Both of these enzymes are distributed throughout all the kingdoms and generally adopted for modification of organoleptic properties. When it comes to non-lipase lipolytic enzymes, one such class of enzymes is currently identified in Actinobacteria. These are known as cholesterol esterases. Lipases were initially reported in *Streptomyces* genus, and recently, a lipase of the arylesterase group able to hydrolyze specifically phthalate esters to a free phthalic acid and simple *n*-alcohols was isolated from a *Rhodococcus* (*Nocardia*) *erythropolis.* Parallel studies on *S. lavendulae* have also indicated to produce cholesterol esterases. When it comes to phospholipases, they are classified into four groups A (*S. cinnamomeus*), B (*S. hiroshimensis*), C (*S. griseus*), and D (*S. chromofuscus*). Variants like phospholipase D have a clinical importance because of their involvements in the determination of serum choline phospholipids. Study conducted by Sutto-Ortiz found novel microbial phospholipases from actinomycetes (Streptomyces, Micromonospora). Production of lipases by Actinobacteria was found in some other studies as well, and some of them are phospholipase 2 by *S. violaceoruber*, and phospholipase 1 and phospholipase B by Streptomyces genus (McMahon et al. 2012; Takemori et al. 2012; Matsumoto et al. 2013).

18.16 Other Enzymes

Apart from the main industrial enzymes discussed, there is a whole bunch of other different types of enzymes, which are being adopted for industrial usage. Some of these include dextranases, aminoacylases, β-*N*-acetyl-D-glucosaminidases, 1, 3-α and 1, 3-β glucanases, *N*-acetylmuramidase, neuraminidases, peptide hydrolases, L-asparaginase, and penicillin amidase. Although these enzymes have potential implementations in different industries, the scale of production is comparatively low (Wink et al. 2017) (Table 18.3).

18.17 Role of Actinobacteria in Agriculture, Crop Protection, and Livestock Industries

A great role of Actinobacteria has been identified in many aspects in agriculture, crop protection, and livestock industries. Plant growth promoters (PGPs) are natural fertilizers, especially designed for accelerating the growth of the plants without disturbing their natural physiological actions. Most of the PGPs were, however, previously limited to plant-based natural products, and recently, there has been a trend shift where plant beneficial symbiotic bacteria have been prioritized in the sustainable agriculture as an alternative to chemical fertilizers. Soil-borne Actinobacteria play a lead role in modern-day PGP, and their spore-forming/biocontrol properties have made them an excellent choice for PGP development. Since most of the epiphytic, endophytic, and rhizosphere Actinobacteria are capable of nitrogen fixation, phosphate solubilization, siderophore production, production of

Table 18.3 Other actinobacterial enzymes of industrial importance

Enzyme	Microorganisms
Aminoacylase	*S. olivaceus, S. roseiscleroticus, S. sparsogenes*
	S. tuirus, S. olivaceus, Streptoverticillium spp.
	Mycobacterium smegmatis
β-*N*-Acetyl-D-glucosaminidase	*Streptomyces griseus*
1, 3-α and 1, 3-β glucanase	*S. murinus, Streptomyces* spp. K.27-4
	S. cellulosae, S. werraensis, S. chartreusis
	Actinomyces levoris, S. globisporus
Neuraminidase	*Actinomyces naeslundii, A. viscosus*
	S. griseus, Mycobacterium spp., *Nocardia* spp.
	S. sapporonensis, Streptomyces fradiae
	S. rimosus, Thermomonospora spp.
L-Asparaginase	*Mycobacterium bovis, M. tuberculosis*
	Pseudonocardia endophytica VUK-10
	Streptomyces spp. WS3/1, *S. acrimycini* NGP
	Streptomyces spp. (SS7), *S. halstedii*
Penicillin amidase	*Mycobacterium, Nocardia, Streptomyces* spp.

antifungal metabolites, phytohormones, and volatile organic compound, they are currently being studied in different formulations of biofertilizers. Actinomyces species, already being the major fraction of the soil microbiota, are capable of enhancing the nutrient mobility of the soil, and *Streptomyces* are responsible for majority of this work. Apart from *Streptomyces*, species like *Nocardia*, *Micromonospora*, and *Streptosporangium* spp. play a supportive role in the soil behaving as potential PGP microbes. PGP Actinobacteria behave as biological nitrogen fixers (BNFs) converting gaseous nitrogen into ammonium and nitrate forms. This cuts off the requirement of synthetic fertilizers and ensures optimal soil health in the long run. Recent studies show that symbiotic Actinomyces species like *Frankia* spp. are highly effective BNFs that are capable of providing the plants with necessary nitrogen through root invasion. Other endophytic actinobacterial species like *Agromyces*, *Arthrobacter*, *Micromonospora*, *Corynebacterium*, *Propionibacterium*, *Mycobacterium*, and *Streptomyces* are also used as BNF. Phosphorous is another major nutrient necessary for plant growth. Although phosphorous is found both in the organic and inorganic forms, only 0.1% is available as soluble phosphorous to be absorbed by plants. *Streptomyces*, *Rhodococcus and Arthrobacter*, *Gordonia* and *Micromonospora* spp., seed, and field inoculums have proved the capability to solubilize soil phosphorus by PGP Actinobacteria. *Micromonospora endolithica, Micromonospora aurantiaca,* and *S. griseus* are now being used as phosphorous solubilizers in bean and wheat cultivations, respectively. Furthermore, the ability of Actinobacteria to produce organic acids (citric acid, gluconic acid, oxalic acid, lactic acid, malic acid, succinic acid, and propionic acid), antibiotic production (nigericin, streptomycin, etc.), volatile oil production (methyl vinyl products), and enzymes (proteases, chitinase, etc.), and ability to create a competitive microbial microenvironment (via siderophore production, extermination of pathogens) are other plus points when they are being formulated as all in one biofertilizer. Additionally, siderophore production by Actinobacteria aids in iron absorption of plants because of chelation effects that covert insoluble hydroxides and oxy-hydroxides into chelated soluble compounds. Catechols, desferrioxamines, coelichelin, hydroxamates, and heterobactin siderophores produced by *Streptomyces, Micrococcus, Microbacterium, Kocuria, Corynebacterium, Arthrobacter, Rhodococcus,* and *Nocardia* spp. play a major role serving this purpose. In addition, they behave as plant-protective agents attributed to their capability to trap iron from phytopathogens. For instance, *Streptomyces* spp. have proven effective in controlling *Rhizoctonia solani* infections, while *Streptomyces* species living in symbiosis with medicinal plants such as *S. diastaticus, S. fradiae,* and *S. collinus* are capable of controlling *Alternaria solani, Sclerotium rolfsii,* and *Fusarium oxysporum* infections. Non-Streptomyces Actinobacteria are also capable of exerting biocontrol effects, i.e., *Actinoplanes* spp. against *Pythium* damping-off and *Microbispora* spp. against *Gaeumannomyces* graminis var. *tritici* in wheat. Also, the antibiotics produced by Actinobacteria are now adopted for the production of plant protective compounds (i.e., Plantomycin®, Kasumin™, Bio-Mycin/Omycin, PH-D®, Endorse™, AFFIRM^{WDG}, Mycostop™, Micro108®). Furthermore, research indicates that production of geldanamycin by *S. violaceusniger* YCED9 and *S. hygroscopicus* was effective in protection of lettuce and potato against *P. ultimum* and scab, respectively, while secondary metabolites of *Streptomyces* spp. were able to induce biosynthesis of polyketides including orsellinic acid, lecanoric acid, cathepsin K inhibitors, S-9775A, and S-9775B in *Aspergillus nidulans* (Crawford Khan et al. 1997; Shimizu 2011; Hayat et al. 2020).

In symbiotic situations, Actinobacteria are capable of assisting the plant growth via production of components that degrade virulence factors and to increase the resistance to drought, radiation, and soil heavy metal. In recent years, plant growth-promoting rhizosphere (PGPR) bacteria were tested for their impact as pretreatments on propagules of different food crops. A trial done using *S. griseus* as seed treatments of barley, oat, wheat, and carrot indicated a significant increment in grain yield, dry foliage weight, tiller number, and advanced head emergence in cereal crops. Carrot yields were increased by about an average of 16% compared to its control treatments, and the sizes of carrot were also increased. In another research which adopted test isolates of *S. pulcher, S. canescens,* and *Streptomyces citreofluorescens*, as seed treatments for controlling *Fusarium* spp., *Verticillium* spp., early blight and bacterial canker of tomato, provided promising results by producing high yields of tomato with healthy fruits.

Recent researches prove that extracellular enzyme secretion by Actinobacteria also supports indirectly for plant growth through acceleration of the nutrient cycle, attributed to the saprophytic nature of most Actinobacteria (*Streptomyces* spp., *Nocardia* spp., etc.). Enzyme cocktails secreted by Actinobacteria include lipase, amylase, peroxidase, xylanase, chitinase, keratinases, pectinase, cellulase, and proteases, which coverts complex organic polymer forms of nutrients into simple mineralized forms. Some actinobacterial species are capable of mobilizing certain elements such as silicon, iron, zinc, selenium, boron, and molybdenum, which are essential micronutrients for plant growth. Moreover, *Mesorhizobium* spp. and *Pseudomonas* spp., which are two rhizosphere bacterial species, have proven capable of enhancing the iron (Fe) uptake by *Cicer arietinum* and certain *Streptomyces* spp. are capable of increasing the iron and zinc uptake of leguminous plants up to 40%. Additionally, there is evidence of several plant hormones including indole-3-acetic acid (IAA) and other auxins, gibberellic acid, and cytokinin have been identified in most Actinobacteria. Initially, only the genus *Streptomyces* was known for IAA production, and this includes *S. violaceus, Streptomyces scabies, S. griseus, S. exfoliatus*, and *S. coelicolor* and *S. lividans* (Manulis et al. 1994; Sathya et al. 2016; Nafis et al. 2019).

In most of the agricultural systems worldwide, chemical control of weed/insects has been the method of choice for ages. However, the alarmingly growing levels of resistance and adaptations toward these chemicals request for a sustainable biological technique, which is viable in the long run. Although researchers have exploited the potential of plants, fungi, and generic bacteria, the group Actinobacteria was not in the scope till recent years. Streptomyces group of Actinobacteria have proven to produce several commercially important herbicidal compounds (i.e., *Streptomyces saganonensis* produce herbicides/herbimycins that controls monocotyledonous and dicotyledonous weeds, anisomycin—another herbicide obtained from *Streptomyces toyocaensis*, provided the chemical basis for the development of synthetic commercial herbicides such as methoxyphenone, which is used in controlling crabgrass, bialaphos (broad-spectrum herbicide) derived from *S. viridochromogenes* and *S. hygroscopicus* also produce carbocyclic coformycin and hydantocidin). Some variants of *Streptomyces* with the potential are reported to produce phytotoxins against *Echinochilora crusgalli,* while some other *Streptomyces* are reported to produce phytotoxins controlling the growth of *Cyperus rotundus*. Other non-*Streptomyces* species including *Nocardia, Nocardiopsis, and Micromonospora* also produce necrotic phytotoxins belonging to ethylene-inducing peptide (NEP)-like family of proteins. Additionally, several different insecticides/nematicides are also being extracted from Actinobacteria. Spinosad is a biorational insecticide produced by *Saccharopolyspora spinosa,* which is used in controlling coleopteran, dipteran, lepidopteran, and thysanopteran pests. Avermectins like Milbemectin from *S. avermitilis* is used to control mites in tea and pome fruits. Ivermectin is also another avermectin used as a broad-spectrum insecticide/antiparasitic/nematocidal compound, whereas valinomycin and piericidins produced by *Streptomyces* spp. are also used as another two important insecticides. Actinobacteria also play an important role in the field of livestock production. With the introduction of antibiotics as therapeutic compounds, there had been extensive researches focused on using antibiotics as antimicrobial growth promoters (AGPs). AGPs were first discovered in a research trial, which fed chickens with a fermentation offal from the chlortetracycline production of *S. aureofaciens*. In this trial, enhanced growth rates, low morbidity levels, and high feed performances were observed and since then several AGP compounds were discovered (i.e., monensin produced by *Streptomyces cinnamonensis*, salinomycin produced by *S. albus*, narasin produced by *Streptomyces aureofaciens*, tylosin produced by *Streptomyces fradiae*, avilamycin produced by *S. viridochromogenes*, streptogramins produced by *S. virginiae*/*S. pristinaspiralis*, efrotomycin produced by *Nocardia lactamdurans* and lasalocid). Aquatic bioresource production schemes also extensively use the metabolites of Actinobacteria in order to improve the aquacultural environment quality, as biocontrol agents, and as food biomass for the nutrition for the cultured organisms. Attributed to the symbiotic roles the Actinobacteria play and their material conversion capabilities, ability to tolerate salt and different pH values, and production of probiotics, they have been identified as excellent all in one solution to be incorporated into aquacultural systems. *Bifidobacterium* spp. are used to enhance metamorphosis and survival rate from nauplius to commercial juvenile shrimp, *Micrococcus luteus* is used against furunculosis, *S. fradiae* is used to reduce dissolved ammonia levels, and *Streptomyces* species are used against vibriosis (Fabre et al. 1988; Heisey et al. 1988; Yassir et al. 2002; Aftabuddin et al. 2013; Zhou et al. 2009; Sekizawa and Takematsu 2013; Hamedi et al. 2015a, b; Saxena 2015).

18.18 Role of Actinobacteria in Bioremediation

Actinobacteria are ecologically vital, attributed to their role in recycling biomaterials, degradation of complex organic polymers, and production of different biomolecules as metabolic by-products. Majority of the problematic contaminants are xenobiotic compounds that are usually organic in nature. Bioconcentration of most of these compounds elevated exponentially with the dawn of industrialization. Actinobacteria have proven to have not only the ability to grow at the presence of xenobiotics, but also to biodegrade them into less toxic compounds. The organic acids and different extracellular enzymes, such as laccases, ligninases, peroxidase, aminoacylpropane-1-carboxylase, and peroxidases secreted by these Actinobacteria, are very important in degradation of xenobiotics (Table 18.4).

When it comes to inorganic pollutants, major focus is given for non-biodegradable heavy metals. Accumulation of heavy metal ions has been proven to induce toxicity symptoms due to bioconcentration in humans and animals. These adverse effects usually take place by altering the conformation structure of nucleic acids, proteins, and interference with oxidative phosphorylation. The bioremediation process for heavy metals could be either direct or indirect. Direct reduction in heavy metals adopts metal reductase enzymes, although the efficiency is comparatively low this is considered as the method of choice for groundwater decontamination while indirect methods involved in using sulfate-reducing bacteria to react and precipitate the metals. Actinomycetes are used for biosorption and bioaccumulation, biologically catalyzed immobilization, and biologically catalyzed solubilization of heavy metals. From the six classes of the phylum Actinobacteria, metal-tolerant microorganisms have only been detected within the class Actinobacteria and two other exception from the class *Acidimicrobiia* (Raskin et al. 1994; Onwurah and Nwuke 2004; Meena et al. 2005; Kyrikou and Briassoulis 2007; Yao et al. 2008; Agbor et al. 2011; Amoroso et al. 2013; Alvarez et al. 2017; Chaturvedi and Khurana 2019). The following table summarizes the actinobacterial species adopted in bioremediation of different heavy metals (Table 18.5).

Table 18.4 Organic pollutant converting actinobacterial species

Actinobacteria	Pollutant
Alcaligenes spp.	Polychlorinated biphenyls (PCB)
Actinomadura spp.	Polyhydroxybutyrate (PHB), polylactic acid (PLA), poly(*cis*-1,4-isoprene)
Anoxybacillus spp.	Azo-dye, aromatic hydrocarbons, petroleum chemicals, oils, pentachlorophenol
Arthrobacter spp.	Benzene, polycyclic aromatics, phenoxyacetate
Arthrobacter spp.	*S*-Triazine (atrazine), organochlorines (α, β, γ, hexachlorocyclohexane), organochlorines (α, β, endosulfan), urea (diuron)
Brevibacillus borstelensis	Polythene and related polymers
Dietzia spp.	Disodium terephthalate, aniline, petroleum hydrocarbons and crude oils, long-chain *n*-alkane, prestige oil spill
Excellospora japonica	Poly(tetra methylene succinate) PTMH and polycarbonate
E. viridilutea	Poly(tetra methylene succinate) PTMH and (PC) polycarbonate, PCL (poly-e-caprolactone)
Georgenia daeguensis	4-Chlorophenol
Kocuria rosea	DDT
Microbispora rosea subsp. *Aerate*	Poly(tetra methylene succinate) PTMH and polycarbonate
Mycobacterium spp.	Benzene, cycloparaffins
Nocardioides spp.	2,4-Dichlorophenol and 2,4,5-trichlorophenol
Nocardia spp.	Naphthalene, alkylbenzenes, phenoxyacetate, poly(*cis*-1,4-isoprene)
Pseudomonas spp.	Polychlorinated biphenyls (PCBs)
Pseudonocardia thermophila	Acrylonitrile
Rhodococcus spp.	*S*-Triazine (Atrazine), organochlorines (α, β, endosulfan)
Streptomyces spp.	Pyrethroids (deltamethrin), latex and natural rubber, poly(tetra methylene succinate) PTMH and PCL (poly-e-caprolactone), organochlorines (lindane), phenol and benzoate, polyhydroxybutyrate (PHB), halogenated hydrocarbons, phenoxyacetate
Thermobifida alba	Terephthalic acid
Thermomonospora curvata	Poly(*cis*-1,4-isoprene)
Janibacter spp.	Pentachlorophenol
Gordonia spp.	Organophosphorus (chlorpyrifos)
Xanthomonas spp.	Polycyclic hydrocarbons

Table 18.5 Actinobacterial species adopted in bioremediation of heavy metals and bioplastics

Heavy metals	Actinobacteria
Cadmium (II)	*Actinomyces turicensis* AL36Cd, *Corynebacterium kutscheri* FL108Hg, *Rhodococcus* spp., *Nocardia* spp., *Propionibacterium freudenreichii shermanii* JS, *Bifidobacterium breve* Bbi99/E8, *Frankia* spp., *Amycolatopsis* spp., *Saccharothrix* spp., *Arthrobacter rhombi*-RE, *S. rimosus, S. lunalinharesii, S. zinciresistens, Streptomyces* CCNWNQ0016T, *Streptomyces* spp. F4, Promicromonospora sp. UTMC 2243
Chromium (VI/III)	*Microbacterium liquefaciens, Streptomyces griseus, Streptomyces* spp. VITSVK9, *Streptomyces* spp. 3M, *Streptomyces* sp. R22, *Streptomyces* spp. *thermocarboxydus* NH50, *Flexivirga alba ST13, Streptomyces* sp. C35, *Streptomyces* sp. M40, *Streptomyces* sp. M46, *Streptomyces* sp. MC3, *Streptomyces* sp. MC2, *Streptomyces* sp.MC1, *Streptomyces rimosus, Streptomyces* spp. MS2, *Streptomyces griseus* NCIM 2020, *Streptomyces* sp.VITSVK9, *Streptomyces* spp., *Streptomyces* sp. AB3. *Streptomyces* sp. AB5A, *Streptomyces* sp. A160, *Streptomyces* sp. A161, *Streptomyces* sp. A164, *Streptomyces flavovirens* ON3, *Streptomyces flavovirens* M4, *Streptomyces zinciresistens, Streptomyces lunalinharesii, Streptomyces acrimycini* NGP, *Streptomyces albogriseolus* NGP, *Streptomyces variabilis* NGP, *Streptomyces rimosus, Streptomyces flavovirens* ON3, *Streptomyces flavovirens* M4
Lead (II)	*Streptomyces viridochromogenes, Streptomyce srimosus, Streptomyces plumbiresistens, Propionibacterium freudenreichii shermanii, Bifidobacterium breve* Bbi99/E8JS
Nickel (II)	*Streptomyces rimosus, Streptomyces acidiscabies* E13
Mercury (II)	*Streptomyces coelicolor* M130, *Streptomyces lividans* 1326, *Streptomyces lividans* strain 8, *Streptomyces espinosus* strain 5, *Streptomyces* CHR 3, *Streptomyces* CHR 28
Polylactic acid (PLA)	*Actinomadura, Amycolatopsis, Kibdelosporangium, Micromonospora, Nonomuraea*, Pseudonocardia, *Saccharothrix, Streptoalloteichus, Streptomyces, Thermomonospora,* and *Thermopolyspora*
Polyhydroxybutrate (PHB)	*Streptomyces* spp., *Nocardioides* spp., *Rhodococcus, Arthrobacter*

18.19 Miscellaneous Applications of Actinobacteria

Despite the major applications of Actinobacteria, there are several other potentials of actinobacterial bioprocesses/biosynthetic products that could be and currently are being leveraged for the betterment of other miscellaneous industries. Actinobacteria are being used for optimizing the biofuel synthesis processes. Attributed to the array of degradative enzymes and the ability to withstand harsh environmental conditions and chemical stresses, many Actinobacteria are used for the biocatalysis of bioethanol production. In a recent study which used a genetically modified version of *S. coelicolor* A3 harboring synthetic genes *adhB* and *pdc*, encoding an alcohol dehydrogenase and a pyruvate decarboxylase respectively was proven capable of elevating bioethanol synthesis. Furthermore, research is being carried out to identify whether microbial oils of Actinobacteria could be adopted as feedstocks for biodiesel production. Actinobacterial bioprocesses are also adopted for the synthesis of a sundry of amino and organic acids, vitamins, and probiotics. Currently, Actinobacteria such as *Propionibacterium freudenreichii, Rhodopseudomonas protamicus, S. olivaceus, Propionibacterium shermanii, Micromonospora* spp., *Nocardia gardneri,* and *Nocardia* spp. are adopted for the commercial production of vitamin B_{12}. As another example, *Corynebacterium glutamicum* is capable of producing vitamin D-pantothenate, while genetically modified variants of *Corynebacterium ammoniagenes are* used for riboflavin (vitamin B_2) production. *Corynebacterium glutamicum,* on the other hand, is the workhorse behind the commercial synthesis of many amino acids. For instance, majority of the L-glutamate, methionine, L-threonine, L-tryptophan, L-valine, isoleucine, and L-lysine comes from the actinobacterial genus *Corynebacteria* and to a lesser extent from genera *Brevibacterium* and *Arthrobacter* (Koizumi et al. 2000; Falentin et al. 2010; Saraf and Hastings 2010; Hardter et al. 2012).

Several studies claim the Actinobacteria are also capable of exerting probiotic activity in many animal models. For instance, *Streptomyces* species have been proven to have a positive in vivo effect on controlling white spot syndrome virus and vibriosis in black tiger shrimp. Actinobacterial probiont, *S. werraensis,* is known to secrete a heat-stable, acidic pH-resistant, low molecular weight aggregative peptide pheromone compound that supports its probiotic function, and this property is currently being researched for developing novel probiotics. Actinobacteria are also known to produce a range of biosurfactants, which can be applied in various areas, such as the nutrient, cosmetic, textile, varnish, and pharmaceutical. One such famous biosurfactant would be the lipopeptide antibiotic daptomycin, developed and marketed as cubicin by cubist pharmaceuticals.

Actinobacteria play a pivotal role in the gut microbiota of most of the animals although they represent only a small percentage of the intestinal microbial community. Research evidence shows that they engage actively in maintaining gut homeostasis and therefore could be adopted as effective therapeutic probiotics. Actinobacteria are capable of functioning as growth promoters, production of inhibitory compounds, agents supporting the improvement of nutrient digestion, immunity boosters, growth regulators, and components that sequester nutrients (Defoirdt et al. 2007). Animal gut Actinobacteria is composed of *Bifidobacteria* spp., *Propionibacteria* spp., *Corynebacteria* spp., and *Streptomyces* spp. out of which *Bifidobacteria* spp. making up the major fraction. *Bifidobacteria* have been proven effective in the maintenance of intestinal barrier functions, increasing tight junction expression and mucin biosynthesis, catabolism, and provision of energy to gut liner cells. Species like Bacteroides, *Bifidobacterium,* and *Ruminococcus* and *Roseburia* have been identified to aid in the process of utilization of macro-polysaccharides, oligosaccharides, fibers, fatty acids, and unabsorbed sugars. They also support for downstream metabolism of proteins, lipids, and absorption/biosynthesis of vitamins including vitamin K, B12, iron, calcium, and magnesium. Bifidobacteria have been identified to stimulate intraepithelial lymphocytes, the production of mucosal immunoglobulins, and the promotion of a tolerogenic immune response. Furthermore, the secondary metabolites of actinobacterial species in general are known to function as AGPs (antibiotic growth promoters) while acting as biological prophylactics against pathogenic bacteria. For instance, the production of bacteriocins (Desriac et al. 2010), siderophores (Lalloo et al. 2010), enzymes (protease, amylase, lipase), hydrogen peroxide, and organic acids (Sugita et al. 1997) by *Aeromicrobium, Dietzia, Marinispora, Marinophilus, Salinispora, Solwaraspora, Salinibacterium, Kocuria, Williamsia, and Verrucosispora, Actinomyces, Actinopolyspora, Micromonospora, Micropolyspora, Nocardia, Rhodococcus, Streptomyces, Streptosporangium*, and *Streptoverticillium* has accepted probiotic application ranging from ruminant farming to aquaculture. Recent studies show that *Streptomyces* species are known to exert an antibiotic effect on pathogenic *Vibrio* spp. through siderophore action and by secreting quorum sensing inhibitors leading to retardation of biofilm formation of pathogenic *Vibrio* spp. (Iwatsuki et al. 2008). *Streptomyces* probionts have been proven effective against WSSV in in vitro *Penaeus monodon* test models (Jenifer et al. 2015). Although Actinobacteria are employed as probiotics in various test settings, it is, however, vital to elucidate the underlying molecular mechanisms of probiotic effects and possible synergistic/antagonistic effects when used in formulations with other microorganisms.

Actinobacteria are also used as indicators of organoleptic qualities such as odor and flavor, for instance, secondary metabolites, geosmin and 2-methylisoborneol (MIB), are two of the determinants of water quality and indicators of the presence of trivial compounds, such as acetic acid, acetaldehyde, ethyl alcohol, isobutyl alcohol, isobutyl acetate, and ammonia. These chemical agents are isolated for synthesizing organoleptic modifiers like *trans*-1,10-dimethyl-*trans*-9-decalol (geosmin), 1,2,7,7-tetramethyl-2-norbornanol 6-ethyl-3isobutyl-2-pyrone(mucidone), 2-isobutyl-3-methoxypyrazine or 2-isopropyl-3-methoxypyrazine (produced mostly by *Streptomyces* spp.), 2-methylisoborneol (produced by *Actinomadura* spp.), 6-methyl-5-hepten-2-one (produced by *Thermoactinomyces* spp.), and dimethyl trisulfide (produced by *Pseudonocardia* spp., *Saccharomonospora* spp., *Thermoactinomyces* spp., *Thermomonospora* spp.). Furthermore, marine Actinobacterial compounds are being used as biocidal agents for circumventing the problems of biofouling in aquatic infrastructure. Due to biodegradable nature, safe and highly effective biocidal profiles, most of the conventional anti-biofouling agents such as copper tributyltin, irgarol, diuron, and chlorothalonil are nowadays being replaced from compounds such as diketopiperazines, diterpene, extracted from *Streptomyces* spp. and fungicidal extracts derived from *S. praecox, S. fradiae, S. filamentosus*, and *S. albidoflavus*. From another perspective, attributed to the toxic nature of most of the synthetic dyes used in various industries, actinobacterial

Table 18.6 Actinobacterial species adopted in production of nanoparticles

Actinobacteria	Nanoparticle produced
S. naganishii (MA7), *S. glaucus, S. hygroscopicus, S. albidoflavus, Nocardiopsis* sp. MBRC-1, *Rhodococcus* spp., *S. rochei*	Silver/5–10 nm
S. viridogens	Gold/18–20 nm
Thermomonospora spp.	Gold/30–60 nm
S. aureofaciens, Thermoactinomyces spp.	Gold/5–50 nm
Nocardia farcinica	Gold/15–20 nm
Streptomyces sp. LK3	Silver/5 nm
Gordonia amarae	Gold/15–40 nm

pigments have gained its popularity as replacement dye options. For instance, anthracycline glycosides such as rhodomycin derived from *Synodontis violaceus*, phenoxazinones like actinomycin extracted from *Streptomyces* spp., III undecylprodigiosin and IV metacycloprodigiosin belonging to prodigiosin group of dyes and produced by *S. longispororuber,* and granaticin (classified as naphthoquinone) biosynthesized by *S. litmocidin*, natural green, red, and brown pigments produced by *S. torulosus* are excellent examples for these of replacement options that have made their way into commercialization. Actinobacteria are also becoming increasingly popular in the fields of gold biotechnology where they are being adopted as nano-factories for production of nanoparticles. This is mainly attributed to the, safe, low cost, nontoxic, and polydispersive properties of the nanoparticles produced by Actinobacteria. *Streptomyces, Arthrobacter, Rhodococcus, Thermomonospora,* and *Nocardiopsis* genera are the most important in this field of application, and these nanoparticles are used to produce antibacterial, antifungal, anticancer, antibiofouling, antimalarial, antiparasitic, and antioxidant products downstream (Table 18.6 and Fig. 18.1).

18.20 Conclusion and Future Perspectives

To date, a sundry of biologically active secondary metabolites with medical, biotechnological, agricultural, and industrial applications have been isolated from Actinobacteria. Attributed to their extreme species diversity and the scarcity of research work, they still remain the biggest prospect in obtaining compounds of vital importance and process for the welfare and well-being of humans and the environment. Despite the advancements in the fields of microbiology and biotechnology, actinobacterial resources will continue to be of interest to humankind, providing sustainable and environmentally friendly solutions.

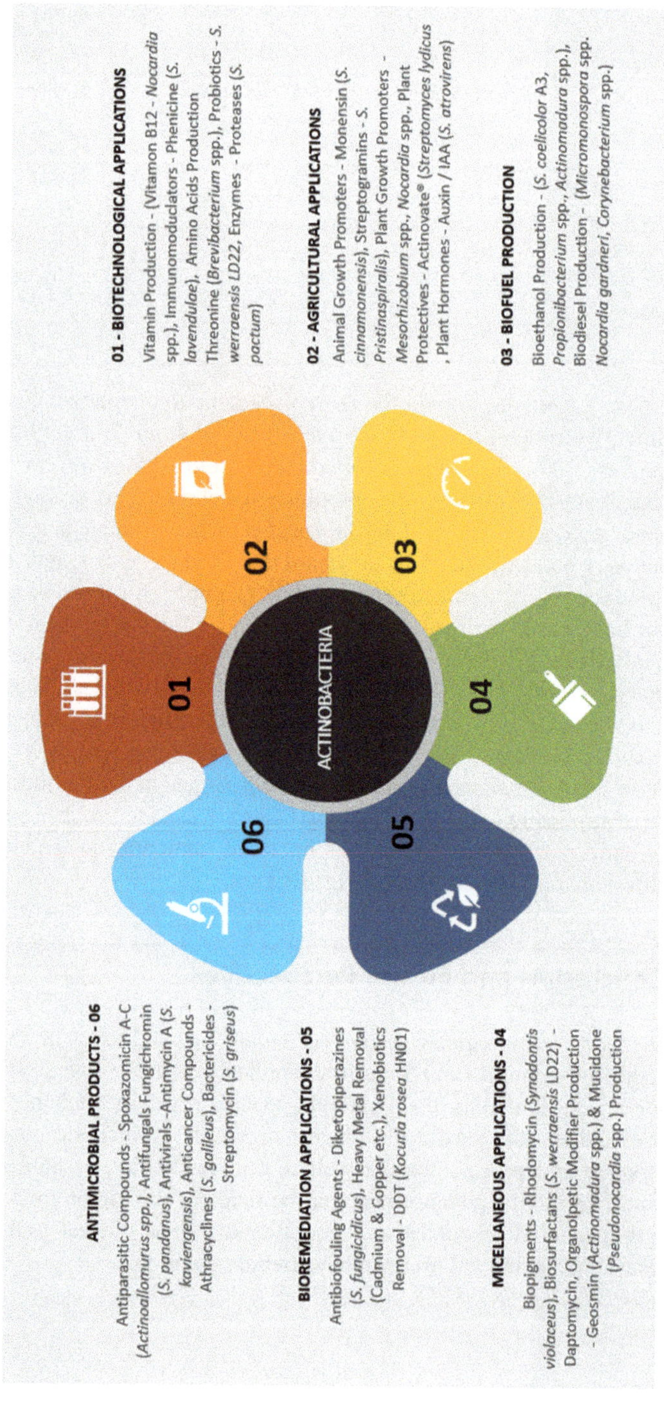

Fig. 18.1 Role of Actinobacteria

References

Aftabuddin S, Kashem MA, Kader MA, Sikder MNA, Hakim MA (2013) Use of Streptomyces fradiae and Bacillus megaterium as probiotics in the experimental culture of tiger shrimp Penaeus monodon (Crustacea, Penaeidae). Aquac Aquar Conserv Legis 6(3):253–267

Agbor VB, Cicek N, Sparling R, Berlin A, Levin DB (2011) Biomass pretreatment: fundamentals toward application. Biotechnol Adv 29(6):675–685

Alvarez A, Saez JM, Costa JSD, Colin VL, Fuentes MS, Cuozzo SA, Benimeli CS, Polti MA, Amoroso MJ (2017) Actinobacteria: current research and perspectives for bioremediation of pesticides and heavy metals. Chemosphere 166:41–62

Amoroso MJ, Benimeli CS, Cuozzo SA (eds) (2013) Actinobacteria: application in bioremediation and production of industrial enzymes. CRC Press

Arasu MV, Al-Dhabi NA, Saritha V, Duraipandiyan V, Muthukumar C, Kim S-J (2013) Antifeedant, larvicidal and growth inhibitory bioactivities of novel polyketide metabolite isolated from Streptomyces sp. AP-123 against Helicoverpa armigera and Spodoptera litura. BMC Microbiol 13(1):1–6

Bae M, Chung B, Oh K-B, Shin J, Oh D-C (2015) Hormaomycins B and C: new antibiotic cyclic depsipeptides from a marine mudflat-derived Streptomyces sp. Mar Drugs 13(8):5187–5200

Berezin V, Abdukhakimova D, Trenozhnikova L, Bogoyavlenskiy A, Turmagambetova A, Issanov A, Azizan A (2019) Antiviral activities of extremophilic actinomycetes extracts from Kazakhstan's unique ecosystems against influenza viruses and paramyxoviruses. Virol J 16(1):1–16

Brandelli A (2008) Bacterial keratinases: useful enzymes for bioprocessing agroindustrial wastes and beyond. Food Bioprocess Technol 1(2):105–116

Bürstner N, Roggo S, Ostermann N, Blank J, Delmas C, Freuler F, Gerhartz B et al (2015) Gift from nature: cyclomarin A kills mycobacteria and malaria parasites by distinct modes of action. ChemBioChem 16(17):2433–2436

Chakraborty S, Khopade A, Biao R, Jian W, Liu X-Y, Mahadik K, Chopade B, Zhang L, Kokare C (2011) Characterization and stability studies on surfactant, detergent and oxidant stable α-amylase from marine haloalkaliphilic Saccharopolyspora sp. A9. J Mol Catal B Enzym 68(1):52–58

Charan RD, Schlingmann G, Janso J, Bernan V, Feng X, Carter GT (2004) Diazepinomicin, a new antimicrobial alkaloid from a marine Micromonospora sp. J Nat Prod 67(8):1431–1433

Chaturvedi S, Khurana SMP (2019) Importance of actinobacteria for bioremediation. In: Plant biotechnology: progress in genomic era. Springer, Singapore, pp 277–307

Chen Y, Liu R-H, Li T-X, Huang S-S, Kong L-Y, Yang M-H (2017) Enduspeptides A-F, six new cyclic depsipeptides from a coal mine derived Streptomyces sp. Tetrahedron 73:527–531

Chen M-H, Chang S-S, Dong B, Yu L-Y, Wu Y-X, Wang R-Z, Jiang W, Gao Z-P, Si S-Y (2018) Ahmpatinin i Bu, a new HIV-1 protease inhibitor, from Streptomyces sp. CPCC 202950. RSC Adv 8(10):5138–5144

Cheng C, MacIntyre L, Abdelmohsen UR, Horn H, Polymenakou PN, Edrada-Ebel RA, Hentschel U (2015) Biodiversity, anti-trypanosomal activity screening, and metabolomic profiling of actinomycetes isolated from Mediterranean sponges. PLoS One 10(9):e0138528

Christopherson MR, Suen G, Bramhacharya S, Jewell KA, Aylward FO, Mead D, Brumm PJ (2013) The genome sequences of Cellulomonas fimi and "Cellvibrio gilvus" reveal the cellulolytic strategies of two facultative anaerobes, transfer of "Cellvibrio gilvus" to the genus Cellulomonas, and proposal of Cellulomonas gilvus sp. nov. PLoS One 8(1):e53954

Crawford Khan NI, Filonow AB, Singleton LL (1997) Augmentation of soil with sporangia of Actinoplanes spp. for biological control of Pythium damping-off. Biocontrol Sci Technol 7(1):11–22

De Azeredo LAI, Freire DMG, Soares RMA, Leite SGF, Coelho RRR (2004) Production and partial characterization of thermophilic proteases from Streptomyces sp. isolated from Brazilian cerrado soil. Enzyme Microb Technol 34(3–4):354–358

Deepika L, Kannabiran K (2010) Isolation and characterization of antagonistic actinomycetes from marine soil. J Microb Biochem Technol 2:001–006

Defoirdt T, Boon N, Sorgeloos P, Verstraete W, Bossier P (2007) Alternatives to antibiotics to control bacterial infections: luminescent vibriosis in aquaculture as an example. Trends Biotechnol 25(10):472–479

Desriac F, Defer D, Bourgougnon N, Brillet B, Le Chevalier P, Fleury Y (2010) Bacteriocin as weapons in the marine animal-associated bacteria warfare: inventory and potential applications as an aquaculture probiotic. Mar Drugs 8(4):1153–1177

Esawy MA (2007) Isolation and partial characterization of extracellular keratinase from a novel mesophilic Streptomyces albus AZA. Res J Agric Biol Sci 3(6):808–817

Fabre B, Armau E, Etienne G, Legendre F, Tiraby G (1988) A simple screening method for insecticidal substances from actinomycetes. J Antibiot 41(2):212–219

Falentin H, Deutsch SM, Parayre-Breton S, Dherbecourt J, Jan G, Thierry A, Maillard MB et al (2010) The complete genome of Propionibacterium freudenreichii CIRM 1 a hardy actinobacteria with food and probiotic applications. In: 3rd International Symposium on Propionibacteria and Bifidobacteria: Dairy and Probiotic, vol 149, p np

Fernandes TAR, da Silveira WB, Passos FML, Zucchi TD (2014) Laccases from Actinobacteria—what we have and what to expect. Adv Microbiol 2014

Galm U, Hager MH, Van Lanen SG, Ju J, Thorson JS, Shen B (2005) Antitumor antibiotics: bleomycin, enediynes, and mitomycin. Chem Rev 105(2):739–758

Gasmi M, Kitouni M, Carro L, Pujic P, Normand P, Boubakri H (2019) Chitinolytic actinobacteria isolated from an Algerian semi-arid soil: development of an antifungal chitinase-dependent assay and GH18 chitinase gene identification. Ann Microbiol 69(4):395–405

Gohain A, Manpoong C, Saikia R, De Mandal S (2020) Actinobacteria: diversity and biotechnological applications. In: Recent advancements in microbial diversity, pp 217–231

Grasso LL, Martino DC, Alduina R (2016) Production of antibacterial compounds from Actinomycetes. In: Actinobacteria-basics and biotechnological applications, pp 177–198

Hamedi J, Dehhaghi M, Mohammdipanah F (2015a) Isolation of extremely heavy metal resistant strains of rare Actinomycetes from high metal content soils in Iran. Int J Environ Res 9(2)

Hamedi J, Moghimi H, Papiran R, Mohammadipanah F (2015b) Screening of phytotoxic activity and nlp genes from rhizosphere actinomycetes. Ann Microbiol 65(1):527–532

Hardter U, Luzhetska M, Ebeling S, Bechthold A (2012) Ethanol production in Actinomycetes after expression of synthetic adhB and pdc. Open Biotechnol J 6(1):13–16

Hassan HM, Degen D, Jang KH, Ebright RH, Fenical W (2015) Salinamide F, new depsipeptide antibiotic and inhibitor of bacterial RNA polymerase from a marine-derived Streptomyces sp. J Antibiot 68(3):206–209

Hassan HM, Boonlarppradab C, Fenical W (2016) Actinoquinolines A and B, anti-inflammatory quinoline alkaloids from a marine-derived Streptomyces sp., strain CNP975. J Antibiot 69(7):511–514

Hayat S, Ashraf A, Aslam B, Asif R, Muzammil S, Zahoor MA, Waseem M et al (2020) Actinobacteria: potential candidate as plant growth promoters. In: Plant stress physiology. IntechOpen

Heisey RM, Huang J, Mishra SK, Keller JE, Miller JR, Putnam AR, D'Silva TDJ (1988) Production of valinomycin, an insecticidal antibiotic, by Streptomyces griseus var. flexipertum var. nov. J Agric Food Chem 36(6):1283–1286

Huang X-H, Nishida H, Tomoda H, Tabata N, Shiomi K, Yang D-J, Takayanagi H, Omura S (1995) Terpendoles, novel AC AT inhibitors produced by Albophoma yamanashiensis. J Antibiot 48(1):5–11

Igarashi Y, Iida T, Oku N, Watanabe H, Furihata K, Miyanouchi K (2012) Nomimicin, a new spirotetronate-class polyketide from an actinomycete of the genus Actinomadura. J Antibiot 65(7):355–359

Iwatsuki M, Uchida R, Yoshijima H, Ui H, Shiomi K, Matsumoto A, Takahashi Y, Abe A, Tomoda H, Ōmura S (2008) Guadinomines, type III secretion system inhibitors, produced by Streptomyces sp. K01-0509. J Antibiot 61(4):222–229

Jakubiec-Krzesniak K, Rajnisz-Mateusiak A, Guspiel A, Ziemska J, Solecka J (2018) Secondary metabolites of actinomycetes and their antibacterial, antifungal and antiviral properties. Pol J Microbiol 67(3):259

Janardhan A, Kumar AP, Viswanath B, Gopal DVRS, Narasimha G (2018) Antiviral and larvicidal properties of novel bioactive compounds produced from marine actinomycetes. Russ J Mar Biol 44(5):424–428

Jenifer JSCA, Donio MBS, Michaelbabu M, Vincent SGP, Citarasu T (2015) Haloalkaliphilic Streptomyces spp. AJ8 isolated from solar salt works and its' pharmacological potential. AMB Express 5(1):1–12

Kane SP, Hanes SD (2017) Unexplained increases in serum vancomycin concentration in a morbidly obese patient. Intens Crit Care Nurs 39:55–58

Karmakar M, Ray RR (2011) Current trends in research and application of microbial cellulases. Res J Microbiol 6(1):41–53

Kawase T, Saito A, Sato T, Kanai R, Fujii T, Nikaidou N, Miyashita K, Watanabe T (2004) Distribution and phylogenetic analysis of family 19 chitinases in Actinobacteria. Appl Environ Microbiol 70(2):1135–1144

Kim SH, Shin Y, Lee SH, Oh WK, Oh KB, Lee SK, Shin J, Oh DC (2015) The secondary metabolites from halophilic actinomycetes from a solar saltern in Korea. Planta Med 81(11): PQ17

Kino T, Hatanaka H, Hashimoto M, Nishiyama M, Goto T, Okuhara M, Kohsaka M, Aoki H, Imanaka H (1987) FK-506, a novel immunosuppressant isolated from a Streptomyces I. Fermentation, isolation, and physico-chemical and biological characteristics. J Antibiot 40(9):1249–1255

Koizumi S, Yonetani Y, Maruyama A, Teshiba S (2000) Production of riboflavin by metabolically engineered Corynebacterium ammoniagenes. Appl Microbiol Biotechnol 53(6):674–679

Kondratyuk TP, Park E-J, Yu R, Van Breemen RB, Asolkar RN, Murphy BT, Fenical W, Pezzuto JM (2012) Novel marine phenazines as potential cancer chemopreventive and anti-inflammatory agents. Mar Drugs 10(2):451–464

Korkmaz HATİCE, Unaldi MN, Aslan B, Coral G, Arikan BURHAN, Dincer SADIK, Colak O (2003) Keratinolytic activity of Streptomyces strain BA7, a new isolate from Turkey. Ann Microbiol 53:85–93

Kudo N, Wolff B, Sekimoto T, Schreiner EP, Yoneda Y, Yanagida M, Horinouchi S, Yoshida M (1998) Leptomycin B inhibition of signal-mediated nuclear export by direct binding to CRM1. Exp Cell Res 242(2):540–547

Kuhad RC, Gupta R, Singh A (2011) Microbial cellulases and their industrial applications. Enzyme Res 2011:280696

Kumar Y, Goodfellow M (2008) Five new members of the Streptomyces violaceusniger 16S rRNA gene clade: streptomyces castelarensis sp. nov., comb. nov., Streptomyces himastatinicus sp. nov., Streptomyces mordarskii sp. nov., Streptomyces rapamycinicus sp. nov. and Streptomyces ruanii sp. nov. Int J Syst Evol Microbiol 58(6):1369–1378

Kyrikou I, Briassoulis D (2007) Biodegradation of agricultural plastic films: a critical review. J Polym Environ 15(2):125–150

Lacombe-Harvey M-È, Brzezinski R, Beaulieu C (2018) Chitinolytic functions in actinobacteria: ecology, enzymes, and evolution. Appl Microbiol Biotechnol 102(17):7219–7230

Lalloo R, Moonsamy G, Ramchuran S, Görgens J, Gardiner N (2010) Competitive exclusion as a mode of action of a novel Bacillus cereus aquaculture biological agent. Lett Appl Microbiol 50(6):563–570

Lamilla C, Pavez M, Santos A, Hermosilla A, Llanquinao V, Barrientos L (2017) Bioprospecting for extracellular enzymes from culturable Actinobacteria from the South Shetland Islands, Antarctica. Polar Biol 40(3):719–726

Lawson PA (2018) The phylum Actinobacteria. In: The bifidobacteria and related organisms. Academic, pp 1–8

Lee J-G, Yoo I-D, Kim W-G (2007) Differential antiviral activity of benzastatin C and its dechlorinated derivative from Streptomyces nitrosporeus. Biol Pharm Bull 30(4):795–797

Lee M-J, Kong D, Han K, Sherman DH, Bai L, Deng Z, Lin S, Kim E-S (2012) Structural analysis and biosynthetic engineering of a solubility-improved and less-hemolytic nystatin-like polyene in Pseudonocardia autotrophica. Appl Microbiol Biotechnol 95(1):157–168

Lee S-H, Shin HJ, Kim D-Y, Shim D-W, Kim T-J, Ye S-K, Won H-S, Koppula S, Kang T-B, Lee K-H (2013) Streptochlorin suppresses allergic dermatitis and mast cell activation via regulation of Lyn/Fyn and Syk signaling pathways in cellular and mouse models. PLoS One 8(9):e74194

Lee D-S, Yoon C-S, Jung Y-T, Yoon J-H, Kim Y-C, Hyuncheol O (2016) Marine-derived secondary metabolite, Griseusrazin A, suppresses inflammation through heme oxygenase-1 induction in activated RAW264. 7 macrophages. J Nat Prod 79(4):1105–1111

Liu X, Gan M, Dong B, Zhang T, Li Y, Zhang Y, Fan X et al (2013) 4862F, a new inhibitor of HIV-1 protease, from the culture of Streptomyces I03A-04862. Molecules 18(1):236–243

Manulis S, Shafrir H, Epstein E, Lichter A, Barash I (1994) Biosynthesis of indole-3-acetic acid via the indole-3-acetamide pathway in Streptomyces spp. Microbiology 140(5):1045–1050

Maskey RP, Helmke E, Kayser O, Fiebig HH, Maier A, Busche A, Laatsch H (2004) Anti-cancer and antibacterial trioxacarcins with high anti-malaria activity from a marine Streptomycete and their absolute stereochemistry. J Antibiot 57(12):771–779

Matsumoto Y, Mineta S, Murayama K, Sugimori D (2013) A novel phospholipase B from Streptomyces sp. NA 684–purification, characterization, gene cloning, extracellular production and prediction of the catalytic residues. FEBS J 280(16):3780–3796

McCarthy AJ, Peace E, Broda P (1985) Studies on the extracellular xylanase activity of some thermophilic actinomycetes. Appl Microbiol Biotechnol 21(3):238–244

McMahon MD, Guan C, Handelsman J, Thomas MG (2012) Metagenomic analysis of Streptomyces lividans reveals host-dependent functional expression. Appl Environ Microbiol 78(10):3622–3629

Meena AK, Mishra GK, Rai PK, Rajagopal C, Nagar PN (2005) Removal of heavy metal ions from aqueous solutions using carbon aerogel as an adsorbent. J Hazard Mat 122(1–2):161–170

Minotto E, Milagre LP, Oliveira MT, Van Der Sand ST (2014) Enzyme characterization of endophytic actinobacteria isolated from tomato plants. J Adv Sci Res 5(2)

Mitra P, Chakrabartty PK (2005) An extracellular protease with depilation activity from Streptomyces nogalator. J Sci Ind Res 64:978–983

Moreira KA, Cavalcanti MTH, Duarte HS, Tambourgi EB, de Melo EHM, Silva VL, Porto ALF, de Lima Filho JL (2001) Caracterização parcial de proteases extracelulares de Streptomyces clavuligerus usando um meio de cultura econômico. Braz J Microbiol 32(3):215–220

Morosoli R, Bertrand J-L, Mondou F, Shareck F, Kluepfel D (1986) Purification and properties of a xylanase from Streptomyces lividans. Biochem J 239(3):587–592

Nafis A, Raklami A, Bechtaoui N, El Khalloufi F, El Alaoui A, Glick BR, Hafidi M, Kouisni L, Ouhdouch Y, Hassani L (2019) Actinobacteria from extreme niches in morocco and their plant growth-promoting potentials. Diversity 11(8):139

Niladevi KN, Prema P (2008) Effect of inducers and process parameters on laccase production by Streptomyces psammoticus and its application in dye decolourization. Bioresour Technol 99(11):4583–4589

Oh D-C, Gontang EA, Kauffman CA, Jensen PR, Fenical W (2008) Salinipyrones and pacificanones, mixed-precursor polyketides from the marine actinomycete Salinispora pacifica. J Nat Prod 71(4):570–575

Onwurah INE, Nwuke C (2004) Enhanced bioremediation of crude oil-contaminated soil by a Pseudomonas species and mutually associated adapted Azotobacter vinelandii. J Chem Technol Biotechnol 79(5):491–498

Padilla MA, Rodrigues RAF, Bastos JCS, Martini MC, de Souza Barnabé AC, Kohn LK, Uetanabaro APT et al (2015) Actinobacteria from termite mounds show antiviral activity against

bovine viral diarrhea virus, a surrogate model for hepatitis C virus. Evid Based Complement Alternat Med 2015:745754
Palaniyandi SA, Yang SH, Suh J-W (2014) Cellulase production and saccharification of rice straw by the mutant strain Hypocrea koningii RSC1. J Basic Microbiol 54(1):56–65
Prudhomme J, McDaniel E, Ponts N, Bertani S, Fenical W, Jensen P, Le Roch K (2008) Marine actinomycetes: a new source of compounds against the human malaria parasite. PLoS One 3(6): e2335
Pulgar G, Maria E, Saadeddin A (2014) The cellulolytic system of Thermobifida fusca. Crit Rev Microbiol 40(3):236–247
Puttaswamygowda GH, Olakkaran S, Antony A, Purayil AK (2019) Present status and future perspectives of marine actinobacterial metabolites. In: Recent developments in applied microbiology and biochemistry. Academic, pp 307–319
Raja A, Prabakarana P (2011) Actinomycetes and drug-an overview. Am J Drug Discov Dev 1(2):75–84
Raskin I, Kumar NPBA, Dushenkov S, Salt D (1994) Bioconcentration of heavy metals by plants. Curr Opin Biotechnol 5:285–290
Ratnakomala S, Perwitasari U (2020) The amylase production by Actinobacteria isolated from rumen fluid. In: IOP Conference Series: Earth and Environmental Science, vol 439, no 1, p 012019. IOP Publishing
Sacramento DR, Coelho RRR, Wigg MD, de Toledo Luna Linhares LF, dos Santos MGM, de Azevedo Soares Semêdo LT, da Silva AJR (2004) Antimicrobial and antiviral activities of an actinomycete (Streptomyces sp.) isolated from a Brazilian tropical forest soil. World J Microbiol Biotechnol 20(3):225–229
Salwan R, Sharma V (2018) The role of actinobacteria in the production of industrial enzymes. In: New and future developments in microbial biotechnology and bioengineering. Elsevier, pp 165–177
Sanglier JJ, Haag H, Huck TA, Fehr T (1993) Novel bioactive compounds from actinomycetes: a short review (1988–1992). Res Microbiol 144(8):633–642
Santos JD, Vitorino I, Reyes F, Vicente F, Lage OM (2020) From ocean to medicine: pharmaceutical applications of metabolites from marine bacteria. Antibiotics 9(8):455
Saraf M, Hastings A (2010) Biofuels, the role of biotechnology to improve their sustainability and profitability. In: Biodiversity, biofuels, agroforestry and conservation agriculture, pp 123–148
Sathya A, Vijayabharathi R, Srinivas V, Gopalakrishnan S (2016) Plant growth-promoting actinobacteria on chickpea seed mineral density: an upcoming complementary tool for sustainable biofortification strategy. 3 Biotech 6(2):1–6
Sato S, Iwata F, Yamada S, Katayama M (2012) Neomaclafungins A–I: oligomycin-class macrolides from a marine-derived actinomycete. J Nat Prod 75(11):1974–1982
Saxena S (2015) Agricultural applications of microbes. In: Applied microbiology. Springer, New Delhi, pp 37–54
Sekizawa Y, Takematsu T (2013) How to discover new antibiotics for herbicidal use. In: Natural Products: Proceedings of the 5th International Congress of Pesticide Chemistry, Kyoto, Japan, 29 August–4 September 1982, p 261. Elsevier
Shata HMAH, Farid MAF (2012) Optimization of extraction parameters for keratinase recovery from fermented feather under solid state fermentation by Streptomyces sp. NRC 13S. J Appl Biol Chem 55(3):149–156
Shimizu M (2011) Endophytic actinomycetes: biocontrol agents and growth promoters. In: Bacteria in agrobiology: plant growth responses. Springer, Berlin, Heidelberg, pp 201–220
Shivlata L, Satyanarayana T (2017) Actinobacteria in agricultural and environmental sustainability. In: Agro-environmental sustainability. Springer, Cham, pp 173–218
Solecka J, Zajko J, Postek M, Rajnisz A (2012) Biologically active secondary metabolites from Actinomycetes. Open Life Sci 7(3):373–390
Stackebrandt E, Schumann PETER (2006) Introduction to the taxonomy of actinobacteria. Prokaryotes 3:297–321

Sugita H, Matsuo N, Hirose Y, Iwato M, Deguchi Y (1997) Vibrio sp. strain NM 10, isolated from the intestine of a Japanese coastal fish, has an inhibitory effect against Pasteurella piscicida. Appl Environ Microbiol 63(12):4986–4989

Suthindhiran K, Jayasri MA, Dipali D, Prasar A (2014) Screening and characterization of protease producing actinomycetes from marine saltern. J Basic Microbiol 54(10):1098–1109

Syed DG, Lee JC, Li W-J, Kim C-J, Agasar D (2009) Production, characterization and application of keratinase from Streptomyces gulbargensis. Bioresour Technol 100(5):1868–1871

Takahashi Y, Nakashima T (2018) Actinomycetes, an inexhaustible source of naturally occurring antibiotics. Antibiotics 7(2):45

Takasuka TE, Book AJ, Lewin GR, Currie CR, Fox BG (2013) Aerobic deconstruction of cellulosic biomass by an insect-associated Streptomyces. Sci Rep 3(1):1–10

Takemori D, Yoshino K, Eba C, Nakano H, Iwasaki Y (2012) Extracellular production of phospholipase A2 from Streptomyces violaceoruber by recombinant Escherichia coli. Protein Expr Purif 81(2):145–150

Techapun C, Poosaran N, Watanabe M, Sasaki K (2003) Thermostable and alkaline-tolerant microbial cellulase-free xylanases produced from agricultural wastes and the properties required for use in pulp bleaching bioprocesses: a review. Process Biochem 38(9):1327–1340

ul Hassan SS, Shaikh AL (2017) Marine actinobacteria as a drug treasure house. Biomed Pharmacother 87:46–57

Wink J, Kroppenstedt RM, Seibert G, Stackebrandt E (2003) Actinomadura namibiensis sp. nov. *International journal of systematic and evolutionary microbiology* 53(3):721–724

Wink J, Mohammadipanah F, Hamedi J (eds) (2017) Biology and biotechnology of actinobacteria. Springer International Publishing, Berlin

Yang J, Yang Z, Yin Y, Rao M, Liang Y, Ge M (2016) Three novel polyene macrolides isolated from cultures of *Streptomyces lavenduligriseus*. J Antibiot (Tokyo) 69:62–65

Yao J, Tian L, Wang Y, Djah A, Wang F, Chen H, Su C et al (2008) Microcalorimetric study the toxic effect of hexavalent chromium on microbial activity of Wuhan brown sandy soil: an *in vitro* approach. Ecotoxicol Environ Saf 69(2):289–295

Yassin AF (2014) The family Actinomycetaceae. In: The prokaryotes: actinobacteria, vol 4, pp 41–79

Yassir RY, Adel ME, Azze A (2002) Use of probiotic bacteria as growth promoters, antibacterial and the effect on physiological parameters of Oreochromis niloticus. J Fish Dis 22:633–642

Zhou Q, Li K, Jun X, Bo L (2009) Role and functions of beneficial microorganisms in sustainable aquaculture. Bioresour Technol 100(16):3780–3786

CPSIA information can be obtained
at www.ICGtesting.com
Printed in the USA
LVHW061132250422
716828LV00031B/82

9 789811 661310